Wetlands

This book is printed on acid-free paper. ∞

Copyright © 2000 by John Wiley & Sons, Inc. All rights reserved.

Published simultaneously in Canada.

No part of this publication may be reproduced, stored in a retrieval system or transmitted in any form or by any means, electronic, mechanical, photocopying, recording, scanning or otherwise, except as permitted under Sections 107 or 108 of the 1976 United States Copyright Act, without either the prior written permission of the Publisher, or authorization through payment of the appropriate per-copy fee to the Copyright Clearance Center, 222 Rosewood Drive, Danvers, MA 01923, (978) 750-8400, fax (978) 750-4744. Requests to the Publisher for permission should be addressed to the Permissions Department, John Wiley & Sons, Inc., 605 Third Avenue, New York, NY 10158-0012, (212) 850-6011, fax (212) 850-6008, E-Mail: PERMREQ@WILEY.COM.

This publication is designed to provide accurate and authoritative information in regard to the subject matter covered. It is sold with the understanding that the publisher is not engaged in rendering professional services. If professional advice or other expert assistance is required, the services of a competent professional person should be sought.

Library of Congress Cataloging-in-Publication Data:

ISBN: 0-471-29232-X

Printed in the United States of America.

10 9 8 7 6 5 4 3 2

**William J. Mitsch and
James G. Gosselink
Columbus, Ohio
June 2000**

Wetlands

Third Edition

William J. Mitsch

Professor
The Ohio State University
Columbus, Ohio

James G. Gosselink

Professor Emeritus
Louisiana State University
Baton Rouge, Louisiana

John Wiley & Sons, Inc.
New York | Chichester | Weinheim | Brisbane | Singapore | Toronto

To Rebecca, Jane, Mary Cecilia, and especially Ruthmarie (Mitsch).
To Jena, Leila, Patrick, and especially Jean (Gosselink).

Contents

Part 5   Wetland Management

Preface

This is the third edition—the one we swore we would never do. However, we succumbed, not to economic forces, but to the goal of making this wetland textbook the best edition yet. You will be the judge of that, but we are very pleased with our effort. You will notice a new look with this edition, in terms of both the larger format and the professionally drawn illustrations. We believe that the illustrations are just as important as the text in conveying the science and management of wetlands. Therefore, we painstakingly redrew almost every illustration to be easily read and added quite a few new illustrations. We also tried to make the illustrations consistent throughout the book.

In essence, this is a brand new book in terms of the text, too. There are an estimated 700 new citations in this book since the 1993 edition. However, we removed 400 or so that referred to material now so well understood as to need no citation or that did not contribute any more to the modern view of wetlands. The last edition had about 1,500 citations; we now have about 1,800 citations, with almost 40 percent of them new. No chapter escaped our heavy editing pen. The general outline of five sections remains similar to previous editions, but we added three chapters, one to the introductory section and two to the last section on wetland management.

First and most important, we have noticed over the past 7 years since we wrote the second edition that wetlands are receiving much more international attention. Through efforts such as the Ramsar Convention and well-attended international wetland conferences, interest in protecting wetlands has exploded around the world. We have also noticed an increased attention to wetlands in concert with increasing attention to biodiversity and ecosystem services around the world. Therefore, in addition to the visible new features of this textbook, we believe that this third edition is also much *more international*. We added a chapter (Chapter 3) specifically on wetlands of the world outside of North America. We complemented that with a new chapter (Chapter 4) on wetlands of North America with much more attention paid to Canadian wetlands. The international flavor of the book is evident in almost every chapter.

We noticed that there was so much information on the legal and policy aspects of wetland protection, both in the United States and elsewhere, that we included a separate chapter on *wetland laws and protection* (Chapter 18). Since the turbulent times for wetlands in the United States in the early 1990s, things have settled down politically, but we present the history of U.S. wetland policy so we will remember the good and not repeat the bad.

We paid a lot more attention in this third edition to the art and science of *wetland creation and restoration,* both for habitat replacement and for water pollution control. Those subjects are now covered in two chapters rather than one as in the previous edition. It is astounding what we have learned about the building and restoring of wetlands over the past decade, but it is equally formidable as to what we still do not know about these subjects and how we are as likely to fail in our attempts to design and build wetlands as we are to succeed. We relied in both of these chapters on well-documented case studies that have taught us a lot about wetland building.

Other areas in which the book has been enhanced, and we think improved, over previous editions include:

- Several more *wetland definitions,* including a new one from a U.S. National Research Council panel that met for two years to try to define what constitutes a wetland

- An *expanded hydrology chapter,* which provides many more examples of wetland hydroperiods and approaches for determining a wetland's hydrologic budget

- An introduction to the *new nomenclature of hydric soils*

- A much better picture of *gaseous emissions from wetlands,* particularly methane, and the implications of these gases for global climate change

- A summary of the revisit that scientists made to the subject of *coastal marsh decomposition and outwelling,* 20 years after the theories were first developed on these subjects

- A better description of the *global extent of mangroves* and a more detailed discussion of the important *role of crabs in mangrove swamps,* particularly for biodiversity, leaf litter burial, and energy transfers

- A discussion of the recent literature on the *importance of hurricanes on mangrove forest composition and function*

- The inclusion of other forested wetlands, particularly *red maple swamps* and *white cedar swamps* in the freshwater swamp chapter and *black spruce swamps* in the peatland chapter

- A significant rewrite of the peatland chapter, with particular attention paid to the wide array of *peatland classification systems and nomenclature*

- A better description of *gas flow in vascular wetland plants,* both as avenues of greenhouse gases and as adaptations for plant survival in anoxic soils

- Additional discussion of *animal adaptations to wetlands,* including behavioral adaptations and mutualism and commensalism

- Additional discussion of *new theories of wetland development,* including functional guilds, gap models, the centrifugal organization concept, and ecosystem-level theories such as pulse stability, self-design, and ecosystem engineers

- An *updated look at tidal freshwater marshes* as three distinct types rather than one

- An inclusion of plant vegetation zonation for *sub-Saharan African marshes* and *inland saline marshes* of the midwestern United States in the freshwater marsh chapter and *European salt marshes* in the tidal salt marsh chapter

- Additional discussion of *European reed die-back* in the freshwater marsh chapter

- More attention to *nongame birds and amphibians* in freshwater marshes and wetlands in general

- Discussion in several chapters of the *effects of wetlands on nutrients and the effects of excessive nutrients on wetlands*

- A new section on *carbon budgets for northern peatlands*

- Consistent *conceptual ecosystem models* for each of the seven ecosystem chapters that illustrate both the similarities and the differences among these wetland types

- More emphasis on *semi-arid riparian wetlands* typical of states in the western United States

- A better contrast, with illustrations, of the *flood pulse concept* and the *river continuum concept* of riverine systems

- A better description of both the role of wetlands in *global climate change* and the role of climate change in affecting wetlands

- A detailed discussion of the *hydrogeomorphic classification* as both a way to classify wetlands and a way to compare the functions of wetlands

- A revisit of the *problems and paradoxes of quantifying wetland values* based on a paper we published recently in *Ecological Economics*

- Updated information on *peat harvesting, fisheries harvests,* and *aquaculture* in relation to wetlands

- An update on the *legal status of wetland protection* in the United States, particularly as it relates to delineation and mitigation policies

- A better summary of the *Ramsar Convention* and an update to the year 2000, including a summary of how many wetlands are enrolled in this program and a new map showing the location of Ramsar wetland sites around the world

- Several *case studies of proposed and actual wetland creation and restoration* on a large scale in the United States

- A summary of *two decades of wetlands creation and restoration* for habitat replacement in the United States, and measures of how successful we have been at it

- Information on simple models used to design *wastewater (treatment) wetlands,* design criteria of how to build these wetlands, and updated information on what they cost to build and operate

- Details, including photos and maps, of some of the most *well-known experimental wetlands* in the United States, including the Experimental Cypress Domes (Florida), the Des Plaines River Demonstration Project (Illinois), Houghton Lake Treatment Wetland (Michigan), and the Olentangy River Wetland Research Park (Ohio)
- A new section in the last chapter on *wetland delineation*

We also added a comprehensive glossary and three appendices with useful information: a list of all the wetland books we could find; a less comprehensive, but nevertheless useful, list of permanent web sites for finding the latest information on wetlands around the world; and a table of useful conversion factors.

As has always been the case, we could not have completed this project without a tremendous amount of help from others. In fact, during the preparation of this edition of the book, it was all too tempting to leave the manuscript typing for a minute and send off an e-mail, asking for assistance. Our friends, colleagues, and employees always came through. We hope we do not miss anybody who made a significant contribution, but here we list as many as we can recall: Aimlee Laderman, Yale University; Scott Bridgham, University of Notre Dame; Philippe Gerbeaux and Ferne McKenzie, New Zealand Department of Conservation; Chris Tanner and Long Nguyen, NIWA, New Zealand; Robert Twilley, Southwestern Louisiana University; John Day, Charles Sasser, Elaine Evers, Sam Meyers, and Jenneke Visser, Louisiana State University; Andy Clewell, Quincy, Florida; Donald Hey and Kathleen Paap, The Wetland Initiative, Chicago; Bob Kadlec, Chelsea, Michigan; Mike Chimney and Tom Fontaine, South Florida Water Management District; Jenny Davis, Murdoch University, Western Australia; Tim Daniel, Jack Henry, and Kim Baker, Ohio Department of Natural Resources; Kathy Ewel, U.S. Forest Service, Hawaii; Ken Mavuti, University of Nairobi, Kenya; Robert Gough, Rosebud, South Dakota; Virginie Bouchard, Kim Hornung, Janice Gilbert, Changwoo Ahn, Eric Lohan, and Naiming Wang, The Ohio State University; Jos Verhoeven, University of Utrecht; and Douglas Wilcox, U.S. Geological Survey, Ann Arbor, Michigan. We give a special thanks to Mike Weinstein and Lisa Young from New Jersey Sea Grant for making available an advance copy of the 1998 symposium on salt marshes, which will be published almost simultaneously with our book. We would like to give a particular thanks to Ann Mischo and Ron McLean from Ohio State University's Technical Services Department for the splendid artwork, that helps so much in making this book both attractive and usable.

We also appreciate the professional effort on the part of our editors, Neil Levine, Joel Stein, and Jennifer Mazurkie, at John Wiley & Sons. They hounded us with professional aplomb to get this book done in an ambitious schedule after author footdragging for a year or more. Jane Kinney, formerly of Van Nostrand Reinhold, which was bought by John Wiley & Sons after we signed the contract, deserves a lot of credit for talking us into agreeing to this third edition. Frances Andersen did a wonderful job of copyediting the manuscript.

We want to make particular mention of two wetland ecologists who died since our last edition. Fran Heliotis of George Mason University died much too early after some excellent contributions to wetland ecology. And Bill Niering of Connecticut College, a hero to both of us, died in late summer 1999, just after a session of what he did so well—teaching. With Bill Niering, it certainly is the case that we could see further because we were standing on the shoulders of a giant.

We appreciate the patience, help, and tolerance of our spouses, Ruthmarie and Jean, and our children who saw us a little less frequently while we stayed in our respective bunkers working on this edition.

William J. Mitsch   James G. Gosselink
Columbus, Ohio   Rock Island, Tennessee

June 2000

Part 1

Introduction

Wetlands: Human History,
Use, and Science

Wetlands are a major feature of the landscape in almost all parts of the world. Although many cultures have lived among and even depended on wetlands for centuries, the modern history of wetlands is fraught with misunderstanding and fear. Wetlands have disappeared at alarming rates throughout the developed and developing worlds. Now their many values are being recognized, and wetland protection is the norm in many parts of the world. Wetlands have properties that are not adequately covered by present terrestrial and aquatic ecology, suggesting that there is a case to be made for wetland science as a unique discipline encompassing many fields, including terrestrial and aquatic ecology, chemistry, hydrology, and engineering. Wetlands are unique because of their hydrologic conditions and their role as ecotones between terrestrial and aquatic systems. Wetland management, as the applied side of wetland science, requires an understanding of the scientific aspects of wetlands balanced with legal, institutional, and economic realities. As interest in wetlands grows, so do professional societies that are concerned with wetlands, as well as the amount of journals and literature about wetlands.

Wetlands are among the most important ecosystems on Earth. In the great scheme of things, the swampy environment of the Carboniferous period produced and preserved many of the fossil fuels on which we now depend. In more recent biological and human time periods, wetlands have been valuable as sources, sinks, and transformers of a multitude of chemical, biological, and genetic materials. Although the value of wetlands for fish and wildlife protection has been known for several decades, some of the other benefits have been identified only recently.

Wetlands are sometimes described as "the kidneys of the landscape" because they function as the downstream receivers of water and waste from both natural and human sources. They stabilize water supplies, thus ameliorating both floods and drought. They have been found to cleanse polluted waters, protect shorelines, and recharge groundwater aquifers.

Wetlands also have been called "biological supermarkets" because of the extensive food chain and rich biodiversity that they support. They play major roles in the landscape by providing unique habitats for a wide variety of flora and fauna. Now that we have become concerned about the health of our entire planet, wetlands are being described by some as carbon dioxide sinks and climate stabilizers on a global scale.

These valuable uses of wetlands are now being recognized and translated into wetland protection laws, regulations, and management plans. Wetlands have been drained, ditched, and filled throughout history but never as quickly or as effectively as was undertaken in countries such as the United States beginning in the mid-1800s. Since then, more than half of the nation's original wetlands have been drained.

More recently, wetlands have become the *cause célèbre* for conservation-minded people and organizations throughout the world, in part because they have been disappearing at alarming rates and in part because their disappearance represents an easily recognizable loss of natural areas to economic "progress." Many scientists, engineers, lawyers, and regulators are now finding it both useful and necessary to become specialists in wetland ecology and wetland management in order to understand, preserve, and even reconstruct these fragile ecosystems. This book is for these aspiring wetland specialists, as well as for those who would like to know more about the structure and function of these unique ecosystems. It is a book about wetlands—how they work and how we manage them.

Human History and Wetlands

There is no way to estimate the impact humans have had on the global extent of wetlands except to observe that, in developed and heavily populated regions of the world, the impact has ranged from significant to total. However, the importance of wetland environments to the development and sustenance of cultures throughout human history is unmistakable. Since early civilization, many cultures have learned to live in harmony with wetlands and have benefited economically from surrounding wetlands (Nicholas, 1998), whereas other cultures quickly drained the landscape. The ancient Babylonians, Egyptians, and the Aztec in what is now Mexico developed specialized systems of water delivery involving wetlands. Mexico City, in fact, is the site of a wet-land/lake that disappeared during the past 400 years as a result of human influence (Fig. 1-1). Major cities of the world, such as Chicago and Washington, DC, in the United States and Christchurch, New Zealand, and parts of Paris, France, stand on sites that were once part wetlands. Many of the large airports (in Boston, New Orleans, and J. F. Kennedy in New York, to name a few) are situated on former wetlands.

Coles and Coles (1989) refer to the people who live in proximity to wetlands and whose culture is linked to them as *wetlanders*. For example, the Camarguais of

Figure 1-1 A plan of present-day Mexico City, c. 1556, by Giovanni Battista Ramusio, *El conquistador anónimo.* (*From "Mexico—Esplendores de treinta siglos," reprinted by the Metropolitan Museum of Art. New York, and Amigos de las Artes de Mexico, Los Angeles; reprinted by permission*)

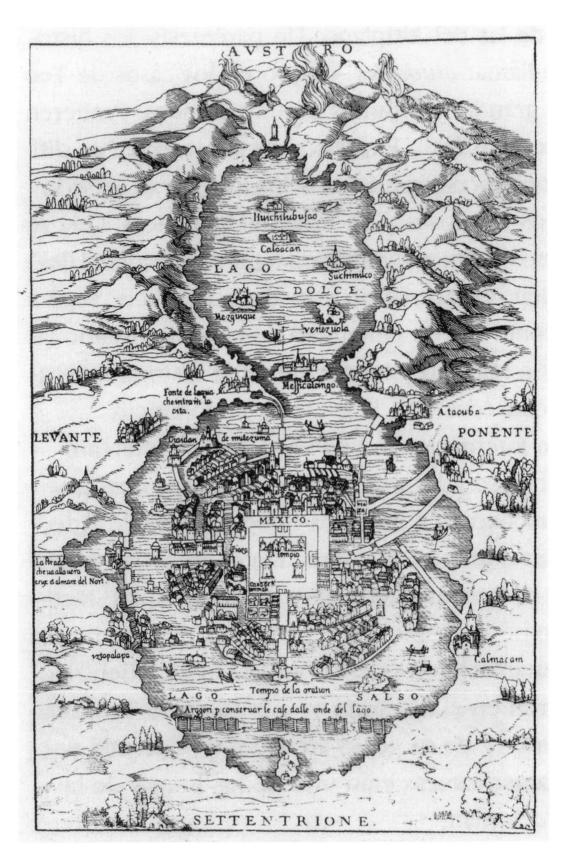

5

southern France (Fig. 1-2), the Cajuns of Louisiana (Fig. 1-3), the Marsh Arabs of southern Iraq (Fig. 1-4), and some Far Eastern cultures (Fig. 1-5) have lived in harmony with wetlands for hundreds of years. Native Americans in North America have, for centuries, harvested and reseeded wild rice (*Zizania aquatica*) along the littoral zone of lakes and streams (Fig. 1-6). The Sokaogon Chippewa in Wisconsin have a saying that "wild rice is like money in the bank." Domestic wetlands such as rice paddies feed an estimated half of the world's population (Fig. 1-7). Countless other plant and animal products are harvested from wetlands throughout the world. Many aquatic plants besides rice such as Manchurian wild rice (*Zizania latifolia*) are harvested as vegetables in China (Fig. 1-8). Cranberries are harvested from bogs, and the industry continues to thrive today in North America (Fig. 1-9). The Russians, Finns, Estonians, and Irish, among other cultures, have mined their peatlands for centuries, using peat as a source of energy (Figs. 1-10 and 1-11). *Sphagnum* peat is now harvested for horticultural purposes throughout the world. In southwestern New Zealand, for example, surface *Sphagnum* has been harvested since the 1970s for export as a potting medium (Fig. 1-12).

Wetlands can be an important source of protein. The production of fish in shallow ponds or rice paddies developed several thousands of years ago in China and Southeast Asia, and crayfish harvesting is still practiced in the wetlands of Louisiana and the Philippines. Shallow lakes and wetlands are an important provider of protein in many parts of sub-Saharan Africa (Fig. 1-13). Coastal marshes in northern Europe, the British Isles, and New England were used for centuries and are still used today for grazing of animals and hay production (Fig. 1-14), and coastal mangroves are harvested for timber, food, and tannin in many countries throughout Indo-Malaysia, East Africa, and Central and South America. Reeds and even the mud from coastal and inland marshes have been used for wall construction, fence material, and thatching for roofs in Europe, Iraq, Japan, and China (Fig. 1-15).

Figure 1-2 The Camargue region of southern France in the Rhone River delta is a historically important wetland region in Europe where Camarguais have lived since the Middle Ages. (*Photograph by Tom Nebbia, Horseshoe, North Carolina, reprinted by permission*)

Figure 1-3 A Cajun lumberjack camp in the Atchafalaya Swamp of coastal Louisiana. American Cajuns are descendants of the French colonists of Acadia (present-day Nova Scotia, Canada), who moved to the Louisiana delta in the last half of the 18th century and flourished within the bayou wetlands. (*Photograph courtesy of the Louisiana Collection, Tulane University Library, New Orleans, reprinted by permission*)

Figure 1-4 The Marsh Arabs of southern Iraq have lived for centuries on artificial islands in marshes at the confluence of the Tigris and Euphrates Rivers. (*Photograph by Nik Wheeler, Los Angeles, reprinted by permission*)

Figure 1-5 Interior wetlands in Weishan County, Shandong Province, China, where approximately 60,000 people live amid wetland–canal systems and harvest aquatic plants for food and fiber. (*Photograph by W. J. Mitsch*)

Figure 1-6 Native American "ricers" from the Sokaogon Chippewa Reservation poling and "knocking" wild rice (*Zizania aquatica*) as they have for hundreds of years on Rice Lake in Forest County, Wisconsin. (*Photograph by R. P. Gough, reprinted by permission*)

8

Figure 1-7 Rice production and water buffalo are supported by the wetland environment in many Asian countries such as at this location in Thailand. (*Photograph by Philip Moore, reprinted by permission*)

Figure 1-8 Wetland plants such as *Zizania latifolia* are harvested and sold in markets such as this one in Suzhou, Jiangsu Province, China. This and several other aquatic plants are cooked and served as vegetables in China. (*Photograph by W. J. Mitsch*)

Figure 1-9 Cranberry wet harvesting is done by flooding bogs in several regions of North America. The cranberry plant (*Vaccinium macrocarpon*) is native to the bogs and marshes of North America and was first cultivated in Massachusetts. It is now also an important fruit crop in Wisconsin, New Jersey, Washington, Oregon, and parts of Canada. (*Reproduced with the permission of Ocean Spray Cranberries, Inc., Lakeville-Middleboro, Massachusetts*)

Figure 1-10 The harvesting of peat or "turf" as a fuel has been a tradition in several parts of the world as shown by this scene of "turf carts" on Moneystown Bog in County Wicklow, Ireland, around 1990. (*Photograph by J. M. Synge; reprinted by permission courtesy John Millington Synge Trustees and Board of Trinity College, Dublin*)

Figure 1-11 Modern-day peat mining on a large scale in Estonia, near Tartu. (*Photograph by W. J. Mitsch*)

Figure 1-12 *Sphagnum* moss harvesting in Westland, South Island, New Zealand, for gardens and potting of plants. (*Photograph by C. Pugsley, copyright New Zealand Department of Conservation, Wellington, reprinted by permission*)

Figure 1-13 Humans use the wetlands of sub-Saharan Africa for sustenance as with this man fishing for lungfish (*Protopterus aethiopicus*) in Lake Kanyaboli, western Kenya. (*Photograph by K. M. Mavuti, reprinted by permission*)

Figure 1-14 Grazing of restored sea and coastal marshes by sheep is common in many parts of Europe such as this scene along the western coastline of Schleswig-Holstein, Germany. (*Photograph by W. J. Mitsch*)

12

Figure 1-15 A ''wetland house'' in the Ebro River delta region on the Mediterranean Sea, Spain. The walls are made from wetland mud and the roof is thatched with reed grass and other wetland vegetation. (*Photograph by W. J. Mitsch*)

Literary References to Wetlands

With all of these valuable uses, not to mention the aesthetics of a landscape in which water and land often provide a striking panorama, one would expect wetlands to be revered by humanity; this has certainly not always been the case. Wetlands have been depicted as sinister and forbidding, and as having little economic value throughout most of history. For example, in the *Divine Comedy*, Dante describes a marsh of the Styx in Upper Hell as the final resting place for the wrathful:

> Thus we pursued our path round a wide arc of that ghast pool,
> Between the soggy marsh and arid shore,
> Still eyeing those who gulp the marish [marsh] foul.
>
> —Dante Alighieri

Centuries later, Carl Linnaeus, crossing the Lapland peatlands, compared that region to that same Styx of Hell:

> Shortly afterwards began the muskegs, which mostly stood under water; these we had to cross for miles; think with what misery, every step up to our knees. The whole of this land of the Lapps was mostly muskeg, *hinc vocavi Styx*. Never can the priest so describe hell, because it is no worse. Never have poets been able to picture Styx so foul, since that is no fouler.
>
> —Carl Linnaeus, 1732

In the 18th century, an Englishman who surveyed the Great Dismal Swamp on the Virginia–North Carolina border and is credited with naming it described the wetland as

[a] horrible desert, the foul damps ascend without ceasing, corrupt the air and render it unfit for respiration. . . . Never was Rum, that cordial of Life, found more necessary than in this Dirty Place.

—Colonel William Byrd III (1674–1744), "Historie of the Dividing Line Betwixt Virginia and North Carolina" in *The Westover Manuscripts,* written 1728–1736, Petersburg, VA; E. and J. C. Ruffin, printers, 1841, 143 pp.

Even those who study and have been associated with wetlands have been belittled in literature:

Hardy went down to botanise in the swamp, while Meredith climbed towards the sun. Meredith became, at his best, a sort of daintily dressed Walt Whitman: Hardy became a sort of village atheist brooding and blaspheming over the village idiot.

—G. K. Chesterton (1874–1936), Chapter 2 in *The Victorian Age in Literature,* Henry Holt and Company, New York, 1913

The English language is filled with words that suggest negative images of wetlands. We get *bogged down* in detail; we are *swamped* with work. Even the mythical *bogeyman,* the character featured in stories that frighten children in many countries, may be associated with European bogs. Grendel, the mythical monster in one of the oldest surviving pieces of Old English literature and Germanic epic, *Beowulf,* comes from the peatlands of present-day northern Europe:

Grendel, the famous stalker through waste places, who held the rolling marshes in his sway, his fen and his stronghold. A man cut off from joy, he had ruled the domain of his huge misshapen kind a long time, since God had condemned him in condemning the race of Cain.

—Beowulf, translated by William Alfred, *Medieval Epics,* The Modern Library, New York, 1993

Hollywood has continued the depiction of the sinister and foreboding nature of wetlands and their inhabitants, in the tradition of Grendel, with movies such as the classic *Creature from the Black Lagoon* (1954), a comic-book-turned-cult-movie, *Swamp Thing* (1982), and its sequel *Return of the Swamp Thing* (1989). Even Swamp Thing, the man/monster depicted in Figure 1-16, evolved in the 1980s from a feared creature to a protector of wetlands, biodiversity, and the environment.

Wetland Conservation

Prior to the mid-1970s, the drainage and destruction of wetlands were accepted practices in the United States and were even encouraged by specific government policies. Wetlands were replaced by agricultural fields and by commercial and residential development (Fig. 1-17). Had those trends continued, the resource would be in danger of extinction. Only through the combined activities of hunters and anglers, scientists and engineers, and lawyers and environmentalists has the case been made for wetlands as a valuable resource whose destruction has serious economic as well as ecological and aesthetic consequences for the nations of the world (see Chapter 16). This increased level of respect in U.S. public policy was first reflected in activities such as the sale of federal "duck stamps" to waterfowl hunters that began in 1934; other countries such as New Zealand have followed suit (Fig. 1-18). Approximately 1.8 million ha of wetlands were preserved as waterfowl habitat by the U.S. duck stamp program alone through 1995. The federal government now supports a variety of

Figure 1-16 The sinister image of wetlands, especially swamps, is often promoted in popular media such as Hollywood movies and comic books, although the man-turned-plant "Swamp Thing" is now a hero as he fights injustice and even toxic pollution. (*Swamp Thing © 2000 DC Comics. All rights reserved. Used with permission.*)

Figure 1-17 Agricultural and urban development has transformed half of the wetlands in the United States to uplands since presettlement times. For example, in the Everglades in southern Florida, some of the landscape has been transformed from **a.** swamps and sloughs typical of the southern Everglades to **b.** high-density housing surrounding artificial ponds. (*a—Photograph by F. Sklar, reprinted from Mitsch et al., 1994a, p. 34, with permission from Elsevier Science;* **b**—*Photograph by W. J. Mitsch*)

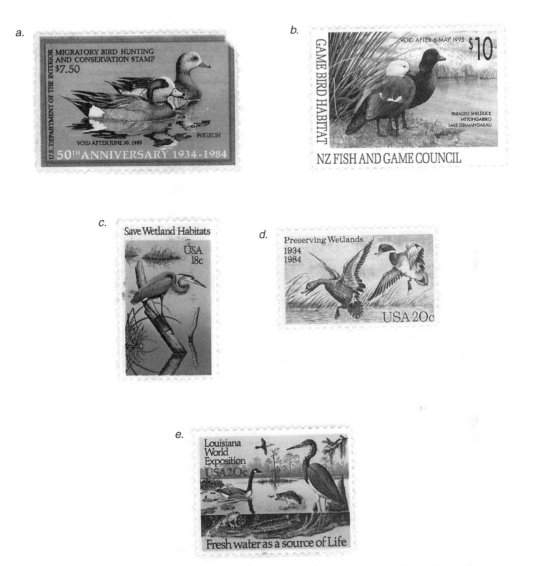

Figure 1-18 Interest in wetlands has been reflected in stamps such as: **a.** "Duck Stamps," which must be purchased by any waterfowl hunter over the age of 16 in the United States; the revenue has been used to acquire and protect wetlands used for hunting; **b.** a similar duck stamp for hunting of game birds in New Zealand; **c.** a wetland stamp as part of a habitat preservation commemorative stamp set (1979); **d.** a wetland stamp honoring the 50th anniversary of the duck stamp (1984); and **e.** a wetland panorama for the Louisiana World Exposition of 1984.

other wetland protection programs. Individual states have also enacted wetland protection laws or have used existing statutes to preserve these valuable resources. That interest, which blossomed in the late 1970s in the United States, has now spread around the world and international programs such as the Ramsar Convention have arisen out of nongovernmental organizations (NGOs) dedicated to preserving wetlands. However, as long as wetlands remain more difficult to stroll through than a forest and more difficult to cross by boat than a lake, they will remain a misunderstood ecosystem to many people.

Wetland Science and Wetland Scientists

Even after the ecological and economic benefits of wetlands were determined and became widely appreciated, wetlands have remained an enigma to scientists. They are difficult to define precisely, not only because of their great geographical extent, but also because of the wide variety of hydrologic conditions in which they are found. Wetlands are usually found at the interface of terrestrial ecosystems, such as upland forests and grasslands, and aquatic systems such as deep lakes and oceans (Fig. 1-19a),

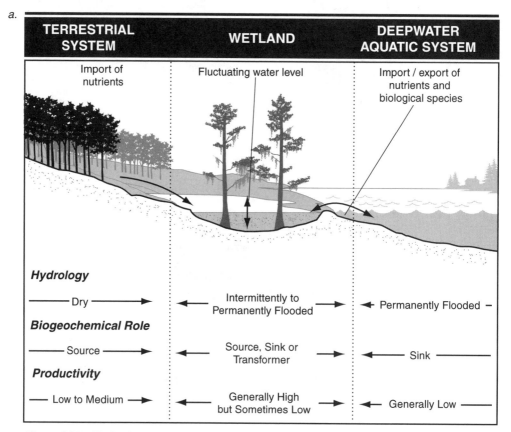

Figure 1-19 Wetlands are often located as   **a.** ecotones between dry terrestrial systems and permanently flooded deepwater aquatic systems such as rivers, lakes, estuaries, or oceans. They can also be found in the landscape as   **b.** isolated basins with little outflow and no adjacent deepwater system.

b.

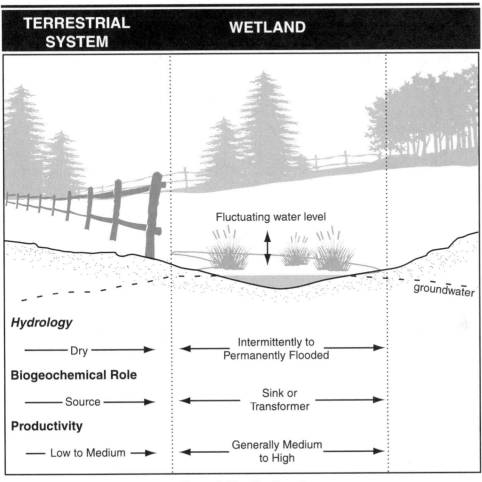

Figure 1-19 *(Continued).*

making them different from each yet highly dependent on both. They are also found in seemingly isolated situations, where the nearby aquatic system is often a groundwater aquifer (Fig. 1-19b). Because wetlands combine attributes of both aquatic and terrestrial ecosystems but are neither, they have fallen between the cracks of the scientific disciplines of terrestrial and aquatic ecology.

A specialization in the study of wetlands is often termed *wetland science* or *wetland ecology,* and those who carry out such investigations are called *wetland scientists* or *wetland ecologists.* The term *mire ecologist* has also been used. Some have suggested that the study of all wetlands be termed *telmatology* (*telma* being Greek for bog), a term originally coined to mean "bog science" (Zobel and Masing, 1987). No matter what the field is called, it is apparent that there are several good reasons for treating wetland ecology as a distinct field of study:

1. Wetlands have unique properties that are not adequately covered by present ecological paradigms and by fields such as limnology, estuarine ecology, and terrestrial ecology.

2. Wetland studies have begun to identify some common properties of seemingly disparate wetland types.

3. Wetland investigations require a multidisciplinary approach or training in a number of fields not routinely studied or combined in university academic programs.

4. There is a great deal of interest in formulating sound policy for the regulation and management of wetlands. These regulations and management approaches need a strong scientific underpinning integrated as wetland ecology.

A growing body of evidence suggests that the unique characteristics of wetlands—standing water or waterlogged soils, anoxic conditions, and plant and animal adaptations—may provide some common ground for study that is neither terrestrial ecology nor aquatic ecology. Wetlands provide opportunities for testing "universal" ecological theories and principles involving succession and energy flow, which were developed for aquatic or terrestrial ecosystems. For example, wetlands provided the setting for the successional theories of Clements (1916) and the energy flow approaches of Lindeman (1942). They also provide an excellent laboratory for the study of principles related to transition zones, ecological interfaces, and ecotones.

Wetlands have often been described as being ecotones, that is, transition zones between uplands such as forests and farmlands and deepwater aquatic systems such as rivers, lakes, and estuaries (see, e.g., Fig. 1-19a). This niche in the landscape allows wetlands to function as organic exporters or inorganic nutrient sinks. Also, this transitional position often leads to high biodiversity in wetlands, which "borrow" species from both aquatic and terrestrial systems. Rather than being simply ecotones, wetlands are ecosystems unto themselves. They have some characteristics of deepwater systems such as algae, benthic invertebrates, nekton, anoxic substrate, and water movement. On the other hand, they also have vascular plant flora similar in structure to those found in uplands. Some wetlands, because of their connections to both upland and aquatic systems, have the distinction of being among the most productive ecosystems on Earth.

Our knowledge of different wetland types such as those discussed in this book is, for the most part, isolated in distinctive literatures and scientific circles. One set of literature deals with coastal wetlands, another with forested wetlands and freshwater marshes, and still another with peatlands. Some investigators have analyzed the properties and functions common to all wetlands. This is probably one of the most exciting areas for wetland research because there is so much to be learned. Comparisons of wetland types have shown, for example, the importance of hydrologic flowthrough for the maintenance and productivity of these ecosystems. The anoxic biochemical processes that are common to all wetlands provide another area for comparative research and pose many questions: What are the roles of different wetland types in local and global biochemical cycles? How do the activities of humans influence these cycles in various wetlands? What are the synergistic effects of hydrology, chemical inputs, and climatic conditions on wetland biological productivity? How can plant and animal adaptations to anoxic stress be compared in various wetland types?

The true wetland ecologist must be an ecological generalist because of the number of sciences that bear on those ecosystems. Wetland flora and fauna are often uniquely adapted to a substrate that may vary from submerged to dry. Emergent plant species support both aquatic benthic animals and terrestrial insects. Because hydrologic conditions are so important in determining the structure and function of the wetland ecosystems, a wetland scientist should be well versed in surface and groundwater hydrology. The shallow-water environment means that chemistry—particularly for water, sediments, soils, and water–sediment interactions—is an important science. Similarly, questions about wetlands as sources, sinks, or transformers of chemicals require investigators to be versed in a number of biological and chemical techniques. The identification of wetland vegetation and animals requires botanical and zoological skills, and backgrounds in microbial biochemistry and soil science contribute significantly to the understanding of the anoxic environment. Understanding adaptations of wetland biota to the flooded environment requires backgrounds in both biochemistry and physiology. If wetland scientists are to become more involved in the management of wetlands, some engineering techniques, particularly for wetland hydrologic control or wetland creation, need to be learned. Wetlands are seldom, if ever, self-sustaining systems. Rather, they interact strongly with adjacent terrestrial and aquatic ecosystems. Hence, a holistic view of these complex landscapes can be achieved only through an understanding of the principles of ecology, especially those that are part of ecosystem and landscape ecology and systems analysis. Finally, if wetland management involves the implementation of wetland policy, then training in the legal and policy-making aspects of wetlands is warranted.

Many scientists now study wetlands. Only a relatively few pioneers, however, investigated these systems in any detail prior to the 1960s. Most of the early scientific studies dealt with classical botanical surveys or investigations of peat structure. A number of early scientific studies of peatland hydrology were also produced, particularly in Europe and Russia. Later, investigators such as Chapman, Teal, Sjörs, Gorham, Eugene and H. T. Odum, and their colleagues and students began to use modern ecosystem approaches in wetland studies (Table 1-1). Several research centers devoted

Table 1-1 Some pioneer researchers in wetland ecology and representative citations for their early work

Wetland Type and Researcher	Country	Representative Citations
COASTAL MARSHES/MANGROVES		
Valentine J. Chapman	New Zealand	Chapman (1938, 1940)
John Henry Davis	USA	Davis (1940, 1943)
John M. Teal	USA	Teal (1958, 1962)
Lawrence R. Pomeroy	USA	Pomeroy (1959)
Eugene P. Odum	USA	E. P. Odum (1961)
D. S. Ranwell	UK	D. S. Ranwell (1972)
PEATLANDS/FRESHWATER WETLANDS		
C. A. Weber	Germany	Weber (1907)
Herman Kurz	USA	Kurz (1928)
A. P. Dachnowski-Stokes	USA	Dachnowski-Stokes (1935)
R. L. Lindeman	USA	Lindeman (1941, 1942)
Eville Gorham	UK/USA	Gorham (1956, 1961)
Hugo Sjörs	Sweden	Sjörs (1948, 1950)
G. Einar Du Rietz	Sweden	Du Rietz (1949, 1954)
P. D. Moore/D. J. Bellamy	UK	Moore and Bellamy (1974)
S. Kulczynski	Poland	Kulczynski (1949)
R. S Clymo	UK	Clymo (1963, 1965)

to the study of wetlands were established in the United States, including the Sapelo Island Marine Institute in Georgia; the Center for Coastal, Energy, and Environmental Resources at Louisiana State University; the Center for Wetlands at the University of Florida; and the Pacific Estuarine Research Laboratory at San Diego State University. In addition, a professional society now exists, the *Society of Wetland Scientists,* which has among its goals to provide a forum for the exchange of ideas within wetland science and to develop wetland science as a distinct discipline. INTECOL, the *International Association of Ecology,* sponsors a major international wetland conference every four years somewhere in the world.

Wetland Managers and Wetland Management

Just as there are wetland scientists who are uncovering the processes that determine wetland functions and values, so too are there those who are involved, by choice or by vocation, in some of the many aspects of wetland management. These individuals, whom we call *wetland managers,* are engaged in activities that range from waterfowl production to wastewater treatment. They must be able to balance the scientific aspects of wetlands with a myriad of legal, institutional, and economic constraints to provide optimum wetland management. The management of wetlands has become increasingly important in many countries because government policy and wetland regulation seek to reverse historic wetland losses in the face of continuing draining or encroachment by agricultural enterprises and urban expansion. The simple act of being able to identify the boundaries of wetlands has become an important skill for a new type of wetland technician called a *wetland delineator.*

Private organizations such as *Ducks Unlimited, Inc.,* and *The Nature Conservancy* have protected wetlands by purchasing thousands of hectares of wetlands throughout North America. Through a treaty known as the *Ramsar Convention* and an agreement jointly signed by the United States and Canada in 1986 called the *North American Waterfowl Management Plan,* wetlands are now being protected primarily for their waterfowl value on an international scale. In 1988, a federally sponsored *National Wetlands Policy Forum* (1988) in the United States raised public and political awareness of wetland loss and recommended a policy of "no net loss" of wetlands. This recommendation has stimulated widespread interest in wetland restoration and creation to replace lost wetlands and "no net loss" has remained the policy of wetland protection in the United States since the late 1980s.

Subsequently, a National Research Council report in the United States (NRC, 1992) called for the fulfillment of an ambitious goal of gaining 4 million ha of wetlands by the year 2010 largely through the reconversion of crop and pasture land. Wetland creation for specific functions is an exciting new area of wetland management that needs trained specialists and may eventually stem the tide of loss and lead to an increase in this important resource. Another National Research Council report (NRC, 1995) reviewed the scientific basis for wetland delineation and classification, particularly as it related to the regulation of wetlands in the United States at that time.

The Wetland Scientific Literature

The increasing interest and emphasis on wetland science and management has been demonstrated by a veritable flood of books, reports, scientific studies, and conference proceedings, most in the last two decades of the 20th century (see Appendix A). The journal citations in this book are only the tip of the iceberg of the literature on wetlands, much of which has been published since the mid-1980s. Two journals, *Wetlands* and *Wetlands Ecology and Management,* are now published to disseminate scientific and management papers on wetlands, and several other scholarly journals frequently publish papers on wetlands. Dozens of wetland meeting proceedings and journal special issues have been published from conferences on wetlands held throughout the world. Beautifully illustrated popular books and articles with color photographs have been developed on wetlands by Niering (1985), Littlehales and Niering (1991), Mitchell et al. (1992), Kusler et al. (1994), and Rezendes and Roy (1996) on North American wetlands; by McComb and Lake (1990) on Australian wetlands; and by Finlayson and Moser (1991) and Dugan (1993) on wetlands of the world.

Government agencies and NGOs around the world have contributed significantly to the wetland literature and to our understanding of wetland functions and values. In the United States, the *U.S. Fish and Wildlife Service* has been involved in the classification and inventory of wetlands and has published a series of community profiles on various regional wetlands. The *U.S. Environmental Protection Agency* has been interested in the impact of human activity on wetlands, and in wetlands as possible systems for the control of water pollution. Along with the *U.S. Army Corps of Engineers,* the U.S. Environmental Protection Agency, especially through its Office of Wetlands, Oceans, and Watersheds (OWOW), the U.S. Fish and Wildlife Service, and the *Natural Resources Conservation Service* now are the primary wetland management agencies in the United States.

Wetland management organizations such as the *Association of State Wetland Managers* and the *Society of Wetland Scientists* focus on disseminating information on wetlands, particularly in North America. The *International Union for the Conservation of Nature and Natural Resources* (IUCN) and the *Ramsar Convention,* both based in Switzerland, have developed a series of publications on wetlands of the world. *Wetlands International* is the world's leading nonprofit organization concerned with the conservation of wetlands and wetland species. It comprises a global network of governmental and nongovernmental experts working on wetlands. Activities are undertaken in more than 120 countries worldwide. The headquarters for its Africa, Europe, Middle East (AEME) branch is located in Wageningen, The Netherlands; its Americas branch is headquartered in Montreal, Canada.

Recommended Readings

Kusler, J., W. J. Mitsch, and J. S. Larson. 1994. Wetlands. Scientific American 270(1):64–70.

Maltby, E. 1986. Waterlogged Wealth: Why Waste the World's Best Wet Places? Earthscan, Washington, DC. 200 pp.

Tiner, R. W. 1998. In Search of Swampland: A Wetland Sourcebook and Field Guide. Rutgers University Press, New Brunswick, NJ. 264 pp.

Williams, M., ed. 1990. Wetlands: A Threatened Landscape. Basil Blackwell, Oxford. 419 pp.

Definitions of Wetlands

Wetlands have many distinguishing features, the most notable of which are the presence of standing water for some period during the growing season, unique soil conditions, and organisms, especially vegetation, adapted to or tolerant of saturated soils. Wetlands are not easily defined, however, especially for legal purposes, because they have a considerable range of hydrologic conditions, because they are found along a gradient at the margins of well-defined uplands and deepwater systems, and because of their great variation in size, location, and human influence. Terms such as swamp, marsh, fen, and bog have been used in common speech for centuries and are frequently used and misused today. Formal definitions have been developed by several federal agencies in the United States, by scientists in Canada and the United States, and through an international treaty known as the Ramsar Convention. These definitions include considerable detail and are used for both scientific and management purposes. No absolute answer to "What is a wetland?" should be expected, but legal definitions involving wetland protection are becoming increasingly comprehensive.

The most common questions that the uninitiated ask about wetlands are "Now, what exactly is a wetland?" or "Is that the same as a swamp?" These are surprisingly good questions, and it is not altogether clear that they have been answered completely by wetland scientists and managers. Wetland definitions and terms are many and are often confusing or even contradictory. Nevertheless, definitions are important both for the scientific understanding of these systems and for their proper management.

In the 19th century, when the drainage of wetlands was the norm, a wetland definition was unimportant because it was considered desirable to produce uplands from wetlands by draining them. In fact, the word "wetland" did not come into common use until the mid-20th century. One of the first references to the word was in the publication *Wetlands of the United States* (Shaw and Fredine, 1956). Before that time, wetlands were referred to by the many common terms that developed in the 19th century and before, such as swamp, marsh, bog, fen, mire, and moor (see

Chapter 3). Even as the value of wetlands was being recognized in the early 1970s, there was little interest in precise definitions until it was realized that a better accounting of the remaining wetland resources was needed and definitions were necessary to achieve that inventory. When national and international laws and regulations pertaining to wetland preservation began to be written in the late 1970s, the need for precision became even greater as individuals recognized that definitions were having an impact on what they could or could not do with their land. The definition of a wetland, and by implication its boundaries (referred to as "delineation" in the United States), became important when society began to recognize the value of these systems (see Chapter 16) and began to translate that recognition into laws to protect itself from further wetland loss (see Chapter 18). However, just as an estimate of the boundary of a forest or a desert or a grassland is based on scientifically defensible criteria, so too should the definition of wetlands be based on scientific measures to as great a degree as possible. What society chooses to do with wetlands, once the definition has been chosen, remains a political decision.

Distinguishing Features of Wetlands

We can easily identify a coastal salt marsh, with its great uniformity of cordgrass and its maze of tidal creeks, as a wetland. A cypress swamp, with majestic trees festooned with Spanish moss and standing in knee-deep water, provides an unmistakable image of a wetland. A northern sphagnum bog, surrounded by tamarack trees that quake as people trudge by, is another easily recognized wetland. All of those sites have several features in common. All have shallow water or saturated soil, all accumulate organic plant material that decomposes slowly, and all support a variety of plants and animals adapted to the saturated conditions. Wetland definitions, then, often include three main components:

1. Wetlands are distinguished by the presence of water, either at the surface or within the root zone.
2. Wetlands often have unique soil conditions that differ from adjacent uplands.
3. Wetlands support vegetation adapted to the wet conditions (*hydrophytes*) and, conversely, are characterized by an absence of flooding-intolerant vegetation.

This three-level approach to the definition of wetlands is illustrated in Figure 2-1. Climate and geomorphology define the degree to which wetlands can exist, but the starting point is the *hydrology,* which, in turn, affects the *physicochemical environment,* including the soils, which, in turn, determines with the hydrology what and how much *biota,* including vegetation, is found in the wetland. This model is reintroduced and discussed in more detail in Chapter 5.

The Difficulty of Defining Wetlands

Although the concepts of shallow water or saturated conditions, unique wetland soils, and vegetation adapted to wet conditions are fairly straightforward, combining these three factors to obtain a precise definition is difficult because of a number of characteristics that distinguish wetlands from other ecosystems yet make them less easy to define:

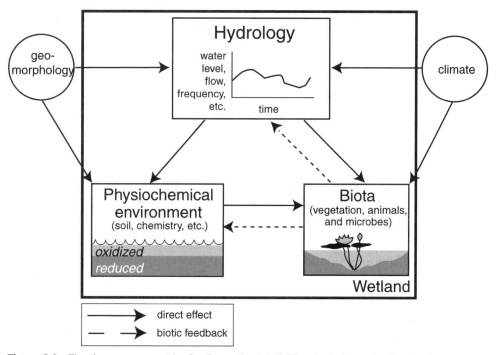

Figure 2-1 The three-component basis of a wetland definition: hydrology, physicochemical environment, and biota. From these components, the current approach to defining wetlands based on three indicators—hydrology, soils, and vegetation—is based. Note that these three components are not independent and that there is significant feedback from the biota. (*After NRC, 1995*)

1. Although water is present for at least part of the time, the depth and duration of flooding vary considerably from wetland to wetland and from year to year. Some wetlands are continually flooded, whereas others are flooded only briefly at the surface or even just below the surface. Similarly, because fluctuating water levels can vary from season to season and year to year in the same wetland type, the boundaries of wetlands cannot always be determined by the presence of water at any one time.

2. Wetlands are often located at the margins between deep water and terrestrial uplands and are influenced by both systems. This ecotone position has been suggested by some as evidence that wetlands are mere extensions of either the terrestrial or the aquatic ecosystem or both, and have no separate identity. Most wetland scientists, however, see emergent properties in wetlands not contained in either upland or deepwater systems.

3. Wetland species (plants, animals, and microbes) range from those that have adapted to live in either wet or dry conditions (*facultative*) to those adapted to only a wet environment (*obligate*), making difficult their use as wetland indicators.

4. Wetlands vary widely in size, ranging from small prairie potholes of a few hectares in size to large expanses of wetlands several hundreds of square kilometers in area. Although this range in scale is not unique to wetlands, the question of scale is important for their conservation. Wetlands can be lost in large parcels or, more commonly, one small piece at a time in a process called *cumulative loss*. Are wetlands better defined functionally on a large scale or in small parcels?

5. Wetland location can vary greatly, from inland to coastal wetlands and from rural to urban regions. Whereas most ecosystem types, for example, forests or lakes, have similar ecosystem structure and function, there are great differences among different wetland types such as coastal salt marshes, inland pothole marshes, and forested bottomland hardwoods.

6. Wetland condition, or the degree to which the wetland is influenced by humans, varies greatly from region to region and from wetland to wetland. In rural areas, wetlands are likely to be associated with farmlands, whereas wetlands in urban areas are often subjected to the impact of extreme pollution and altered hydrology associated with housing, feeding, and transporting a large population. Many wetlands can easily be drained and turned into dry lands by human intervention; similarly, altered hydrology or increased runoff can cause wetlands to develop where they were not found before. Some animals such as beavers, muskrats, and alligators can play a role in developing wetlands. Because wetlands are so easily disturbed, it is often difficult to identify them after such disturbances; this is the case, for example, with wetlands that have been farmed for a number of years.

As described by R. L. Smith (1980), "Wetlands are a half-way world between terrestrial and aquatic ecosystems and exhibit some of the characteristics of each." They form part of a continuous gradient between uplands and open water. As a result, the upper and the lower limits of wetland excursion are arbitrary boundaries in any definition. Consequently, few definitions adequately describe all wetlands. The problem of definition usually arises at the edges of wetlands, toward either wetter or drier conditions. How far upland and how infrequently should the land flood before we can declare that it is not a wetland? At the other edge, how far can we venture into a lake, pond, estuary, or ocean before we leave a wetland? Does a floating mat of vegetation define a wetland? What about a submerged bed of rooted vascular vegetation?

The frequency of flooding is the variable that has made the definition of wetlands particularly controversial. Some classifications include seasonally flooded bottomland hardwood forests, whereas others exclude them because they are dry for most of the year. Because wetland characteristics grade continuously from aquatic to terrestrial, there is no single, universally recognized definition of a wetland. This lack has caused confusion and inconsistencies in the management, classification, and inventorying of wetland systems, but, considering the diversity of types, sizes, locations, and conditions of wetlands in this country, inconsistencies should be no surprise.

Formal Definitions

Precise wetland definitions are needed for two distinct interest groups: (1) wetland scientists and (2) wetland managers and regulators. The wetland scientist is interested in a flexible yet rigorous definition that facilitates classification, inventory, and research. The wetland manager is concerned with laws or regulations designed to prevent or control wetland modification and, thus, needs clear, legally binding definitions. Because of these differing needs, different definitions have evolved for the two groups. The discrepancy between the regulatory definition of *jurisdictional wetlands* and other definitions in the United States has meant, for example, that maps developed for wetland inventory purposes cannot be used for regulating wetland development. This is a source of considerable confusion to regulators and landowners.

Scientific Definitions

Early U.S. Definition: Circular 39 Definition

One of the earliest definitions of the term *wetlands,* one that is still used today by both wetland scientists and managers, was presented by the U.S. Fish and Wildlife Service in 1956 in a publication that is frequently referred to as Circular 39 (Shaw and Fredine, 1956):

> The term "wetlands". . . refers to lowlands covered with shallow and sometimes temporary or intermittent waters. They are referred to by such names as marshes, swamps, bogs, wet meadows, potholes, sloughs, and river-overflow lands. Shallow lakes and ponds, usually with emergent vegetation as a conspicuous feature, are included in the definition, but the permanent waters of streams, reservoirs, and deep lakes are not included. Neither are water areas that are so temporary as to have little or no effect on the development of moist-soil vegetation.

The Circular 39 definition (1) emphasized wetlands that were important as water-fowl habitats and (2) included 20 types of wetlands that served as the basis for the main wetland classification used in the United States until the 1970s (see Chapter 21). It thus served the limited needs of both the wetland manager and the wetland scientist.

U.S. Fish and Wildlife Service Definition

Perhaps the most comprehensive definition of wetlands was adopted by wetland scientists in the U.S. Fish and Wildlife Service in 1979, after several years of review. The definition was presented in a report entitled *Classification of Wetlands and Deepwater Habitats of the United States* (Cowardin et al., 1979):

> Wetlands are lands transitional between terrestrial and aquatic systems where the water table is usually at or near the surface or the land is covered by shallow water . . . . Wetlands must have one or more of the following three attributes: (1) at least periodically, the land supports predominantly hydrophytes; (2) the substrate is predominantly undrained hydric soil; and (3) the substrate is nonsoil and is saturated with water or covered by shallow water at some time during the growing season of each year.

This definition was significant for its introduction of a number of important concepts in wetland ecology. It was one of the first definitions to introduce the concepts of *hydric soils* and *hydrophytes,* and it served as the impetus for scientists and managers to define these terms more accurately (NRC, 1995). Designed for the scientist as well as the manager, it is broad, flexible, and comprehensive and includes descriptions of vegetation, hydrology, and soil. It has its main utility in scientific studies and inventories and generally has been more difficult to apply to the management and regulation of wetlands. It is still frequently accepted and employed in the United States today and was, at one time, accepted as the official definition of wetlands by India. Like the Circular 39 definition, this definition serves as the basis for a detailed wetland classification and an updated and comprehensive inventory of wetlands in the United States. The classification and inventory are described in more detail in Chapter 21.

Canadian Wetland Definitions

Canadians, who deal with vast areas of inland northern peatlands, have developed a specific national definition of wetlands. At a workshop of the Canadian National

Wetlands Working Group, Zoltai (1979) defined wetlands as "areas where wet soils are prevalent, having a water table near or above the mineral soil for the most part of the thawed season, supporting a hydrophytic vegetation."

Tarnocai (1979), at the same workshop, presented the definition used in the Canadian Wetland Registry, an inventory and data bank on Canadian wetlands. That definition is similar to two definitions formally published in the book *Wetlands of Canada* by the National Wetlands Working Group (1988). First, Zoltai (1988) defined a wetland as:

> land that has the water table at, near, or above the land surface or which is saturated for a long enough period to promote wetland or aquatic processes as indicated by hydric soils, hydrophytic vegetation, and various kinds of biological activity which are adapted to the wet environment.

Zoltai (1988) notes that "wetlands include waterlogged soils where in some cases the production of plant materials exceeds the rate of decomposition." He describes the wet and dry extremes of wetlands as:

- shallow open waters, generally less than 2 m; and
- periodically inundated areas only if waterlogged conditions dominate throughout the development of the ecosystem.

Tarnocai et al. (1988) offered a slightly reworded definition in that same publication as the basis of the Canadian wetland classification system. That definition, repeated by Zoltai and Vitt (1995) and Warner and Rubec (1997) in later years, is the official definition of wetlands in Canada:

> land that is saturated with water long enough to promote wetland or aquatic processes as indicated by poorly drained soils, hydrophytic vegetation and various kinds of biological activity which are adapted to a wet environment.

These definitions emphasize wet soils, hydrophytic vegetation, and "various kinds" of other biological activity. The distinction between "hydric soils" in the Zoltai definition and "poorly drained soils" in the current, more accepted definition probably is a reflection of using hydric soils exclusively to define wetlands (see Chapter 6).

U.S. National Academy of Sciences Definition

In the early 1990s, amid renewed regulatory controversy in the United States as to what constitutes a wetland, the U.S. Congress asked the private nonprofit National Academy of Sciences to appoint a committee through its principal operating agency, the National Research Council, to undertake a scientific review of the scientific aspects of wetland characterization. Seventeen committee members met for the first time in September 1993. The committee was charged with considering: (1) the adequacy of the existing definition of wetlands; (2) the adequacy of science for evaluating the hydrologic, biological, and other ways that wetlands function; and (3) regional variation in wetland definitions. The report produced by that committee two years later was entitled *Wetlands: Characteristics and Boundaries* (NRC, 1995) and included yet another scientific definition, referred to as a "reference definition" in that it was meant to stand "outside the context of any particular agency, policy or regulation" (NRC, 1995):

A wetland is an ecosystem that depends on constant or recurrent, shallow inundation or saturation at or near the surface of the substrate. The minimum essential characteristics of a wetland are recurrent, sustained inundation or saturation at or near the surface and the presence of physical, chemical, and biological features reflective of recurrent, sustained inundation or saturation. Common diagnostic features of wetlands are hydric soils and hydrophytic vegetation. These features will be present except where specific physiochemical, biotic, or anthropogenic factors have removed them or prevented their development.

Although little formal use has been made of this definition, it remains the most comprehensively developed scientific wetland definition. It uses the terms *hydric soils* and *hydrophytic vegetation,* as did the early U.S. Fish and Wildlife Service definition, but indicates that they are "common diagnostic features" rather than absolute necessities in designating a wetland.

An International Definition

The International Union for the Conservation of Nature and Natural Resources (IUCN) at the Convention on Wetlands of International Importance Especially as Waterfowl Habitat, better known as the *Ramsar Convention,* adopted the following definition of wetlands (Navid, 1989; Finlayson and Moser, 1991):

areas of marsh, fen, peatland or water, whether natural or artificial, permanent or temporary, with water that is static or flowing, fresh, brackish, or salt including areas of marine water, the depth of which at low tide does not exceed 6 meters.

This definition, which was adopted at the first meeting of the convention in Ramsar, Iran, in 1971, states that wetlands may incorporate riparian and coastal zones adjacent to the wetlands and islands or bodies of marine water deeper than 6 m at low tide lying within the wetlands. This definition does not include vegetation or soil and extends wetlands to water depths of 6 m or more, well beyond the depth usually considered wetlands in the United States and Canada. As predicted by Navid (1989), this definition has been interpreted to include "a wide variety of habitat types including rivers, coastal areas, and even coral reefs." The rationale for such a broad definition of wetlands "stemmed from a desire to embrace all the wetland habitats of migratory water birds" (Scott and Jones, 1995). A description of Ramsar's current program and activities is included in Chapter 18.

Legal Definitions

When protection of wetlands began in earnest in the mid-1970s in the United States, there arose an almost immediate need for precise definitions that were based as much on closing legal loopholes as on science. Two such definitions have developed in U.S. agencies—one for the U.S. Army Corps of Engineers to enforce its legal responsibilities with a "dredge-and-fill" permit program in the Clean Water Act and the other for the U.S. Natural Resources Conservation Service to administer wetland protection under the so-called "swampbuster" provision of the Food Security Act. Both agencies were parties to an agreement in 1993 to work together on administering a unified policy of wetland protection in the United States. Yet the two separate definitions remain.

U.S. Army Corps of Engineers Definition

A U.S. government regulatory definition of wetlands is found in the regulations used by the U.S. Army Corps of Engineers for the implementation of a dredge-and-fill permit system required by Section 404 of the 1977 Clean Water Act Amendments. The latest version of that definition is given as follows:

> The term "wetlands" means those areas that are inundated or saturated by surface or ground water at a frequency and duration sufficient to support, and that under normal circumstances do support, a prevalence of vegetation typically adapted for life in saturated soil conditions. Wetlands generally include swamps, marshes, bogs, and similar areas. (33 CFR 328.3(b); 1984)

This definition replaced a 1975 definition that stated that "those areas that normally are characterized by the prevalence of vegetation that *requires* saturated soil conditions for growth and reproduction" (42 *Fed. Reg.* 3712X, July 19, 1977; italics added) because the Corps of Engineers found that the old definition excluded "many forms of truly aquatic vegetation that are prevalent in an inundated or saturated area, but that do not require saturated soil from a biological standpoint for their growth and reproduction." The words "normally" in the old definition and "that under normal circumstances do support" in the new definition were intended "to respond to situations in which an individual would attempt to eliminate the permit review requirements of Section 404 by destroying the aquatic vegetation . . ." (quotes from 42 *Fed. Reg.* 37128, July 19, 1977). The need to revise the 1975 definition illustrates how difficult it has been to develop a legally useful definition that also accurately reflects the ecological reality of a wetland site.

This legal definition of wetlands has been debated in the courts in several cases, some of which have become landmark cases. In one of the first court tests of wetland protection, the Fifth Circuit of the U.S. Court of Appeals ruled in 1972, in *Zabel v. Tabb,* that the U.S. Army Corps of Engineers has the right to refuse a permit for filling of a mangrove wetland in Florida. In 1975, in *Natural Resources Defense Council v. Callaway,* wetlands were included in the category "waters of the United States," as described by the Clean Water Act. Prior to that time, the Corps of Engineers regulated dredge-and-fill activities (Section 404 of the Clean Water Act) for navigable waterways only; since that decision, wetlands have been legally included in the definition of waters of the United States. In 1985, the question of regulation of wetlands reached the U.S. Supreme Court for the first time. The Court upheld the broad definition of wetlands to include groundwater-fed wetlands in *United States v. Riverside Bayview Homes, Inc.* (Want, 1994). In that case, the Supreme Court affirmed that the U.S. Army Corps of Engineers had jurisdiction over wetlands that were adjacent to navigable waters, but it left open the question as to whether it had jurisdiction over nonadjacent wetlands (NRC, 1995).

Food Security Act Definition

In December 1995, the U.S. Department of Agriculture, through its Soil Conservation Service [now known as the Natural Resources Conservation Service (NRCS)], was brought into the arena of wetland definitions and wetland protection by means of a provision known as "swampbuster" in the 1985 Food Security Act (see Chapter 18). On agricultural land in the United States that, prior to December 1995, had been exempt from regulation, wetlands were now protected. As a result of this *swampbuster provision*, a definition, known as the NRCS or Food Security Act definition, was included in the Act (16 CFR 801(a)(16); 1985):

> The term "wetland" except when such term is part of the term "converted wetland" means land that—
>> (A) has a predominance of hydric soils;
>> (B) is inundated or saturated by surface or ground water at a frequency and duration sufficient to support a prevalence of hydrophytic vegetation typically adapted for life in saturated soil conditions; and
>> (C) under normal circumstances does support a prevalence of such vegetation.
>
> For purposes of this Act and any other Act, this term shall not include lands in Alaska identified as having high potential for agricultural development which have a predominance of permafrost soils.

The emphasis on this agriculture-based definition is on hydric soils. The omission of wetlands that do not have hydric soils, while not invalidating this definition, makes it less comprehensive than some others, for example, the NRC (1995) definition. The most curious feature of this definition is its wholesale exclusion of the largest state in the United States from the definition of wetlands. The exclusion of Alaskan wetlands that have a high potential for agriculture makes this definition even less of a scientific and more of a regulatory or even political definition. There is no scientific distinction between the characteristics of Alaskan wetlands and wetlands in the rest of the United States except for climatic differences and the presence of permafrost under many but certainly not all Alaskan wetlands (NRC, 1995).

Jurisdictional Wetlands

Since 1989, the term *jurisdictional wetland* has been used for legally defined wetlands in the United States to delineate those areas that are under the jurisdiction of Section 404 of the Clean Water Act or the swampbuster provision of the Food Security Act. The Army Corps of Engineers' definition cited previously emphasizes only one indicator, vegetative cover, to determine the presence or absence of a wetland. It is difficult to include soil information and water conditions in a wetland definition when its main purpose is to determine jurisdiction for regulatory purposes and there is little time to examine the site in detail. The Food Security Act definition, on the other hand, includes hydric soils as the principal determinant of wetlands.

It is likely that most of the wetlands that are considered "jurisdictional wetlands" by the preceding two definitions fit the scientific definition of wetlands. It is also just as likely that some types of wetlands, particularly those that have less chance of developing hydric soil characteristics or hydrophytic vegetation (e.g., riparian wetlands), would not be identified as jurisdictional wetlands with the legal definitions. And, of course, excluding Alaskan wetlands "having high potential for agricultural development" from the Food Security Act definition has no scientific basis at all but is a political decision.

Those who delineate wetlands are interested in a definition that allows the rapid identification of a wetland and the degree to which it has been or could be altered. They are interested in the delineation of wetland boundaries; establishing boundaries is facilitated by defining the wetland simply, according to the presence or absence of certain species of vegetation or aquatic life or the presence of simple indicators such as hydric soils. Several U.S. federal manuals spelling out specific methodologies for identifying jurisdictional wetlands were written or proposed in the 1980s and early

1990s. The manuals differed, however, in the prescribed ways these three criteria are proved in the field. The first of these manuals, written in 1987, is now the one most widely used to field-identify wetlands in the United States. These aspects of wetland management are discussed in more detail in Chapter 18. All three manuals indicated that the three criteria for wetlands, namely, wetland hydrology, wetland soils, and hydrophytic vegetation, must be present. However, as illustrated in Figure 2-1, these three variables are not independent; strong evidence of long-term wetland hydrology, for example, should almost ensure that the other two variables are present. Furthermore, there are potentially other indicators of the physicochemistry and biota beyond hydric soils and hydrophytic vegetation that may one day serve as useful indicators of wetlands.

Choice of a Definition

A wetland definition that will prove satisfactory to all users has not yet been developed because the definition of wetlands depends on the objectives and the field of interest of the user. Different definitions can be formulated by the geologist, soil scientist, hydrologist, biologist, ecologist, sociologist, economist, political scientist, public health scientist, and lawyer. This variance is a natural result of the differences in emphasis in the definer's training and a result of the different ways in which individual disciplines deal with wetlands. For ecological studies and inventories, the 1979 U.S. Fish and Wildlife Service definition has been and should continue to be applied to wetlands in the United States. When wetland management, particularly regulation, is necessary, the U.S. Army Corps of Engineers' definition, as modified, is probably most appropriate. Just as important as the precision of the definition of a wetland, however, is the consistency with which it is used. That is the difficulty we face when science and legal issues meet, as they often do, in resource management questions such as wetland conservation versus wetland drainage. Applying a comprehensive definition in a uniform and fair way requires a generation of well-trained wetland scientists and managers armed with a fundamental understanding of the processes that are important and unique to wetlands.

Recommended Readings

Bosselman, F. P. 1996. Limitations inherent in the title to wetlands at common law. Stanford Law Journal 15:247–337.

Environmental Defense Fund and World Wildlife Fund. 1992. How Wet Is a Wetland? The Impact of the Proposed Revisions to the Federal Wetlands Delineation Manual. Environmental Defense Fund and World Wildlife Fund, Washington, DC. 175 pp.

National Research Council (NRC). 1995. Wetlands: Characteristics and Boundaries. National Academy Press, Washington, DC. 306 pp.

Wetlands of the World

The extent of the world's wetlands is generally thought to be from 7 to 9 million km², or about 4 to 6 percent of the land surface of the Earth. The loss of wetlands in the world is difficult to determine, but it is probably on the order of 50 percent in the lower 48 states of the United States, with much higher rates of loss in Europe and parts of Australia, Canada, and Asia. The historical terminology of wetlands has been confusing and often contradictory. Terms such as billabong, bog, bottomland, carr, fen, lagoon, mangal, marsh, mire, moor, muskeg, playa, pocosin, pothole, reedswamp, slough, swamp, and vernal pool in the English language give great confusion to the categorization of wetlands of the world. We describe a number of wetlands of particular interest from around the world with regard to their extent, uniqueness, and other unusual attributes. They include coastal and inland deltas, riverine wetlands, salt marshes and mangroves, freshwater marshes, and peatlands. Almost all are impacted by human activities, yet most remain functional ecosystems.

The Global Extent of Wetlands

Wetlands include the swamps, bogs, marshes, mires, fens, and other wet ecosystems found throughout the world. They are found on every continent except Antarctica and in every clime, from the Tropics to the tundra (Fig. 3-1). Any estimate of the extent of wetlands in the world is difficult and depends on the definition used (see Chapter 2 for formal definitions and Chapter 4 for classifications used in this book), as well as the pragmatic difficulty of quantifying wetlands in aerial and satellite images that are now the most common sources of data.

Based on several estimates, the extent of the world's wetlands is generally thought to be from 7 to 9 million km² (Table 3-1), or about 4 to 6 percent of the land surface of the Earth. Maltby and Turner (1983), based on the work of Russian geographers, estimated that more than 6.4 percent of the land surface of the world, or 8.6 million km², is wetland (Table 3-1). Almost 56 percent of this estimated total wetland area is found in tropical (2.6 million km²) and subtropical (2.1 million km²) regions.

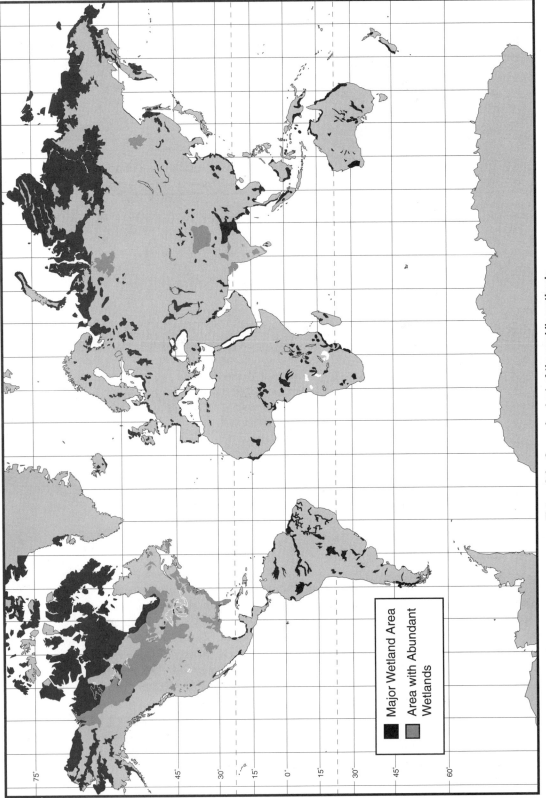

Figure 3-1 General extent of the world's wetlands.

Major Wetland Area

Area with Abundant Wetlands

Table 3-1 Comparison of several estimates of extent of wetlands in the world by climatic zone

| Zone[a] | Wetland Area ($\times 10^6$ km^2) | | | |
	Maltby and Turner (1983)[b]	Matthews and Fung (1987)	Aselmann and Crutzen (1989)	Gorham (1991)
Polar/boreal	2.8	2.7	2.4	3.5
Temperate	1.0	0.7	1.1	—
Subtropical/tropical	4.8	1.9	2.1	—
Rice paddies	—	1.5	1.3	—
Total wetland area	8.6	6.8	6.9	—

[a] Definitions of polar, boreal, temperate, and tropical vary among studies.
[b] Based on Bazilevich et al. (1971).
Source: Mitsch et al. (1994) and Mitsch (1996, 1998b).

Wetlands occupy 1 million km^2 in subboreal regions (called temperate in Table 3-1), 2.6 million km^2 in boreal regions, and 0.2 million km^2 in polar regions.

Other estimates of the global extent of wetlands have been developed from studies of the role that wetlands play in global biogeochemical cycles, especially with regard to greenhouse gases. Using global digital databases (1 degree resolution), Matthews and Fung (1987) estimated that there are 5.3 million km^2 of wetlands in the world, with a higher percentage of wetlands being boreal and a far lower percentage of wetlands being subtropical and tropical than those estimated by Maltby and Turner (1983). Aselmann and Crutzen (1989, 1990) estimated that there are 5.6 million km^2 of natural wetlands in the world, with a higher amount and percentage of wetlands in the temperate region than given in either of the earlier estimates. They used regional wetland surveys and monographs rather than maps, which Matthews and Fung (1987) used to make their estimate. These two research groups estimated the coverage by rice paddies but did not include this in their total wetland area. Matthews et al. (1991) estimated 1.5 million km^2 of rice paddies, whereas Aselmann and Crutzen (1989) estimated 1.3 million km^2 of rice paddies. By including rice fields, their estimates of the extent of the world's wetlands are 6.8 and 6.9 million km^2, respectively (Table 3-1). The Matthews–Fung estimate divided wetlands into five classes: (1) forested bogs, (2) nonforested bogs, (3) forested swamps, (4) nonforested swamps, and (5) alluvial formations. They estimated that about half of the world's natural wetlands are in the boreal region between 50 and 70° N; 90 percent of the wetlands in this region was defined by them as bogs (Matthews, 1990). Aselmann and Crutzen's (1989, 1990) wetlands categories included: (1) bogs, (2) fens, (3) swamps, (4) marshes, and (5) floodplains. Bogs and fens accounted for about 60 percent of the world's wetlands (3.35 km^2), an estimate that is very close to the 3.46 million km^2 of northern boreal and subarctic peatlands given by Gorham (1991). Aselmann and Crutzen (1989) described bogs and fens as also occurring in both temperate (40–50° N) and tropical latitudes. Both Matthews and Fung (1987) and Aselmann and Crutzen (1989) showed a much lower extent of wetlands in tropical and subtropical regions than did Maltby and Turner (1983), although definitions of zones differ. By way of comparison, a total of 750,000 km^2 of wetlands has been registered with the Convention on Wetlands of International Importance as of 2000. This represents about 8 to 10 percent of the world's total wetlands.

A more recent estimate of the world's wetlands was developed by the U.S. Department of Agriculture (USDA), which classified wetlands as: (1) inland, (2) riparian or ephemeral, (3) organic, (4) salt affected, and (5) permafrost affected (Eswaran and Reich, 1996). Using a different cartographic technique, the USDA determined that 13.7 percent (18.8 million km^2) of the Earth's surface is wetland, more than twice the estimates given in Table 3-1. About 14 percent of Canada, a wetland-rich country, was estimated to be wetland by Canada's National Wetlands Working Group (1988) (see Chapter 4). Therefore, it is unlikely that the USDA estimate is correct, as this would assume that the Canadian average should be used for the entire world.

Wetland Losses

The rate at which wetlands are being lost on a global scale is unknown. There are too many vast areas of wetlands where accurate records have not been kept, and many wetlands in the world were drained centuries ago. It is probably safe to assume that we are still losing wetlands at a fairly rapid rate globally and that we have perhaps lost as much as 50 percent of the original wetlands on the face of the Earth (Dugan, 1993). There are a number of areas where the loss rate has been documented (Table 3-2). Threats to wetlands take many forms (Williams, 1990; Dugan, 1993; Table 3-3). The estimate of about 50 percent loss of wetlands since European settlement in the lower 48 states is fairly accurate (see Chapter 4). Likewise, the 90 percent loss of wetlands in New Zealand is well documented. Several regions of the world, for

Table 3-2 Percentage loss of wetlands in various geographic locations in the world

Location	Percentage Loss	Reference
NORTH AMERICA		
United States	53	Dahl (1990)
Canada		National Wetlands Working Group
Atlantic tidal and salt marshes	65	(1988)
Lower Great Lakes–St. Lawrence River	71	
Prairie potholes and sloughs	71	
Pacific coastal estuarine wetlands	80	
AUSTRALASIA		
Australia	>50	Australian Nature Conservation
Swan Coastal Plain	75	Agency (1996)
Coastal New South Wales	75	
Victoria	33	
River Murray basin	35	
New Zealand	>90	Dugan (1993)
Philippine mangrove swamps	67	Dugan (1993)
CHINA	60	Lu (1995)
EUROPE	>90	Estimate

Source: Mitsch (1998b).

Table 3-3 Causes of wetland losses and degradation in the world[a]

Cause	Estuaries	Floodplains	Freshwater Marshes	Lakes/Littoral Zone	Peatlands	Swamp Forest
HUMAN ACTIONS—DIRECT						
Agriculture, forestry, mosquito control drainage	xx	xx	xx	x	xx	xx
Stream channelization and dredging; flood control	x		x			
Filling—solid-waste disposal; roads; development	xx	xx	xx	x		
Conversion to aquaculture/mariculture	xx					
Dikes, dams, seawall, levee construction	xx	x	x	x		
Water pollution—urban and agricultural	xx	xx	xx	xx		
Mining of wetlands of peat and other materials	x	x		xx	xx	xx
Groundwater withdrawal		x	xx			
HUMAN ACTIONS—INDIRECT						
Sediment retention by dams and other structures	xx	xx	xx			
Hydrologic alteration by roads, canals, etc.	xx	xx	xx	xx		
Land subsidence due to groundwater and other resource extraction	xx	xx	xx			
NATURAL CAUSES						
Subsidence	x			x	x	x
Sea-level rise	xx					xx
Drought	xx	xx	xx	x	x	x
Hurricanes and other storms	xx	x			x	x
Erosion	xx	x			x	
Biotic effects		xx	xx	xx		

[a] xx, common and important cause of wetland loss and degradation; x, present but not a major cause of wetland loss and degradation. Blank indicates that effect is generally not present except in exceptional situations
Source: Dugan (1993).

example, Europe and parts of Australia, Canada, and China, have lost considerably more, and there are probably other regions that have lost less. The loss rate of 60 percent from China in Table 3-2 is based on the present estimate of 250,000 km^2 of natural wetlands in the country out of a total of 620,000 km^2, including artificial wetlands (rice paddies, etc.) (Lu, 1995).

With over 70 percent of the world's population living on or near coastlines, coastal wetlands have long been destroyed through a combination of excessive harvesting, hydrologic modification and seawall construction, coastal development, pollution, and other human activities. Likewise, inland wetlands have been continually affected, particularly through hydrologic modification and agricultural and urban development. The loss of inland wetlands is the result of a number of causes, the most notable being drainage for agriculture, forestry, and mosquito control; filling for residential, commercial, and industrial development; filling for solid-waste disposal; and mining of peat (Table 3-3).

The propensity in the East was not to drain valuable wetlands entirely, as has been done in the West, but to work within the aquatic landscape, albeit in a heavily managed way. Dugan (1993) makes the interesting comparison between *hydraulic civilizations* (European in origin) that controlled water flow through the use of dikes, dams, pumps, and drainage tile, partially because water was only seasonally plentiful, and *aquatic civilizations* (Asian in origin) that better adapted to their surroundings of water-abundant floodplains and deltas and took advantage of nature's pulses such as flooding. It is because the former approach of controlling nature rather than working with it is so dominant around the world today that we find such high losses of wetlands worldwide.

Wetland Terms and Types

A number of common terms such as swamp, marsh, and mire have been used over the years to describe different types of wetlands (Table 3-4). The history of the use and misuse of these words has often revealed a decidedly regional or at least continental origin. Although the lack of standardization of terms is confusing, many of the old terms are rich in meaning to those familiar with them. They often bring to mind vivid images of specific kinds of ecosystems that have distinct vegetation, animals, and other characteristics. Each of the terms has a specific meaning to some, and many are still widely used by both scientists and laypersons alike. A *marsh* is known by most as a herbaceous plant wetland. A *swamp,* on the other hand, has woody vegetation, either shrubs or trees. There are subtle differences among marshes. A marsh with significant (>30 cm) standing water throughout much of the year is often called a *deepwater marsh.* A shallow marsh with waterlogged soil or shallow standing water is sometimes referred to as a *sedge meadow* or a *wet meadow.* Intermediate between a marsh and a meadow is a *wet prairie.* Several terms are used to denote peat-accumulating systems. The most general term is *peatland,* which is generally synonymous with *moor* and *muskeg.* There are many types of peatlands, however, the most general being *fens* and *bogs.*

Within the international scientific community, these common terms do not always convey the same meaning relative to a specific type of wetland. In fact, some languages have no direct equivalents for certain kinds of wetlands. The word *swamp* has no direct equivalent in Russian because there are few forested wetlands there. On the other hand, *bog* can easily be translated because bogs are a common feature of the Russian landscape. The word *swamp* in North America clearly refers to a wetland

Table 3-4 Common terms used for various wetland types in the world

Billabong—Australian term for a riparian wetland that is periodically flooded by the adjacent stream or river.

Bog—A peat-accumulating wetland that has no significant inflows or outflows and supports acidophilic mosses, particularly *Sphagnum*.

Bottomland—Lowland along streams and rivers, usually on alluvial floodplains, that is periodically flooded. When forested, it is called a bottomland hardwood forest in the southeastern and eastern United States.

Carr—Term used in Europe for forested wetlands characterized by alders (*Alnus*) and willows (*Salix*). Cumbungi swamp—Cattail (*Typha*) marsh in Australia.

Fen—A peat-accumulating wetland that receives some drainage from surrounding mineral soil and usually supports marshlike vegetation.

Lagoon—Term frequently used in Europe to denote a deepwater enclosed or partially opened aquatic system, especially in coastal delta regions.

Mangal—Same as mangrove.

Mangrove—Subtropical and tropical coastal ecosystem dominated by halophytic trees, shrubs, and other plants growing in brackish to saline tidal waters. The word "mangrove" also refers to the dozens of tree and shrub species that dominate mangrove wetlands.

Marsh—A frequently or continually inundated wetland characterized by emergent herbaceous vegetation adapted to saturated soil conditions. In European terminology, a marsh has a mineral soil substrate and does not accumulate peat. See also Tidal freshwater marsh and salt marsh.

Mire—Synonymous with any peat-accumulating wetland (European definition); from the Norse word "myrr." The Danish and Swedish word for peatland is now "mose."

Moor—Synonymous with peatland (European definition). A highmoor is a raised bog; a lowmoor is a peatland in a basin or depression that is not elevated above its perimeter. The primitive sense of the Old Norse root is "dead" or barren land.

Muskeg—Large expanse of peatlands or bogs; particularly used in Canada and Alaska.

Oxbow—Abandoned river channel, often developing into a swamp or marsh.

Pakihi—Peatland in southwestern New Zealand dominated by sedges, rushes, ferns, and scattered shrubs. Most pakihi form on terraces or plains of glacial or fluvial outwash origin and are acid and exceedingly infertile.

Peatland—A generic term of any wetland that accumulates partially decayed plant matter (peat).

Playa—An arid- to semiarid-region wetland that has distinct wet and dry seasons. Term used in the southwest United States for marshlike ponds similar to potholes, but with a different geologic origin.

Pocosin—Peat-accumulating, nonriparian freshwater wetland, generally dominated by evergreen shrubs and trees and found on the southeastern Coastal Plain of the United States. The term comes from the Algonquin for "swamp on a hill."

Pothole—Shallow marshlike pond, particularly as found in the Dakotas and central Canadian provinces, the so-called "prairie pothole" region.

Raupo swamp—Cattail (*Typha*) marsh in New Zealand.

Reedmace swamp—Cattail (*Typha*) marsh in the UK.

Reedswamp—Marsh dominated by *Phragmites* (common reed); term used particularly in Europe.

Riparian ecosystem—Ecosystem with a high water table because of proximity to an aquatic ecosystem, usually a stream or river. Also called bottomland hardwood forest, floodplain forest, bosque, riparian buffer, and streamside vegetation strip.

Salt marsh—A halophytic grassland on alluvial sediments bordering saline water bodies where water level fluctuates either tidally or nontidally.

Sedge meadow—Very shallow wetland dominated by several species of sedges (e.g., *Carex, Scirpus, Cyperus*).

Slough—An elongated swamp or shallow lake system, often adjacent to a river or stream. A slowly flowing shallow swamp or marsh in the southeastern United States (e.g., cypress slough). From the Old English word "sloh" meaning a watercourse running in a hollow.

Swamp—Wetland dominated by trees or shrubs (U.S. definition). In Europe, forested fens and wetlands dominated by reed grass (*Phragmites*) are also called swamps (see Reedswamp).

Tidal freshwater marsh—Marsh along rivers and estuaries close enough to the coastline to experience significant tides by nonsaline water. Vegetation is often similar to nontidal freshwater marshes.

Table 3-4 *(Continued).*

Vernal pool—Shallow, intermittently flooded wet meadow, generally typical of Mediterranean climate with dry season for most of the summer and fall. Term is now used to indicate wetlands temporarily flooded in the spring throughout the United States.

Wad (pl. Wadden)—Unvegetated tidal flat originally referring to the northern Netherlands and northwestern German coastline. Now used throughout the world for coastal areas.

Wet meadow—Grassland with waterlogged soil near the surface but without standing water for most of the year.

Wet prairie—Similar to a marsh, but with water levels usually intermediate between a marsh and a wet meadow.

---

dominated by woody plants—shrubs or trees. In Europe, *reedswamps* are dominated by reed grass (*Phragmites*), a dense-growing but nonwoody plant. In Africa, what would be called a *marsh* in the United States is referred to as a *swamp*. A cutoff meander of a river is called a *billabong* in Australia (Shiel, 1994) and an *oxbow* in North America.

Even common and scientific names of plants and animals can become confusing on a global scale. *Typha* spp., a cosmopolitan wetland plant, is called *cattail* in America, *reedmace* in the United Kingdom, *cumbungi* in Australia, and *raupo* in New Zealand. It is also referred to as *bulrush* in New Zealand. True bulrush is still called *Scirpus* spp. in North America and *Schoenoplectus* spp. in much of the rest of the world. *Scirpus fluviatilis* (river bulrush) is *Bolboschoenus fluviatilis* in much of the rest of the world. The Great Egret in North America is *Casmerodius albus,* whereas the Great Egret in Australia is *Ardea alba*. To further complicate matters, the Australian version of the Great Egret is called *Egretta alba* in New Zealand and is not called an egret at all but a White Heron.

Table 3-5 illustrates the confusion in terminology that occurs because of different regional or continental uses of terms for similar types of wetlands. In North American terminology, nonforested inland wetlands are often casually classified either as peat-forming, low-nutrient acid bogs or as marshes. European terminology, which is much older, is also much richer and distinguishes at least four different kinds of freshwater wetlands—from mineral-rich reed beds, called reedswamps, to wet grassland marshes, to fens, and, finally, to bogs or moors. To some (e.g., Moore, 1984), all of these wetland types are considered *mires*. According to others, mires are limited to peat-building wetlands. The European classification is based on the amount of surface

Table 3-5   Comparison of terms used to describe similar inland nonforested freshwater wetlands

North American terminology	◄———— Marsh or fen ————►	◄——— Bog ———►	
European terminology	◄ Swamp ► ◄ Marsh ► ◄ Fen ►	◄——— Bog ———►	
Characteristics			
Vegetation	◄ Reeds ► ◄ Grasses and sedges ►	◄——— Mosses ———►	
Hydrology	◄——— Rheotrophic ———►	◄——— Ombrotrophic ———►	
Soil	◄——— Mineral ———►	◄——— Peat ———►	
pH	◄——— Roughly neutral ———►	◄——— Acid ———►	
Trophic state	◄ Eutrophic ► ◄ Mesotrophic ►	◄——— Oligotrophic ———►	

water and nutrient inflow (rheotrophy), type of vegetation, pH, and peat-building characteristics.

Two points can be made about the use of common terms in classifying wetland types: First, the physical and biotic characteristics grade continuously from one of these wetland types to the next; hence, any classification based on common terms is, to an extent, arbitrary. Second, the same term may refer to different systems in different regions. The common terms continue to be used, even in the scientific literature; we simply suggest that they be used with caution and with an appreciation for an international audience.

Major Wetland Regions of the World

Some of the wetland areas around the world, unique for their extent, variety, and historical importance, are briefly discussed in this chapter and are located in Figure 3-2. North American wetlands shown in the figure are discussed in the next chapter.

Central and South America

Central American Freshwater Tidal Marshes

Rivers on the Pacific Coast side of Central America are shorter and more seasonal than their counterparts on the Caribbean side of the isthmus (Dugan, 1993). As a result of this and the prevailing climate, wetlands near the Pacific Coast tend to be very seasonal with wet summers and very dry winters. One of the most important wetlands in this Central American setting for many years was a seasonal, freshwater marsh at the Palo Verde National Park in Costa Rica. The 500-ha tidal freshwater marsh receives rainwater and overflow water from the Tempisque River during the wet season, which, in turn, discharges to the Gulf of Nicoya about 20 km downstream of the wetland on Costa Rica's Pacific Coast. The marsh dries out almost completely by March during the dry season. It was habitat for about 60 resident and migratory birds (McCoy and Rodriguez, 1994). Thousands of migrating Black-bellied Whistling Ducks and Blue-winged Teal and hundreds of Northern Shoveler, American Wigeon, and Ring-necked Ducks visited the wetland during the dry season. More recently, the marsh has been almost completely taken over by cattail (*Typha domingensis*), which covered 95 percent of the marsh by the late 1980s. This is a common problem with marsh managers throughout the world, where clonal dominants such as *Typha* tend to choke off any other vegetation and make a poor habitat for many waterfowl and other birds. Curiously, the diversity of birds was partially maintained because of cattle grazing, which was permitted until 1980. Site managers have recently tried to reintroduce cattle grazing, burning, disking, below-water mowing, and mechanical crushing to control the *Typha*. The only method that was consistently successful was crushing the cattails (McCoy and Rodriguez, 1994).

The Orinoco River Delta

The Orinoco River delta of Venezuela (Fig. 3-3) was explored by Columbus during one of his early voyages. It covers 36,000 km² and is dominated along its brackish shoreline by magnificent mangrove forests (Fig. 3-4). The Orinoco Delta economy is based on cattle ranching, with the cattle being shipped out during the high-water season, as well as on cacao production and palm heart canning. The delta's indigenous

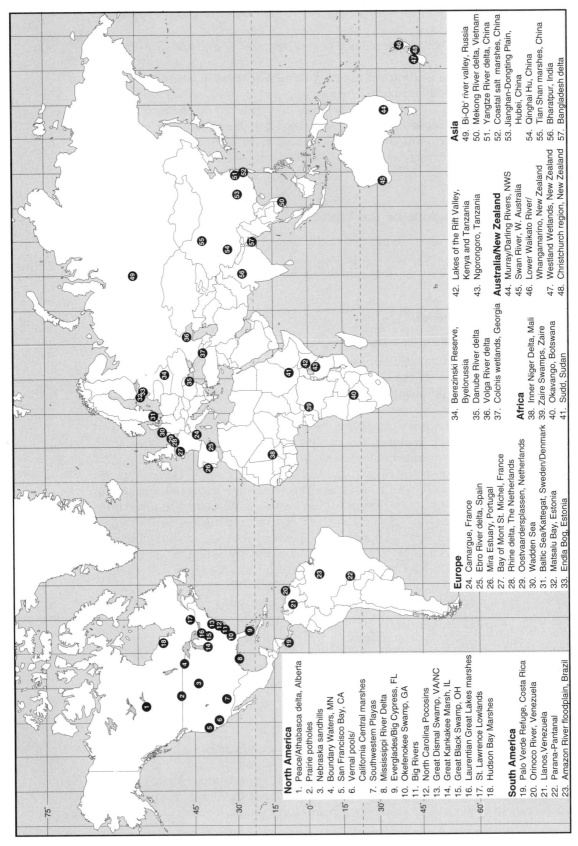

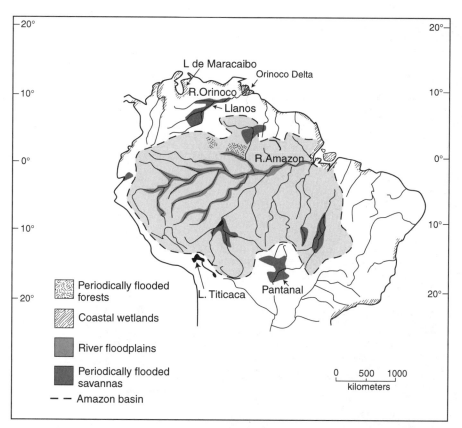

Figure 3-3 Major wetland areas of tropical South America. (*After Junk, 1993*)

population practices subsistence farming and fishing and exports salted fish to the population centers bordering the region (Dugan, 1993). Although some regions are protected and some conservation efforts have been made by government and industry, grazing and illegal hunting have been detrimental to the area's flora and fauna.

The Llanos

The western part of the Orinoco River basin in western Venezuela and northern Colombia (Fig. 3-3) is a very large (450,000 km^2) sedimentary basin called the Llanos. This region represents one of the largest inland wetland areas of South America (Dugan, 1993). The Llanos has a winter wet season coupled with a summer dry season, which causes it to be a wetland dominated by savanna grasslands and scattered palms rather than floodplain forests typical of the Orinoco Delta (Junk, 1993). The region is an important wading-bird habitat and is rich with such animals as the caiman (*Caiman* sp.) and the red piranha (*Serrasalmus nattereri*) (Dugan, 1993).

Figure 3-2 Location of major international wetlands described in this book. Significant North American wetlands are described in Chapter 4; all others are discussed or mentioned in this chapter.

Figure 3-4   Mangroves of the Orinoco River delta in Venezuela. (*Reprinted from Mitsch et al., 1994a, Fig. 15, p. 19, with permission from Elsevier Science*)

The Pantanal

One of the largest regional wetlands in the world is the Gran Pantanal of the Paraguay–Paraná River basin and Mato Grosso and Mato Grosso do Sul, Brazil (Por, 1995), located almost exactly in the geographic center of South America (Fig. 3-3). The wetland complex is 140,000 km², four times the size of the Florida Everglades with about 131,000 km² of that area flooded annually (Hamilton et al., 1996). The annual period of flooding (called the *cheia*) from March through May supports luxurious aquatic plant and animal life and is followed by a dry season (called the *seca*) from September through November when the Pantanal reverts to vegetation typical of dry savannas. There are also specific terms for the period of rising waters (*enchente*) from December through February and the period of declining waters (*vazante*) from June through August (Heckman, 1994). There is also an asynchronous pattern to flooding in the Pantanal; while maximum rainfall and upstream flows occur in January, water stage does not peak until May in the downstream reaches (Hamilton et al., 1996).

Just as in the Everglades cycle of wet and dry seasons, the biota spread across the landscape during the wet season and concentrate in fewer wet areas in a food chain frenzy during the dry season. Even though the Pantanal is one of the least known regions of the globe, it is legendary for its bird life (Fig. 3-5). Por (1995) reports that the Pantanal is potentially home to about 650 to 700 species of birds. There are 13 species of herons and egrets, 3 stork species, 6 ibis and spoonbill species, 6 duck species, 11 rail species, and 5 kingfisher species. Wetland birds also include the Anhinga and the magnificent symbol of the Pantanal, the Jabiru, the largest flying bird of the

Figure 3-5 The seasonally flooded Pantanal of South America is a haven to 650 to 700 species of birds. Shown here are egrets, herons, and the jabiru (*Jabiru mycteria*), intermixed with cattle, during the dry period. (*Photograph by W. J. Mitsch*)

Western Hemisphere. In addition, the wetland supports abundant populations of the *jacaré,* or caimin (Fig. 3-6), relative of the North American crocodile, and the largest rodent the capybara (*Hydrochoerus hydrochaeris*).

The threats to the Pantanal are many, but, until recently, there was a semibalance between human use of the Pantanal region, particularly for cattle ranching during the dry season, and the ecological functions of the region. The ecological health of the Pantanal, however, is in a state of developmental uneasiness. Some of the rivers are polluted with metals, particularly mercury, from gold-mining activity and by agrochemicals from farms. Although the Pantanal provides tourist revenues, it is also the site of illegal wildlife trafficking and cocaine smuggling. In such a vast and remote wetland, law enforcement is physically difficult and prohibitively expensive. The Hidrovia Project, an international project to develop the Paraguay and Paraná Rivers into a 3,340-km waterway, was agreed to by five South American countries in 1988 at the First International Meeting for the Development of the Paraguay–Paraná Waterway in Campo Grande, Brazil. Since then, there has been a flurry of activity by

Figure 3-6   The *jacaré,* or caimin (*Caimin yacare*), of the Pantanal. (*Photograph by W. J. Mitsch*)

both developers and environmentalists on this project. The political and economic will to see this project completed is as great as the uncertainties of the effects that this massive project will have on the Pantanal (Gottgens et al., in press).

The Amazon

Vast wetlands are found along many of the world's rivers, well before they reach the sea, especially in tropical regions. The Amazon River in Brazil is one of the best examples (Fig. 3-3). It is considered one of the world's major rivers, with a flow that results in about one-sixth to one-fifth of all the fresh water in the world (Junk, 1993). Deforestation from development threatens many Amazon aquatic ecosystems and has great social ramifications for people displaced in the process. Some of the floodplain forested wetlands of the Amazon, which are estimated to cover about 300,000 km², undergo flooding with flood levels reaching 5 to 15 m or more (Fig. 5-3n). During the flood season, it is possible to boat around the canopy of trees (Fig. 3-7).

Europe

Mediterranean Sea Deltas

The saline deltaic marshes of the mostly tideless Mediterranean Sea are among the most biologically rich in Europe. The Rhone River delta created France's most important wetland, the Camargue (Fig. 3-8; see also Fig. 1-2), an expanse of wetlands centered around the 9,000-ha Étang du Vaccarès. This land is home to the free-roaming horses celebrated in literature and film; here, too, is a species of bull that inhabited Gaul several

Figure 3-7 When the Amazon River is flooded annually, it is possible to boat around the treetops. (*Photograph by W. Junk, reprinted by permission*)

Figure 3-8   **The Camargue of the Rhone River delta in southern France is highly affected by a Mediterranean climate of hot, dry summers and cool, wet winters.** (*Photograph by W. J. Mitsch; reprinted from Mitsch et al., 1994a, p. 13, Fig. 6, with permission from Elsevier Science*)

thousand years ago before being driven south by encroaching human settlements. The Camargue is also home to one of the world's 25 major flamingo nesting sites, and France's only such site. The sense of mystery and the feeling for space and freedom pervading the Camargue are linked, of course, with the gypsies, who have gathered at Les Saintes-Maries-de-la-Mer since the 15th century, as well as with the Camarguais cowboys, the *gardians,* who ride their herds over the lands (Fig. 1-2). Aquatic plants and plant communities differ distinctly from those of northern Europe or tropical Africa, as the landscape transitions from dune to lagoon, to marshland, to grassland, and then to forest. Current set-aside agricultural policies in Europe call for restoration of some of the rice fields in the Camargue (Mesléard et al., 1995).

A principal delta on the Spanish Mediterranean coast is the Ebro Delta (Fig. 3-9), located halfway between Barcelona and Valencia and fed by the Ebro River, which flows hundreds of kilometers through arid landscape to the sea. The delta itself, covered with extensive and ancient rice paddies, also has salt marshes dominated by several species of *Salicornia* and other halophytes. Lagoons are populated with a wide variety of avian species. Some restoration of rice paddies to *Phragmites* marshes has been attempted in the delta (Comin et al., 1997).

Rhine River Delta

The Rhine River is a highly managed river and a major transportation artery in Europe. The Netherlands, which comes from *Nederland,* meaning "low country," is, essentially, the Rhine River delta, and although the Dutch language did not even have a word for "wetlands," the English word was adopted in the 1970s. The Netherlands is one of the most hydraulically controlled locations on Earth (Fig. 3-10). It is

Figure 3-9   The Ebro River delta in Spain is being restored into a deltaic wetland on the Mediterranean coast. At one time, the area was extensively managed for rice production. Freshwater systems are now partially dominated by *Phragmites*, a plant that is considered desirable in Europe. The saline part of the delta is dominated by *Salicornia* salt marshes that thrive despite the absence of any significant tide. (*Photograph by W. J. Mitsch*)

currently estimated that 16 percent of the Netherlands is wetland; the Dutch have warmed to the idea of the importance of wetlands and have registered 7 percent of the country as internationally important wetlands with the Ramsar Convention on Wetlands of International Importance (Wolff, 1993). Today, several governmental initiatives are designed to encourage some water to enter, or at least remain, on the lands (Fig. 3-11), in great contrast to earlier Dutch traditions of controlling water in this close-to-sea-level environment. Earlier in this century, thousands of hectares were reclaimed from the Zuiderzee; today, some of these areas are reverting back to wetlands. For example, beginning in 1968, the Oostvaardersplassen in the Flevoland Polder, originally created as a site for industrial development, was artificially flooded in order to create a wildlife sanctuary. The 5,600-ha site is now a habitat for birds such as herons, cormorants, and spoonbills (250 bird species have been recorded there, 90 of which have bred there), as well as for Konik horses, descended from the original Tarpan wild horses of Western Europe (Fig. 3-12). Cattle have been crossbred from Scottish, Hungarian, and Camarguais breeds in an effort to recreate the original oxen of Europe. The Oostvaardersplassen is now one of the most popular places in the Netherlands for bird watching, and this wetland, less than 30 years old, has become a national treasure.

Coastal Marshes, Mud Flats, and Bays of Northern Europe

Extensive salt marshes and mud flats are found along the Atlantic Ocean and the North Sea coastlines of Europe from the Mira Estuary in Portugal to the Wadden

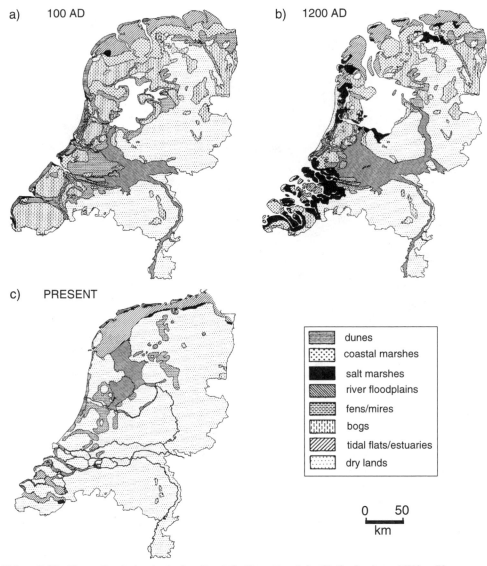

Figure 3-10 The estimated extent of wetlands in the present-day Netherlands and Rhine River delta in: **a.** 100 A.D.; **b.** 1200 A.D.; and **c.** present day. (*From Wolff, 1993, pp. 3–5, Figs. 1–3, reprinted with kind permission from Kluwer Academic Publishers*)

Sea of the Netherlands, Germany, and Denmark. These marshes contrast with the extensive salt marshes of North America, which stretch from the Bay of Fundy in Canada to southern Florida and the Gulf of Mexico, in dominant vegetation, tidal inundation, and sediment transport. One of the better known coastal wetland areas in France is at the Normandy–Brittany border near the world-famous abbey of Mont-St.-Michel, perched atop a promontory in a bay of the English Channel and accessible to pilgrims and tourists by day only until the tides turn it into an island. Some of the

Figure 3-11 The Rhine River near Arnhem, the Netherlands, is once again being allowed to flood the floodplain, with seasonal grazing by cattle and other domestic animals between flood seasons. (*Photograph by W. J. Mitsch; reprinted from Mitsch et al., 1994a, p. 15, Fig. 8, with permission from Elsevier Science*)

Figure 3-12 Konik horses (descended from the Tarpan wild horses of Western Europe) are among the unusual features of the Oostvaardersplassen, one of the largest and best known created wetlands in the Netherlands. It was originally designed for industrial development and is now one of the Netherlands' best birding locations. (*Photograph by W. J. Mitsch; reprinted from Mitsch et al., 1994a, p. 15, Fig. 9, with permission from Elsevier Science*)

most extensive salt marshes of Europe are found surrounding the abbey (Fig. 3-13). There has been a 60 percent drainage of coastal marshes since the beginning of the century in this region, but now coastal wetlands are better protected even though sheep grazing is still commonly practiced on these marshes. At nearby Le Vivier-sur-Mer, mussels are grown on *bouchots* (mussel beds created by sinking poles into the mud flats) in the shelter of a 30-km dike built in the 11th century.

The Wadden Sea, making up over 8,000 km² of shallow water, extensive tidal mud flats, marsh, and sand, is considered by some to be Western Europe's most important coastal wetlands. Over the past five centuries or more, drainage of the coastal land created hundreds of square kilometers of arable land as the local residents created more and more land out of the sea (Fig. 3-14). The wetlands extend for more than 500 km along the coasts of Denmark (10 percent), Germany (60 percent), and the Netherlands (30 percent), supporting North Sea fisheries, including almost 50 percent of the North Sea's sole, 60 percent of brown shrimp, 80 percent of plaice, and nearly all herring for some part of their life cycle (Maltby, 1986; Dugan, 1993).

Numerous bays surround the Baltic Sea and adjacent seas in northern Europe, many with extensive wetlands despite the fact that most of the rivers that feed these brackish seas are relatively small. Matsalu Bay, a water meadow and reed marsh in northwestern Estonia, has been known for years as a very important bird habitat. The wetland covers about 500 km² with much of that designated the Matsalu State Nature Preserve. Great numbers of birds, as many as 300,000 to 350,000, including swans, mallards, pintails, coots, geese, and cranes, stay in the Matsalu wetland during migration in the spring.

Figure 3-13   **Upper salt marsh at Mont-St.-Michel Bay, Normandy, France, with Mont-St.-Michel abbey in the background. (*Photograph by A. Mauxion, reprinted by permission*)**

Figure 3-14 Ancient drainage ditch near several kilometers from the current shoreline of the Wadden Sea in Schleiswig-Holstein, Germany. By developing drainage ditches, the Dutch and Germans encroached gradually on the Wadden Sea over the centuries, producing more arable land. (*Photograph by W. J. Mitsch*)

In many bays adjacent to southern Sweden, both in the Baltic Sea and in the Kattegat, there are significant eutrophication problems due to excessive nitrogen from agricultural practices and due to the loss of wetlands, drainage of streams, and destruction of riparian zones. For example, Lanholm Bay, a small (300 km²) bay in southwestern Sweden on the Kattegat, experienced significant blooms of *Cladophora* in the 1970s, which has affected fisheries and tourism in the region. This was followed by the development of a much more extensive low-oxygen condition (hypoxia) due to excessive plankton production in large parts of the Kattegat (22,000 km²) itself in the 1980s (Fleischer et al., 1994). A goal of 50 percent reduction of nitrogen loading has been proposed for streams entering the coastal waters of Sweden, and restored wetlands are being considered as a cost-effective measure for reducing the nitrogen loads (Fleischer et al., 1991, 1994; Kessler and Jansson, 1994).

Inland Deltas

There are a great number of important wetlands around the world that form not as coastal deltas, but as inland deltas or coastal marshes along large bodies of brackish and freshwater systems. There are several significant inland deltas in southeastern Europe. The 6,000-km² Danube River delta, one of the largest and most natural European wetlands (IUCN, 1993), has been degraded by drainage and by activities related to agricultural development, gravel extraction, and dumping. The Danube Delta is the home of floating marshes of *Phragmites communis,* representing—in biological terms—an endpoint in development. On the edge of the Caspian Sea, the Volga River forms one of the world's largest inland deltas (19,000 km²), a highly "braided" delta over 120 km in length and spreading over 200 km at the sea's edge. The most extensive wetland area occurs in the delta of the Caspian Sea as the sea declined in water level, creating extensive *Phragmites* marshes and water lotus (*Nelumbo nucifera*) beds (Finlayson and Volz, 1994; Fig. 3-15). Seventy percent of the world's sturgeon comes from the Caspian Sea (Jones, 1993c), and the delta is a wintering site in mild winters for water birds and a major staging area for a broad variety of water bird, raptor, and passerine species. A series of dams destroyed the river's natural hydrology, and heavy industrial and agricultural pollution, as well as sea-level decline in the Caspian Sea, are making an impact.

Yet another lowland inland delta is the Colchis wetlands of eastern Georgia (Fig. 3-16), a 13,000-km² region of subtropical alder (*Alnus glutinosa, A. barbata*) swamps and sedge–rush–reed marshes created by tectonic settling plus backwaters from the

Figure 3-15 Lotus bed in the Volga Delta, Russia. (*Photograph by C. M. Finlayson, reprinted by permission*)

Figure 3-16 The Colchis wetlands of western Georgia near the eastern tip of the Black Sea. This is the setting for the mythological adventures of Jason and the Argonauts, who sought the Golden Fleece. Shown here is ecologist H. T. Odum trudging through the wetland. (*Photograph by W. J. Mitsch*)

rivers discharging into the eastern Black Sea. This wetland is found in an area of great mythological interest because it is supposedly where Jason and the Argonauts (the Greek story of Argonautica as told by Apollonius) "hid their ship in a bed of reeds" (Grant, 1962) as they attempted to claim the Golden Fleece from the King of Colchis. These wetlands must indeed be long-lived!

European Peatlands

Gorham (1991) estimated that 3.46 million km^2 of northern boreal and subarctic peatlands exist, probably well over half the world's wetlands. A good portion of these peatlands are found in the Old World, where peatlands spread across a significant portion of Ireland, Scandinavia, Finland, northern Russia, and many of the former Soviet republics. The Endla Bog in Estonia (Fig. 3-17) and the Berezinski Bog in Byelorussia (Fig. 3-18) are both examples of peatlands that have been protected as nature preserves and are in seminatural states. The 76,000-ha Berezinski reservation

Figure 3-17 The Endla Bog in central Estonia. (*Photograph by W. J. Mitsch*)

Figure 3-18 Forested (*Pinus*) peatland in Berezinski Biospheric Reserve, Byelorussia. (*Photograph by W. J. Mitsch; reprinted from Mitsch et al., 1994a, p. 28, Fig. 27, with permission from Elsevier Science*)

in northeastern Byelorussia is over half peatland and predominantly forested peatland dominated by pine (*Pinus*), birch (*Betula*), and black alder (*Alnus*).

Africa

An abundance of wetlands is found in sub-Saharan Africa (Fig. 3-19). The vast size of some of these major wetlands is far beyond what the Western world experiences, with examples such as the Zaire Swamps (200,000 km^2), the Inner Niger Delta of Mali (320,000 km^2 when flooded), the Sudd of the Upper Nile (>30,000 km^2 when flooded), and the Okavango (16,000 km^2) (Denny, 1993; Dugan, 1993).

Okavango Delta

One of the great seasonally pulsed inland deltas of the world, the Okavango Delta forms at the convergence of the Okavango River and the sands of the Kalahari in

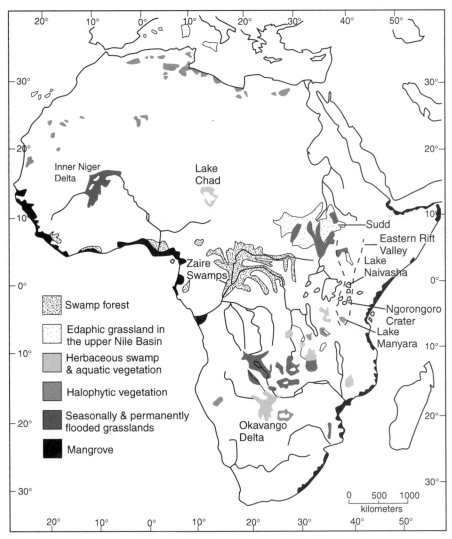

Figure 3-19 **Map of major wetland areas of Africa. (*After Denny, 1993*)**

Botswana, Africa. It has a web of channels, islands, and lagoons supporting crocodiles, elephants, and buffalo, and more than 400 bird species. Tilapia spawn in the Okavango Delta. Many of northern Botswana's diverse tribes find refuge there. The majority of the inhabitants depend on the delta's resources. Like so many other deltas, however, the Okavango is threatened by the increased burning and clearing associated with crop production. Tourism is an issue here, too, as in many other wetland sites; it is the largest single employer in Maun, located on the edge of the delta (Maltby, 1986). Maun also benefits economically from the wetland's water lily tubers, bulrush roots, palm hearts, and palm wine, made from the sap of the *Hyphaene* palm. Fencing, roofing, and wall materials are also derived from the wetlands.

East Africa Tropical Marshes

Several wetlands that form around tropical lakes in Africa are typical of what are called "swamps" in Old World usage of the word but would be "marshes" in New World terminology (see Wetland Terms and Types earlier in this chapter). These highly productive wetland margins tend to be dominated by tropical species of cattail (*Typha domingensis*) and papyrus (*Cyperus papyrus*), often with mats of floating plants (*Eichhornia crassipes* and *Salvinia molesta*). Many lakes and wetlands are found along the 6,500-km Rift Valley of eastern Africa. Not far from Nairobi on the floor of the Rift Valley, Lake Naivasha (Fig. 3-20) is one of the most studied tropical lakes in East Africa (Gaudet, 1977, 1979; Harper, 1992; Harper et al., 1995). The area provides a home for nearly the entire range of ducks and herons found in eastern Africa (Finlayson and Moser, 1991). Vegetation changes in Lake Naivasha have been caused recently by a combination of water-level fluctuations, the introduction of crayfish (*Procambarus clarkii*), and the physical effects of floating rafts of *Eichhornia crassipes* (Harper et al., 1995). Other vast wetlands are found in the Rift Valley in northern Tanzania, including the shorelines of Lake Manyara (Fig. 3-21) and the wetlands of Ngorongoro Crater (Fig. 3-22). Two swamps, Mandusi Swamp and Gorigor Swamp, and one lake, Lake Makat, are found in the caldera. The abundant wildlife of Ngorongoro was summarized by the East African/German conservationist Bernhard Grzimek who stated, "It is impossible to give a fair description of the size and beauty of the Crater, for there is nothing with which one can compare it. It is one of the wonders of the World" (Hanby and Bygott, 1998).

Australia/New Zealand

Billabongs

Australia's wetlands are distinctive for their seasons of general dryness due to high evaporation rates and low rainfall (McComb and Lake, 1990). Wetlands do occur on the Australian mainland, but only where the accumulation of water is possible, generally on the eastern and western portions of the continent. Thus, there are not many permanent wetlands; most are intermittent and seasonal. Furthermore, because of the high evaporation rates, saline wetlands and lakes are not uncommon (McComb and Lake, 1990). A particular feature in eastern Australia is the *billabong* (Shiel, 1994), a semipermanent pool that develops from an overflowing river channel (Fig. 3-23). Although found throughout Australia, the best concentration of billabongs is along the Murray and Darling Rivers in southeastern Australia. There are about 1,400 wetlands representing 32,000 ha in four watersheds alone in New South Wales

Figure 3-20 Tropical wetlands of the Rift Valley in eastern Africa support papyrus (*Cyperus papyrus*) and floating mats of water hyacinths (*Eichhornia crassipes*) and *Salvinia molesta*. Many lakes and wetlands are found along the 6,500-km rift including Lake Naivasha in Kenya shown here. (*Photograph by K. Mavuti, reprinted by permission*)

Figure 3-21 Wildlife is abundant in the Rift Valley lakes and wetlands. This photo, showing wildebeests, monkeys, and Yellow-billed Storks (*Ibis ibis*), is along Lake Manyara, Tanzania, one of the southernmost lakes along the Rift Valley. (*Photograph by W. J. Mitsch*)

Figure 3-22 Waterfowl in the wetlands of the Ngorongoro Crater, northern Tanzania, including Yellow-billed Duck (*Anas undulata*), Red-billed Duck (*A. erythrorhynchos*), and Egyptian Goose (*Alopochen aegyptiaca*). (*Photograph by W. J. Mitsch*)

Figure 3-23 A billabong of New South Wales, Australia, showing bulrushes, river red gum in the background, and invasive *Salvinia molesta* on the water's surface.

(McComb and Lake, 1990). Billabongs support a variety of aquatic plants, are a major habitat for birds and fish, and are often surrounded by one of many species of eucalyptus, especially the river red gum *Eucalyptus camaldulensis* (Fig. 3-23). The billabongs serve as refuges for aquatic animals during the dry season when the rivers come close to drying (Finlayson and Moser, 1991).

Western Australia

The Mediterranean-type climate of southwestern Australia does, in fact, favor a wide variety of wetlands, which are especially important to the waterfowl that are separated from the rest of the continent by vast expanses of desert. Swamps are numerous, and many can be found inland or just above the saline wetlands of the tidal rivers and bays of the Swan Coastal Plain, near Perth (Fig. 3-24). Nevertheless, it is estimated that 75 percent of the wetlands in the Swan Coastal Plain in southwestern Australia have been lost (Chambers and McComb, 1994).

New Zealand

For a small country, New Zealand has a wide variety of wetland types. In fact, there is one location in North Island where all seven types of wetlands described in this book (see Chapter 4) can, in theory, be seen from one general location in a helicopter. However, New Zealand has lost 90 percent of its wetlands, amounting to an estimated 311,300 ha.

The western region of South Island, called Westland, is sometimes humorously called "Wetland" because of the enormous amount of rain it receives (2–10 m annually) due to its location between the Tasman Sea to the west and the Southern

Figure 3-24   Freshwater wetland in the Swan Coastal Plain, Western Australia. (*Photograph by J. Davis, reprinted by permission*)

Alps to its east. It is also, not surprisingly, the location of a great variety of coastal wetlands. Grand Kahikatea (*Dacrycarpus dacrydiodes*) or "white pine" forested wetlands, reminiscent of the bald cypress swamps of the southeastern United States (Fig. 3-25), are found throughout Westland and also on North Island. *Pakihi* (peatlands) are found on both North Island and South Island. One of the largest wetlands in North Island is Whangamarino Wetland (Fig. 3-26), a 7,300-ha peatland and seasonally flooded swamp adjacent to the Waikato River, New Zealand's largest river (Clarkson, 1997; Shearer and Clarkson, 1998). Management issues facing this and other peatlands in the area are reduced inundation by the river, silt deposition from agricultural development, increased fire frequency over presettlement times, and invading willows and other exotics. Flax (*Phormium tenax*) swamps and raupo (*Typha orientalis*) marshes are also common in New Zealand. Willows (*Salix* spp.) are generally considered undesirable woody invaders to many of these wetlands.

Asia

Central Russian Bi-Ob' River Valley

The Bi-Ob' region of central Russia is a large floodplain on the Ob River between Kazakhstan to the south and the Ob River's estuary to the north on the Kara Sea. This valley of channels, floodplain lakes, and river distributaries is actually an inland delta caused more by decreased sea levels than by deposited sediments. The region has been described as "the largest single breeding area for waterfowl in Eurasia" (Dugan, 1993).

Figure 3-25 Kahikatea Swamp in the background with Okarito Lagoon in the foreground in western New Zealand. Kahikatea is locally called "white pine." These forests once dominated both coastlines of New Zealand. (*Photograph by W. J. Mitsch*)

Figure 3-26   Peatland in the lower Waikato River basin, about 60 km south of Aukland, New Zealand. Circular ponds with earthen paths are hunting ponds. (*Photograph by W. J. Mitsch*)

Southeast Asian River Deltas

Over 80 percent of Asian wetlands are located in seven countries: Indonesia, China, India, Papua New Guinea, Bangladesh, Myanmar, and Vietnam. The diversity of Asia's wetlands is reflected in its intertidal mud flats, swamp forests, natural lakes, open marshes, arctic tundra, and mangrove forests (recognized as one of the most productive ecosystems in the world—yielding over 70 direct and indirect uses of the forest or its products—but now threatened by logging). The snowfields and glaciers of the Himalayas are the birthplace of many of the world's well-known rivers, including the Ganges, the Indus, the Mekong, and the Yangtze. The Mekong, Southeast Asia's longest river, begins in the Tibetan Plateau, enters its lower basin at the boundary of Myanmar, Laos, and Thailand, and then flows to the ocean through one of the world's great deltas. The basin catchment area is more than 600,000-km^2 and includes Laos, Cambodia, Thailand, and Vietnam. There has been little coordination among these countries concerning its management, especially with regard to the extensive wetlands in the Mekong Delta region. Problems stemming from devegetation and drainage during the war years have been exacerbated by more recent efforts at agricultural intensification, urbanization, industrialization, and dam and reservoir construction (Beilfuss and Barzen, 1994). Even drained soil became acidic (pH < 3) when sulfur-rich soils oxidized, making them unsuitable for agriculture. Current research is focused on the mangrove wetlands of the Mekong Delta, especially because they are an important source of fuel and medicine. Forty-seven percent of the total fish catch from Mekong wetlands contributes between 50 and 70 percent of the protein needs of the 20 million people living in the delta (Maltby, 1986). Restoration of a freshwater portion of the Mekong Delta, known as the Dong Thap Muoi ("Plain of Reeds"), is currently underway with the help of the International Crane Foundation (Beilfuss and Barzen, 1994).

Chinese Wetlands

The total area of wetlands in China ranks as Asia's highest (Lu, 1990; Chen, 1995) with an estimated 625,000 km² of wetlands, 250,000 km² of which are natural with the rest artificial wetlands (rice paddies, fish ponds, etc.). Natural wetlands thus comprise about 2.5 percent of the country (Lu, 1995). Few of the wetlands are preserved in semipristine conditions as is done in the West as habitat alone; most wetlands in China provide fish, cattle, grain, duck, and other food, as well as habitat and recreation benefits, in a symbiotic relationship between humans and nature. However, an estimated 40 percent of the wetlands in China that have been designated as wetlands of international importance are "under moderate to severe threat from conversion to agricultural land, increased siltation due to catchment degradation, pollution, over-fishing, and hunting" (Parish and Elliott, 1990). Many of the important wetlands of China are found in the delta regions of the Changjiang (Yangtze) River (Fig. 3-27), the Zhujiang (Pearl) River, and the Liaohe River. Because these regions are among the most populated in the world, very few natural wetlands remain, having been converted to rice paddies or fish ponds. In many cases, wetlands and reed (*Phragmites*) fields are connected hydrologically with fish ponds and rice paddies to enhance food and fiber production (Ma et al., 1993). There are also extensive inland wetlands associated with the Yangtze River, particularly in the Jianghan–

Figure 3-27 Aquaculture site in the reeds of the Taihu (Tai Lake) shoreline in the Yangtze River delta of Eastern China. (*Photograph by W. J. Mitsch; reprinted from Mitsch et al., 1994a, p. 17, Fig. 11, with permission from Elsevier Science*)

Dongting Plain in the middle of the river valley near Wuhan, Hubei Province. This approximately 10,000-km^2 area of former marshes and lakes has been extensively drained and diked, yet consistently suffers crop damage due to excessive water. Integrating these backwater areas with the Yangtze River as they once were is probably impossible, but converting wet areas from rice and other crops to wetland crops such as lotus (*Nelumbo nucifera*) and wild rice stem (*Zizania latifolia*) has been suggested as a viable "ecological" approach (Bruins et al., 1998).

In contrast to the marshes found on the European and eastern North American coastlines, many of China's coastal marshes are of more recent origin. As described by Chung (1994), both *Spartina anglica* and *S. alterniflora* have been introduced in China in an effort to restore the eastern coastline. After 30 years of experiments and full-scale projects, many of these salt marsh plantations have proved to be very successful, both ecologically and economically. Almost 40,000 ha have been restored by Professor Chung-Hsin Chung and his colleagues at Nanjing University for accretion for agricultural use, for mitigation of saline soils, as a source of green manure and animal fodder, and for seashore stabilization (Qin et al., 1998).

China has several inland wetland areas in the western mountain and desert regions as part of the more than 250,000 km^2 of wetlands in the country. The sources of several major rivers can also be found in the Qinghai–Tibetan Plateau—the Yellow, the Yangtze, the Indus, and the Ganges—along with high-altitude lakes and bogs. Most of the plateau's larger lakes are saline, and, at 458,000 ha, Qinghai Hu (Lake) is the largest (Fig. 3-28). As this whole area is experiencing desiccation, the lakes are shrinking, most recently at a rate of 12 cm/yr in depth. Nevertheless, these wetlands

Figure 3-28   A bird island in Qinghai Hu, western China. This lake is in an arid region of China and is the largest saltwater lake in the country. (*Photograph by J. Lu, reprinted from Mitsch et al., 1994a, p. 33, Fig. 33, with permission from Elsevier Science*)

are habitat for millions migratory and resident birds comprising over 160 species (Lu, 1990). One of the world's great mountain lakes, Issyk Kul, is a brackish wetland lying in a basin of the Tianshan mountain chain along the China–Kazakhstan border (Fig. 3-29). Some of these wetlands, while important havens for migratory waterfowl, are also important for regional fisheries and for supplying reeds used in the manufacture of paper (Lu, 1990).

Indian Freshwater Marshes

The world's second most populous nation is India. It is slightly more than one-third the size of the United States, but has more than three times as many people. Droughts, soil erosion, overgrazing, and desertification are common. Agriculture employs two-thirds of the labor force, based in and around the alluvial plains and coastal zones on 55 percent of the land. The wetlands are under intense pressure for farm expansion, water control, and urbanization. Flooding cycles on alluvial valleys have been aggravated by these developments, resulting in "natural" disasters to humans and habitat alike. A few conservation wetlands remain under moderate protection, sometimes as remnants of the formerly upper-class lands. Keoladeo National Park in Bharatpur (Fig. 3-30) is an example, where the hunting reserve is now a protected area of international significance (Prasad et al., 1996). About 850 ha of the park are wetlands. The local economy benefits from tourism and also collects or illegally harvests products from the area. The protected wildlife heritage includes migratory species from northern Asia. In all, more than 350 species of birds, 27 mammals, 13 amphibians, 40 fish, and 90 wetland flowering plants are found in the park (Prasad et al., 1996).

Figure 3-29 River and wetlands in the desert areas of Xinjiang, China. The river is fed from the Tianshan Mountains. (*Photograph by J. Lu, reprinted by permission*)

Figure 3-30 Keoladeo National Park, Bharatpur, India, during flooding season. (*Reprinted from Mitsch et al., 1994a, p. 32, Fig. 33, with permission from Elsevier Science*)

Recommended Readings

Dugan, P. 1993. Wetlands in Danger. Michael Beasley, Reed International Books, London. 192 pp.

Finlayson, C. M., and A. G. van der Valk, eds. 1995. Classification and Inventory of the World's Wetlands. Special Issue of Vegetatio 118:1–192.

Finlayson, M., and M. Moser, eds. 1991. Wetlands. Facts on File, Oxford. 224 pp.

McComb, A. J., and P. S. Lake, 1990. Australian Wetlands. Angus and Robertson, London. 258 pp.

Mitsch, W. J., ed. 1994. Global Wetlands: Old World and New. Elsevier, Amsterdam. 967 pp.

Whigham, D. F., D. Dykyjová, and S. Hejny, eds. 1993. Wetlands of the World I: Inventory, Ecology, and Management. Kluwer Academic Publishers, Dordrecht, The Netherlands. 768 pp.

Wetlands of North America

In this book, we classify wetlands as seven major types: tidal salt marshes, tidal freshwater marshes, mangrove wetlands, freshwater marshes, peatlands, freshwater swamps, and riparian wetlands. Wetlands are numerous and diverse in North America, but drastic changes have occurred since presettlement times. An estimated 53 percent of the original wetlands in the conterminous United States has been lost because of drainage and other human activities. On a regional basis, the greatest wetland losses occurred in the lower Mississippi alluvial plain and the prairie pothole region of the north central states. Estimates of original wetlands in the lower 48 states vary around 89 million ha. The most accurate estimate of present wetland area is 42 million ha, of which approximately 80 percent are inland wetlands and 20 percent are coastal. Current measures of the wetland area in Alaska are about 71 million ha. By contrast, Canada has an estimated 127 million ha of wetlands, more than the U.S. total (including Alaska). Although Canada has not suffered the same rate of loss as has the lower 48 states, high wetland loss rates of 70 to 90 percent are common in the most populated southern Ontario. Overall, with 2.4 million km^2, North America has about 30 percent of the world's wetlands. Historical and ecological characteristics of some important regional wetlands of the United States and Canada, several of which have been threatened or eliminated by human development, are discussed here.

North America has always had an abundance and a diversity of wetlands. The wetlands that exist now, however, may represent only a fraction of those seen by early settlers and pioneers as they advanced across the continent about 200 years ago. Salt marshes and mangroves probably represented inhospitable barriers to many of the explorers and settlers as they arrived in the New World. Soggy marshes and wet meadows must have evoked both trepidation and awe in the first settlers as they passed through the hills and mountains of the East and entered the level prairie of the Midwest. Peatlands and prairie potholes had to be a common sight to those who settled in the north plains in what is now Michigan, Wisconsin, Minnesota, the Dakotas, and the central Canadian provinces. The early explorers and settlers on the lower Mississippi River or rivers in the southeastern Coastal Plain must have marveled at the mammoth structures of many of the old-growth cypress swamps and bottomland hardwood forests that lined those rivers. All of those wetlands undoubtedly played an important role in the development of North America. In the early period of settlement, wetlands were accepted as part of the landscape, and there was little desire or capability to change their hydrologic conditions to any great degree. As described by Shaw and Fredine (1956):

> The great natural wealth that originally made possible the growth and development of the United States included a generous endowment of shallow-water and waterlogged lands. The original inhabitants of the New World had utilized the animals living among these wet places for food and clothing, but they permitted the land to essentially remain unchanged.

As the westward movement slowed and towns, villages, cities, and farms began to expand on a regional scale, wetlands were increasingly viewed as wastelands that should be drained for reasons as varied as disease control and agricultural expansion. There followed an unfortunate yet understandable period in the country's growth, when wetland drainage and destruction became the accepted norm. The Swamp Land Acts of 1849, 1850, and 1860 set the tone for that period by encouraging drainage in 15 of the interior and West Coast states. Coastal wetlands in the populated Northeast were drained and dissected to accommodate urban development. Many old-growth forested bottomlands and swamps in the Southeast were lumbered for their valuable products, reflecting little concern for their regeneration for the enjoyment and use of future generations. From the middle of the 19th century to the middle of the 20th century, the United States and large parts of Canada went through a period in which wetland removal was not questioned. Indeed, it was considered the proper thing to do. Despite all of this wetland loss and degradation, a great wealth and diversity of wetland types remains in North America.

Wetland Types Used in This Book

The wetland classification system used by many scientists around the world is the hierarchical classification developed by the U.S. Fish and Wildlife Service as part of the U.S. National Wetlands Inventory (Cowardin et al., 1979). This classification is formal and all-encompassing, but it is also much too complex to use conveniently as a way of organizing this book. (The formal wetland classification scheme is described more fully in Chapter 21.) In this book, we describe seven major types of wetlands divided into two major groups: (1) coastal—salt marsh, tidal freshwater marsh, mangrove, and (2) inland—freshwater marsh, peatland, freshwater swamp, and riparian ecosystem (Table 4-1). These classes of wetlands are generally recognizable ecosystems about which extensive research literature is available. Regulatory agencies also deal with these systems, and management strategies and regulations have been developed for these wetland types. It is recognized that there are other types of wetlands such as inland saline marshes, scrub–shrub swamps, and even red maple swamps that may "fall between the cracks" in this simple wetland classification, but the seven classes cover most wetlands currently found in North America. Collectively, these categories

Table 4-1 Wetland types described in this book with estimated area in the United States, including Alaska

Type of Wetland	Area in United States ($\times 10^6$ ha)	Book Chapter
Coastal wetlands		
Tidal salt marshes	1.9	9
Tidal freshwater marshes	0.8	10
Mangrove wetlands	0.5	11
Inland wetlands		
Freshwater marshes	27	12
Peatlands	55	13
Freshwater swamps ⎫	25	⎧14
Riparian systems ⎭		⎩15
Total	111	

encompass about 111 million ha of wetlands in the United States and 127 million ha of wetlands in Canada.

Coastal Wetlands

Several types of wetlands in coastal areas are influenced by alternate floods and ebbs of tides. Near coastlines, the salinity of the water approaches that of the ocean, whereas, further inland, the tidal effect can remain significant even when the salinity is that of fresh water. Coastal wetlands include tidal salt marshes, tidal freshwater marshes, and mangrove swamps. The total area of wetlands considered as coastal or estuarine wetlands in the United States, including Alaska, is approximately 3.2 million ha, with about 1.9 million ha as salt marsh and 0.5 million ha as mangrove (Table 4-2). Alaskan estuarine wetlands were estimated to cover 0.86 million ha (Hall et al., 1994), with about 16.9 percent of these wetlands (0.146 million ha) vegetated and thus presumably salt marsh. The vast majority of estuarine wetlands in Alaska were classified by Hall et al. (1994) as "nonvegetated" and are, therefore, not discussed in this book.

Table 4-2 Estimated area of coastal wetlands in the United States ($\times 1,000$ ha)

	Salt Marsh[a]	Freshwater Tidal Marsh[b]	Mangrove[b]	Total
Atlantic Coast	669	400		1,069
Gulf of Mexico	1,011	362	506	1,879
Pacific Coast	49	57		106
Alaska[c]	146			146
Total	1,875	819	506	3,200

[a] Watzin and Gosselink (1992).
[b] Field et al. (1991). Mangroves are estimated as forested wetlands in Ten Thousand Islands Drainage Area, southwest Florida.
[c] Hall et al. (1994).

Tidal Salt Marshes

Salt marshes (Fig. 4-1) are found throughout the world along protected coastlines in the middle and high latitudes. On the eastern coast of the United States, salt marshes are often dominated by the grass *Spartina alterniflora* in the low intertidal zone, *Spartina patens* and the rush *Juncus* in the upper intertidal zone. Plants and animals in these systems have adapted to the stresses of salinity, periodic inundation, and extremes in temperature. Salt marshes are most prevalent in the United States along the eastern coast from Maine to Florida and on into Louisiana and Texas along the Gulf of Mexico. They are also found in narrow belts on the West Coast of the United States and along much of the coastline of Alaska. The area and type of coastal wetland depend on where the inland boundary of the coast is drawn. The estimated total area of salt marshes in the United States is 1.9 million ha, most (53 percent) of which are found along the Gulf of Mexico (Table 4-2).

Tidal Freshwater Marshes

Inland from the tidal salt marshes but still close enough to the coast to experience tidal effects (Fig. 4-2), these wetlands, dominated by a variety of grasses and by annual and perennial broad-leaved aquatic plants, are found primarily along the Middle and South Atlantic coasts and along the coasts of Louisiana and Texas. Estimates of tidal freshwater wetlands in the United States range from 164,000 ha along the Atlantic Coast (W. E. Odum et al., 1984) to 820,000 ha for the conterminous United States (Field et al., 1991; Table 4-2). The uncertainty in the estimate relates to where the line is drawn between tidal and nontidal areas in determining how much of the extensive Gulf Coast freshwater wetlands are tidal. Tidal freshwater marshes can be described as intermediate in the continuum from coastal salt marshes to freshwater

Figure 4-1   Tidal salt marsh.

Figure 4-2   Tidal freshwater marsh.

marshes. Because they are tidally influenced but lack the salinity stress of salt marshes, tidal freshwater marshes have often been reported to be very productive ecosystems, although a considerable range in their productivity has been measured.

Mangrove Wetlands

The tidal salt marsh is replaced by the mangrove swamp (Fig. 4-3) in subtropical and tropical regions of the world. The word *mangrove* refers both to the wetland itself and to the salt-tolerant trees that dominate those wetlands. In the United States, mangrove wetlands are limited primarily to the southern tip of Florida, although small mangrove stands are scattered as far north as Louisiana and Texas. Approximately 287,000 to 500,000 ha (Table 4-2) of these wetlands exist in the United States, a small fraction of the 14 million ha of mangroves estimated to be found worldwide (Finlayson and Moser, 1991). In Florida, mangrove wetlands are generally dominated by the red mangrove tree (*Rhizophora*) and the black mangrove tree (*Avicennia*). Like the salt marsh, the mangrove swamp requires protection from the open ocean and occurs in a wide range of salinity and tidal influence.

Inland Wetlands

On an areal basis, most of the wetlands of the United States and Canada are not located along the coastlines but are found in interior regions (these wetlands are called "nontidal" in coastal regions to distinguish them from coastal wetlands; no such term is used by inland wetland scientists). Frayer et al. (1983) estimated that

Figure 4-3 Mangrove wetland.

32 million ha or about 80 percent of the total wetlands in the lower 48 states are inland. Our estimate is that 38 million ha or about 93 percent of the total wetlands in the lower 48 states are inland (Tables 4-2 and 4-3). If we include Alaska, there are a total of 107 million ha of inland wetlands in the United States (Table 4-3). It is difficult to put these wetlands into simple categories. Our simplified scheme divides them into two groups that are decidedly regional—northern peatlands and southern swamps, and into two other categories that are found throughout the climatic zones—freshwater marshes and riparian ecosystems. These divisions roughly parallel the divisions that persist in both the scientific literature and the specializations of wetland scientists.

Freshwater Marshes

This category includes a diverse group of wetlands characterized by (1) emergent soft-stemmed aquatic plants such as cattails, arrowheads, pickerelweed, reeds, and several other species of grasses and sedges; (2) a shallow-water regime; and (3) generally shallow peat deposits (Fig. 4-4). These wetlands are ubiquitous in North America

Table 4-3   Estimated area of inland wetlands in the United States (×1,000 ha)

	Peatlands[a]	Marshes	Swamps	Total
Lower 48 states	3,700	9,932	25,138	38,771
Alaska	51,754	17,004	[b]	68,758
Total	55,454	26,936	25,138	107,529

[a] Peatlands of the lower 48 states estimated to be 3.7 million ha (Minnesota has 2.7 million ha alone).
[b] All 5.4 million ha of palustrine forested wetlands in Alaska are assumed to be peatland.
Source: Hall et al. (1994).

and are estimated to cover about 7 to 10 million ha in the United States (Table 4-3). Major regions where marshes dominate include the prairie pothole region of the Dakotas, the Great Lake coastal marshes, and the Everglades of Florida. They occur in isolated basins, as fringes around lakes, and along sluggish streams and rivers.

Peatlands

As defined here, peatlands include the deep peat deposits of the boreal regions of the world (Fig. 4-5). In the United States, these systems are limited primarily to Wisconsin, Michigan, Minnesota, and the glaciated Northeast, although similar peat deposits, called pocosins, are found on the Coastal Plain of the Southeast. There are also mountaintop bogs in the Appalachian Mountains of West Virginia such as those found in the Canaan Valley. Minnesota, containing an estimated 2.7 million ha, has the largest peatland area in the United States (Glaser, 1987). The much more extensive peatlands of Alaska and Canada cover an estimated 52 million ha and 111 million ha, respectively (Zoltai, 1988; Hall et al., 1994). Bogs and fens, the two major types

Figure 4-4   Inland freshwater marsh.

Figure 4-5   Peatland (*Photograph by C. D. A. Rubec, reprinted by permission*).

of peatlands, occur as thick peat deposits in old lake basins or as blankets across the landscape. Many of these lake basins were formed by the last glaciation, and the peatlands are considered to be a late stage of a "filling-in" process. There is a wealth of European scientific literature on this wetland type, much of which has influenced the more recent North American literature on the subject. Bogs are noted for their nutrient deficiency and waterlogged conditions and for the biological adaptations to these conditions such as carnivorous plants and nutrient conservation.

Freshwater Swamps

These are freshwater forested wetlands of the southeastern United States that have standing water for most, if not all, of the growing season (Fig. 4-6). These swamps occur under a variety of nutrient and hydrologic conditions and are normally dominated by various species of cypress (*Taxodium*) and gum/tupelo (*Nyssa*). These wetlands can occur as isolated cypress domes fed primarily by rainwater or as alluvial swamps that are flooded annually by adjacent streams and rivers. In the northern parts

Figure 4-6   Freshwater swamp.

of the lower 48 states, red maple (*Acer rubrum*) swamps are common, although there are no trees quite as adapted to flooding anywhere in North America as are *Taxodium* and *Nyssa*.

Riparian Ecosystems

Extensive tracts of riparian wetlands, which occur along rivers and streams, are occasionally flooded by those bodies of water but are otherwise dry for varying portions of the growing season (Fig. 4-7). Riparian forests and freshwater swamps combined constitute the most extensive class of wetlands in the United States, covering from 22 to 25 million ha (Dahl and Johnson, 1991; Hall et al., 1994). In the southeastern United States, riparian ecosystems are referred to as bottomland hardwood forests. They contain a diverse vegetation that varies along gradients of flooding frequency. Riparian wetlands also occur in arid and semiarid regions of the United States, where they are often a conspicuous feature of the landscape in contrast with the surrounding arid grasslands and desert. Riparian ecosystems are generally considered to be more productive than the adjacent uplands because of the periodic inflow of nutrients, especially when flooding is seasonal rather than continuous.

Figure 4-7 Riparian forested wetland.

The Status of North American Wetlands

The major wetland regions of North America are shown in Figure 4-8. Regional diversity and the lack of unanimity about the definition of a wetland, as described in Chapters 2 and 3, make it difficult to inventory the wetland resources in general. Nevertheless, attempts have been made to find out how many wetlands there are in the United States and Canada and to determine the rate at which they are changing.

United States Wetlands

Based on a review of several studies that have been made of wetland trends in the United States, two general statements can be made: (1) Estimates of the area of wetlands in the United States, while they vary widely, are becoming more accurate; and (2) most studies indicate a rapid rate of wetland loss in the United States, at least prior to the mid-1970s. Historical estimates of wetland area in the 48 conterminous United States are given in Table 4-4. The numbers vary widely for several reasons. First, the purposes of the inventories varied from study to study. Early wetland censuses, for example, Wright (1907) and Gray et al. (1924), were undertaken to identify lands suitable for drainage for agriculture. Later inventories of wetlands (Shaw and Fredine, 1956) were concerned with waterfowl protection. Only within the last decade have wetland inventories considered all of the values of these ecosystems. Second, the definition and classification of wetlands varied with each study, ranging from simple terms to complex hierarchical classifications. Third, the methods available for estimating wetlands changed over the years or varied in accuracy. Remote sensing

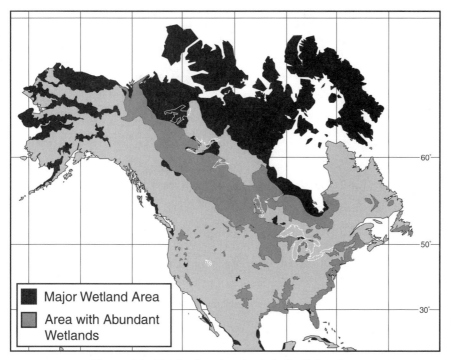

Figure 4-8 **Distribution of wetlands in North America.**

from aircraft and satellites is one example of a technique for wetland studies that was not generally available or used before the 1970s. Early estimates, in contrast, were often based on fragmentary records. Fourth, in a number of instances, the borders of geographical or political units changed between censuses, leading to gaps or overlaps in data. Finally, significant drainage of wetlands has occurred since the early estimates were made at the beginning of the century.

Presettlement Wetland Estimates

Roe and Ayres (1954) included drained wet soils along with swamps and marshes to estimate 87 million ha of wetlands in presettlement times in what are now the lower 48 states. This estimate often has been compared with existing wetland acreage to determine losses. The U.S. Congress Office of Technology Assessment interpreted prior wetland and wet soil data to suggest a range of 60 to 75 million ha, although the authors admit that the estimates "are limited by the lack of good data on the amount of land that has been drained or otherwise reclaimed and the relationship between wetlands and wet soils" (Office of Technology Assessment, 1984). Two other analyses (see Dahl, 1990) give estimates close to Roe and Ayres' original figure of 87 million ha. Given the uncertainty of the original records, these are likely to be as definitive as is possible.

Present-Day Wetland Estimates

One of the first wetland surveys in the United States based on wetland habitat values rather than on values for agriculture was carried out in 1954 and published by the

Table 4-4 **Estimates of wetland area in the United States at different times**

Date	Wetland Area (×10[6] ha)[a]	Reference
Presettlement	87	Roe and Ayres (1954)
Presettlement	86.2	USDA estimate, in Dahl (1990)
Presettlement	89.5	Dahl (1990)
1906	32[b]	Wright (1907)
1922	37 (total)	Gray et al. (1924)
	3 (tidal)	
	34 (inland)	
1940	39.4[c]	Whooten and Purcell (1949)
1954	30.1[d] (total)	Shaw and Fredine (1956)
	3.8 (coastal)	
	26.3 (inland)	
1954	43.8 (total)	Frayer et al. (1983)
	2.3 (estuarine)	
	41.5 (inland)	
1974	40.1 (total)	Ibid.; Tiner (1984)
	2.1 (estuarine)	
	38.0 (inland)	
mid-1970s	42.8[e] (total)	Dahl and Johnson (1991)
	2.2 (estuarine)	
	40.6 (inland)	
mid-1980s	41.8[e] (total)	Dahl and Johnson (1991)
	2.2 (estuarine)	
	39.6 (inland)	

[a] For 48 conterminous states unless otherwise noted.
[b] Does not include tidal wetlands or eight public land states in West.
[c] Outside of organized drainage enterprises.
[d] Only included wetlands important for waterfowl.
[e] Based on estimates of NWI classes for vegetated estuarine and palustrine wetlands.

U.S. Fish and Wildlife Service two years later as Circular 39 (Shaw and Fredine, 1956). This survey, which relied on a classification scheme of 20 wetland types (described in Chapter 21), estimated that the nation had 30.1 million ha of wetlands that were important to waterfowl. Of this area, 25.7 million ha were identified as inland freshwater wetlands, 0.6 million ha as inland saline wetlands, 2.1 million ha as coastal saline wetlands, and 1.6 million ha as coastal freshwater wetlands. The major shortcoming of the Circular 39 survey was that it considered only wetlands that were important for waterfowl and thus failed to consider a large portion of wetlands in the United States. Nevertheless, it is still referred to today and represents a benchmark summary of waterfowl wetlands for the 48 conterminous states.

A later assessment of wetland abundance in the United States, called the National Wetland Trends Study (NWTS), was conducted by the U.S. Fish and Wildlife Service (Frayer et al., 1983). This study used statistical analysis of wetland data derived from detailed mapping for the National Wetlands Inventory to estimate the total coverage of wetlands in the lower 48 states. The results indicated that, in the mid-1950s, there were 43.8 million ha of wetlands in the United States, about 45 percent more than the estimate by the Circular 39 study (Shaw and Fredine, 1956). This difference reflects the broader mandate of the NWTS study in contrast to the Circular 39 study, which focused only on wetlands that were important to waterfowl. The NWTS study also estimated that, by the mid-1970s, wetlands in the United States had decreased to 40.1 million ha. Of these totals, all were inland wetlands except for 2.3 million ha of estuarine wetlands in the 1950s and 2.1 million ha of estuarine wetlands in the 1970s. More recent Fish and Wildlife Service analyses of mid-1970s and mid-1980s data (Dahl and Johnson, 1991) show higher estimates for the 1970s and a continuing slow decline to about 41.8 million ha in the mid-1980s.

State-by-State Distribution

The extent of wetlands by state from estimates by the National Wetlands Inventory in the mid-1980s is shown in Table 4-5. Several states in the midwestern United States (Illinois, Indiana, Iowa, Kentucky, Missouri, and Ohio) plus California all had wetland losses of more than 80 percent. Collectively, seven states in the Mississippi River basin (Indiana, Illinois, Iowa, Minnesota, Missouri, Ohio, and Wisconsin) have had about 18.6 million ha of land artificially drained (Zucker and Brown, 1998), principally for agricultural production; these states collectively show a loss of 14.1 million ha of wetlands during the past 200 years (Table 4-5), or 30 percent of the wetland loss of the entire conterminous United States. States with high densities of wetlands—Minnesota, Illinois, Louisiana, and Florida—had among the highest losses of total area of wetlands—2.6, 2.8, 3.0, and 3.8 million ha, respectively.

Including Alaska in wetlands estimates of the United States changes the national numbers altogether. The estimate by Dahl (1990) of 69 million ha of wetlands in Alaska alone is more than the total estimate for the lower 48 states (42 million ha). Hall et al. (1994), in a more detailed study four years later, estimated that there are 71 million ha of wetlands in Alaska. The inclusion of Alaska in wetland surveys of the United States more than doubles the U.S. wetland area.

Wetland Losses

Overall, about 53 percent of the wetlands in the conterminous United States were lost from the 1780s to the 1980s (Table 4-6). Wetland area decreased from 89.5 million ha to 42.2 million ha. Shaw and Fredine (1956) gave early evidence of the magnitude of wetland loss in the United States by demonstrating that there was a 46 percent loss of wetlands in seven states that were covered by the Swamp Land Act of 1850 from 1850 to 1950. In a study of wetland losses in the United States, Frayer et al. (1983) estimated a net loss from the 1950s to the 1970s of more than 3.7 million ha (a loss rate of approximately 0.4 percent yr^{-1}), or an average annual loss of 185,000 ha. This loss represents a wetland area equivalent to the combined size of Massachusetts, Connecticut, and Rhode Island. Freshwater marshes and forested wetlands were hardest hit. They disappeared at rates of 7 and 4.8 percent per decade, respectively.

Wetland losses have continued into the 1980s and 1990s despite the enactment of strong wetland protection laws in the mid-1970s in the United States. However, the loss rates appear to be slightly slower now. An overall rate of about 0.25 percent yr^{-1} was reported for the United States from the 1970s to 1980s (Table 4-6). Tiner (1998) estimated that wetlands were still being lost at a rate of 47,000 ha yr^{-1} in the United States, or about 0.1 percent yr^{-1}, 10 years after the federal laws protecting wetlands went into effect. Losses from the 1980s through the end of the century have been more difficult to interpret because it was about this time that strong wetland protection laws took effect in the United States. While it has been hard to document,

Table 4-5   Surveys of area of wetlands by state in the United States[a]

State	Original Wetlands Circa 1780 (×1,000 ha)	National Wetlands Inventory, mid-1980s (×1,000 ha)	Change (percent)[b]
Alabama	3,063	1,531	−50
Alaska	68,799	68,799	−0.1
Arizona	377	243	−36
Arkansas	3,986	1,119	−72
California	2,024	184	−91
Colorado	809	405	−50
Connecticut	271	70	−74
Delaware	194	90	−54
Florida	8,225	4,467	−46
Georgia	2,769	2,144	−23
Hawaii	24	21	−12
Idaho	355	156	−56
Illinois	3,323	508	−85
Indiana	2,266	304	−87
Iowa	1,620	171	−89
Kansas	340	176	−48
Kentucky	634	121	−81
Louisiana	6,554	3,555	−46
Maine	2,614	2,104	−19
Maryland	668	178	−73
Massachusetts	331	238	−28
Michigan	4,533	2,259	−50
Minnesota	6,100	3,521	−42
Mississippi	3,995	1,646	−59
Missouri	1,960	260	−87
Montana	464	340	−27
Nebraska	1,178	771	−35
Nevada	197	96	−52
New Hampshire	89	81	−9
New Jersey	607	370	−39
New Mexico	291	195	−33
New York	1,037	415	−60
North Carolina	4,488	2,300	−44
North Dakota	1,994	1,008	−49
Ohio	2,024	195	−90
Oklahoma	1,150	384	−67
Oregon	915	564	−38
Pennsylvania	456	202	−56
Rhode Island	42	26	−37
South Carolina	2,596	1,885	−27
South Dakota	1,107	720	−35
Tennessee	784	318	−59
Texas	6,475	3,080	−52
Utah	325	226	−30
Vermont	138	89	−35
Virginia	748	435	−42
Washington	546	380	−31
West Virginia	54	41	−24
Wisconsin	3,966	2,157	−46
Wyoming	809	506	−38
Total wetlands	158,395	111,060	−30
Total "lower 48"	89,491	42,240	−53

[a] Dahl (1990).
[b] From original wetlands (1780s) to mid-1980s.

Table 4-6 **Estimates of wetland losses in the conterminous United States**

Period	Wetland Loss		Reference
	×1,000 ha	Percent	
TOTAL WETLANDS			
Presettlement–1980s	47,300	53	Dahl (1990)
1922–1954		0.2/yr	Zinn and Copeland (1982)
1954–1970s		0.50–0.65/yr	Ibid.
1950s–1970s	3,700 (185/yr)	8.5 (0.4/yr)	Frayer et al. (1983)
1970s–1980s	1,057	2.5	Dahl and Johnson (1991)
~1985	47/yr	0.1/yr	Tiner (1998)
SOUTHERN BOTTOMLAND HARDWOOD FORESTS			
1883–1991	6,500	77	The Nature Conservancy (1992)
1940–1975	2,300 (65/yr)	16 (0.45/yr)	Turner et al. (1981)
1960–1975	2,600 (175/yr)	18 (1.2/yr)	Ibid.
COASTAL WETLANDS			
1922–1954	260 (8.1/yr)	6.5 (0.2/yr)	Gosselink and Baumann (1980)
1954–1974	370 (19/yr)	9.9 (0.5/yr)	Ibid.
1950s–1970s	146 (7.3/yr)	—	Tiner (1984)
1970s–1980s[a]	29 (2.9/yr)	1.7 (0.15/yr)	Dahl and Johnson (1991)
LOUISIANA COASTAL WETLANDS			
1958–1974	10.8/yr	0.86/yr	Dunbar et al. (1992)
1983–1990	6.6/yr	—	

[a] Vegetated estuarine emergent wetland.

wetland losses have been at least partially offset by a major effort in wetland restoration and creation during this period.

Specific Wetland Losses

Several studies have described losses of particular types of wetlands. Gosselink and Baumann (1980) estimated that a loss rate of coastal wetlands of 8,100 ha yr^{-1} between 1922 and 1954 accelerated to a loss rate of 19,000 ha yr^{-1} between 1954 and the 1970s (Table 4-6). By comparison, the U.S. Fish and Wildlife Service estimated that 7,300 ha yr^{-1} of estuarine wetlands were lost from the 1950s to the 1970s, and 2,900 ha yr^{-1} from the mid-1970s to the mid-1980s. After New Jersey enacted its Wetlands Act of 1970, losses decreased from 1,300 ha yr^{-1} to less than 20 ha yr^{-1} (Tiner, 1998). Thus, for coastal wetlands, loss rates declined as public understanding of the value of wetlands increased and wetland protection laws and regulations became more stringent.

In another example of coastal wetland loss, Louisiana experienced an average annual loss rate of 10,800 ha yr^{-1} from 1958 to 1974 (Dunbar et al., 1992) as relative sea level increased due to land subsidence, sediment starvation, altered hydrology, and a number of other human activities (Baumann and Turner, 1990; Turner, 1997). This rate of wetland loss along the Gulf of Mexico represented a very high 0.86

percent yr^{-1}. The loss rate declined to 6,600 ha yr^{-1} from 1983 to 1990 (Dunbar et al., 1992). Oil and gas extraction activities in the inshore coastal marshes peaked in the 1970s and canal dredging for access to drill sites and for navigation fell precipitously thereafter. It is likely that this change, along with more stringent environmental control of permits for and design of wetland-related activities, has contributed substantially to the reduced loss rate.

Bottomland hardwood forests of the lower Mississippi River alluvial floodplain provide another sobering example. These forests covered approximately 8.5 million ha prior to European settlement. Now fewer than 2 million ha remain, more than 95 percent of them in Louisiana, Mississippi, and Arkansas (The Nature Conservancy, 1992). The 23 percent that have survived are seriously fragmented and have lost many of their original functions (Fig. 4-9).

Wetland Conversions

By themselves, estimates of net wetland losses provide an incomplete picture of the dynamics of change. A more complete picture would show that human activities converted millions of hectares of wetlands from one class to another. Through these conversions, some wetland classes increased in area at the expense of other types (Fig. 4-10). Considering the period from the mid-1970s to the mid-1980s, swamps and forested riparian wetlands suffered the greatest loss, 1.4 million ha. Although 800,000 ha were converted to agricultural and other land uses, large areas were converted to other wetland types: 292,000 ha to marshes, 195,000 ha to scrub and shrubs, and 32,000 ha to nonvegetated wetlands. Although shrub wetlands lost 208,000 ha to agriculture and other nonwetland uses, this was almost offset by the conversion of forested wetlands to shrubs, leaving a net loss of 65,000 ha. The net gain of 89,000 ha of marshes occurred despite a loss of 213,000 ha to agriculture and other land uses because 320,000 ha of swamps and shrub wetlands changed to marshes (Dahl and Johnson, 1991). In this example, most of the scrub–shrub wetlands are probably areas recently cut over for their timber. Thus, the trends reflect extensive logging, with the logged areas subsequently cleared for agriculture or changing to marshes because hydrologic alterations associated with logging were inimitable to tree regrowth. In this report, based on the National Wetlands Inventory classification scheme, scrub–shrub wetlands were considered a separate class of palustrine wetlands; in this book, we consider them with swamps and riparian wetlands.

Canadian Wetlands

Wetland Extent

Canada has about three times the area of wetlands than is found in the lower 48 states of the United States. By one estimate, Canada has about 127.2 million ha of wetlands, or about 14 percent of the country (Zoltai, 1988). Most of that area (111.3 million ha) is defined as peatland. The greatest concentration of Canadian wetlands can be found in the provinces of Manitoba and Ontario. The National Wetlands Working Group (1988), which provided a particularly comprehensive description of major regional wetlands in Canada, estimated that there were 22.5 million ha and 29.2 million ha, respectively, of wetlands in these two provinces, or about 41 percent of the total wetlands of Canada. Much of this total is boreal forested peatlands as

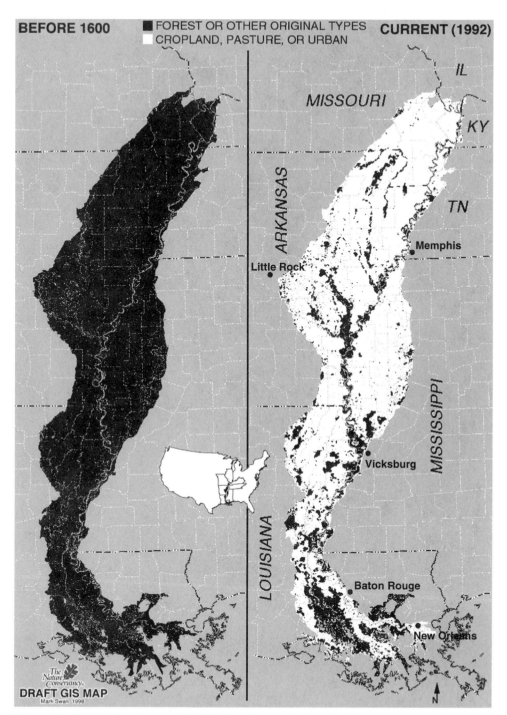

Figure 4-9 Historical and present distribution of bottomland wetland forests in the Mississippi River floodplain. (*Courtesy of The Nature Conservancy*)

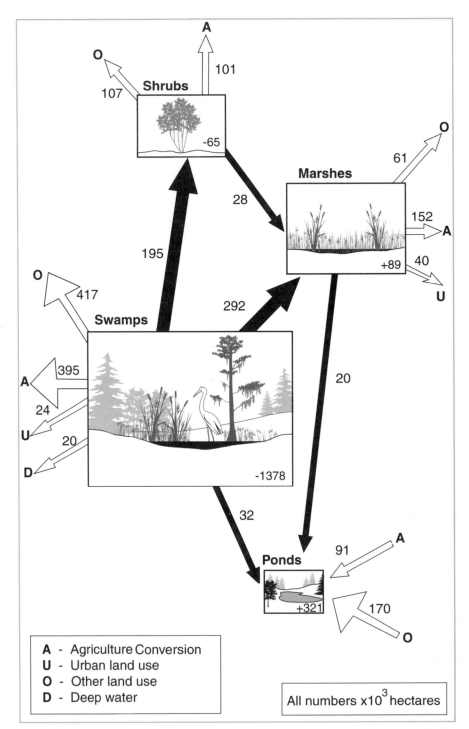

Figure 4-10 Wetland conversion in the conterminous United States, mid-1970s to mid-1980s. The figure shows how misleading the net change figures are. For example, although there was a net gain in freshwater marshes (89,000 ha), it occurred along with a loss of about 1,378,000 ha of swamps, some of which were converted to freshwater marshes. (*After Dahl and Johnson, 1991*)

bogs and fens, but there are also many deltaic and shoreline marshes and floodplain swamps in the region (Zoltai et al., 1988).

Wetland Losses

Because of the vastness of Canada and its wetlands, and because the low-population regions have had less impact on wetland loss than the coastal and southern regions of Canada, there has been no attempt to summarize the loss of wetlands in Canada to one number as has been the case for the conterminous United States. However, some estimates do exist for the more populated regions of Canada (Table 4-7). There has been a loss of 65 percent of coastal marshes in the Atlantic region, 68 percent of all wetlands in southern Ontario, and 70 percent of wetlands in the prairie pothole region (Rubec, 1994). The most extensive wetland loss has occurred in southern Ontario, Canada's most populated region, particularly from Windsor on the west toward and past Toronto on the east, where 80 to more than 90 percent wetland loss is common. Further north to Quebec City, Quebec, and further west to Thunder Bay, Ontario, loss rates are less (Table 4-7). Few data are available on the conversion of wetlands to other uses in rural areas, even in eastern Canada. However, studies have suggested that 32 percent of the tidal marshes along the St. Lawrence Estuary have been converted to agricultural use and that, on the St. Lawrence River between Cornwall and Quebec, there was a 7 percent loss in wetland area from 1950 to 1978 alone (National Wetlands Working Group, 1988).

North American Total

In total, about 2.4 million km² of wetlands are found in the United States and Canada combined, with more than half of that area of wetlands (1.3 million km²) in Canada. Collectively, North American wetlands represent about one-third of the world's known wetlands.

Table 4-7 **Wetland losses near major urban centers in Quebec and Ontario, Canada**

Urban Center Region (UCR)	UCR Area (km²)	Estimated Presettlement Wetland Area (km²)	Wetland Area in 1981 (km²)	Percentage Loss
Chicoutimi–Jonquiere (QC)	398	114	22	81
Montreal (QC)	3,148	1,945	228	88
Quebec City (QC)	1,000	405	168	58
Hamilton (ON)	863	102	18	82
Kitchener (ON)	1,482	101	63	37
London (ON)	972	27	7.7	70
Oshawa (ON)	191	18	4.9	73
Ottawa–Hull (ON,QC)	2,864	868	297	66
St. Catherines–Niagara (ON)[a]	1,282	522	119	77
Sudbury (ON)	1,174	148	80	45
Thunder Bay (ON)	529	104	73	30
Toronto (ON)	2,732	172	23	87
Windsor (ON)	403	422	11	97

[a] 1976 data.
Source: National Wetlands Working Group (1988).

Some Major Regional Wetlands of North America

Many regions in the United States and Canada support, or once supported, large, contiguous wetlands or many smaller and more numerous wetlands. These are called regional wetlands in this book. They are often large, heterogeneous wetland areas such as the Okefenokee Swamp in Georgia and Florida that defy categorization as one type of wetland ecosystem. Regional wetlands can also be large expanses of a single class of wetlands such as the prairie pothole region of the Dakotas and Minnesota in the United States and Manitoba, Saskatchewan, and Alberta in Canada, or the pocosins of North Carolina. Some regional wetlands such as the Great Dismal Swamp on the Virginia–North Carolina border have been drastically altered since presettlement times, and others such as the Great Kankakee Marsh of northern Indiana and Illinois and the Great Black Swamp of northwestern Ohio have almost disappeared as a result of extensive drainage programs. Still others such as the Peace–Athabasca Delta in Alberta, Canada, remain somewhat isolated and minimally impacted by humans.

Each of these regional wetland areas has had a significant influence on the culture and development of its region, and many have benefited from major investigations by wetland scientists. These studies have taught us much about wetlands and have identified much of their intrinsic value. Descriptions of some of the regional wetlands, located in Figures 3-2 and 4-11, are given here. The reader is referred to Chapters 9 through 15 and to the scientific literature cited throughout this book for a more complete discussion of the ecology of these wetlands. For those who wish a more complete description of some of these wetlands, Littlehales and Niering (1991) have combined to produce beautifully illustrated descriptions of North American wetlands and Dugan (1993) presents a significant section on North American wetlands.

The Everglades and Big Cypress Swamp

The southern tip of Florida, from Lake Okeechobee southward to the Florida Bay, harbors one of the unique regional wetlands in the world. The region encompasses three major types of wetlands in its 34,000-km² area: the Everglades, the Big Cypress Swamp, and the coastal mangroves and Florida Bay (Fig. 4-12a and b). The water that passes through the Everglades on its journey from Lake Okeechobee to the sea is often called a river of grass, although it is often only centimeters in depth and 80 km wide. The Everglades is dominated by sawgrass (*Cladium jamaicense*), which is actually a sedge, not a grass. The expanses of sawgrass, which can be flooded by up to a meter of water in the wet season (summer) and burned in a fire in the dry season (winter/spring), are interspersed with deeper water sloughs and tree islands, or *hammocks,* that support a vast diversity of tropical and subtropical plants, including hardwood trees, palms, orchids, and other air plants. To the west of the sawgrass Everglades is the Big Cypress Swamp, called big because of its great expanse, not because of the size of the trees. The swamp is dominated by cypress interspersed with pine flatwoods and wet prairie. It receives about 125 cm of rainfall per year but does not receive major amounts of overland flow as the Everglades does. The third major wetland type, mangroves, forms impenetrable thickets where the sawgrass and cypress swamps meet the saline waters of the coastline.

Numerous popular books and articles, including the classic *The Everglades: River of Grass* by Marjory Stoneman Douglas (1947), have been written about the Everglades

Figure 4-11 Location of some of the regional wetlands in North America as described in this chapter.

and its natural and human history. Davis (1940, 1943) gives some of the earliest and best descriptions of the plant communities in southern Florida. The findings of several years of ecological studies that have taken place in the Everglades have been summarized by Gleason (1974, 1984), Gunderson and Loftus (1993), and Davis and Ogden (1994).

Since about half of the original Everglades has been lost to agriculture (the Everglades Agricultural Area) in the north and to urban development in the east (Fig. 4-12b), concern for the remaining Everglades has been extended to the quality and quantity of water delivered to the Everglades through a series of canals and water conservation areas. The Everglades is currently the site of the largest proposed restoration effort in the United States. The project includes the expertise of all major federal and state environmental agencies and universities in the region, as well as a commitment of $8 billion by the federal government and the state of Florida. The comprehensive restoration blueprint includes plans for improving the water quality as it leaves

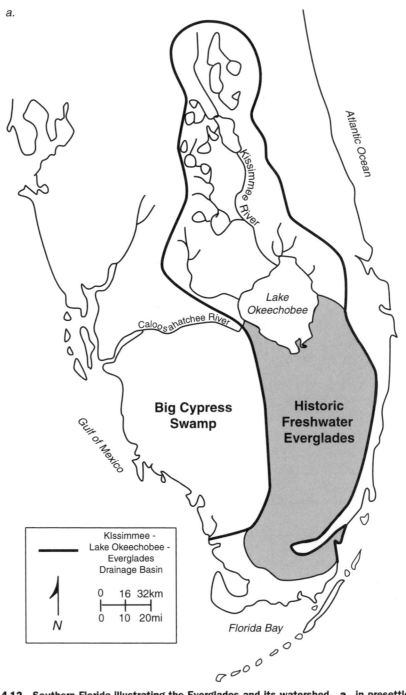

a.

Atlantic Ocean

Kissimmee River

Lake
Okeechobee

Caloosahatchee River

Gulf of Mexico

**Big Cypress
Swamp**

**Historic
Freshwater
Everglades**

KIssimmee -
Lake Okeechobee -
Everglades
Drainage Basin

0	16	32km
0	10	20mi

N

Florida Bay

Figure 4-12 Southern Florida illustrating the Everglades and its watershed **a.** in presettlement times and **b.** in present conditions with cities, drainage works, water conservation areas (WCAs), and the Everglades Agricultural Area (EAA). (*After Light and Dineen, 1994*)

b.

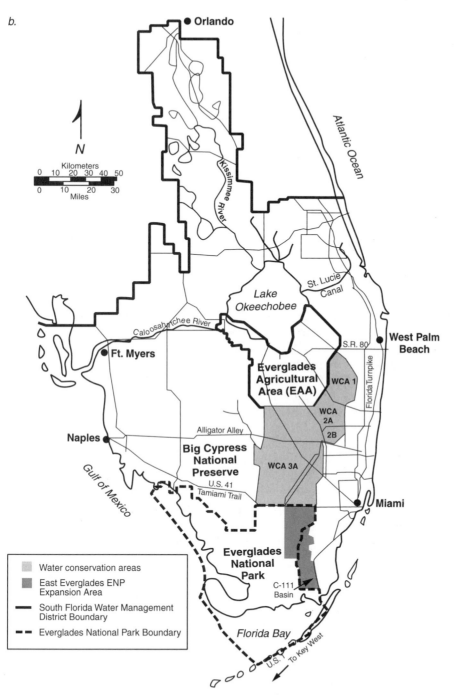

Figure 4-12 *(Continued).*

the agricultural areas and for modifying the hydrology to conserve and restore habitat for declining populations of wading birds such as the wood stork and the white ibis (Walters et al., 1992). North of the Everglades, there is a renewed effort to restore the ecological functions of the Kissimmee River, including many of its backswamp areas (Dahm, 1995). This river feeds Lake Okeechobee, which, in turn, spills over to the Everglades.

The Okefenokee Swamp

The Okefenokee Swamp on the Atlantic Coastal Plain of southeastern Georgia and northeastern Florida is a 1,750-km^2 mosaic of several different types of wetland communities. It is believed to have been formed during the Pleistocene or later when ocean water was impounded and isolated from the receding sea by a sand ridge (now referred to as the Trail Ridge) that kept water from flowing directly toward the Atlantic. The swamp forms the headwaters of two river systems: the Suwannee River, which flows southwest through Florida to the Gulf of Mexico, and the St. Mary's River, which flows southward and then eastward to the Atlantic Ocean.

Much of the swamp is now part of the Okefenokee National Wildlife Refuge, established in 1937 by Congress. The Okefenokee is named for an Indian word meaning "Land of Trembling Earth" because of the numerous vegetated floating islands that dot the wet prairies. Six major wetland communities comprise the Okefenokee Swamp: (1) pond cypress forest, (2) emergent and aquatic bed prairie, (3) broad-leaved evergreen forest, (4) broad-leaved shrub wetland, (5) mixed cypress forest, and (6) black gum forest (Auble et al., 1982). Pond cypress (*Taxodium distichum* var. nutans), black gum (*Nyssa sylvatica* var. biflora), and various evergreen bays (e.g., *Magnolia virginiana*) are found in slightly elevated areas where water and peat deposits are shallow. Open areas, called prairies, include lakes, emergent marshes of *Panicum* and *Carex,* floating-leaved marshes of water lilies (e.g., *Nuphar* and *Nymphaea*), and bladderwort (*Utricularia*). Fires that actually burn peat layers are an important part of this ecosystem and have recurred in a 20- to 30-year cycle when water levels became very low (Cypert, 1961). Many believe that the open prairies represent early successional stages, maintained by burning and logging, of what would otherwise be a swamp forest.

A more complete description of the Okefenokee Swamp is contained in a book compendium edited by Cohen et al. (1984) and in numerous papers (e.g., Wright and Wright, 1932; Cypert, 1961, 1972; Schlesinger and Chabot, 1977; Schlesinger, 1978; Auble et al., 1982; Bosserman, 1983a, b; Patten, 1994a).

The Pocosins

Pocosins are evergreen shrub bogs found on the Atlantic Coastal Plain from Virginia to northern Florida. These wetlands are particularly dominant in North Carolina, where an estimated 3,700 km^2 remained undisturbed or only slightly altered in 1980, whereas 8,300 km^2 were drained for other land uses between 1962 and 1979 (Richardson et al., 1981). The word *pocosin* comes from the Algonquin phrase for "swamp on a hill." In successional progression and in nutrient-poor acid conditions, pocosins resemble bogs typical of much colder climes and, in fact, were classified as bogs in the 1954 National Wetlands Survey (Shaw and Fredine, 1956). Richardson et al. (1981) described the typical pocosin ecosystem in North Carolina as being dominated by evergreen shrubs and pine (*Pinus serotina*). Pocosins are found "growing on water-logged, acid, nutrient poor, sandy or peaty soils located on broad, flat topographic plateaus, usually removed from large streams and subject to periodic burning" (Richardson et al., 1981). Draining and ditching for agriculture and forestry have affected pocosins in North Carolina. Proposed peat mining and phosphate mining could cause serious losses of these wetlands.

Summaries of the ecological, economic, and legal aspects of pocosin management are included in *Pocosin Wetlands,* edited by Richardson (1981), and in a review paper by Richardson (1983). Descriptions of the extent and phytosociology of these wetlands have been published by Wells (1928), Woodwell (1956), Wilson (1962), and Kologiski (1977); see also Chapter 13.

The Great Dismal Swamp

The Great Dismal Swamp is one of the northernmost "southern" swamps on the Atlantic Coastal Plain and one of the most studied and romanticized wetlands in the United States. The swamp covers approximately 850 km^2 in southeastern Virginia and northeastern North Carolina near the urban sprawl of the Norfolk–Newport News–Virginia Beach metropolitan area. It once extended over 2,000 km^2 (F. P. Day, 1982). The swamp has been severely affected by human activity during the past 200 years. Draining, ditching, logging, and fire played a role in diminishing its size and altering its ecological communities (Berkeley and Berkeley, 1976). The Great Dismal Swamp was once primarily a magnificent bald cypress–gum swamp that contained extensive stands of Atlantic white cedar (*Chamaecyparis thyoides*). Although remnants of those communities still exist today, much of the swamp is dominated by red maple (*Acer rubrum*), and mixed hardwoods are found in drier ridges (F. P. Day, 1982). In the center of the swamp lies Lake Drummond, a shallow, tea-colored, acidic body of water. The source of water for the swamp is thought to be underground along its western edge as well as surface runoff and precipitation. Drainage occurred in the Great Dismal Swamp as early as 1763 when a corporation called the Dismal Swamp Land Company, which was owned in part by George Washington, built a canal from the western edge of the swamp to Lake Drummond to establish farms in the basin. In general, that effort, like several others in the ensuing years, failed. Timber companies, however, found economic reward in the swamp by harvesting the cypress and cedar for shipbuilding and other uses. One of the last timber companies that owned the swamp, the Union Camp Corporation, gave almost 250 km^2 of the swamp to the federal government to be maintained as a national wildlife refuge (Dabel and Day, 1977).

An extensive literature describes the history and ecology of the Great Dismal Swamp (e.g., Whitehead, 1972; Berkeley and Berkeley, 1976). The structural and functional characteristics of the vegetation of the swamp have been described in several studies from Old Dominion University (Dabel and Day, 1977; Day and Dabel, 1978; McKinley and Day, 1979; Atchue et al., 1982, 1983; Gomez and Day, 1982; Train and Day, 1982; Day, 1982, 1987; Day et al., 1988; Laderman, 1989; Powell and Day, 1991). Also, at least one book, *The Great Dismal Swamp* (Kirk, 1979), describes the ecological and historical aspects of this important wetland. The extent and management of Atlantic white cedar, a dominant species in the Great Dismal Swamp, are presented by Sheffield et al. (1998).

The Big Rivers of the South Atlantic Coast

The Atlantic Coastal Plain, extending from North Carolina to the Savannah River in Georgia, is a land dominated by forested wetlands and marshes and cut by large rivers that drain the Piedmont and cross the Coastal Plain in a northwest–southeast direction to the ocean. These rivers include the Roanoke, Chowan, Little Pee Dee, Great Pee Dee, Lynches, Black, Santee, Congaree, Altamaha, Cooper, Edisto, Combahee, Coosawhatchie, and Savannah, as well as a host of smaller tributaries. Extensive bottomland hardwood forests and cypress swamps line these rivers and spread into the lowlands between them. Interspersed among these forests are hundreds of Carolina Bays, small elliptical-shaped lakes of uncertain origin surrounded by or overgrown with marshes and forested wetlands (Knight et al., 1984; Lide et al., 1995). The origin of these lake–wetland complexes, of which there are more than 500,000 along the eastern Coastal Plain, has been suggested to be meteor showers, wind, or ground-water flow (D. C. Johnson, 1942; H. T. Odum, 1951; Prouty, 1952; Savage, 1983). Along the coast, freshwater tides on the lower rivers formerly overflowed extensive forests, but many of these were cleared in the early 1800s to establish rice plantations. Most of the rice plantations have since been abandoned, and the former fields are now extensive freshwater marshes that have become a paradise for ducks and geese. The estuaries at the mouths of the rivers support the most extensive salt marshes on the Southeast Coast.

In 1825, Robert Mills wrote of Richland County, South Carolina: "What clouds of miasma, invisible to sight, almost continually rise from these sinks of corruption, and who can calculate the extent of its pestilential influence?" (quoted by Dennis, 1988). At that time, only 10,000 ha of the 163,000-ha county were being cultivated. Almost all the rest was a vast, untouched swamp. Our appreciation of these swamps has changed dramatically since that time, and parts of this swamp are now the Congaree Swamp National Monument and the Francis Beidler Forest; the latter includes the world's largest virgin cypress–tupelo (*Taxodium–Nyssa*) swamp and is now an Audubon sanctuary. Both preserves contain extensive stands of cypress more than 500 years old that escaped the logger's ax in the late 1800s. Descriptions of the wetlands of this region were written by Savage (1956, 1983), Cely (1974), Gaddy (1978), Wharton (1978), Porcher (1981), Sharitz and Gibbons (1982), and Dennis (1988).

The Prairie Potholes

A significant number of small wetlands, primarily freshwater marshes, are found in a 780,000-km^2 region in the states of North Dakota, South Dakota, and Minnesota and in the Canadian provinces of Manitoba, Saskatchewan, and Alberta (Fig. 4-13). These wetlands, called *prairie potholes,* were formed by glacial action during the Pleistocene. This region is considered one of the most important wetland regions in the world because of its numerous shallow lakes and marshes, its rich soils, and its warm summers (Weller, 1994). Wet-and-dry cycles are a natural part of the ecology of these prairie wetlands. In fact, many of the prairie potholes might not exist if there were no periodic dry periods. In some cases, dry periods of 1 to 2 years every 5 to 10 years are required to maintain emergent marshes. Another feature of this wetland region is the occasional presence of saline wetlands and lakes caused by high evapo-transpiration/precipitation ratios. Salinities as high as 370 parts per thousand (ppt) have been recorded for some hypersaline lakes in Saskatchewan. It is estimated that

Figure 4-13 Oblique aerial view of prairie pothole wetlands, showing many small ponds surrounded by wetland vegetation, in the middle of large agricultural fields. (*File photograph, U.S.
Fish and Wildlife Service, Northern Prairie Wildlife Research Center, Jamestown, North Dakota*)

50 to 75 percent of all the waterfowl originating in North America in any given year
comes from this region. More than half of the original 80,000 km² of wetlands in
the prairie pothole region may have been drained or altered, primarily for agriculture.
An estimated 500 km² of prairie pothole wetlands in North Dakota, South Dakota,
and Minnesota were lost between 1964 and 1968 alone (Weller, 1994). However,
major efforts to protect the remaining prairie potholes are progressing. Thousands
of square kilometers of wetlands have been purchased under the U.S. Fish and Wildlife
Service Waterfowl Production Area program in North Dakota alone since the early
1960s. The Nature Conservancy and other private foundations have also purchased
many wetlands in the region.

Much of the early literature on prairie potholes has centered on their role in the
breeding and feeding of migratory waterfowl and other nongame species (e.g., Stewart
and Kantrud, 1973; Kantrud and Stewart, 1977). A wetland classification was developed by Stewart and Kantrud (1971, 1972) for the prairie pothole region. The social
and economic implications of draining prairie pothole wetlands were described by
Leitch (1986, 1989), Leitch and Danielson (1979), and Leitch and Ekstrom (1989).
The ecology of the marshes of this region is discussed in books written by Weller
(1994) and Galatowitsch and van der Valk (1994) and edited by van der Valk (1989).

The Nebraska Sandhills

South of the prairie pothole region is an irregular-shaped region of 51,800 km² in
northern Nebraska described as "the largest stabilized dune field in the Western
Hemisphere" (Novacek, 1989). These Nebraska sandhills represent an interesting

and sensitive coexistence of wetlands, agriculture, and a very important aquifer-recharge area. The area was originally mixed-grass prairie composed of thousands of small wetlands in the interdunal valleys. Much of the region is now used for farming and rangeland agriculture, and many of the wetlands in the region have been preserved even though the vegetation is often harvested for hay or grazed by cattle. The Ogallala Aquifer is an important source of water for the region and is recharged to a significant degree through overlying dune sands and to some extent through the wetlands. It has been estimated that there are 558,000 ha of wetlands in the Nebraska sandhills, many of which are interconnected wet meadows or shallow lakes that contain water levels determined by both runoff and regional water table levels (LaBaugh, 1986; Winter, 1986). The wetlands in the region have been threatened by agricultural development, especially pivot irrigation systems that cause a lowering of the local water tables despite increased wetland flooding in the vicinity of the irrigation systems. Like the prairie potholes to the north, the Nebraska sandhill wetlands are important breeding grounds for numerous waterfowl, including about 2 percent of the Mallard breeding population in the north-central flyway (Novacek, 1989).

The Great Kankakee Marsh

For all practical purposes, this wetland no longer exists, although, until about 100 years ago, it was one of the largest marsh–swamp basins in the interior United States. Located primarily in northwestern Indiana and northeastern Illinois, the Kankakee River basin is 13,700 km^2 in size, including 8,100 km^2 in Indiana, where most of the original Kankakee Marsh was located. From the river's source to the Illinois line, a direct distance of only 120 km, the river originally meandered through 2,000 bends along 390 km, with a nearly level fall of only 8 cm/km. Numerous wetlands, primarily wet prairies and marshes, remained virtually undisturbed until the 1830s, when settlers began to enter the region. The naturalist Charles Bartlett (1904) described the wetland as follows:

> More than a million acres of swaying reeds, fluttering flags, clumps of wild rice, thick-crowding lily pads, soft beds of cool green mosses, shimmering ponds and black mire and trembling bogs—such is Kankakee Land. These wonderful fens, or marshes, together with their wide-reaching lateral extensions, spread themselves over an area far greater than that of the Dismal Swamp of Virginia and North Carolina.

The Kankakee region was considered a prime hunting area until the wholesale draining of the land for crops and pasture began in the 1850s. The Kankakee River and almost all of its tributaries in Indiana were channelized into a straight ditch in the late 19th century and early 20th century. In 1938, the Kankakee River in Indiana was reported to be one of the largest drainage ditches in the United States; the Great Kankakee Marsh was essentially gone by then.

Some historical and scientific literature exists for the Great Kankakee Marsh. Early accounts of the region were given by Bartlett (1904) and Meyer (1935). Scientists have studied riparian forested wetlands in the Illinois portion of the wetland, where some preservation was achieved (Mitsch et al., 1979b, c; Mitsch and Rust, 1984). More recently, there has been a major effort to restore parts of the Great Kankakee Marsh in northwestern Indiana.

The Great Black Swamp

Another vast wetland of the Midwest that has practically ceased to exist is the Great Black Swamp in what is now northwestern Ohio, although several wetlands remain as marshes managed for waterfowl at the western end of Lake Erie. The Great Black Swamp (Fig. 4-14) was once a combination of marshland and forested swamps that extended about 160 km long and 40 km wide in a southwesterly direction from the lake and covered an estimated 4,000 km². The bottom of an ancient extension of Lake Erie, the Black Swamp was named for the rich, black muck that developed in areas where drainage was poor as a result of several ridges that existed perpendicular to the direction of the flow to the lake (Herdendorf, 1987). There are numerous accounts of the difficulty that early settlers and armies (especially during the War of 1812) had in negotiating this region, and few towns of significant size have developed in the location of the original swamp. One account of travel through the region in the late 1700s suggested that "man and horse had to travel mid-leg deep in mud" for three days just to cover a distance of only 50 km (Kaatz, 1955). As with many other wetlands in the Midwest, state and federal drainage acts led to the rapid drainage of this wetland until little of it was left by the beginning of the 20th century. Only one small example of an interior forested wetland and several coastal marshes (about 150 km²) remain of the original western Lake Erie wetlands. Descriptions of some of these remaining coastal wetlands of Lake Erie are presented in Prince and D'Itri (1985), Herdendorf (1987), Krieger et al. (1992), Mitsch (1989, 1992b), Mitsch et al. (1994, 2000b), Özesmi and Mitsch (1997), and Gottgens et al. (1998).

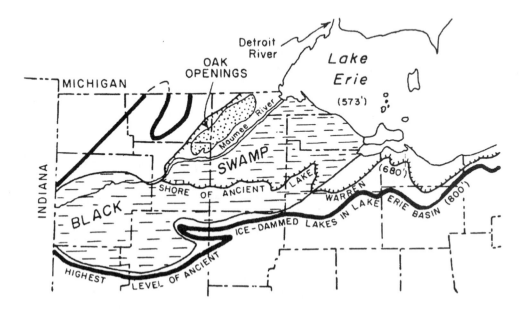

Figure 4-14 The Great Black Swamp as it probably existed 200 years ago in northwestern Ohio. Essentially none of this 4,000-km² wetland remains. (*After Forsyth, 1960*)

The Mississippi River Delta

As the Mississippi River reaches the last phase of its journey to the Gulf of Mexico in southeastern Louisiana, it enters one of the most wetland-rich regions of the world. The total area of marshes, swamps, and shallow coastal lakes covers more than 36,000 km². As the Mississippi River distributaries reach the sea, forested wetlands give way to freshwater marshes and then to salt marshes. The salt marshes are some of the most extensive and productive in the United States, and depend on the influx of fresh water, nutrients, and organic matter from upstream swamps. The total amount of freshwater and saltwater wetlands is decreasing at a rapid rate in coastal Louisiana, amounting to a total wetland loss of between 66 km² yr⁻¹ (Dunbar, 1990; Dunbar et al., 1992) and 90 km² yr⁻¹ (National Wetlands Research Center, Lafayette, LA) attributed to both natural and artificial causes (Dunbar, 1990; Dunbar et al., 1992).

Characteristic of the riverine portion of the delta, the Atchafalaya River, a distributary of the Mississippi River, serves as both a flood-relief valve for the Mississippi River and a potential captor of its main flow. The Atchafalaya Basin by itself is the third-largest continuous wetland area in the United States and contains 30 percent of all the remaining bottomland forests in the entire lower Mississippi alluvial valley (The Nature Conservancy, 1992). The river passes through this narrow 4,700-km² basin for 190 km, supplying water for 1,700 km² of bottomlands and cypress–tupelo swamps and another 260 km² of permanent bodies of water (Hern et al., 1980). The Atchafalaya Basin, contained within a system of artificial and natural levees, has had a controversial history of human intervention. Its flow is controlled by structures located where it diverges from the Mississippi River main channel, and it has been dredged for navigation and to prevent further infilling of the basin by Mississippi River silt. It has been channelized for oil and gas production. The old-growth forests were logged at the turn of the century, and the higher lands are now in agricultural production.

Another frequently studied wetland area in the delta is the Barataria Bay estuary in Louisiana, an interdistributary basin of the Mississippi River that is now isolated from the Mississippi River by a series of flood-control levees. This basin, 6,500 km² in size, contains 700 km² of wetlands, including cypress–tupelo swamps, bottomland hardwood forests, marshes, and shallow lakes (Conner and Day, 1976).

Currently, the U.S. Army Corps of Engineers, in cooperation with other federal and state agencies, as well as university personnel and private environmental and engineering firms, is designing and implementing a comprehensive strategy for conservation and restoration of the delta. This is being funded to the tune of millions of dollars per year under the federal Coastal Wetlands Planning, Protection and Restoration Act of 1990, but proposed major diversions of fresh water and silt from the Mississippi River into the coastal wetlands and bays are estimated to cost billions of dollars.

Gosselink et al. (1998) provide a recent review of the history and ecology of the delta. Numerous studies have been published on the delta wetlands and freshwater swamps (Penfound and Hathaway, 1938; Montz and Cherubini, 1973; Conner and Day, 1976, 1982, 1987; J. W. Day et al., 1977; Conner et al., 1981; Kemp and Day, 1984) and coastal marshes (Gosselink, 1984; Conner and Day, 1987). The land loss in coastal Louisiana was most recently described by Britsch and Kemp (1990), Dunbar (1990), and Dunbar et al. (1992).

San Francisco Bay

One of the most altered and most urbanized wetland areas in the United States is San Francisco Bay in northern California. The marshes surrounding the bay covered more than 2,200 km^2 when the first European settlers arrived. Almost 95 percent of these marshes have since been destroyed.

The ecological systems that make up San Francisco Bay range from deep, open water to salt and brackish marshes. The salt marshes are dominated by Pacific cordgrass (*Spartina foliosa*) and pickleweed (*Salicornia virginica*), and the brackish marshes support bulrushes (*Scirpus* spp.) and cattails (*Typha* spp.). Soon after the beginning of the Gold Rush in 1849, the demise of the bay's wetlands began. Industries such as agriculture and salt production first used the wetlands, clearing the native vegetation and diking and draining the marsh. At the same time, other marshes were developing in the bay as a result of rapid sedimentation. The sedimentation was caused primarily by upstream hydraulic mining. Sedimentation and erosion continue to be the greatest problems encountered in the remaining tidal wetlands.

Canada's Central and Eastern Province Wetlands

The marshes of this region, especially those along the Great Lakes in Ontario and in the St. Lawrence lowlands of Ontario and Quebec, are important habitats for migratory waterfowl (Glooschenko and Grondin, 1988). Several of the notable wetlands in the region include Long Point and Point Pelee on Lake Erie, the St. Clair National Wildlife Area on Lake St. Clair along the Great Lakes in southern Ontario, and the Cap Tourmente National Wildlife Area and Lac St.-François along the upper St. Lawrence River in eastern Ontario and southwestern Quebec. Cap Tourmente, a 2,400-ha tidal freshwater marsh complex located about 50 km northeast of Quebec City, Quebec, was the first wetland in Canada designated as a Ramsar site in international importance. It consists of both intertidal mud flats and freshwater marshes, as well as nontidal marshes, swamps, shrub swamps, and peatlands. The Cap Tourmente freshwater tidal marshes are subjected to heavy tidal flooding, with tidal amplitudes of 4.1 m at mean tides and 5.8 m during spring tides. The Cap Tourmente National Wildlife Area has a wide range of communities, including 400 ha of tidal marsh, 100 ha of coastal meadow, 700 ha of agricultural land, and 1,198 ha of forest. The *Scirpus americanus* marshes of the St. Lawrence such as those found at Cap Tourmente are restricted to the freshwater tidal portion of the river with only 4,000 ha remaining in the entire region. Although increasing numbers of greater snow geese have led to a depletion of *Scirpus* rhizomes, which may eventually cause a deterioration of the marshes at Cap Tourmente, the snow geese remain one of the notable features of this wetland during the migratory season.

Marshes of this region are considered the least stable class of wetlands and are significantly changed by variations in rainfall, flooding patterns, lake levels, and herbivory from one year to the next. The marshes along the Great Lakes are generally diked and heavily managed to buffer them from the year-to-year fluctuations in lake levels. This temperate region in Canada also has a considerable number of hardwood forested swamps dominated by red and sugar maples (*Acer rubrum* and *A. saccharinum*) and ash (*Fraxinus* spp.). Without human intervention, these swamps are quite stable; however, logging has been frequent, including clear-cutting. A clear-cut swamp is often replaced by a marsh and the successional pattern starts all over again.

The peatlands of northern Ontario and Manitoba are extensive regions that are used less by waterfowl and more by a wide variety of mammals, including moose, wolf, beaver, and muskrat. Wild rice (*Zizania palustris*), a common plant in littoral zones of boreal lakes, is often harvested for human consumption (Glooschenko and Grondin, 1988). Some of the boreal wetlands are mined for peat that is used for horticultural purposes or fuel. Fens of the region can be quite stable and are fairly common; bogs are also stable in this region, but less common. Radiocarbon dating of the bottom peat layers in Quebec bogs suggests that they began as fens between 9,000 and 5,500 years ago. Once formed, open bogs are quite stable, and forested bogs are even more stable, although they can revert to open bogs if fire occurs.

Hudson–James Bay Lowlands

A large wetland complex is found in northern Ontario and Manitoba and the eastern Northwest Territories, wrapping around the southern shore of the Hudson Bay (Fig. 4-5) and its southern extension, James Bay. These Hudson–James Bay lowlands are part of the vast subarctic wetland region of Canada, which stretches from the Hudson Bay northwestward to the northwestern corner of Canada and into Alaska and covers 760,000 km^2 of Canada (Zoltai et al., 1988). One of the largest and best described managed wetland sites in this region is the 24,000-km^2 Polar Bear Provincial Park in northern Ontario. Two additional sanctuaries of note are located in the southern James Bay: the Hannah Bay Bird Sanctuary and the Moose River Bird Sanctuary, which total 250 km^2. The region is dominated by extensive areas of mud flats, intertidal marshes, and supertidal meadow marshes, which grade into peatlands, interspersed with small lakes, thicket swamps, forested bogs and fens, and open bogs, fens, and marshes away from the shorelines. A band of the high subarctic wetland region along the southern shore of the bay is dominated by sedges (*Carex* spp.), cotton grasses (*Eriophorum* spp.), and clumps of birches (*Betula* spp.). The more southerly low subarctic wetland region is made up of low, open bogs, sedge–shrub fens, moist sedge-covered depressions, and open pools and small lakes separated by ridges of peat, lichen–peat-capped hummocks, raised bogs, and beach ridges. Even though the tidal range from the Hudson Bay is small, the gradual slope of land allows much tidal inundation of flats that vary from 1 to 5 km in width. Low-energy coasts with wide coastal marshes occur in the southern James Bay; high-energy coasts with sand flats and sand beaches are found along the Hudson Bay shoreline itself. Isostatic rebound following glacial retreat has resulted in the emergence of land from the bay at a rate of 1.2 m per century for the past 1,000 years, the greatest rate of glacial rebound in North America.

The coastal marshes, intertidal sand flats, and river mouths of the Hudson–James Bay lowlands serve as breeding and staging grounds for a large number of migratory waterfowl, including the Lesser Snow Goose, which was once in danger of disappearing but is now flourishing; Canada Goose; Black Duck; Pintail; Green-winged Teal; Mallard; American Wigeon; Shoveler; and Blue-winged Teal. The western and south-western coasts of the Hudson and James Bays form a major migration pathway for many shorebird species as well, including Red Knot, Short-billed Dowitcher, Dunlin, Greater Yellowlegs, Lesser Yellowlegs, Ruddy Turnstone, and Black-bellied Plover. The wetlands of Polar Bear Provincial Park itself provide nesting habitat for Red-throated, Arctic, and Common Loons; American Bittern; Common and Red-breasted Merganser; Yellow Rail; Sora; Sandhill Crane; and several gulls and terns.

Peace–Athabasca Delta

The Peace–Athabasca Delta in Alberta, Canada, is the largest freshwater inland boreal delta in the world and is relatively undisturbed by humans. It actually comprises three deltas: the Athabasca River delta (1,970 km²), the Peace River delta (1,684 km²), and the Birch River delta (168 km²). It is one of the most important waterfowl nesting and staging areas in North America and is the staging area for breeding ducks and geese on their way to the MacKenzie River lowlands, Arctic river deltas, and Arctic islands. The major lakes of the delta are very shallow (0.6–3.0 m) and have a thick growth of submerged and emergent vegetation during the growing season. The delta consists of very large flat areas of deposited sediments with some outcropping islands of the granitic Canadian Shield. The site has the following types of wetlands: emergent marshes, mud flats, fens, sedge meadows, grass meadows, shrub–scrub wetlands, deciduous forests of balsam (*Populus balsamifera*) and birch (*Betula* spp.), and coniferous forests dominated by white and black spruce (*Picea glauca* and *P. mariana*). Owing to the shallow water, high fertility, and relatively long growing season for that latitude, the area is an abundant food source of particular importance during drought years on the prairie potholes to the south. All four major North American flyways cross the delta, with the most important being the Mississippi and central flyways.

Studies in the early 1970s showed that water levels on the delta required regulating to mitigate the effects of the Bennett Dam, and weirs were subsequently constructed at Riviere des Rochers and Revillon Coupe. The dam, located upstream on the Peace River in British Columbia, was constructed in 1967 and caused a significant drop in water flow to the delta, resulting in insufficient water levels to fill the numerous perched basins in the area (Healey, 1994). The effects of the reduced water flow as a result of this dam construction have been mitigated by the construction of weirs on the Peace River tributaries, which have nearly restored natural summer peak water levels in the delta. The amplitude of seasonal and annual fluctuations, however, is still less than under the natural water flow regime prior to the construction of the dam (Healey, 1994). Between 1976 and 1986, about 30 to 40 percent of the delta experienced severe drying and woody vegetation was rapidly colonizing.

At least 215 species of birds, 44 species of mammals, 18 species of fish, and thousands of species of insects and invertebrates are found in the delta. Up to 400,000 birds use this wetland in the spring, and more than one million birds in autumn. Waterfowl species recorded in the delta area include Lesser Snow Goose, White-fronted Goose, Canada Goose, Tundra Swan, all four species of the loon, all seven species of North American grebe, and 25 species of duck. The world's entire population of the endangered Whooping Crane nests in the northern part of the delta area. The site also contains the largest undisturbed grass and sedge meadows in North America, which support an estimated 10,000 wood and plains buffalo.

Recommended Readings

Dahl, T. E. 1990. Wetlands Losses in the United States, 1780s to 1980s. U.S. Department of the Interior, Fish and Wildlife Service, Washington, DC. 21 pp.

Field, D. W., A. J. Reyer, P. V. Genovese, and B. D. Shearer. 1991. Coastal Wetlands of the United States. National Oceanic and Atmospheric Administration and U.S. Fish and Wildlife Service, Washington, DC. 59 pp.

Hall, J. V., W. E. Frayer, and B. O. Wilen. 1994. Status of Alaska Wetlands. U.S. Fish and Wildlife Service, Alaska Region, Anchorage, AK. 32 pp.

Littlehales, B., and W. A. Niering. 1991. Wetlands of North America. Thomasson-Grant, Charlottesville, VA. 160 pp.

National Wetlands Working Group. 1988. Wetlands of Canada. Ecological Land Classification Series, No. 24. Environment Canada, Ottawa, Ontario, and Polyscience Publications, Montreal, Quebec. 452 pp.

Niering, W. A. 1985. Wetlands. Alfred A. Knopf, New York. 638 pp.

Watzin, M. C., and J. G. Gosselink. 1992. Coastal Wetlands of the United States. Louisiana Sea Grant College Program, Baton Rouge, LA, and U.S. Fish and Wildlife Service, Lafayette, LA. 15 pp.

Part 2

The Wetland Environment

Chapter 5

Wetland Hydrology

Hydrologic conditions are extremely important for the maintenance of a wetland's structure and function. Hydrologic conditions affect many abiotic factors, including soil anaerobiosis, nutrient availability, and, in coastal wetlands, salinity. These, in turn, determine the biota that develop in a wetland. Finally, completing the cycle, biotic components are active in altering the wetland hydrology and other physicochemical features. The hydroperiod, or hydrologic signature of a wetland, is the result of the balance between inflows and outflows of water (called the water budget), the wetland basin geomorphology, and the subsurface conditions. The hydroperiod can have dramatic seasonal and year-to-year variations, yet it remains the major determinant of wetland processes. The major components of a wetland's water budget include precipitation, evapotranspiration, and overbank flooding in riparian wetlands; surface flows, groundwater fluxes, and tides in coastal wetlands. Simple determinations of hydroperiod, water budget, and turnover time in wetland studies can contribute to a better understanding of wetland function. Hydrology affects species composition and richness, primary productivity, organic accumulation, and nutrient cycling in wetlands. In general, productivity is high in wetlands that have high flowthrough of water and nutrients or in wetlands with pulsing hydroperiods. Decomposition in wetlands is slower in anaerobic standing water than it is under dry conditions. Although many wetlands are organic exporters, this cannot be generalized even within one wetland type. Nutrient cycling is enhanced by hydrology-mediated inputs, and nutrient availability is often changed by reduced conditions in wetland substrates.

The hydrology of a wetland creates the unique physicochemical conditions that make such an ecosystem different from both well-drained terrestrial systems and deepwater aquatic systems. Hydrologic pathways such as precipitation, surface runoff, groundwater, tides, and flooding rivers transport energy and nutrients to and from wetlands. Water depth, flow patterns, and duration and frequency of flooding, which are the result of all of the hydrologic inputs and outputs, influence the biochemistry of the soils and are major factors in the ultimate selection of the biota of wetlands. Biota ranging from microbial communities to vegetation to waterfowl are all constrained or enhanced by hydrologic conditions. An important point about wetlands—one that is often missed by ecologists who begin to study these systems—is this: *Hydrology is probably the single most important determinant of the establishment and maintenance of specific types of wetlands and wetland processes.* An understanding of rudimentary hydrology should be in the repertoire of any wetland scientist.

The Importance of Hydrology in Wetlands

Wetlands are transitional between terrestrial and open-water aquatic ecosystems. They are transitional in terms of spatial arrangement, for they are usually found between uplands and aquatic systems (see Fig. 1-19a). They are also transitional in the amount of water they store and process, and in other ecological processes that result from the water regime. Wetlands form the aquatic boundary of the habitats of many terrestrial plants and animals; they also form the terrestrial edge for many aquatic plants and animals. Hence, small changes in hydrology can result in significant biotic changes.

A conceptual model describing the fundamental role of hydrology in wetlands is shown in Figure 5-1. The starting point for the *hydrology* of a wetland is the climate and basin geomorphology. All things being equal, wetlands are more prevalent in cool or wet climates than in hot or dry climates. Cool climates have less water loss from the land via evapotranspiration, whereas wet climates have excess precipitation. The second important factor is the geomorphology of the landscape and basin. Steep terrain tends to have fewer wetlands than flat or gently sloping landscapes. Isolated basins have different potential for wetlands than do tidal-fed or river-fed environments. When climate, basin geomorphology, and hydrology are considered as one unit, it is referred to as a wetland's *hydrogeomorphology*. The hydrology of a wetland directly modifies and changes its *physicochemical environment* (chemical and physical properties), particularly oxygen availability and related chemistry such as nutrient availability, pH, and toxicity (e.g., the production of hydrogen sulfide). Hydrology also transports sediments, nutrients, and even toxic materials into wetlands, thereby further influencing the physicochemical environment. Except in nutrient-poor wetlands such as bogs, water inputs are the major source of nutrients to wetlands. Hydrology also causes water outflows from wetlands that often remove biotic and abiotic material such as dissolved organic carbon, excessive salinity, toxins, and excess sediments and detritus. Some modifications in the physicochemical environment, such as the buildup of sediments, can modify the hydrology by changing the basin geometry or affecting the hydrologic inflows or outflows (pathway A in Fig. 5-1).

Modifications of the physiochemical environment, in turn, have a direct impact on the *biota* in the wetland. When hydrologic conditions in wetlands change even slightly, the biota may respond with massive changes in species composition and richness and in ecosystem productivity. Biota such as emergent aquatic plants adapt to the anoxia in the sediments, although most vascular plant species are excluded by the anoxia. The level of nutrients in the sediments determines productivity and which species will dominate that productivity. Animals adapted to shallow water and this vegetation cover will, in turn, flourish. Microbes able to metabolize in anoxic conditions dominate the reduced sediments, while aerobic microorganisms survive in a thin layer of oxidized sediments and in the water column if oxygen is present there. When

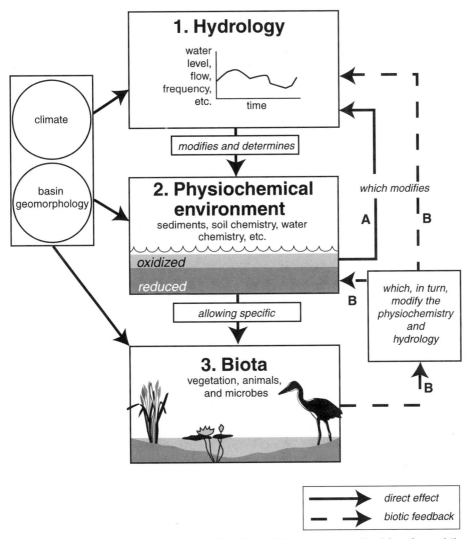

Figure 5-1 Conceptual diagram illustrating the effects of hydrology on wetland function and the biotic feedbacks that affect wetland hydrology. Pathways A and B are feedbacks to the hydrology and physicochemistry of the wetland.

hydrologic patterns remain similar from year to year, a wetland's biotic structural and functional integrity may persist for many years.

Biotic Control of Wetland Hydrology

Just as many other ecosystems exert feedback (cybernetic) control of their physical environments, wetland biota are not passive to their hydrologic conditions. Feedback loop B in Figure 5-1 shows that the biotic components of wetlands, in turn, can control the hydrology and chemistry of their environment through a variety of mechanisms.

Microbes, in particular, catalyze virtually all chemical changes in wetland soils and, thus, control nutrient availability to plants and even the production of phytotoxins such as sulfides. Plants, animals, and microbes that use these essential biological feedback mechanisms have been formally recognized in the ecological literature as *ecosystem engineers* (Jones et al., 1994, 1997; Alper, 1998). Plants cause changes in their physical environment through processes such as peat building, sediment trapping, nutrient retention, water shading, and transpiration. Wetland vegetation influences the hydrologic conditions of the physicochemical environment by binding sediments to reduce erosion, by trapping sediments, by interrupting water flows, and by building peat. Accumulated sediments and organic matter, in turn, interrupt water flows and can eventually decrease the duration and frequency by which the wetlands are flooded. Bogs build peat to the point at which they are no longer influenced at the surface by the inflow of mineral waters. Some trees in some southern swamps save water by their deciduous nature, their seasonal shading, and their relatively slow rates of transpiration. In more temperate climates, trees that invade shallow marshes and vernal pools can decrease water levels during the growing season by increasing transpiration, thus allowing even more woody plants to take over. Removal of these trees in what appears to be a dry forest sometimes surprisingly causes standing water and marsh vegetation to reappear (Golet et al., 1993).

Several animals are particularly noted for their contributions to hydrologic modifications and subsequent changes in wetlands (Fig. 5-2). The exploits of beavers (*Castor canadensis*) in much of North America in both creating and destroying wetland habitats are well known. They build dams on streams, backing up water across great expanses and creating wetlands where none existed before. In colonial times, beaver populations covered the entire American continent north of Mexico before they were drastically reduced by fur trappers. Oaks and Whitehead (1979) suggested that they have been an important causal force in the creation of the Great Dismal Swamp of Virginia and North Carolina. They are also considered a factor contributing to altered global carbon biogeochemistry (Naiman, 1988; Johnston and Naiman, 1990; Johnston, 1994).

The impact of beavers in presettlement times in North America could have been dramatic. Hey and Philippi (1995) estimated that a population of 40 million beavers could have accounted for 207,000 km^2 of beaver ponds (wetlands) in the upper Mississippi and Missouri River basins before European trappers entered the region and that, with the demise of the beaver, only 1 percent of those beaver ponds exist today.

Muskrats (*Ondatra zibethicus*) burrow through wetlands, changing flow patterns and sometimes water levels directly. They harvest large amounts of emergent vegetation for their food and to build winter lodges, thereby opening up large areas of marshes. Geese, especially Canada geese (*Branta canadensis*) and several varieties of Snow geese (*Chen* spp.), cause *eat-outs*, or major wetland vegetation removal by herbivory, in many parts of the world (Middleton, 1999). Newly planted wetlands are particularly susceptible to Canada geese eat-outs in North America. By removing vegetation cover, these herbivores reset the successional status of the wetlands and, thus, have a major impact on wetland hydrology.

The American alligator (*Alligator mississipiensis*) is known for its role in the Florida Everglades in constructing "gator holes" that serve as oases for fish, turtles, snails, and other aquatic animals during the dry season. In all of these cases, the biota of the ecosystem have contributed to their own survival, to the survival of other

Figure 5-2 Two animals, **a.** beaver and **b.** alligator, can significantly modify the hydrology and subsequent chemical and physical properties of wetlands. (*Photographs by M. Liptak and D. M. Dennis, reprinted by permission*)

species, and to the elimination of others by influencing the ecosystem's hydrology and other physical characteristics.

Studies of Wetland Hydrology

Many early wetland investigations that dealt with hydrology explored the relationships between hydrologic variables (usually water depth) and wetland productivity or species composition. Many case studies of the hydrology of wetlands have been published (Table 5-1); fewer studies have described in detail the hydrologic characteristics within specific wetland types. An exception to this has been the study of northern peatlands, for which a wealth of literature exists, including work from the former Soviet Union, the British Isles, and North America, particularly the Lake Agassiz region of northern Minnesota. Some of the more notable hydrology studies for other types of wetlands in the United States have included salt marshes in New England (Hemond and Fifield, 1982), cypress swamps in Florida (Heimburg, 1984), and large-scale wetland complexes at the Okefenokee Swamp in Georgia (Rykiel, 1984; Hyatt and Brook, 1984). A major contribution to the understanding of wetland hydrology resulted from papers published from a 1989 symposium on wetland hydrogeology held at the 28th International Geological Congress in Washington, DC (Winter and Llamas, 1993).

Wetland Hydroperiod

The *hydroperiod* is the seasonal pattern of the water level of a wetland and is like a hydrologic signature of each wetland type. It defines the rise and fall of a wetland's surface and subsurface water. It characterizes each type of wetland, and the constancy of its pattern from year to year ensures a reasonable stability for that wetland. The hydroperiod is an integration of all inflows and outflows of water, as influenced by physical features of the terrain and by proximity to other bodies of water. Many terms are used to describe qualitatively a wetland's hydroperiod. Table 5-2 gives several definitions that have been suggested by the U.S. Fish and Wildlife Service. For wetlands that are not subtidal or permanently flooded, the amount of time that a wetland is in standing water is called the *flood duration,* and the average number of times that a wetland is flooded in a given period is known as the *flood frequency.* Both terms are used to describe periodically flooded wetlands such as coastal salt marshes and riparian wetlands.

Some typical hydroperiods for a diverse set of wetlands are shown in Figure 5-3. A coastal salt marsh has a hydroperiod of semidiurnal flooding and dewatering superimposed on a twice-monthly pattern of spring and ebb tides, as illustrated in Figure 5-3a. Wetlands along coastlines often show some of this same spring-and-ebb pulsing (Fig. 5-3b), whereas others reflect seasonal water level changes of freshwater inflows and the water levels of the ocean itself (Fig. 5-3c). The hydroperiods of coastal wetlands along the Laurentian Great Lakes vary considerably, depending on whether pumps and water management are used or whether the marshes are open to the seasonal patterns of river flows and lake levels (Fig. 5-3d). Water levels for interior wetlands such as the prairie potholes of North America vary considerably from year to year (see the next section) with differences depending on the variability of the climate (Fig. 5-3e) and whether the marsh is affected by groundwater, which tends to make water levels less seasonally variable (Fig. 5-3f). Some of the

Table 5-1 Selected publications on wetland hydrology

Type of Study	References
PEATLAND STUDIES	
Minnesota peatland hydrology	Heinselman (1963, 1970)
	Bay (1967, 1969)
	Boelter and Verry (1977)
	Verry and Boelter (1979)
	Siegel (1983, 1988b)
	Siegel and Glaser (1987)
	Glaser et al. (1990, 1997a, b)
Alaska wetlands	Ford and Bedford (1987)
	Siegel (1988c, d)
Lake Michigan shoreline dune peatlands	Wilcox et al. (1986)
	Shedlock et al. (1993)
	Doss (1993)
Pocosins/Carolina Bays, North Carolina	Skaggs et al. (1991)
	Lide et al. (1995)
Russian peatland hydrology	Romanov (1968)
	Ivanov (1981)
Permafrost peatlands of North America	Woo and Winter (1993)
SPECIFIC STUDY SITES	
Florida cypress dome	Heimburg (1984)
	Ewel and Smith (1992)
Okefenokee Swamp, Georgia	Rykiel (1984)
	Hyatt and Brook (1984)
South Florida wetland hydrology	Duever (1988, 1990)
New England wetlands	O'Brien (1988)
Salt marsh hydrology—eastern coastline of the United States	Hemond and Fifield (1982)
	Nuttle and Hemond (1988)
	Harvey and Odum (1990)
	Nuttle and Harvey (1995)
	Harvey and Nuttle (1995)
Prairie pothole wetlands	Winter (1989)
	Winter and Rosenberry (1995)
Bottomland hardwood forests—southern United States	Gonthier (1996)
Constructed wetland hydrology—midwestern United States	Hensel and Miller (1991)
	Hey et al. (1994a)
	Hunt et al. (1996)
	Koreny et al. (1999)
Wastewater treatment wetlands	Kadlec (1989)
	Kadlec and Knight (1996)
	Werner and Kadlec (1996)
	Walker (1998)
Wetland delineation criteria	Skaggs et al. (1994)
GENERAL STUDIES OF WETLAND HYDROLOGY	
	Linacre (1976)
	Gosselink and Turner (1978)
	Carter et al. (1979)
	LaBaugh (1986)
	Carter (1986)
	Carter and Novitzki (1988)
	Winter (1988, 1992)
	Siegel (1988a)
	Winter and Llamas (1993)
	Doss (1995)

Table 5-2 Definitions of wetland hydroperiods

TIDAL WETLANDS

Subtidal—permanently flooded with tidal water
Irregularly exposed—surface exposed by tides less often than daily
Regularly flooded—alternately flooded and exposed at least once daily
Irregularly flooded—flooded less often than daily

NONTIDAL WETLANDS

Permanently flooded—flooded throughout the year in all years
Intermittently exposed—flooded throughout the year except in years of extreme drought
Semipermanently flooded—flooded during the growing season in most years
Seasonally flooded—flooded for extended periods during the growing season, but usually no
  surface water by end of growing season
Saturated—substrate is saturated for extended periods during the growing season, but standing
  water is rarely present
Temporarily flooded—flooded for brief periods during the growing season, but water table is
  otherwise well below surface
Intermittently flooded—surface is usually exposed with surface water present for variable periods
  without detectable seasonal pattern

Source: After Cowardin et al. (1979).

most seasonally variable wetlands are the vernal pools of central California, where surface water essentially disappears in this Mediterranean-type climate for all but four or five months (Fig. 5-3g). Cypress domes in central Florida have standing water during the wet summer season and dry periods in the late autumn and early spring (Fig. 5-3h). Low-order riverine wetlands such as the alluvial swamps in the southeastern United States respond sharply to local rainfall events rather than to general seasonal patterns (Fig. 5-3i). The hydroperiods of many bottomland hardwood forests and swamps in colder climates have distinct periods of surface flooding in the winter and early spring due to snow and ice conditions followed by spring floods but otherwise have a water table that can be a meter or more below the surface (Fig. 5-3j and k). Peatlands in cooler climates can have hydroperiods with little pronounced seasonal fluctuation, as in the fen from North Wales in Figure 5-3l. If peatlands such as the pocosins of North Carolina are located in regions of warm summers, significant patterns of seasonal water-level change will occur (Fig. 5-3m). The most dramatic hydroperiods result from high-order rivers that are more influenced by seasonal patterns of precipitation throughout a large watershed than by local precipitation, leading to a more predictable and seasonally distinct hydroperiod. For example, the annual fluctuation of water in the tropical floodplain forests along the Amazon River is a predictable seasonal pattern that can include a seasonal fluctuation in water level of 5 to 10 m due to flooding of upstream rivers (Fig. 5-3n).

Year-to-Year Fluctuations

The hydroperiod, of course, is not the same each year but varies statistically according to climate and antecedent conditions. Great variability can be seen from year to year

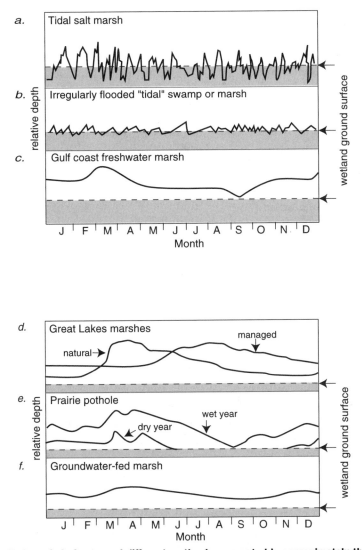

Figure 5-3 Hydroperiods for several different wetlands, presented in approximately the same relative scale: **a.** tidal salt marsh, Rhode Island; **b.** irregularly flooded "tidal" swamp or marsh; **c.** Gulf Coast freshwater marsh, Louisiana; **d.** Great Lakes marshes, northern Ohio (natural and managed); **e.** prairie pothole marsh with little groundwater flow (dry and wet years); **f.** groundwater-fed prairie pothole marsh; **g.** vernal pool, California; **h.** subtropical cypress dome, Florida; **i.** alluvial swamp, North Carolina; **j.** bottomland hardwood forest, northern Illinois; **k.** mineral soil swamp, Ontario, Canada; **l.** rich fen, North Wales; **m.** pocosin or Carolina Bay, North Carolina; **n.** tropical floodplain forest, Amazon River, Manaus, Brazil. (*After Nixon and Oviatt, 1973; Mitsch et al., 1979b; Gilman, 1982; Junk, 1982; Zedler, 1987; Mitsch et al., 1989; van der Valk, 1989; Brinson, 1993b; Woo and Winter, 1993*)

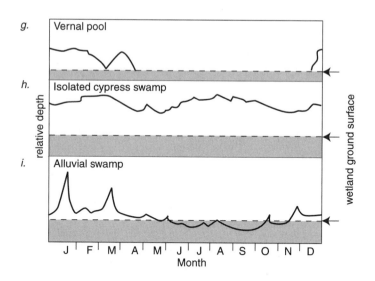

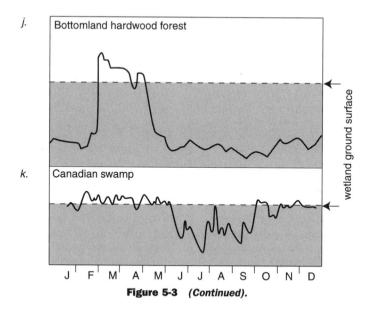

Figure 5-3 *(Continued).*

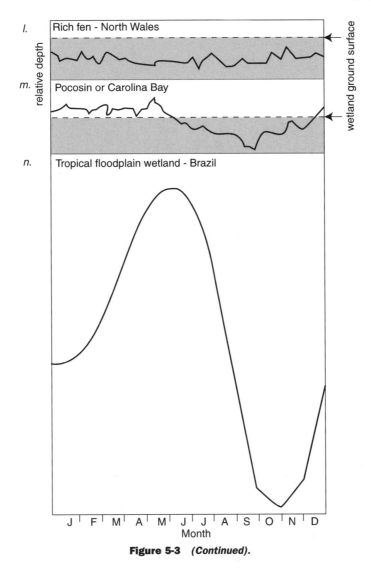

Figure 5-3 *(Continued).*

for some wetlands, as illustrated in Figure 5-4 for a prairie pothole regional wetland in Canada and for the Big Cypress Swamp region of south Florida. In the pothole region, a wet–dry cycle of 10 to 20 years is seen; spring is almost always wetter than fall but depths vary significantly from year to year (Fig. 5-4a). Figure 5-4b illustrates cases of an even seasonal rainfall pattern for the Big Cypress Swamp between 1957 and 1958, which caused a fairly stable hydroperiod throughout the year, and a significant dry season during 1970 and 1971, which caused the hydroperiod to vary about 1.5 m between high and low water.

Pulsing Water Levels

Water levels in most wetlands are generally not stable but fluctuate seasonally (high-order riparian wetlands), daily or semi-daily (types of tidal wetlands), or unpredictably

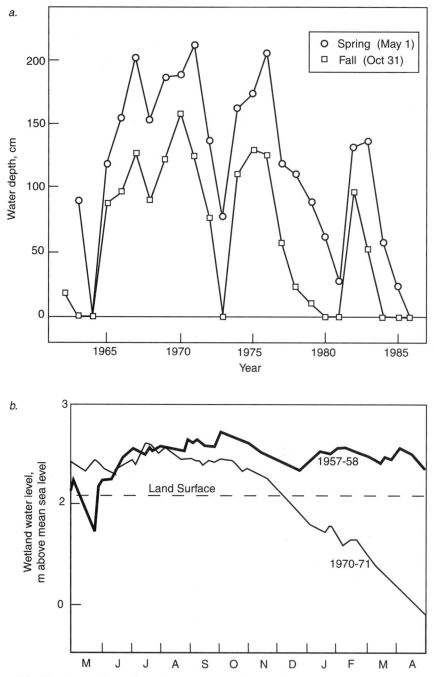

Figure 5-4 Year-to-year fluctuations in wetland water levels in two regions: **a.** spring and fall water depths for 1962–1986 in shallow open-water wetlands in the prairie pothole region of southwestern Saskatchewan, Canada, and **b.** wet and dry season hydrographs for the Big Cypress Swamp region of the Everglades, southwestern Florida. (*a. after Kantrud et al., 1989; Millar, 1971; b. after Freiberger, 1972; Duever, 1988*)

(wetlands in low-order streams and coastal wetlands with wind-driven tides). In fact, wetland hydroperiods that show the greatest differences between high and low water levels such as those seen in riverine wetlands are often caused by flooding "pulses" that occur seasonally or periodically (Junk et al. 1989). These pulses nourish the riverine wetland with additional nutrients and carry away detritus and waste products. Pulse-fed wetlands are often the most productive wetlands and are the most favorable for exporting materials, energy, and biota to adjacent ecosystems (see Specific Effects of Hydrology on Wetlands later in this chapter). Despite this obvious fact, many wetland managers, especially those who manage wetlands for waterfowl, often attempt to control water levels by isolating formerly open wetlands with dikes and pumps (Mitsch, 1992b). Fredrickson and Reid (1990) stated that "Because the goal of many [wetland] management scenarios is to counteract the effects of seasonal and long-term droughts, a general tendency is to restrict water level fluctuations in managed wetlands. This misconception is based on the fact that most wetland wildlife requires water for most stages in their life cycles." Kushlan (1989) suggested that, because the avian fauna that use wetlands often possess adaptations to fluctuating water levels, the active manipulation of water levels may be appropriate in artificially managed wetlands. A seasonally fluctuating water level, then, is the rule, not the exception, in most wetlands. Unfortunately, the retaining dikes that enable the control of water levels in managed marshes also tend to restrict water flows through the managed area, thus reducing both the range of fluctuation and the water renewal rate.

The Wetland Water Budget

The hydroperiod, or hydrologic state of a given wetland, can be summarized as being a result of the following factors:

1. The balance between the inflows and outflows of water
2. The surface contours of the landscape
3. Subsurface soil, geology, and groundwater conditions

The first condition defines the water budget of the wetland, whereas the second and the third define the capacity of the wetland to store water. The general balance between water storage and inflows and outflows, illustrated in Figure 5-5, is expressed as

$$\frac{\Delta V}{\Delta t} = P_n + S_i + G_i - ET - S_o - G_o \pm T \qquad (5.1)$$

where V = volume of water storage in wetlands
$\Delta V/\Delta t$ = change in volume of water storage in wetland per unit time, t
P_n = net precipitation
S_i = surface inflows, including flooding streams
G_i = groundwater inflows
ET = evapotranspiration
S_o = surface outflows
G_o = groundwater outflows
T = tidal inflow $(+)$ or outflow $(-)$

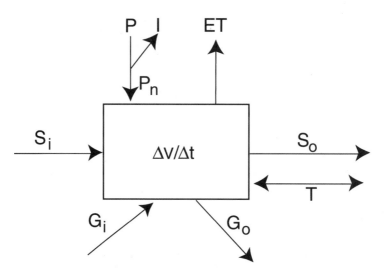

Figure 5-5 Generalized water budget for a wetland with corresponding terms as in Equation 5-1. *P*, precipitation; *ET*, evapotranspiration; *I*, interception; *P$_n$*, net precipitation; *S$_i$*, surface inflow; *S$_o$*, surface outflow; *G$_i$*, groundwater inflow; *G$_o$*, groundwater outflow; $\Delta V/\Delta t$, change in storage per unit time; *T*, tide or seiche.

The average water depth, *d*, at any one time, can further be described as

$$d = \frac{V}{A} \tag{5.2}$$

where A = wetland surface area

Each of the terms in Equation 5-1 can be expressed in terms of depth per unit time (e.g., cm/yr) or in terms of volume per unit time (e.g., m^3/yr.)

Examples of Water Budgets

Equation 5.1 serves as a useful summary of the major hydrologic components of any wetland water budget. Examples of hydrologic budgets for several wetlands are shown in Figure 5-6. The terms in the equation vary in importance according to the type of wetland observed; furthermore, not all terms in the hydrologic budget apply to all wetlands (Table 5-3). There is a large variability in certain flows, particularly in surface inflows and outflows, depending on the openness of the wetlands. An alluvial cypress swamp in southern Illinois received a gross inflow of floodwater from one flood that was more than 50 times the gross precipitation for the entire year (Fig. 5-6a). Even the net surface inflow from that flood (the water left behind after the flooding river receded) was three times the precipitation input for the entire year. Surface and groundwater inflows to a coastal Lake Erie marsh in Ohio were estimated to be almost 20 times the precipitation for a major part of a drought year (Fig. 5-6b), and tides contributed 10 times the precipitation to a black mangrove swamp in Florida (Fig. 5-6c). In contrast to these inflow-dominated wetlands, surface inflow is approximately equal to the precipitation inflow in the prairie pothole marshes of North Dakota (Fig. 5-6d); considerably less than the precipitation for the Okefenokee Swamp in Georgia (Fig. 5-6e) and a rich fen in North Wales (Fig. 5-6f); and essentially nonexistent in the upland Green Swamp of central Florida (Fig. 5-6g), a bog in Massachusetts (Fig.

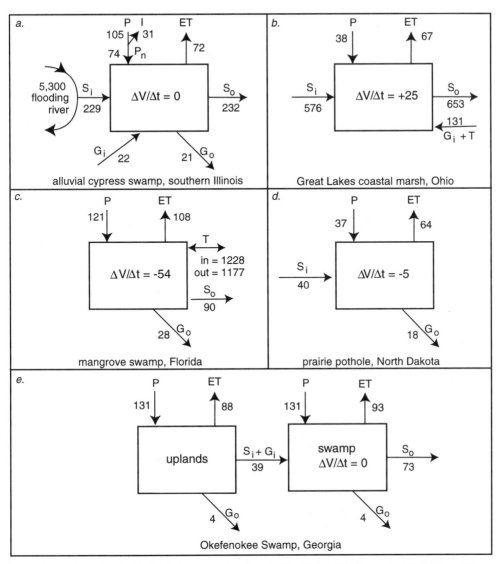

Figure 5-6 Annual water budgets for several wetlands. See Figure 5-5 for symbol definitions. All values are expressed in centimeters per year (cm/yr) except **b.**, which is March–September only. (*After Mitsch, 1979; Mitsch and Reeder, 1992; Twilley, 1982; Shjeflo, 1968; Rykiel, 1984; Gilman, 1982; Pride et al., 1966; Hemond, 1980; Richardson, 1983*)

5-6h), and a pocosin wetland of North Carolina (Fig. 5-6i). In most of these examples, the change in storage is small or zero, indicating that the water level at the end of the study period (usually an annual cycle) is close to where it was at the beginning of the study period.

Residence Time

A generally useful concept of wetland hydrology is that of the *renewal rate* or *turnover rate* of water, defined as the ratio of throughput to average volume within the system:

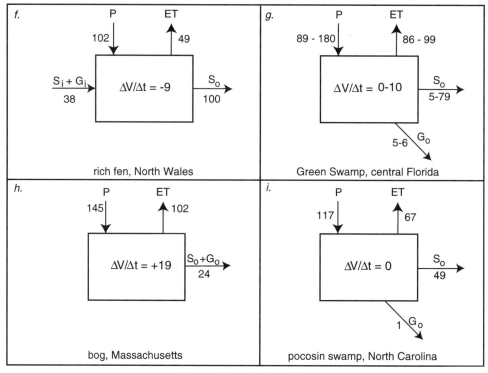

Figure 5-6 *(Continued)*.

Table 5-3 Major components of hydrologic budgets for wetlands

Component	Pattern	Wetlands Affected
Precipitation	Varies with climate, although many regions have distinct wet and dry seasons	All
Surface inflows and outflows	Seasonally, often matched with precipitation pattern or spring thaw; can be channelized as streamflow or nonchannelized as runoff; includes river flooding of alluvial wetlands	Potentially all wetlands except ombrotrophic bogs; riparian wetlands, including bottomland hardwood forests and other alluvial wetlands, are particularly affected by river flooding
Groundwater	Less seasonal than surface inflows and not always present	Potentially all wetlands except ombrotrophic bogs and other perched wetlands
Evapotranspiration	Seasonal with peaks in summer and low rates in winter. Dependent on meterorological, physical, and biological conditions in wetlands	All
Tides	One to two tidal periods per day; flooding frequency varies with elevation	Tidal freshwater and salt marshes; mangrove swamps

$$t^{-1} = \frac{Q_t}{V} \tag{5.3}$$

where t^{-1} = renewal rate (time^{-1})
      Q_t = total inflow rate (volume/time)
      V = average volume of water storage in wetland

Few measurements of renewal rates have been made in wetlands, although it is a frequently used parameter in limnological studies. Chemical and biotic properties are often determined by the openness of the system, and the renewal rate is an index of this since it indicates how rapidly the water in the system is replaced. The reciprocal of the renewal rate is the *turnover time* or *residence time* (t, sometimes called *detention time* by engineers for constructed wetlands), which is a measure of the average time that water remains in the wetland. Recent evidence, however, suggests that the theoretical residence time, as calculated by Equation 5-3, is often much longer than the actual residence time of water flowing through a wetland, because of nonuniform mixing (Kadlec and Knight, 1996; Werner and Kadlec, 1996; see Chapter 20). Because there are often parts of a wetland where waters are stagnant and not well mixed, the theoretical residence time (t) estimate should be used with caution when estimating the hydrodynamics of wetlands.

Precipitation

Wetlands occur most extensively in regions where *precipitation*, a term that includes rainfall and snowfall, is in excess of losses such as evapotranspiration and surface runoff (Fig. 5-7). The dividing line between excess precipitation in the eastern United States and precipitation deficit in the western part of the country is generally the Mississippi River. Wetland-rich regions such as the eastern provinces of Canada have 50 to 60 cm/yr of excess precipitation (precipitation less evaporative losses), whereas regions of the southwestern United States have precipitation deficits of 100 cm or more and generally few wetlands. Exceptions to this generality occur in coastal salt marshes fed by tides or in arid regions where riparian wetlands depend more on river flow than on local precipitation.

The fate of precipitation that falls on wetlands with forested, shrub, or emergent vegetation is shown in Figure 5-8. When some of the precipitation is retained by the vegetation cover, particularly in forested wetlands, the amount that actually passes through the vegetation to the water or substrate below is called *throughfall*. The amount of precipitation that is retained in the overlying vegetation canopy is called *interception*. Interception depends on several factors, such as the total amount of precipitation, the intensity of the precipitation, and the character of the vegetation, including the stage of vegetation development, the type of vegetation (e.g., deciduous or evergreen), and the strata of the vegetation (e.g., tree, shrub, or emergent macrophyte). The percentage of precipitation that is intercepted in forests varies between 8 and 35 percent. One review cites a median value of 13 percent for several studies of deciduous forests and 28 percent for coniferous forests (Dunne and Leopold, 1978). The water budget in Figure 5-6a illustrates that 29 percent of precipitation in a forested wetland was intercepted by a canopy dominated by *Taxodium distichum* a deciduous conifer.

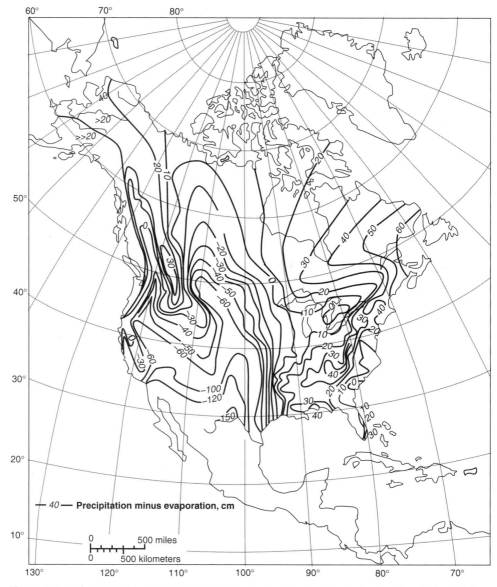

Figure 5-7   Annual precipitation less open-water evaporation (cm) in North America. (*After Winter and Woo, 1990; Woo and Winter, 1993*)

Little is known about the interception of precipitation by emergent herbaceous macrophytes, but it probably is similar to that measured in grasslands or croplands. Essentially, in those systems, interception at maximum growth can be as high as that in a forest (10–35 percent of gross precipitation). An interesting hypothesis about interception and the subsequent evaporation of water from leaf surfaces is that, because the same amount of energy is required whether water evaporates from the surface of a leaf or is transpired by the plant, the evaporation of intercepted water is not "lost" because it may reduce the amount of transpiration loss that occurs (Dunne and

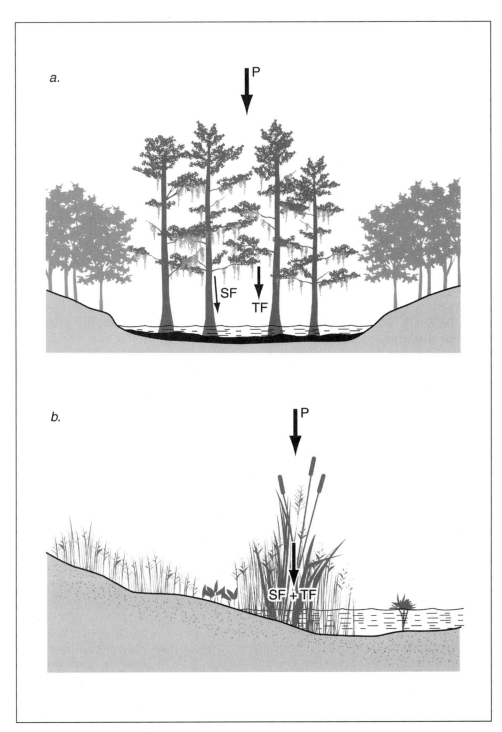

Figure 5-8 Fate of precipitation in **a.** a forested wetland and **b.** a marsh. *P*, precipitation; *TF*, throughfall; *SF*, stemflow.

Leopold, 1978). This argues that wetlands with high and low interception may be similar in overall water loss to the atmosphere.

Another term related to precipitation, *stemflow*, refers to water that passes down the stems of the vegetation (Fig. 5-8). This flow is generally a minor component of the water budget of a wetland. For example, Heimburg (1984) found that stemflow was, at maximum, 3 percent of throughfall in cypress dome wetlands in north-central Florida.

These terms are related in a simple water balance as follows:

$$P = I + TF + SF \tag{5.4}$$

where P = total precipitation
I = interception
TF = throughfall
SF = stemflow

The total amount of precipitation that actually reaches the water's surface or substrate of a wetland is called the net precipitation (P_n) and is defined as

$$P_n = P - I \tag{5.5}$$

Combining Equations 5-4 and 5-5 yields the most commonly used form for estimating net precipitation in wetlands

$$P_n = TF + SF \tag{5.6}$$

Surface Flow

Watersheds and Runoff

The percentage of precipitation that becomes surface flow depends on a number of variables, with climate being the most important (Fig. 5-9). Humid cool regions such as the Pacific Northwest, western British Columbia, and the northeastern Canadian provinces have 60 to 80 percent of precipitation converted to runoff. In the arid southwestern United States, less than 10 percent of the already low precipitation becomes runoff. This difference is related, in large part, to the higher temperatures in the arid Southwest, which translate into higher evapotranspiration rates, greater soil moisture deficits, and higher soil infiltration rates than in the Northeast. Even though runoff in arid regions is small relative to that in humid areas, it does contribute streamflow, which is an important part of a riparian wetland's water budget. Wetlands can be receiving systems for surface water flows (*inflows*), or surface water streams can originate in wetlands to feed downstream systems (*outflows*). Surface outflows are found in many wetlands that are located in the upstream reaches of a watershed. These wetlands are often important water flow regulators for downstream rivers. Some wetlands have surface outflows that develop only when their water stages exceed a critical level.

Wetlands are subjected to surface inflows of several types. *Overland flow* is non-channelized sheet flow that usually occurs during and immediately following rainfall or a spring thaw, or as tides rise in coastal wetlands. A wetland influenced by a drainage basin may receive channelized *streamflow* during most or all of the year. Wetlands are often an integrated part of a stream or river, for example, as instream freshwater marshes or riparian bottomland forests. Wetlands that form in wide shallow expanses

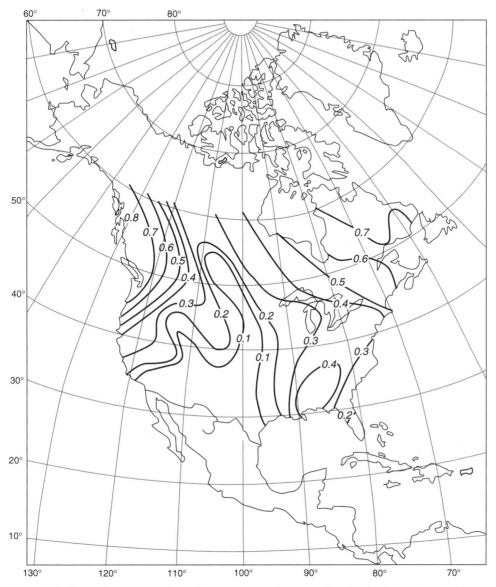

Figure 5-9 Ratio of mean annual runoff to mean annual precipitation for North America. A higher number indicates a higher percentage of precipitation that becomes runoff and streamflow. (*After Hare, 1980; Woo and Winter, 1993*)

of river channels or floodplains adjacent to them are greatly influenced by the seasonal streamflow patterns of the river. Wetlands can also receive surface inflow from seasonal or episodic pulses of flood flow from adjacent streams and rivers that may otherwise not be connected hydrologically with the wetland. Coastal saline and brackish wetlands are also significantly influenced by freshwater runoff and streamflow (in addition to tides) that contribute nutrients and energy to the wetland and often ameliorate the effects of soil salinity and anoxia.

Surface inflow from a drainage basin into a wetland is usually difficult to estimate without a great deal of data. Nevertheless, it is often one of the most important sources of water in a wetland's hydrologic budget. The direct runoff component of streamflow refers to rainfall during a storm that causes an immediate increase in streamflow. An estimate of the amount of precipitation that results in direct runoff, or *quickflow,* from an individual storm can be determined from the following equation:

$$S_i = R_p P A_w \tag{5.7}$$

where S_i = direct surface runoff into wetland (m^3 per storm event)
  R_p = hydrologic response coefficient
  P = average precipitation in watershed (m)
  A_w = area of watershed draining into wetland (m^2)

This equation states that the flow is proportional to the volume of precipitation ($P \times A_w$) on the watershed feeding the wetland in question. The values of R_p, which represent the fraction of precipitation in the watershed that becomes direct surface runoff, range from 4 to 18 percent for small watersheds in the eastern United States (R. Lee, 1980). As suggested by Figure 5-9, R_p increases with latitude. Otherwise, slope and type of vegetation appear to have little influence on R_p in a watershed with a mature forest cover. As the following paragraph suggests, land use and soil type can strongly influence runoff.

While Equation 5-7 predicts the volume of direct runoff caused by a storm event, in some cases wetland scientists and managers might be interested in calculating the peak runoff (*flood peak*) into a wetland caused by a specific rainfall event. Although this is generally a difficult calculation for large watersheds, a formula with the unlikely name of the *rational runoff method* is a widely accepted and useful way to predict peak runoff for watersheds less than 80 ha in size. The equation is given by

$$S_{i(pk)} = 0.278 C I A_w \tag{5.8}$$

where $S_{i(pk)}$ = peak runoff into wetland (m^3/s)
  C = rational runoff coefficient (see Table 5-4)
  I = rainfall intensity (mm/h)
  A_w = area of watershed draining into wetland (km^2)

The coefficient C, which ranges from 0 to 1 (Table 5-4), depends on the upstream land use. Concentrated urban areas have a coefficient ranging from 0.5 to 0.95, and rural areas have lower coefficients that greatly depend on soil type, with sandy soils lowest ($C = 0.1$–0.2) and clay soils highest ($C = 0.4$–0.5).

Channelized Streamflow

Channelized streamflow into and out of wetlands is described simply as the product of the cross-sectional area of the stream (A) and the average velocity (v) and can be determined through stream velocity measurements in the field:

$$S_i \text{ or } S_o = A_x v \tag{5.9}$$

where S_i, S_o = surface channelized flow into or out of wetland (m^3/s)
  A_x = cross-sectional area of stream (m^2)
  v = average velocity (m/s)

Table 5-4   **Values of the rational runoff coefficient *C* used to calculate peak runoff**

	C
URBAN AREAS	
Business areas: high-value districts	0.75–0.95
neighborhood districts	0.50–0.70
Residential areas: single-family dwellings	0.30–0.50
multiple-family dwellings	0.40–0.75
suburban	0.25–0.40
Industrial areas: light	0.50–0.80
heavy	0.60–0.90
Parks and cemeteries	0.10–0.25
Playgrounds	0.20–0.35
Unimproved land	0.10–0.30
RURAL AREAS	
Sandy and gravelly soils: cultivated	0.20
pasture	0.15
woodland	0.10
Loams and similar soils: cultivated	0.40
pasture	0.35
woodland	0.30
Heavy clay soils; shallow soils over bedrock:	
cultivated	0.50
pasture	0.45
woodland	0.40

Source: Dunne and Leopold (1978).

The velocity can be determined in a number of ways, ranging from hand-held velocity meter readings taken at various locations in the stream cross section to the floating-orange technique where the velocity of a floating orange or similar fruit (which is 90 percent or more water and therefore floats but just beneath the water surface) is timed as it goes downstream. If a continuous or daily record of streamflow is needed, then a *rating curve* (Fig. 5-10), a plot of instantaneous streamflow (as estimated using Eq. 5-9) versus stream elevation or stage, is useful. If this type of rating curve is developed for a stream (the basis of most hydrologic streamflow gauging stations operated by the U.S. Geological Survey), then a simple measurement of the stage in the stream can be used to determine the streamflow. Since hydrographs generally assume a constant water gradient, caution should be taken in using this approach for streams flowing into wetlands to ensure that no "backwater effect" of the wetland's water level will affect the stream stage at the point of measurement.

When an estimate of surface flow into or out of a riverine wetland is needed and no stream velocity measurements are available, the *Manning equation* can often be used if the slope of the stream and a description of the surface roughness are known:

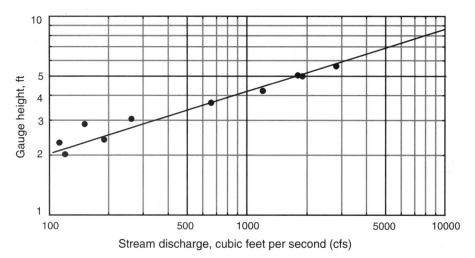

Figure 5-10 Rating curve for streamflow determination as a function of stream stage. 100 cfs = 2.832 m³/s. (*After Dunne and Leopold, 1978*)

$$S_i \text{ or } S_o = \frac{A_x R^{2/3} s^{0.5}}{n} \tag{5.10}$$

where n = roughness coefficient (Manning coefficient; see Table 5-5)
R = hydraulic radius (m) [cross-sectional area divided by the wetted perimeter; this is an estimate of the relative portion of the stream cross section (and, hence, flow volume) in contact with the stream bed]
s = channel slope (dimensionless)

The equation states that flow is proportional to stream cross section, as modified by the roughness of the stream bed and the proportion of flow in contact with that bed. Although the potential exists for their use in wetland studies, the roughness coefficients given in Table 5-5 and the Manning equation (Eq. 5.10) have not been used very often. The relationship is particularly useful for estimating streamflow where velocities are too slow to measure directly, and to estimate flood peaks from high-water marks on ungauged streams. These circumstances are common in wetland studies.

Table 5-5 Roughness coefficients (*n*) for manning equation used to determine streamflow in natural streams and channels

Stream Conditions	Manning Coefficient, *n*
Straightened earth canals	0.02
Winding natural streams with some plant growth	0.035
Mountain streams with rocky stream bed	0.040–0.050
Winding natural streams with high plant growth	0.042–0.052
Sluggish streams with high plant growth	0.065
Very sluggish streams with high plant growth	0.112

Source: Chow (1964) and R. Lee (1980).

Floods and Riparian Wetlands

A special case of surface flow occurs in wetlands that are in floodplains adjacent to rivers or streams and are occasionally flooded by those rivers or streams. These ecosystems are often called *riparian wetlands* (Chapter 15). The flooding of these wetlands varies in intensity, duration, and number of floods from year to year, although the probability of flooding is fairly predictable. In the eastern and midwestern United States and in much of Canada, a pattern of winter or spring flooding caused by rains and sudden snowmelt is often observed (Fig. 5-11). When river flow begins to overflow onto the floodplain, the streamflow is referred to as *bankfull discharge*. A hydrograph of a stream that flooded its riparian wetlands above bankfull discharge is shown in Figure 5-12. Leopold et al. (1964) showed a remarkable consistency in the hydrographs of riparian streams in the United States, in that they tend to overflow their banks at intervals between 1 and 2 years for bankfull discharge, with an average of approximately 1.5 years (Fig. 5-13). The *recurrence interval* is the average interval between the recurrence of floods at a given or greater depth. The inverse of the recurrence interval is the average probability of flooding in any one year. Figure 5-13 indicates that a stream in the midwestern and southern United States will overflow its banks onto the adjacent riparian forest with an average recurrence interval of 1.5 years (or a probability of 1/1.5, or 67 percent, of overbank flooding in any one year). Stated another way, these rivers, on average, overflow their banks in 2 out of every 3 years. Figure 5-13 also illustrates that twice bankfull discharge occurs at recurrence intervals of approximately 5 years; this flow, however, results in only a 40 percent greater river

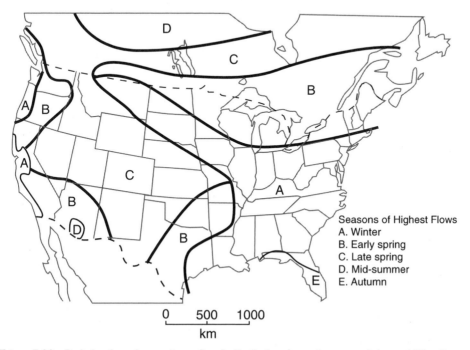

Figure 5-11 Periods of maximum streamflow in North American streams and rivers. (*After Beaumont, 1975*)

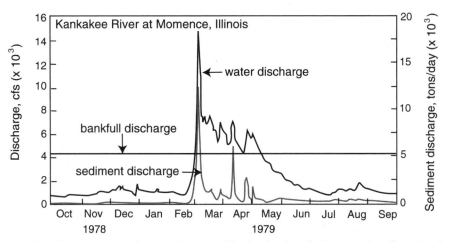

Figure 5-12 River hydrograph from northeastern Illinois, showing discharge and sediment load of the river and discharge at which a riparian wetland is flooded (bankfull discharge). 1,000 cfs = 28.32 m³/s. (*After Bhowmik et al., 1980*)

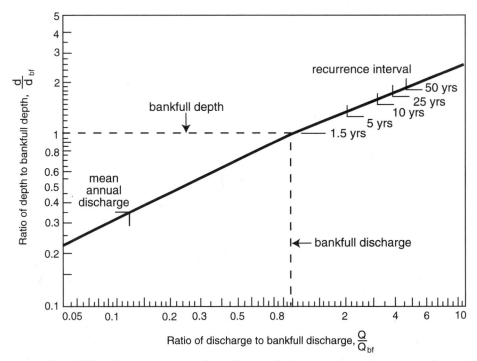

Figure 5-13 Relationships among streamflow (discharge), stream depth, and recurrence interval for streams and rivers in the midwestern and southern United States. *Q*, stream discharge; *Q*bf, bankfull discharge; *d*, stream depth; *d*bf, bankfull depth (depth of river with floodplain is initially flooded). (*After Leopold et al., 1964*)

132

depth over bankfull depth on the floodplain. This predictable relationship suggests that in natural stream systems the size of a stream channel is related to the hydraulic energy that scours the stream bed.

Weir Flow Measurement

When it is confined to a channel, surface outflow from a wetland can be determined with the general equations for surface flow (see Eqs. 5-9 and 5-10). When a continuous record is desirable, a rating curve related to stream stage, as shown in Figure 5-10, can be developed. When a weir or other control structure is used (Fig. 5-14), the outflow of a wetland can also be estimated to be a function of the water level in the wetland itself according to the equation:

$$S_o = xL^y \tag{5.11}$$

where S_o = surface outflow
L = wetland water level above a control structure crest
(level at which flow just begins)
x, y = calibration coefficients

If a control structure such as a rectangular or V-notched weir is used to measure the outflow from a wetland, standard equations of the form of Equation 5-11 can be obtained from water measurement manuals (e.g., U.S. Department of Interior, 1984). Care should be taken to calibrate standard weir equations with actual measurements of streamflow and water level.

Figure 5-14 V-notched weir used for measuring surface water flow in small streams.

Groundwater

Recharge and Discharge Wetlands

Groundwater can heavily influence some wetlands, whereas in others it may have hardly any effect at all. The influence of wetland recharge and discharge on groundwater resources has often been cited as one of the most important attributes of wetlands, but it does not hold for all wetland types; nor is there sufficient experience with site-specific studies to make many generalizations (Siegel, 1988a). Groundwater inflow results when the surface water (or groundwater) level of a wetland is lower hydrologically than the water table of the surrounding land (called a *discharge wetland* by geologists, who generally view their water budget from a groundwater, not a wetland, perspective). Wetlands can intercept the water table in such a way that they have only inflows and no outflows, as shown for a prairie marsh in Figure 5-15a. Another type of discharge wetland, called a *spring* or *seep* wetland, is often found at the base of steep slopes where the groundwater surface intersects the land surface (Fig. 5-15b). This type of wetland can be an isolated low point in the landscape; more often, it discharges excess water downstream as surface water or as groundwater, as shown in the riparian wetland in Figure 5-15c.

When the water level in a wetland is higher than the water table of its surroundings, groundwater will flow out of the wetland (called a *recharge wetland;* Fig. 5-15d). When a wetland is well above the groundwater of the area, the wetland is referred to as being *perched* (Fig. 5-15e). This type of wetland, also referred to as a "surface water depression wetland" by Novitzki (1979), loses water only through infiltration into the ground and through evapotranspiration. Tidally influenced wetlands often have significant groundwater inflows that can reduce soil salinity and keep the wetland soil wet even during low tide (Fig. 5-15f).

A final type of wetland, one that is fairly common, is little influenced by or has little influence on groundwater. Because wetlands often occur where soils have poor permeability, the major source of water can be restricted to surface water runoff, with losses occurring only through evapotranspiration and other surface outflows. This type of wetland often has fluctuating hydroperiods and intermittent flooding (e.g., some prairie potholes; Fig. 5-3e), and standing water is dependent on seasonal surface inflows. If, on the other hand, such a wetland were to be influenced by groundwater, its water level would be better buffered against dramatic seasonal changes or at least it would be semipermanently flooded (Fig. 5-3f; Winter, 1988).

Darcy's Law

The flow of groundwater into, through, and out of a wetland is often described by *Darcy's law,* an equation familiar to groundwater hydrologists. This law states that the flow of groundwater is proportional to (1) the slope of the piezometric surface (the hydraulic gradient) and (2) the hydraulic conductivity, or *permeability,* the capacity of the soil to conduct water flow. In equation form, Darcy's law is given as

$$G = kA_x s \qquad (5.12)$$

where G = flow rate of groundwater (volume per unit time)
k = hydraulic conductivity or permeability (length per unit time)
A_x = groundwater cross-sectional area perpendicular to the direction of flow
s = hydraulic gradient (slope of water table or piezometric surface)

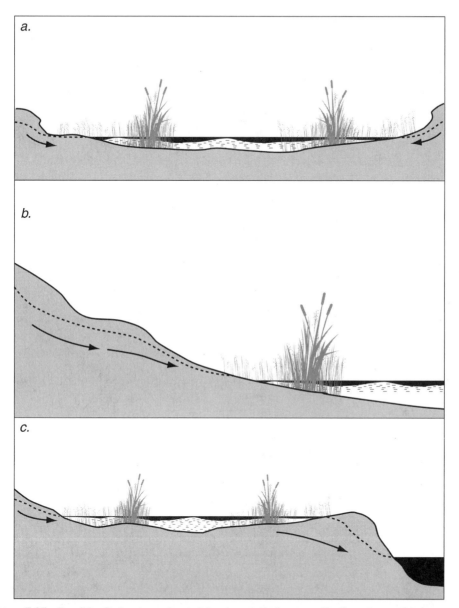

Figure 5-15   Possible discharge–recharge interchanges between wetlands and groundwater systems including   **a.** marsh as a depression receiving groundwater flow (''discharge wetland''); **b.** groundwater spring or seep wetland or groundwater slope wetland at the base of a steep slope; **c.** floodplain wetland fed by groundwater;   **d.** marsh as a ''recharge wetland'' adding water to groundwater;   **e.** perched wetland or surface water depression wetland;   **f.** groundwater flow through a tidal wetland.

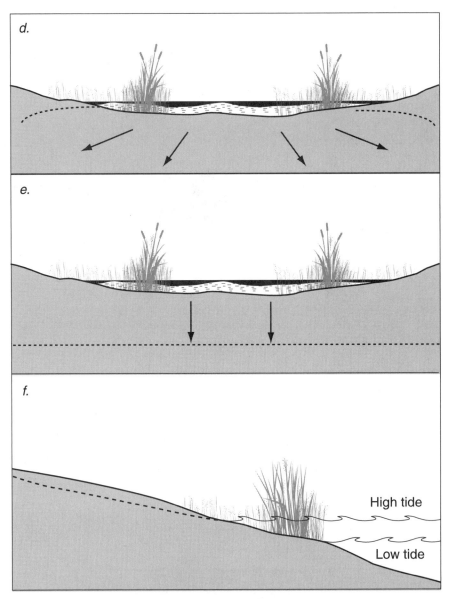

Figure 5-15 *(Continued).*

Despite the importance of groundwater flows in the budgets of many wetlands, there is a poor understanding of groundwater hydraulics in wetlands, particularly in those that have organic soils. The hydraulic conductivity can be predicted for some peatland soils from their bulk density or fiber content, both of which can easily be measured (Fig. 5-16). In general, the conductivity of organic peat decreases as the fiber content decreases through the process of decomposition. Water can pass through fibric, or poorly decomposed, peats a thousand times

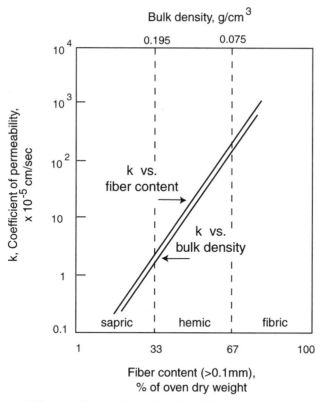

Figure 5-16 **Permeability of peatland soil as a function of fiber content and bulk density. (*After Verry and Boelter, 1979*)**

faster than it can through more decomposed sapric peats. The type of plant material that makes up the peat is also important. Peat composed of the remains of grasses and sedges such as *Phragmites* and *Carex,* for example, is more permeable than the remains of most mosses, including sphagnum. The hydraulic conductivity of peat can vary over several orders of magnitude, showing a range almost as great as the range for mineral soil between clay ($k = 5 \times 10^{-7}$ cm/s) and sand ($k = 5 \times 10^{-2}$ cm/s) (Table 5-6). There has been disagreement over the appropriate methods for measuring hydraulic conductivity in wetlands and about whether Darcy's law applies to flow through organic peat (Hemond and Goldman, 1985; Kadlec, 1989).

Evapotranspiration

The water that vaporizes from water or soil in a wetland (*evaporation*), together with moisture that passes through vascular plants to the atmosphere (*transpiration*), is called *evapotranspiration.* The meteorological factors that affect evaporation and tran-

Table 5-6 **Typical hydraulic conductivity for wetland soils compared with other soil materials**

Wetland or Soil Type	Hydraulic Conductivity, k (cm/s $\times$ 10^{-5})	Reference
NORTHERN PEATLANDS		
Highly humified blanket bog, UK	0.02–0.006	Ingram (1967)
Fen, Russia		
Slightly decomposed	500	Romanov (1968)
Moderately decomposed	80	
Highly decomposed	1	
Carex fen, Russia		
0–50 cm deep	310	Romanov (1968)
100–150 cm deep	6	
North American peatlands (general)		
Fibric	>150	Verry and Boelter (1979)
Hemic	1.2–150	
Sapric	<1.2	
COASTAL SALT MARSH		
Great Sippewissett Marsh, Massachusetts (vertical conductivity)		Hemond and Fifield (1982)
0–30 cm deep	1.8	
High permeability zone	2,600	
Sand–peat transition zone	9.4	
NONPEAT WETLAND SOILS		
Cypress dome, Florida		
Clay with minor sand	0.02–0.1	Smith (1975)
Sand	30	
Okefenokee Swamp watershed, Georgia	3.4–834	Hyatt and Brook (1984)
MINERAL SOILS (general)		
Clay	0.05	Linsley and Franzini (1979)
Limestone	5.0	
Sand	5000	

Source: Partially after Rycroft et al. (1975).

spiration are similar as long as there is adequate moisture, a condition that almost always exists in most wetlands. The rate of evapotranspiration is proportional to the difference between the vapor pressure at the water surface (or at the leaf surface) and the vapor pressure in the overlying air. This is described in a version of *Dalton's law:*

$$E = cf(u)\,(e_{\mathrm{w}} - e_{\mathrm{a}}) \tag{5.13}$$

where E = rate of evaporation
c = mass transfer coefficient
$f(u)$ = function of wind speed, u
e_{w} = vapor pressure at surface, or saturation vapor pressure at wet surface
e_{a} = vapor pressure in surrounding air

Evaporation and transpiration are enhanced by the same meteorological conditions such as solar radiation or surface temperature that increase the value of the vapor pressure at the evaporating surface and by factors such as decreased humidity or increased wind speed that decrease the vapor pressure of the surrounding air. This equation assumes an adequate supply of water for capillary movement in the soil or for access by rooted plants. When the water supply is limited (not a frequent occurrence in wetlands), evapotranspiration is limited as well. Transpiration can also be physiologically limited in plants through the closing of leaf stomata despite adequate moisture during periods of stress such as anoxia.

Direct Measurement of Wetland Evapotranspiration

Pan Evaporation

Several direct measurement techniques can be used in wetlands to determine evapotranspiration. The classical reference method is the measurement of evaporation from a water-filled pan (Fig. 5-17), usually by measuring the weight loss, by measuring the volume required to replace lost water over a period of time, or by measuring the drop in water level. This is generally considered a measurement of potential evaporation, since the evaporating surface is saturated. The method is tedious and the results often poorly correlated with actual evaporation from vegetated surfaces, because the transpiration, unsaturated soils, winds, and shading effects of the plant canopy all influence the rate, often in unknown ways (see the following discussion). On the other hand, pan evaporation provides a reference evaporation rate for comparison with other techniques. Furthermore, since wetland soils tend to be saturated most of

Figure 5-17   **Standard evaporation pan used for estimating evaporation and transpiration.**

the time, the pan method may be more accurate for wetlands than for terrestrial environments.

Diurnal Method

Over fairly uniform areas, it is possible to determine evapotranspiration from heat and water balances through the plant canopy (Hsu et al., 1972). Evapotranspiration from wetlands has also been calculated from measurements of the increase in water vapor in air flowing through vegetation chambers (S. L. Brown, 1981) and from observing the diurnal cycles of groundwater or surface water in wetlands (Mitsch et al., 1977; Heimburg, 1984; Ewel and Smith, 1992). This latter method, shown in Figure 5-18, can be calculated as follows:

$$ET = S_y(24h \pm s) \tag{5.14}$$

where ET = evapotranspiration (mm/day)
$\quad S_y$ = specific yield of aquifer (unitless)
$\qquad$ = 1.0 for standing-water wetlands
$\qquad$ < 1.0 for groundwater wetlands
$\quad h$ = hourly rise in water level from midnight to 4:00 A.M. (mm/h)
$\quad s$ = net fall (+) or rise (−) of water table or water surface in one day

The pattern assumes active "pumping" of water by vegetation during the day and a constant rate of recharge equal to the midnight-to-4:00-A.M. rate. This method also assumes that evapotranspiration is negligible around midnight and that the water table around this time approximates the daily mean. The water level is usually at or

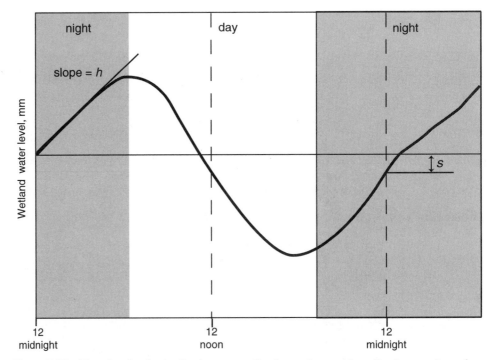

Figure 5-18 Diurnal water fluctuation in some wetlands can be used to estimate evapotranspiration as in Equation 5-14.

near the root zone in many wetlands, a necessary condition for this method to measure evapotranspiration accurately (Todd, 1964).

Empirical Estimates of Wetland Evapotranspiration

Thornthwaite Equation

Evapotranspiration can be determined with any number of empirical equations that use easily measured meteorological variables. One of the most frequently used empirical equations for evapotranspiration from terrestrial ecosystems, which has been applied with some success to wetlands, is the *Thornthwaite equation* for potential evapotranspiration (Chow, 1964):

$$ET_i = 16(10T_i/I)^a \qquad (5.15)$$

where ET_i = potential evapotranspiration for month i (mm/month)
T_i = mean monthly temperature (°C)
I = local heat index = $\sum_{i=1}^{12} (T_i/5)^{1.514}$
$a = (0.675 \times I^3 - 77.1 \times I^2 + 17{,}920 \times I + 492{,}390) \times 10^{-6}$

As an example, this equation was used to determine evapotranspiration from the Okefenokee Swamp in Georgia by Rykiel (1984). For the 26-year period examined in that study, average evapotranspiration ranged from 21 mm/month in December to 179 mm/month in July. However, Kadlec et al. (1988) tested the Thornthwaite equation on wetland evapotranspiration in Michigan and Nevada and found it to underpredict actual evapotranspiration, especially in the arid Nevada site.

Penman Equation

A second empirical relationship that has had many applications in hydrologic and agricultural studies but relatively few in wetlands is the *Penman equation* (Penman, 1948; Chow, 1964). This equation, based on both Dalton's law and the energy budget approach, is given as

$$ET = \frac{\Delta H + 0.27 E_a}{\Delta + 0.27} \qquad (5.16)$$

where ET = evapotranspiration (mm/day)
Δ = slope of curve of saturation vapor pressure versus mean air temperature (mmHg/°C)
H = net radiation (cal/cm²-day)
$\quad = R_t (1 - a) - R_b$
R_t = total shortwave radiation
a = albedo of wetland surface
R_b = effective outgoing longwave radiation = $f(T^4)$
E_a = term describing the contribution of mass transfer to evaporation
$\quad = 0.35(0.5 + 0.00625u)(e_w - e_a)$
u = wind speed 2 m above ground (km/day)
e_w = saturation vapor pressure of water surface at mean air temperature (mmHg)
e_a = vapor pressure in surrounding air (mmHg)

The Penman equation was compared with the pan evaporation (multiplied by a factor of 0.8) and other methods at natural enriched fens in Michigan and constructed wetlands in Nevada by Kadlec et al. (1988). They found that the Penman equation, like the Thornthwaite equation, generally underpredicted evapotranspiration from the Michigan wetland but agreed within a few percent with other measurement techniques for the Nevada wetlands.

Hammer and Kadlec Equation

Another empirical relationship for describing summer evapotranspiration using solar energy was developed by Scheffe (1978) and was described by Hammer and Kadlec (1983). The equation, which was used individually for sedge, willow, leatherleaf, and cattail vegetation covers, is

$$ET = a + bR_t + cT_a + dH_r + eu \qquad (5.17)$$

where a, b, c, d, e = correlation coefficients

R_t = incident shortwave radiation (measured by pyranograph)

T_a = air temperature

H_r = relative humidity

u = wind speed

The equation gives estimates that are better than some more frequently used evapotranspiration relationships. When the results of using this model were compared to actual measurements, the radiation term was shown to dominate (Hammer and Kadlec, 1983).

Because of the many meteorological and biological factors that affect evapotranspiration, none of the many empirical relationships, including the Thornthwaite, Penman, and Hammer and Kadlec equations, is entirely satisfactory for estimating wetland evapotranspiration. Lee (1980) cautions that there is "no reliable method of estimating evapotranspiration rates based on simple weather-element data or potential evapotranspiration." Nevertheless, these equations of potential evapotranspiration offer the most cost-effective first approximations for estimating water loss. Furthermore, when applied to wetlands, which are only rarely devoid of an adequate water supply, they may be more reliable than their applications to upland terrain, where evapotranspiration can be limited by a lack of soil water.

Effects of Vegetation on Wetland Evapotranspiration

A question about evapotranspiration from wetlands that does not elicit a uniform answer in the literature is, "Does the presence of wetland vegetation increase or decrease the loss of water compared to that which would occur from an open body of water?" Data from individual studies are conflicting. Obviously, the presence of vegetation retards evaporation from the water surface, but the question is whether the transpiration of water through the plants equals or exceeds the difference (Kadlec et al., 1988; Kadlec, 1989). Eggelsmann (1963) found evaporation from bogs in Germany to be generally less than that from open water except during wet summer months. In studies of evapotranspiration from small bogs in northern Minnesota, Bay (1967) found it to be 88 percent to 121 percent of open-water evaporation. Eisenlohr (1976) reported 10 percent lower evapotranspiration from vegetated prairie potholes than from nonvegetated potholes in North Dakota. Hall et al. (1972) estimated that a stand of vegetation in a small New Hampshire wetland lost 80 percent more water than did the open water in the wetland. In a forested pond cypress dome in north-central Florida, Heimburg (1984) found that swamp evapotranspiration was about 80 percent of pan evaporation during the dry season (spring and fall) and as low as 60 percent of pan evaporation during the wet season (summer). S. L. Brown (1981) found that transpiration losses from pond cypress wetlands were lower than evaporation from an open-water surface even with adequate standing water.

In the arid West, it has been a longstanding practice to conserve water for irrigation and other uses by clearing riparian vegetation from streams. In this environment where groundwater is often well below the surface but within the rooting zone of deep-rooted plants, trees "pump" water to the leaf surface and actively transpire even when little evaporation occurs at the soil surface.

The conflicting measurements and the difficulty of measuring evaporation and evapotranspiration led Linacre (1976) to conclude that neither the presence of wetland vegetation nor the type of vegetation had major influences on evaporation rates, at least during the active growing season. Bernatowicz et al. (1976) also found little difference in evapotranspiration among several species of vegetation. The general unimportance of plant species variation on overall wetland water loss is probably a reasonable conclusion for most wetlands, although it is clear that the type of wetland ecosystem and the season are important considerations. Ingram (1983), for example, found that fens have about 40 percent more evapotranspiration than do treeless bogs and that evaporation from the bogs is less than potential evapotranspiration in the summer and greater than potential evapotranspiration in the winter.

In some cases the type of vegetation in the wetland does matter. When trees are removed from some forested swamps where the soil is hydric but there is little surface flooding, standing water may return and, with it, herbaceous marsh vegetation. This resets a hydrologic succession; woody plants are able to reinvade the marsh during dry years and reestablish the site back to a forested wetland. This pattern is illustrated in Chapter 12.

Tides

The periodic and predictable tidal inundation of coastal salt marshes, mangroves, and freshwater tidal marshes is a major hydrologic feature of these wetlands. The tide acts as a stress by causing submergence, saline soils, and soil anaerobiosis; it acts as a subsidy by removing excess salts, reestablishing aerobic conditions, and providing nutrients. Tides also shift and alter the sediment patterns in coastal wetlands, causing a uniform surface to develop.

Typical tidal patterns for several coastal areas of the United States are shown in Figure 5-19a. Seasonal as well as diurnal patterns exist in the tidal rhythms. Annual variations of mean monthly sea level are as great as 25 cm (Fig. 5-19b). Tides also have significant bimonthly patterns because they are generated by the gravitational pull of the moon and, to a lesser extent, the sun. When the sun and the moon are in line and pull together, which occurs almost every two weeks, *spring tides,* or tides of the greatest amplitude, develop. When the sun and the moon are at right angles, *neap tides,* or tides of least amplitude, occur. Spring tides occur roughly at full and new moons, whereas neap tides occur during the first and third quarters.

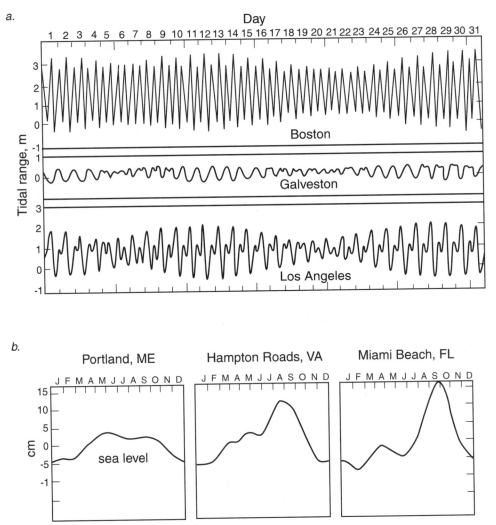

Figure 5-19 Patterns of tides: **a.** daily tides for a month and **b.** seasonal changes in mean monthly sea level for several locations in North America. (*After Emery and Uchupi, 1972*)

Tides vary more locally than regionally. The primary determinant is the coastline configuration. In North America, tidal amplitudes vary from less than 1 m along the Texas Gulf Coast to several meters in the Bay of Fundy in Canada. Tidal amplitude can actually increase as one progresses inland in some funnel-shaped estuaries. Typically, on a rising tide, water flows up tidal creek channels until the channels are bankfull. It overflows first at the upstream end, where tidal creeks break up into small creeklets that lack natural levees. The overflowing water spreads back downstream over the marsh surface. On falling tides, the flows are reversed. At low tides, water continues to drain through the natural levee sediments into adjacent creeks because these sediments tend to be relatively coarse; in the marsh interior, where sediments are finer, drainage is poor and water is often impounded in small depressions in the marsh.

Specific Effects of Hydrology on Wetlands

The effects of hydrology on wetland structure and function can be described with a complicated series of cause-and-effect relationships. A conceptual model of the general effects of hydrology in wetland ecosystems was shown in Figure 5-1. The effects are shown to be primarily on the chemical and physical aspects of the wetlands, which, in turn, influence the biotic components of the ecosystem. The biotic components, in turn, have a feedback effect on hydrology. Several principles underscoring the importance of hydrology in wetlands can be elucidated from studies that have been conducted to date. These principles, discussed later, are as follows:

1. Hydrology leads to a unique vegetation composition but can limit or enhance species richness.

2. Primary productivity and other ecosystem functions in wetlands are often enhanced by flowing conditions and a pulsing hydroperiod and are often depressed by stagnant conditions.

3. Accumulation of organic material in wetlands is controlled by hydrology through its influence on primary productivity, decomposition, and export of particulate organic matter.

4. Nutrient cycling and nutrient availability are both significantly influenced by hydrologic conditions.

Species Composition and Richness

Hydrology is a two-edged sword for species composition and diversity in wetlands. It acts as a limit or a stimulus to species richness, depending on the hydroperiod and physical energies. At a minimum, the hydrology acts to select water-tolerant vegetation in both freshwater and saltwater conditions and to exclude flood-intolerant species. Of the thousands of vascular plants that are on Earth, relatively few have adapted to waterlogged soils. (These adaptations are discussed in more detail in Chapter 7.) Although it is difficult to generalize, many wetlands that sustain long flooding durations have lower species richness in vegetation than do less frequently flooded areas. Waterlogged soils and the subsequent changes in oxygen content and other chemical conditions significantly limit the number and the types of rooted plants that can survive in this environment. McKnight et al. (1981), describing the effects of water on species composition in a riparian wetland, stated that "In general, as one goes from the hydric [wet] to the more mesic [dry] bottomland sites, the possible combinations or mixtures of species increases."

In general, species richness, at least in the vegetation community, increases as flowthrough increases (Table 5-7). Flowing water can be thought of as a stimulus to diversity, probably caused by its ability to renew minerals and reduce anaerobic conditions. Hydrology also stimulates diversity when the action of water and transported sediments creates spatial heterogeneity, opening up additional ecological niches. When rivers flood riparian wetlands or when tides rise and fall on coastal marshes, erosion, scouring, and sediment deposition sometimes create niches that allow diverse habitats to develop. On the other hand, flowing water can also create a relatively uniform surface that might enable monospecific stands of *Typha* or *Phragmites* to dominate a freshwater marsh or *Spartina* to dominate a coastal marsh. Keddy (1992) likened water-level fluctuations in wetlands to fires in forests. They eliminate

Table 5-7 Relationship between hydrologic regime and species richness in northern Minnesota peatlands

		Species Present					
	Tree	Shrub	Field Herbs	Grasses and Ferns	Ground Layer	Total	Flow Conditions
1. Rich swamp forest	6	16	28	11	10	71	Good surface flow; minerotrophic
2. Poor swamp forest	3	14	17	12	5	51	Downstream from 1; not adapted to strong water flow
3. Cedar string bog and fen	3	10	10	12	4	39	Better drainage than 2
4. Larch string bog and fen	3	9	9	12	4	37	Similar to 3; sheet flow
5. Black spruce feather moss forest	2	9	2	2	10	25	Gentle water flow on semiconvex template
6. Sphagnum bog	2	8	2	1	7	20	Isolated; little standing water
7. Sphagnum heath	2	6	2	2	5	17	Wet, soggy, and on convex template

Source: After Gosselink and Turner (1978) and Heinselman (1970).

one growth form of vegetation (e.g., woody plants) in favor of another (e.g., herbaceous species) and allow regeneration of species from buried seeds (see Chapter 8).

Primary Productivity

In general, the "openness" of a wetland to hydrological fluxes is probably one of the most important determinants of potential primary productivity. For example, peatlands that have flowthrough conditions (fens) have long been known to be more productive than stagnant raised bogs (see Chapter 13). A number of studies have found that wetlands in stagnant (nonflowing) or continuously deep water have low productivities, whereas wetlands that are in slowly flowing strands or are open to flooding rivers have high productivities.

This relationship between hydrology and ecosystem primary productivity has been investigated most extensively for forested wetlands. A general relationship was developed by Mitsch and Ewel (1979) for cypress productivity as a function of hydrology in Florida. That study concluded that

> Cypress–hardwood associations, found primarily in riverine and flowing strand
> systems, have the most productive cypress trees. The short hydroperiod favors both
> root aeration during the long dry periods and elimination of water-intolerant
> species during the short wet periods. The continual supply of nutrients with the
> flooding river system conditions may be a second important factor in maintaining
> these high productivities.

Productivity was found to be low under both continually flooded conditions and drained conditions. S. L. Brown (1981) found that much of the variation in biomass productivity of cypress wetlands in Florida could be explained by the variation in nutrient inflow, as measured by phosphorus. Productivity was lowest when nutrients were brought into the system solely by precipitation and was highest when large amounts of nutrients were passed through the wetlands by flooding rivers. Brown suggested that rather than there being a simple relationship between wetland productivity and hydrology, there is a more complex relationship among hydrology, nutrient inputs, and wetland productivity, decomposition, export, and nutrient cycling. Hydrology, then, also influences wetland productivity by being the main pathway through which nutrients are transported to many wetlands.

The influence of hydrologic conditions on freshwater marsh productivity is less certain. If peak biomass or similar measures are used as indicators of marsh productivity, some studies have shown the classical stimulation of vegetation along the water's edge (W. E. Odum et al., 1984; Gosselink, 1984), whereas other studies have indicated a higher macrophyte productivity in sheltered, nonflowing marshes than in wetlands open to flowing conditions or coastal influences. For example, consistently higher macrophyte biomass was found in wetlands isolated from surface fluxes with artificial dikes than in wetlands open to coastal fluxes along Lake Erie (Table 5-8). Several explanations are possible: (1) The coastal fluxes may also be serving as a stress as well as a subsidy on the macrophytes; (2) the open marshes may be exporting a significant amount of their productivity; and (3) the diked wetlands have more predictable hydroperiods. Conversely, a study of the influence of flowthrough conditions on water column primary productivity of constructed marshes found that, after two years of experimentation, water column (phytoplankton and submerged aquatics) productivity was higher in high-flow wetlands than in low-flow wetlands (Fig. 5-20). While macro-

Table 5-8 Selected macrophyte measurements at peak biomass from diked (hydrologically isolated) and undiked wetlands of Ohio's coastal Lake Erie

Measure of Vegetation Structure	Average ± Standard Error	
	Diked Wetlands ($n = 6$)	Undiked Wetlands ($n = 4$)
Biomass, g dry weight/m^2	897 ± 277	473 ± 149
# species/plot[a]	1.7 ± 0.3	1.4 ± 0.3
# stems/m^2	597 ± 211	241 ± 59

[a] Only species greater than 10 percent by weight per plot. Plots were 0.5 m^2 randomly placed in each wetland (3 to 6 per wetland).
Source: Mitsch (1992b) and Mitsch et al. (1994).

phyte productivity may take many years to respond to the difference in hydrology, water column productivity, which is mainly due to attached and planktonic algae, responds relatively quickly to changing hydrologic conditions.

Saltwater tidal wetlands subject to frequent tidal action are generally more productive than those that are only occasionally inundated. A comparison of several Atlantic Coast salt marshes, for example, showed a direct relationship between tidal range (as a measure of water flux) and end-of-season peak biomass of *Spartina alterniflora* (Fig. 5-21). Apparently, vigorous tides increase the nutrient subsidy and cause a flushing of toxic materials such as salt with vigorous tidal fluxes. Freshwater tidal wetlands may be even more productive than saline tidal wetlands because they receive the energy and nutrient subsidy of tidal flushing while avoiding the stress of saline soils.

Despite the overwhelming evidence of the influence of hydrology on wetlands, some investigators have cautioned against always ascribing a direct linkage between hydrologic variables and wetland productivity. Richardson (1979) stated that "a

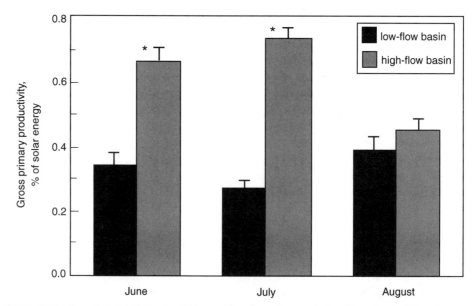

Figure 5-20 Aquatic primary productivity as a function of average flowthrough conditions in constructed marshes in the Midwestern United States. (*After Cronk and Mitsch, 1994*)

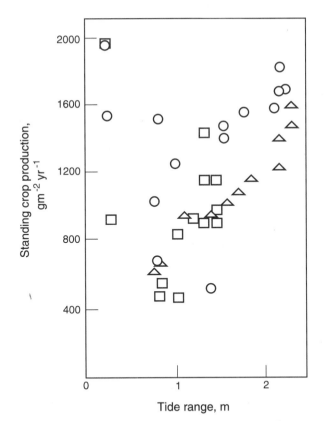

Figure 5-21 Production of *Spartina alterniflora* versus mean tidal range for several Atlantic Coast salt marshes. Different symbols indicate different data sources. (*After Steever et al., 1976*)

definitive statement about the influence of water levels on net primary productivity for all wetland types is impossible, since responses of individual species to water fluctuations vary." Water levels, however, are not the same as flowthrough of water. The higher the flowthrough, the higher is the input of associated nutrients and allochthonous energy. Although individual species vary inconsistently in their responses to water levels and hydrology, ecosystem-level responses may be more consistent.

Organic Accumulation and Export

Wetlands can accumulate excess organic matter as a result of either increased primary productivity (as described previously) or decreased decomposition and export. Notwithstanding the discrepancies from short-term litter decomposition studies, peat accumulates to some degree in all wetlands as a result of these processes. In fact, wetlands worldwide are major carbon sinks in the biosphere (see Chapter 16). The effects of hydrology on decomposition pathways are even less clear than the effects on primary productivity discussed previously. Probably the lack of agreement among the many studies published on the subject (cf. Brinson, 1977; W. E. Odum and Heywood, 1978; Chamie and Richardson, 1978; Deghi et al., 1980; Brinson et al., 1981a; van der Valk et al., 1991) is due to the complexity of the decomposition process. In general, decomposition of organic detritus requires electron donors (usually oxygen, but alternate chemicals such as sulfate or nitrate may be effective under anoxic conditions), moisture, inorganic nutrients, and microorganisms capable of metabolizing in the specific environment concerned. The observed rate of organic decomposition is also influenced by the ambient temperature and by the activity of macrodetritivores that shred the plant remains and/or repackage it as bacterially inoculated fecal pellets. Hydrology modifies many of these variables; for example, moisture depends on the flooding regime, flowing water carries oxygen and nutrients, while in stagnant water oxygen is rapidly depleted and nutrients are transformed to more or less available forms. Given this complexity, it is not surprising that the results of short-term *in situ* decomposition studies often disagree. (See Chapter 6 and the Ecosystem Function sections of Chapters 9 through 15 for further details.)

The importance of hydrology for organic carbon export is obvious. A generally higher rate of export is to be expected from wetlands that are open to the flowthrough of water. Riparian wetlands often contribute large amounts of organic detritus to streams, including macrodetritus such as whole trees. For many years, salt marshes and mangrove swamps were considered major exporters of their production [e.g., 45 percent estimated by Teal (1962) for a salt marsh; 28 percent measured by Heald (1969) for a mangrove swamp], but the generality of this concept is not accepted by coastal ecologists (Nixon, 1980; see also Chapters 6 and 9). Hydrologically isolated wetlands such as northern peatlands have much lower organic export. For example, Bazilevich and Tishkov (1982) found that only 6 percent of the net productivity of a fen in Russia was exported by surface and subsurface flows.

Nutrient Cycling

Nutrients are carried into wetlands by the hydrologic inputs of precipitation, river flooding, tides, and surface and groundwater inflows. Outflows of nutrients are controlled primarily by the outflow of water. These hydrologic/nutrient flows are also important determinants of wetland productivity and decomposition (see previous sections). Intrasystem nutrient cycling is generally, in turn, tied to pathways such as primary productivity and decomposition. When productivity and decomposition rates are high, as in flowing water or pulsing hydroperiod wetlands, nutrient cycling is rapid. When productivity and decomposition processes are slow, as in isolated ombrotrophic bogs, nutrient cycling is also slow.

The hydroperiod of a wetland has a significant effect on nutrient transformations, on the availability of nutrients to vegetation, and on loss from wetland soils of nutrients that have gaseous forms (see Chapter 6). Thus, nitrogen availability and loss are affected in wetlands by the reduced conditions that result from waterlogged soil. Typically, a narrow oxidized surface layer develops over the anaerobic zone in wetland soils, causing a combination of reactions in the nitrogen cycle—nitrification and denitrification—that may result in substantial losses of dinitrogen gas to the atmosphere. Furthermore, ammonium nitrogen is usually the form of nitrogen most available to plants in wetland soils, because the anaerobic environment favors the reduced ionic form over the nitrate common in agricultural soils.

Flooding of wetland soil, by altering both the pH and the redox potential of the soil, influences the availability of other nutrients. The pH of both acid and alkaline soils tends to converge on a pH of 7 when they are flooded (see Chapter 6). The redox potential, a measure of the intensity of oxidation or reduction of a chemical or biological system, indicates the state of oxidation (and, hence, availability) of several nutrients. Phosphorus is known to be more soluble under anaerobic conditions, due, in part, to the hydrolysis and reduction of ferric and aluminum phosphates to more soluble compounds. The availability of major ions such as potassium and magnesium and several trace nutrients such as iron, manganese, and sulfur is also affected by hydrologic conditions in the wetlands.

Techniques for Wetland Hydrology Studies

It is curious that so little attention has been paid to hydrologic measurements in wetland studies, despite the importance of hydrology in ecosystem function. A great deal of information can be obtained with only a modest investment in supplies and equipment. A diagram summarizing many of the hydrology measurements typical for developing a wetland's water budget is given in Figure 5-22. Water levels can be recorded continuously with a water-level recorder or during site visits with a staff gauge. With records of water level, all of the following hydrologic parameters can be determined: hydroperiod, frequency of flooding, duration of flooding, and water depth. Water-level recorders can also be used to determine the change in storage in a water budget, as in Equation 5-1.

Evapotranspiration measurements are more difficult to obtain, but there are several empirical relationships such as the Thornthwaite equation that use meteorological variables. Evaporation pans can also be used to estimate total evapotranspiration from wetlands, although pan coefficients are highly variable. Evapotranspiration of continuously flooded nontidal wetlands can also be determined by monitoring the diurnal water-level fluctuation as described in Figure 5-18.

Precipitation or throughfall or both can be measured by placing a statistically adequate number of rain gauges in random locations throughout the wetland or by utilizing weather station data. Surface runoff to wetlands can usually be determined as the increase in water level in the wetland during and immediately following a storm after net precipitation has been subtracted. Weirs can be constructed on more permanent streams to monitor surface water inputs and outputs.

Groundwater flows are usually the most difficult and most costly hydrologic flows to measure accurately. In some cases, clusters of shallow monitoring wells, placed around a wetland, will help indicate the direction of groundwater flow and the slope of the water or hydraulic gradient as required in Equation 5-12. The wells are called *piezometers* when they are only partially screened and thus measure the piezometric head of an isolated part of the groundwater rather than being screened through the entire length of the well and thus measuring the surface water aquifer. Piezometers can be installed by professional well-drilling companies or, for low-budget installations, can generally be installed with augers or as well points. Estimates of permeability or hydraulic conductivity are then required to quantify the flows. Permeability can be estimated through *in situ* pump tests using the wells (Todd, 1964) or through laboratory analysis of intact soil cores. The variability of results among different hydraulic conductivity measuring techniques suggests that caution should be made in using these numbers (Koreny et al., 1999). If a wetland is a perched or a recharge wetland, seepage can be estimated either through a water budget approach (e.g., subtracting evapotranspiration losses from water-level decreases when there are no other inflows or outflows) or by using half-barrel seepage meters. Other methods available to measure groundwater flows in wetlands include the use of stable isotopes,

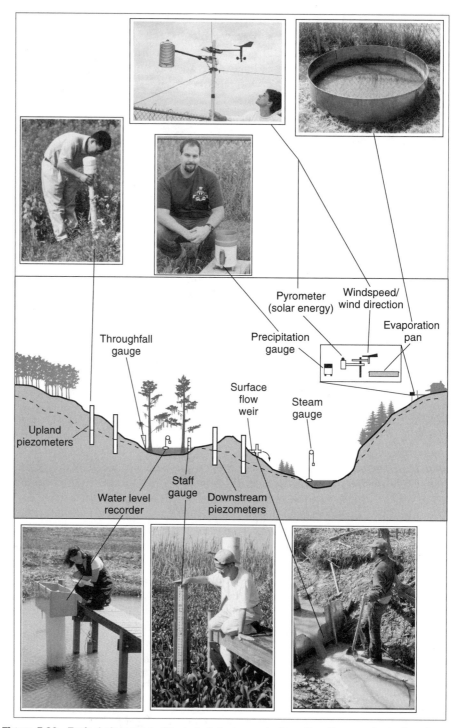

Figure 5-22 Typical placement of hydrology instruments for estimating a water budget for a floodplain wetland.

152

generally $^{18}O/^{16}O$ or $^{2}H/^{1}H$, because of the propensity of the lighter isotope in each case to evaporate more readily, allowing water to be "tagged" according to its source (Hunt et al., 1996). Groundwater flow models have also been used to estimate the flow of groundwater into and out of wetlands with some success (Hunt et al., 1996; Koreny et al., 1999).

The uncertainty in the scientific literature concerning many wetland processes (e.g., the rates of organic matter decomposition discussed earlier) is often closely related to unquantified hydrologic parameters. Thus, careful attention to quantification of pertinent hydrologic parameters in wetland research studies is virtually certain to improve our understanding of the ecological processes that control wetlands.

Recommended Readings

Carter, V. 1986. An overview of the hydrologic concerns related to wetlands in the United States. Canadian Journal of Botany 64:364–374.

Glaser, P. H., D. I. Siegel, E. A. Romanowicz, and Y. P. Shen. 1997. Regional linkages between raised bogs and the climate, groundwater, and landscape in north-western Minnesota. Journal of Ecology 85:3–16.

LaBaugh, J. W. 1986. Wetland ecosystem studies from a hydrologic perspective. Water Resources Bulletin 22:1–10.

Siegel, D. I., and P. H. Glaser. 1987. Groundwater flow in the bog/fen complex, Lost River Peatland, northern Minnesota. Journal of Ecology 75:743–754.

Winter, T. C., and M. R. Llamas, eds. 1993. Hydrogeology of Wetlands. Special Issue of Journal of Hydrology 141:1–269.

Chapter 6

Wetland Biogeochemistry

Wetland biogeochemical cycles feature a combination of chemical transformations and chemical transport processes not shared by many other ecosystems. Wetland soils, known as hydric soils, are formed when oxygen is cut off due to the presence of water, causing reduced conditions. They can be organic soils or mineral soils. Hydric mineral soils can be identified through redox concentrations, redox depletions, and reduced matrices. Transformations of nitrogen, sulfur, iron, manganese, and carbon occur as a result of these anaerobic conditions. Some transformations cause toxic conditions, as with the production of hydrogen sulfide, whereas others, such as denitrification and methanogenesis, cause a loss of chemicals to the atmosphere. Many of the transformations are mediated by microbial populations that are adapted to the anaerobic environment. Chemicals are hydrologically transported to wetlands through precipitation, surface flow, groundwater, and tides. Wetlands dominated only by precipitation are generally nutrient poor, whereas concentrations of chemicals flowing into wetlands from the other three sources are highly variable. Wetlands can be sources, sinks, or transformers of chemicals. Not all wetlands are nutrient sinks, nor are the patterns consistent from season to season or from year to year. Wetlands are often chemically coupled to adjacent ecosystems by the export of organic materials, although the direct effects on adjacent ecosystems have been difficult to quantify. Although wetlands are similar to terrestrial and aquatic ecosystems in that they can be high-nutrient or low-nutrient systems, there are several differences, particularly in the importance of sediment storage of nutrients and in the functioning of the vegetation in the cycle of different nutrients.

The transport and transformation of chemicals in ecosystems, known as *biogeochemical cycling*, involve a great number of interrelated physical, chemical, and biological processes. The unique and diverse hydrologic conditions in wetlands (discussed in the previous chapter) markedly influence biogeochemical processes. These processes result not only in changes in the chemical forms of materials but also in the spatial movement of materials within wetlands, as in water–sediment exchange and plant uptake, and with the surrounding ecosystems, as in organic exports. These processes, in turn, determine overall wetland productivity. The interrelationships among hydrology, biogeochemistry, and response of wetland biota were summarized in Figure 5-1.

The biogeochemistry of wetlands can be divided into (1) intrasystem cycling through various transformation processes and (2) the exchange of chemicals between a wetland and its surroundings. Although few transformation processes are unique to wetlands, the permanent or intermittent flooding of these ecosystems causes certain processes to be more dominant in wetlands than in either upland or deep aquatic ecosystems. For example, while *anaerobic*, or oxygenless, conditions are sometimes found in other ecosystems, they prevail in wetlands. The soils in wetlands are characterized by waterlogged conditions during part or all of the year, which produce reduced conditions, which, in turn, have a marked influence on several biochemical transformations unique to anaerobic conditions.

This intrasystem cycling, along with hydrologic conditions, influences the degree to which chemicals are transported to or from wetlands. An ecosystem is considered biogeochemically *open* when there is an abundant exchange of materials with its surroundings. When there is little movement of materials across the ecosystem boundary, it is biogeochemically *closed*. Wetlands can fall into either category. For example, wetlands such as bottomland forests and tidal salt marshes have a significant exchange of minerals with their surroundings through river flooding and tidal exchange, respectively. Other wetlands such as ombrotrophic bogs and cypress domes have little material exchange except for gaseous matter that passes into or out of the ecosystem. These latter systems depend more on intrasystem cycling than on throughput for their chemical supplies.

Wetland Soils

Types and Definitions

Wetland soils are both the medium in which many of the wetland chemical transformations take place and the primary storage of available chemicals for most wetland plants. They are often described as *hydric soils*, defined by the U.S. Department of Agriculture's Natural Resources Conservation Service (NRCS, 1998) as "soils that formed under conditions of saturation, flooding or ponding long enough during the growing season to develop anaerobic conditions in the upper part." Wetland soils are of two types: (1) *mineral soils* or (2) *organic soils*. Nearly all soils have some organic material; but when a soil has less than 20 to 35 percent organic matter (on a dry-weight basis), it is considered a mineral soil.

Organic soils and organic soil materials are defined under either of two conditions of saturation:

1. Soils are saturated with water for long periods or are artificially drained and, excluding live roots, (a) have 18 percent or more organic carbon if the mineral fraction is 60 percent or more clay, (b) have 12 percent or more organic carbon if the mineral fraction has no clay, or (c) have a proportional content of organic carbon between 12 and 18 percent if the clay content of the mineral fraction is between 0 and 60 percent (Fig. 6-1); or

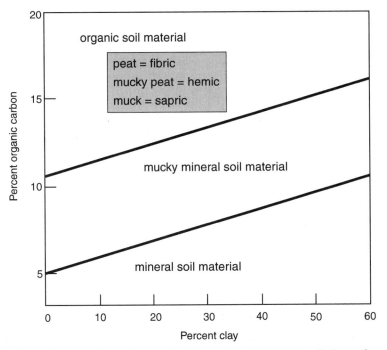

Figure 6-1 Percentage organic carbon required for a soil material to be called organic, mucky mineral soil, or mineral soil material. (*After NRCS, 1998*)

2. Soils are never saturated with water for more than a few days and have 20 percent or more organic carbon.

For an estimate of organic carbon when organic matter content is known,

$$\%C_{org} = \%OM/2 \tag{6.1}$$

where $\%C_{org}$ = percentage of organic carbon
$\%OM$ = percentage of organic matter

Any soil material that is not included in the preceding definition is considered mineral soil material. Where mineral soils occur in wetlands such as in some freshwater marshes or riparian forests, they generally have a soil profile made up of horizons, or layers. The upper layer of wetland mineral soils is often organic peat composed of partially decayed plant materials.

Although the preceding definition of organic soil is applicable to many types of wetlands, particularly to northern peatlands, *peat,* a generic term for relatively undecomposed organic soil, is not usually that strictly defined. Most peats contain less than 20 percent unburnable inorganic matter (and therefore usually contain more than 80 percent burnable organic material, which is about 40 percent organic carbon). Some soil scientists, however, allow up to 35 percent unburnable inorganic matter (approximately 33 percent organic carbon), and commercial operations sometimes allow 55 percent unburnable material (22 percent organic carbon).

Organic soils are different from mineral soils in several physicochemical features other than the percentage of organic carbon (Table 6-1):

Table 6-1 Comparison of mineral and organic soils in wetlands

	Mineral Soil	Organic Soil
Organic content (percent)	Less than 20 to 35	Greater than 20 to 35
Organic carbon (percent)	Less than 12 to 20	Greater than 12 to 20
pH	Usually circumneutral	Acid
Bulk density	High	Low
Porosity	Low (45–55%)	High (80%)
Hydraulic conductivity	High (except for clays)	Low to high[a]
Water holding capacity	Low	High
Nutrient availability	Generally high	Often low
Cation exchange capacity	Low, dominated by major cations	High, dominated by hydrogen ion
Typical wetland	Riparian forest, some marshes	Northern peatland

[a]See Chapter 5.

1. *Bulk density and porosity.* Organic soils have lower bulk densities and higher water-holding capacities than do mineral soils. Bulk density, defined as the dry weight of soil material per unit volume, is generally 0.2 to 0.3 g/cm^3 when the organic soil is well decomposed, although peatland soils composed of sphagnum moss can be extremely light, with bulk densities as low as 0.04 g/cm^3. By contrast, mineral soil bulk density generally ranges between 1.0 and 2.0 g/cm^3. Bulk density is low in organic soils because of their high porosity, or percentage of pore spaces. Peat soils generally have at least 80 percent pore spaces and are, thus, 80 percent water by volume when flooded. Mineral soils generally range from 45 to 55 percent total pore space, regardless of the amount of clay or texture.

2. *Hydraulic conductivity.* Both mineral and organic soils have wide ranges of possible hydraulic conductivities; the latter depends on their degree of decomposition (see Table 5-6 and Fig. 5-16). Organic soils may hold more water than mineral soils, but, given the same hydraulic conditions, they do not necessarily allow water to pass through more rapidly.

3. *Nutrient availability.* Organic soils generally have more minerals tied up in organic forms unavailable to plants than do mineral soils. This follows from the fact that a greater percentage of the soil material is organic. This does not mean, however, that there are more total nutrients in organic soils; very often, the opposite is true in wetland soils. For example, organic soils can be extremely low in bioavailable phosphorus or iron content—enough to limit plant productivity.

4. *Cation exchange capacity.* Organic soils have a greater cation exchange capacity, defined as the sum of exchangeable cations (positive ions) that a soil can hold. Figure 6-2 summarizes the general relationship between organic content and cation exchange capacity of soils. Mineral soils have a cation exchange capacity that is dominated by the major metal cations (Ca^{2+}, Mg^{2+}, K^+, and Na^+). As organic content increases, both the percentage and the amount of exchangeable hydrogen ions increase. For sphagnum moss peat, the high cation capacity is caused by long-chain polymers of uronic acid (Clymo, 1983).

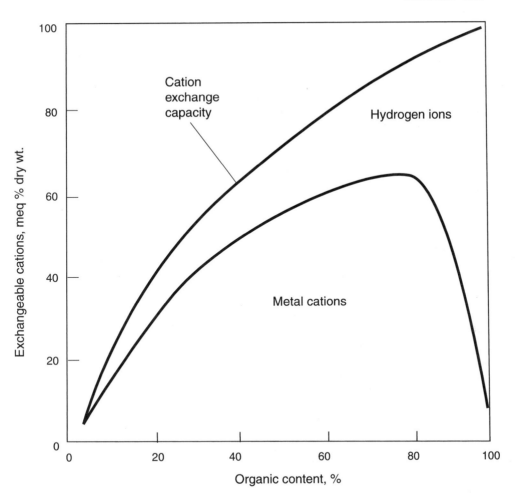

Figure 6-2 Relationship between cation exchange capacity and organic content for wetland soils. For low organic content (mineral soils), the cation exchange capacity is saturated by metal cations; when organic content is high, the exchange capacity is dominated by hydrogen ions. (*After Gorham, 1967*)

Organic Wetland Soil

Organic soil is composed primarily of the remains of plants in various stages of decomposition and accumulates in wetlands as a result of the anaerobic conditions created by standing water or poorly drained conditions. Two of the more important characteristics of organic soil, including soils commonly termed peat and *muck,* are the botanical origin of the organic material and the degree to which it is decomposed (Clymo, 1983). Several of the properties that have been discussed, including bulk density, cation exchange capacity, hydraulic conductivity, and porosity, are often dependent on these characteristics. Therefore, it is often possible to predict the range of the physical properties of an organic soil if the origin and state of decomposition can be observed in the field or laboratory.

Botanical Origin

The botanical origin of the organic material can be (1) mosses, (2) herbaceous material, and (3) wood and leaf litter. For most northern peatlands, the moss is usually *Sphagnum*, although several other moss species can dominate if the peatland is receiving inflows of mineral water. Organic soils can originate from herbaceous grasses such as reed grass (*Phragmites*), wild rice (*Zizania*), and salt marsh cordgrass (*Spartina*), or from sedges such as *Carex* and *Cladium*. Organic soils can also be produced in freshwater marshes by plant fragments from a number of nongrass and nonsedge plants, including cattails (*Typha*) and water lilies (*Nymphaea*). In forested wetlands, the peat can be a result of woody detritus or leaf material or both. In northern peatlands, the material can originate from birch (*Betula*), pine (*Pinus*), or tamarack (*Larix*), and in southern deepwater swamps, the organic horizon can be composed of material from cypress (*Taxodium*) or water tupelo (*Nyssa*) trees.

Decomposition

The state of decomposition, or humification, of wetland soils is the second key characteristic of organic peat. As decomposition proceeds, albeit at a very slow rate in flooded conditions, the original plant structure is changed physically and chemically until the resulting material little resembles the parent material. As peat decomposes, bulk density increases, hydraulic conductivity decreases, and the quantity of larger (>1.5 mm) fiber particles decreases as the material becomes increasingly fragmented. Chemically, the amount of peat "wax," or material soluble in nonpolar solvents, and lignin increase with decomposition, whereas cellulose compounds and plant pigments decrease (Clymo, 1983). When some wetland plants such as salt marsh grasses die, the detritus rapidly loses a large percentage of its organic compounds through leaching. These readily soluble organic compounds are thought to be easily metabolized in adjacent aquatic systems.

Classification and Characteristics

Organic soils (*histosols*) are classified into four groups, three of which are considered hydric soils:

1. *Saprists* (muck). Two-thirds or more of the material is decomposed, and less than one-third of plant fibers are identifiable.
2. *Fibrists* (peat). Less than one-third of material is decomposed, and more than two-thirds of plant fibers are identifiable.
3. *Hemists* (mucky peat or peaty muck). Conditions fall between saprist and fibrist soil.
4. *Folists*. Organic soils caused by excessive moisture (precipitation > evapotranspiration) that accumulate in tropical and boreal mountains; these soils are not classified as hydric soils as saturated conditions are the exception rather than the rule.

Organic soil is generally dark in color, ranging from the dark black soils characteristic of mucks such as those found in the Everglades in Florida to the dark brown color of partially decomposed peat from northern bogs.

Mineral Wetland Soil

Mineral soils, when flooded for extended periods, develop certain characteristics that allow for their identification. These characteristics are collectively called *redoximorphic features,* defined as features formed by the reduction, translocation, and/or oxidation of iron and manganese oxides (Vepraskas, 1995).

The development of redoximorphic features in mineral soils is mediated by micro-biological processes. The rate at which they are formed depends on three conditions, all of which must be present:

1. Sustained anaerobic conditions
2. Sufficient soil temperature (5°C is often considered "biological zero," below which much biological activity ceases or slows considerably)
3. Organic matter, which serves as a substrate for microbial activity

Reduced Matrices and Redox Depletions

One characteristic of many hydric mineral soils that are semipermanently or perma-nently flooded is the development of black, gray, or sometimes greenish or blue–gray color as the result of a process known as *gleying.* This process, also known as *gleization,* is the result of the chemical reduction of iron (see Iron and Manganese Transforma-tions later in this chapter). When soils are not saturated with water, iron (ferric = Fe^{3+}) oxides are the principal chemicals that give the soil its typical red, brown, yellow, or orange color. Manganese (Mn^{3+} or Mn^{4+}) oxides give the soil a black color. When soils are flooded and become reduced, the iron is reduced to a soluble form of iron (ferrous = Fe^{2+}) and the manganese is reduced to its soluble manganous (Mn^{2+}) form. These soluble forms of iron and manganese can be leached out of the soil, leaving the natural (gray or black) color of the parent sand, silt, or clay, called the matrix. A similar term used to describe these reduced soils is *redox depletions*—iron is reduced and then depleted from the soil matrix. In a similar manner, *clay depletions* occur when clay is selectively removed along root channels after iron and manganese oxides have been depleted, only to redeposit as clay coatings on soil particles below the clay depletions (Vepraskas, 1995).

Oxidized Rhizosphere

Another characteristic of some mineral wetland soils is the presence of an *oxidized rhizosphere* (also called *oxidized pore linings*) that results from the capacity of many hydrophytes to transport oxygen through above-ground stems and leaves to below-ground roots (Fig. 6-3). Excess oxygen, beyond the root's metabolic needs, diffuses from the roots to the surrounding soil matrix, forming deposits of oxidized iron along small roots. When a wetland soil is examined, these oxidized rhizosphere deposits can often be seen as thin traces through an otherwise dark matrix.

Redox Concentrations

Mineral soils that are seasonally flooded, particularly by alternate wetting and drying, develop spots of highly oxidized materials called *mottles* or *redox concentrations* (Fig. 6-4). Mottles and redox concentrations are orange/reddish-brown (because of iron) or dark reddish-brown/black (because of manganese) spots seen throughout an other-wise gray (gleyed) soil matrix and suggest intermittently exposed soils with spots of iron and manganese oxides in an otherwise reduced environment. Mottles are relatively insoluble, enabling them to remain in soil long after it has been drained.

Modern Nomenclature

A revised set of terms defining redoximorphic features has been devised by soil scientists to describe indicators of hydric soils, or more properly, to identify an *aquic condition—*

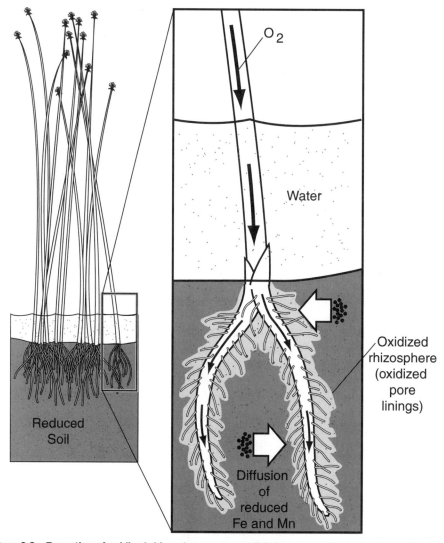

Figure 6-3 Formation of oxidized rhizospheres, or pore linings, around the roots of a wetland plant due to the transport of excess oxygen by wetland plants to their roots. When the plant dies, pore linings of iron and manganese oxides remain. (*After Vepraskas, 1995*)

the condition in which soils are saturated with water, are reduced, and display redoximorphic features (Vepraskas, 1995). The term *aquic condition* was introduced in the early 1990s to better reconcile field techniques that used soil colors (e.g., iron reduction or oxidation) with the former term *aquic moisture regime*—any soil that was saturated with water and chemically reduced such that no dissolved oxygen was present. The redoximorphic features that can be used to identify aquic conditions are (Vepraskas, 1995):

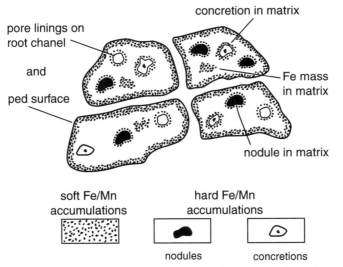

Figure 6-4 Different kinds of redox concentrations, or mottles, in soil peds (soil macroparticles), including nodules and concretions, iron masses in soil matrix (also called reddish mottles), and pore linings on root channel (also called oxidized rhizospheres). (*After Vepraskas, 1995*)

1. *Redox concentrations.* Accumulation of iron and manganese oxides (formerly called mottles) in at least three different structures (Fig. 6-4):
    a. *Nodules and concretions.* Firm to extremely firm irregularly shaped bodies with diffuse boundaries
    b. *Masses.* Formerly called "reddish mottles"
    c. *Pore linings.* Formerly included "oxidized rhizosphores" (Fig. 6-3)
2. *Redox depletions.* Low-chroma (≤ 2) bodies with high values (≥ 4) including:
    a. *Iron depletions.* Sometimes called "gray mottles" or "gley mottles"; these are low-chroma bodies
    b. *Clay depletions.* Contain less iron, manganese, and clay than adjacent soils
3. *Reduced matrices.* Low-chroma soils (because of presence of Fe^{2+} *in situ* that change color if exposed to air and iron is oxidized to Fe^{3+}

Mineral Hydric Soil Determination

In practice, the determination of whether a mineral soil is a hydric soil is a complicated process, but it is often done by determining soil color relative to a standard color chart called the Munsell soil color chart (Fig. 6-5). Soils that contain *low chromas* (as indicated by the color chips on the left-hand side of the color chart in Fig. 6-5b) indicate hydric soils. Soils that contain bright reds, browns, yellows, or oranges are nonhydric. In general, a chroma of 2 or less on the Munsell color chart is necessary for a soil to be classified as a hydric soil. These color charts are commonly used in the United States to identify the presence of hydric soils for the delineation of wetlands.

a.

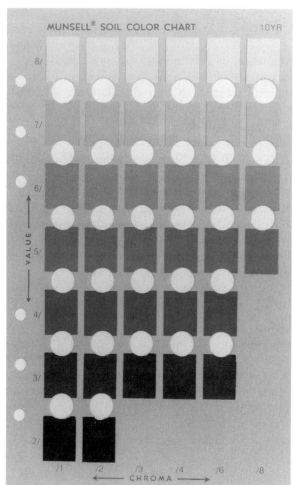

b.

Figure 6-5 Hydric soils can be identified by comparing the soil color with standard soil color charts such as the Munsell® Soil Color Chart shown here: **a.** the book and **b.** a sample page in the book. The hue, given in the upper right-hand corner of the chart (10YR in this case) indicates the relation to standard spectral colors such as red (R) or yellow (Y). The value notation (vertical scale) indicates the soil lightness (darker with lower value) and the chroma (horizontal scale) indicates the color strength or purity, with grayer soils to the left. Chromas of 2 or less generally indicate hydric soils. (*Courtesy of GretagMacbeth, New Windsor, New York; reprinted by permission*)

Chemical Transformations in Wetlands

Oxygen and Redox Potential

When soils, whether mineral or organic, are inundated with water, anaerobic conditions usually result. When water fills the pore spaces, the rate at which oxygen can diffuse through the soil is drastically reduced. Diffusion of oxygen in an aqueous solution has been estimated at 10,000 times slower than oxygen diffusion through a porous medium such as drained soil (Greenwood, 1961; Gambrell and Patrick, 1978). This low diffusion rate leads relatively quickly to anaerobic, or reduced, conditions, with the time required for oxygen depletion on the order of several hours to a few days after inundation begins (Fig. 6-6). The rate at which the oxygen is depleted depends on the ambient temperature, the availability of organic substrates for microbial respiration, and sometimes the chemical oxygen demand from reductants such as ferrous iron. The resulting lack of oxygen prevents plants from carrying out normal aerobic root respiration and strongly affects the availability of plant nutrients and toxic materials in the soil. As a result, plants that grow in anaerobic soils generally have a number of specific adaptations to this environment (see Chapter 7).

It is not always true that oxygen is totally depleted from the soil water of wetlands. There is usually a thin layer of oxidized soil, sometimes only a few millimeters thick,

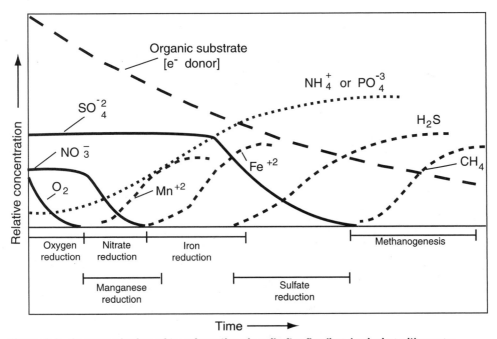

Figure 6-6 Sequence in time of transformations in soil after flooding, beginning with oxygen depletion and followed by nitrate and then sulfate reduction. Increases are seen in reduced manganese (manganous), reduced iron (ferrous), hydrogen sulfide, and methane. Note the gradual decrease in organic substrate (electron donor) and increases in available ammonium (NH_4^+) and phosphate (PO_4^{3-}) ions. The graph can also be interpreted as relative concentrations with depth in wetland soils. (*After Reddy and D'Angelo, 1994*)

at the surface of the soil at the soil–water interface (Fig. 6-7). The thickness of this oxidized layer is directly related to:

1. The rate of oxygen transport across the atmosphere–surface water interface
2. The small population of oxygen-consuming organisms present
3. Photosynthetic oxygen production by algae within the water column
4. Surface mixing by convection currents and wind action (Gambrell and Patrick, 1978)

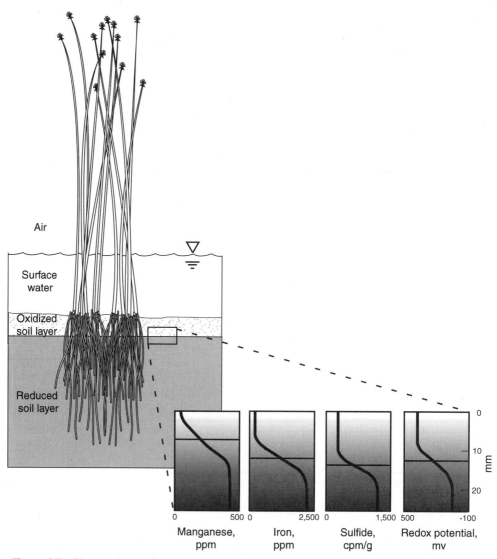

Figure 6-7 Characteristics of many wetland soils showing a shallow oxidized soil layer over a reduced soil layer. Also shown are soil profiles of reduced forms of manganese (sodium acetate–extractable manganese), iron (ferrous iron), and sulfur (sulfide), and redox potential. (*After Patrick and Delaune, 1972; Gambrell and Patrick, 1978*)

Even though the deeper layers of the wetland soils remain reduced, this thin oxidized layer is often very important in the chemical transformations and nutrient cycling that occur in wetlands. Oxidized ions such as Fe^{3+}, Mn^{4+}, NO_3^-, and $SO_4^=$ are found in this microlayer, whereas the lower anaerobic soils are dominated by reduced forms such as ferrous and manganous salts, ammonia, and sulfides. Because of the presence of oxidized ferric iron (Fe^{3+}) in the oxidized layer, the soil surface often is a brown or brownish-red color, in contrast to the bluish-gray to greenish-gray color of the reduced gleyed sediments, dominated by ferrous iron (Fe^{2+}).

Redox potential, or oxidation–reduction potential, a measure of the electron pressure (or availability) in a solution, is often used to quantify further the degree of electrochemical reduction of wetland soils. *Oxidation* occurs not only during the uptake of oxygen but also when hydrogen is removed (e.g., $H_2S \rightarrow S^{2-} + 2H^+$) or, more generally, when a chemical gives up an electron (e.g., $Fe^{2+} \rightarrow Fe^{3+} + e^-$). *Reduction* is the opposite process of releasing oxygen, gaining hydrogen (hydrogenation), or gaining an electron.

Redox potential can be measured in wetland soils and is a quantitative measure of the tendency of the soil to oxidize or reduce substances. When based on a hydrogen scale, redox potential is referred to as E_H and is related to the concentrations of oxidants {ox} and reductants {red} in a redox reaction by the *Nernst equation:*

$$E_H = E^0 + 2.3[RT/nF]\log[\{ox\}/\{red\}] \tag{6.2}$$

where E^0 = potential of reference (mV)
 R = gas constant = 81.987 cal deg^{-1} mol^{-1}
 T = temperature (K)
 n = number of moles of electrons transferred
 F = Faraday constant = 23,061 cal/mole-volt

Redox potential can be measured with a platinum electrode, easily constructed in the laboratory (Fig. 6-8). Several papers discuss constructing platinum electrodes (Faulkner et al., 1989; Swerhone et al., 1999), using microplatinum electrodes (Meijer and Avnimelech, 1999), or using steel rods (Bridgham et al., 1991) to indicate flooding and/or redox conditions in pond and wetland soils. Electric potential in units of millivolts (mV) is measured relative to a hydrogen electrode ($H^+ + e \rightarrow H$) or to a calomel reference electrode. As long as free dissolved oxygen is present in a solution, the redox potential varies little (in the range of +400 to +700 mV). However, it becomes a sensitive measure of the degree of reduction of wetland soils after oxygen disappears, ranging from +400 mV down to −400 mV.

As organic substrates in a waterlogged soil are oxidized (donate electrons), the redox potential drops as a sequence of reductions (electron gains) takes place (Fig. 6-6). Because organic matter is one of the most reduced of substances, it can be oxidized when any number of terminal electron acceptors is available, including O_2, NO_3^-, Mn^{2+}, Fe^{3+}, or $SO_4^=$. Rates of organic decomposition are most rapid in the presence of oxygen and slower for electron acceptors such as nitrates and sulfates (Fig. 6-9).

The oxidation of organic substrate is described by the following equation, which illustrates the organic substrate as an electron (e^-) donor (Fig. 6-6):

$$[CH_2O]_n + nH_2O \rightarrow nCO_2 + 4ne^- + 4nH^+ \tag{6.3}$$

Various chemical and biological transformations take place as coupled oxidation (e^- donor)–reduction (e^- acceptor) reactions. Equations 6-3 and 6-4 make one such

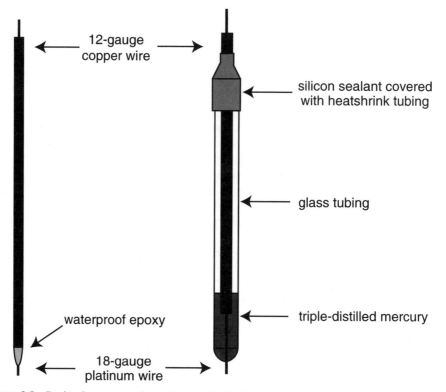

Figure 6-8 Design for constructing redox electrodes for measurement of redox potential in wetlands. (*After Faulkner et al., 1989*)

coupled reaction. These occur in a predictable sequence (Fig. 6-6), within predictable redox ranges to provide electron acceptors for this oxidation or decomposition (Table 6-2). The first and most common is through aerobic oxidation when oxygen itself is the terminal electron acceptor at a redox potential of between 400 and 600 mV:

$$O_2 + 4e^- + 4H^+ \rightarrow 2H_2O \qquad (6.4)$$

One of the first reactions that occur in wetland soils after they become anaerobic (i.e., the dissolved oxygen is depleted) is the reduction of NO_3^- (nitrate) first to NO_2^- (nitrite) and ultimately to N_2O or N_2; nitrate becomes an electron acceptor at a redox potential of approximately 250 mV:

$$2NO_3 + 10e^- + 12H^+ \rightarrow N_2 + 6H_2O \qquad (6.5)$$

As the redox potential continues to decrease, manganese is transformed from manganic to manganous compounds at about 225 mV:

$$MnO_2 + 2e^- + 4H^+ \rightarrow Mn^{2+} + 2H_2O \qquad (6.6)$$

Iron is transformed from ferric to ferrous form at about +100 to −100, while sulfates are reduced to sulfides at −100 to −200 mV:

$$Fe(OH)_3 + e^- + 3H^+ \rightarrow Fe^{2+} + 3H_2O \qquad (6.7)$$

$$SO_4^= + 8e^- + 9H^+ \rightarrow HS^- + 4H_2O \qquad (6.8)$$

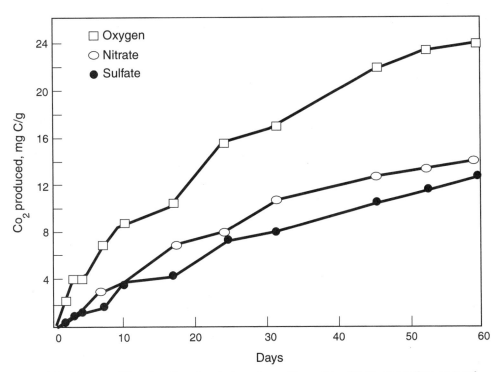

Figure 6-9 Decomposition of sediment organic carbon with oxygen, nitrate, and sulfate as terminal electron acceptors. The conditions were maintained at redox potentials of 312, 115, and −123 mV, respectively, for oxygen, nitrate, and sulfate. (*After Reddy and Graetz, 1988*)

Finally, under the most reduced conditions, the organic matter itself or carbon dioxide becomes the terminal electron acceptor below −200 mV, producing low-molecular-weight organic compounds and methane gas, as, for example,

$$CO_2 + 8e^- + 8H^+ \rightarrow CH_4 + 2H_2O \qquad (6.9)$$

These redox potentials are not precise thresholds, as pH and temperature are also important factors in the rates of transformation. These major chemical transformations

Table 6-2 Oxidized and reduced forms of several elements and approximate redox potentials for transformation

Element	Oxidized Form	Reduced Form	Approximate Redox Potential for Transformation (mV)
Nitrogen	NO_3^- (nitrate)	N_2O, N_2, NH_4^+	250
Manganese	Mn^{4+} (manganic)	Mn^{2+} (manganous)	225
Iron	Fe^{3+} (ferric)	Fe^{2+} (ferrous)	+100--100
Sulfur	$SO_4^=$ (sulfate)	$S^=$ (sulfide)	−100--200
Carbon	CO_2 (carbon dioxide)	CH_4 (methane)	Below −200

and others related to the nitrogen, iron, manganese, sulfur, and carbon cycles are discussed next.

pH

Wetland soils and overlying waters occur over a wide range of pH. Organic soils in wetlands are often acidic, particularly in peatlands in which there is little groundwater inflow. On the other hand, mineral soils often have more neutral or alkaline conditions. There are specific connections between redox potential and pH because the specific redox potential at which chemicals are stable in either reduced or oxidized states is pH dependent and can be shown on redox–pH stability diagrams (see, e.g., Stumm and Morgan, 1996). The general consequence of flooding previously drained soils is to cause alkaline soils to decrease in pH and acid soils to increase in pH (Fig. 6-10), the former stemming from the buildup of CO_2 and then carbonic acid, the latter stemming from the reduction of ferric iron hydroxides, as shown in Equation 6-7 (Ponnamperuma, 1972). This neutral pH is generally in the range of 6.7 to 7.2. This convergence to neutrality is usually what can be expected to happen with soil pH when lands that had previously been drained become flooded, as occurs in the construction of wetlands. For some organic soils high in iron content, submergence does not always increase pH (Fig. 6-10). Peat soils often remain acidic during submergence through the slow oxidation of sulfur compounds near the surface, producing sulfuric acid, and the production of humic acids and selective cation exchange by sphagnum mosses.

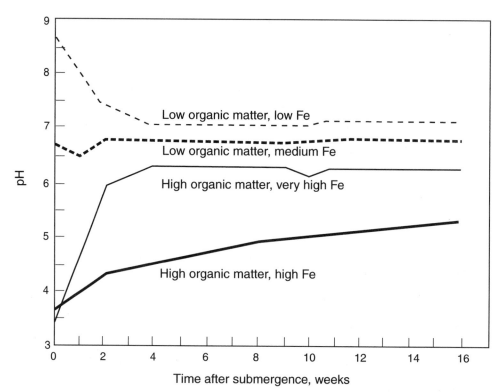

Figure 6-10 Changes in pH of soils of different organic and iron content after flooding. (*After Ponnamperuma, 1972*)

Nitrogen Transformations

Nitrogen appears in a number of oxidation states in wetlands, several of which are important in a wetland's biogeochemistry. Nitrogen is often the most limiting nutrient in flooded soils, whether the flooded soils are in natural wetlands or on agricultural wetlands such as rice paddies. Nitrogen is considered one of the major limiting factors in coastal waters, making the nitrogen dynamics in coastal wetlands particularly significant. Because of the presence of anoxic conditions in wetlands, microbial denitrification of nitrates to gaseous forms of nitrogen in wetlands and their subsequent release to the atmosphere remain one of the more significant ways in which nitrogen is lost from the lithosphere and hydrosphere to the atmosphere. Nitrates serve as one of the first terminal electron acceptors in wetland soils in this situation after the disappearance of oxygen (Table 6-2), making them an important chemical in the oxidation of organic matter in wetlands. Because humans have essentially doubled the amount of nitrogen that enters the land-based nitrogen cycle through fertilizer manufacturing, increased use of nitrogen-fixing crops, and fossil fuel burning (Vitousek et al., 1997), the ability of wetlands to serve as "sinks" for nitrogen is now being widely investigated.

Nitrogen transformations in wetlands (Fig. 6-11) involve several microbiological processes, some of which make the nutrient less available for plant uptake. The

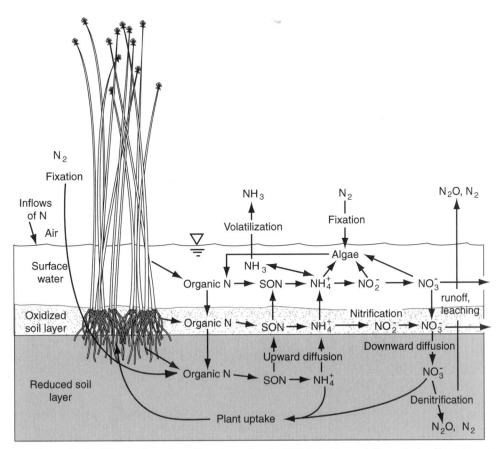

Figure 6-11 Nitrogen transformations in wetlands. SON indicates soluble organic nitrogen.

ammonium ion (NH_4^+), with a nitrogen oxidation state of -3, is the primary form of mineralized nitrogen in most flooded wetland soils, although much nitrogen can be tied up in organic forms in highly organic soils. The presence of an oxidized zone over the anaerobic or reduced zone is critical for several of the pathways.

Nitrogen mineralization refers to a series of biological transformations that converts organically bound nitrogen to ammonium nitrogen as the organic matter is being decomposed and degraded. This pathway occurs under both anaerobic and aerobic conditions and is often referred to as *ammonification*. Typical formulas for the mineralization of a simple soluble organic nitrogen (SON) compound, urea, are given as

$$NH_2CONH_2 + H_2O \rightarrow 2NH_3 + CO_2 \tag{6.10}$$

$$NH_3 + H_2O \rightarrow NH_4^+ + OH^- \tag{6.11}$$

Once the ammonium ion (NH_4^+) is formed, it can take several possible pathways. It can be absorbed by plants through their root systems or by anaerobic microorganisms and converted back to organic matter. Under high-pH conditions (pH > 8), a common occurrence in marshes with excessive algal blooms, the ammonium ion can be converted to NH_3, which is then released to the atmosphere through *volatilization*. The ammonium ion can also be immobilized through ion exchange onto negatively charged soil particles. Because of the anaerobic conditions in wetland soils, ammonium would normally be restricted from further oxidation and would build up to excessive levels were it not for the thin oxidized layer at the surface of many wetland soils. The gradient between high concentrations of ammonium in the reduced soils and low concentrations in the oxidized layer causes an upward diffusion of ammonium, albeit very slowly, to the oxidized layer. In this aerobic environment, ammonium nitrogen can be oxidized through the process of *nitrification* in two steps by *Nitrosomonas* sp.:

$$2NH_4^+ + 3O_2 \rightarrow 2NO_2^- + 2H_2O + 4H^+ + energy \tag{6.12}$$

and by *Nitrobacter* sp.:

$$2NO_2^- + O_2 \rightarrow 2NO_3^- + energy \tag{6.13}$$

Nitrification can also occur in the oxidized rhizosphere of plants where adequate oxygen is often available to convert the ammonium nitrogen to nitrate nitrogen (Reddy and Graetz, 1988).

Nitrate (NO_3), as a negative ion rather than the positive ammonium ion, is not subject to immobilization by negatively charged soil particles and is, thus, much more mobile in solution. If it is not assimilated immediately by plants or microbes (*assimilatory nitrate reduction*) or is lost through groundwater flow stemming from its rapid mobility, it has the potential to undergo *dissimilatory nitrogenous oxide reduction*, a term that refers to several pathways of nitrate reduction. The most prevalent are reduction to ammonia and *denitrification*. Denitrification, carried out by microorganisms under anaerobic conditions, with nitrate acting as a terminal electron acceptor, results in the loss of nitrogen as it is converted to gaseous nitrous oxide (N_2O) and molecular nitrogen (N_2):

$$C_6H_{12}O_6 + 4NO_3 \rightarrow 6CO_2 + 6H_2O + 2N_2 \tag{6.14}$$

Denitrification has been documented as a significant path of loss of nitrogen from most kinds of wetlands, including salt marshes, freshwater marshes, forested wetlands, and rice paddies. Denitrification is inhibited in acid soils and peat and is, therefore, thought to be of less consequence in northern peatlands (Etherington, 1983). As illustrated in Figure 6-11, the entire process occurs after (1) ammonium nitrogen diffuses to the aerobic soil layer, (2) nitrification occurs, (3) nitrate nitrogen diffuses back to the anaerobic layer, and (4) denitrification, as described in Equation 6-14, occurs. The diffusion rates of the ammonium ion to the aerobic soil layer and the nitrate ion to the anaerobic layer are governed by the concentration gradients of the ions. There is generally a steep gradient of ammonium between the anaerobic and aerobic layers. Nevertheless, because nitrate diffusion rates in wetland soils are seven times faster than ammonium diffusion rates, ammonium diffusion and subsequent nitrification appear to limit the entire process of nitrogen loss by denitrification (Reddy and Patrick, 1984; Reddy and Graetz, 1988).

Nitrogen fixation results in the conversion of N_2 gas to organic nitrogen through the activity of certain organisms in the presence of the enzyme nitrogenase. It may be the source of significant nitrogen for some wetlands. Nitrogen fixation, which is carried out by certain aerobic and anaerobic bacteria and blue–green algae, is favored by low-oxygen tensions because nitrogenase activity is inhibited by high oxygen. In wetlands, nitrogen fixation can occur in overlying waters, in the aerobic soil layer, in the anaerobic soil layer, in the oxidized rhizosphere of the plants, and on the leaf and stem surfaces of plants (Reddy and Graetz, 1988). Bacterial nitrogen fixation can be carried out by nonsymbiotic bacteria, by symbiotic bacteria of the genus *Rhizobium,* or by certain actinomycetes. Whitney et al. (1975, 1981) and Teal et al. (1979) documented the fact that bacterial fixation was the most significant pathway for nitrogen fixation in salt marsh soils. On the other hand, both nitrogen-fixing bacteria and nitrifying bacteria are virtually absent from the low-pH peat of northern bogs (Moore and Bellamy, 1974). Cyanobacteria (blue–green algae), as nonsymbiotic nitrogen fixers, are also frequently found in waterlogged soils of wetlands and can contribute significant amounts of nitrogen. For example, rates of 2.5 g N m^{-2} yr^{-1} for blue–green algae nitrogen fixation compared to only 0.01 to 0.05 g N m^{-2} yr^{-1} by heterotrophic bacteria were found in flooded soils in Louisiana (Buresh et al., 1980a). Nitrogen fixation by blue–green algae is also important in northern bogs and rice cultures, which are often too acidic to support large bacterial populations (Etherington, 1983).

Iron and Manganese Transformations

Below the reduction of nitrate on the redox potential scale comes the reduction of manganese and iron (see Eqs. 6-6 and 6-7). Iron and manganese are found in wetlands primarily in their reduced forms (ferrous and manganous, respectively; Table 6-2), and both are more soluble and more readily available to organisms in those forms. Manganese is reduced slightly before iron on the redox scale, but otherwise it behaves similarly to iron. The direct involvement of bacteria in the reduction of MnO_2 (Eq. 6-6) has been questioned by some, although several experiments have shown the generation of energy by the bacterial reduction of oxidized manganese (Laanbroek, 1990).

Iron can be oxidized from the ferrous to the insoluble ferric form by chemosynthetic bacteria in the presence of oxygen:

$$4Fe^{2+} + O_2(aq) + 4H^+ \rightarrow 4Fe^{3+} + 2H_2O \qquad (6.15)$$

Although this reaction can occur nonbiologically at neutral or alkaline pH, microbial activity has been shown to accelerate ferrous iron oxidation by a factor of 10^6 in coal mine drainage water (Singer and Stumm, 1970). It is thought that a similar type of bacterial process exists for manganese.

Iron bacteria are thought to be responsible for the oxidation to insoluble ferric compounds of soluble ferrous iron that originated in anaerobic groundwaters in northern peatland areas. These "bog-iron" deposits form the basis of the ore that has been used in the iron and steel industry. Iron in its reduced ferrous form causes a gray–green coloration (gleying) of mineral soils [$Fe(OH)_2$] instead of the normal red or brown color in oxidized conditions [$Fe(OH)_3$]. This gives a relatively easy field check on the oxidized and reduced layers in a mineral soil profile (see Wetland Soils earlier in this chapter).

Iron and manganese in their reduced forms can reach toxic concentrations in wetland soils. Ferrous iron, diffusing to the surface of the roots of wetland plants, can be oxidized by oxygen leaking from root cells, immobilizing phosphorus and coating roots with an iron oxide, and causing a barrier to nutrient uptake (Gambrell and Patrick, 1978).

Sulfur Transformations

Sulfur occurs in several different states of oxidation in wetlands, and like nitrogen, it is transformed through several pathways that are mediated by microorganisms (Fig. 6-12). Sulfur is rarely present in such low concentrations that it is limiting to plant or animal growth in wetlands. The release of the reduced form of sulfur, sulfide, when wetland sediments are disturbed causes the odor familiar to those who carry out research in wetlands—the smell of rotten eggs. On the redox scale, sulfur compounds are the next major electron acceptors after nitrates, iron, and manganese, with reduction occurring at about -100 to -200 mV on the redox scale (see Table 6-2). The most common oxidation states (valences) for sulfur in wetlands are:

Form	Valence
$S^=$ (sulfide)	-2
S (elemental sulfur)	0
S_2O_3 (thiosulfate)	$+2$
$SO_4^=$ (sulfate)	$+6$

Sulfate reduction can take place as *assimilatory sulfate reduction* in which certain sulfur-reducing obligate anaerobes such as *Desulfovibrio* bacteria utilize the sulfates as terminal electron acceptors in anaerobic respiration:

$$4H_2 + SO_4^= \rightarrow H_2S + 2H_2O + 2OH^- \tag{6.16}$$

This sulfate reduction can occur over a wide range of pH, with highest rates prevalent near neutral pH.

There have been a few measurements of the rate at which hydrogen sulfide is produced in and released from wetlands, and those measurements have ranged from 0.004 to 2.6 g S m^{-2} yr^{-1} (Table 6-3). Although the data in Table 6-3 vary over a wide range, it can probably be safely generalized that saltwater wetlands have higher rates of sulfide emission per unit area than do freshwater wetlands where sulfate ions

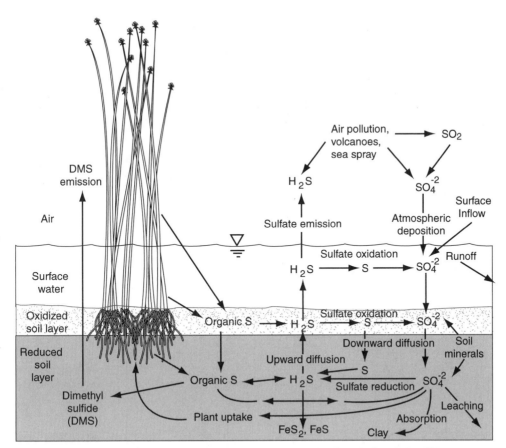

Figure 6-12 Sulfur transformations in wetlands. DMS indicates dimethyl sulfide.

are much less abundant (~2700 mg/L in sea water; ~10 mg/L in fresh water). Sulfur can also be released to the atmosphere as organic sulfur compounds, especially as dimethyl sulfide (DMS), $(CH_3)_2S$, and this flux is thought by some to be as important as or more important than H_2S emissions from some wetlands (Faulkner and Richardson, 1989). The general concensus, however, is that most DMS comes from oceans as a product of decomposing phytoplankton cells and that the most important loss of sulfur from terrestrial freshwater wetland systems is H_2S (Schlesinger, 1991).

Sulfides can be oxidized by both chemoautotrophic and photosynthetic microorganisms to elemental sulfur and sulfates in the aerobic zones of some wetland soils. Certain species of *Thiobacillus* obtain energy from the oxidation of hydrogen sulfide to sulfur, whereas other species in this genus can further oxidize elemental sulfur to sulfate. These reactions are summarized as follows:

$$2H_2S + O_2 \rightarrow 2S + 2H_2O + energy \qquad (6.17)$$

and

$$2S + 3O_2 + 2H_2O \rightarrow 2H_2SO_4 + energy \qquad (6.18)$$

Table 6-3 Biogenic emissions of H₂S from wetlands

Wetland Type	Sampling Location	Emission Rates (g S m⁻² yr⁻¹)			Source
		Mean	Min.	Max.	
FRESHWATER					
Marsh	North Carolina	0.60	0.08	1.27	Aneja et al. (1981)
Marsh	Florida	0.08	0.05	0.11	Castro and Dierberg (1987)
Swamp	Ivory Coast	2.63[a]	—	—	Delmas et al. (1980)
Cypress swamp	Florida	0.004	0.001	0.006	Castro and Dierberg (1987)
Cypress swamp	North Carolina, New York, Georgia	—	0.001	0.16	Adams et al. (1981)
SALTWATER					
Salt marsh	North Carolina	0.15	0.04	0.41	Aneja et al. (1981)
Salt marsh	Florida, Texas, North and South Carolina, Delaware, Louisiana		0.02	601.6	Adams et al. (1981)
Salt marsh	New York	0.55	0.00	41.5	Hill et al. (1978)
Salt marsh	Massachusetts	2.16	1.26	5.04	Steudler and Peterson (1984)
Salt marsh (*Juncus*)	Florida	0.008	0.003	0.015	Castro and Dierberg (1987)
Mangrove swamp	Florida	0.11	0.00	—	Castro and Dierberg (1987)

[a]Waterlogged soil only.
Source: Castro and Dierberg (1987).

Photosynthetic bacteria such as the purple sulfur bacteria found in salt marshes and mud flats are capable of producing organic matter in the presence of light according to the following equation:

$$CO_2 + H_2S + light \rightarrow CH_2O + S \qquad (6.19)$$

This reaction uses hydrogen sulfide as an electron donor rather than H_2O but is otherwise similar to the more traditional photosynthesis equation. This reaction often takes place under anaerobic conditions where hydrogen sulfide is abundant but at the surface of sediments where sunlight is also available.

Sulfide Toxicity

Hydrogen sulfide, which is characteristic of anaerobic wetland sediments, can be toxic to rooted higher plants and microbes, especially in saltwater wetlands where the concentration of sulfates is high. The negative effects of sulfides on higher plants have been described by Ponnamperuma (1972) as attributable to a number of causes, including the following:

1. The direct toxicity of free sulfide as it comes in contact with plant roots
2. The reduced availability of sulfur for plant growth because of its precipitation with trace metals
3. The immobilization of zinc and copper by sulfide precipitation

In wetland soils that contain high concentrations of ferrous iron (Fe^{2+}), sulfides can combine with iron to form insoluble ferrous sulfides (FeS), thus reducing the toxicity of the free hydrogen sulfide (Gambrell and Patrick, 1978). Ferrous sulfide gives the black color characteristic of many anaerobic wetland soils; one of its common mineral forms is pyrite, FeS_2, the form of sulfur commonly found in coal deposits.

Carbon Transformations

The major processes of carbon transformation under aerobic and anaerobic conditions are shown in Figure 6-13. Photosynthesis and aerobic respiration, of course, dominate the aerobic horizons (aerial and aerobic water and soil) with H_2O as the major electron donor in photosynthesis and oxygen as the terminal electron acceptor in respiration:

$$6CO_2 + 12H_2O + light \rightarrow C_6H_{12}O_6 + 6O_2 + 6H_2O \qquad (6.20)$$

$$C_6H_{12}O_6 + 6O_2 \rightarrow 6CO_2 + 6H_2O + 12e^- + energy \qquad (6.21)$$

The degradation of organic matter by aerobic respiration (Eq. 6-21) is fairly efficient in terms of energy transfer. However, because of the anoxic nature of wetlands, anaerobic processes, less efficient in terms of energy transfer, occur in close proximity to aerobic processes. Two of the major anaerobic processes are fermentation and methanogenesis. The *fermentation* of organic matter, also called *glycolysis* for the substrate involved, occurs when organic matter itself is the terminal electron acceptor in anaerobic respiration by microorganisms and forms various low-molecular-weight acids and alcohols and CO_2. Examples are lactic acid

$$C_6H_{12}O_6 \rightarrow 2CH_3CH_2OCOOH \qquad (6.22)$$
$$\text{(lactic acid)}$$

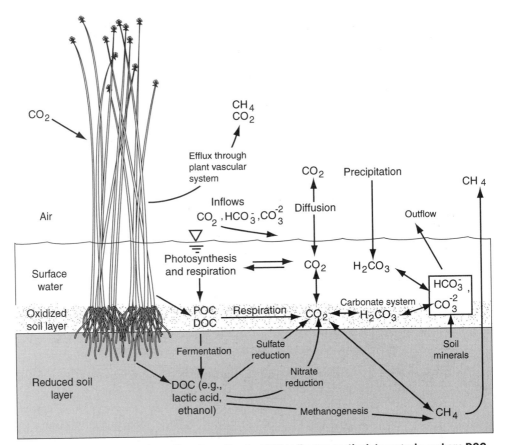

Figure 6-13 **Carbon transformations in wetlands. POC indicates particulate organic carbon; DOC indicates dissolved organic carbon.**

and ethanol

$$C_6H_{12}O_6 \rightarrow 2CH_3CH_2OH + 2CO_2 \qquad (6.23)$$
$$\text{(ethanol)}$$

Fermentation can be carried out in wetland soils by either facultative or obligate anaerobes. Although *in situ* studies of fermentation in wetlands are rare, it is thought that "fermentation plays a central role in providing substrates for other anaerobes in sediments in waterlogged soils" (Wiebe et al., 1981). It represents one of the major ways in which high-molecular-weight carbohydrates are broken down to low-molecular-weight organic compounds, usually as dissolved organic carbon, which are, in turn, available to other microbes (Valiela, 1984).

Methanogenesis occurs when certain bacteria (methanogens) use CO_2 as an electron acceptor for the production of gaseous methane (CH_4), as shown in Equation 6-9, or, alternatively, use a low-molecular-weight organic compound such as one from a methyl group:

$$CH_3COO + 4H_2 \rightarrow 2CH_4 + 2H_2O \qquad (6.24)$$

Methane, which can be released to the atmosphere when sediments are disturbed, is often referred to as *swamp gas* or *marsh gas*. Methane production requires extremely reduced conditions, with a redox potential below -200 mV, after other terminal electron acceptors (O_2, NO_3, and $SO_4^=$) have been reduced. Methanogenesis is carried out by methanogens—a group of microbes called the Archaea, a group of prokaryotes that includes several obligate halophiles and thermophiles (Boon, 1999).

The rates of methanogenesis from both saltwater and freshwater wetlands, as well as from domestic wetlands such as rice paddies, have a considerable range (Table 6-4). Comparison of rates of methane production from different studies is difficult, because different methods are used and because the rates depend on both temperature (season) and hydroperiod. Methanogenesis has clear seasonal patterns in temperate-zone wetlands (Fig. 6-14) but the pattern depends on constant flooding. Harriss et al. (1982) noted maximum methane production in a Virginia freshwater swamp between April and May and a net uptake of methane by the wetland during a drought when the wetland soil was exposed to the atmosphere. Wiebe et al. (1981) found methane production to generally peak in late summer in a Georgia salt marsh. Boon and Sorrell (1991) and Sorrell and Boon (1992), in a detailed study of gas production (mostly methane) from billabong sediments in Australia, found clear patterns of higher rates in the summer ($>25°C$) than in winter but no significant difference between rates during the night and during the day (Fig. 6-14).

Most of the methane emission studies to date have been in peatlands (bogs and fens) and freshwater marshes. More studies have been undertaken in the temperate zone compared to boreal or tropical climes (Table 6-4). Researchers in boreal wetlands have found beaver ponds to have much higher methane flux rates than other wetland types (Naiman et al., 1991; Roulet et al., 1992a,b) and neutral fens to have higher rates than acid fens and bogs (Crill et al., 1988). A comparison of methanogenesis between freshwater wetlands (marshes and swamps) and marine wetlands (salt marshes and mangroves) shows that the rate of methane production is higher in the former (up to 500 mg C m^{-2} day^{-1}) than in the latter (up to 100 mg C m^{-2} day^{-1}), apparently because of the lower amounts of sulfate competition for oxidizable substrate in freshwater systems (see Carbon–Sulfur Interactions, p. 182).

Table 6-4 **Ranges of mean methane emission rates (number of sites/treatments) for major wetland types**

	CH$_4$ Emission Rates (mg C m^{-2} day^{-1})		
	Boreal	Temperate	Subtropical/Tropical
FRESHWATER WETLANDS			
Tundra	3.7–1,500 (12)	—	—
Bog	0.7–17 (5)	20–221 (7)	—
Fen	14–325 (11)	3–314 (8)	—
Freshwater marsh	23–80 (2)	0.1–498 (17)	29–443 (7)
Forest swamp	5–66 (2)	7.4–106 (6)	44–144 (7)
Rice paddy	—	10–880 (34)	47–486 (9)
SALTWATER WETLANDS			
Salt marsh	—	0–109 (17)	2.5 (1)
Mangrove	—	—	3–61 (3)

Source: Mitsch and Wu (1995).

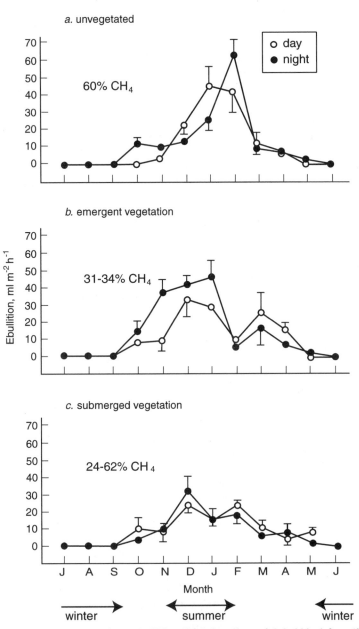

Figure 6-14. Seasonal patterns of gas ebullition (flux of methane-rich bubbles) from three different wetland community types in a floodplain lake (billabong) along the River Murray, New South Wales, Australia: **a.** no vegetation; **b.** beds of the emergent plant *Eleocharis sphacelata;* **c.** beds of the submerged aquatic plant *Vallisneria gigantea.* Methane concentrations were 60 percent of the emissions from the bare area, 31–54 percent of the emergent plant site, and 24–62 percent of the submerged aquatic plant site. (*After Sorrell and Boon, 1992*)

The data in Table 6-4 can be compared with generalizations that have been used to estimate global methane production from wetlands (Table 6-5). Gorham (1991) assumed an average of 77.5 mg C m^{-2} day^{-1} in estimating the global contributions of northern peatlands. Table 6-4 suggests that the means of several studies range from 0.7 to 17 mg C m^{-2} day^{-1} for boreal bogs to a much higher 14 to 325 mg C m^{-2} day^{-1} for the more mineral-rich fens. Thus, depending on the mix of bogs and fens in northern peatlands, Gorham's assumption could be low or high. Matthews and Fung (1987) used cross-latitude assumptions to determine average emission rates of methane per wetland type and found that bogs have the highest rate of methanogenesis (Table 6-5). Aselmann and Crutzen (1989) found completely the opposite and suggested a geometric mean emission rate of 11 mg C m^{-2} day^{-1} in increasing order of bogs, fens, swamps, marshes, and rice paddies (Table 6-5). Fens to have much higher emissions in both boreal and temperate regions than do forested swamps and marshes.

Some of the preceding measurements may have led to integrated annual estimates of methane emission that are too high. Moore and Roulet (1995) suggest that most annual flux measurements in Canada are less than 10 g CH$_4$ m^{-2} yr^{-1} with the primary controlling mechanisms being soil temperature, water table position, or a combination of both. Roulet et al. (1992a), in a detailed field study of 12 minerotrophic wetlands and 3 beaver ponds in the low boreal region of central Canada, estimated a habitat-weighted average emission of 1.6 g CH$_4$ m^{-2} yr^{-1}, which would translate to a daily rate of only 3.35 mg C m^{-2} day^{-1}. If their study is accurate, it suggests that emissions of methane from northern peatlands may be much less than originally estimated. By contrast, Sorrell and Boon (1992) found extremely high annual rates in Australian billabongs of about 16 to 30 g CH$_4$ m^{-2} yr^{-1}.

Gaseous Transport in Plants

With the exception of CO_2 and O_2, gases emitted from wetlands have generally been assumed to (1) emanate from the sediment or soil surface through the water column, (2) bubble or diffuse to the surface in a process called *ebullitive flux* and then exit to the atmosphere, or (3) pass through the vascular system of emergent plants (Boon, 1999). Boon and Sorrell (1995) noted that there were substantially more methane fluxes during the day than during the night in chamber studies of Australian wetlands when wetland plants were included in the chambers. They also noted that there was a discrepancy in chambers between the total methane flux and the amount measured by inverted funnels (which capture the ebullitive flux). As a result, the pressures, flows, and gas concentrations were measured within a dominant wetland plant, *Eleocharis*

Table 6-5 Methane production means (mg C m^{-2} day^{-1}) used for global methane emission estimates

Wetland Type	Matthews and Fung (1987)	Aselmann and Crutzen (1989)
Peatlands (bogs)	150	11
Fens	—	60
Forested swamps	53	63
Marshes	90	190
Riparian wetlands	23	75
Rice paddies	—	230

sphacelata, in both "influx" culms, which could generate high pressures, and "efflux" culms, which could not. Sorrell and Boon (1994) and Sorrell et al. (1994) demonstrated that the methane concentration was three orders of magnitude greater in the efflux culms than in the influx culms (Fig. 6-15). Carbon dioxide concentrations, as expected, were 50 times higher, whereas dissolved oxygen concentrations decreased 20 percent. These studies and others suggest that between 50 and 90 percent of all methane generated from a vegetated wetland comes through the vascular system of emergent plants (Boon, 1999).

Carbon–Sulfur Interactions

The sulfur cycle is important in some wetlands for the oxidation of organic carbon. This is particularly true in most coastal wetlands where sulfur is abundant. In general, methane is found at low concentrations in reduced soils when sulfate concentrations are high (Gambrell and Patrick, 1978; Valiela, 1984). Possible reasons for this phenomenon include (1) competition for substrates that occurs between sulfur and methane bacteria, (2) the inhibitory effects of sulfate or sulfide on methane bacteria, (3) a possible dependence of methane bacteria on products of sulfur-reducing bacteria, and (4) a stable redox potential that does not drop low enough to reduce CO_2 because of an ample supply of sulfate. Other evidence suggests that methane may actually be oxidized to CO_2 by sulfate reducers (Valiela, 1984).

Sulfur-reducing bacteria require an organic substrate, generally of low molecular weight, as a source of energy in converting sulfate to sulfide (Eq. 6-16). The process of fermentation described previously can conveniently supply these necessary low-molecular-weight organic compounds such as lactate or ethanol (see Eqs. 6-22 and 6-23 and Fig. 6-13). The equations for sulfur reduction, also showing the oxidation of organic matter, are given as follows:

$$2CH_3CHOHCOO^- + SO_4^= + 3H^+$$
$$\text{(lactate)} \hspace{4cm} (6.25)$$
$$\rightarrow 2CH_3COO_- + 2CO_2 + 2H_2O + HS^-$$

and

$$CH_3COO^- + SO_4^= \rightarrow 2CO_2 + 2H_2O + HS^- \hspace{1cm} (6.26)$$
$$\text{(acetate)}$$

The importance of this fermentation–sulfur reduction pathway in the oxidation of organic carbon to CO_2 in saltwater wetlands is demonstrated by comparing similar measurements in a New England salt marsh and freshwater lake sediments in Wisconsin (Table 6-6). Fully 54 percent of the carbon dioxide evolution from the salt marsh was caused by the fermentation–sulfur reduction pathway, with aerobic respiration accounting for another 45 percent. Only a small percentage of carbon release (0.7 percent) was caused by methanogenesis. A similar study using the same methods in a salt marsh in Georgia (Howarth and Giblin, 1983) yielded sulfate reduction rates that were one-third of those measured in the New England salt marsh, possibly due to less underground organic productivity in the Georgia marsh.

In anaerobic lake sediments in Wisconsin (similar in function to freshwater marsh sediments), almost 80 percent of the carbon was produced through methanogenesis with a minor contribution due to fermentation followed by sulfate reduction (Table 6-6). Smith and Klug (1981) found that 2.5 times more organic carbon was mineral-

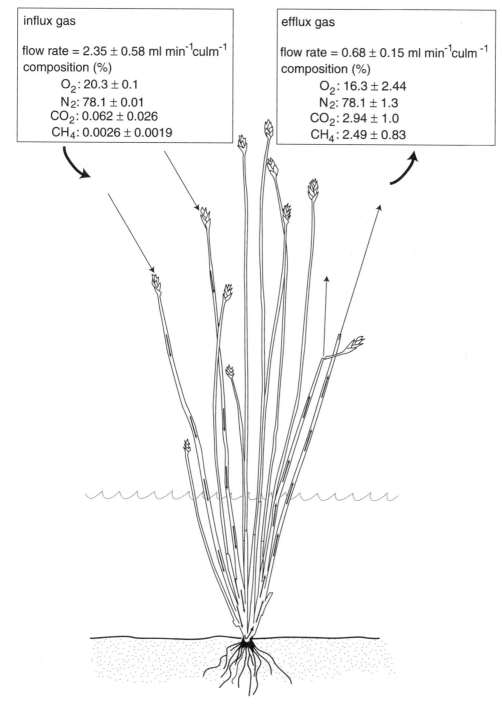

influx gas

flow rate = 2.35 ± 0.58 ml min^{-1}culm^{-1}
composition (%)
O_2: 20.3 ± 0.1
N_2: 78.1 ± 0.01
CO_2: 0.062 ± 0.026
CH_4: 0.0026 ± 0.0019

efflux gas

flow rate = 0.68 ± 0.15 ml min^{-1}culm^{-1}
composition (%)
O_2: 16.3 ± 2.44
N_2: 78.1 ± 1.3
CO_2: 2.94 ± 1.0
CH_4: 2.49 ± 0.83

Figure 6-15 Typical fluxes and concentrations of gases in "influx" and "efflux" of the emergent plant *Eleocharis sphacelata* as measured in Australian wetlands. Fluxes through emergent plants are due to both simple diffusion along a gradient and convective flow of pressurized gas. (*After Sorrell et al., 1994; Boon, 1998*)

Table 6-6   Carbon dioxide release from mineralization of organic matter in a New England salt marsh and Wisconsin freshwater lake sediments

Pathway	Salt Marsh (g C m^{-2} yr^{-1})	Lake Sediment (g C m^{-2} yr^{-1})
Aerobic respiration	361	—
Nitrate reduction	5	8[a]
Fermentation–sulfate reduction	432	61
Methanogenesis	6	254
Total	804	—

[a] Estimated from previous study.
Source: Ingvorsen and Brock (1982) and Howes et al. (1984, 1985).

ized by methanogenesis than through sulfate reduction in a lake in Michigan. In a freshwater billabong in Australia, Boon and Mitchell (1995) demonstrated that methanogenesis accounted for 30 to 50 percent of the total benthic carbon flux and that a major portion of the carbon fixed by plants leaves the wetland via methanogenesis. Estimates of the absolute values of these processes, when compared with those for the salt marsh (Table 6-6), support the generalizations that oxidation of organic carbon by methane production is dominant in freshwater wetlands, whereas oxidation of organic carbon by sulfate reduction is dominant in saltwater wetlands (Capone and Kiene, 1988).

Phosphorus Transformations

Phosphorus (Fig. 6-16) is one of the most important limiting chemicals in ecosystems, and wetlands are no exception. It is a major limiting nutrient in northern bogs, freshwater marshes, and southern deepwater swamps. In other wetlands such as agricultural wetlands and salt marshes, phosphorus is an important mineral, although it is not considered a limiting factor because of its relative abundance and biochemical stability. Phosphorus retention is considered one of the most important attributes of natural and constructed wetlands, particularly those that receive nonpoint source pollution or wastewater [see the ecosystem chapters (Chapters 9–15) and also Chapter 20].

Phosphorus occurs as soluble and insoluble complexes in both organic and inorganic forms in wetland soils (Table 6-7). Inorganic forms include the ions PO_4^{3-}, HPO_4^{2-}, and $H_2PO_4^-$, the predominant form depending on the pH. Phosphorus also has an affinity for calcium, iron, and aluminum, forming complexes with those elements when they are readily available. Phosphorus occurs in a sedimentary cycle (Fig. 6-16) rather than in a gaseous cycle such as the nitrogen cycle. At any one time, a major proportion of the phosphorus in wetlands is tied up in organic litter and peat and in inorganic sediments, with the former dominating peatlands and the latter dominating mineral soil wetlands. The principal inorganic form is orthophosphate. One of the ways in which phosphorus is categorized is by its differential solubility in various chemical extractants (Reddy et al., 1999). One such fractionation scheme, developed by van Eck (1982), uses the categorization given in Table 6-8. The analytical measure of biologically available orthophosphates is sometimes called *soluble reactive phosphorus* (SRP), although the equivalence among soluble reactive phosphorus, exchangeable phosphorus, and orthophosphate is not exact. However, they are often used as indica-

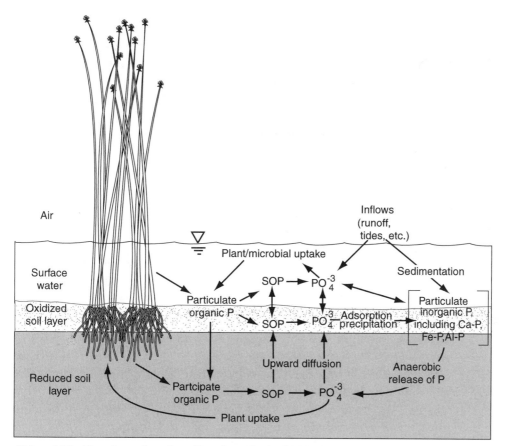

Figure 6-16 Phosphorus transformations in wetlands. SOP indicates soluble organic phosphorus.

Table 6-7 Major types of dissolved and insoluble phosphorus in natural waters

Phosphorus	Soluble Forms	Insoluble Forms
INORGANIC		
	Orthophosphates ($H_2PO_4^-$, HPO_4^{2-}, PO_4^{3-}) Polyphosphates Ferric phosphate ($FeHPO_4^+$)	Clay–phosphate complexes Metal hydroxide phosphates, e.g., vivianite $Fe_3(PO_4)_2$, variscite $Al(OH)_2H_2PO_4$
	Calcium phosphate ($CaH_2PO_4^+$)	Minerals, e.g., apatite [$Ca_{10}(OH)_2(PO_4)_6$]
ORGANIC		
	Dissolved organics, e.g., sugar phosphates, inositol phosphates, phospholipids, phosphoproteins	Insoluble organic phosphorus bound in organic matter

Table 6-8 Categorization of phosphorus in waters based on sequential extraction

Phosphorus Fraction	Extractant
Exchangeable phosphorus	0.5 M NaCl
Labile organic phosphorus and carbonate-bound phosphorus	1 M NH_4Cl
Fe–Al–P	0.1 M NaOH
Ca–Mg–P	0.5 M HCl
Resistant organic phosphorus	Remaining residue

Source: Reddy et al. (1999), quoting van Eck (1982).

tors of the bioavailability of phosphorus. Dissolved organic phosphorus and insoluble forms of organic and inorganic phosphorus are generally not biologically available until they are transformed into soluble inorganic forms.

Although phosphorus is not directly altered by changes in redox potential as are nitrogen, iron, manganese, and sulfur, it is indirectly affected in soils and sediments by its association with several elements, especially iron, that are so altered. Phosphorus is rendered relatively unavailable to plants and microconsumers by:

1. The precipitation of insoluble phosphates with ferric iron, calcium, and aluminum under aerobic conditions

2. The adsorption of phosphate onto clay particles, organic peat, and ferric and aluminum hydroxides and oxides

3. The binding of phosphorus in organic matter as a result of its incorporation into the living biomass of bacteria, algae, and vascular macrophytes

There are three general conclusions about the tendency of phosphorus to precipitate with selected ions: (1) Phosphorus is fixed as aluminum and iron phosphates in acid soils, (2) phosphorus is bound by calcium and magnesium in alkaline soils, and (3) phosphorus is most bioavailable at slightly acidic to neutral pH (Reddy et al., 1999). The precipitation of the metal phosphates and the adsorption of phosphates onto ferric or aluminum hydroxides and oxides are believed to result from the same chemical forces, namely, those involved in the forming of complex ions and salts.

In many surface water wetlands, high algal productivity can pull CO_2 out of the water, shift the whole carbonate equilibrium, and drive the pH as high as 9 or 10 on a diurnal basis. Under these conditions, coprecipitation of phosphorus as it adsorbs onto calcite and precipitates as calcium phosphate can be significant just as precipitation of calcium carbonate is also accelerated (Reddy et al., 1999). Richardson (1985) and Reddy and D'Angelo (1994) suggested that a good indicator of phosphorus sorption on a wetland soil is the amount of oxalate-extractable iron and aluminum in the soil.

The sorption of phosphorus onto clay particles is believed to involve both the chemical bonding of the negatively charged phosphates to the positively charged edges of the clay and the substitution of phosphates for silicate in the clay matrix. This clay–phosphorus complex is particularly important for many wetlands, including riparian wetlands and coastal salt marshes, because a considerable portion of the phosphorus brought into these systems by flooding rivers and tides is brought in sorbed to clay particles. Thus, phosphorus cycling in many mineral soil wetlands tends to follow the sediment pathways of sedimentation and resuspension. Because most wetland macrophytes obtain their phosphorus from the soil, the sedimentation of phosphorus sorbed onto clay particles is an indirect way in which the phosphorus is made available to the biotic components of the wetland. In essence, the plants transform inorganic phosphorus to organic forms that are then stored in organic peat, mineralized by microbial activity, or exported from the wetland (Fig. 6-16).

When soils are flooded and conditions become anaerobic, several changes in the availability of phosphorus result. A well-documented phenomenon in the hypolimnion of lakes is the increase in soluble phosphorus when the hypolimnion and the sediment–water interface become anoxic. In general, a similar phenomenon often occurs in wetlands on a compressed vertical scale. As ferric (Fe^{3+}) iron is reduced to more soluble ferrous (Fe^{2+}) compounds, phosphorus that is in a specific ferric phosphate analytically known as reductant-soluble phosphorus (Gambrell and Patrick, 1978; Faulkner and Richardson, 1989) is released into solution. Other reactions that may be important in releasing phosphorus upon flooding are the hydrolysis of ferric and aluminum phosphates and the release of phosphorus sorbed to clays and hydrous oxides by the exchange of anions (Ponnamperuma, 1972). Phosphorus can also be released from insoluble salts when the pH is changed either by the production of organic acids or by the production of nitric and sulfuric acids by chemosynthetic bacteria. Phosphorus sorption onto clay particles, on the other hand, is highest under acidic to slightly acidic conditions (Stumm and Morgan, 1996).

Chemical Transport into Wetlands

The inputs of materials to wetlands occur through geologic, biologic, and hydrologic pathways. The geologic input from weathering of parent rock, although poorly understood, may be important in some wetlands. Biologic inputs include photosynthetic uptake of carbon, nitrogen fixation, and biotic transport of materials by mobile animals such as birds. Except for gaseous exchanges such as carbon fixation in photosynthesis and nitrogen fixation, however, elemental inputs to wetlands are generally dominated by hydrologic inputs.

Precipitation

Table 6-9 describes typical chemical characteristics of precipitation as measured in several locations in North America. The levels of chemicals in precipitation are variable but very dilute. Relatively higher concentrations of magnesium and sodium are associated with maritime influences, whereas high calcium indicates continental influences. Precipitation tends to contain contaminants at higher concentrations in short storms and when precipitation is infrequent.

Human influence on the chemicals in precipitation has been significant, particularly because of the burning of fossil fuels and subsequent increased concentrations of sulfates and nitrates in the atmosphere, the causes of the familiar acid rain (more properly called acid deposition). The data in Table 6-9 for the chemical characteristics of precipitation for New Hampshire are weighted averages over several years and do not reflect recent trends. Long-term monitoring at this site showed that, after several years of relatively high concentrations of sulfates in precipitation, there has been a steady decrease in the concentrations of sulfates and basic cations. Sulfate concentrations in precipitation decreased approximately 40 percent from about 3.1 mg/L in the late 1960s to 1.9 mg/L in the 1980s, probably due to the decline in the emissions

Table 6-9 Chemical characteristics of bulk precipitation (mg/L)

Chemical	Georgia[a]	Newfoundland[b]	Wisconsin–Minnesota[b]	New Hampshire[c]
Ca^{2+}	0.17	0.8	1.0–1.2	0.13
Mg^{2+}	0.05	—	—	0.04
Na^+	—	5.2	0.2–0.5	0.11
K^+	0.14	0.3	0.2	0.06
NO_3^-	0.26	—	—	1.47
NH_4^+	—	—	—	0.19
Cl^-	—	8.9	0.1	0.40
$SO_4^=$	—	2.2	1.4	2.6
P(soluble)	0.017	—	—	0.01

[a] Schlesinger (1978).
[b] Gorham (1961), as summarized by Moore and Bellamy (1974).
[c] Likens et al. (1985).

of sulfur dioxide in the northeastern United States over that period (Driscoll et al., 1989). Long-term trends of nitrates in precipitation, at least in the eastern and parts of the midwestern United States, did not show similar decreases, because emissions from sources such as automobiles continued to increase over that period. Some wetlands such as northern peatlands or southern cypress domes are fed primarily by precipitation and therefore may be more susceptible to anthropogenic inputs in precipitation, especially where acidic deposition is severe. In fact, well before there was a general concern for their effects on aquatic and terrestrial systems, the possible effects of anthropogenic inputs of acid precipitation to certain wetlands were discussed in biogeochemical studies of English peatlands by Gorham (1956, 1961).

Streams, Rivers, and Groundwater

As precipitation reaches the ground in a watershed, it infiltrates into the ground, passes back to the atmosphere through evapotranspiration, or flows on the surface as runoff. When enough runoff comes together, sometimes combined with groundwater flow, in channelized streamflow, its mineral content is different from that of the original precipitation.

There is not, however, a typical water quality for surface and subsurface flows. Figure 6-17 describes the cumulative frequency of the ionic composition of freshwater streams and rivers in the United States. It shows, for example, the average concentrations of the many ions at the 50 percent line. Average NO_3^- concentrations are about 1 mg/L, whereas the average for Mg^{2+} is about 10 mg/L. The curves demonstrate the wide range over which these chemicals are found and the median values (50 percent line) of these ranges.

The "average" concentration of dissolved materials in the world's rivers is given in Table 6-10. The variability in concentrations of chemicals in runoff and streamflow that enter wetlands is caused by several factors:

1. *Groundwater influence.* The chemical characteristics of streams and rivers depend on the degree to which the water has previously come in contact with underground formations and on the types of minerals present in those formations. Soil and rock weathering, through dissolution and redox reactions, provides major dis-

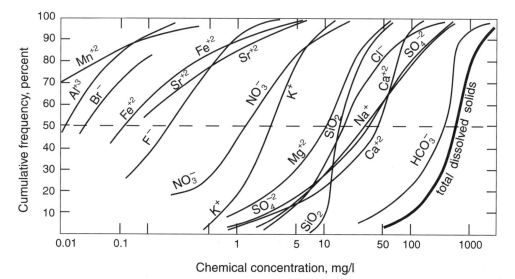

Figure 6-17 Cumulative frequency curves for concentrations of various dissolved minerals in surface waters. Horizontal dashed line indicates median concentrations, 90 percent indicates the 90th percentile concentration, etc. (*After Davis and DeWiest, 1966*)

Table 6-10 Average chemical concentrations (mg/L) of ocean water and river water

Chemical	Sea Water[a]	"Average" River[b]
Na^+	10,773	6.3
Mg^{2+}	1,294	4.1
Ca^{2+}	412	15
K^+	399	2.3
Cl^-	19,344	7.8
SO_4^{2-}	2,712	11.2
HCO_3^-/CO_3^{2-}	142	58.4
B	4.5[c]	0.01[c]
F	1.4[c]	0.1[c]
Fe	<0.01[c]	0.7
SiO_2	<0.1->10[c]	13.1[c]
N	0-0.5[c]	0.2[c]
P	0-0.07[c]	0.02[c]
Particulate Organic Carbon	0.01-1.0[c]	5-10[c]
Dissolved Organic Carbon	1-5[c]	10-20[c]

[a] Riley and Skirrow (1975).
[b] D. A. Livingston (1963).
[c] Burton and Liss (1976).

solved ions to waters that enter the ground. The dissolved materials in surface water can range from a few milligrams per liter, found in precipitation, to 100 or even 1,000 mg/L. The ability of water to dissolve mineral rock depends, in part, on its nature as a weak carbonic acid. The rock being mineralized is also an important consideration. Minerals such as limestone and dolomite yield high levels of dissolved ions, whereas granite and sandstone formations are relatively resistant to dissolution.

2. *Climate.* Climate influences surface water quality through the balance of precipitation and evapotranspiration. Arid regions tend to have higher concentrations of salts in surface waters than do humid regions. Climate also has a considerable influence on the type and extent of vegetation on the land, and it therefore indirectly affects the physical, chemical, and biological characteristics of soils and the degree to which soils are eroded and transported to surface waters.

3. *Geographic effects.* The amounts of dissolved and suspended materials that enter streams, rivers, and wetlands also depend on the size of the watershed, the steepness or slope of the landscape, the soil texture, and the variety of topography. Surface waters that have high concentrations of suspended (insoluble) materials caused by erosion are often relatively low in dissolved substances. On the other hand, waters that have passed through groundwater systems often have high concentrations of dissolved materials and low levels of suspended materials. The presence of upstream wetlands also influences the quality of water entering downstream wetlands (see Coupling with Adjacent Ecosystems later in this chapter).

4. *Streamflow/ecosystem effects.* The water quality of surface runoff, streams, and rivers varies seasonally. It is most common that there is an inverse correlation between streamflow and concentrations of dissolved materials (Hem, 1970; Dunne and Leopold, 1978). During wet periods and storm events, the water is contributed primarily by recent precipitation that becomes streamflow very quickly without coming into contact with soil and subsurface minerals. During low flow, some or much of the streamflow originates as groundwater and has higher concentrations of dissolved materials. This inverse relationship, however, is not always the case, as illustrated by the relationship found in several years of study at the Hubbard Brook, New Hampshire, watershed (Fig. 6-18). There, the small watershed had remarkably similar concentrations of dissolved substances despite a wide range of streamflow, caused by the biotic and abiotic "regulation" of water quality in the small forested watershed and the stream itself (Bormann et al., 1969; Likens et al., 1985).

The relationship between particulate matter and streamflow is often found to be a nonlinear positive relationship (Fig. 6-18). Part of this relationship is based on the fact that streamflow from a small watershed is often low in the growing season, even for relatively heavy precipitation, because of high interception and evapotranspiration. On the other hand, precipitation in the nongrowing season, often combined with saturated soils and little evapotranspiration, creates high streamflow and sediment transport. In one example, Omernik (1977) showed that, for a large sample of streams throughout the United States, concentrations of nitrogen and phosphorus increased with discharge in disturbed watersheds because of increased erosion but decreased with streamflow in natural watersheds presumably because of reduced erosion and increased dilution.

As with all generalizations, the increase in particulate matter concentration with flow is not always the case. For example, in many streams in the midwestern United States, bioturbation stemming from active fish populations (e.g., common carp, *Cypri-*

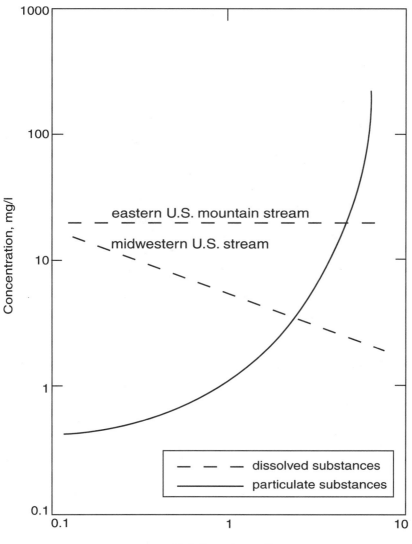

Figure 6-18 Generalized relationships between streamflow and concentrations of dissolved and particulate substances. Data for midwestern U.S. stream is generalization from several studies. Data for a small eastern U.S. watershed is Hubbard Brook, New Hampshire. (*After Likens et al., 1985*)

nus carpio) and summer algal blooms can actually cause sediment concentrations to be higher under the low-flow conditions typical of late summer.

5. *Human effects.* Water that has been modified by humans through, for example, sewage effluent, urbanization, and runoff from farms often drastically alters the chemical composition of streamflow and groundwater that reach wetlands. If drainage is from agricultural fields, higher concentrations of sediments and nutrients and some herbicides and pesticides might be expected. Urban and suburban drainage is often

lower than that from farmland in those constituents, but it may have high concentrations of trace organics, oxygen-demanding substances, and some toxins.

Estuaries

Wetlands such as salt marshes and mangrove swamps are continually exchanging tidal waters with adjacent estuaries and other coastal waters. The quality of these waters differs considerably from that of the rivers described previously. Although estuaries are places where rivers meet the sea, they are not simply places where sea water is diluted with fresh water (J. W. Day et al., 1989). Table 6-10 contrasts the chemical makeup of average river water with the average composition of sea water. The chemical characteristics of sea water are fairly constant worldwide compared with the relatively wide range of river water chemistry. Total salt concentrations typically range from 33 to 37 ppt. Although sea water contains almost every element that can go into solution, 99.6 percent of the salinity is accounted for by 11 ions. In addition to seawater dilution, estuarine waters can also involve chemical reactions when sea and river waters meet, including the dissolution of particulate substances, flocculation, chemical precipitation, biological assimilation and mineralization, and adsorption and absorption of chemicals on and into particles of clay, organic matter, and silt. In most estuaries and coastal wetlands, biologically important chemicals such as nitrogen, phosphorus, silicon, and iron come from rivers, whereas other important chemicals such as sodium, potassium, magnesium, sulfates and bicarbonates/carbonates come from ocean sources (J. W. Day et al., 1989).

Chemical Mass Balances of Wetlands

A quantitative description of the inputs, outputs, and internal cycling of materials in an ecosystem is called an *ecosystem mass balance.* If the material being measured is one of several elements such as phosphorus, nitrogen, or carbon that are essential for life, then the mass balance is called a *nutrient budget.* In wetlands, mass balances have been developed both to describe ecosystem function and to determine the importance of wetlands as sources, sinks, and transformers of chemicals. Extensive literature reviews on the subject of the influence of wetlands on water quality, the net result of these mass balances, have been provided by Nixon and Lee (1986), Johnston et al. (1990), and Johnston (1991).

A general mass balance for a wetland, as shown in Figure 6-19, illustrates the major categories of pathways and storages that are important in accounting for materials passing into and out of wetlands. Nutrients or chemicals that are brought into the system are called *inputs* or *inflows.* For wetlands, these inputs are primarily through hydrologic pathways (described in Chapter 5) such as precipitation, surface water and groundwater inflow, and tidal exchange. Biotic pathways of note that apply to the carbon and nitrogen budgets are the fixation of atmospheric carbon through photosynthesis and the capture of atmospheric nitrogen through nitrogen fixation.

Hydrologic *exports,* or *losses* or *outflows,* are by both surface water and groundwater unless the wetland is an isolated basin that has no outflow such as a northern ombrotrophic bog. The long-term burial of chemicals in the sediments is also considered a nutrient or chemical outflow, although the depth at which a chemical goes from internal cycling to permanent burial is an uncertain threshold. The depth of available chemicals is usually defined by the root zone of vegetation in the wetland. Biologically

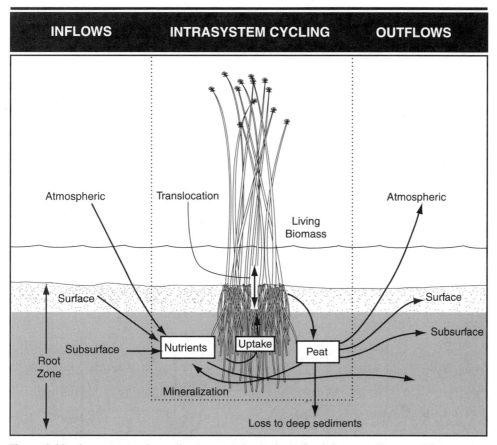

| INFLOWS | INTRASYSTEM CYCLING | OUTFLOWS |

Figure 6-19   **Components of a wetland mass balance, including inflows, outflows, and intrasystem cycling.**

mediated exports to the atmosphere are also important in the nitrogen cycle (denitrification) and in the carbon cycle (respiratory loss of CO_2). The significance of other losses of elements to the atmosphere such as ammonia volatilization and methane and sulfide releases is not well understood, although they are potentially important pathways for individual wetlands, as well as for the global cycling of minerals (see Chemical Transformations in Wetlands earlier in this chapter).

Intrasystem cycling involves exchanges among various *pools,* or *standing stocks,* of chemicals within a wetland. This cycling includes pathways such as litter production, remineralization, and various chemical transformations discussed earlier. The *translocation* of nutrients from the roots through the stems and leaves of vegetation is another important intrasystem process that results in the physical movement of chemicals within a wetland.

Figure 6-20 illustrates in more detail the major pathways and storages that investigators should consider when developing chemical mass balances for wetlands. Major exchanges with the surroundings are shown as exchanges of particulate and dissolved material with adjacent bodies of water (pathways 1 and 2), exchange through groundwater (pathways 7 and 8), inputs from precipitation (pathways 9 and 10), and burial

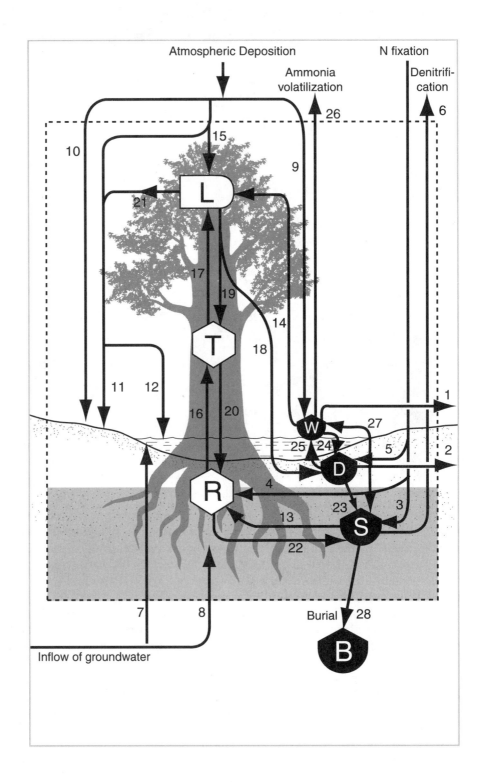

Atmospheric Deposition

N fixation

Ammonia
volatilization

Denitrifi-
cation

26

6

15

10

9

L

21

17

19

14

T

18

11   12

16   20

W

27

25 24

5

1

D

2

R

4

3

23

13

22

S

7   8

Burial   28

Inflow of groundwater

B

194

in sediments (pathway 28). Exchanges specific to a nitrogen mass balance, namely, nitrogen fixation (pathways 3–5), denitrification (pathway 6), and ammonia volatilization (pathway 26), are also shown in the diagram. A number of intrasystem pathways such as stemflow (pathway 12), root sloughing (pathway 22), detritus–water exchanges (pathways 24 and 25), and sediment–water exchanges (pathway 27) can be very important in determining the fate of chemicals in wetlands but are extremely difficult to measure.

Few, if any, investigators have developed a complete mass balance for wetlands that includes measurement of all of the pathways shown in Figure 6-20, but the diagram remains a useful guide to those considering studies of wetlands as sources, sinks, or transformers of chemicals as described later. The chemical balances that have been developed for various wetlands are extremely variable. A few generalizations emerge from these studies:

1. Wetlands serve as sources, sinks, or transformers of chemicals, depending on the wetland type, the hydrologic conditions, and the length of time the wetland has been subjected to chemical loadings. When wetlands serve as sinks for certain chemicals, the long-term sustainability of that situation depends on the hydrologic and geomorphic conditions, the spatial and temporal distribution of chemicals in the wetland, and the ecosystem succession. Wetlands can become saturated in certain chemicals after a number of years, particularly if loading rates are high.

2. Seasonal patterns of nutrient uptake and release are characteristic of many wetlands. In temperate climates, retention of certain chemicals such as nutrients is greatest during the growing season primarily because of higher microbial activity in the water column and sediments and secondarily because of greater macrophyte productivity. For example, in cold temperate climates, distinct seasonal patterns of nitrate retention are evident in many cases, with greater retention during the summer months when warmer temperatures accelerate both denitrification microbial activity and algal and macrophyte growth.

3. Wetlands are frequently coupled to adjacent ecosystems through chemical exchanges that significantly affect both systems. Ecosystems upstream of wetlands are often significant sources of chemicals to wetlands, whereas

---

Figure 6-20 Model of major chemical storages and flows in wetlands. Storages: L, aboveground shoots or leaves; T, stems, branches, perennial above-ground storage; R, roots and rhizomes; W, surface water; D, litter and detritus; S, near-surface sediments; B, deep sediments essentially removed from internal cycling. Flows: 1 and 2 are exchanges of dissolved and particulate matter with adjacent waters; 3–5 are nitrogen fixation in sediments, rhizosphere microflora, and litter; 6 is denitrification; 7 and 8 are groundwater inputs; 9 and 10 are atmospheric inputs (e.g., precipitation); 11 and 12 are throughfall and stemflow; 13 is uptake by roots; 14 is foliar uptake from surface water; 15 is foliar uptake directly from precipitation; 16 and 17 are translocation from roots through stem to leaves; 18 is litterfall; 19 and 20 are translocation of materials from leaves back to stems and roots; 21 is leaching from leaves; 22 is death/decay of roots; 23 is incorporation of detritus into peat; 24 is adsorption from water to detritus; 25 is release from detritus to water; 26 is volatilization of ammonia; 27 is sediment–water exchange; and 28 is long-term burial of sediments. (*After Nixon and Lee, 1986*)

downstream aquatic systems often benefit either from the ability of wetlands to retain certain chemicals or from the export of organic materials.

4. Contrary to popular opinion that all wetlands are highly productive, wetlands can be either highly productive ecosystems rich in nutrients or systems of low productivity caused by a scarce supply of nutrients.

5. Nutrient cycling in wetlands differs from both aquatic and terrestrial ecosystem cycling in temporal and spatial dimensions. For example, more nutrients are tied up in sediments and peat in wetlands than in most terrestrial systems, and deepwater aquatic systems have autotrophic activity more dependent on nutrients in the water column than on nutrients in the sediments.

6. Anthropogenic changes have led to considerable changes in chemical cycling in many wetlands. Although wetlands are quite resilient to many chemical inputs, the capacity of wetlands to assimilate anthropogenic wastes from the atmosphere or hydrosphere is not limitless.

Wetlands as Sources, Sinks, or Transformers of Nutrients

There has been much discussion and research by wetland scientists about whether wetlands are nutrient sources, sinks, or transformers. A wetland is considered a *sink* if it has a net retention of an element or a specific form of that element (e.g., organic or inorganic), that is, if the inputs are greater than the outputs (Fig. 6-21a). If a wetland exports more of an element or material to a downstream or adjacent ecosystem than would occur without that wetland, it is considered a *source* (Fig. 6-21b). If a wetland transforms a chemical from, say, dissolved to particulate form but does not change the amount going into or out of the wetland, it is considered to be a *transformer* (Fig. 6-21c). Part of the interest in the source–sink–transformer question was stimulated by studies that hypothesized the importance of salt marshes as sources of particulate carbon for adjacent estuaries and other studies that suggested the importance of wetlands as sinks for certain chemicals, particularly nitrogen and phosphorus. The two concepts of one wetland being a source and a sink for various materials are not mutually exclusive; a wetland can be a sink for an inorganic form of a nutrient and a source for an organic form of the same nutrient. The desire of wetland scientists to determine conclusively whether wetlands are sources or sinks of nutrients has often been hampered by the imprecise use of the words "source" and "sink" and by the inadequacy of the techniques used to measure the nutrient fluxes in a wetland nutrient budget. All that can be said with certainty is that many wetlands act as sinks for particular inorganic nutrients and many wetlands are sources of organic material to downstream or adjacent ecosystems.

Seasonal Patterns of Uptake and Release

The fact that a wetland is a sink or a source of nutrients on a year-by-year basis suggests nothing about the seasonal differences in nutrient uptake and release. In temperate climates, seasonal patterns of nutrient retention and release or at least differences in retention rates persist, particularly where biological rather than chemical processes dominate. During the growing season, there is a high rate of uptake of nutrients by emergent and submerged vegetation from the water and sediments.

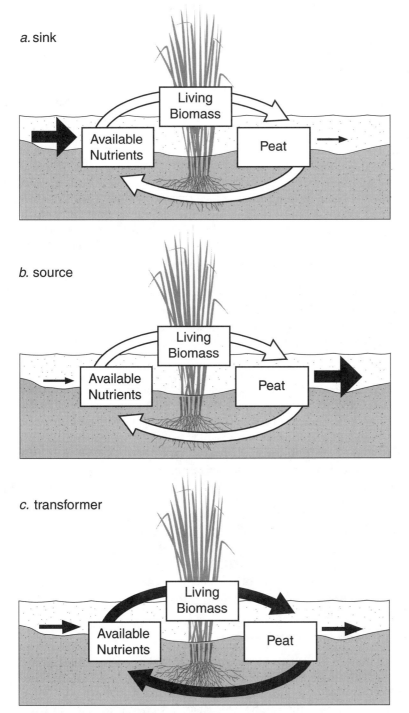

Figure 6-21 Wetlands as   **a.** inorganic nutrient sink;   **b.** source of total nutrients;
c. transformer of inorganic nutrients to organic nutrients.

Increased microbiological immobilization of nutrients and uptake by algae and epiphytes also lead to a retention of inorganic forms of nitrogen and phosphorus. If algae are so productive as to increase the pH, precipitation of nutrients, particularly phosphorus, can occur. By the time the vascular plants die, they have translocated a substantial portion of the nutrient material back to the roots and rhizomes. A substantial portion of the nutrients, however, is lost to the waters through litterfall and subsequent leaching. Cold weather brings on diminished microbial activity in the water column and sediments, slowing down processes. In the spring, excessive runoff coupled with still cooler temperatures leads to high throughflow and less nutrient retention. This generally leads to a diminished retention or even net export of nutrients in the fall through the early spring.

Several studies, particularly on freshwater marshes and salt marshes, have documented seasonal changes of nutrient export that occurred in cool climates. Lee et al. (1975) found that the marshes that they studied act as nutrient sinks during the summer and fall and as nutrient sources in the spring. This pattern had two potential benefits, according to the authors: (1) The problem of lake enrichment downstream of the marshes was decreased in the summer, the time of the most serious algal blooms in lakes; and (2) it might be economically and ecologically reasonable to treat waters for nutrient removal only during the periods of high flow, when the marsh is exporting nutrients, allowing the marsh to be the nutrient removal system when it is acting as a sink. The seasonality of nutrient retention by freshwater marshes was also observed in another Wisconsin riverine marsh by Klopatek (1978), who found that the marsh acted as both a source and a sink of nutrients and that the pattern depended on the hydrologic conditions, the anaerobiosis in the sediments, and the activity of microbes and emergent macrophytes. Simpson et al. (1978) described the movement of inorganic nitrogen and phosphorus in a freshwater tidal marsh in New Jersey and concluded: "It appears almost all habitats of freshwater tidal marshes may be sinks for inorganic N and PO_4–P during the vascular plant growing season and that certain habitats may continually function as sinks." Woodwell and Whitney (1977) found that there was a seasonal shift from uptake of phosphate in the cold months to export of phosphate in the warm months in a New York salt marsh. This pattern is opposite to the seasonal pattern that would be expected if plant uptake were the dominant sink in the growing season. Spieles and Mitsch (2000) describe the seasonal retention of nitrate nitrogen in constructed marshes in central Ohio (Fig. 6-22). Much of this seasonal pattern is attributed to the temperature effects on denitrification. All wetland basins retained 50 to 60 percent of nitrate nitrogen, by mass or concentration, in the summer, but retentions were not nearly as effective in the winter and spring. In fact, one of the three wetlands showed signs of nitrate–nitrogen export in the spring.

Coupling with Adjacent Ecosystems

In the beginning of this chapter, wetlands were described as being either "open" or "closed" to hydrologic transport. For those that are open to export, the chemicals are often transformed from inorganic to organic forms and transported to downstream ecosystems. This connection can be from riparian wetlands to adjacent streams, rivers, and downstream ecosystems; from tidal salt marshes to the estuary; or from in-stream riverine marshes to the river itself. There is considerable evidence that watersheds that drain wetland regions export more organic material but retain more nutrients than do watersheds that do not have wetlands. For example, in Figure 6-23, the slope of the

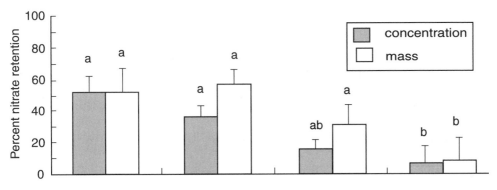

a. River water wetland

b. River water wetland 2

c. Waste water wetland

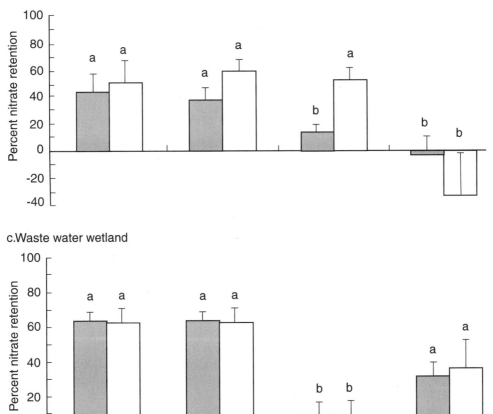

Figure 6-22 Seasonal pattern of nitrate–nitrogen retention in created wetland basins in central Ohio **a. and b.** river-fed wetlands; **c.** municipal wastewater-fed wetland. Similar letter above bars indicates no significant difference ($\alpha = 0.05$). (*After Spieles and Mitsch, 2000*)

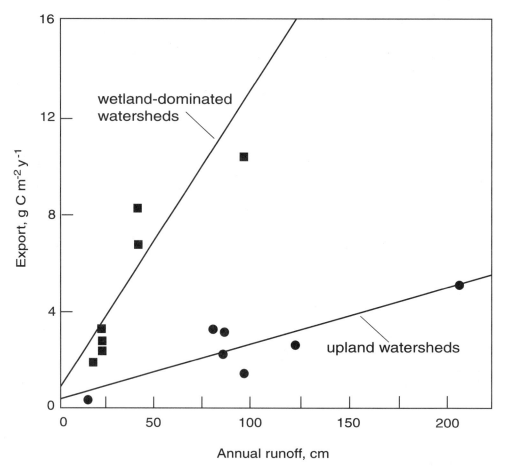

Figure 6-23   **Organic carbon export from wetland-dominated watersheds compared with nonwetland watersheds. (*After Mulholland and Kuenzler, 1979*)**

line for a swamp-draining watershed is much steeper than that for upland watersheds, indicating a much greater organic export for a given runoff. Nixon (1980) summarized total organic export data from several studies of salt marshes in the United States and suggested a general range of 100 to 200 g m^{-2} yr^{-1} carbon export from salt marshes to adjacent estuaries. This is an order of magnitude greater than the export of carbon from freshwater wetland watersheds as shown in Figure 6-23, indicating that coastal marshes, with their high productivity and frequent tidal exchange, may be more consistent exporters of organic matter than freshwater wetlands. The export indicated in Figure 6-23, however, is based on the entire watershed, some of which is not wetlands.

In an evaluation of the cumulative effects of wetlands on stream water quality in a multicounty region in Minnesota, Johnston et al. (1990) found that the area of wetland in the watershed related, through a principal components analysis, only to lower conductivity, chloride, and lead. All are indicators of urban runoff and urban areas generally have fewer wetlands. They also found that lower annual concentrations (and, hence, export) of inorganic suspended solids, fecal coliform, nitrates, and conductivity and flow-weighted ammonium nitrogen and total phosphorus were related to wetland proximity. The authors suggested that "These findings do not necessarily mean that wetlands farther upstream from a sampling station are less important to water quality than proximal wetlands; just that their effects on nutrients are not detectable very far downstream, or are offset by downstream inputs."

The effects of export on adjacent ecosystems have generally been difficult to quantify, although some attempts have been made to establish a cause-and-effect relationship. Figure 6-24 shows the general scatter of data that results when estuarine aquatic productivity is plotted versus the amount of adjacent salt marshes. The graph shows both highly productive estuaries with few adjacent marshes and estuaries of low productivity amid great expanses of marshes, but the general trend is for greater estuarine productivity with more coastal wetlands.

High- and Low-Nutrient Wetlands

There is a misconception in some ecological texts that all wetlands are high-nutrient, highly productive ecosystems. Although many types of wetlands such as tidal marshes and riparian forests have more than adequate supplies of nutrients and, consequently,

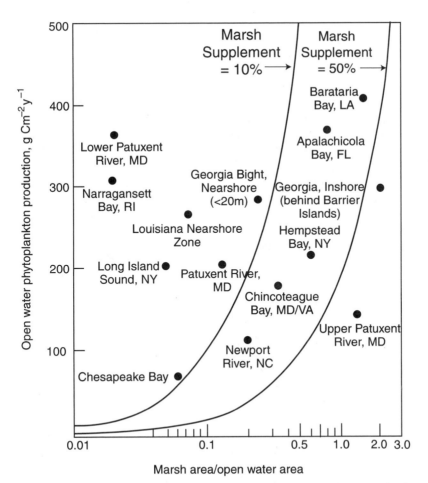

Figure 6-24 Relationship between estuarine primary productivity and relative area of adjacent salt marshes. (*After Nixon, 1980*)

are highly productive, there are many wetland types that have low supplies of nutrients. Table 6-11 lists some of the characteristics of high-nutrient (eutrophic) and low-nutrient (oligotrophic) wetlands. The terms *eutrophic* and *oligotrophic,* usually used to denote the trophic state of lakes and estuaries, are appropriate for wetlands despite the difference in structure. In fact, the use of the terms originated in the classification of peatlands (Hutchinson, 1973) and was later adapted for open bodies of water.

The intrasystem cycling of nutrients in wetlands depends on the availability of nutrients and the degree to which processes such as primary productivity and decomposition are controlled by the wetland environment. The availability of nitrogen and phosphorus is significantly altered by anoxic conditions as discussed earlier in this chapter, but these nutrients do not necessarily limit production or consumption. Ammonium nitrogen is high in most wetland soils, and soluble inorganic phosphorus is often in abundance. Thus, intrasystem cycling is often limited by the effects that hydrologic conditions have on pathways such as primary productivity and decomposition, as discussed in Chapter 5. It is therefore possible to have wetlands with extremely rapid yet open nutrient cycling (e.g., some high-nutrient freshwater marshes) and wetlands with extremely slow nutrient cycling (e.g., low-nutrient bogs).

Comparison of Wetlands with Terrestrial and Aquatic Systems

Wetlands are, in one sense, an "in-between world"—between upland terrestrial ecosystems and deepwater aquatic systems (Fig. 1-19a). Accordingly, they have characteristics of both systems. One of the major differences between wetlands and drier upland ecosystems is that, in the former, more nutrients are tied up in organic deposits and are lost from ecosystem cycling as peat deposits or organic exports. Because wetlands are more frequently open to nutrient fluxes than are upland ecosystems, they may not be as dependent on the recycling of nutrients; wetlands that are not open to these fluxes often have lower productivities and slower nutrient cycling than comparable upland ecosystems.

Wetlands are similar to deep aquatic systems in that most of the nutrients are often permanently tied up in sediments and peat. In most deep aquatic systems, the retention of nutrients in organic sediments is probably longer than in wetlands, although few such comparisons have been made. Wetlands, however, usually involve larger biotic storages of nutrients than do deep aquatic systems, which are primarily

Table 6-11   **Characteristics of high-nutrient (eutrophic) and low-nutrient (oligotrophic) wetlands**

Characteristic	Low-Nutrient Wetland	High-Nutrient Wetland
Inflows of nutrients	Mainly precipitation	Surface water and groundwater
Nutrient cycling	Tight closed cycles; adaptations such as carnivorous plants and nutrient translocations	Loose open cycle; few adaptations to shortages
Wetland as source or sink of nutrients	Neither	Either
Exporter of detritus	No	Usually
Net primary productivity	Low (100–500 g m^{-2} yr^{-1})	High (1,000–4,000 g m^{-2} yr^{-1})
Examples	Bogs; cypress domes	Floodplain wetlands; many coastal marshes

plankton dominated. Aquatic system nutrient cycling in the autotrophic zone of lakes and coastal waters is more rapid than it is in the autotrophic zone of most wetlands. Another obvious difference between wetlands and lakes or coastal waters is that most wetland plants obtain their nutrients from the sediments, whereas phytoplankton depend on nutrients dissolved in the water column. Wetland plants have often been described as "nutrient pumps" that bring nutrients from the anaerobic sediments to the above-ground strata. Phytoplankton in lakes and estuaries can be viewed as being "nutrient dumps" that take nutrients out of the aerobic zone and, through settling and death, deposit the nutrients in the anaerobic sediments. Thus, the plants in these two environments can be viewed as having decidedly different functions in nutrient cycling.

Anthropogenic Effects

Human influences have caused significant changes in the chemical cycling in many wetlands. These changes have taken place as a result of land clearing and subsequent erosion, hydrological modifications such as stream channelization and dams, and pollution. Increased erosion in the uplands leads to increased deposition of sediments in the lowland wetlands such as forested swamps and coastal salt marshes. This increased accumulation of sediments can cause increased biochemical oxygen demand (BOD) and can alter the hydrologic regime of the wetlands in a relatively short time. Stream channelization and dams can lead to a change in the flooding frequency of many wetlands and, thus, alter the inputs of nutrients. Dams generally serve as nutrient traps, retaining materials that would otherwise nourish downstream wetlands. In some cases, stream channelization has led to stream downcutting that ultimately drains wetlands.

Sources of pollution also have had localized effects on nutrient cycling in wetlands. Pollutants such as BOD, toxic materials, oils, trace organics, and metals have frequently been added purposefully or accidentally to wetlands from municipal industrial wastes and urban and rural runoff. The effects of toxic materials on wetland nutrient cycles are poorly understood, although wetlands were often used for the disposal of such materials.

Although wetlands cover less than 7 percent of the Earth's land area, they have a disproportionate influence on global cycles of such elements as carbon (because of the enormous stores of peat in boreal regions), nitrogen, and sulfur. Hence, there has been considerable interest in the role of wetlands in such issues as global climate change (see, e.g., Immirzi et al., 1992; Mitsch and Wu, 1995). This role is discussed in more detail in Chapter 16. The effect that global change has on wetlands is discussed in Chapter 17.

Recommended Readings

Natural Resources Conservation Service (NRCS). 1998. Field Indicators of Hydric Soils in the United States, Version 4.0. G. W. Gurt, P. M. Whited, and R. F. Pringle, eds. USDA, NRCS, Ft. Worth, TX. 30 pp.

Reddy, K. R., and E. M. D'Angelo. 1994. Soil processes regulating water quality in wetlands. In W. J. Mitsch, ed. Global Wetlands: Old World and New. Elsevier, Amsterdam, pp. 309–324.

Reddy, K. R., R. H. Kadlec, E. Flaig, and P. M. Gale. 1999. Phosphorus retention in streams and wetlands: a review. Critical Reviews in Environmental Science and Technology 29:83–146.

Schlesinger, W. H. 1991. Biogeochemistry: An Analysis of Global Change. Academic Press, San Diego. 443 pp.

Stumm, W., and J. J. Morgan. 1996. Aquatic Chemistry: Chemical Equilibria and Rates in Natural Waters, 3rd ed. John Wiley & Sons, New York. 1022 pp.

Vepraskas, M. J. 1995. Redoximorphic Features for Identifying Aquic Conditions. Technical Bulletin 301, North Carolina Agricultural Research Service, North Carolina State University, Raleigh, NC. 33 pp.

Chapter 7

Biological Adaptations to the Wetland Environment

The wetland environment is, in many ways, physiologically harsh. Major stresses are anoxia and the wide salinity and water fluctuations characteristic of an environment that is neither terrestrial nor aquatic. Adaptations to this environment have an energy cost, either because an organism's cells operate less efficiently (conformer) or because the organism expends energy to protect its cells from the external stress (regulator). At the cell level, all organisms have similar adaptations, although unicellular organisms appear to show more novelty. Adaptations of these organisms include the ability to respire anaerobically, to detoxify end products of anaerobic metabolism, to use reduced organic compounds in the sediment as energy sources, and to use mineral elements in the sediment as alternative electron acceptors when oxygen is unavailable. Multicellular plants and animals have a wider range of responses available to them because of the flexibility afforded by the development of organ systems and division of labor within the body, mobility, and complex life-history strategies. To counter anoxia, one important structural adaptation in vascular plants is the development of pore space in the cortical tissues, which allows oxygen to diffuse from the aerial parts of the plant to the roots to supply root respiratory demands. Animals have developed both structural and physiological adaptations to reduced oxygen availability, such as specialized tissues or organ systems, mechanisms to increase the oxygen gradient into the body, better means of circulation, and more efficient respiratory pigment systems. Salt stresses are met in plants and animals with specialized tissues or organs to regulate the internal salt concentration or to protect the rest of the body from the effects of salt (osmoregulators), or with increased metabolic and physiological tolerance to salt at high concentrations (osmoconformers). In motile organisms, behavioral adaptations, such as avoidance, are commonly found. Even in sessile seed-producing plants behavioral adaptations have evolved, which typically concern the timing of seed production, mechanisms of seed distribution, and advantageous germination patterns.

Wetland environments are characterized by stresses that most organisms are ill equipped to handle. Aquatic organisms are not adapted to deal with the periodic drying that occurs in many wetlands. Terrestrial organisms are stressed by long periods of flooding. Because of the shallow water, temperature extremes on the wetland surface are greater than would ordinarily be expected in aquatic environments. However, the most severe stress is probably the absence of oxygen in flooded wetland soils, which prevents organisms from respiring through normal aerobic metabolic pathways. In the absence of oxygen, the supply of nutrients available to plants is also modified, and concentrations of certain elements and organic compounds can reach toxic levels. In coastal wetlands, salt is an additional stress to which organisms must respond. It is not surprising that those plants and animals regularly found in wetlands have evolved functional mechanisms to deal with these stresses. Adaptations can be broadly classified as those that enable the organism to tolerate stress and those that enable it to regulate stress. *Tolerators* (also called *resisters*) have functional modifications that enable them to survive and often to function efficiently in the presence of stress. *Regulators* (alternatively called *avoiders*) actively avoid stress or modify it to minimize its effects.

The specific mechanisms for either tolerating or regulating are many and varied (Table 7-1). In general, unicellular organisms show biochemical adaptations that are also characteristic of the range of cell-level adaptations found in more complex multicellular plants and animals. Vascular plants show both structural and physiological adaptations. Animals have the widest range of adaptations, not only through biochemical and structural means but also by behavioral responses, using to advantage their mobility and their life-history patterns (Table 7-1).

Cell-Level Adaptations

For unicellular organisms that have little mobility, the range of adaptations is limited. Most adaptations of this group are metabolic. Because the metabolism of all living cells is similar, these adaptations are characteristic of cell-level adaptations in general, although some of the bacterial responses to anoxia are beyond anything found in multicellular organisms.

Anoxia

When an organic wetland soil is flooded, the oxygen available in the soil and in the water is rapidly depleted through metabolism by organisms that normally use oxygen

Table 7-1   Types of adaptations of the wetland environment found in different kinds of organisms and communities of organisms

Type of Organisms	Level of Organization	Type of Adaptation
Bacteria/protists	Organelle	Biochemical
	Cell	Physiological/structural
Plant	Tissue	
	Organ	
Animal	Individual	Behavioral
Plant/Animal group	Community	Commensal/mutualistic

as the terminal electron acceptor for oxidation of organic molecules. The rate of diffusion of molecular oxygen through water is orders of magnitude slower than through air, and cannot supply the metabolic demand of submerged soil organisms under most circumstances. When the demand exceeds the supply, dissolved oxygen is depleted, the redox potential in the soil drops rapidly, and other ions (nitrate, manganese, iron, sulfate, and carbon dioxide) are progressively reduced (see Chapter 6 and Table 6-2).

Although some abiotic chemical reduction occurs in the soil, virtually all of these reductions are coupled to microbial respiration. When oxygen concentrations first become limiting, most cells, bacterial or otherwise, use internal organic compounds as electron acceptors. The glycolytic or fermentation pathway of sugar metabolism results in the anaerobic production of pyruvate, which is subsequently reduced to ethyl alcohol, lactic acid, or other reduced organic compounds, depending on the organism. A number of bacteria also have the ability to couple their oxidative-respiratory reactions to the reduction of inorganic ions (other than molecular oxygen) in the surrounding medium, using them as electron acceptors. Many bacterial species are facultative anaerobes, capable of switching from aerobic to anaerobic respiration. Others have become so specialized, however, that they can grow only under anaerobic conditions and rely on specific electron acceptors other than oxygen in order to respire. *Desulfovibrio* is one such genus. It uses sulfate as its terminal electron acceptor, forming sulfides that give the marsh its characteristic rotten-egg odor. Other microbially mediated chemical reactions in anoxic sediments were discussed in Chapter 6.

Most bacteria require organic energy sources. In contrast, nonphotosynthetic autotrophic bacteria are adapted to use reduced inorganic compounds in wetland muds as an energy source for growth. The genus *Thiobacillus,* for example, captures the energy in the sulfide bonds formed by *Desulfovibrio* and, in the process, converts sulfide into elemental sulfur. Members of the genus *Nitrosomonas* oxidize ammonia to nitrite. *Siderocapsa* can capture the energy released in the oxidation of the ferrous ion to the ferric form. In this way, some bacteria not only survive in the anoxic environment of wetland soils but sometimes require it and obtain their metabolic energy from it.

Salt

In a freshwater aquatic or soil environment, the osmotic concentration of the cytoplasm in living cells is higher than that of the surrounding medium. This enables the cells to develop turgor, that is, to absorb water until the turgor pressure of the cytoplasm is balanced by the resistance of their cell membranes and walls. In coastal wetlands, organisms must cope with high and variable external salt concentrations. The dangers of salts are twofold—osmotic and directly toxic. The immediate effect of an increase in salt concentration in a cell's environment is osmotic. If the osmotic potential surrounding the cell is higher than that of the cell cytoplasm, water flows out of the cell and the cytoplasm dehydrates. This is a rapid reaction that can occur in a matter of minutes and may be lethal to the cell. Even the "tightest" membranes leak salts passively so that, in the absence of any active regulation by the cell, inorganic salts gradually diffuse into the cell. Although the absorption of inorganic ions such as Na^+ may relieve the osmotic gradient across the cell membrane, these ions at high concentrations in the cytoplasm are also toxic to most organisms, posing a second threat to survival.

Unicellular organisms have adapted in a number of ways to cope with these twin problems of osmotic shock and toxicity. There is no evidence that cells are able to retain water against an osmotic gradient. Instead, in order to maintain their water potential, the internal osmotic concentration of salt-adapted cells is usually slightly higher than the external concentration. Indeed, the high specific gravities of halophiles—literally "salt-loving" organisms—can be accounted for only by the presence of inorganic salts at high concentrations. Analyses of cell contents show, however, that the balance of specific ions is usually quite different from that of the external solution. For example, potassium is usually accumulated and sodium is usually diluted relative to external concentrations (Table 7-2). Active transport mechanisms that accumulate or excrete ions across cell membranes are universal features of all living cells, and although they depend on a cellular supply of biological energy, there is no evidence that large energy expenditures are needed to maintain the gradients shown in Table 7-2. They can be maintained in the cold, in cells apparently carrying out little metabolism. This suggests that the accumulated ions are loosely bound or complexed within the cell cytoplasm or that the cytoplasmic water has a more ordered structure than external water. As a result, cytoplasmic ions such as potassium and sodium are less free than in external solutions but still free enough to be osmotically and physiologically active (Kushner, 1978).

Although inorganic ions seem to make up the bulk of the osmotically active cell solutes in most halophilic bacteria, in others the internal salt concentration can be substantially lower than the external concentration. In these organisms, the rest of the osmotic activity is supplied by organic compounds. For example, the halophilic green alga *Dunaliella virigus* contains large amounts of glycerol, the concentration varying with the external salt concentration; and certain salt-tolerant yeasts regulate

Table 7-2 Internal ionic concentrations of bacteria growing with NaCl at different concentrations

| | Ion Concentration (M) | | | |
| | External Medium | | Cell Cytoplasm | |
Bacterium	Na$^+$	K^{+a}	Na$^+$	K$^+$
Vibrio costicolaa	1.0	0.004	0.68	0.22
	0.6	0.008	0.50	0.52
	1.0	0.008	0.58	0.66
	1.6	0.008	1.09	0.59
	2.0	0.008	0.90	0.57
Paracoccus halodinitrificans	1.0	0.004	0.31	0.47
Pseudomonas 101	1.0	0.0055	0.90	0.71
	2.0	0.0055	1.15	0.89
	3.0	0.005	1.04	0.67
Marine pseudomonad B-16	0.3	—	0.12	0.37
Unidentified salt-tolerant rod	0.6	0.04	0.05	0.34
	4.4	0.04	0.62	0.58
Halobacterium cutirubrum[b]	3.33	0.05	0.80	5.32

[a] The difference between K$^+$ concentration in different studies may be related to whether the cultures were in a stationary phase or were growing exponentially.
[b] Several workers have reported similar or higher results for K$^+$ in *H. cutirubrum* or *H. halobium*.
Source: Kushner (1978).

internal osmotic concentration with polyols such as glycerol and arabitol. The enzymes of these organisms seem to be salt sensitive, and it has been suggested that the organic compounds act like "compatible solutes" that raise the osmotic pressure without interfering with enzymatic activity (Kushner, 1978).

The steric configurations of enzymes of salt-sensitive organisms are typically modified by salt at high concentrations. In addition, where an enzyme is activated by a specific ion, NaCl can interfere with its activity. The enzymes of true halophiles, in contrast, are able to function normally in the presence of inorganic ions at high concentrations or may even require them.

Vascular Plant Adaptations

Multicellular organization adds another layer of complexity to individuals compared to unicellular organization. This has enabled plants and animals to develop a wider range of adaptations than bacteria to anoxia and to salt. At the same time, some adaptations found in unicellular organisms, such as the ability to use reduced inorganic compounds in the sediment as a source of energy, are not found in multicellular organisms. These adaptations typically develop in specialized tissue and organ systems.

Vascular emergent and floating-leaved wetland plants are sessile; only their roots are in an anoxic or salty environment. Typically, if the roots of a flood-sensitive upland plant are inundated, the oxygen supply rapidly decreases. This shuts down the aerobic metabolism of the roots, impairs the energy status of the cells, and reduces nearly all metabolically mediated activities such as cell extension and division and nutrient absorption. Even when cell metabolism shifts to anaerobic glycolysis, adenosine triphosphate (ATP) production is reduced. Toxic metabolic end products of fermentation may accumulate, causing cytoplasmic acidosis and eventually death (Roberts, 1988). Anoxia is soon followed by pathological changes in the mitochondrial structure, including swelling, the reduction of cristae number, and the development of a transparent matrix. The complete destruction of mitochondria and other organelles occurs within 24 hours (Vartapetian, 1988). Anoxia changes the chemical environment of the root, increasing the availability of reduced minerals such as iron, manganese, and sulfur. These may accumulate to toxic levels in the root (Ernst, 1990). In addition, if the sensitive plants are in a marine environment, they suffer osmotic shock and salt toxicity.

Anoxia

In contrast to flood-sensitive plants, flood-tolerant species (hydrophytes) possess a range of adaptations that enable them either to tolerate stresses or to avoid them. There are several adaptations by hydrophytes that allow them to tolerate anoxia in wetland soils. These adaptations can be grouped into two main categories:

1. Structural (or morphological) adaptations
   a. Aerenchyma
   b. Special organs or responses
      i. Adventitious roots
      ii. Stem elongation

          iii. Lenticels
          iv. Pneumatophores
       c. Pressurized gas flow
   2. Physiological adaptations
       a. Anaerobic respiration
       b. Malate production

Details of these adaptations are discussed next.

Aerenchyma

Virtually all hydrophytes have elaborate structural (or morphological) mechanisms to avoid root anoxia. These responses to flooding are mechanisms that increase the oxygen supply to the plant either by growth into aerobic environments or by enabling oxygen to penetrate more freely into the anoxic zone. The primary plant strategy in response to flooding is the development of air spaces (*aerenchyma*) in roots and stems, which allow the diffusion of oxygen from the aerial portions of the plant into the roots (Fig. 7-1). Aerenchyma development is not extensive in the absence of flooding and is characteristic of flood-tolerant plant species, not flood-sensitive ones (Burdick and Mendelssohn, 1990; Pezeshki et al., 1991; Grosse et al., 1998). In plants with well-developed aerenchyma, the root cells no longer depend on the diffusion of oxygen from the surrounding soil, the main source of root oxygen to terrestrial plants. Unlike the plant porosity of normal mesophytes, which is usually a low 2 to 7 percent of volume, up to 60 percent of the volume of the roots of wetland species consists of pore space. Air spaces are formed either by cell separation during maturation of the root cortex or by cell breakdown. They result in a honeycomb structure. Air spaces are not necessarily continuous throughout the stem and roots. The thin lateral cellular partitions within the aerenchyma, however, are not likely to impede internal gas diffusion significantly (Armstrong, 1975). The same kind of cell lysis and air space

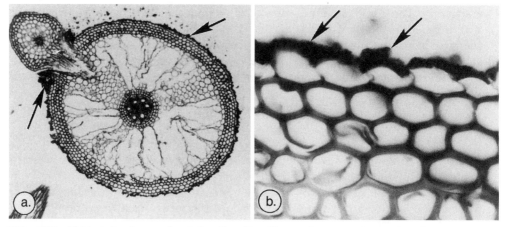

Figure 7-1 Light photomicrographs of *Spartina alterniflora* roots: **a.** cross section of a stream-side root; arrows indicate the presence of red ferric deposits on the root epidermis. ×192; **b.** streamside root cross section showing the presence of similar materials on the external walls of the epidermal cells. ×1,143. Note the extensive pore space (aerenchyma) in the roots. (*From Mendelssohn and Postek, 1982; © 1982 by the Botanical Society of America, reprinted by permission*)

development has been described in submerged stem tissue (Jackson, 1990). Roots of flood-tolerant species such as rice form aerenchyma even in aerated apical cells (Webb and Jackson, 1986), and it is not clear whether this "preadaptation" is ethylene-mediated (Jackson, 1985).

Effectiveness of Aerenchyma: Because flooding interrupts the flow of oxygen from soil to root, a key question is the efficacy of adaptations to correct that interruption. The sufficiency of the oxygen supply to the roots depends on the root permeability (i.e., how leaky the root is to oxygen, which can move out into the surrounding soil), the root respiration rate, the length of the diffusion pathway from the upper parts of the plant, and the root porosity (the pore space volume). Models of gas diffusion show that, under most circumstances, root porosity is the overriding factor governing internal root oxygen concentration (Fig. 7-2). The effectiveness of aerenchyma in supplying oxygen to the roots has been demonstrated in a number of plant species. For example, the root respiration of flood-tolerant *Senecio aquaticus* was only 50 percent inhibited by root anoxia, whereas that of *S. jacobaea,* a flood-sensitive species, was almost completely inhibited. Greater root porosity in the tolerant species was the primary factor that contributed to the difference (Lambers et al., 1978). The most extensively studied flood-tolerant plant is rice. Rice plants grown under continuous flooding develop greater root porosity than unflooded plants, and this maintains the oxygen concentration in the root tissues. When deprived of oxygen, rice root mitochondria degraded in the same way as did flood-sensitive pumpkin plants, showing that the basis of resistance in flooded plants was by the avoidance of root anoxia, not by physiological changes in cell metabolism (Levitt, 1980).

Hormones: Hormonal changes, especially the concentration of ethylene in hypoxic tissues, initiate structural adaptations. Ethylene production is stimulated by flooded conditions (Jackson, 1993; Blom and Voesenek, 1996; Grosse et al., 1998). Ethylene rapidly concentrates in flooded tissues because its diffusion rate in water is approximately 10,000 times slower than that in air. Ethylene stimulates cellulase activity in the cortical cells of a number of plant species, with the subsequent collapse and disintegration of cell walls (Kawase, 1981). Not all of the effects of ethylene occur in the hypoxic tissues; there is strong evidence that ethylene precursors produced in hypoxic tissues diffuse to aerobic tissues, are converted into ethylene, and modify plant response in those aerobic zones (Fig. 7-3; Bradford and Yang, 1980; Jackson, 1988).

Special Organs

In addition to aerenchyma development, ethylene has been reported to stimulate the formation of *adventitious roots,* which develop in both flood-tolerant trees (e.g., *Salix* and *Alnus*) and herbaceous species and flood-intolerant (e.g., tomato) plants just above the anaerobic zone when these plants are flooded (Wample and Reid, 1979; Jackson, 1985). These roots are able to function normally in an aerobic environment.

Another response stimulated by submergence is rapid stem elongation in such aquatic and semiaquatic plants as the floating heart (*Nymphoides peltata*), rice (*Oryza sativa*), and bald cypress (*Taxodium distichum*) (Demaree, 1932; Malone and Ridge, 1983; Raskin and Kende, 1983; Blom et al., 1990; Blom and Voesenek, 1996).

The red mangrove (*Rhizophora* spp.) grows on arched prop roots in tropical and subtropical tidal swamps around the world (see Chapter 11). These prop roots have

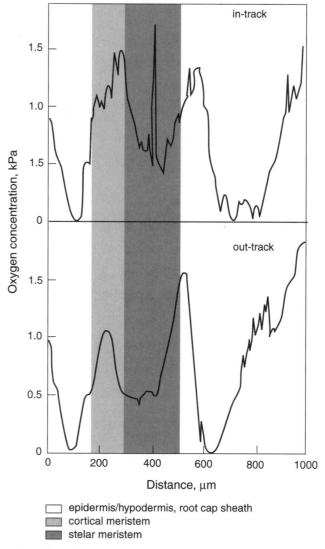

Figure 7-2 Microelectrode polarographic "radial" oxygen profile across primary root of maize (length approximately 90 mm) at an apical position within the zone of the root cap sheathing of the meristem. "In-track" refers to the movement of the electrode from the outside of the root to the central part; "out-track" is the reverse movement. kPa = Boltzman constant × pressure in newton/m² (pascals). (*After Armstrong et al., 1994; Blom and Voesenek, 1996*)

numerous small pores, termed *lenticels,* above the tide level, which terminate in long, spongy, air-filled, submerged roots. The oxygen concentration in these roots, embedded in anoxic mud, may remain as high as 15 to 18 percent continuously, but if the lenticels are blocked, this concentration can fall to 2 percent or less in two days (Scholander et al., 1955). Lenticels in the stems of flood-tolerant species such as *Alnus glutinosa* and *Nyssa sylvatica* also serve as conduits to the aerenchymatous tissue in the stem.

Similarly, the black mangrove (*Avicennia* spp.) produces thousands of *pneumato-*

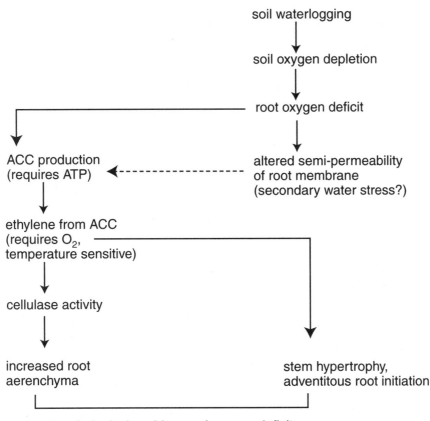

Figure 7-3 Schematic of the development of morphological acclimation characteristics in response to flooding stress. (*After McLeod et al., 1988*)

phores (air roots) about 20 to 30 cm high by 1 cm in diameter, spongy, and studded with lenticels. They protrude out of the mud from the main roots and are exposed during low tides. The oxygen concentration of the submerged main roots has a tidal pulse, rising during low tide and falling during submergence, reflecting the cycle of emergence of the air roots (Scholander et al., 1955). The "knees" of bald cypress (*Taxodium distichum*) are thought to be pneumatophores that improve gas exchange to the root system (see Chapter 14).

Pressurized Gas Flow

Dacey (1980, 1981) first described a particularly interesting adaptation that increases the oxygen supply to the floating-leaved water lily (*Nuphar luteum*). Since then, a similar adaptation of pressurized gas flow from the surface to the rhizosphere has been demonstrated for other floating-leaved species (Grosse and Mevi-Schütz, 1987; Mevi-Schütz and Grosse, 1988; Grosse et al., 1991). Brix et al. (1992) tested 14 emergent plants in southwestern Australia and found significant gas flow (0.2–> 10 cm^3 min^{-1} culm^{-1}) in 8 of the species, including *Baumea articulata, Cyperus involucratus, Eleocharis sphacelata, Schoenoplectus validus, Typha domingensis, T. orientalis,*

Phragmites australis, and *Juncus ingens* (Table 7-3). These results for such a wide variety of plants suggest that internal pressurization and pressurized gas flow may be common to many hydrophytes. They also investigated the diel variation in convective gas flow for *T. domingensis* and found a dramatic diel pattern related to air and subsequent leaf temperatures (Fig. 7-4). Gas flows of 0.1 to 0.2 cm^3 min^{-1} $culm^{-1}$ occurred at night but increased to a rate as high as 3 cm^3 min^{-1} $culm^{-1}$ during the afternoon. The results were interpreted to suggest that humidity-induced pressuriza-tion was the dominant driving force for the gas flow. Furthermore, the pressure produced by this range of plants matched very nicely the approximate depths at which these plants can potentially occur in wetlands (Table 7-3). *Phragmites, Eleocharis,* and the two *Typha* species can grow in water depths up to 2 m; *Schoenoplectus, Juncus,* and *Baumea* are found in water less than 1 m deep; the two *Cyperus* species, *Bolboschoe-nus,* and *Canna* grow in very shallow water to wet soils. Air moves into the internal gas spaces (the lacunar system) of aerial leaves and is forced down through the aerenchyma of the stem into the roots by a slight pressure ($\sim$200–1,300 Pa) generated by a gradient in temperature and water vapor pressure (Brix et al., 1992). Older leaves often lose their capacity to support pressure gradients, and so the return flow of gas

Table 7-3 **Pressurized gas flow in culms or leaves of 13 wetland plants and 1 upland plant in Australia**[a]

Species	n	ΔP_s (Pa)	Flow Rate (cm^3 min^{-1} $culm^{-1}$)
POTENTIALLY DEEPWATER PLANTS			
Phragmites australis	12	573 ± 54	5.3 ± 0.4[b]
Typha orientalis	8	1,070 ± 120	4.4 ± 0.3[b]
Typha domingensis	6	780 ± 140	3.4 ± 0.4[b]
MARGINAL DEPTH (< 1 m) PLANTS			
Juncus ingens	11	222 ± 24	1.2 ± 0.1[c]
Eleocharis sphacelata	10	1,080 ± 86	0.85 ± 0.02
Schoenoplectus validus	9	1,310 ± 124	0.29 ± 0.05
Baumea articulata	16	494 ± 58	0.23 ± 0.06
VERY SHALLOW WATER OR MOIST-SOIL PLANTS			
Cyperus involucratus	11	903 ± 234	0.33 ± 0.09[c]
Canna sp.	5	27 ± 5	0.06 ± 0.01
Myriophyllum papillosum[d]	6	68 ± 12	0.04 ± 0.01
Cyperus eragrostis	8	111 ± 34	0.02 ± 0.01
Ludwigia pelloides[d]	5	57 ± 1	<0.01
Bolboschoenus medianus	15	2 ± 31	<0.01
NOT A TRUE WETLAND PLANT			
Arundo donax	6	1 ± 10	<0.01

[a] Water depths refer to the potential depths that these plants can grow based on other studies and plant size. ΔP_s refers to the static pressure differential in the plant stem. Plants are listed in order of decreasing gas flow rates. Numbers indicate averages ± standard deviations.
[b] Small specimens or leaves had to be removed to get flow rates within measuring range.
[c] Gas flow measured through detached culms.
[d] Creeping, floating plants that grow in shallow water.
Source: Brix et al. (1992).

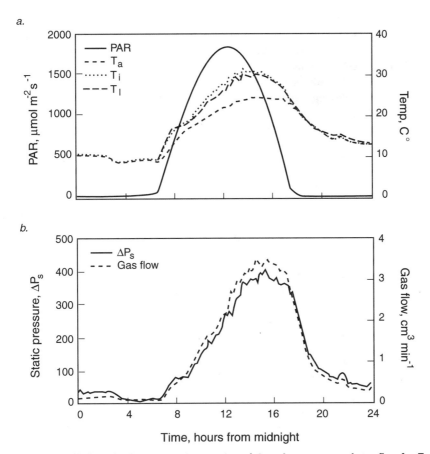

Figure 7-4 Diel variation of solar energy, temperature, internal pressure, and gas flow for *Typha domingensis:* **a.** photosynthetically available solar radiation (PAR), air temperature (T_a), internal leaf temperature (T_i), and leaf surface temperature (T_l); **b.** internal static pressure differential (ΔP_s). (*After Brix et al., 1992*)

from the roots is through the older leaves, which are rich in carbon dioxide and methane from root respiration. Brix (1989) did find that the gas exchange to the rhizosphere through dead culms of *P. australis* was sufficient to maintain aerobic respiration of the plant roots. The dead culms also provided an escape channel for excess CO_2 and CH_4 in the roots.

Grosse and others (Grosse and Schröder, 1984; Schröder, 1989; Grosse et al., 1992) described a similar process in swamp trees, specifically in common alder (*Alnus glutinosa*), the dominant tree species of European floodplain forests and riverine temperate forests. Seedlings and dormant (leafless) trees of flood-tolerant species show enhanced gas transport from aerial shoots to the roots when the shoots are heated by the sun or incandescent light, compared to plants in the dark. Grosse et al. (1998) called this phenomenon "pressurized gas flow" or "thermo-osmotic" gas flow. This phenomenon occurs when a temperature gradient is established between the exterior ambient air and the interior gas spaces in a plant's cortical tissue. A second requirement is a permeable partition between the exterior and interior with pore diameters "similar to or smaller than the 'mean free path length' of the gas molecules in the system (e.g., 70 nm at room temperature and standard barometric pressure" (Grosse et al., 1998). In alder, meristematic tissue in the lenticels forms such a partition. When the surface of the stem is mildly heated by sunlight, the mean free path length of gas molecules in the intercellular spaces of the plant increases, preventing the molecules from moving out through the osmotic barrier in the lenticels. The cooler exterior molecules, however, can still diffuse into the plant. This sets up an internal pressure gradient that forces gas down through the plant stem to the roots. This "thermal pump" is not as effective in moving oxygen to the roots in alder as is extensive aerenchyma tissue. For example, alder seedlings grown two months in flooded soil transported oxygen at eight times the rate of seedlings grown in aerated soil, the difference being due to aerenchyma and lenticel development under flooded conditions. In comparison, the thermo-osmotic effect (in the absence of flooding) led to a fourfold increase in the rate of gas transport. The thermal pump is also not as active in foliated trees as in dormant ones. Therefore, for trees, the adaptation appears to be most effective in enhancing root aeration during seedling establishment in saturated soils before aerenchyma development is accomplished, and in deciduous trees during the dormant season.

Secondary Effects of Root Aeration

Rhizosphere Oxygenation

When anoxia is moderate, the magnitude of oxygen diffusion through many wetland plants into the roots is apparently large enough not only to supply the roots but also to diffuse out, oxidize the adjacent anoxic soil, and produce an oxidized rhizosphere (Teal and Kanwisher, 1966; Howes et al., 1981; Laanbroek, 1990). Through scanning electron microscopy coupled with X-ray microanalysis, Mendelssohn and Postek (1982) showed that the brown deposits found around the roots of *Spartina alterniflora* are composed of iron and manganese deposits formed when root oxygen comes in contact with reduced soil ferrous ions (Fig. 7-1). It has been suggested (Armstrong, 1975; Engelaar et al., 1993) that oxygen diffusion from the roots is an important mechanism that moderates the toxic effects of soluble reduced ions such as manganese in anoxic soil and restores ion uptake and plant growth. These ions tend to be reoxidized and precipitated in the rhizosphere, which effectively detoxifies them. In a similar vein, McKee et al. (1988) determined that soil redox potentials were higher and that pore water sulfide concentrations were three to five times lower in the presence of the aerial prop roots of the red mangrove or the pneumatophores of the black mangrove than in nearby bare mud soils. This suggests the diffusion of oxygen from the mangrove roots into the soil. An interesting possibility is that the root systems of these flood-tolerant plants may modify sediment anoxia enough to allow the survival of nearby nontolerant plants (Ernst, 1990).

The presence of *oxidized rhizospheres* (now called *oxidized pore linings* by soil scientists; see Chapter 6), which form as a result of root oxidation, is an important way in which wetlands can be identified. Long after the plant roots die, residual veins of red and orange, resulting from oxidized iron (Fe^{3+}) deposits, remain in many mineral soils, a tell-tale sign that hydrophytes had been living in the soil. They are used in wetland delineation practices as one indicator that hydric soils and, thus, wetlands are present (Tiner, 1998).

Water Uptake

Plants intolerant to anaerobic environments typically show decreased water uptake despite the abundance of water, probably as a response to an overall reduction of root metabolism. Decreased water uptake results in symptoms similar to those seen under drought conditions: closing of stomata, decreased carbon dioxide uptake, decreased transpiration, and wilting (Mendelssohn and Burdick, 1988). Stomatal closure is mediated by the hormone abscisic acid (Dorffling et al., 1980), the concentrations of which increase in leaf tissues as a result of root flooding. The increased concentration probably results from the reduced export of leaf abscisic acid to the flooded roots (Jackson, 1990). The adaptive advantage of these responses is probably the same as for drought-stricken plants—to minimize water loss and accompanying damage to the cytoplasm. The accompanying depression of the photosynthetic machinery is generally seen as an unavoidable corollary.

Nutrient Absorption

One of the earliest processes affected by anoxia is the absorption by plants of nutrients from the substrate. The availability of many nutrients in the soil is modified by an anoxic environment. In general, flood-intolerant plant species lose the ability to control nutrient absorption because of the tissue energy deficit brought on by anoxia. Although some studies (e.g., Mendelssohn and Burdick, 1988) have concluded that nutrient uptake in most flood-tolerant species appears unchanged, perhaps because the plant can maintain near-normal metabolism, other studies (Grosse and Meyer, 1992) have illustrated that sufficient oxygen is necessary for nutrient uptake. This is illustrated in Figure 7-5, where hydroponically grown alder (*Alnus glutinosa*) seedlings were subjected to light-induced pressurized gas flow. Under anoxic conditions, nitrate, potassium, and phosphate uptakes were reduced to about 20, 30, and 70 percent, respectively, of aerated conditions (Grosse et al., 1998).

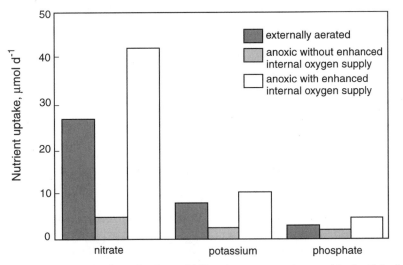

Figure 7-5 **Effect of root anoxia and enhanced internal oxygen supply on nutrient uptake by one-year-old foliated *Alnus glutinosa* (common alder). Trees were grown hydroponically. (*After Grosse and Meyer, 1992; Grosse et al., 1998*)**

The major nutrients that have been studied are those most affected by soil anaerobiosis: nitrogen, phosphorus, iron, manganese, and sulfur.

Nitrogen: In reduced soils, nitrates are replaced by ammonium, although most plants preferentially absorb the oxidized form (nitrate). Despite this change in the sediment supply of available nitrogen, most wetland plants that have been studied are able to maintain normal rates of nitrogen uptake. This ability is related to three factors: first, the possibility that ammonium is oxidized to nitrate in the rhizosphere through radial oxygen loss from the roots; second, the possibility that some wetland species are able to absorb the ammonium directly; and third, the ability of the plant to maintain metabolic activity sufficient to absorb nutrients, which is known to be an energy-requiring process (Morris and Dacey, 1984; Schat, 1984).

Phosphorus: The availability of phosphorus generally increases in waterlogged soils. It is, however, precipitated by iron (which also increases in availability). Generally, studies have shown reduced uptake by flood-intolerant species, probably because of the energy requirement, but no effect or even enhanced uptake by flood-tolerant species.

Iron and Manganese: Because extremely small concentrations of iron and manganese are required by plants, they can reach toxic levels in many environments. Both elements are reduced and become much more available in flooded soils, where, because of their high concentrations, they escape every metabolic control and become concentrated in plant tissues (Ernst, 1990). Wetland plants tolerate these elements by means of several adaptations. First, the oxidized rhizosphere can precipitate and reduce the concentration that reaches the roots. Second, much of the minerals taken up into the tissues can be sequestered in cell vacuoles, in the shoot vascular tissue, or in senescing tissues where they do not influence the metabolism of healthy cytoplasm. Third, many wetland plants appear to have a much higher than average metabolic tolerance for these ions (Ernst, 1990).

Sulfur: Sulfur as sulfide is toxic to plant tissues. The element is reduced to sulfide in anaerobic soils and accumulates to toxic concentrations in salt marshes (Goodman and Williams, 1961; Koch and Mendelssohn, 1989). Although sulfate uptake is metabolically controlled, sulfide can enter the plant without control and is found in elevated concentrations in many flood-adapted species under highly reduced conditions. In experiments with *Spartina alterniflora,* a salt marsh species, and *Panicum hemitomon,* a freshwater marsh species, Koch et al. (1990) reported that the activity of alcohol dehydrogenase (ADH), the enzyme that catalyzes the terminal step in alcohol fermentation, was significantly inhibited by hydrogen sulfide and that this inhibition may help to explain the physiological mechanism of sulfide phytotoxicity often seen in salt marshes. This suppression of an anaerobic pathway by high concentrations of sulfides is in addition to the known suppression of aerobic energy production that occurs in wetland plants (Allam and Hollis, 1972; Havill et al., 1985; Pearson and Havill, 1988). Sulfur tolerance in wetland plants varies widely probably because of the variety of detoxification mechanisms available. These include the oxidation of sulfide to sulfate through root aeration of the rhizosphere; the accumulation of sulfate in the vacuole; the conversion to gaseous hydrogen sulfide, carbon disulfide, and dimethylsulfide and their subsequent diffusive loss; and a metabolic tolerance to elevated sulfide concentrations (Ernst, 1990).

Respiration

The presence of root air space is not a sufficient condition for effective oxygen transport to the root. If oxygen leaks from the cortex into the surrounding sediments before it reaches the root tips where metabolic activity is greatest, it is of little value to the plant. Smits et al. (1990b) reported different patterns of oxygen leakage among several plant species; some showed oxygen leakage along the whole length of the roots, and others (e.g., water lily) leaked oxygen only from the 1-cm apex.

Mendelssohn and Postek (1982) found that the oxidized rhizosphere was only 1/50 as well developed around the roots of inland marsh plants as streamside ones, indicating that root oxygen availability was limited in inland sites. This limitation is probably related to the more anoxic conditions inland, placing greater demand on the ability of the root aerenchyma to deliver oxygen from above ground.

Under conditions of oxygen deprivation, plant tissues respire anaerobically, as described for bacterial cells. In most plants, pyruvate, the end product of glycolysis, is decarboxylated to acetaldehyde, which is reduced to ethanol (Fig. 7-6). Both of these compounds are potentially toxic to root tissues. Flood-tolerant plants often have adaptations to minimize this toxicity. For example, under anaerobic conditions, *S. alterniflora* roots show much increased activity of ADH, the inducible enzyme that

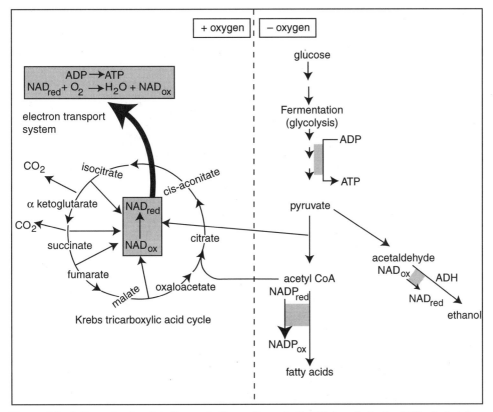

Figure 7-6 **Schematic of metabolic respiration pathway in flood-tolerant plants. ADH, alcohol dehydrogenase; NAD, nicotinamide adenine dinucleotide; NADP, NAD phosphate; subscripts refer to oxidized (ox) and reduced (red) forms.**

catalyzes the reduction of acetaldehyde to ethanol. The increase in the enzyme indicates a switch to anaerobic respiration, and it explains why acetaldehyde does not accumulate in the root tissue. Ethanol does not accumulate, either, although its production is apparently stimulated (Mendelssohn et al., 1982). It diffuses from rice roots during anaerobiosis (Bertani et al., 1980), thus preventing a toxic buildup, and it is probable that this occurs in *Spartina* also.

Another metabolic strategy reduces the production of alcohol by shifting the metabolism to accumulate nontoxic organic acids instead (Fig. 7-6). MacMannon and Crawford (1971) suggested that malic acid accumulation may be a characteristic feature of wetland species. The accumulation of malate cannot easily be interpreted, in part because malate is an intermediate in several metabolic pathways.

The metabolic problem encountered by plants deprived of oxygen is the loss of the electron acceptor that enables normal energy metabolism through ATP formation and use. The metabolic bottleneck in this process is often the electron-accepting coenzyme nicotinamide adenine dinucleotide (NAD), which is reduced in the oxidative steps of carbohydrate metabolism and then reoxidized in the mitochondria by molecular oxygen to yield the biological energy currency ATP. In the absence of oxygen, reduced NAD (NADH) accumulates and "jams" the metabolic system, blocking ATP generation (Fig. 7-6). As described previously, in fermentation, acetaldehyde takes the place of oxygen, reoxidizing reduced NAD. Malate acts in the same way through the tricarboxylic acid cycle. Thus, glycolysis can occur as long as NADH is reoxidized to the oxidized form, NAD. Another route, perhaps the main route of NADH oxidation in vascular plants, may be the oxidation of pyrimidine nucleotides (Jackson and Black, 1993; Blom and Voesenek, 1996). This allows the fermentation pathway to provide a steady source of energy (ATP) to the roots, but it is an inefficient pathway of metabolism. NAD phosphate (NADP) serves a function similar to that of NAD in lipid synthesis. Reduced NADP is oxidized during the synthesis of fatty acids from acetyl coenzyme A and can be recycled in oxidative pentose phosphate metabolism to provide metabolic intermediates and oxidized NADP for more fatty acid synthesis. This cycle does not require oxygen. In some plant species, it has been observed that fatty acids accumulate under hypoxic conditions and this lipid synthesis may even continue (Fox et al., 1988). This has obvious adaptive significance, especially for seed germination and growth under water or in other anaerobic environments.

Intraspecies variations related to metabolic and morphological characteristics point to the genetic basis for plant performance in wetland environments. Smits et al. (1990a) reported a positive correlation between the number of ADH isozymes and ethanol production in the roots of a number of aquatic macrophytes. They proposed that high enzyme polymorphism confers a selective advantage in hypoxic sediments. Similarly, in a flood-tolerant *Echinochloa* species, an extra ADH isozyme not present in aerobic soils was induced under anoxia (Fox et al., 1988). Despite these examples, the evidence relating genetic differentiation of isozymes to adaptive mechanisms of flood tolerance is not at all clear. For example, both *Sporobolus virginicus* and *Spartina patens* have morphologically and genetically distinct forms typically found in environments that have different inundation patterns. However, both forms of *S. virginicus* can adapt physiologically to accommodate anaerobic substrates (Donovan and Gallagher, 1984). Silander (1984) concluded from his study of *S. patens* that the genetic structure of the morphologically distinct forms could be explained only in terms of the simultaneous operation of selection, inbreeding, and migration.

Salt

At the cell level, plants behave toward salt in much the same way as bacteria do, and their adaptive strategies are identical. Vascular plants, however, have also developed adaptations that take advantage of their structural complexity. These include barriers to prevent or control the entry of salts, and organs specialized to excrete salts. In both cases, specialized cells bear most of the burden of the adaptation, allowing the remaining cells to function in a less hostile environment. It has generally been thought that the air space of the root cortex is freely accessible from the rhizosphere; consequently, the endodermis forms the first real barrier to the upward movement of solutes from the soil. This was confirmed by the observation that the roots of plants in high-salt environments often have much higher salt concentrations (and must also have higher salt tolerance) than the leaves do. However, X-ray microanalysis of the salt-resistant *Puccinellia peisonis* shows a decreasing concentration of sodium and an increasing concentration of potassium in the roots from the outer cortex through the endodermis to the stele (Stelzer and Lauchli, 1978). Both the inner cells of the cortex and the passage cells of the stele seem to be barriers to sodium transport, whereas potassium moves through fairly freely. The selectivity for potassium (or the exclusion of sodium) was seen to be common in bacteria also (Table 7-2). In contrast to this picture, Moon et al. (1986) reported that, in the mangrove *Avicennia marina*, the primary barrier to passive salt incursion into the plant apoplast is at the root periderm and exodermis so that the root cortex is protected from high salt concentrations. Uptake through the symplasm is restricted mainly to the terminal third- and fourth-order roots, and the large lateral roots serve primarily as the means of vascular transport and support.

As a result of the filtering out of salt at the root apoplast, the sap of many halophytes is almost pure water. The mangroves *Rhizophora, Laguncularia,* and *Sonneratia* exclude salts almost completely. Their sap concentrations are only about 1 to 1.5 mg NaCl per ml (compared with about 35 mg/ml in sea water). *Avicennia* has a higher sap concentration of about 4 to 8 mg/ml (Scholander et al., 1966), or about 10 percent of external concentration (Moon et al., 1986). When fluid is forced out of the leaves of these species by pressure, it is almost pure distilled water. Thus, both the root and the leaf cell membrane act like ultrafilters.

The leaf cytoplasm must have an osmotic potential higher than the osmotic potential of the sap in order to retain water, and in mangroves, 50 to 70 percent of this osmotic potential is obtained from sodium and chloride ions. Most of the remainder is presumably organic. In *Batis,* a succulent halophyte, NaCl alone makes up 90 percent of the total osmoticum (Scholander et al., 1966).

Some plants that do not exclude salts at the root, or are "leaky" to salt, have secretory organs. The leaves of many salt marsh grasses, for example, characteristically are covered with crystalline salt particles excreted through specialized salt glands embedded in the leaf. These glands do not function passively; instead, they selectively remove certain ions from the vascular tissues of the leaf. In *Spartina,* for example, the excretion is enriched in sodium, relative to potassium. These two mechanisms, salt exclusion and salt secretion, protect the shoot and leaf cells of the plant from high concentrations of salt and presumably maintain an optimum ionic balance between mono- and divalent cations and between sodium and potassium. At the same time, the osmotic concentration of the cells of salt-tolerant plants must be maintained at a level high enough to allow the absorption of water from the root medium. Where the inorganic salt concentration is kept low, organic compounds make up the rest of the osmoticum in the cells.

Photosynthesis

One adaptation that many wetland plant species share with plants in other stressed environments, especially in drought-stressed environments, is the C_4 biochemical pathway of photosynthesis (formally called the Hatch–Slack–Kortschak pathway, after its discoverers). It gets its identity from the fact that the first product of CO_2 incorporation is a four-carbon compound, oxaloacetic acid. This pathway is outlined in simplified form in Figure 7-7, and C_3 plants are compared with C_4 plants in Table 7-4. (The first compound resulting from CO_2 incorporation in C_3 plants is a three-carbon compound, phosphoglyceric acid.) C_3 plants are much more common than C_4 plants. Although water is a universal feature of wetlands, plants in saline wetland habitats have much the same problem of water availability as plants in arid areas do. In both cases, the water potential of the substrate is very low. In arid zones, it is low because the soil is dry; in saline wetlands, it is low because of the salt content. In wetlands, water uptake is accompanied by a mass flow of dissolved salts to the roots; their absorption must be regulated, at an energy cost, by the plant. Therefore, in both environments, mechanisms that reduce water loss (transpiration) provide an adaptive advantage. For a C_3 plant to take up carbon dioxide for photosynthesis, its stomata must be open, and if they are open during the bright hours of the day, water loss is excessive.

Plants that fix carbon by the C_4 pathway can use carbon dioxide more effectively than other plants. They may be able to fix CO_2 in the dark, and are able to withdraw carbon dioxide from the atmosphere until its concentration falls below 20 ppm (as compared with 30 to 80 ppm for C_3 plants). This is achieved by using phosphoenolpyruvate (PEP), which has a high affinity for CO_2, as the carbon dioxide acceptor instead

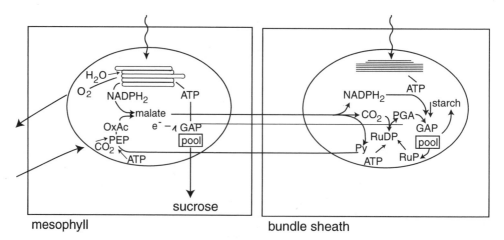

Figure 7-7   A much simplified diagram of CO_2 fixation via the Hatch–Slack–Kortschak pathway in C_4 plants. PEP, phosphoenolpyruvate; OxAc, oxaloacetate; PGA, 3-phosphoglyceric acid; GAP, 3-phosphoglyceraldehyde; RuP, ribulose-5-phosphate; Py, pyruvate. PGA is also produced by carboxylation of C_2 compounds that appear in the pool; the regeneration of PEP from PGA, in which water is given off, is not shown. (*After Larcher, 1995*)

Table 7-4 **Comparison of aspects of photosynthesis of herbaceous C$_3$ and C$_4$ plants**

Photosynthetic Characteristics	C$_3$	C$_4$
Initial CO$_2$ fixation enzyme	PEP carboxylase[a]	RuBP carboxylase[a]
Location of initial carboxylation	Mesophyll	Bundle sheath
Theoretical energy requirement for net CO$_2$ fixation, CO$_2$: ATP : NADPH	1 : 3 : 2	1 : 5 : 2
CO$_2$ compensation concentration (ppm CO$_2$)	30–70	0–10
Transpiration ratio (g H$_2$0 transpired/g dry weight)	450–950	250–350
Optimum day temperature for net CO$_2$ fixation (°C)	15–25	30–47
Response of net photosynthesis to increasing light intensity	Saturation at 1/4 to 1/2 full sunlight	Proportional to or saturation at full sunlight
Maximum rate of net photosynthesis (mg CO$_2$/dm^2 leaf surface/hr)	15–40	40–80
Maximum growth rate (g m^{-2} d^{-1})	19.5	30.3
Dry matter production (g m^{-2} yr^{-1})	2,200	3,860

[a] RuBP, ribulose-bis-phosphate; PEP, phosphoenolpyruvate.
Source: Based on data from Black (1973) and Fitter and Hay (1987).

of the ribulose diphosphate acceptor of the conventional pathway. In addition, the malate formed by this carboxylation is nontoxic and can be stored in the cell until it can be decarboxylated and the released carbon dioxide fixed through the normal C$_3$ pathway (Fig. 7-7). The production of PEP and the C$_4$ metabolism provide a possible pathway for recycling carbon dioxide from cell respiration. In addition, plants using the C$_4$ pathway of photosynthesis have low photorespiration rates and the ability to use efficiently even the most intense sunlight. These differences make C$_4$ plants more efficient than most C$_3$ plants both in their rates of carbon fixation and in the amount of water used per unit of carbon fixed (Table 7-4; Fitter and Hay, 1987). Finally, Armstrong (1975) suggested that water conservation mechanisms in wetland plants have the additional function of reducing the rate at which soil toxins are drawn toward the root. This increases the probability of detoxifying them as they move through the oxidized rhizosphere.

Among the common wetland angiosperms that have been shown to photosynthesize through the C$_4$ pathway are *S. alterniflora, S. townsendii, S. foliosa, Cyperus rotundus, Echinochloa crus-galli, Panicum dichotomiflorum, P. virgatum, Paspalum distichum, Phragmites communis,* and *Sporobolus cryptandrus.* It is obvious that this adaptation is fairly common in the wetland environment. Table 7-5 compares some of the photosynthetic attributes of two salt marsh species, *S. alterniflora,* a C$_4$ plant, and *Juncus roemerianus,* a C$_3$ plant. The tall form of *S. alterniflora* grows along creek banks; the short form grows farther inland in poorly drained soils. Compared with *J. roemerianus, S. alterniflora* has a higher rate of photosynthesis, a lower CO$_2$ compensation concentration (the CO$_2$ concentration in the leaf cellular spaces when photosynthesis is reduced to zero), a lower respiration rate in the light, and a higher temperature optimum. Water use efficiency (Table 7-5) is an important index of the ability of plants to photosynthesize with a minimum water loss, especially in arid or saline environments, where available water is in scarce supply. *S. alterniflora* is almost twice as efficient in this respect as *J. roemerianus.*

Table 7-5 Comparison of photosynthetic characteristics of two wetland plant species: *Spartina alterniflora*, a C$_4$ plant, and *Juncus roemerianus*, a C$_3$ plant

Photosynthetic Characteristics	*Spartina alterniflora*		*Juncus roemerianus*
	Tall Form	Short Form	
Maximum seasonal net photosynthetic rate [mg CO_2/ cm^{-2} s^{-1} (month)]	90 (Sept.)	65 (July)	60 (March)
Photosynthetic light response (Fig. 7-8)	Nonsaturating	Saturating	Nonsaturating
CO_2 compensation concentration (ppm)	12	84	84
Photorespiration at 21% O_2 (mg CO_2 cm^{-2} s^{-1})	6.7	18.2	9.1
(% of photosynthesis)	(11)	(40)	(54)
Temperature optimum (summer) (°C)	30–35	30–35	25
Water use efficiency (mg CO_2/g H_2O)	15	12–15	8–9

Source: Based on data from Giurgevich and Dunn (1978, 1979).

One characteristic in which *J. roemerianus* does not fit the typical C$_3$ plant pattern is its photosynthetic response to light intensity (Fig. 7-8). Both species increase their photosynthetic rates as light intensity increases to full sunlight, although C$_3$ plants typically reach saturation at no more than one-half full intensity. The short form of *S. alterniflora* behaves in a way intermediate to the tall form and the C$_3$ plant. It is not known whether this reflects a switch from C$_4$ to C$_3$ metabolism under conditions of oxygen stress. Although the C$_4$ adaptations suggest some selective advantages for plants in wetlands, such advantages are not enough to displace well-adapted C$_3$ species such as *J. roemerianus.*

Whole Plant Strategies

Many plant species have evolved avoidance or escape strategies by life-history adaptations. The most common of these strategies are the timing of seed production in the non–flood season by either delayed or accelerated flowering (Blom et al., 1990); the production of buoyant seeds that float until they lodge on high, unflooded ground; the germination of seeds while the fruit is still attached to the tree (vivipary), as in the red mangrove (see Chapter 11); the production of a large, persistent seed bank (Voesenek, 1990), and the production of tubers, roots, and seeds that can survive long periods of submergence. In many riparian wetlands, flooding occurs primarily during the winter and early spring, when trees are dormant and much less susceptible to anoxia than they are during the active growing season.

Animal Adaptations

Animals are exposed to the same range of environmental conditions in wetlands as unicellular organisms and plants, but because of their complexity, their adaptations are more varied. The adaptation may be as varied as a biochemical response at the

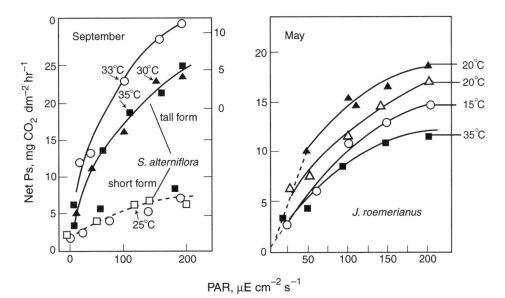

Figure 7-8 Light–photosynthesis curves for two wetland species with different biochemical pathways of photosynthesis at different ambient temperatures: *Spartina alterniflora*, a C₄ plant; and *Juncus roemerianus*, a C₃ plant. PAR, photosynthetically active radiation. (*After Giurgevich and Dunn, 1978, 1979*)

cell level, a physiological response of the whole organism such as a modification of the circulatory system, or a behavioral response such as modified feeding habits. Furthermore, although it is convenient to discuss the specific response mechanisms to individual kinds of environmental stresses, in reality an organism must respond simultaneously to a complex of environmental factors, and it is the success of this integrated response that determines its fate. For example, one possible response to stress is avoidance by moving out of the stress zone. In wetlands, however, that might mean moving from an anoxic zone within the soil to the surface, where temperature extremes and desiccation pose a different set of physiological problems. Thus, the organism's successful adaptations are often compromises that enable it to live with several competing environmental demands.

Anoxia

At the cell level, the metabolic responses of animals to anoxia are similar to those of bacteria. Vertebrates, however, tend to have less ability to adapt to anaerobic conditions than invertebrates. The vertebrates and many invertebrates are limited in anaerobic respiration to glycolysis or to the pentose monophosphate pathway whose dominant end product is lactate. In all chordates, the internal cell environment is closely regulated. As a result, most adaptations are organism-level ones to maintain the internal environment. Vernberg and Vernberg (1972) list six major kinds of adaptations that marine organisms have evolved to control gaseous exchange:

1. Development or modification of specialized regions of the body for gaseous exchange; for example, gills on fish and crustacea, parapodia on polychaetes

2. Mechanisms to improve the oxygen gradient across a diffusible membrane; for example, by moving to oxygen-rich environments or by moving water across the gills by ciliary action

3. Internal structural changes such as increased vascularization, a better circulation system, or a stronger pump (the heart)

4. Modification of respiratory pigments to improve oxygen-carrying capacity

5. Physiological adaptations, including shifts in metabolic pathways and heart pumping rates

6. Behavioral patterns such as decreased locomotor activity or closing a shell during low oxygen stress

All but the behavioral adaptations involve modifications of tissue and organ systems and associated physiological changes. Adaptations in this list are not mutually exclusive. Many animals combine several different types of adaptations.

Examples of different kinds of adaptations are numerous. We give a few here to illustrate their diversity. Crabs inhabit a wide range of marine habitats. The number and the total volume of gills of crabs living on land are less per unit of body weight than those of aquatic species. In addition, the gills of some intertidal crabs have become highly sclerotized so that the gill leaves do not stick together when the crab is out of water (Vernberg and Vernberg, 1972). Tube-dwelling amphipods apparently can function efficiently with low oxygen supplies. At saturated oxygen tensions, they exhibit an intermittent rhythm of ventilation. At low tide, when the oxygen in their burrows drops to very low levels, they ventilate continuously but do not hyperventilate as free-swimming amphipods do. Because of the resistance of their tubes to water flow, hyperventilation would be energetically expensive for tube-dwelling amphipods (Vernberg, 1981).

Many marine animals associated with anoxic wetland soils have high concentrations of respiratory pigments or pigments that have unusually high affinities for oxygen or both. These include the nematode (*Enoplus communis*), the Atlantic bloodworm (*Glycera dibranchiata*), the clam (*Mercenaria mercenaria*), and even the land crab (*Carooma quannumi*) (Vernberg and Vernberg, 1972; Vernberg and Coull, 1981).

Fiddler crabs (*Uca* spp.) illustrate the complex behavioral and physiological patterns to be found in the intertidal zone. These crabs are active during low tides, feeding daily when the marsh floor is exposed. (Incidentally, this pattern of activity is based on an innate lunar rhythm, not on a direct sensing of low water levels. When transported miles from the ocean, fiddlers continue to be active at the time that low tide would occur in their new location.) When the tide rises, they retreat to their burrows, where the oxygen concentration can become very low because fiddler crabs apparently do not pump water in their burrows. Not only are these species relatively resistant to anoxia, but also their critical oxygen tension (i.e., the tension below which respiratory activity is reduced) is low, 0.01 to 0.03 atm for inactive and 0.03 to 0.08 atm for active crabs. They can continue to consume oxygen down to a level of 0.004 atm (Vernberg and Vernberg, 1972). When oxygen levels get very low in the burrows, the crabs simply become inactive and consume very little oxygen. They may remain that way for several tidal cycles without harm.

Intertidal bivalves close their valves tightly or loosely when the tide recedes. Widdows et al. (1979) found that four different bivalves had lower respiration rates in air (valves closed) than in water. All could respire anaerobically, but the accumulation of end products of anaerobic respiration depended on how tightly their shells were closed and, thus, how much oxygen they received. The tolerance to anoxia may change during the life of an organism. The larvae of fiddler crabs, which are planktonic, are much more sensitive to low oxygen than are the burrowing adults. An interesting but rather unusual adaptation is that of a gastrotrich (*Thiodasys sterreri*), which is reported to be able to use sulfide as an energy source under extreme anaerobiosis (Maguire and Boaden, 1975).

Salt

Like their responses to oxygen stress, the major mechanisms of adaptation by animals to salt involve control of the body's internal environment. Most simple marine animals are *osmoconformers;* that is, their internal cell environment follows closely the osmotic concentration of the external medium. In animals that have greater body complexity, however, *osmoregulation,* that is, control of internal osmotic concentration, is the rule. This is particularly true of animals that inhabit the upper intertidal zone, where they are exposed to widely varying salinities and to prolonged periods of desiccation. *Euryhaline* organisms can tolerate wide fluctuations in salinity. *Stenohaline* organisms, on the other hand, survive within fairly narrow osmotic limits. Most marsh organisms must be euryhaline, but they can be either osmoconformers or osmoregulators.

Figure 7-9 illustrates the imperfect osmoregulation found in penaeid shrimp. The hemolymph concentration of a perfect osmoconformer would follow the solid line of isotonicity (internal concentration equal to ambient concentration). In contrast, a perfect osmoregulator would have a constant internal concentration, which would be illustrated by a horizontal line on the graph. The brown shrimp (*Penaeus aztecus*) is intermediate between these two positions. The internal environment varies much less than the external medium does, but it is not constant. At low external salt concentrations, the shrimp is hyperosmotic, usually achieved by concentrating sodium and chloride ions. This condition indicates a water potential gradient into the organism. Regulation at high ambient salt concentration is hypo-osmotic. In this case, the water potential gradient is directed out of the animal, which means that dehydration would occur if the body covering were not, to some extent, impervious to water movement. Animals that possess the ability for hypo-osmotic regulation must have some mechanism to lower the osmotic concentration of the body. This is accomplished through special regulatory organs and organ systems, chiefly renal organs (kidneys or more primitive nephridia, antennal glands), gills, salt-secretory, nasal, or rectal glands, and the specialized excretory functions of the gut. These organs are able to move ions across cell membranes against the concentration gradient (at an energy cost to the animal), concentrating them in some excretory product such as urine.

Figure 7-9 also illustrates the differences in adaptation to salinity changes in species of crabs that inhabit different environments. The aquatic species *Cancer* is an osmoconformer. Species that are submerged most of the time but are subject to wider osmotic fluctuations (*Hemigrapsus* and *Pachygrapsus*) are imperfect osmoregulators. The other species (including *Uca,* a common marsh crab) are from the high intertidal zone. They are excellent osmoregulators, possessing adaptations obviously useful in controlling the variable salinity and frequent desiccation of their habitat. Regulation in these species is controlled both by differences in exoskeleton permeability and by specialized excretory organs. The exoskeleton of terrestrial crabs is less permeable to water and salt than those of semiterrestrial species, which, in turn, are less permeable

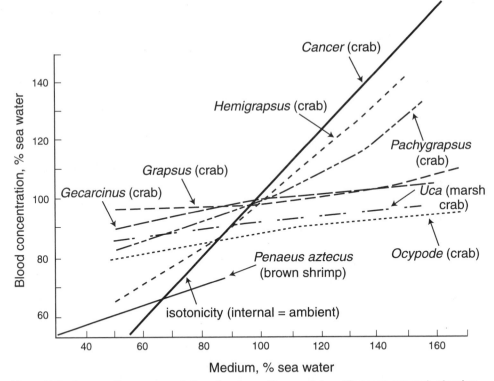

Figure 7-9 Comparative osmoregulation of crabs and brown shrimp (*Penaeus aztecus*), showing different degrees of adaptation to the marine environment. Line of perfect osmoconformity (isotonicity) is also shown for comparison. A horizontal line would indicate perfect osmoregulation. Of the crab species, *Cancer* is an aquatic species; *Hemigrapsus* and *Pachygrapsus* are low intertidal zone species; others are terrestrial or high intertidal zone inhabitants. (*After Gross, 1964; Bishop et al., 1980*)

than subtidal crabs. The antennal glands seem to control the concentrations of specific ions in the hemolymph. Osmoregulation, however, is controlled by the gills and the posterior diverticulum of the alimentary canal (Vernberg and Vernberg, 1972). The complexity of the adaptations is illustrated by the permeability of the foregut of the land crab (*Gecarcinus lateralis*) to both water and salt. This permeability varies with time and environmental circumstance and is under neuroendocrine control (Mantel, 1968).

Feeding

As with reproductive adaptations, the broad range of animal feeding responses closely reflects their habitats. Adaptations of feeding appendages, for example, seem to be more closely related to feeding habits than to taxonomic relationships. Many organisms that exist in marsh sediments are adapted for the direct absorption of dissolved organic compounds from their environment. For example, infaunal polychaetes can supply a major portion of their energy requirements from the rich supply of dissolved amino acids present in their environment, but epifaunal species are unable to take advantage of amino acids at concentrations typical of their environment (Vernberg, 1981).

Many mud-dwelling organisms have one or more adaptations to selective feeding on microscopic particles by means of pseudopods, cilia, mucus, and setae, or they may ingest substrate unselectively. Sikora (1977) suggested that the appendages of many macrobenthic organisms (shrimp, crabs) are adapted to feeding on microscopic meio-benthic organisms and that these latter organisms are major intermediaries in the marsh/estuary food chain.

Behavioral Adaptations

As might be expected, adaptations to specific habitats involve virtually every facet of an organism's existence. We have focused primarily on the metabolic and structural responses of the individual to stresses of the wetland environment. In terms of species survival, howevers, reproduction is equally important. In an evolutionary sense, a species strategy must be to produce reproductively active offspring at minimum energy cost. For infauna in the marsh sediment, where mobility is restricted, this is often accomplished by direct contact of organisms for fertilization and by direct growth of offspring through the elimination of larval stages or by larvae that remain in place. Epibenthic organisms in a fluctuating environment, in contrast, often produce great numbers of pelagic larvae that are widely distributed by currents and tides. Reproductive behavior is complex, and it is not always easy to see any adaptive significance in the responses that have been observed. For example, the subtidal clam *Rangia cuneata* requires a salinity shock (of about 5 ppt up or down) to release its gametes, even though the female may be gravid more than half of the year. When salinities remain constant, for example, when an area is impounded, the clam eventually dies out. The intertidal crab *Sesarma cinereum* requires low estuarine salinity for larval development. The fourth zoeal stage, in particular, is sensitive to salinity, and best development occurs at 26.7 ppt. In the succeeding megalops stage, however, it can withstand a wide range of salinities and temperatures (Costlow et al., 1960).

In addition to developing strategies for reproduction, many animals have adapted their lifestyles to take seasonal advantage of the wetland ecotone. For example, bottom-land forests that flood during winter and drain in the spring are exceptionally high in productivity. While the forest floor is inundated in the early spring, many fish and shellfish species move up into the floodplain from the adjacent river to feed and to spawn. They move with the water's edge as floodwaters recede, feeding on abundant detritus and benthic invertebrates (Lambou, 1990). Terrestrial animals follow the receding water line for the same reasons. Deer, turkey, bear, migrating songbirds, and other species take advantage of the acorns and other hard seeds and nuts available in fall and early winter when surrounding uplands have lost much of their habitat value (Gosselink et al., 1990b).

Mutualism and Commensalism

The close interactions among members of an ecological community reflect the high degree of adaptability of members of the community, not only to their physical environment, but also to their biological environment. This chapter has documented adaptations to the physical and chemical wetland environment, but positive interactions among organisms are also predicted to play a significant role in ecosystem dynamics, especially in marginal or stressed environments such as wetlands (Bertness and Callaway, 1994; Bertness and Hacker, 1994). Documentations of these effects in wetland environments are relatively few. Many appear to involve nutrients that are limiting in these environments. For example, Grosse et al. (1990) reported increased levels of nitrogen fixation by the symbiotic fungus *Frankia alni,* presumably because of thermal pumping of oxygen through and out of the root system of common alder into the rhizosphere.

Ellison et al. (1996) reported a mutualistic interaction between root-fouling sponges (*Todania ignis* and *Haliclona implexiformis*) and the red mangroves (*Rhizophora mangle*) on which they grow. Fine adventitious mangrove rootlets ramify throughout the sponges. They absorb dissolved ammonium from the sponges, which stimulates additional root growth. The sponges also protect the roots from isopod attack. Mangrove roots, in turn, provide the only hard substrate for sponges in this habitat, and they stimulate sponge growth by leaking carbon.

These examples of the positive interactions among wetland species point to an extremely interesting line of neglected research that may lead to important new insights into the complexity of mutualistic adaptations in wetland ecosystems, and their importance, not only to the organisms involved, but also to the energetic dynamics of the entire community.

Recommended Readings

Grosse, W., H. B. Büchel, and S. Lattermann. 1998. Root aeration in wetland trees and its ecophysiological significance. In A. D. Laderman, ed. Coastally Restricted Forests. Oxford University Press, New York, pp. 293–305.

Larcher, W. 1995. Physiological Plant Ecology, 3rd ed. Springer-Verlag, New York. 522 pp.

Vernberg, F. J., and W. B. Vernberg, eds. 1981. Functional Adaptations of Marine Organisms. Academic Press, New York.

Wetland Ecosystem Development

Wetland ecosystems have traditionally been considered transitional seres between open lakes and terrestrial forests. The accumulation of organic material from plant production was seen to build up the surface until it was no longer flooded and could support flood-tolerant terrestrial forest species (autogenic succession). An alternative hypothesis is that the vegetation found at a wetland site consists of species adapted to the particular environmental conditions of that site (allogenic succession). Observed zonation patterns, in this view, reflect underlying environmental gradients rather than autogenic successional patterns. Present evidence seems to suggest that both allogenic and autogenic forces act to change wetland vegetation and that the idea of a regional terrestrial climax is inappropriate. Several models have been described for wetland plant development, including the functional guild model, the environmental sieve model, the gap dynamic model, and the centrifugal organization concept.

If one looks at ecosystem attributes as indices of ecosystem maturity, wetlands appear to be mature in some respects and young in others. Generally, productivity is high, some production is exported, and mineral cycles are open, all indications of young systems. On the other hand, most wetlands accumulate much structural biomass in peat, all wetlands are detrital systems, spatial heterogeneity is generally high, and life cycles are complex. These properties indicate maturity. Although the input of water and nutrients varies in different types of wetlands by as much as five orders of magnitude, ecosystem response in terms of productivity, biomass, and nutrient storage varies by only a factor of 2 to 4. Wetland ecosystem response is controlled by stores of soil organic material that stabilize the flooding pattern and provide a steady source of nutrients to the plants, minimizing the impact of external supplies. The strategy for ecosystem development in wetlands includes concepts such as pulse stability whereby the wetland maintains itself because of and not in spite of natural pulses such as tides and floods, and self-design whereby the wetland develops according to the multitude of propagules that come its way. Open systems are, thus, more biologically diverse. In self-design, the presence and survival of species due to the continuous introduction of them and their propagules is the essence of the successional and functional development of an ecosystem. At landscape scales, observed patterns of wetlands, aquatic and upland habitats, reflect a complex and dynamic interaction of physical (allogenic) and biotic (autogenic) forces acting on the geomorphic template of the landscape.

Wetland Plant Development

The beginning and subsequent development of a plant community is characterized by the initial conditions at the site and by subsequent events, including the availability of viable seeds or other propagules, appropriate environmental conditions for germination and subsequent growth, and replacement by plants of the same or different species as site conditions change in response to both abiotic and biotic factors. The concept of succession, that is, the replacement of plant species in an orderly sequence of development, in particular, has exerted a strong influence on plant ecology throughout this century. This concept has a long history. It was first clearly enunciated by Clements (1916) and applied to wetlands by the English ecologist W. H. Pearsall in 1920 and by an American, L. R. Wilson, in 1935. E. P. Odum (1969) adapted and extended the ideas of those early ecologists to include ecosystem properties such as productivity, respiration, and diversity.

The classical use of the term *succession* involves three fundamental concepts: (1) Vegetation occurs in recognizable and characteristic *communities;* (2) community change through time is brought about by the biota (i.e., changes are *autogenic*); and (3) changes are linear and *directed* toward a mature stable *climax* ecosystem (E. P. Odum, 1971).

Although the classical concept of succession has been a dominating paradigm of great importance in plant ecology, it is presently in disarray. As early as 1917, Gleason enunciated an *individualistic* hypothesis to explain the distribution of plant species. His ideas have developed into the *continuum* concept (Whittaker, 1967; McIntosh, 1980), which holds that the distribution of a species is governed by its response to its environment (*allogenic succession*). Because each species responds differently to its environment, no two occupy exactly the same zone. The observed invasion/replacement sequence is also influenced by the chance occurrence of propagules at a site. The result is a continuum of overlapping sets of species, each responding to subtly different environmental cues. In this view, no communities exist in the sense used by Clements, and although ecosystems change, there is little evidence that this is directed or that it leads to a particular climax.

Autogenic versus Allogenic Processes

A key issue in discussions of ecosystem development is whether the biota on a site determine their own future by modifying their own environment, or if the development of an ecosystem is simply a response to the external environment. In the classical view of succession, wetlands are considered transient stages in the *hydrarch development* of a terrestrial forested climax community from a shallow lake (Fig. 8-1). In this view, lakes gradually fill in as organic material from dying plants accumulates and minerals are carried in from upslope. At first, change is slow because the source of organic material is single-celled plankton. When the lake becomes shallow enough to support

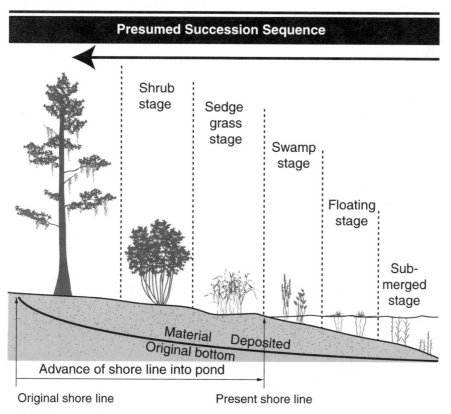

Figure 8-1 Diagram of classical hydrarch succession at the edge of a pond. (*After Wilson and Loomis, 1967*)

rooted aquatic plants, however, the pace of organic deposition increases. Eventually, the water becomes shallow enough to support emergent marsh vegetation, which continues to build a peat mat. Shrubs and small trees appear. They continue to transform the site to a terrestrial one, not only by adding organic matter to the soil but also by drying it through enhanced evapotranspiration. Eventually, a climax terrestrial forest occupies the site (see, e.g., W. S. Cooper, 1913). The important point to note in this description of hydrarch succession is that most of the change is brought about by the plant community itself as opposed to externally caused environmental changes.

How realistic is this concept of succession? It is certainly well documented that forests do occur on the sites of former lakes (Larsen, 1982), but the evidence that the successional sequence leading to these forests was autogenic is not clear. Because peat building is crucial to filling in a lake and its conversion to dry land, key questions involve the conditions for peat accumulation and the limits of that accumulation. Peat underlies many wetlands, often in beds 10 m or more deep. Several scientists (McCaffrey, 1977; Delaune et al., 1983b) have shown that, in coastal marshes, it has accumulated and is still accumulating at rates varying from less than 1 to about 15 mm/yr. Most of this accumulation seems to be associated with rising sea levels (or submerging land). By contrast, northern inland bogs accumulate peat at rates of 0.2 to 2 mm/yr (see Chapter 13).

In general, accumulation occurs only in anoxic sediment. When organic peats are drained, they rapidly oxidize and subside, as farmers who cultivate drained marshes have discovered. As the wetland surface accretes and approaches the water surface or at least the upper limit of the saturated zone, peat accretion in excess of subsidence must cease. It is hard to see how this process can turn a wetland into a dry habitat that can support terrestrial vegetation unless there is a change in hydrologic conditions that lowers the water table. For example, Cushing (1963) used paleoecological techniques to show that most of the peatlands in the Lake Agassiz plain (Minnesota and south-central Canada) formed during the Mid-Holocene (beginning about 4,000 years ago) during a moist climatic period when surface water levels rose about 4 m (see Chapter 13).

Wetlands are in the center of this dispute about the importance of autogenic versus allogenic processes because of their transitional nature. In addition to being *seres,* wetlands are often described as being *ecotones,* that is, transitional spatial gradients between adjacent aquatic and terrestrial environments. Thus, wetlands can be considered transitional in both space and time. As ecotones, wetlands usually interact strongly to varying (allogenic) forcing functions from both ends of the ecotone. These forces may push a wetland toward its terrestrial neighbor if, for example, regional water levels fall, or toward its aquatic neighbor if water levels rise. Alternately, plant production of organic matter may raise the level of the wetland, resulting in a drier environment in which different species succeed. Since these environmental changes can be subtle, it is often difficult to determine whether the observed ecosystem response is autogenic or allogenic. Without careful measurements, the causes of the response are often obscure.

A particularly well-researched example illustrates the problem. In the early 20th century, Cowles (1899, 1901, 1911) and Shelford (1907, 1911, 1913) studied ponds of different ages in the Indiana dunes region along the southern shore of Lake Michigan. The ponds were of different ages and were thought to represent an autogenic successional sequence. Along this age gradient, the young ponds were deep and dominated by aquatic vegetation. Older ponds were shallower and supported emergent vascular plants along their borders. The oldest ponds were shallowest and contained the most "terrestrial" vegetation. This sequence was interpreted as evidence of classical autogenic succession.

Wilcox and Simonin (1987) and Jackson et al. (1988) revisited the Indiana dunes ponds. Using modern quantitative methods of ordination, they found the same progression of plant species from young to old ponds, supporting the sequence observed by earlies workers. In addition to the present vegetation, they also examined pollen and macrofossils in the sediments of a 3,000-year-old pond to determine whether the sediments support the presumed successional sequence found in the modern-day chronosequences. The pollen and macrofossil data older than 150 B.P. consisted of a diverse assemblage of submersed, floating-leaved, and emergent macrophyte groups. After 150 B.P., the data indicated a major and rapid vegetation change, which the authors attributed to post-European settlement, such as railroad construction and forest clearing.

To further evaluate the historical changes in the Indiana dunes ponds, Singer et al. (1996) examined the sediment pollen and macrofossil record of aquatic and emergent plants in one of the old Indiana dunes pond sites and compared it with the regional terrestrial pollen record (of airborne pollen found in the same cores). The latter tracks long-term climate changes in the region. If aquatic and emergent paleotaxonomic remains showed changes in species dominance that mirrored the terrestrial pollen record, then the changes in the ponds could be attributed to regional climate change rather than to autogenic processes. From their 10,000-year record, Singer et al. (1996) determined that historic changes in pond vegetation did, indeed, correspond to regional climate change (Fig. 8-2). Between 10,000 and 5700 B.P., the sampled area was a shallow lake; the regional climate was mesic (a pine/oak/elm terrestrial assemblage). A rapid increase in oak and hickory pollen around 5700 B.P. signaled a regional climate shift to a drier environment. At the same level in the sediment record, the pond macrofossil record showed a rapid shift to a peat-forming marsh environment. After about 3000 B.P., modest increases in beech and birch pollen suggested a trend toward a cooler, moister climate. The concomitant pond vegetation remained dominated by emergents, but transitions among several taxa suggest that water-level fluctuations and occasional fires were characteristic of the period.

These three studies, taken together, provide a fuller, more complex picture of plant development than the autogenic succession process proposed in Cowles' and

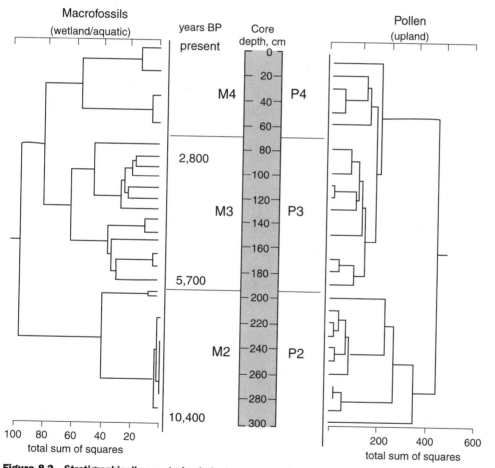

Figure 8-2 Stratigraphically constrained cluster diagrams for macrofossil and pollen data from an Indiana dunes pond in northern Indiana adjacent to Lake Michigan. Macrofossil zonation is based on presence/absence of aquatic and wetland taxa; pollen zonation is based on percentages of selected upland pollen types. Close individual taxa indicate close occurrence in fossil record. Thus the macrofossil record indicates 3 different groups of organisms. (*After Singer et al., 1996*)

Shelford's earlier studies. The picture that emerges is one of an interaction between allogenic and autogenic processes, with allogenic forces driving the development of the biotic system, but modified by autogenic processes. Over the 10,000-year span of the fossil record, changes in the plant assemblage correlated well with regional climate change. However, during the same period, the lake was slowly filling with organic sediments, first 100 cm of gyttja, characteristic of open freshwater systems; then 200 cm of fibrous peat, characteristic of vascular aquatic plants. During the period of a slow climate shift to a less xeric environment after approximately 3000 B.P., the pond environment remained a marsh, although the species assemblage changed. This fits well with the idea that the organic sediments moderated the climatic influence on the local water levels. Finally, during the modern period after about 150 B.P., human activities, which probably altered water levels locally, resulted in rapid vegetation changes.

The Community Concept and the Continuum Idea

The Indiana dunes ponds example is only a small part of the extensive literature concerning questions about plant and ecosystem development in wetlands. The idea of the community is particularly strong in wetland literature. Historic names for different kinds of wetlands—marshes, swamps, carrs, fens, bogs, reedswamps—often used with the name of a dominant plant (*Sphagnum* bog, leatherleaf bog, cypress swamp)—signify our recognition of distinctive associations of plants that are readily recognized and at least loosely comprise a community. One reason these associations are so clearly identified is that zonation patterns in wetlands often tend to be sharp, having abrupt boundaries that call attention to vegetation change and, by implication, the uniqueness of each zone. The plant community is central to the historic idea of succession because the mature climax resulting from succession was presumed to be a predictable group of plant species, with each group dependent on the regional climate.

Although wetland communities were historically identified qualitatively, the application of objective statistical clustering techniques supports the community idea, at least in some instances. For example, the classical syntaxonomical treatment of European *Spartina* communities (a semiquantitative analysis of vegetation stands based on dominants and observed similarities) resulted in the classification of these marshes into a number of subassociations (Beeftink and Gehu, 1973). A numerical classification of the same areas, based on similarity ratios and a statistical clustering technique, identified virtually the same subassociations (Kortekaas et al., 1976; Fig. 8-3). It is important to notice that different degrees of clustering occur with these data. In Figure 8-3, three main groups are associated with the dominance of three different *Spartina* species: *S. maritima, S. alterniflora,* and *S. townsendii*. These three groups break down further into various subassociations. The decision about what level of similarity, if any, identifies a community is entirely arbitrary.

The identification of a community is also, to some extent, a conceptual issue that is confused by the scale of perception. Field techniques are adequate to describe the vegetation in an area and its variability. However, its homogeneity—one index of community—may depend on size. For example, Louisiana coastal marshes have been classified into four zones, or communities, based on the dominant vegetation (Chabreck, 1972). If the size of the sampling area is large enough, any sample within one of these zones will always identify the same species. If smaller grids are used, however, differences appear within a zone. The intermediate marsh zone is dominated on a broad scale by *S. patens;* but aerial imagery shows patterns of vegetation within the

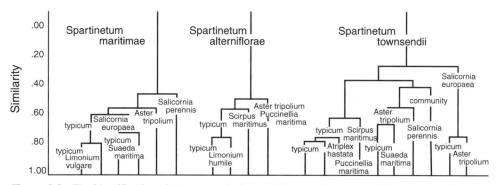

Figure 8-3 The identification of plant associations within *Spartina* communities using statistical clustering techniques. These are virtually the same plant associations identified by the less quantitative classical syntaxonomical treatment of the data. (*After Kortekaas et al., 1976*)

zone, and intensive sampling and cluster analysis of the vegetation reveal at least five subassociations that are characteristic of intermediate marshes. Is the intermediate marsh a community? Are the subassociations communities? Or is the community concept a pragmatic device to reduce the bewildering array of plants and possible habitats to a manageable number of groups within which there are reasonable similarities of ecological structure and function?

Supporters of the continuum concept would argue that the scale dependence of plant associations illustrates that individual species are simply responding to subtle environmental cues, implying little, if anything, about "communities," and that plant zonation simply indicates an environmental gradient to which individual species are responding. The reason zonation is so sharp in many wetlands, they argue, is that environmental gradients are "ecologically" steep and groups of species have fairly similar tolerances that tend to group them on these gradients. Figure 8-3 can easily be interpreted to support this contention, for it shows that the similarity level between two groups of plants is never more than 0.85 (1.00 indicates identity) and may be as low as 0.50. Figure 8-4 shows the distribution of swamp trees and submersed aquatic vegetation along an ordination axis. Although the species overlap, the distribution of each seems to be distinct, leaving no reflection of a community. The idea that each species is found where the environment is optimal for it makes perfect sense to ecophysiologists and autecologists, who interpret the success of a species in terms of its environmental adaptation.

One major difference between classical community ecologists and proponents of the continuum idea is the greater emphasis put on allogenic processes by the latter. In wetlands, abiotic environmental factors often seem to overwhelm biotic forces. Under these circumstances, the response of the vegetation is determined by these abiotic factors. Hydrologic conditions, for example, were described in Chapter 5 as having a particular significance for wetland structure and function. In coastal areas, plants can do little to change the tidal pulse of water and salt. Tidal energy may be modified by vegetation as stems create friction that slows currents or as dead organic matter accumulates and changes the surface elevation. These effects are limited, however, by the overriding tides. These wetlands are often in dynamic equilibrium with the abiotic forces, an equilibrium that E. P. Odum (1971) called *pulse stability*.

In the low-energy environment of northern peatlands, in contrast to tidal marshes, hydrologic flows can be dramatically changed by biotic forces, resulting in distinctive

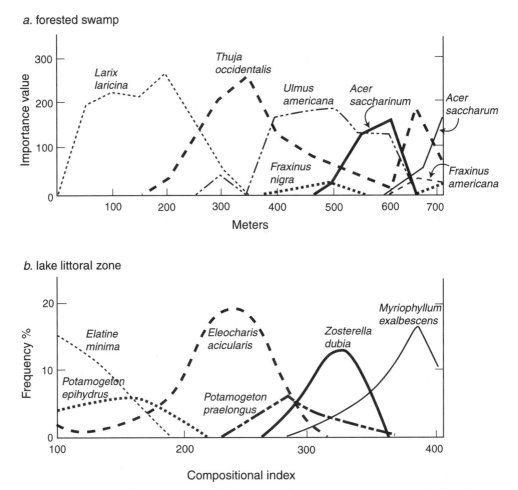

Figure 8-4 Examples of gradient analysis of wetlands in **A.** a forested swamp and **B.** submersed aquatic vegetation in the littoral zone of Wisconsin lakes. (*After van der Valk, 1982, based on original data from Beschel and Webber, 1962; Curtis, 1959*)

patterned landscapes (Glaser and Wheeler, 1980; Glaser et al., 1981; Glaser, 1983c; Siegel, 1983; Foster et al., 1983; Rochefort et al., 1990; see Chapter 13). Thus, changes in wetlands may be autogenic but are not necessarily directed toward a terrestrial climax. In fact, wetlands in dynamically stable environmental regimes seem to be extremely stable, contravening the central idea of succession. Walker (1970) found from pollen profiles that the successional sequence in northern peatlands was variable: There were reversals and skipped stages that may have been influenced by the dominant species first reaching a site (Fig. 8-5). A bog, not some type of terrestrial forest, was the most common endpoint in most of the sequences described.

Linear Directed Change

If plant species development on a site is determined by allogenic processes and is, therefore, simply a response to environmental forcing processes, then the successional concept of linear directed change makes little sense. Although the scientific literature

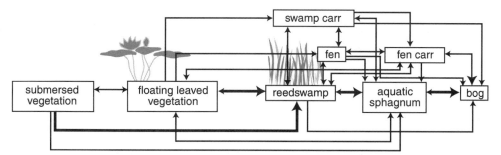

Figure 8-5 Successional sequences reconstructed from stratigraphic and palynological studies of postglacial British peatlands. Thicker lines indicate the more common transitions. (*After Walker, 1970*)

is replete with schematic diagrams showing the expected successional sequence from wetland to terrestrial forest, most of these are based on observed zonation patterns (or chronosequences), assuming that these spatial patterns presage the temporal pathway of change.

However, paleoecological analyses of soil profiles (such as those discussed earlier for the Indiana dunes ponds; Fig. 8-2) provide the best evidence to evaluate the concept. These records, mostly from northern peat bogs, suggest two generalizations: (1) In some sites, the present vegetation has existed for several thousands of years (Redfield, 1972); and (2) climatic change and glaciation had major impacts on plant species composition and distribution; generally, bogs expanded during warm, wet periods and contracted during cool, drier periods, although the influence of local topographic, drainage, and other site conditions often masked regional climatic shifts (Dopson et al., 1986; Casparie and Streefkerk, 1992). Pollen sequences, however, are generally consistent across Europe and North America, indicating a response to similar global climate shifts (McIntosh, 1985). West (1964), as quoted in McIntosh (1985), wrote:

> We may conclude that our present plant communities have no long history in the Quaternary, but are merely temporary aggregations under given conditions of climate, other environmental factors, and historical factors.

Seed Banks

Many recent studies have documented the role of chance in the development of plant communities, especially in the early stages. In this respect, studies of seed banks and their role in the introduction and invasion of plant species have been important.

If the development of plants on a site can be explained only in terms of the response of individual species to local conditions, the previous history of the site is important because it determines what propagules are present for future invasion. This—the sediment *seed bank*—has been found to be extremely variable—both in space and in time. Pederson and Smith (1988) made the following generalizations about marsh seed banks:

1. Freshwater marshes with drawdowns produce the greatest number of seeds. This has also been confirmed by Siegley et al. (1988) and Welling et al. (1988b).

2. Seed banks are dominated by the seeds of annual plants and flood-intolerant species. Areas that contain emergent plants have greater seed densities than mud flats. Perennials generally produce fewer seeds that have shorter viability than annuals. They are more likely to reproduce by asexual means such as rhizomes. Saline zones produce few seeds. The salt marsh is an example of a perennial-dominated system in which most reproduction is asexual.

3. Seed distribution decreases exponentially with the depth of the sediment.

4. Water is a major factor in seed banks. Seeds are concentrated along drift lines. The kinds of seeds produced depend on the flooding regime—by submergents when deep flooded, by emergents when periodically flooded, and by flood-intolerant annuals during drawdowns.

The germination of seedlings from a seed bank is similarly influenced by many factors that vary in space and time. Welling et al. (1988a) stated that differences in environmental conditions seem to have less impact on the distribution of seedlings along an elevation gradient than the distribution of seeds. Nevertheless, environmental factors such as flooding, temperature, soil chemistry, soil organic content, pathogens, nutrients (Gerritsen and Greening, 1989; Willis and Mitsch, 1995; Wijte and Gallagher 1996a), and allelopathy have been shown to influence recruitment. Water, in particular, is a critical variable because most seeds require moist but not flooded conditions for germination and early seedling growth. As a result of this restrictive moisture requirement, it is common to find even-aged stands of trees at low elevations in riparian wetlands, reflecting seed germination during relatively uncommon years when water levels were unusually low during the spring and summer.

Postrecruitment processes play a major role in the distribution of adult plants at a site, leading to plant assemblages that cannot be predicted from the seed bank alone (Welling et al., 1988a; Wijte and Gallagher, 1996b). Thus, in coastal areas where the dominant plant species, *Spartina alterniflora*, occurs in large monotypic stands, it is often the pioneer species and remains dominant throughout the life of the marsh. In contrast, in tidal and nontidal freshwater marshes, the seed bank is much larger and richer, and the first species to invade a site may later be replaced by other species. For example, in one study primary succession on delta islands along the Louisiana coast is characterized by willows (*Salix* spp.) on the higher portions of the intertidal zone and arrowheads (*Sagittaria* spp.) on lower elevations (Shaffer et al., 1992). After 15 years, willows enriched with many additional understory species still dominated the high ground, whereas the arrowheads at intermediate elevations had been replaced by a rich mix of annual and perennial grasses and had died out at low elevations, leaving bare mud flats.

Models of Wetland Community Development

Functional Guild Model

Historically, although the community concept has been of immense value in ecology, it has been criticized for being imprecise and not subject to accurate predictive models for ecological communities. A number of ecologists have addressed this problem in different ways. One approach is to describe communities in terms of functional guilds that can be defined by measurable traits. A *guild* is defined as a group of functionally similar species in a community (Pianka, 1983). The guild concept has at least three important advantages over the generalized community concept: (1) It collapses the large number of species in a community to a manageable subset; (2) it defines guilds in terms of measurable functional properties; and (3) it enables the prediction of what guilds will be found given specific environmental conditions. The guild concept is not new. Its use is well established, for example, in the study of birds and mammals (Simberloff and Dayan, 1991). Boutin and Keddy (1993) illustrated a functional classification of wetland plants, and Keddy (1992a) described how a similar approach could be used to predict the plant species (or guilds) in a specific habitat. In the Boutin–Keddy study, 27 functional traits (Table 8-1) of 43 wetland plant species from a number of different environments in the northeastern United States were

Table 8-1   Traits measured on wetland plant species for functional guild classification

A. TRAITS MEASURED ON 1-YEAR-OLD PLANTS IN THE GARDEN

1	Life span:
	1 = annuals
	2 = facultative annuals (100% flowering)
	3 = partly facultative annuals (>50<100 % flowering)
	4 = perennials (<50% flowering)
2	Percentage flowering first year
3	Final height or highest height (cm)
4	Rate of shoot extension (cm/day): $\dfrac{\log_n \text{height at day 94} - \log_n \text{height at day 36}}{\text{Day 94} - \text{Day 36}}$
5	Total biomass at harvest (g)
6	Above-ground biomass (g)
7	Below-ground biomass (g)
8	Ratio below-ground/above-ground biomass
9	Photosynthetic area (cm^2); includes leaves and green stems
10	Photosynthetic area/total biomass (cm^2/g)
11	Photosynthetic area/total volume occupied by a plant (cm^2/ml) measured by displacement of water in graduated cylinder
12	Total biomass/total volume (g/ml)
13	Total number of tillers or shoots
14	Crown cover (cm^2): $((D_1 + D_2)/4)^2$
	where D_1 = first measure of crown diameter
	D_2 = second measurement at right angle to first
15	Stem diameter at ground level (cm)
16	Depth to below-ground system (cm)
17	Diameter of below-ground system, i.e., rhizome or main roots (cm)
18,19	Shortest (18) and longest (19) distances between two shoots or tillers (measure of degree of clumping of aerial stems) (cm)

B. TRAITS MEASURED ON PLANTS IN NATURAL WETLANDS (ADULT TRAITS)

20	Total height (cm)
21	Total number of tillers or shoots
22	Stem diameter at ground level (cm)
23, 24	Shortest (23) and longest (24) distances between two shoots or tillers (cm)
25	Diameter of below-ground system, i.e., rhizome or main roots (cm)
26	Depth to below-ground system (cm)

C. TRAIT MEASURED UNDER GREENHOUSE CONDITIONS

27	Relative growth rate (RGR) (day^{-1}) between days 10 and 30

Source: Boutin and Keddy (1993).

measured and the results interpreted by cluster analyses. Figure 8-6 summarizes the results, which groups the species according to their traits into seven guilds, ranging from fast-growing obligate annuals that flowered in the first year to clonal species with deep roots that spread vegetatively. This exercise reduced 43 individual species to 7 groups or guilds within which the species are functionally similar. Most of the guilds appear to fall along a continuum of life histories adjusted to different light regimes, which is consistent with the results of other studies.

The traits of these guilds can then be used to predict their presence in defined habitats. For example, Keddy (1992a) described assembly and response rules by which a community of plants can be predicted from a list of species and their traits and an environmental filter can be developed to delete those species that, because of their traits, will not be found in the community. The same procedure can be used to predict community species composition as a result of changes in the environmental filter. Figure 8-7 illustrates the procedure, which is explained with an example from van der Valk (1981). From a total pool of freshwater wetland species, those that cannot pass the first filter, for example, a shallow-water emergent environment, are deleted. If the marsh is later continually flooded, most of the adults will die from submergence or grazing by muskrats, as predicted from deletion rules based on flood tolerance of adults. New vegetation then arises from buried propagules based on the regeneration requirements of species (addition rules). The result is a new set of species at time $t + 1$.

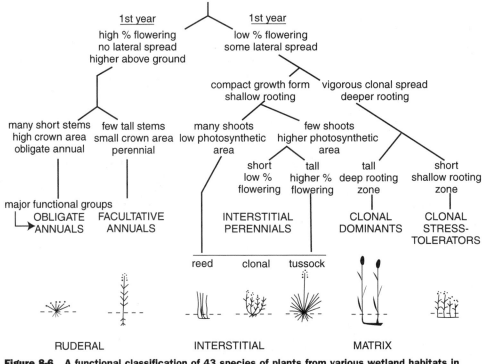

Figure 8-6 A functional classification of 43 species of plants from various wetland habitats in eastern North America, based on 27 plant traits displayed in Table 8-1. (*After Boutin and Keddy, 1993*)

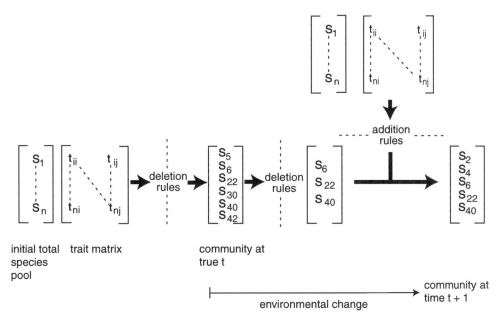

Figure 8-7 General procedure for response rules to delete and add species to a community after environmental perturbation. Species composition at time *t* is determined from an initial total list and trait matrix of all species, according to deletion rules specified by an environmental filter. Subsequent changes to the environmental filter result in deletion and/or addition to determine the plant composition at time *t* +1. See discussion in text. (*After Keddy, 1992a*)

Environmental Sieve Model

van der Valk's (1981) environmental sieve model, also a Gleasonian model (Fig. 8-8), is similar to Keddy's model in several ways. The presence and the abundance of each species depend on its life history and its adaptation to the environment of a site. In van der Valk's model, all plant species are classified into life-history types, based on potential life span, propagule longevity, and propagule establishment requirements. Each life-history type has a unique set of characteristics and, thus, potential behavior in response to controlling environmental factors such as water-level changes. These environmental factors comprise the "environmental sieve" in van der Valk's model. As the environment changes, so does the sieve and, hence, the species present. L. M. Smith and Kadlec (1985) tested the model's ability to predict species composition in a fresh marsh after a fire and were satisfied with the qualitative results.

Gap Dynamic Model

Chen and Twilley (1998) used a gap dynamic computerized model to simulate the growth and composition of a mangrove forest in south Florida (Fig. 8-9). This model is an individual-based model that tracks the growth of each tree in a forest gap of defined size, based on species-specific life-history traits and limitations of resource availability on the individual. This approach has been used extensively to model temperate and boreal forests and to simulate terrestrial forest dynamics (Shugart et al., 1992). In general, it is useful in forest studies because of the impossibility of obtaining long-term data in most sites.

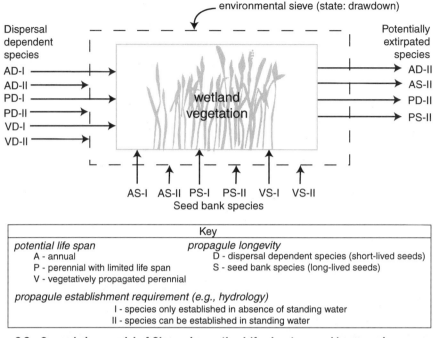

Figure 8-8 General sieve model of Gleasonian wetland (freshwater marsh) succession proposed by van der Valk. (*After van der Valk, 1981*)

Only three mangrove species occur in south Florida, although they can form diverse ecological types—riverine, fringe, basin, overwash, hammock, and dwarf mangroves (see Chapter 11). Simulations indicated that the initial conditions of the forest following disturbance, recruitment, and nutrient availability acted in combination to control mangrove regeneration. The model also predicted that *Laguncularia racemosa* (white mangrove) would dominate the early stages of forest development because it can resprout after trunk destruction and because it grows faster than the other two species in fertile and low-salinity conditions (Fig. 8-9a). However, in 500-year simulations, the early dominance of *L. racemosa* was replaced by *Avicennia germinans* (black mangrove), even when the recruitment rate of the latter was reduced to half that of the other species (Fig. 8-9b).

Centrifugal Organization Concept

A number of other models of community change have been developed, although few have been applied to wetlands. Grime (1979) proposed that changes in species composition and richness of herbaceous plants was related to the gradients of disturbance and stress factors, which reduced biomass and determined which functional plant strategies would work best. Tilman (1982) suggested that competition among plants controlled community plant distribution, with each species limited by a different ratio of resources and spatial heterogeneity of the resources.

Wisheu and Keddy (1992) combined aspects of both Grime's and Tilman's models to propose a model of centrifugal organization of plant communities (Fig. 8-10a). Centrifugal organization describes the distribution of species and vegetation types

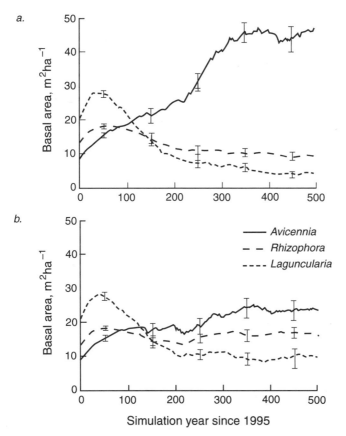

Figure 8-9 Projected basal area of three mangrove species at the lower Shark River estuary, Florida, simulated with a model under different recruitment rates: **a.** equal recruitment rate for the three mangrove species; **b.** low black mangrove (*Avicennia germinans*) recruitment rate. (*After Chen and Twilley, 1998*)

along standing-crop gradients caused by combinations of environmental constraints. Wisheu and Keddy (1992) summarize the concept as follows:

> Gradients radiate outwards from a single core habitat to many different peripheral habitats. The assumed mechanism is a competitive hierarchy where weaker competitors are restricted to the peripheral end of the gradient as a result of a trade-off between competitive ability and tolerance limits. The benign ends of the gradients comprise a core habitat which is dominated by the same species. At the peripheral end of each axis, species with specific adaptations to particular sources of adversity occur.

Wisheu and Keddy propose that the core habitat in wetlands has low disturbance and high fertility and is dominated by species that form dense canopies such as *Typha* in the eastern United States (Fig. 8-10b). Peripheral habitats represent different kinds and combinations of stresses (infertility, disturbance) and support distinctive plant associations. The model allows one to predict how changes in gradients and, hence, peripheral habitats will change community composition. In the case of the *Typha*-core centrifugal model shown in Figure 8-10b, ice scouring, infertile sandy soils,

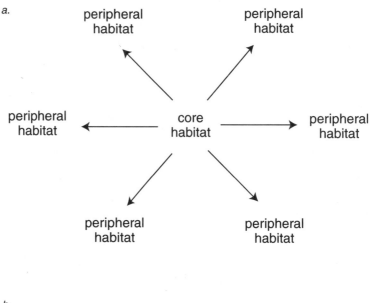

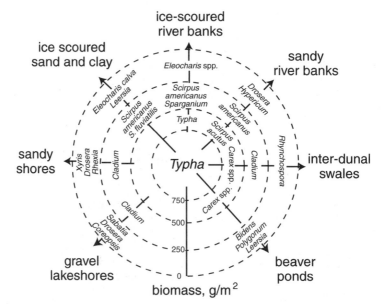

Figure 8-10 Centrifugal organization models illustrating **a.** transitions from core habitat to peripheral habitats along resource or stress gradients (general model), and **b.** freshwater wetland pattern for eastern North America, where large, leafy species such as cattail (*Typha* spp.) occupy the core habitat, while several different species and communities occupy peripheral habitats stressed by infertile sand, ice scouring, and beaver activity. (*After Wisheu and Keddy, 1992*)

flooding by beavers, and open shorelines are among the stresses that shift communities to less productive, albeit possibly more diverse, assemblages.

Ecosystem-Level Processes

So far in this chapter, we have discussed vegetational changes in wetlands. We summarize this discussion with a statement by Niering (1989):

> Traditional successional concepts have limited usefulness when applied to wetland dynamics. Wetlands typically remain wet over time exhibiting a wetland aspect rather than succeeding to upland vegetation. Changes that occur may not necessarily be directional or orderly and are often not predictable on the long term. Fluctuating hydrologic conditions are the major factor controlling the vegetation pattern. The role of allogenic factors, including chance and coincidence, must be given new emphasis. Cyclic changes should be expected as water levels fluctuate. Catastrophic events such as floods and droughts also play a significant role in both modifying yet perpetuating these systems.

E. P. Odum (1969) described the maturation of ecosystems as a whole (as distinct from communities and species) in an article entitled "The Strategy of Ecosystem Development." Immature ecosystems, he observed, are characterized, in general, by high production to biomass ($P : B$) ratios; an excess of production over community respiration ($P : R$ ratio > 1); simple, linear, grazing food chains; low species diversity; small organisms; simple life cycles; and open mineral cycles. In contrast, mature ecosystems such as old-growth forests tend to use all their production to maintain themselves and, therefore, have $P : R$ ratios about equal to one and little, if any, net community production. Production may be lower than in immature systems, but the quality is better; that is, plant production tends to be high in fruits, flowers, tubers, and other materials that are rich in protein. Because of the large structural biomass of trees, the $P : B$ ratio is small. Food chains are elaborate and detritus based, species diversity is high, the space is well organized into many different niches, organisms are larger than in immature systems, and life cycles tend to be long and complex. Nutrient cycles are closed; nutrients are efficiently stored and recycled within the ecosystem.

It is instructive to see how wetland ecosystems fit into this scheme of ecosystem development. Do their ecosystem-level characteristics fit the classical view that all wetlands are immature transitional seres? Or do they resemble the mature features of a terrestrial forest? For the wetland ecosystems covered in this book, Table 8-2 displays an evaluation of the system attributes discussed by E. P. Odum (1969). For comparison, we have included a generalized developing (immature) and a mature ecosystem from Odum's article. The quantitative values are very rough because they represent means that reflect wide variation and were derived from incomplete data presented in the ecosystem chapters of this book. Nevertheless, the table provides the following interesting insights:

1. Wetland ecosystems have properties of both immature and mature ecosystems. For example, nearly all of the nonforested wetlands have $P : B$ ratios intermediate between developing and mature systems and $P : R$ ratios greater than one. Primary production tends to be very high compared with most terrestrial ecosystems. These attributes are characteristic of immature ecosystems. On the other hand, all of the ecosystems are detrital based, with complex food webs characteristic of mature systems.

Table 8-2 Ecosystem attributes of wetlands compared with E. P. Odum's[a] successional attributes

	Community Energetics				Community Structure		
Ecosystem Type	P : R Ratio	P : B Ratio	Net Primary Productivity (g C m^{-2} day^{-1})	Food Chains	Total Organic Matter (kg/m^2)	Species Diversity	Spatial Heterogeneity
Developing	<1 or >1	High (2–5)	High (~2–3)	Linear, grazing	Small (<2)	Low	Poorly organized
Mature	1	Low (<0.1)	Low (~1)	Weblike, detritus	Large (~20)	High	Well organized
INLAND WETLANDS							
Northern peatlands and bogs	>1	0.1	0.8 (0.2–1.4)	Weblike, detritus	7.8 (1.2–16)	Low	Well organized
Freshwater marshes	>1	1.2	3.9 (0.7–8.2)	Weblike, detritus	0.75–2.3	High	Well organized
Swamp forests	1.3 (1.1–1.5)	0.07 (0.015–0.09)	1.2 (0.5–1.9)	Weblike, detritus	22.6 (7.4–4.5)	Fairly low	Well organized
Riparian forests	≥1	0.06	1.4	Weblike, detritus	17.4 (10–29)	High	Well organized
COASTAL WETLANDS							
Salt marshes	1.5	2	2.2 (0.45–5.7)	Weblike, detritus	1.1	Low	Well organized
Tidal freshwater marshes	>1	1.2	1.9	Weblike, detritus	1.1 (0.4–2.3)	Fairly low	Well organized
Mangroves	1.9 (0.7–3.3)	—	3 (0–7.5)	Weblike, detritus	11 (1–29)	Plants: low Animals: high	Well organized

Ecosystem Type	Life History		Nutrient Cycles		Selection Pressure	
	Organism Size	Life Cycle	Mineral Cycles	Role of Detritus	Growth Form	Production
Developing[a]	Small	Short, simple	Open	Unimportant	r	Quantity
Mature[a]	Large	Long, complex	Closed	Important	K	Quality
INLAND WETLANDS						
Northern peatlands and bogs	Small to large	Long	Closed	Important	K	Quality?
Freshwater marshes	Fairly small	Short, complex	Closed	Important	K?	Quality
Swamp forests	Plants: large Animals: small	Long, simple Short	Open	Important	Plants: K Animals: r	Quantity
Riparian forests	Plants: large Animals: small to large	Long Short to long	Open to closed	Important	K	Quality
COASTAL WETLANDS						
Salt marshes	Small	Short, complex	Open	Important	r	Quantity
Tidal freshwater marshes	Small	Short, complex	Open	Important	r?	Quantity
Mangroves	Plants: large Animals: small	Long, simple Short, complex	Open	Important	Plants: K Animals: r?	Quantity

[a] E. P. Odum (1969, 1971).

2. E. P. Odum (1971) used live biomass as an index of structure or "information" within an ecosystem. Hence, a forested ecosystem is more mature in this respect than a grassland. This relationship is reflected in the high $P : B$ ratios (immature) of nonforested wetlands and the low $P : B$ ratios (mature) of forested wetlands. In a real sense, however, peat is a structural element of wetlands because it is a primary autogenic factor modifying the flooding characteristic of a wetland site. If peat were included in biomass, herbaceous wetlands would have the high biomass and low $P : B$ ratios characteristic of more mature ecosystems. For example, a salt or fresh marsh has a live peak biomass of less than 2 kg/m². However, the organic content of a meter depth of peat (peats are often many meters deep) beneath the surface is on the order of 45 kg/m². This is comparable to the above-ground biomass of the most dense wetland or terrestrial forest. As a structural attribute of a marsh, peat is an indication of a maturity far greater than the live biomass alone would signify.

3. Mineral cycles vary widely in wetlands, ranging from extremely open riparian systems in which surface water (and nutrients) may be replaced thousands of times each year to bogs in which nutrients are derived from precipitation alone and are almost quantitatively retained. An open nutrient cycle is a juvenile characteristic of wetlands, directly related to the large flux of water through these ecosystems. On the other hand, even in a system as open as a salt marsh that is flooded daily, about 80 percent of the nitrogen used by vegetation during a year is recycled from mineralized organic material (Delaune and Patrick, 1979).

4. Spatial heterogeneity is generally well organized in wetlands along allogenic gradients. The sharp, predictable zonation patterns and abundance of land–water interfaces are examples of this spatial organization. In forested wetlands, vertical heterogeneity is also well organized. This organization is an index of mature ecosystems. In most terrestrial ecosystems, however, the organization results from autogenic factors in ecosystem maturation. In wetlands, most of the organization seems to result from allogenic processes, specifically hydrologic and salinity gradients created by slight elevation changes across a wetland. Thus, the "maturity" of a wetland's spatial organization consists of a high level of adaptation to prevailing microhabitat differences.

5. Life cycles of wetland consumers are usually relatively short but are often exceedingly complex. The short cycle is characteristic of immature systems, although the complexity is a mature attribute. Once again, the complexity of the life cycles of many wetland animals seems to be as much an adaptation to the physical pattern of the environment as to the biotic forces. A number of animals use wetlands only seasonally or only during certain life stages. For example, small marsh fish and shellfish make daily excursions into wetlands during high tides, retiring to adjacent ponds during ebb tides. Many fish and shellfish species migrate from the ocean to coastal wetlands to spawn or for use as a nursery. Waterfowl use northern wetlands to nest and southern wetlands to overwinter, migrating thousands of miles between the two areas each year.

The Strategy of Wetland Ecosystem Development

In the previous sections, we showed that wetlands possess attributes of both immature and mature systems and that both allogenic and autogenic processes are important. In this section, we suggest that in all wetland ecosystems there is a common theme: Development insulates the ecosystem from its environment. At the level of individual species, this occurs through genetic (structural and physiological) adaptations to anoxic sediments and salt (see Chapter 7). At the ecosystem level, it occurs primarily through peat production, which tends to stabilize the flooding regime and shifts the main source of nutrients to recycled material within the ecosystem. In forests, shading is important in regeneration following disturbance.

Turnover Rates and Productivity

The intensity of water flow over and through a wetland can be described by the water renewal rate (t^{-1}), the ratio of throughflow to the volume stored on the site (see Chapter 5). In wetlands, t^{-1} varies by five orders of magnitude (Fig. 8-11), ranging from about one per year in northern bogs to as much as 7,500 per year in low-lying riparian forests. The nutrient input (except nitrogen fixation) to a site follows closely the water renewal rate because nutrients are carried to a site by water. The amount of nitrogen delivered to a wetland site, for example, also varies by five orders of magnitude, ranging from less than 1 g m^{-2} yr^{-1} in a northern bog to perhaps 10,000 g m^{-2} yr^{-1} in a riparian forest (Fig. 8-11 and Table 8-3). Not all of this nitrogen is available to plants in the ecosystem because, in extreme cases, it is flowing through much faster than it can be immobilized, but these figures indicate the potential nutrient supply to the ecosystem.

In spite of the extreme variability in these outside (allogenic) forces, wetland ecosystems are remarkably similar in many respects. Total stored biomass, including peat to 1 m depth, ranges from 40 to 60 kg m^{-2}—less than twofold. Soil nitrogen similarly varies only about threefold, from about 500 to 1,500 g m^{-2}.

Net primary production, a key index of ecosystem function, varies by only a factor of about four. Mean values for different ecosystem types are usually in the range of 600 to 2,000 g m^{-2} yr^{-1} (Table 8-3). Thus, although it has been shown in a number of studies of individual species (e.g., *Spartina alterniflora;* Steever et al., 1976) or ecosystems (e.g., cypress swamps; Conner et al., 1981) that production is directly proportional to the water renewal rate, when different wetland ecosystems that constitute greatly different water regimes are compared, the relationship breaks down. The apparent contradiction may be explained primarily by the role of stored nutrients, especially nitrogen, within the ecosystem. As the large store of organic nitrogen in the sediment (Table 8-3) mineralizes, it provides a steady source of inorganic nitrogen for plant growth. In most wetland ecosystems, "new" nitrogen is not adequate or is barely adequate to supply the plants' demands but the demands are small in comparison to the amounts of stored nitrogen in the sediments (Table 8-3). As a result, most of the nitrogen demand is satisfied by recycling, even in systems as open as salt marshes (Delaune and Patrick, 1979). External nitrogen provides a subsidy to this basic supply. Therefore, growth is often apparently limited by the mineralization rate, which, in turn, is strongly temperature and hydroperiod dependent. Temperatures during the growing season are uniform enough to provide a similar nitrogen supply to plants in different wetland systems, except probably in northern bogs. There, the low temperature and short growing season limit mineralization and restrict nutrient input. The combination of the two factors limits productivity.

Thus, as wetland ecosystems develop, they become increasingly insulated from the variability of the environment by storing nutrients. Often, the same process that stores nutrients, that is, peat accumulation, also reduces the variability of flooding, further stabilizing the system. The surface of marshes, in general, is built up by the deposition of peats and waterborne inorganic sediments. As the elevation increases,

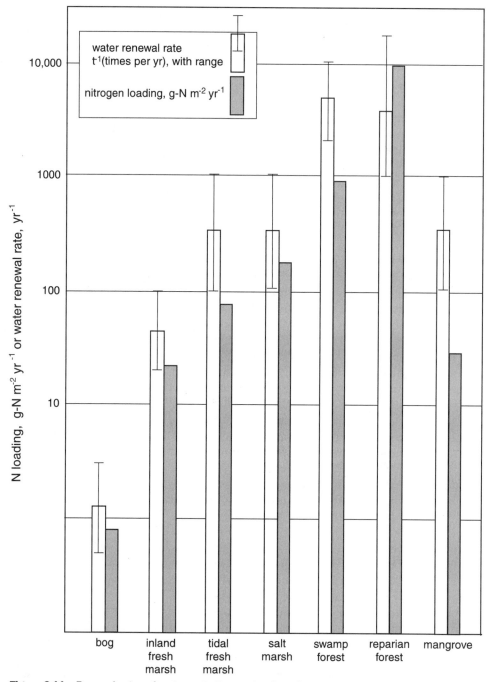

Figure 8-11 Renewal rates of water and nitrogen loading of major wetland types discussed in this book in the ecosystem chapters (Chapters 9–15).

Table 8-3 Comparison of primary productivity and nitrogen dynamics in major wetland types

Wetland Type[a]	Net Primary Production ($g\ m^{-2}\ yr^{-1}$) (range)	Total Biomass (kg/m^2)	Soil Nitrogen ($g\ N\ m^{-2}\ yr^{-1}$)	Nitrogen Loading ($g\ N\ m^{-2}\ yr^{-1}$)	Plant Nitrogen Uptake ($g\ N\ m^{-2}\ yr^{-1}$)	Ratio N Throughput Soil Store (yr^{-1}) (col 4/col 3)	Ratio N Throughput Uptake (col 4/col 5)
Northern bog	560 (153–1,943)	53	500	0.8	9	0.002	0.09
Inland fresh marsh	1980 (1,070–2,860)	46	1,600	22	48	0.01	0.46
Tidal fresh marsh	1,370 (780–2,100)	46	1,340	75	54	0.06	1.4
Salt marsh	1,950 (330–3,700)	46	1,470	30–100	25	0.02–0.07	1.2–4.0
Swamp forest	870 (390–1,780)	52	1,300	900	14	0.7	64
Riparian forest	1,040 (750–1,370)	37	900	10,000	17	11	600
Mangrove	1,500 (0–4,700)	60	1,400	30	24	0.02	1.2
Range, all wetlands	(560–1,980)	37–60	500–1,470	0.8–10.000	9–54	0.0002–11	0.09–600

[a] Values are rough averages with large variability, based on data presented in Chapters 9–15.

flooding becomes less frequent and sediment input decreases. In the absence of overriding factors, coastal wetland marshes in time reach a stable elevation somewhere around local mean high water. The surfaces of riparian wetlands similarly rise until they become only infrequently flooded. Northern bogs grow by peat deposition above the water table, stabilizing at an elevation that maintains saturated peat by capillarity. Prairie potholes may be exceptions to these generalizations. They appear to be periodically "reset" by a combination of herbivore activity and long-term precipitation cycles and to achieve stability only in some cyclic sense.

Pulse Stability

In contrast to the lack of evidence for community succession to a stable set of species, which was summarized earlier in a statement from Niering (1989), perhaps the concept of a progression toward a mature ecosystem (E. P. Odum, 1969) has greater merit. The attributes of a mature stable ecosystem place it in dynamic equilibrium with its environment, and although individual species may come and go, a mature ecosystem is stable in the sense that it has built-in mechanisms (species diversity, nutrient storage, and recycling) that resist short-term environmental fluctuations. In fact, E. P. Odum et al. (1995) suggest that natural processes pulse regularly and the mature ecosystem responds in a pulsing steady state. In wetlands, examples of this phenomenon include salt marshes, tidal freshwater marshes, riverine forests, and seasonally flooded freshwater marshes, all of which are functionally similar despite marked differences in species composition, diversity, and community structure. Odum et al. (1995) suggest that natural pulses such as tides pump energy into ecosystems and enhance productivity. Biotic events are geared to and take advantage of these pulses, for example, the influx of small fish into flooded marshes to feed during high tide or the capturing of young fish in backwater oxbows and billabongs during flooding, with the captured fish serving as food for wading birds during periods of low water. This concept is referred to as *pulse stability.*

Self-Organization and Self-Design

Self-design (Mitsch and Wilson, 1996; Metzker and Mitsch, 1997; Mitsch et al., 1998) and the related concept of self-organization (H. T. Odum, 1989) are important concepts in wetland ecosystem development. Most wetland ecosystems are continually open to atmospheric, hydrologic, and biotic inputs of propagules of plants, animals, and microbes. *Self-organization,* as discussed by Odum (1989), manifests itself in both microcosms and newly created ecosystems "showing that after the first period of competitive colonization, the species prevailing are those that reinforce other species through nutrient cycles, aids to reproduction, control of spatial diversity, population regulation, and other means" (Odum, 1989). *Self-design* relies on the self-organizing ability of ecosystems; natural processes (wind, rivers, tides, biotic inputs, etc.) contribute to species introduction; selection of those species that will dominate from this gene inflow is then nature's manifestation of ecosystem design. In self-design, the presence and survival of species due to the continuous introduction of them and their propagules is the essence of the successional and functional development of an ecosystem. This can be thought of as analogous to the continuous production of mutations necessary for evolution to proceed. In the context of ecosystem restoration and creation, self-design means that, if an ecosystem is open to allow "seeding," through human or natural means, of enough species' propagules, the system itself will optimize its design by selecting for the assemblage of plants, microbes, and animals that is best adapted to the existing conditions. It is an important process to be investigated, particularly in view of the interest in restoring and creating wetlands (see Chapter 19). In contrast to the self-design approach, the wetland restoration approach that is more commonly used today involves the introduction of organisms (often plants), the survival of which becomes the measure of success of the restoration. This has sometimes been referred to as the "designer wetland" approach (Mitsch, 1998c; van der Valk, 1998). This latter approach, while understandable because of the natural human tendency to control events, may be less sustainable than an approach that relies more on nature.

In a multiyear, whole-ecosystem experiment in newly constructed freshwater marsh basins in central Ohio, Mitsch et al. (1998) describe how 2,500 individuals of 12 plant species were introduced to one wetland basin while the other remained an unplanted control, essentially testing the self-design capabilities of nature with and without human intervention. Both basins had identical inflows of river water and hydroperiods. For the first 3 years of the wetland experiment, 17 different biotic and abiotic functional indicators of wetland function were measured (Table 8-4). After

Table 8-4 Indicators of ecosystem function for first 3 years of two newly constructed 1-ha experimental freshwater marsh basins at the Olentangy River Wetland Research Park, Columbus, OH. Wetland 1 (W1) was planted at the beginning of year 1; Wetland 2 (W2) remained as an unplanted control

Indicator of Ecosystem Function	Year 1	Year 2	Year 3
Hydrology		Essentially identical[a]	
Geomorphology		Essentially identical[a]	
Macrophytes			
1. Percentage cover	W1 = W2 ~ 0	W1 > W2	W1 = W2
2. Species richness	W1 > W2	W1 > W2	W1 > W2
Microphytes			
3. Species richness	W1 = W2	W1 > W2	W1 > W2
4. Aquatic metabolism	W1 = W2	W1 < W2	W1 = W2
Water quality changes			
5. Temperature	W1 > W2	W1 < W2	W1 < W2
6. Turbidity	W1 = W2	W1 < W2	W1 = W2
7. Dissolved oxygen	W1 < W2	W1 < W2	W1 = W2
8. pH	W1 < W2	W1 < W2	W1 = W2
9. Conductivity	W1 = W2	W1 > W2	W1 = W2
10. Redox	W1 = W2	W1 > W2	W1 > W2
Nutrient retention			
11. Total phosphorus	W1 < W2	W1 = W2	W1 = W2
12. Soluble reactive phosphorus	W1 = W2	W1 > W2	W1 = W2
13. Nitrate + nitrite	W1 = W2	W1 = W2	W1 > W2
Benthic invertebrate diversity			
14. Shannon–Weaver index	W1 = W2	W1 < W2	W1 = W2
15. "Clean water" species richness	W1 < W2	W1 > W2	W1 = W2
Bird use			
16. Abundance	W1 = W2	W1 > W2	W1 = W2
17. Species richness	W1 = W2	W1 > W2	W1 = W2
Functional equality (%)[b]	65	12	71

[a] Basins were essentially identical in morphology and hydrology by design.
[b] Percentage of indicators that are similar or identical in both wetland basins.
Source: Mitsch et al. (1998).

only 3 years, there was convergence of wetland function of the planted and unplanted basins with 71 percent of the functional indicators essentially the same in the two basins. This convergence in year 3 followed the second year where only 12 percent of the indicators were similar. Most important, hundreds of taxa, both aquatic and terrestrial, were continually introduced to these wetland basins, primarily because the wetland basins were hydrologically open systems, and many of them survived. In 3 years, over 50 species of macrophytes, 130 genera of algae, over 30 taxa of aquatic invertebrates, and dozens of bird species found their way naturally to the wetlands. This continual introduction of species, whether introduced through flooding and other abiotic and biotic pathways, appeared to have a much longer-lasting effect in development of these ecosystems than the few species of plants that were introduced to one of the wetlands in the beginning.

Ecosystem Engineers

A new way to describe the importance of autogenic successional processes involves the recently introduced term *ecosystem engineer*—a term that is used to describe organisms that have dramatic and important effects on an ecosystem (Jones et al., 1997; Alper, 1998). [This concept, also discussed in Chapter 5, should not be confused with the newly developing field of ecological engineering (Mitsch, 1993, 1998a).] In wetlands, examples of ecosystem engineers could be muskrats and beavers, both of which can have dramatic effects on vegetation cover and ecosystem hydrology in freshwater marshes. In these cases, the biota do show a dramatic feedback to the many features of the wetland (e.g., water levels or vegetation productivity). One could argue that these ecosystem engineers "set back" succession to an earlier stage; an alternative argument is that they are part of the ecosystem development and that their behavior and its effects should be both expected and appreciated as normal ecosystem behavior.

Landscape Patterns

Many large wetland landscapes develop predictable and often complex patterns of aquatic, wetland, and terrestrial habitats or ecosystems. In high-energy environments, these patterns appear to reflect abiotic forces, but they are largely controlled by biotic processes in low-energy environments. At the high-energy end of the spectrum, the microtopography and sediment characteristics of mature floodplains—complex mosaics of river channels, natural levees, back swamps, abandoned first and second terrace flats, and upland ridges—reflect the flooding pattern of the adjacent river (Gosselink et al., 1990a,b; see Chapter 15). The vegetation responds to the physical topography and sediments with typical zonation patterns. Salt marshes similarly develop a characteristic pattern of tidal creeks, creekside levees, and interior flats that determine the zonation pattern and vigor of the vegetation (Fig. 8-12a; see also Chapter 9).

At the low-energy end of the spectrum, the characteristic pattern of strings and flarks stretching for miles across northern peatlands appears to be primarily controlled by biotic processes (Glaser, 1987; Rochefort et al., 1990; see Chapter 13). Similarly, in many freshwater marshes, herbivores can be major actors in the development of landscape patterns (Fig. 8-12b). In actuality, both physical (climatic, topographic, hydrologic) and biotic (production rates, root binding, herbivory, peat accumulation) processes combine in varying proportions and interact to produce observed wetland landscape patterns.

Figure 8-12 Landscape patterns in wetlands: **a.** the physically controlled pattern of tidal creeks in a Louisiana salt marsh; **b.** a muskrat ''eat-out'' in a brackish marsh on the Louisiana coast. Note the high density of muskrat houses.

Recommended Readings

Grime, H. H. 1979. Plant Strategies and Vegetation Processes. John Wiley & Sons, New York.

Keddy, P. A. 1992. Assembly and response rules: two goals for predictive community ecology. Journal of Vegetation Science 3:157–164.

Odum, E. P. 1969. The strategy of ecosystem development. Science 164:262–270.

Tilman, D. 1988. Plant Strategies and the Dynamics and Structure of Plant Communities. Princeton University Press, Princeton, NJ. 376 pp.

van der Valk, A. G. 1981. Succession in wetlands: A Gleasonian approach. Ecology 62:688–696.

Part **3**

Coastal Wetland Ecosystems

C h a p t e r **9**

Tidal Salt Marshes

The salt marsh, distributed worldwide along coastlines in middle and high latitudes, is a complex ecosystem in dynamic balance with its surroundings. These marshes flourish wherever the accumulation of sediments is equal to or greater than the rate of land subsidence and where there is adequate protection from destructive waves and storms. The important physical and chemical variables that determine the structure and function of the salt marsh include tidal flooding frequency and duration, soil salinity, soil permeability, and nutrient limitation, particularly by nitrogen. The vegetation of the salt marsh, primarily salt-tolerant grasses and rushes, develops in identifiable zones in response to these and possibly other factors. Mud and epiphytic algae are also often an important component of the autotrophic community. The heterotrophic communities are dominated by detrital food chains, with the grazing food chain being much less trophically significant.

Salt marshes are among the most productive ecosystems in the world. Many measurements of the net primary productivity of salt marshes suggest that regional differences are related to available solar energy and, to some extent, to available nutrient imports by large rivers. Below-ground production and anaerobic decomposition are major processes in the overall energy balance. Sulfur transformations in this environment assume much of the role of oxygen, so that a major portion of the below-ground flow of organic energy cycles through reduced sulfur compounds. The decomposition of dead vegetation in salt marshes, at or near the surface, is carried out by fungi and bacteria, which enhance the protein content of the detrital mixture for other marsh estuarine organisms. Salt marshes are outwelling systems that export organic energy to the estuary and coastal ocean. The pathway of energy outwelling is complex, and may occur passively through coastal currents, by the migration of marine transient species, and by trophic relays of food energy sources originating on the marsh. Nutrients are exported in organic matter, but when inorganic nutrients are included in the budget, nutrient cycle measurements show few consistent patterns.

Beeftink (1977a) defined a salt marsh as a "natural or semi-natural halophytic grassland and dwarf brushwood on the alluvial sediments bordering saline water bodies whose water level fluctuates either tidally or non-tidally." Salt marshes, dominated by rooted vegetation that is alternately inundated and dewatered by the rise and fall of the tide, appear from afar to be vast fields of grass of a single species. In reality, salt marshes have a complex zonation and structure of plants, animals, and microbes, all tuned to the stresses of salinity fluctuations, alternate drying and submergence, and extreme daily and seasonal temperature variations. A maze of tidal creeks with plankton, fish, nutrients, and fluctuating water levels crisscrosses the marsh, forming conduits for energy and material exchange with the adjacent estuary. Studies of a number of different salt marshes have found them to be highly productive and to support the spawning and feeding habits of many marine organisms. Thus, salt marshes and tropical mangrove swamps throughout the world form an important interface between terrestrial and marine habitats.

Salt marshes have been studied extensively. For many years, the standard text on salt marshes was *Salt Marshes and Salt Deserts of the World* written by V. J. Chapman (1960). The ecosystem approach to wetlands was pioneered by scientists at the University of Georgia Marine Laboratory on Sapelo Island, Georgia, and early interest in this ecosystem was aroused by a symposium at that laboratory in 1958 (Ragotzkie et al., 1959) and continued by many researchers (see, e.g., Pomeroy and Wiegert, 1981). The proceedings of a major symposium on salt marshes, commemorating the 40th anniversary of the Sapelo Island symposium, has been published (Weinstein and Kreeger, 2000). A less technical early description of salt marshes was written by Teal and Teal (1969). A number of thorough reviews of regional coastal saline marshes were commissioned by the U.S. Fish and Wildlife Service: Northeast Atlantic Coast marshes (Teal, 1986), New England high salt marshes (Nixon, 1982), Southeast Atlantic Coast marshes (Wiegert and Freeman, 1990), Northeast Gulf Coast irregularly flooded salt marshes (Stout, 1984), Gulf Coast deltaic marshes (Gosselink, 1984), southern California coastal salt marshes (Zedler, 1982a), and San Francisco Bay tidal marshes (Josselyn, 1983). Additional useful summaries include Florida coastal wetlands (Montague and Wiegert, 1990), New England salt marshes (Bertness, 1992), and a book by Chabreck (1988). Descriptions of European salt marshes are provided by Ranwell (1972), Beeftink (1977a, b), Adam (1990), Allen and Pye (1992), and LeFeuvre and Dame (1994), whereas reviews of Australian salt marshes are provided by Bridgewater and Cresswell (1993), Cresswell and Bridgewater (1998), Latchford (1998), and Adam (1994, 1998).

Geographical Extent

Salt marshes are found in the middle and high latitudes along intertidal shores throughout the world (Fig. 9-1a). They are replaced by mangrove swamps along coastlines in tropical and subtropical regions (between 25° N and 25° S latitude; see Chapter 11). The distribution of salt marshes in United States is shown in Figure 9-1b.

Salt marshes can be narrow fringes on steep shorelines or expanses of several kilometers wide. They are found near river mouths, in bays, on protected coastal plains, and around protected lagoons. Different plant associations dominate different coastlines, but the ecological structure and function of salt marshes is similar around the world. Based on the work of Chapman, the world's salt marshes can be divided into the following major geographical groups:

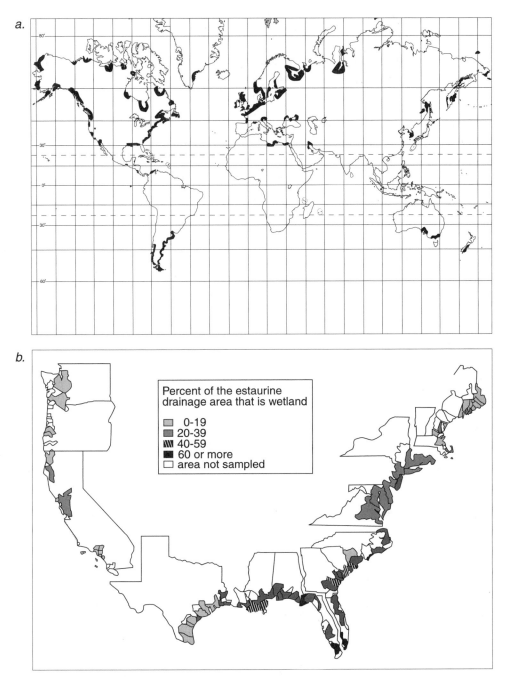

Figure 9-1 Distribution of **a.** salt marshes of the world (*from Chapman, 1977*) and **b.** wetlands in coastal drainage areas of the United States (*from Field et al., 1991*). Figure 9-1b includes freshwater tidal marshes and mangroves, as well as tidal salt marshes.

1. *Arctic.* This group includes marshes of northern Canada, Alaska, Greenland, Iceland, northern Scandinavia, and Russia. Probably the largest extent of marshes in North America, as much as 300,000 km^2, occurs along the southern shore of the Hudson Bay. These marshes, influenced by ice, extreme low temperatures, a positive water balance, and numerous inflowing streams, can generally be characterized as brackish rather than saline (Ewing and Kershaw, 1986). Various species of the sedge *Carex* and the grass *Puccinellia phryganodes* often dominate. Parts of the southwestern coast of Alaska are dominated by species of *Salicornia* and *Suaeda*.

2. *Northern Europe.* This group includes marshes along the west coast of Europe from the Iberian Peninsula to Scandinavia, including Great Britain and the Baltic Sea coast. Most of the western European coastal environment is characterized by a moderate climate with sufficient precipitation but high salinities toward the southern extremes. Dominant species include *Puccinellia maritima, Juncus gerardi, Salicornia* spp., and *Spartina anglica* and *S. townsendii.* The west coast of Great Britain and parts of the Scandinavian and Baltic Sea coasts, where substrates are dominated by sand and salinities are low, are populated by *Festuca rubra, Agrostis stolonifera, Carex paleacea, Juncus bufonius, Desmoschoenus bottanica,* and *Scirpus* spp. The muddy coast of the English Channel is dominated by *Spartina townsendii.* Salt marshes in northern Europe are often characterized by a lack of vegetation in the intertidal zone, in contrast to North American marshes.

3. *Mediterranean.* This group includes the arid, rocky-to-sandy, high-salinity coasts of the Mediterranean Sea. The salt marshes are dominated by low shrubby vegetation, *Arthrocnemum, Limonium, Juncus* spp., and the halophyte *Salicornia* spp.

4. *Eastern North America.* These marshes, mostly dominated by *Spartina* and *Juncus* species, are found along the eastern coasts of the United States and Canada and the Gulf Coast of the United States. This group is further divided into three subgroups:

a. *Bay of Fundy.* River and tidal erosion is high in the soft rocks of this region, producing an abundance of reddish silt. The tidal range, as exemplified at the Bay of Fundy, is large, leading to a few marshes in protected areas and considerable depth of deposited sediments. *Puccinellia americana* dominates the lower marsh, and *Juncus balticus* is found on the highest levels.

b. *New England.* Marshes are built mainly on marine sediments and marsh peat, and there is little transport of sediment from the hard-rock uplands. These marshes range from Maine to New Jersey and are dominated by *Spartina alterniflora* in the low marsh with *S. patens,* mixed with *Distichlis spicata* in the high marsh.

c. *Coastal Plain.* These marshes extend southward from New Jersey along the southeastern coast of the United States to Texas along the Gulf of Mexico. Major rivers supply an abundance of silt from the recently elevated Coastal Plain. The tidal range is relatively small. The marshes are laced with tidal creeks. Mangrove swamps replace salt marshes along the southern tip of Florida. Because of the extensive delta marshes built by the Mississippi River, the Gulf Coast contains about 60 percent of the coastal salt and fresh marshes of the United States (Fig. 9-1b; Field et al., 1991). Dominant species are *Spartina alterniflora, S. patens, Juncus roemerianus,* and *Distichlis spicata.*

5. *Western North America.* Compared with the Arctic and the eastern coast of North America, salt marshes are far less developed along the western coasts of the United States and Canada because of the geomorphology of the coastline. On this

rugged coast with its Mediterranean-type climate is found a narrow belt of *Spartina foliosa,* often bordered by broad belts of *Salicornia* and *Suaeda. Spartina alterniflora* is a nonnative invasive plant in coastal marshes north of California.

6. *Australasia.* Salt marshes are frequently found in river deltas along the temperate coastlines of eastern Asia, Australia, and New Zealand on the Pacific Ocean, Indian Ocean, and Tasman Sea.

a. *Eastern Asia.* The coasts of China, Japan, Russia, and Korea are generally rugged and uplifted, with moderate precipitation but limited marsh development. These marshes are dominated by *Triglochin maritima, Limonium japonicum, Salicornia,* and *Zoysia macrostachya.* Major areas of salt marsh restoration have occurred on China's eastern coastline, owing to the introduction of *Spartina anglica* and *S. alterniflora.*

b. *Australia.* This group also includes New Zealand and Tasmania. It is characterized by high rainfall and geographic isolation. Cosmopolitan species in Australian salt marshes include *Sporobolus virginicus, Sarcocornia quinqueflora,* and *Suaeda australis.* However, invasion of salt marshes of Australia and New Zealand by several species is common. In New Zealand and other temperate-zone salt marshes of the region, *Spartina anglica* is a major invasive species. Even with less rainfall and a clearly defined seasonal pattern of wet and dry on the western coast of Australia, salt marshes can be found, particularly around Shark Bay and the Peel–Harvey estuaries. Unlike the general case around much of the world, a majority of the salt marshes of Australia are found in tropical regions (Adam, 1998).

7. *South America.* South American coasts too far south and too cold for mangroves are rugged and geographically isolated. They are dominated by unique species of *Spartina, Limonium, Distichlis, Juncus, Heterostachys,* and *Allenrolfea.*

8. *Tropics.* Although mangroves generally dominate tropical coastlines, salt marshes are found in the Tropics on high-salinity flats that mangroves cannot tolerate. *Spartina* spp. and the halophytic genera *Salicornia* and *Limonium* often dominate.

Geomorphology

The physical features of tides, sediments, freshwater inputs, and shoreline structure determine the development and extent of salt marsh wetlands within their geographical range. Coastal salt marshes are predominantly intertidal; that is, they are found in areas at least occasionally inundated by high tide but not flooded during low tide. A gentle, rather than steep, shoreline slope allows for tidal flooding and the stability of the vegetation. Adequate protection from wave and storm energy is also a physical requirement for the development of salt marshes. Sediments that build salt marshes originate from upland runoff, marine reworking of the coastal shelf sediments, or organic production within the marsh itself.

Although a number of different patterns of development can be identified, salt marshes can be classified broadly into those that were formed from reworked marine sediments on marine-dominated coasts and those that were formed in deltaic areas where the main source of mineral sediment is riverine. The former type is typical of most of the world's coastlines. Deltaic marshes develop mainly where large rivers debouch onto low-energy coasts. Figure 9-2 shows the delta plain area of the world's largest rivers. These rivers span the world's latitudes and climatic zones. In the Tropics,

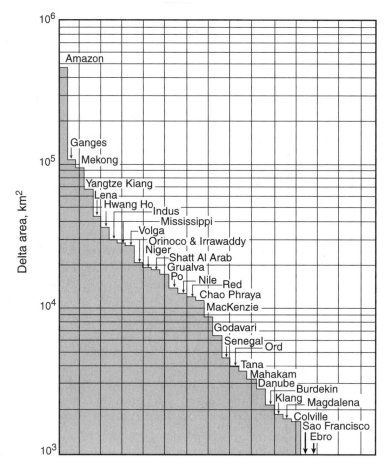

Figure 9-2 The area of deltaic plains of selected major river systems of the world. (*From Coleman and Roberts, 1989*)

the tidally influenced wetlands of these deltas are mangroves. Above 25° latitude, mangroves give way to saline marshes. In North America, large deltas are restricted to the coasts of the South Atlantic and the Gulf of Mexico. The Mississippi River deltaic marshes are the major example of this type of development and are the most extensive coastal marshes in the United States.

Marine-Dominated Marsh Development

On marine-dominated coasts, salt marsh development requires sufficient shelter to ensure sedimentation and to prevent excessive erosion from wave action (Beeftink, 1977a). Marshes can develop at the mouths of estuaries where sediments are deposited by the river and behind spits and bars that offer protection from waves and longshore currents:

1. *Shelter of Spits, Offshore Bars, and Islands.* Salt marshes will form along coastlines only where a bar, a neck of land (called a spit), or an island acts to trap sediment on its lee side and protects the marsh from the full forces of the open sea. The most

extensive examples of this type of coastal salt marsh have developed behind outer barrier reefs along the Georgia–Carolina coast. In their early development, sedimentation is rapid, both vertically and laterally. The resulting marshes are low in the intertidal zone and are usually dominated by *Spartina alterniflora*. The drainage pattern is well developed, and there is a pronounced meandering and erosion of tributaries at the headward end. As the marsh matures, sedimentation is primarily vertical because lateral deposition is balanced by erosion at the seaward ends of the marsh. As the elevation of the marsh approaches the highest excursion of the tide, sediment deposition slows and an equilibrium is attained between deposition and erosion. At maturity, the lower streamside area is a monotypic stand of *S. alterniflora,* the rest a high marsh dominated by a mixture of short *S. alterniflora, Salicornia* spp., *Distichlis spicata,* and *Juncus roemerianus.* As the marsh continues to age, high marsh becomes increasingly dominant. The extensive drainage system is slowly filled so that surface flows dominate and tidal forces become increasingly weak. When this occurs in areas of excess rainfall, fresh marsh and terrestrial plant species start to invade (Frey and Basan, 1985; Wiegert and Freeman, 1990).

   2. *Protected Bays and Estuaries.* Several large bays such as the Chesapeake Bay, the Hudson Bay, the Bay of Fundy, and the San Francisco Bay are adequately protected from storms and waves so that they can support extensive salt marshes. These salt marshes have features of both marine and deltaic origins. They occur on the shores of estuaries where shallow water and low gradients lead to river sediment deposition in areas protected from destructive wave action. Tidal action must be strong enough to maintain salinities above about 5 ppt; otherwise, the salt marsh will be replaced by reeds, rushes, and other freshwater aquatic plants.

River-Dominated Marsh Development

Major rivers carrying large sediment loads build marshes into shallow estuaries or out onto the shallow continental shelf where the ocean is fairly quiet. The size of a delta increases with the size of the inflowing river's drainage basin and its discharge, but is modified by such factors as the slope of the ocean shelf into which the river drains and the tidal range. Steep slopes have high wave energy. Under these circumstances, the delta periphery is smoothed out and the delta does not protrude from the adjacent shoreline. In contrast, in coasts with shallow slopes and low wave energies, deltas can build out onto the continental shelf. These deltas tend to have long shorelines relative to the straight-line width of the delta. The interaction of river discharge and tidal energy determines the salinity of the delta wetlands (Fig. 9-3), with fresh river water reducing salinities and tidal action extending the zone of marine–riverine interactions (Coleman and Roberts, 1989).

   Figure 9-4 shows the Holocene sequence of delta lobes that constructed the present Mississippi River delta. Typically, the first marshes developing on newly deposited sediments are dominated by freshwater species. However, the river course shifts through geologic time as the delta lobe is extended and the river loses efficiency. The abandoned marshes, no longer supplied with fresh river water, become increasingly marine influenced. In the Mississippi River delta, these marshes undergo a 5,000-year cycle of growth as fresh marshes, transition to salt marshes, and, finally, degradation back to open water under the influence of subsidence and marine transgression. During the last stage, the seaward edges of the marshes are reworked into barrier islands and spits in the same way as coastal marshes on the Atlantic Coast.

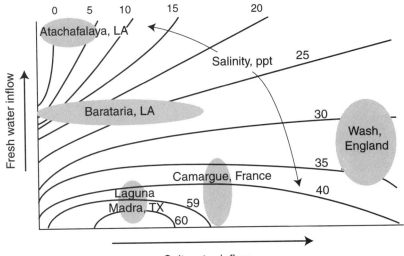

Figure 9-3 A conceptual diagram of the average salinities of salt marshes as related to the tidal range (saltwater inflow) and freshwater supply (river inflow).

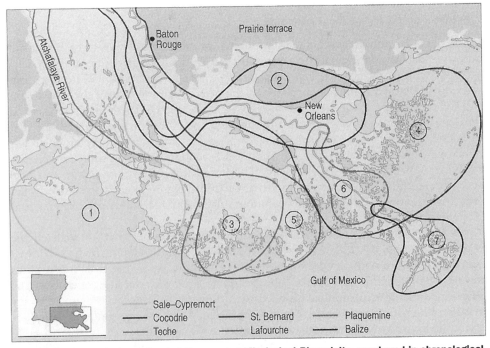

Figure 9-4 Lobes that constructed the present Mississippi River delta, numbered in chronological order of formation. Light shading shows extent of current coastal marsh and bottomland forest. (*From Gosselink et al., 1998, modified from Kolb and Van Lopik, 1958*)

Marsh Stability

The long-term stability of a salt marsh is determined by the relative rates of two processes: (1) sediment accretion on the marsh (including the production and the deposition of peat by growing plants), which causes it to expand outward and grow upward in the intertidal zone; and (2) coastal submergence caused by rising sea level and marsh surface subsidence (Fig. 9-5). These two processes are, to some extent, self-regulating; as a marsh subsides, it is inundated more frequently and, thus, receives more sediment and stores more peat (because the substrate is more anoxic and organic deposits degrade more slowly). Conversely, if a marsh accretes faster than it is submerging, it gradually rises out of the intertidal zone, is flooded less frequently, receives less sediment, and oxidizes more peat.

Local conditions have an overriding control over the balance achieved between accretion and submergence. The amount and type of sediment particles in the water column are determined by their source, whether coarse or fine, organic or inorganic, marine or riverine. The amplitude and frequency of the tide as well as the volume and seasonal variation of river flows onto the coast determine current speeds and, hence, the capacity to carry sediments. The morphology of the area determines the pattern of flow, erosion, and deposition.

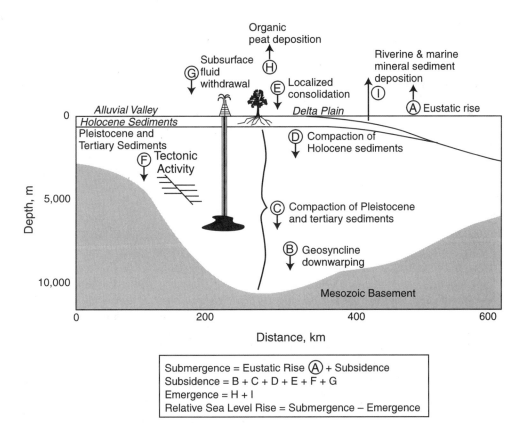

Figure 9-5 Factors controlling relative sea level in the Mississippi River delta plain. (*After Penland et al., 1988*)

Although physical processes probably dominate in coastal marshes, the effects of biota can also be significant. This is especially true when one considers the formation of peat, which is almost entirely caused by the *in situ* production of organic matter by marsh plants. Indeed, several studies have concluded that the vertical accretion of organic marshes is determined by below-ground organic production. Mineral sediments are important, apparently because they stimulate production and perhaps buffer the soil redox potential, rather than directly controlling accretion rates (Bricker-Urso et al., 1989; Callaway et al., 1997; Anisfeld et al., 1999). In addition to *in situ* organic production, the biota of the marsh control their physical environment in several other ways (O'Neil, 1949; Frey and Basan, 1985; Wiegert and Freeman, 1990): (1) Emergent grass dampens wind-generated waves, changing the sediment transport capacity of flooding water compared to open-water areas; (2) stems and leaves slow the water velocity, thus promoting sediment deposition; (3) changes caused by plants in the salinity of surrounding waters may influence the deposition of clays; (4) roots and rhizomes increase the stability of the sediment and its resistance to hydraulic erosion; (5) algal, bacterial, and diatom films help trap fine sediments; (6) colonial animals influence deposition and sediment structure; for example, oyster colonies directly modify the flow of water over them, and the dense concentration of fiddler crab burrows directly influences sediment permeability; (7) macroinvertebrates trap enormous quantities of suspended detritus, depositing it as feces or pseudofeces; and (8) grazing by waterfowl and mammals such as the nutria may completely denude an area, exposing it to tide and wind-driven erosive forces.

A few examples will serve to illustrate the wide diversity of developmental paths in a salt marsh.

Gulf Coast of North America

Sea level has been quite stable throughout the world for the last 5,000 years. Along the northern Gulf Coast, however, submergence is currently rapid, mostly because of the subsidence of the surface by compaction of deltaic sediments and downwarping of the older Pleistocene surface. There, where tidal energy is low and Mississippi River sediments are no longer supplied to the coastal marshes (they are channeled directly into deep offshore waters), accretion is not keeping up with submergence and salt marshes are degrading rapidly. Delaune et al. (1983b) showed that the percentage of open water in a Gulf coastal marsh was directly related to the rate of coastal submergence over an 85-year period. The average rate of coastal submergence was 1.2 cm/yr compared to a marsh accretion (sediment deposition) rate of 0.66 to 0.78 cm/yr. The accumulated aggradation deficit was closely related to wetland loss. Deteriorating salt marshes along this coast receive most of their sediments from the reworking of marine and estuarine deposits during severe storms, especially hurricanes (Table 9-1), but stable marshes depend more on riverine input during spring floods (Baumann et al., 1984). Both inland and streamside salt marshes showed a net loss of elevation in this subsiding landscape, with a net gain only on streamside marshes when hurricane inputs are included (Table 9-1).

When vertical marsh accretion is unable to keep up with the combined effects of subsidence and sea-level rise, marsh plant growth is stressed by increased flooding. This leads to a positive feedback loop on marsh degradation in which reduced primary production reduces root formation, which, in turn, decreases the rate of vertical accretion and increases the duration and depth of flooding, further reducing plant

Table 9-1 Rates of sediment accumulation of salt marshes

| | Rate (cm/100 years) | | | |
Vegetation Zone	Accretion	Subsidence	Net Accumulation	Age (years)
GULF COAST (Louisiana)[a]				
Spartina alterniflora–S. patens				
Barataria Bay (deteriorating)				
Streamside				
With hurricanes (4 yr)	150	123	27	
Without hurricanes (3 yr)	110	123	−13	
Inland				
With hurricanes	90	123	−33	
Without hurricanes	60	123	−63	
Four League Bay (stable)				
Streamside, no hurricanes	130	85	45	
Inland, no hurricanes	56	85	−29	
NORTH ATLANTIC COAST (Massachusetts)[b]				
Spartina alterniflora	61	30	31	490
S. patens–Distichlis spicata	38	30	8	600
Juncus gerardi	32	30	2	1,200
SOUTHERN ENGLAND (Dorset)[c]				
Mudflats	340–390	10.50		
Spartina anglica zone	11–65	10.50		
Halimione zone	29–61	10.50		
High marsh	17–33	10.50		

[a]Data from Baumann et al. (1984).
[b]Data from Chapman (1960).
[c]Data from Gray (1992) and Allen (1992).

production (Delaune et al., 1994). The net result of this cycle is plant senescence, followed rapidly by substrate loss and the formation of a shallow pond.

In two independent studies of a highly organic salt marsh, Delaune et al. (1994) and Day et al. (1994) described the rapid collapse of the peat surface when the vegetation died. The marsh surface decreased about 8 to 10 cm over a period of one winter (Day et al., 1994) or 2 years (Delaune et al., 1994). This surface collapse was not attributed to erosion, as the stubble of the dead plants remained on the surface. However, the death of the plants' root systems led, subsequently, to a shift from the dominance of biological processes that preserve the marsh root mat (live roots protect the substrate from erosion) to physical processes that led to erosion.

It is not clear what causes the rapid marsh surface collapse subsequent to plant senescence. In the examples cited previously, Day et al. (1994) suggested that, following plant death, the formerly living root mass is rapidly decomposed by anaerobic sulfate-reducing bacteria. Evidence for this high rate of decomposition was the development of a white film from the chemoautotrophic bacterium, *Beggiatoa,* in the dying marshes. This bacterium uses the reduced sulfate product, hydrogen sulfide, as an energy source. Rapid root decomposition would lead to collapse of the structural components of the marsh substrate.

From their study of the same general marsh system, Delaune et al. (1994) suggested that surface collapse was caused by death of the living root network in the upper layers of the soil and loss of the root membrane integrity that allows the accumulation of gases in the root aerenchyma. Most of the substrate volume (about 90 percent) is occupied by pore space (Nyman et al., 1990; Gosselink and Sasser, 1995). Part of this is water filled; the rest gas filled. If much of the gas volume is related to intact live roots, which seems a reasonable assumption since wetland species are known to develop aerenchyma (see Chapter 7), the death of these roots might lead to loss of membrane integrity and gas volume and, hence, to substrate collapse.

North Atlantic Coast

Along the North Atlantic Coast, the processes of accretion and submergence are apparently close to a dynamic equilibrium. There, the sea level has been rising at between only 1 and 3 mm/yr for the past several thousand years (Teal, 1986). Accretion in many marshes has been somewhat faster. Redfield (1965, 1972) gave a detailed description of the development of this kind of salt marsh in New England during the past 4,000 years in a general model of peat and sediment accumulation in the presence of a continually slowly rising sea level (Fig. 9-6). As the sea level rises ($HW_0 \rightarrow HW_3$), the marsh extends inland and the upland is covered by marsh peat. At the same time, if sediments accumulate beyond the lower limit of marsh vegetation (called the *thatch line*) at a rate greater than the rise in sea level, the intertidal marsh migrates seaward to maintain its critical elevation. The upper (high) marsh may also develop seaward over the old intertidal peat. The expansion of the marsh in this example occurs not primarily because of a large mineral sediment source and relatively high tidal energy compared to the Gulf Coast but because peat formation and aggradation exceed a submergence rate only a fraction of that experienced in Louisiana (Table 9-1).

By examining old maps and by determining the depth of marsh peat and the sedimentation rate (which can be estimated from radioactive markers such as ^{210}Pb and ^{137}Cs), scientists can measure the age of salt marshes. Such studies indicate that the oldest present-day salt marshes were formed during the last 3,000 to 4,000 years. In one study of a New England salt marsh (Table 9-1), it was found that the lower (seaward) marsh, dominated by *Spartina alterniflora*, accumulated sediments at a much greater rate than did the more inland upper marsh dominated by *Juncus gerardi*.

Subarctic North America

A third example of marsh development is that of the northern part of the North American continent, which is emerging slowly as the land rises at a rate of about

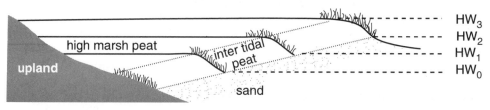

Figure 9-6 Development of a New England salt marsh. HW_0, HW_1, HW_2, and HW_3 refer to successive high water levels as sea level rises. (*From Redfield, 1965*)

1 cm/yr in response to the melting of the ice sheet that covered the land during the last ice age (Martini et al., 1980). As a result, in subarctic marshes such as those along the southern shore of the Hudson Bay, the sea is retreating, shallow flats are being exposed and invaded by salt marsh species, and the whole marsh is expanding outward.

These examples show how delicately poised the salt marsh is geomorphically. A change in the rate of sea-level rise or sedimentation of as little as a millimeter or two per year can determine whether a marsh will expand, retreat, degrade, or remain stable.

Hydrology

Tidal energy represents a subsidy to the salt marsh that influences a wide range of physiographic, chemical, and biological processes, including sediment deposition and scouring, mineral and organic influx and efflux, flushing of toxins, and the control of sediment redox potential. These physical factors, in turn, influence the species that occur on the marsh and their productivity. The lower and upper limits of the marsh are generally set by the tide range. The lower limit is set by the depth and the duration of flooding and by the mechanical effects of waves, sediment availability, and erosional forces (Chapman, 1960). The upland side of the salt marsh generally extends to the limit of flooding on extreme tides, normally between mean high water and extreme high water of spring tides (Beeftink, 1977a). Based on marsh elevation and flooding characteristics, the marsh is often divided into two zones, the upper marsh (*high marsh*) and the intertidal lower marsh (*low marsh*). The upper marsh is flooded irregularly and can experience at least 10 days of continuous exposure to the atmosphere, whereas the lower marsh is flooded almost daily, and there are never more than 9 continuous days of exposure (Table 9-2). This classification is simplistic, as will be clear from examples discussed in the Ecosystem Structure section. Nevertheless, it is a first approximation that has considerable practical value. In the Gulf Coast marshes of the United States, the terms *streamside* and *inland* generally replace low and high marsh, respectively, because in these flat, expansive marshes the streamside levees are actually the highest marsh elevations.

Tidal Creeks

A notable physiographic feature of salt marshes, especially low marshes, is the development of tidal creeks in the marsh itself. These creeks develop, as do rivers, "with minor irregularities sooner or later causing the water to be deflected into definite channels" (Chapman, 1960). Redfield (1972) suggested that these tidal creeks had already developed on sand flats before they were encroached upon by advancing intertidal peat. The creeks serve as important conduits for material and energy transfer

Table 9-2 Hydrologic demarcation between low marsh and high marsh in salt marshes

| Marsh | Submergences | | Maximum Period of Continuous Exposure (days) |
	Per Day in Daylight	Per Year	
High marsh	<1	<360	≥10
Low marsh	>1.2	>360	≤9

Source: Chapman (1960).

between the marsh and its adjacent body of water. A tidal creek has salinity similar to that of the adjacent estuary or bay, and its water depth varies with tide fluctuations. Its microenvironments include different vegetation zones along its banks that have aquatic food chains important to the adjacent estuaries. Because the flow in tidal channels is bidirectional, the channels tend to remain fairly stable; that is, they do not meander as much as streams that are subject to a unidirectional flow. As marshes mature and sediment deposition increases elevation, however, tidal creeks tend to fill in and their density decreases (Fig. 9-7).

Sediments

The sediment source and tidal current patterns determine the sediment characteristic of the marsh. Salt marsh sediments can come from river silt, from organic productivity in the marsh itself, or from reworked marine deposits. As a tidal creek rises out of its banks, water flowing over the marsh slows and drops its coarser grained sediment load near the stream edge, creating a slightly elevated streamside levee. Finer sediments drop out farther inland. This gives rise to the well-known "streamside" effect, characterized by the greater productivity of grasses along tidal channels than inland, a result

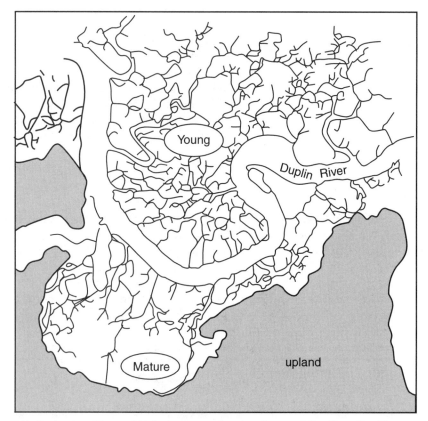

Figure 9-7 Drainage patterns of typical young and mature lagoonal *Spartina alterniflora* marshes in the Duplin River drainage, Doboy Sound, Georgia. (*After Wiegert and Freeman, 1990, and Wadsworth, 1979*)

of the slightly larger nutrient input, higher elevation, and better drainage. Figure 9-8 shows inorganic sediment concentrations for a salt marsh in the low-energy environment of the Louisiana coast, during flood and ebb tides. Sediments are being deposited as the water moves inland from the creek, but, in this example, there appears to be little, if any, resuspension during ebb tide. Water velocity on the marsh is about 15 percent of that in the adjacent creek. The source of mineral sediments is not as important for the productivity of the marsh as elevation, drainage, and organic content, all of which are determined by local hydrologic factors.

Pannes

A distinctive feature of salt marshes is the occurrence of pannes (pans). The term *panne* is used to describe bare, exposed, or water-filled depressions in the marsh (Wiegert and Freeman, 1990), which may have different sources. In the higher reaches of the marsh, inundated by only the highest tides, pannes, also called *sand barrens* (Frey and Basan, 1985), appear to form where evaporation concentrates salts in the

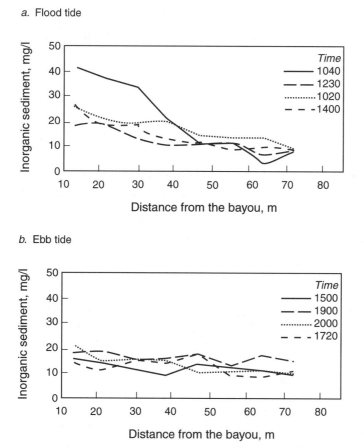

Figure 9-8 Inorganic suspended sediment concentration versus distance from the creek bank during   a. flood tide and   b. ebb tide. Data collected May 16–17, 1991, in a coastal Louisiana salt marsh. (*After Wang et al., 1993*)

substrate, killing the rooted vegetation. These exposed barrens are often covered by thin films of blue–green algae. *Mud barrens* are naturally occurring depressions in the marsh that are intertidal and retain water even during low tide. These pannes are often barren of vascular vegetation or support submerged or floating vegetation because of the continuous standing water and the elevated salinities when evaporation is high. They are continually forming and filling due to shifting sediments and organic production. The vegetation that develops in a mud panne, for example, wigeon grass (*Ruppia* sp.), is tolerant of salt at high concentrations in the soil water. Relatively permanent ponds are formed on some high marshes and are infrequently flooded by tides. Because of their shallow depth and their support of submerged vegetation, pannes are used heavily by migratory waterfowl. Pannes are a common feature due to human intervention, occurring where free tidal movement has been blocked by roads or levees, where spoil deposits have elevated a site, or where soil excavation, for example, for highway construction, has occurred in a marsh.

Chemistry

The development and zonation of vegetation in the salt marsh are influenced by several chemical factors. Three of the most important are the soil water salinity, which is linked with tidal flooding frequency; the availability of nutrients, particularly macronutrients such as nitrogen; and the degree of anaerobiosis, which controls the pathway of decomposition and nutrient availability.

Salinity

A dominant factor in the productivity and species selection of the salt marsh is the salinity of the overlying water and the soil water. The salinity in the marsh soil water depends on several factors:

1. *Frequency of Tidal Inundation.* The lower marsh soils that are flooded frequently tend to have a fairly constant salinity approximating that of the flooding sea water. On the other hand, the upper marsh that is only occasionally flooded experiences long periods of exposure that may lead to either higher or lower salt concentrations.

2. *Rainfall.* Frequent rainfall tends to leach the upper soil in the high marsh of its salts. Frequent periods of drought, on the other hand, lead to higher salt concentrations in the soil.

3. *Tidal Creeks and Drainage Slopes.* The presence of tidal creeks and steep slopes that drain away saline water can lead to lower soil water salinity than that which would occur under poorly drained conditions.

4. *Soil Texture.* Silt and clay materials tend to reduce drainage rates and retain more salt than does sand.

5. *Vegetation.* The vegetation itself has an influence on soil salinity. Evaporation of water from the marsh surface is reduced by vegetation cover, but transpiration is increased. The net effect depends on the type of vegetation and the environmental setting. Salt marsh vegetation also changes the ion balance in soils when roots take up ions selectively from the surrounding soil solution.

6. *Depth to Water Table.* When groundwater is close to the surface, soil water salinity fluctuations are less.

7. *Freshwater Inflow.* The inflow of fresh water in rivers, as overland flow, or in groundwater tends to dilute the salinity in both the salt marsh and the surrounding estuary. The early spring flood periods along much of the Atlantic and Gulf coasts of the United States lead to significant reductions in the salinity of downstream coastal marshes.

8. *Fossil Salt Deposits.* The presence of fossil salt deposits in the substrate can increase salt concentrations in the root zone, as occurs in the Hudson and James Bays (Price and Woo, 1988).

The interaction of these factors on marsh salinity is illustrated by two examples. The first, as described in Figure 9-3, relates to the average salinity of marshes world-wide, as influenced by the relative size of the freshwater source (here shown as river inflow, although local rainfall contributes to the freshwater input) to the marine input as indicated by the range of the tide. In areas that experience a large tide range (e.g., the Wash, England), marshes tend to approximate the ambient marine water salinity even though rainfall may be significant. In coastal marshes adjacent to large rivers, on the other hand (e.g., the north coast of the Gulf of Mexico), fresh water dilutes marine sources and the marshes are brackish or even fresh. Extreme salinities are found in subtropical areas such as the Texas Gulf Coast, where rivers and rainfall supply little fresh water and tides have a narrow range so that flushing is reduced. As a result, marine water is concentrated by evapotranspiration, often to double seawater strength or even higher.

A second example (Fig. 9-9) shows the development of a lateral salinity gradient as a function of the flooding frequency in an Atlantic Coast salt marsh. Near the adjacent tidal creek, frequent tidal inundation keeps sediment salinity at or below sea strength. As the marsh elevation increases, the inundation frequency decreases and

Vegetation zones	Tidal creek	*Spartina alterniflora*		Salt flats	*Salicornia, Batis*	*Juncus*
		Tall	Short			
Frequency of flooding, %	100	80-100	40-80	5-10	4-8	2-5
Interstitial salinity, ppt	20	23	33	127	41	24.5

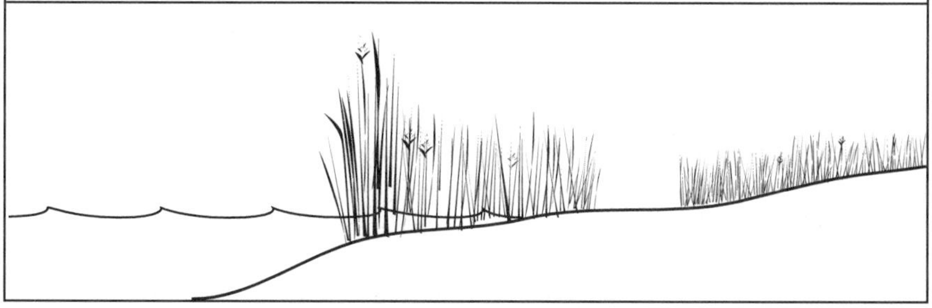

Figure 9-9   The relation of a salt flat's interstitial salinity and its vegetation. (*After Wiegert and Freeman, 1990, and Antlfinger and Dunn, 1979*)

the finer sediments drain poorly. At the elevation shown in Figure 9-9 as salt flats, infrequent spring tides bring in salt water that is concentrated by evaporation. Flushing is not frequent enough to remove these salts, and so they accumulate to lethal levels. Above this elevation, tidal flooding is so infrequent that salt input is restricted, and flushing by rainwater is sufficient to prevent salt accumulation. In this way, the salt gradient set up by the interaction of marsh elevation, tides, and rain often controls the general zonation pattern of vegetation and its productivity. Within the salt marsh zone itself, however, all plants are salt tolerant, and it is misleading to account for plant zonation and productivity on the basis of salinity alone. Salinity, after all, is the net result of many hydrodynamic factors, including slope and elevation, tides, rainfall, freshwater inputs, and groundwater. Thus, when *Spartina* flourishes in the intertidal zone, it is also responding to tides that reduce the local salinity, remove toxic materials, supply nutrients, and modify soil anoxia. All of these factors collectively contribute to different productivities and different growth forms in the intertidal and high marshes.

Nutrients

Salt marshes usually occur where sea water mixes with fresh water coming from land runoff, streams, rivers, groundwater, and precipitation. The degree of mixing of fresh water and oceanic water determines, to a large extent, the elemental composition, including salinity, of water flowing over the marsh. The mixing zone is broad: Salt marsh salinities range from as low as 10 ppt to as high as 60 ppt or more. Most salt marshes experience salinities between the lower limit and ocean salinity (36 ppt), depending on the proportion of fresh water in the mix. Thus, the concentration of most elements needed by plants for growth also varies widely, depending on their source and the location of a salt marsh in the mixing zone. Most macronutrient cations—calcium, magnesium, and potassium, in particular, and the anion sulfate—are derived primarily from ocean water, whereas silicon, phosphate, iron, zinc, and copper are more concentrated in freshwater sources. Nitrogen is about equally concentrated in sea and river water, and because it has a volatile phase, it is both gained and lost from the atmosphere (see Chapter 6).

Nitrogen

The availability of nutrients, particularly nitrogen and phosphorus, in the salt marsh soil is important for the productivity of the salt marsh ecosystem. A number of studies (e.g., Valiela and Teal, 1974; Smart and Barko, 1980) have shown that salt marsh vegetation is primarily nitrogen limited. Typical concentrations of total nitrogen and ammonium nitrogen in a salt marsh soil transect (Fig. 9-10) show that ammonium nitrogen, the primary form available to marsh vegetation, is only a small percentage (less than 1 percent) of the total nitrogen in the marsh soil. This is typical of organic wetland soils in general. Mendelssohn (1979), however, found ammonium to be the dominant form of available inorganic nitrogen in salt marsh interstitial water in North Carolina by one to two orders of magnitude over nitrate nitrogen. This results from the near anaerobic conditions usually present in the soil water, which preclude the buildup of nitrate nitrogen.

Another interesting feature of Figure 9-10 is the increase in ammonium in the inland direction. Although salt marsh plant growth rates are generally limited by the supply of inorganic nitrogen, in this example ammonium concentrations are high precisely where growth is poorest. A number of scientists have observed this phenome-

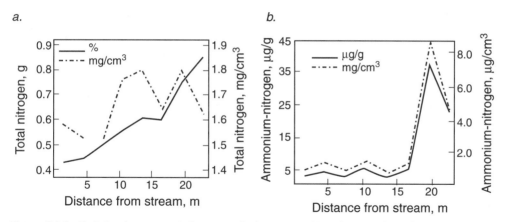

Figure 9-10   Variation in   **a.** total nitrogen and   **b.** extractable soil ammonium nitrogen with distance inland from tidal stream in Louisiana salt marsh. (*After Buresh et al., 1980b*)

non and have tried to understand it. It appears that other, nonnutrient, stresses related to the poor drainage and low redox potentials limit the ability of inland plants to assimilate ammonium nitrogen. Morris (1984) determined that under anaerobic conditions the rates of ammonium absorption by the roots of *Spartina alterniflora* and *S. patens* decreased by 60 and 40 percent, respectively. Absorption rates were further inhibited by high salinities in *S. patens*, but not in *S. alterniflora*. These results suggest that lowered ammonium absorption in the highly anaerobic inland marshes leads to accumulation in the substrate and that, along the creek bank, available nitrogen is kept at low levels by the actively growing plants (Mendelssohn, 1979; Mendelssohn et al., 1982).

The increase in total nitrogen on a dry-weight basis in Figure 9-10 results from a buildup of peat on the streamside–inland gradient (most soil nitrogen is organic); however, soil mass is related primarily to the mineral content. Thus, on a volumetric basis (mg N cm^{-3}), nitrogen is much less variable across the transect, and no trend is observed. It has been argued that, in soils of variable density, density-based measurements of nutrient elements correspond better to the volume invaded by plant roots than mass-based units. Thus, although studies have shown little relationship between soil nutrient concentrations (dry-weight basis) and plant biomass, there tends to be a strong positive correlation between above-ground vegetation biomass and soil nutrient *density* (volumetric basis). In one study by Delaune et al. (1979), only soil carbon concentration was related (negatively) to plant biomass. In contrast, when the nutrient content per unit volume of soil was examined, sodium, potassium, calcium, magnesium, total nitrogen, and extractable phosphorus were all positively correlated with above-ground plant biomass.

Phosphorus

Phosphorus is a nutrient that often limits plant growth, but in salt marsh soils it accumulates in high concentrations and apparently does not limit growth. For example, the marsh sediments along the Georgia coast were estimated to contain enough phosphorus to supply the marsh vegetation for several hundred years (Pomeroy et al., 1972).

Iron

Other plant nutrients have also been suggested as possibly limiting growth in salt marshes. D. A. Adams (1963) found that soluble iron concentrations in the marsh soil water were highest in the more productive zones of the marsh. In fact, several micronutrients, including iron and manganese, are available in high concentrations in marsh soils because of the reducing conditions, and they are more likely to be in toxic concentrations rather than limiting. Iron, for example, is found in *Spartina* tissues at concentrations of about 10 times those in most crop plants.

Sulfur

Sulfur is an interesting marsh soil chemical because of its toxicity, acid-forming properties, and ability to store energy from organic sources. Sea water contains abundant sulfate. When this ion encounters the anoxic marsh soil, it is reduced by soil bacteria to sulfide, which, in turn, can form insoluble pyrites with iron. Hydrogen sulfide (which is responsible for the characteristic rotten-egg odor of salt marsh sediments) is extremely toxic to plants and is probably responsible, at least in part, for the poor performance of inland marsh plants. In addition, when exposed to air, sulfides can be reoxidized to sulfate, forming sulfuric acid, with a resulting drop in soil pH. It has been suggested that local sediment drying (and oxidation) and the subsequent increase in acidity in the soil may account for some patchy death of plants in the marsh (Cooper, 1974). When salt marshes underlain with clays are drained for agricultural production, the soil sulfides oxidize to sulfuric acid, and the soils become too acidic to support crop growth. This is the frequently observed "cat clay" phenomenon.

An intriguing property of sulfide is its ability to act as a storage compound for biologically fixed energy. In the anaerobic soil environment, certain bacteria can use sulfate as an electron acceptor, reducing it to sulfide as they oxidize the organic substrate. A significant portion of the organic energy is transferred to the energy-rich sulfide radical. These sulfide compounds can later be reoxidized in the presence of oxygen, and the stored energy can be used to fuel the growth of sulfur-oxidizing bacteria. It is estimated that as much as 50 to 70 percent of the energy of net primary production flows through reduced inorganic sulfur compounds (see Chapter 6). Most of the stored sulfides are reoxidized on an annual basis by oxygen diffusing into the soil from marsh grass roots. A small percentage of the soluble sulfides, however, may be exported from the marsh in pore water. Other sulfides enter into the sediment food chain directly. For example, the colorless sulfur bacteria (*Beggiatoa* spp.) on marsh and intertidal flats oxidize soluble reduced sulfur compounds to elemental sulfur, storing the energy in bacterial biomass. The bacterial mats are easily suspended by waves and currents and are recycled into the water column (Grant and Bathmann, 1987).

Ecosystem Structure

The salt marsh ecosystem has diverse biological components that include vegetation and animal and microbe communities in the marsh and plankton, invertebrates, and fish in the tidal creeks, pannes, and estuaries. The discussion here will be limited to the biological structure of the marsh itself.

Vegetation

The vegetation of salt marshes can be divided into zones that are related to the upper and lower marshes described previously but that also reflect regional differences. Figure 9-11 shows a typical New England vegetation zonation pattern from streamside to upland. The intertidal zone or low marsh next to the estuary, bay, or tidal creek is dominated by the tall form of *S. alterniflora* Loisel (smooth cordgrass). In the high marsh, *S. alterniflora* gives way to extensive stands of *S. patens* (saltmeadow cordgrass) mixed with *Distichlis spicata* (spikegrass) and occasional patches of the shrub *Iva frutescens* (marsh elder) and various forbs. Beyond the *S. patens* zone and at normal high tide, *Juncus gerardi* (blackgrass) forms pure stands. At the upper edge of a marsh inundated only by spring tides, two groups of species are common, depending on the local rainfall and temperature. Where rainfall exceeds evapotranspiration, salt-tolerant species give way to less tolerant species such as *Panicum virgatum* (switchgrass), *Phragmites australis* (common reed), *Limonium carolinianum* (sea lavender), *Aster* spp. (asters), and *Triglochin maritima* (arrow grass). On the southeastern New England coast where evapotranspiration may exceed rainfall during the summer, salts can accumulate in these upper marshes, and salt-tolerant halophytes such as *Salicornia* spp. (saltwort) and *Batis maritima* flourish. Bare areas with salt efflorescence are common. Other features of New England salt marshes include well-flushed mosquito ditches lined with tall *S. alterniflora* and salt pannes containing short-form *S. alterniflora*.

Characteristic patterns found in other salt marshes are shown in Figure 9-12. South of the Chesapeake Bay along the Atlantic Coast, salt marshes typical of the Coastal Plains appear (Fig. 9-12a). These marshes are similar in zonation to those in

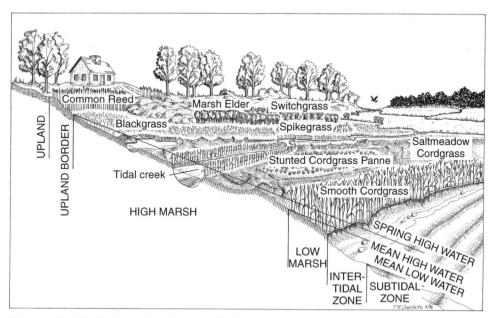

Figure 9-11   Idealized zonation of communities on a typical North Atlantic salt marsh. The location of the different plant associations is strongly influenced by small differences in elevation above the mean high water level. (*After Dreyer and Niering, 1995, reprinted by permission of Tom Ouellette, Vernon, CT*)

a. Southeast Atlantic coast salt marsh

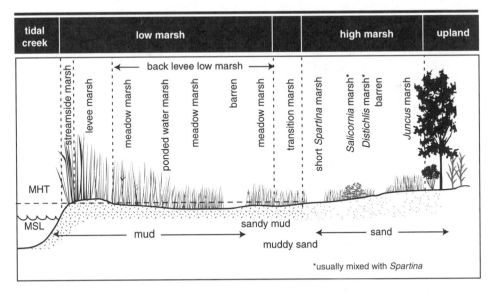

b. Eastern Gulf of Mexico coast salt marsh

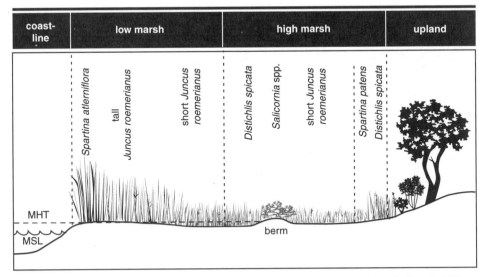

Figure 9-12 Zonation of vegetation in typical salt marshes: **a.** Southeast Atlantic Coast. (*After Wiegert and Freeman, 1990*) **b.** Eastern Gulf of Mexico (*After Montague and Wiegert, 1990*). **c.** Northern France (*After LeFeuvre and Dame, 1994*).

c. European salt marsh

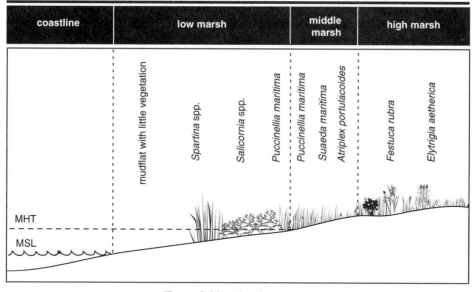

Figure 9-12 *(Continued).*

New England except that (1) tall *S. alterniflora* often forms only in very narrow bands along creeks, (2) the short form of *S. alterniflora* occurs more commonly in the wide middle zone, and (3) *Juncus roemerianus* (black rush) replaces *J. gerardi* in the high marsh. At maturity, low and high marsh areas are approximately equal. The low marsh is almost entirely *S. alterniflora,* tall on the creek bank, and shorter behind the natural levee as elevation gradually increases in an inland direction. It may contain small vegetated or unvegetated ponds and mud barrens. The high marsh is much more diverse, containing short *S. alterniflora* intermixed with associations of *Distichlis spicata, Juncus roemerianus,* and *Salicornia* spp.

Along the Mississippi and northwest Florida coasts, *J. roemerianus* is found in extensive monocultures (Fig. 9-12b). There is often a fringe of *S. alterniflora* along the seaward margin, followed, in an inland direction, by large areas of tall and short *J. roemerianus.* Mixtures of *S. patens* and *D. spicata* line the marsh on the landward edge, and *Salicornia* spp. can be found in small areas such as berms where salt accumulates. Along the northern Gulf Coast, *S. patens* is the dominant species. It occurs in a broad zone inland of the more salt-tolerant *S. alterniflora.* For example, Chabreck (1972) estimated that *S. patens* dominated more than 200,000 ha of coastal marsh in Louisiana.

In Europe, a totally different salt marsh is found, at least compared to the eastern United States marshes (Fig. 9-12c). One of the most notable features is that the intertidal zone between high tide and mean high tide is sparsely covered if it is vegetated at all in Europe, whereas it is dominated by *S. alterniflora* in the United States (Lefeuvre and Dame, 1994). So much of what would be called the low marsh in Europe is, in fact, a mud flat or sparsely vegetated. In Europe, the salt marshes that are studied are mostly between mean high tide and spring tide. The cordgrass found in Europe is generally *S. anglica* or *S. townsendii* and it is found in a relatively narrow band.

Salt Marsh Plant Species

S. alterniflora is a stiff, leafy perennial grass that can grow to 3 m in height. It has two growth forms, tall and short, that are found in different parts of the marsh. The tall form (100–300 cm) generally occurs adjacent to tidal creeks and in the extreme low portions of the intertidal marsh. The short form (17–80 cm) is found away from the creeks in the upper marsh, generally in areas of greater salinity. Since the differences in growth form, underground reserves, and root profiles remained for at least 9 years when plants were grown together in common garden plots, there appears to be a strong genetic control of the morphology and physiology of the two growth forms (Gallagher et al., 1988).

The genus *Spartina* is native to North America. *S. alterniflora,* however, was introduced in Britain and France about 1860 or 1870 and hybridized with *S. maritima,* a European species, forming a sterile species, *S. townsendii.* This sterile hybrid later produced a fertile amphidiploid (i.e., the chromosome number of the original hybrid was doubled). *S. anglica,* the name given to this fertile hybrid, is extremely vigorous and is rapidly expanding along the European coast. Since the 1960s, both *S. anglica* and *S. alterniflora* have been used extensively for reclaiming China's east coast along the China Sea (Chung, 1982, 1983, 1985, 1989). *Spartina*'s introduction there has led to increased arable land, the stabilization of the coastline, decreased soil salinity, and the production of animal fodder, green manure, and fuel (Chung, 1989).

It is of some interest that *S. alterniflora* harbors within its culm vascular spaces an ascosporogenous yeast, *Pichia spartinae.* The yeast is able to use ethanol, plant lipids, and organic acids as its sole carbon sources and amines as its nitrogen sources. It apparently can mobilize, transport, and cycle micronutrients, including iron and possibly zinc, copper, cadmium, nickel, and manganese, from deposits in the rhizosphere. The leakage of oxygen from plant roots oxidizes and immobilizes iron and other micronutrients in this zone, making them unavailable to vascular plants. *P. spartinae* also produces the plant auxin, indoleacetic acid, which is a universal growth regulator in vascular plants. These properties suggest an endosymbiotic relationship between the two species in which *Spartina* provides an energy source for the yeast, which, in turn, may be active in micronutrient absorption and hormonal and metabolic regulation of its host (Nakamura et al., 1991; Catallo et al., 1996).

A congener, *S. patens,* is found widely along the Atlantic and Gulf coasts of North America, from Quebec to Florida and west to Texas. It is also found in the Yucatan Peninsula, Cuba, the West Indies, southern Europe, and northern Africa. It dominates brackish marshes along the Gulf Coast of the United States. Salt tolerance varies over its range, and Hester et al. (1996) reported that these differences are carried in different genotypes.

Distichlis spicata (spikegrass) is another common salt marsh plant that is widely distributed along coastal and inland saline wetlands in the United States. The plant is particularly tolerant of high soil salinity, and it may serve as a pioneer species in salt marsh development.

Benthic Algae

Beneath the grass swards that characterize the salt marsh lies the zone of *edaphic* (benthic) microalgae. The term *edaphic* refers to organisms living on or in the soil. Miller et al. (1996) and MacIntyre et al. (1996) refer to this as the "secret garden," invisible except for the golden-brown sheen of diatoms migrating to the sediment surface in response to the sun's early rays. Nevertheless, the algal flora on and in the sediments of open creek banks and grass-shaded marshes are a diverse and functionally important component of the salt marsh ecosystem. The flora are dominated by small pennate diatoms of the genera *Navicula, Nitzschia,* and *Amphora* (Table 9-3). Studies of New England, southern California, and European salt marshes have also identified extensive populations of cyanobacteria (formerly called blue–green algae or Cyanophyceae), including coccoid, colonial, and filamentous genera. In addition, eukaryotic genera may be periodically abundant, sometimes forming green or yellow–green mats on salt marsh creek banks (Table 9-3).

Round (1960) and Sullivan (1975, 1977, 1978) reported that most diatoms in salt marshes were tolerant of wide salinity ranges and were found across the salt marsh from the low creek bank to high inland marshes. Structural differences were detected beneath different emergent vascular species; differences in assemblages were related to differences in elevation, soil temperature, surface sediment moisture, ammonium concentrations, canopy height, and interactions between diatoms and cyanobacteria in warmer months and green algae in cooler months. Nevertheless, Sullivan (1978) hypothesized the existence of a "single basic edaphic diatom community indigenous to Atlantic and Gulf coast salt marshes."

Although diatoms dominate Atlantic and Gulf Coast marshes, cyanobacteria are dominant in hypersaline marshes in southern California, especially under canopies of *Spartina foliosa* and *Jaumea carnosa* (Sullivan and Currin, 2000).

Consumers

Salt marshes, whose features are characteristic of both terrestrial (aerobic) and aquatic (anoxic) environments, provide a harsh environment for consumers. Salt is an addi-

Table 9-3 Dominant edaphic algal species of salt marshes

Taxonomic Group	Life Form	Dominant Genera
Diatoms	Pennate, biraphid	*Navicula*
		Nitzschia
		Amphora
Cyanobacteria	Coccoid	*Chroococcus*
		Anacystis
	Colonial	*Merismopedia*
	Filamentous	*Anabaena*
		Nostoc
		Calothrix
	Oscillatoria	*Lyngbya*
Eukaryotes	Coenocytic filamentous	*Vaucheria*
Xanthophytes (yellow–green algae)		
Euglenoids		*Euglena*
Chlorophytes (green algae)		*Rhizoclonium*
		Enteromorpha

Source: Blum (1968), Carter (1932, 1933 a,b), Drum and Webber (1966), Polderman (1975, 1978), Polderman and Polderman-Hall (1980), Sullivan (1975, 1977, 1978), Webber (1967, 1968), Williams (1962), and Zedler (1980, 1982 a,b).

tional stress with which they must contend. In addition, the variability of the environment through time is extreme. The dominant plant food source for marsh consumers
is generally a marsh grass, which is usually limited in its nutritional value. Considering
all these limitations, the number of consumers in the salt marsh is perhaps surprisingly diverse.

Relatively little is known about the faunal populations of salt marshes. Species
composition and population size of resident species, however, appear remarkably
similar from site to site, especially across the southeastern United States, where most
of the tidal marshes are found. Descriptions of faunal populations of U.S. salt marshes
are presented in Davis and Gray (1996), Daiber (1977, 1982), Montague et al.
(1981), Pfeiffer and Wiegert (1981), and Montague and Wiegert (1990) and in the
series of wetland community profiles published by the U.S. Fish and Wildlife Service
(Nixon, 1982; Zedler, 1982a; Josselyn, 1983; Gosselink, 1984; Teal, 1986; Wiegert
and Freeman, 1990). Stout (1984) and Heard (1982) are useful sources for *Juncus*-
dominated marshes.

It is convenient to classify consumers according to the type of marsh habitat they
occupy, although the animals, especially in the higher trophic levels, move from one
habitat to another. The marsh can be divided into three major habitats: an aerial
habitat—the above-ground portions of the macrophytes, which is seldom flooded; a
benthic habitat—the marsh surface and lower portions of the living plants; and an
aquatic habitat—the marsh pools and creeks (Fig. 9-13). Our modification of this
classification presented by Montague and Wiegert (1990) is reflected in the following discussion.

Aerial Habitat

The aerial habitat is similar to a terrestrial environment and is dominated by insects
and spiders that live in and on the plant leaves. This is the grazing portion of the salt

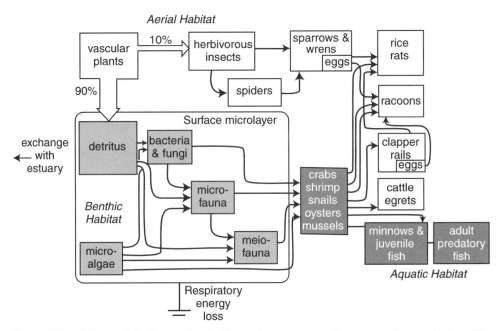

Figure 9-13 **Salt marsh food web, showing the major producer and consumer groups of the aerial
habitat, benthic habitat, and aquatic habitat.** (*After Montague and Wiegert, 1990*)

marsh food web (Fig. 9-13). The most common leaf-chewing organisms in salt marshes in the eastern United States are the arthropod *Orchelimum* (Smalley, 1960), the weevil *Lissorhoptrus,* and the squareback crab *Sesarma* (Pfeiffer and Wiegert, 1981). In addition, there are abundant sap-sucking insects (*Prokelisia marginata, Delphacodes detecta*) that ingest material translocated through the plant's vascular tissue or empty the contents of mesophyll cells. Some idea of the diversity of species occupying this habitat is shown in Table 9-4, which compares the number of herbivorous insect species associated with *S. alterniflora, S. patens,* and *S. foliosa* (a West Coast species). Numerous carnivorous insects are also found in this habitat. Pfeiffer and Wiegert (1981) list 81 species of spiders and insects in North Carolina, South Carolina, and Georgia *Spartina* marshes.

Salt marshes support large populations of wading birds, including egrets, herons, willets, and even woodstorks and Roseate Spoonbills. Coastal marshes also support vast populations of migratory waterfowl, including the Mallard (*Anas platyrhynchos*), American Wigeon (*A. americana*), Gadwall (*A. strepena*), Redheads (*Aythya americana*), and Teals (*Anas discors* and *A. crecca*). Black Duck (*A. rubripes*) is a permanent resident in many marshes, as are a number of songbirds.

A number of birds, including the Marsh Wren (*Cistothorus palustris*) and the Seaside Sparrow (*Ammodramus maritimus*), Laughing Gulls (*Larus atricilla*), and Forster's (*Sterna forsteri*) and Common Terns (*S. hirundo*), feed and nest in the marsh grasses (Burger, 1991). Wrens feed primarily on insects (Kale, 1965), and the sparrows apparently feed on the marsh surface, eating worms, shrimp, small crabs, grasshoppers, flies, and spiders. The Clapper Rail (*Rallus longirostris*) is another permanent marsh resident, feeding primarily on cutworm moths and small crabs. Many nonpermanent insectivorous birds forage in the salt marsh periodically, entering from adjacent fresher marshes, beaches, and upland habitats or migrating through. These include the Sharp-tailed Sparrow (*Ammodramus caudacutus*), swallows (*Tachycineta bicolor, Hirundo rustica,* and *Stelgidopteryx serripennis*), Red-winged Blackbirds (*Agelaius phoeniceus*), and various gulls.

Migratory waterfowl use coastal marshes extensively, mostly as wintering grounds, but also as stopover areas during fall and spring migrations. On the Gulf Coast of the United States, most species appear to favor fresh marshes, but move to adjacent salt marshes in early spring (Abernethy and Gosselink, 1988). In some areas, geese or duck flocks numbering in the hundreds of thousands denude coastal marshes

Table 9-4 Comparisons of the number of herbivorous species occupying various insect orders associated with three intertidal species of *Spartina*

	S. alterniflora	S. patens	S. foliosa
Orthoptera	6	10	0
Thysanoptera	3	8	5
Hemiptera	11	11	1
Homoptera	15	22	7
Coleoptera	19	18	10
Lepidoptera	5	—	2
Diptera	20	16	11
Hymenoptera	29	22	6
Total herbivorous insect species	108	107	42

Source: Pfeiffer and Wiegert (1981).

(Smith and Odum, 1981; Smith, 1983; Miller et al., 1996; Evers et al., 1998). Repeated and intense herbivory, especially when followed by high water levels, salinity extremes, or extended drought, may result in the formation of mud flats or shallow open-water ponds (Miller et al., 1996; Baldwin and Mendelssohn, 1998).

Benthic Habitat

Probably less than 10 percent of the above-ground primary production of the salt marsh is grazed by aerial consumers. Most plant biomass dies and decays on the marsh surface, and its energy is processed through the detrital pathway. The primary consumers are microbial fungi and bacteria. These organisms, in turn, are preyed upon by meiofauna in the decaying grass, the surface microfilm of the marsh, and the decaying bases of plant shoots. Most of these microscopic organisms are protozoa, nematodes, harpacticoid copepods, annelids, rotifers, and larval stages of larger invertebrates (Fenchel, 1969). The larger invertebrates on the marsh surface are of two groups, foragers (deposit feeders) and filter feeders. In a general sense, they are considered aquatic because most have some kind of organ to filter oxygen out of water. Foragers include polychaetes, gastropod mollusks such as *Littorina irrorata* and *Melampus bidentatus*, and crustaceans such as *Uca* spp., the blue crab (*Callinectes sapidus*), and amphipods. These organisms browse on the sediment surface, ingesting algae, detritus, and meiofauna. The filter feeders such as the ribbed mussel (*Geukensia demissus*) and the oyster (*Crassostrea virginica*) filter particles out of the water column.

Aquatic Habitat

Animals classified as aquatic overlap with those in the benthic habitat. For convenience, we include in this group higher trophic levels (mostly vertebrates) and migratory organisms that are not permanent residents of the marsh. Few fish species are permanent residents of the marsh. Most feed along the marsh edges and in small, shallow marsh ponds and move up into the marsh on high tides. Werme (1981) found 30 percent of silverside (*Menidia extensa*) and mummichog (*Fundulus heteroclitus*) in a North Atlantic estuary up in the marsh at high tide. Ruebsamen (1972) reported that common fish in small salt marsh ponds in Louisiana are sheepshead minnow (*Cyprinodon variegatus*), diamond killifish (*Adinia xenica*), tidewater silverside (*Menidia beryllina*), gulf killifish (*Fundulus grandis*), and sailfin molly (*Poecilia latipinna*). Shrimp (*Penaeus* spp.) and blue crabs (*Callinectes sapidus*) are also common. Most other species use the marsh intermittently for shelter and for food but range widely. Many fish and shellfish spawn offshore or upstream and, as juveniles, migrate into the salt marsh, which offers an abundant food supply and shelter. They are concentrated along the edges of the marsh (Zimmerman et al., 1990). As subadults, they migrate back into the estuary or offshore. This group of migratory organisms includes more than 90 percent of the commercially important fish and shellfish of the southeastern Atlantic and Gulf coasts (Table 9-5).

Mammals

Two mammals in North American salt marshes deserve attention because of their impact on salt marshes. The muskrat (*Ondatra zibethicus*) is native to North America; the coypu or nutria (*Myocastor coypus*) is an exotic species introduced from South America. Both prefer fresh marshes but are also found in salt marshes. In Louisiana, the muskrat appears to have been displaced by the nutria from its preferred freshwater habitat into saline marshes. Both mammals are voracious herbivores that consume

Table 9-5 Value of U.S. landings of commercial fish and shellfish species, 1998. About one-half of the value of the fish catch is from estuarine-dependent species, nearly all of which use salt and brackish marshes for food and shelter during some part of their lives

Species	Landings (metric tons)	Value (× $1,000)	Percentage of Total
Estuarine-dependent species			
Crab, blue	101,675	184,250	
Flounder—all	17,622	58,748	
Menhaden—Atlantic	774,631	103,950	
Oyster—eastern	10,807	66,297	
Shrimp—brown, pink, white	116,313	505,860	
Other—striped bass, Atlantic croaker,			
drum, mullet, seatrout	20,226	28,340	
Subtotal		947,451	50
Marine species			
Lobster—>90% American	38,800	275,600	
Scallops	5,500	75,100	
Squid, longfin	18,880	32,141	
Tuna—all	4,267	27,085	
Subtotal		409,926	22
Other		521,540	28
Total		$1,878,917	

Source: National Marine Fisheries Service.

plant leaves and shoots during the growing season and dig up tubers during the winter. They destroy far more vegetation than they ingest and are responsible for "eat-outs" that degrade large areas of marsh (Fig. 8-12). These areas recover extremely slowly, especially in the subsiding environment of the northern Gulf Coast. In European marshes, it is common to have domestic animals (e.g., cattle, sheep, or goats) grazing in coastal salt marshes. This grazing has a profound effect on the plant communities and zonation that develops in these marshes.

Ecosystem Function

Major points that have been demonstrated in several studies about the functioning of salt marsh ecosystems include the following:

1. Annual gross and net primary productivity of macrophytes is high in much of the salt marsh—almost as high as in subsidized agriculture. This high productivity is a result of subsidies in the form of tides, nutrient import, and abundance of water that offset the stresses of salinity, widely fluctuating temperatures, and alternate flooding and drying.

2. Although the biomass of edaphic algae is small, algal production can sometimes be as high as or higher than that of the community's

macrophytes, especially in hypersaline marshes such as those of southern California.

3. Direct grazing of vascular plant tissue is a minor energy flow in the salt marsh, but grazing on edaphic and epiphytic algae is a significant source of high-quality food energy for meio- and macroinvertebrates.

4. Fungi and bacteria are primary consumers that break down and transform indigestible plant cellulose (detritus) into protein-rich microbial biomass for consumers. This detrital pathway is a major flow of energy utilization in the salt marsh.

5. Salt marshes have been shown at times to be both sources and sinks of nutrients, particularly nitrogen.

These and other points will be discussed next.

Primary Productivity

Vascular Plants

Tidal marshes are among the most productive ecosystems in the world, producing annually up to 80 metric tons per hectare of plant material ($8,000$ g m^{-2} yr^{-1}) in the southern Coastal Plain of North America. The three major autotrophic units of the salt marsh include the marsh grasses, the mud algae, and the phytoplankton of the tidal creeks. Extensive studies of the net primary production have been conducted in salt marshes, especially along the Atlantic and Gulf coasts of the United States. A comparison of some of the measured values of net above-ground and below-ground production is given in Table 9-6. Above-ground production varies widely, from as little as 94 g m^{-2} yr^{-1} in a Mediterranean salt marsh to a high of $3,700$ g m^{-2} yr^{-1} in a Georgia study. Below-ground production is difficult to measure, and estimates vary widely from 220 to $2,500$ g m^{-2} yr^{-1} (streamside) and 420 to $6,200$ g m^{-2} yr^{-1} inland (Good et al., 1982).

Several major generalizations about primary productivity in salt marshes can be made:

1. Above-ground productivity of *Spartina* is often higher along creek channels and in low or intertidal marshes than in high marshes because of the increased exposure to tidal and freshwater flow. These conditions also produce the taller forms of *Spartina,* as discussed earlier.

2. In the United States, there is generally greater productivity in the southern Coastal Plain salt marshes than in those farther north because of the greater influx of solar energy and the longer growing season but possibly also because of the nutrient-rich sediments carried by the rivers of that region. For example, White et al. (1978) argued that the productivity of salt marshes in Louisiana and Mississippi may be higher than Atlantic Coast marshes because of the higher nutrient import from the Mississippi River.

3. The wide divergence in measurements of net primary productivity among sites is as much caused by the methods used for measurement as by true differences in productivity. A number of investigators have discussed the variations obtainable with different methods used for calculating productivity. For example, estimates of *S. alterniflora* production, calculated

Table 9-6 A comparison of net primary productivity estimates of different salt marsh plant species in the same region

Species	Net Primary Production (g m^{-2} yr^{-1})[a]		Source
LOUISIANA	Above ground	Below ground	
Distichlis spicata	1,162–1,291		White et al. (1978)
Juncus roemerianus	1,806–1,959		
Spartina alterniflora	1,473–2,895		
Spartina patens	1,342–1,428		
Distichlis spicata	1,967		Hopkinson et al. (1980)
Juncus roemerianus	3,295		
Spartina alterniflora	1,381		
Spartina cynosuroides	1,134		
Spartina patens	4,159		
ALABAMA			
Juncus roemerianus	3,078	7,578	Stout (1978)
Spartina alterniflora	2,029	6,218	
MISSISSIPPI			
Juncus roemerianus	1,300		de la Cruz (1974)
Distichlis spicata	1,072		
Spartina alterniflora	1,089		
Spartina patens	1,242		

[a] The methods are comparable in each study.

from data from the same plots, varied from 1,470 to 2,900 g m^{-2} yr^{-1} in one study (White et al., 1978) and from 830 to 2,700 g m^{-2} yr^{-1} in another study (Kaswadji et al., 1990). Linthurst and Reimold (1978) found as much as a sixfold range in above-ground production in Maine, based solely on the method of calculation of production from the raw data, whereas Shew et al. (1981) reported a fivefold difference in North Carolina.

4. Below-ground production is sizable—often greater than aerial production (Table 9-6). Under unfavorable soil conditions, plants seem to put more of their energy into root production. Hence, roots : shoot ratios seem to be generally higher inland than at streamside locations.

5. A comparison of different salt marsh vascular species (Table 9-6) shows that none comes out the winner in a productivity sweepstakes.

Algae

The productivity of edaphic algae was summarized by Sullivan and Currin (2000). Annual benthic algal production, as measured in a number of studies, ranges from 28 g C m^{-2} yr^{-1} in a Gulf Coast *Juncus roemerianus* marsh to 314 g C m^{-2} yr^{-1} in a southern California *Jaumea carnosa* marsh (Table 9-7). Benthic algal production increases in a southerly direction along the Atlantic Coast, but is lowest on the Gulf Coast. Much of the algal production on the East and West coasts of the United States occurs when the overstory plants are dormant.

Table 9-7 **Comparison of annual benthic microalgal production and ratio of annual benthic microalgal to vascular plant net aerial production (BMP/VPP) in different salt marshes of the United States**

State	Algal Productivity (g C m^{-2} yr^{-1})	BMP/VPP (%)	Reference
Massachusetts	105	25	Van Raalte (1976)
Delaware	61–99	33	Gallagher and Daiber (1974)
South Carolina	98–234	12–58	Pinckney and Zingmark (1993)
Georgia	200	25	Pomeroy (1959)
Georgia	150	25	Pomeroy et al. (1981)
Mississippi	28–151	10–61	Sullivan and Moncreiff (1988)
Texas	71	8–13	Hall and Fisher (1985)
California	185–341	76–140	Zedler (1980)

Source: Sullivan and Currin (2000).

On the Atlantic and Gulf coasts, the productivity of algae is 10 to 60 percent of vascular plant productivity. Zedler (1980), however, found that algal net primary productivity in southern California ranges from 75 to 140 percent of vascular productivity. She hypothesized that the arid and hypersaline conditions of southern California favor algal growth over vascular plant growth.

Limits to Primary Production

We have alluded to factors that control production throughout this chapter. These factors have probably been studied more in salt marshes than in other wetland systems. In a general sense, plant production is a response to light (energy), water, nutrients, and toxins (negatively). Marsh plants grow in full sunlight, they appear to have a limitless water supply, and the sedimentary minerals on which they grow are generally rich in nutrients. As a result, one would expect uniformly high production rates. This is not true. Instead, the salt marsh environment sets limits on production that relate to all three parameters. Plant species undoubtedly differ genetically in productivity potential. However, most studies of environmental controls of production in salt marshes have been carried out on *Spartina alterniflora* and the following discussion focuses on that species as a surrogate for salt marsh production in general.

On a regional scale, primary production limits are set by differences in climate, tidal range, and the source of sedimentary material. Turner (1976) demonstrated that production generally declined with increasing latitude and related this phenomenon to the shorter length of the growing season and lower temperatures in the north (Fig. 9-14).

Beyond latitudinal differences, regional tidal amplitude differences influence primary production. One of the oldest paradigms of salt marsh ecology is the "tidal subsidy" (E. P. Odum, 1980). The hypothesis states that tides, by flushing salts and other toxins out of the marsh and by bringing in nutrients, stimulate marsh growth. Steever et al. (1976) verified this hypothesis by showing that the peak standing biomass of *S. alterniflora* along the New England coast was directly related to the tide range (see Chapter 5).

Variations in productivity on the local scale result from complex interactions of soil anoxia, soluble sulfide, and salinity (Mendelssohn and Morris, 2000). The complex interrelationships among vascular plant growth in the salt marsh and flooding, soil drainage, and salinity are illustrated in Figure 9-15. Although water appears plentiful,

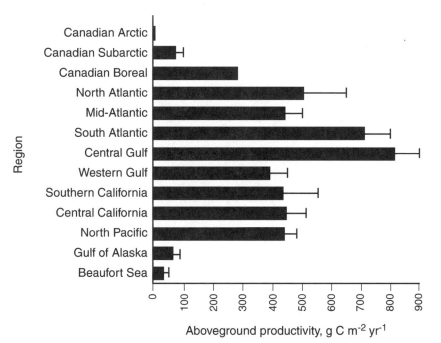

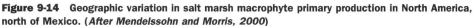

Figure 9-14 Geographic variation in salt marsh macrophyte primary production in North America, north of Mexico. (*After Mendelssohn and Morris, 2000*)

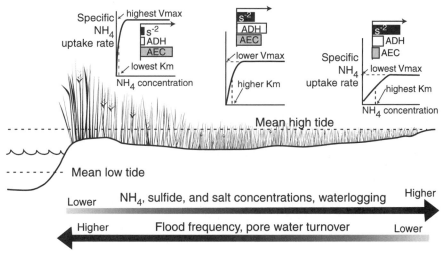

Figure 9-15 Factors controlling the growth rate of *Spartina alterniflora* across the marsh from creek edge to inland. (*After Mendelssohn and Morris, 2000*)

the concentration of dissolved salt makes the salt marsh environment similar in many respects to a desert: The "normal" water gradient is from plant to substrate. To overcome the osmotic influence of salt, plants must expend energy to increase their internal osmotic concentration in order to take up water. As a result, numerous studies confirm that plant growth is progressively inhibited by increasing salt concentrations in the soil. This is true even for the salt-tolerant species of the salt marsh, and the salinity effects may be subtle. For example, Morris et al. (1990) showed that the year-to-year variation in marsh production at a single site on the East Coast was correlated with the mean summer water level, which they equated with soil salinity (soil salinity was inversely correlated with the frequency of marsh flooding in this study).

Another factor limiting production is the degree of anaerobiosis of the substrate (see Chapter 7). Vascular plants, even those that have developed adaptations to anaerobic conditions, grow best in aerobic soils. Many effects of anaerobiosis have been documented: reduced energy availability as the aerobic respiratory pathway is blocked, reduced nutrient uptake (Morris, 1984), the accumulation of toxic sulfides in the substrate (Koch and Mendelssohn, 1989), and changes in the availability of nutrients (Patrick and Delaune, 1972; see Chapter 7). Salt inhibition and oxygen depletion frequently occur together. *Spartina* grows shorter in the inland marsh because its drainage is poor and, hence, oxygen deficits are severe. Salt may concentrate in this environment.

The primary result of poor drainage in inland salt marshes, however, is apparently a dramatically lower soil redox potential, which, in turn, leads to elevated sulfide concentrations. Although *S. alterniflora* is able to mitigate the toxic effects of sulfide to some extent through its ability to transport oxygen through the root system to the rhizosphere (see Chapter 7) and by the enzymatic oxidation of sulfides (R.W. Lee et al., 1999), its growth is inhibited when the interstitial soluble sulfide concentration exceeds 1 mM sulfide (Bradley and Dunn, 1989; Koch et al., 1990).

Fate of Primary Production

Decomposition

Since Teal's seminal publication on energy flow in the salt marsh system (Teal, 1962), salt marshes have been considered detritus systems. Almost three-quarters of the primary production in the salt marsh ecosystem is broken down by bacteria and fungi. In his study of energy flow within the salt marsh environment, Teal (1962) estimated that 47 percent of the total net primary productivity was lost through respiration by microbes. It was largely assumed that the rich secondary productivity of estuaries was fueled by a detritus food web. In recent years, with the development of such new techniques as multiple stable isotope fractionation, these early assumptions have been questioned, and a refined and quite different picture of decomposition and secondary production has emerged. With a few exceptions such as salt marshes in Mediterranean-type climates, primary production is dominated by emergent spermatophytes, usually grasses. When they senesce, the soluble organic contents are rapidly flushed from their tissues. This labile soluble organic matter from both living and decomposing salt marsh vegetation (which may be as much as 25 percent of the initial dry weight of the dying grass) is an important energy source for microorganisms in the marsh and the adjacent estuary (Wilson et al., 1986; Newell and Porter, 2000). The remaining 75 percent of dead vegetation biomass is largely composed of refractory structural lignocellulose that is indigestible by all but a few metazoans. Our ideas about the fate of this vegetation biomass have changed rapidly in the past 10 years. Some key conclusions about the process of decomposition are:

1. The initial secondary producers, or decomposers, on epibenthic marsh grass stems are ascomycetous fungi. These fungi may reach a biomass equal to 3 (summer) to 28 (winter) percent of live *Spartina alterniflora* standing crop. Most of this biomass occurs in standing dead grass or on the marsh surface (Table 9-8). In South Atlantic coastal marshes, fungal productivity is 10 times greater in winter than in summer. In contrast, most of the bacterial biomass is found in the sediment surface microlayer. Productivity of bacteria is twice fungal productivity in summer, but only one-tenth as great in winter. The conversion efficiency of grass biomass to fungal biomass can be as high as 50 percent (Newell and Porter, 2000).

2. There appear to be at least three decomposer groups. (a) Fungi are the major decomposers of the epibenthic standing dead grass. (b) Aerobic bacteria in the surface microlayer decompose the decayed grass leaf shoots that are shredded by gastropods and amphipods and fall to the marsh surface. (c) Anaerobic bacteria, a third group of decomposers in deeper anoxic sediments, are able to use electron acceptors other than oxygen to metabolize. Prime among these are sulfate reducers, which may oxidize a major proportion of the underground senescent root and rhizome biomass.

During the decomposition process, the nitrogen content of the grass/fungal/bacterial brew increases. This is due, in part, to the low C : N ratio of bacterial decomposers compared to raw grass tissue. It was assumed for many years that nitrogen enrichment made the decaying plant material a nutritionally better food supply for consumers (E. P. Odum and de la Cruz, 1967); in more recent studies, however, it was determined that much of the nitrogen is bound in refractory compounds in the decaying grass (Teal, 1986). The nutritious bacterial population is kept at low concentrations by metazoan grazing.

Consumption

These recent discoveries about the decomposition process in marsh macrophytes have led to a reevaluation of the source of energy for the abundant consumer population found in tidal marshes and their associated tidal creeks. The emerging picture is quite different in detail than earlier models such as that of Teal (1962). Although much of the change consists of elaboration and clarification of the detrital process, a major

Table 9-8 Estimates of living fungal biomass (g/m² marsh surface) in standing–decaying *Spartina alterniflora* shoot parts and of total bacteria in the sediment surface microlayer, averaged over tall- to short-form stands in three marsh watersheds

	Summer	Winter
Fungal crop in dead blades	5	23
Fungal crop in dead sheaths + stems	<u>14</u>	<u>29</u>
Sum of fungal crop	19	52
Bacterial crop on dead shoot parts	0.02	0.07
Bacterioplankton, 0.5 m depth flooding tide	0.25	0.15
Bacteria in sediment to 20-cm depth	44	44

Source: Newell and Porter (2000).

shift has been toward a much greater role for algae, both phytoplankton and especially edaphic algae, as major flows of energy in the salt marsh food web. Figure 9-13 illustrates some of these changes in the way the food web might now be structured (Kreeger and Newell, 2000; Newell and Porter, 2000). Vascular plants are still the major source of organic carbon, but few metazoans can assimilate this cellulose-rich material. Hence, direct grazers are limited to several species of herbaceous insect, which collectively consume less than 10 percent of plant production. The ribbed mussel, *Geukensia demissa,* is an exception to this generalization. It has been shown to assimilate aseptic detrital cellulose with an efficiency of up to 15 percent. Kreeger and Newell (2000) suggest that the mussel must either possess endogenous cellulases or contain a vigorous gut flora capable of cellulose breakdown.

Decay begins with fungal decomposition of the aerial parts of the senescent vascular plants. Epiphytic algae growing on the lower parts of the grass culms are also a part of this detrital brew. The complex is ingested and shredded by shredder gastropod snails, such as *Littorina irrorata,* and perhaps amphipods. In a microcosm experiment, the snails had the capacity to ingest 7 percent of their weight of naturally decayed leaves per day and assimilate them with an efficiency of about 50 percent. The epiphytic algae are also ingested and assimilated by amphipods and other organisms grazing on the dead leaf surfaces.

The finely shredded grass/fungal/algal material that falls to the marsh surface is infected by aerobic bacteria that continue the process of decomposition. Also part of this mixture is the algal community, largely diatoms, growing on the marsh surface. This complex is consumed by benthic meiofaunal and macrofaunal deposit feeders. Primary among the meiofauna are nematodes; also feeding on the surface are harpacticoid copepods, amphipods, polychaetes, turbellarians, ostracods, foraminifera, and gastroliths. The larger consumers in this group include fiddler crabs, snails, polychaetes, oligochaetes, and some bivalves.

Finally, some of the finely decomposed organic material on the surface microlayer is periodically suspended by winds and currents, where it mixes with the phytoplankton growing in the water. For example, as much as 25 percent of the suspended algae have been found to be edaphic species (MacIntyre and Cullen, 1995). This sestonic mix of bacterial/organic fragments, free-living bacteria, and algae, are consumed by suspension feeders, especially benthic suspension feeders such as bivalve mollusks and oligochaete annelids. Also active are zooplankton, although they probably do not process as much material as the benthic bivalves.

The meio- and macrofauna feeding on algae, fungi, and bacteria, are, in turn, consumed by animals in the higher trophic levels (Fig. 9-13).

Three important insights resulted from the research highlighted previously:

1. Algae are important components of the salt marsh food web, so much so that Kreeger and Newell (2000) stated:

> We question the paradigm that salt marshes have "detritus-based food webs" (Odum, 1980), considering that the bulk of secondary production by metazoans could actually be linked to primary production by the microphytobenthos rather than through either direct (herbivory) or indirect (detrivory) linkages to primary production by vascular plants.

Multiple stable isotope studies showed conclusively that *Spartina*-derived material does not provide the primary diet for a number of metazoans studied.

2. Most metazoans are probably omnivorous, using different carbon sources seasonally as they are available. Intensive studies of the ribbed mussel (summarized in Kreeger and Newell, 2000) showed that no single component of marsh seston can supply sufficient organic carbon to support the adult mussel (in a marsh in Delaware). The available food sources and probable diet of the mussel change seasonally. Indirect use of vascular plant material through ingestion of microheterotrophic bacteria and protists is the most valuable carbon resource in all seasons except summer, when phytoplankton represent the largest resource (Fig. 9-16). In spring and summer, microheterotrophs can supply about 50 percent of the mussel's carbon demands, in fall and winter all of their carbon requirements. Seasonally, phytoplankton could account for 36 to 100 percent of the mussel's carbon requirements. Benthic microalgae can contribute between 10 and 53 percent. Detrital cellulose was estimated to supply only 0 to 9 percent of carbon demands. These figures give some indication of the omnivorous nature of a single species. Rather than the exception, omnivory is probably the rule.

3. Nitrogen requirements of metazoan consumers may be more critical than organic carbon (energy) requirements. From this point of view, the nitrogen-poor vascular plants cannot supply the needs of their consumers. The microheterotrophic microbial and protist population represents the greatest source of nitrogen for the

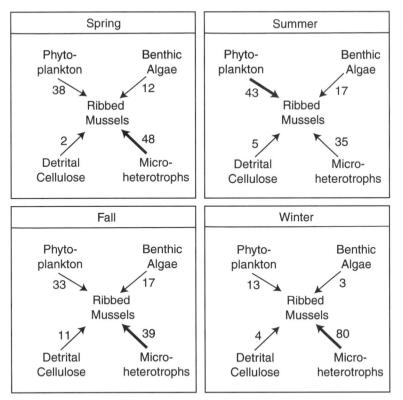

Figure 9-16 Relative food value of phytoplankton, benthic microalgae, detrital cellulose, and microheterotrophs in balancing the carbon demands of ribbed mussel *Geukensia demissa* during the spring, summer, fall, and winter. Values given are percentages of the total carbon potentially assimilated by mussels. (*After Kreeger and Newell, 2000*)

ribbed mussel throughout the year (Kreeger and Newell, 2000). The second most important nitrogen source is phytoplankton, followed by benthic microalgae. In this study, the sum of these three sources almost exactly balanced the nitrogen demands during the spring and fall, suggesting that benthic microalgae could be an essential component of the mussel's diet.

Organic Export

A central paradigm of salt marsh ecology has long been the "outwelling hypothesis," which was first enunciated by E. P. Odum in 1968, based, in part, on a salt marsh energy flow analysis presented by John Teal at the first salt marsh conference, held in 1958 at the University of Georgia Marine Laboratory on Sapelo Island, Georgia (published as Teal, 1962). Odum described salt marshes as "primary production pumps" that feed large areas of adjacent waters (Odum, 1968), and he compared the flow of organic material and nutrients from salt marshes to the *upwelling* of deep ocean water, which supplies nutrients to some coastal waters. Teal (1962) hypothesized that salt marshes exported organic material and energy primarily as detritus from the marsh surface. In the intervening years, there have been many attempts to measure this export. Nixon (1980) summarized studies to 1980, and Childers et al. (2000) completed a more recent update. Teal's (1962) energy flow analysis estimated that about 45 percent of net primary production was exported from the salt marsh. Nixon's summary agreed in that most studies showed an export of dissolved and particulate material, in an amount that could account for about 10 to 50 percent of phytoplankton production in coastal and estuarine waters.

Childers et al. (2000) pointed out that the original hypothesis was ambiguous in that it equated salt marsh export to coastal ocean import. In reality, the flows from a salt marsh are into nearby tidal creeks, and fluxes to the coastal ocean depend on the geomorphology of the estuary and the distance from the marsh to the coast. Hence, salt marshes interact with nearby tidal creeks and the inner estuary, which, in turn, exchange flows with the greater estuarine basin, which, finally, interacts with the coastal ocean (Fig. 9-17). Failure to take these spatial factors into account has

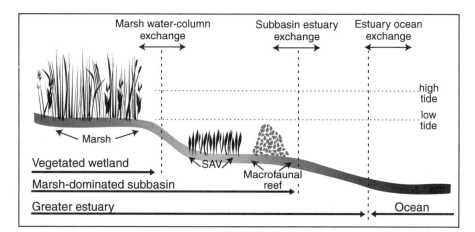

Figure 9-17   **Hierarchy of estuarine–coastal landscape that includes estuarine subbasins nested within the greater estuary and vegetated wetlands nested within both.** (*After Childers et al., 2000*)

made it difficult to compare studies and is one source of the lack of agreement in study results.

The evidence for outwelling rests on more than organic flux data, such as reported in the studies summarized by Nixon (1980) and Childers (1994). Hopkinson (1985) reported that water column respiration offshore of the Georgia barrier islands exceeded *in situ* production; that is, the zone was heterotrophic, implying that organic matter was being imported from the inshore estuaries and marshes. Turner et al. (1979) reported that offshore, within 10 km of the coastal estuaries, primary productivity measurements were often 10 times greater than that farther offshore. They attributed this high productivity to outwelling of nutrients from the estuaries.

Other evidence of outwelling comes from fishery studies. Turner (1977) found a close correlation worldwide between commercial yield of shrimp (which are harvested both in the estuary and offshore) and the area of estuarine intertidal vegetation. Teal and Howes (2000) analyzed fish catch statistics dating back to 1880 from the Long Island Sound, New York, and determined that fish catch was closely related to marsh edge length. Since edge length is an index of accessibility to the marsh, the result implicated salt marsh production in commercial fishery catch.

Several general factors affect the outwelling hypothesis. First, material and energy usually flow from concentrated hot spots to lower concentration areas. Salt marshes are hot spots of production, so it is logical to expect an outwelling of production and food energy (E. P. Odum, 2000). Second, outwelling can be expected to be modified by the geomorphology of the estuary and the location of a salt marsh in the estuary. Thus, open estuaries with salt marshes close to the coast are expected to export more material than estuaries with small coastal passes and distant marshes (W. E. Odum et al., 1979). Finally, salt marshes and coastal estuaries are pulsing systems, with daily tidal variation, seasonal variations in rainfall and river flow, and periodic severe storms. Extreme events often lead to import or export that overwhelms the normal daily fluxes.

Energy Flow

The energy budget study of a Georgia salt marsh by Teal (1962) remains a classic attempt to quantify energy fluxes in the salt marsh, although many of his values would be modified today (Fig. 9-18). Gross primary productivity was calculated to be 6.1 percent of incident sunlight energy, verifying the observation that the salt marsh is one of the most productive ecosystems in the world. Only 1.4 percent of the incident light, however, was converted into organic material available to other organisms. Herbivorous insects, mostly plant hoppers (*Prokelisia*) and grasshoppers (*Orcheli-mum*), consumed only 4.6 percent of the *Spartina* net productivity. The rest of the *Spartina* net productivity and that of the mud benthic algae passed through the detrital–algal food chain or was exported to the adjacent estuary. An estimated 45 percent of the net production was exported from the marsh into the estuary in this study. As we saw in the previous section, this estimate is unrealistically high.

Figure 9-13, which summarizes a number of more recent studies, amplifies the details of Teal's energy flow diagram. The flows are unquantified, but show in much more detail the complexity of the salt marsh system. The decomposer and invertebrate consumer portions of the food web were discussed in the previous section. The figure also illustrates the connections to vertebrates that not only feed and nest or burrow on the marsh, but also export organic energy to the estuary and coastal ocean. This energy transfer likely occurs through changes in both diet and habitat use of fish as

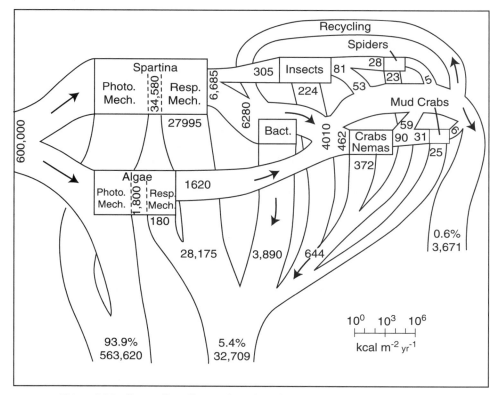

Figure 9-18 Energy flow diagram for a Georgia salt marsh. (*After Teal, 1962*)

they mature (Smith et al., 2000) and through trophic "relays" as marsh-inhabiting minnows are consumed by migratory nektonic predators. Thus, invertebrates on and near the edge of the marsh form the primary link between marsh-associated production and fish consumers. They transform detrital microbes and microalgae into biomass available to fish. This organic material is transferred to the greater estuary and offshore by fish migrations and by trophic transfers to predatory fish, which move between the estuary and the coastal ocean (Cicchetti and Diaz, 2000; Deegan et al., 2000). The magnitude of these live organism energy flows has not been quantified, but indications are that their magnitude in the marsh energy budget is significant.

Nutrient Budgets

Salt marshes have the longest history of nutrient budget studies, have been the subject of the most comprehensive studies of nutrient dynamics, and probably have generated the most controversy about the source–sink question. A critical evaluation of 20 years of research on the role of salt marshes in nutrient cycles was presented by Nixon (1980) with the following general conclusion:

> On the basis of very little evidence, marshes have been widely regarded as strong
> terms (sources or sinks) in coastal marine nutrient cycles. The data we have
> available so far do not support this view. In general, marshes seem to act as
> nitrogen transformers, importing dissolved oxidized inorganic forms of nitrogen

and exporting dissolved and particulate reduced forms. While the net exchanges are too small to influence the annual nitrogen budget of most coastal systems, it is possible that there may be a transient local importance attached to the marsh–estuarine nitrogen flux in some areas. Marshes are sinks for total phosphorus, but there appears to be a remobilization of phosphate in the sediments and a small net export of phosphate from the marsh.

Various studies that led to the preceding summary had depicted salt marshes as either sources or sinks of nutrients. The salt marsh was described as a source of phosphorus for the adjacent estuary in studies in Delaware (Reimold and Daiber, 1970; Reimold, 1972), Georgia (Gardner, 1975), and Maryland (Stevenson et al., 1977). Studies of the Flax Pond Marsh in Long Island, New York (Woodwell and Whitney, 1977; Woodwell et al., 1979), found a net input of organic phosphorus and a net discharge of inorganic phosphate. Nitrogen budgets of the same marsh (Woodwell et al., 1979) and of the Great Sippewissett Marsh in Massachusetts (Valiela et al., 1978) indicated seasonal changes in nitrogen species but, within the accuracy of the measurements, a probable balance between inputs and outputs of nitrogen. Whitney et al. (1981), in a summary of many years of research on Sapelo Island, Georgia, described that marsh as generally balanced with regard to phosphorus and a net sink of nitrogen, primarily because denitrification greatly exceeded nitrogen fixation. In some studies, for example, Valiela et al. (1973), salt marshes have been shown to be nutrient sinks when fertilized with sewage sludge.

Nutrients are carried into salt marshes by precipitation, surface water, groundwater, and tidal exchange. Because many salt marshes appear to be net exporters of organic material (with incorporated nutrient elements), a nutrient budget would be expected to show that the marsh is a net sink for inorganic nutrients (to balance the nutrient budget). Recent studies have shown that this is not always the case.

One of the most ambitious studies of nutrient dynamics in the salt marsh ecosystem was carried out at the Great Sippewissett Marsh in Massachusetts (Valiela et al., 1978; Teal et al., 1979; Kaplan et al., 1979). It is presented here because it shows clearly the many possible nitrogen sources and sinks. A summary diagram of the nitrogen budget developed from this study is given in Figure 9-19. Valiela et al. (1978) estimated the amount of nitrogen in groundwater inputs, precipitation, and tidal exchanges in the marsh. Nitrogen from groundwater entering the marsh primarily as nitrate nitrogen (NO_3–N) and from tidal exchange represented the major fluxes of nitrogen to the salt marsh. Precipitation contributed significantly less nitrogen, mostly as NO_3–N and dissolved organic nitrogen (DON). Nitrogen fixation by bacteria was significant (Teal et al., 1979), whereas that of blue–green algae was much less (Carpenter et al., 1978). Kaplan et al. (1979) found denitrification to be high in the salt marsh, particularly in muddy creek bottoms and in the short *Spartina* marsh. Tidal exchange was found to result in a net export of nitrogen, mostly in the form of DON. The inflows and outflows roughly balanced, but upstream surface water inflow was not quantified. Most marshes deposit nitrogen in unavailable peat sediments. This also was not quantified, although it is important to the structural stability of the marsh.

The issue of whether there is a net uptake or a net discharge of nutrients by the salt marsh was addressed by Nixon (1980; see the previous quotation). Childers et al. (2000) added further insight to his conclusion that, in general, net fluxes from the salt marsh were not very large components of estuarine nutrient cycles. Childers et al. (2000) addressed the question of marsh export to the inner estuary, as determined from eight studies of estuaries in the southeastern United States, where fluxes were

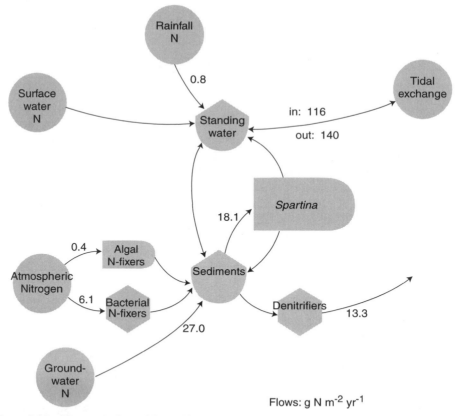

Figure 9-19 **Nitrogen budget of Great Sippewissett Marsh, Massachusetts. (*Data from Valiela et al., 1978; Teal et al., 1979; Kaplan et al., 1979; Valiela and Teal, 1979*)**

measured in flumes erected to funnel water on and off the marsh. That is, these studies investigated marsh fluxes, uncomplicated by questions about what happened to the material in the adjacent estuary. Nutrient and organic matter fluxes were found to be positively correlated with tidal range. Export from the marsh occurred in microtidal situations, changing to import at higher tidal ranges. The switch from export to import consistently occurred at about the 1-m tidal range. It is interesting that suspended sediment fluxes behaved oppositely. Suspended sediments were exported at tidal ranges less than 1 m, but imported at greater tidal ranges. The results for organic constituents appear to contradict the outwelling hypothesis, but Childers et al. (2000) suggest that the data are misleading because, at high tidal energies, plants and sediments would tend to take up increased amounts of nutrients from the inundating water column, and would also increase well-documented subsurface export flows, enriched in nutrients, from the marsh to adjacent tidal creeks. Childers (1994) further estimated the magnitude of these subsurface return flows for tidal ranges greater than 1 m, using a dynamic nutrient budget approach (Childers et al., 1993). Using conservative figures for subsurface flux, Childers estimated that inorganic nitrogen influx from the adjacent water column was greater than subsurface return flow, but phosphorus loss through subsurface flows was three times surface inflows (Table 9-9). This is ecologically reasonable, since inflowing inorganic nitrogen is rapidly

Table 9-9 Concentrations of inorganic nitrogen and phosphorus in net tidal (surface) inflows and benthic (subsurface) effluxes from salt marshes, estimated by a dynamic nutrient budget approach

Flux	Ammonium Nitrogen (mg N/L)	Nitrate/Nitrite Nitrogen (mg N/L)	Soluble Reactive Phosphorus (mg P/L)
Net tidal inflow into salt marsh	2.5	0.58	0.05
Subsurface advection from salt marsh	1.3	0.01	0.16

Source: Childers et al. (1993).

taken up by both sediments and plants, where it is reduced to organic forms and/or lost by denitrification.

Ecosystem Models

Several conceptual and mathematical models have been developed to describe the structure and function of the salt marsh ecosystem. Teal's energy flow diagram (Fig. 9-18), the diagram of a salt marsh food web (Fig. 9-13), and the nitrogen budget of the Great Sippewissett Marsh (Fig. 9-19) represent conceptual models of the salt marsh and contain valuable information about the flow of energy and materials in the marsh. A conceptual model that integrates many of the concepts in these and other models of salt marshes is illustrated in Figure 9-20.

These and other data sets have led to a number of simulation models of salt marshes, including those by Williams and Murdock (1972), Reimold (1974), Wiegert et al. (1975), Hopkinson and Day (1977), Zieman and Odum (1977), Wiegert and Wetzel (1979), Wiegert et al. (1981), Morris (1982), Chalmers et al. (1985), Morris and Bowden (1986), and Wiegert (1986). Reimold (1974) expanded on the phosphorus model of a *Spartina* salt marsh developed by Pomeroy et al. (1972). The model has five phosphorus compartments, including water, sediments, *Spartina*, detritus, and detrital feeders. The model was used to simulate the effects of perturbations such as *Spartina* harvesting on the ecosystem and to help design subsequent field experiments. The simulations demonstrated that *Spartina* regrowth and the resulting phosphorus in the water depended on the time of year that the harvesting of the marsh grass took place.

Morris and Bowden (1986) developed a more detailed model of the sedimentation of exogenous and endogenous organic and inorganic matter, decomposition, above- and below-ground biomass and production, and nitrogen and phosphorus mineralization in an Atlantic Coast salt marsh. They found that the model calculations of nitrogen and phosphorus export were sensitive to small changes in below-ground production and the fraction of refractory organic matter, both parameters for which there is little good verification. The model demonstrates that mature marshes with deep sediments recycle proportionally more of the nutrients required for plant growth than do young marshes.

Wiegert and his associates (Wiegert et al., 1975; Wiegert and Wetzel, 1979; Wiegert et al., 1981; Chalmers et al., 1985; Wiegert, 1986) have put considerable effort into the development of a simulation model for Sapelo Island, Georgia, salt marshes. The model has 14 major compartments and traces the major pathways of carbon in the ecosystem through *Spartina,* algae, grazers, decomposers, and several

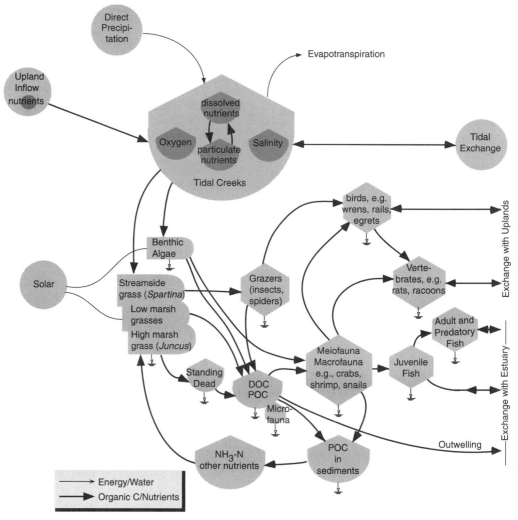

Figure 9-20 **Conceptual model of a tidal salt marsh.**

compartments of abiotic carbon storage. The model was originally constructed to answer three questions (Wiegert et al., 1981):

1. Is the Georgia salt marsh a potential source of carbon for the estuary, or is it a sink for carbon from offshore?

2. What organism groups are most responsible for the processing of carbon in the salt marsh?

3. What parameters are important (but poorly known) for the proper modeling of the salt marsh?

One of the most important revelations from this model-building process was the demonstration of the importance of the tidal export coefficient. It consistently gave export values on the order of $1,000$ g C m^{-2} yr^{-1}, well above the field measurements obtained by other researchers such as E. P. Odum and de la Cruz (1967). The model results led to intensive investigations conducted during the next 5 years of the seasonal variation in particulate and dissolved organic carbon concentrations in the river adjacent to the marsh (Chalmers et al., 1985). This resulted in a more sophisticated model and a revised hypothesis about the dynamics of carbon flux in which material is moved off the marsh in the guts of migratory feeding fish and birds or cycled from the marsh to the upper ends of tidal creeks by local rainfall at low tide and then redeposited on the marsh when tides rise. This is an excellent example of the use of a model as an intellectual tool to drive field research.

Sea-Level-Rise Coastal Models

Some applications of ecological modeling to coastal marshes have been to investigate, on a regional scale, the impacts of future sea-level rise due to global warming. A spatial cell-based simulation model developed by Park and colleagues (Park et al., 1991; J. K. Lee et al., 1991), called SLAMM (Sea Level Affecting Marshes Model), predicted that a 1-m sea-level rise in the next century could lead to a 26 to 82 percent loss of U.S. coastal wetlands, depending on the protection afforded. Most of the loss would occur in the Southeast, especially in Louisiana (Titus et al., 1991). J. K. Lee et al. (1991) predicted a 40 percent loss of wetlands along the northeastern Florida coastline with a 1-m sea-level rise, mostly as low salt marsh.

Recommended Readings

Allen, J. R. L., and K. Pye, eds. 1992. Saltmarshes: Morphodynamics, Conservation and Engineering Significance. Cambridge University Press, Cambridge, UK. 184 pp.

Day, J. W., Jr., C. A. S. Hall, W. M. Kemp, and A. Yáñez-Arancibia. 1989. Estuarine Ecology. John Wiley & Sons, New York. 558 pp.

Gosselink, J. G. 1984. The Ecology of Delta Marshes of Coastal Louisiana: A Community Profile. U.S. Fish and Wildlife Service, Biological Services. FWS/OBS-84/09, Washington, DC. 134 pp.

Pomeroy, L. R., and R. G. Wiegert, eds. 1981. The Ecology of a Salt Marsh. Springer-Verlag, New York. 271 pp.

Weinstein, M. P., and D. A. Kreeger, eds. 2000. Concepts and Controversies in Tidal Marsh Ecology. International Symposium: Concepts and Controversies in Tidal Marsh Ecology, Cumberland College, Vineland, NJ, April 5–9, 1998. Kluwer Academic Publishers, Dordrecht, The Netherlands.

Chapter 10

Tidal Freshwater Marshes

Freshwater coastal wetlands are unique ecosystems that combine many features of both salt marshes and freshwater inland marshes. They act in many ways like salt marshes, but the biota reflect the increased diversity made possible by the reduction of the salt stress found in salt marshes. Plant diversity is high, and more birds use these marshes than any other marsh type. Because they are inland from the saline parts of the estuary, they are often close to urban centers. This makes them more prone to human impact than coastal salt marshes. Tidal freshwater marshes can be grouped into three types: mature marshes, usually with tidal ranges of 1 to 2 m, such as those found on the Atlantic Coast of the United States; mature coastal floating marshes, such as those found on the northern coast of the Gulf of Mexico; and newly emergent marshes formed in prograding deltas, similar to the new marshes in the active deltas of the Mississippi River. Mature tidal freshwater marshes, whether anchored or floating, have large peat reserves. Despite the sometimes large tidal fluxes, these marshes have closed nutrient cycles that depend little on inputs from outside the system. In contrast, the newly emergent marshes are open systems with extremely low sediment organic content. They depend on flooding water to supply their nutrient needs.

This chapter is concerned with freshwater marshes that are close enough to coasts to experience significant tides but, at the same time, are above the reach of oceanic salt water. This set of circumstances usually occurs where precipitation is high or fresh river water runs to the coast, and where the morphology of the coast amplifies the tide as it moves inland. Tidal freshwater marshes are interesting because they receive the same "tidal subsidy" as coastal salt marshes but without the salt stress. One would expect, therefore, that these ecosystems might be very productive and also more diverse than their saltwater counterparts. As tides attenuate upstream, the marshes assume more of the character of inland freshwater marshes (Chapter 12). The distinction between tidal and inland freshwater marshes is not clear-cut because on the coast they form a continuum (Fig. 10-1). In this chapter, we include the tidally dominated

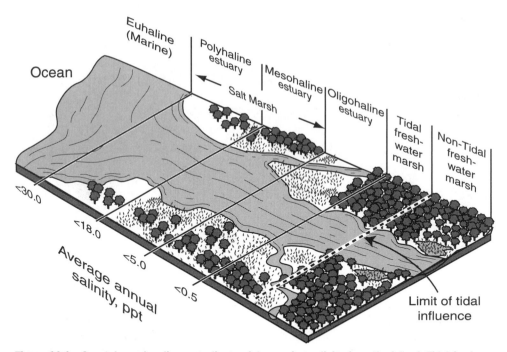

Figure 10-1 **Coastal marshes lie on gradients of decreasing salinity from the inland. Tidal fresh-water marshes still experience tides but are above the salt boundary. Further inland, marshes experience neither salt nor tides.**

systems of the Atlantic Coast and the extensive coastal freshwater marshes of the northern Gulf of Mexico, even though the latter are influenced more by wind tides than by lunar tides.

Coastal freshwater marshes are probably not as well studied or as well understood as other coastal wetlands. Several reviews and studies comparing tidal freshwater marshes with tidal salt and nontidal fresh marshes include Good et al. (1978), Simpson et al. (1983), W. E. Odum et al. (1984), Bowden (1984, 1987), W. E. Odum (1988), and Sasser (1994). These are drawn on heavily in the following discussion.

To illustrate the range of variation in these systems, we draw examples from three quite different types of tidal freshwater marshes. Figure 10-2 illustrates conceptually these three types of marshes, with major flows of water, energy, and nutrients highlighted.

1. *Mature Marshes.* Most of the available information concerns mature (approximately 500 years old with a well-developed peat substrate) marshes, particularly those on the Atlantic Coast of the United States.

2. *Floating Marshes.* On the northern Gulf Coast, many of these mature marshes have broken loose from the underlying mineral substrate and float on the water surface.

3. *New Marshes in Prograding Deltas.* The third type, also from the northern Gulf Coast, is marshes developing on emerging land where major rivers are building new deltas.

a. Mature Marsh

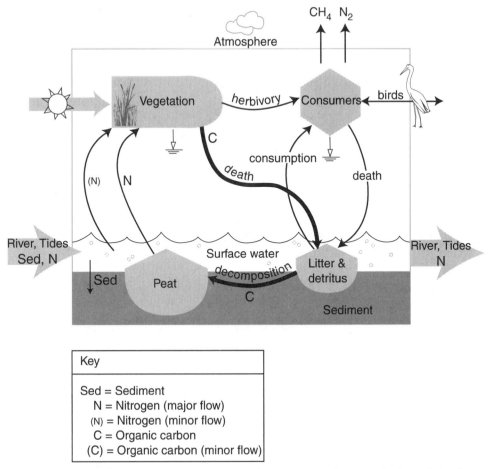

Figure 10-2 **Conceptual models of sediments, nutrients, and organic carbon in three kinds of tidal freshwater marshes:** **a.** **mature marsh;** **b.** **floating marsh;** **c.** **new marshes on emerging delta spays.**

Although these types of tidal marshes share many attributes, in some ways they respond very differently to the same environmental forces.

Geographical Extent

The physical conditions for tidal freshwater marsh development are adequate rainfall or river flow to maintain fresh conditions, a flat gradient from the ocean inland, and a significant tidal range. These conditions occur where major rivers debouch into coastal waters. In the United States, this is predominantly along the Atlantic and northern Gulf coasts. A number of rivers bring fresh water to the coast, precipitation is moderate and fairly evenly spread throughout the year, and the broad coastal plain is flat and deep. The morphology of the system is such that tidal water is often constricted as it moves inland, resulting in amplification of the tidal range, typically from 0.5 to 2 m. Along the northern Gulf Coast, the tidal range is small—less than

b. Floating Marsh

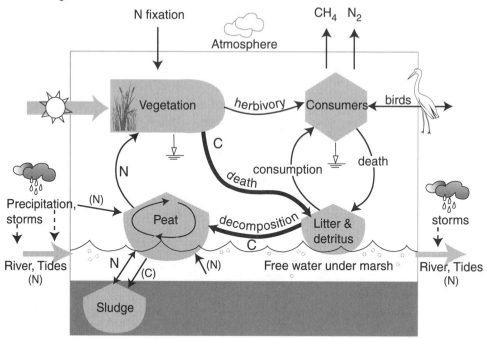

c. New Emergent Marsh

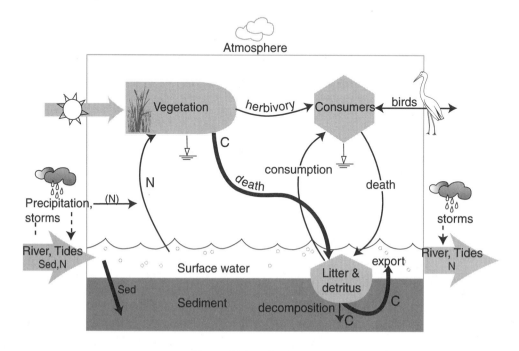

Figure 10-2 *(Continued).*

0.5 m at the coast—and this range attenuates inland. Nevertheless, because the land slope is so slight, freshwater marshes as far inland as 80 km experience some lunar tides, although these are overridden by wind tides and storm runoff.

W. E. Odum et al. (1984) estimated that Atlantic Coast tidal freshwater marshes cover about 164,000 ha (Fig. 10-3). Most of these are along the Mid-Atlantic Coast, perhaps one-half in New Jersey. This area estimate does not include upstream nontidal marshes. The Gulf of Mexico coastal freshwater marshes are concentrated in Louisiana, where they cover about 383,000 ha (Gosselink et al., 1998). Not all of these marshes are tidal.

Geomorphology and Hydrology

Coastal freshwater wetlands occur on many different kinds of substrates, but, in spite of regional differences, their recent geological history is similar (W. E. Odum et al., 1984). Contemporary coastal marshes are recent (Holocene) in origin. They lie in river valleys that were cut during Pleistocene periods of lowered sea levels. When the sea level rose after the last glaciation (15,000–5000 B.P.), freshwater coastal marshes expanded rapidly as drowned river systems were inundated and filled with sediment. Except on the northern Gulf of Mexico, these marshes are probably still expanding because of the recent increased soil runoff associated with forest clearing, agriculture, and other human activities. On the Gulf Coast, the rapid rate of subsidence is degrading mature fresh tidal marshes, while, at the same time, the Mississippi River and its major tributary (the Atchafalaya River) are building new tidal freshwater marshes elsewhere. A vertical section through a present-day tidal freshwater marsh might show a sequence of sediments built on top of an eroded Pleistocene surface cut during a glacial period of lowered sea level. The sediments might include varying layers of riverine, estuarine, and marsh sediments, capped by recent tidal freshwater marsh sediments varying in thickness from less than 1 m to more than 10 m (W. E. Odum et al., 1984).

In Atlantic Coast tidal freshwater marshes, vigorous 2-m semidiurnal tides and accompanying strong currents create a typical elevation gradient from tidal streams into adjoining marshes. Odum et al. (1984) and Metzler and Rosza (1982) described similar cross-sectional profiles, with elevation increasing slowly from the stream edge to adjacent upland areas. The low edge marsh is usually geologically younger than the mature high marsh. A common feature is a slightly elevated levee along the margin of the creek, where overflowing water deposits much of its sediment load. This corresponds to the streamside levee common in salt marshes. The elevation decreases away from the tidal channels, but Simpson et al. (1983) stated that this occurs primarily where the marshes have been impounded.

On the Atlantic Coast, high marshes are flooded to a depth of 30 cm for up to 4 hours on each tidal cycle; the low marshes are flooded deeper and may have standing water for 9 to 12 hours on each cycle. The elevation and flooding differences result in gradients of soil physical and chemical properties and in plant zonation patterns that are a consistent feature of freshwater tidal marshes.

Along the northern Gulf Coast, the weak lunar tide is often minor compared to wind tides. Winds from the east and southeast, which dominate the weather patterns during the summer, blow water into the upper (inland) reaches of coastal estuaries. Conversely, during periods of north winds, common during the winter, water is blown out of the estuaries and abnormally low water periods result. These wind tides are

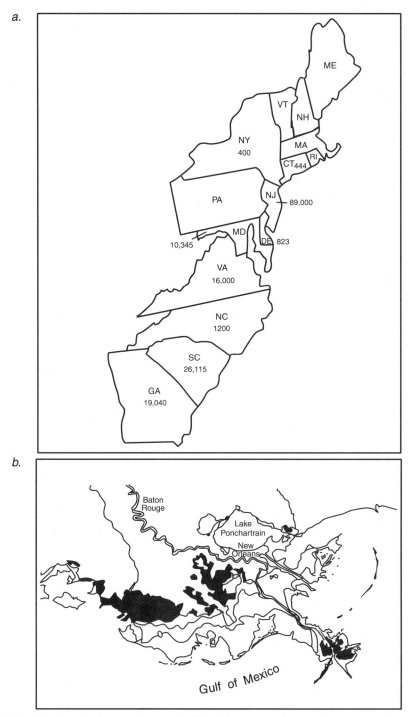

Figure 10-3 Tidal freshwater marshes in the United States: **a.** extent in hectares along the
East Coast and **b.** location in Louisiana, on the northern coast of the Gulf of Mexico. (*a. after
W. E. Odum et al., 1984; b. after Gosselink, 1984*)

responsible for water-level changes of as much as a meter and sometimes more when heavy rains accompany summer storms. Thus, the inundation pattern of Gulf Coast freshwater tidal marshes is not a regular pulse but an irregular pattern of flooding and drying, with a longer duration than lunar tides.

Another difference between Gulf Coast and Atlantic Coast tidal freshwater marshes is their elevation profiles. Because they tend to be very wide—often tens of kilometers—mature Gulf Coast marshes do not show a perceptible increase from low to high marsh. When the marshes occur at the edges of lakes, there is often a small natural levee caused by sediment washover from the lake, but, otherwise, the elevation differences and vegetation patterns do not occur in distinct linear zones. Instead, associations of plant species occur in apparently random patches, and there is considerable intergrading around the edges. An exception are the young, emerging islands of the Mississippi and Atchafalaya River deltas in Louisiana. These are closer to the coast, have a more regular pattern of tidal inundation, and exhibit a profile similar to that of Atlantic Coast marshes (Fig. 10-4a).

With sea level rising along most of the world's coasts, it is appropriate to consider whether tidal freshwater marshes are submerging or whether vertical accretion is keeping pace with sea-level rise. In the new marshes of the Mississippi River delta, high sedimentation rates exceed land submergence, especially during years with major spring floods (Roberts and Van Heerden, 1992). Hence, the marshes are expanding. Floating tidal marshes, to a great extent, are immune to the problem of rising water levels in that they are floating on the surface (Fig. 10-4b). Orson et al. (1990) described the dynamics of accretion in a tidal freshwater marsh in the Delaware River estuary (Fig. 10-5). That small near-urban marsh was an oak swamp until about 1940. The precolonial sediment accumulation rate was a low 0.04 cm yr^{-1}. After colonists arrived around 1800 and diked the river, the sedimentation rate increased to 0.12 cm yr^{-1} despite the lack of tides, perhaps due to land clearing. In 1940, dike failure restored the tidal regime and the present *Zizania, Bidens, Sagittaria, Peltandra, Amaranthus,* and *Polygonum* marsh developed rapidly. The sedimentation rate increased to 1.67 cm yr^{-1} from 1954 to 1966, during a period of unusually frequent storms. Since then, it has averaged 0.97 cm yr^{-1}. This is four times the current rate of sea-level rise (0.267 cm yr^{-1}). The authors speculate that this rapid accretion will continue until the marsh surface reaches an elevation approximating mean high tide.

Biogeochemistry

Typical ranges of chemical parameters in both water and sediments of tidal freshwater marshes are shown in Figure 10-6 and Table 10-1. Freshwater coastal marsh sediments are generally fairly organic. Along the Delaware River, they contain 14 to 40 percent organic material (Whigham and Simpson, 1975); in Virginia, 20 to 70 percent (W. E. Odum, 1988). On the Gulf Coast, the mature marsh sediments, especially in floating marshes, are nearly pure organic peat (Chabreck, 1972; Sasser et al., 1991), whereas in the newly emergent delta lobes the sediments are predominantly fine sands and silts with as little as 0.5 percent organic content (Johnson et al., 1985). The sediments are anaerobic except for a thin surface layer. This condition is reflected in the absence of

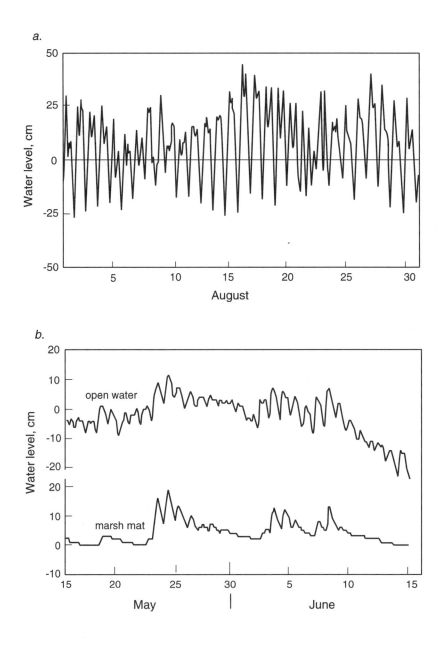

Figure 10-4 Hydrographs from different types of tidal freshwater marshes. **a.** A newly formed marsh in the Atchafalaya Delta, Louisiana. This coast has a diurnal tide with an amplitude of less than 0.5 m. **b.** A floating marsh in the coastal wetlands of the northern Gulf of Mexico. Note the trace of the marsh surface, which tracks water depth. The lunar tidal signal is present but damped. The tailing off of water level during the last week is in response to a frontal system. In contrast to these hydrographs, an Atlantic Coast tidal freshwater marsh hydrograph has a semi-diurnal tide with a range reaching 1 m or more.

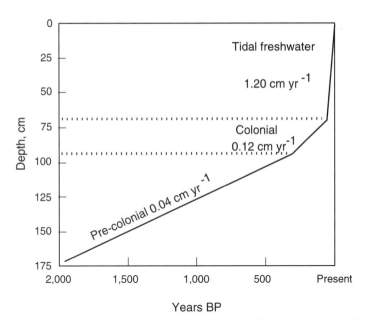

Figure 10-5 Rates of sediment accumulation through time in a tidal freshwater marsh in the Delaware River estuary on the Atlantic Coast of the United States. (*After Orson et al., 1990*)

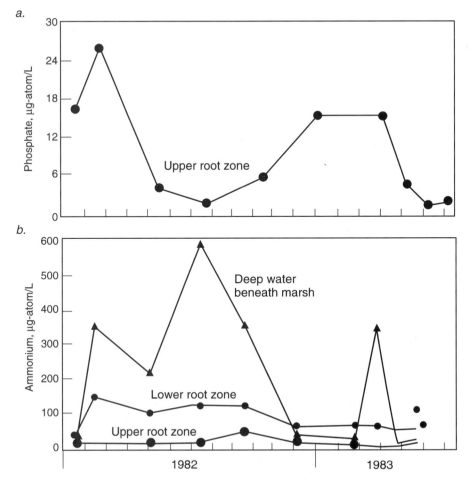

Figure 10-6 Concentration of **a.** phosphate and **b.** ammonium in interstitial water of tidal freshwater marsh sediments in Louisiana. (*After Sasser et al., 1991*)

Table 10-1 Chemistry of water and sediments in selected tidal freshwater marshes in eastern and southern United States

	North River, Massachusetts[a]	Delaware River, New Jersey[b]	James River, Virginia[c]	Barataria Bay, Louisiana[d]
Tidal range (m)	~0.5	3	<1	Low, mostly wind tide
WATER				
Dissolved oxygen (mg/L)	—	4–13	7–12	2–8
$NO_2 + NO_3$ (µg N/L)	100–1,100	40–300	500–1,600	40–370
NH_3 (µg N/L)	0–2,000	40–80	460–470	0–2,780
Reactive P (µg P/L)	10–160	5–20	30–40	35–340
Total P (µg P/L)	—	5–50	160–180	217
SEDIMENTS (top 20 cm)				
pH	—	6.2–7.7	—	6.3
Organic carbon (% of dry weight)	32	14–40	20–50	6–68
Total Kjeldahl N (% of dry weight)	1.7	1.0–1.6	1.5 ± 0.8 (s.d.)	1.5–1.8
Total P (% of dry weight)	0.16	0.12–0.35	0.7	0.09

[a]Tidal marsh; Bowden (1984) and Bowden et al. (1991).
[b]Hamilton and Woodbury Creek marsh; Whigham et al. (1980).
[c]Herring Creek marsh; Adams (1978), Lunz et al. (1978), and W. E. Odum et al. (1984).
[d]Gulf Coast fresh marsh; Chabreck (1972), Hatton (1981), and Ho and Schneider (1976).

nitrate. Ammonium is present in the winter but is reduced to low levels in the summer by plant uptake. Total nitrogen levels are closely related to the organic content, for almost all of the sediment nitrogen is bound in organic form. Phosphorus is more variable (Table 10-1). The cation exchange capacity (CEC) in James River marshes is 40 to 67 meq/100 g dry weight, which is high but typical for a highly organic soil (Wetzel and Powers, 1978). Soil acidity is generally close to neutral.

The water flooding the marsh varies in chemical composition according to the season and the source of water (Fig. 10-7). Recorded concentrations of nutrients in the water have been high, possibly because most of the detailed chemical studies reported for tidal freshwater wetlands have been from more or less polluted sites.

Ecosystem Structure

Macrophytes

Elevation differences across a freshwater tidal marsh correspond to different plant associations (Fig. 10-7). These associations are not discrete enough to call communities, and the species involved change with latitude. Nevertheless, they are characteristic enough to allow some generalizations.

Mature Marshes

On the Atlantic Coast of the United States (Fig. 10-7a and b), submerged vascular plants such as *Nuphar advena* (spatterdock), *Elodea* spp. (waterweed), *Potamogeton* spp. (pondweed), and *Myriophyllum* spp. (water milfoil) grow in the streams and permanent ponds. The creek banks are scoured clean of vegetation each fall by the strong tidal currents, and they are dominated during the summer by annuals such as *Polygonum punctatum* (water smartweed), *Amaranthus cannabinus* (water hemp), and *Bidens laevis* (bur marigold). The natural stream levee is often dominated by *Ambrosia trifida* (giant ragweed). Behind this levee, the low marsh is populated with broad-leaved monocotyledons such as *Peltandra virginica* (arrow arum), *Pontederia cordata* (pickerelweed), and *Sagittaria* spp. (arrowhead).

Typically, the high marsh has a diverse population of annuals and perennials. W. E. Odum et al. (1984) called this the "mixed aquatic community type" in the Mid-Atlantic region. Leck and Graveline (1979) described a "mixed annual" association in New Jersey, Caldwell and Crow (1992) the vegetation of a tidal freshwater marsh in Massachusetts. Generally, the areas were dominated early in the season by perennials such as arrow arum. A diverse group of annuals— *Bidens laevis, Polygonum arifolium* (tear-thumb) and other smartweeds, *Pilea pumila* (clearweed), *Hibiscus coccineus* (rose mallow), *Acnida cannabina,* and others—assumed dominance later in the season. In addition to these associations, there are often almost pure stands of *Zizania aquatica* (wild rice), *Typha* spp. (cattail), *Zizaniopsis miliacea* (giant cutgrass), and *Spartina cynosuroides* (big cordgrass). In the northern Gulf of Mexico, arrowheads (*Sagittaria* spp.) replace arrow arum (*Peltandra* spp.) and pickerelweed (*Pontedaria cordata*) at lower elevations. Visser et al. (1998) described three vegetation associations in this area: (1) Bulltongue (*Sagittaria lancifolia*) occurs with co-dominants maidencane (*Panicum hemitomon*) and spikerush (*Eleocharis* spp.). Commonly the ferns *Thelypteris palustris* and *Osmunda regalis,* wax myrtle (*Myrica cerifera*), and pennywort (*Hydroco-*

a.

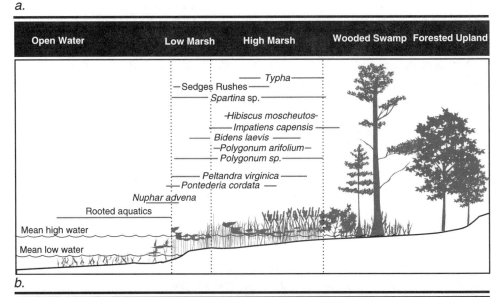

b.

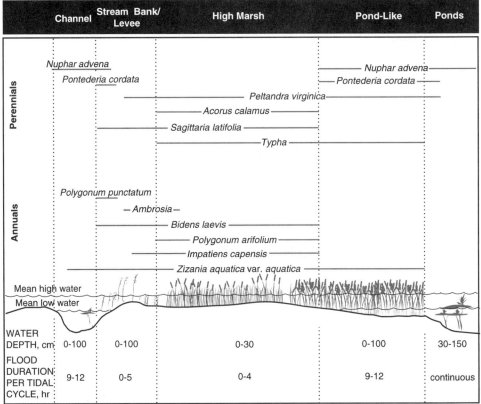

Figure 10-7 **Cross sections across typical freshwater tidal marshes, showing elevation changes and typical vegetation: a. and b. Atlantic Coast marshes; c. new marsh in the Atchafalaya Delta, Louisiana. (***a. After W. E. Odum et al., 1984; b. after Simpson et al., 1983; c. after Gosselink et al., 1998***)**

c.

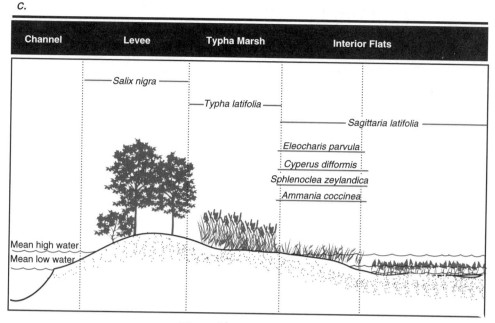

Figure 10-7 *(Continued).*

tyl spp.) are also present. Fifty-two different species occur in this association. (2) A maidencane-dominated association is widespread across the delta, and includes 55 species. (3) Cutgrass (*Zizaniopsis miliacea*) occurring with co-dominant maidencane is relatively uncommon. It includes 20 other species.

Interestingly, rising sea level and/or surface subsidence on both the Gulf Coast and the Atlantic Coast has resulted in vegetation shifts. Although the previously dominant species are still present, in a Chesapeake Bay tidal freshwater wetland, for example, the oligohaline species *Spartina cynosuroides,* which was not among the dominant species in 1974, is now second in peak biomass and fourth in importance value (Perry and Hershner, 1999). Similarly, Visser et al. (1999) reported that the maidencane association (see the preceding discussion) has decreased from 51 percent coverage of the tidal wetlands of Terrebonne Bay (in the Mississippi River delta) in 1968 to only 14 percent in 1992. It has been replaced by *Eleocharis baldwinii*–dominated marshes, which were uncommon in 1968 (3 percent coverage) but in 1992 covered 42 percent of the area.

Floating Marshes

The floating marshes in the tidal reaches of the northern Gulf of Mexico are similar to the nontidal riverine and lacustrine marshes found extensively around the globe. Large expanses of floating marshes are also found in the Danube Delta (*Phragmites communis* marshes; Pallis, 1915), along the lower reaches of the Sud in Africa (papyrus swamps; Beadle, 1974), in South America (floating meadows in lakes of the *varzea;* Junk, 1970), and in Tasmania (floating islands in the Lagoon of Islands; Tyler, 1976). Floating marshes have also been reported in Germany, the Netherlands (Verhoeven, 1986), England (Wheeler, 1980), and North Dakota and Arkansas in the United States (Eisenlohr, 1972; Huffman and Lonard, 1983). In Louisiana, floating marshes are usually floristically diverse, but different stands are dominated by *Panicum hemitomon* with ferns and vines such as *Vigna luteola* and *Ipomoea sagittata; Sagittaria lancifolia* with *Eleocharis* spp., *Panicum dichotomiflorum, Bacopa monnieri,* and *Spartina patens;* and *Eleocharis baldwinii* and *Eleocharis parvula* with *Ludwigia leptocarpa, Phyla nodiflora,* and *Bidens laevis* (Sasser et al., 1996). The marsh substrate is composed of a thick organic mat, entwined with living roots, that rises and falls (all year or seasonally) with the ambient water level (Swarzenski et al., 1991). This type of marsh is interesting in a successional sense because it appears to be an endpoint in development; it is freed from normal hydrologic fluctuation and mineral sediment deposition. Hence, in the absence of salinity intrusions, it appears to support a remarkably stable community (Sasser et al., 1995; see also Chapter 9).

New Marshes

The active deltas of the Mississippi and Atchafalaya Rivers are the sites of the largest newly prograding coastal deltas in the continental United States. Fresh tidal marshes (Fig. 10-7c) that have formed in the last 25 years on emergent islands are dominated on the natural levees by black willow (*Salix nigra*), and the extensive back-island mud flats by common three-square sedge (*Scirpus deltarum*) or by arrowhead (*Sagittaria latifolia*), with areas of cattail (*Typha latifolia*) and a seasonally variable annual/ perennial mix in between. An elephant ear (*Colocasia esculenta*) community flourishes on the water's edge (Johnson et al., 1985; White, 1993).

Seed Banks

The species composition of a tidal freshwater marsh does not appear to depend on the availability of seed in particular locations. Seeds of most species are found in almost all habitats, although the most abundant seed reserves are generally from species found in that marsh zone (Whigham and Simpson, 1975; Leck and Simpson, 1987, 1995; Baldwin et al., 1996). They differ, however, in their ability to germinate under the local field conditions and in seedling survival. Flooding is one of the main controlling physical factors. Many of the common plant species seem to germinate well even when submerged, for example, *Peltandra virginica* and *Typha latifolia,* whereas others such as *Impatiens capensis* (Leck, 1979), *Cuscuta gronovii,* and *Polygonum arifolium* show reduced germination. (Leck and Graveline, 1979; Leck and Simpson, 1987). Competitive factors also play a role: Arrow arum and cattail, for example, produce chemicals that inhibit the germination of seed (McNaughton, 1968; Bonasera et al., 1979); and shading by existing plants is apparently responsible for the inability of arrow arum plants to become established anywhere except along the marsh fringes (Whigham et al., 1979). Some species (*Impatiens capensis, Bidens laevis,* and *Polygonum arifolium*) are restricted to the high marsh because the seedlings are not tolerant of extended flooding (Simpson et al., 1983). Seed bank strategies differ in different zones of the marsh. The seeds of most of the annuals in the high marsh germinate each spring so that there is little carryover in the soil. In contrast, perennials tend to maintain seed reserves. The seeds of most species, however, appear to remain in the soil for a restricted period. In one study, 31 to 56 percent of the seeds was present only in surface samples, and 29 to 52 percent germinated only in sediment samples taken in early spring (Leck and Simpson, 1987). The complex interaction of all these factors has not been elucidated to the extent that it is possible to predict what species will be established where on the marsh.

Algae

In addition to vascular plants, phytoplankton and epibenthic algae abound in freshwater tidal marshes, but relatively little is known about them. In one study of Potomac River marshes, diatoms (Bacillariophytes) were the most common phytoplankton, with green algae (Chlorophytes) comprising about one-third of the population and blue–green algae (Cyanobacteria) present in moderate numbers (Lippson et al., 1979). The same three taxa accounted for most of the epibenthic algae. Indeed, many of the algae in the water column are probably entrained by tidal currents off the bottom. In a study of New Jersey tidal freshwater marsh soil algae, Whigham et al. (1980) identified 84 species exclusive of diatoms. Growth was better on soil that was relatively mineral and coarse compared with growth on fine organic soils. Shading by emergent plants reduced algal populations in the summer months. In nontidal freshwater marshes, algae epiphytic on emergent plants and litter made important contributions to invertebrate consumers (Campeau et al., 1994). Algal biomass is probably two to three orders of magnitude less than peak biomass of the vascular plants (Wetzel and Westlake, 1969), but the turnover rate is much more rapid.

Consumers

Invertebrates

Coastal freshwater wetlands are used heavily by wildlife. The consumer food chain is predominantly detrital, and benthic invertebrates are an important link in the food web. Bacteria and protozoa decompose litter, gaining nourishment from the organic material. It appears unlikely that these microorganisms concentrate in large enough numbers to provide adequate food for macroinvertebrates. Meiobenthic organisms, primarily nematodes, comprise most of the living biomass of anaerobic sediments (J. P. Sikora et al., 1977). They probably crop the bacteria as they grow, packaging them in bite-sized portions for slightly larger macrobenthic deposit feeders. In coastal freshwater marshes, the microbenthos is composed primarily of amoebae (Thecamoebinids, a group of amoebae with theca, or tests). This is in sharp contrast to more saline marshes in which foraminifera predominate (Ellison and Nichols, 1976). The slightly larger macrobenthos is composed of amphipods, especially *Gammarus fasciatus,* oligochaete worms, freshwater snails, and insect larvae. Copepods and cladocerans are abundant in the tidal creeks. The Asiatic clam (*Corbicula fluminaea*), a species introduced into the United States earlier in the century, has spread throughout the coastal marshes of the southern states and as far north as the Potomac River (W. E. Odum et al., 1984). Caridean shrimp, particularly *Palaemonetes pugio,* are common, as are freshwater shrimp, *Macrobrachium* spp. The density and diversity of these benthic organisms are reported to be low compared with those in nontidal freshwater wetlands, perhaps because of the lack of diverse bottom types in the tidal reaches of the estuary. No species are found exclusively in tidal freshwater systems. Instead, those found there appear to have a wide range (Diaz, 1977).

Nekton

Coastal freshwater marshes are important habitats for many nektonic species that use the area for spawning and year-round food and shelter and as a nursery zone and juvenile habitat. Fish of coastal freshwater marshes can be classified into five groups (Fig. 10-8). Most of them are freshwater species that spawn and complete their lives within freshwater areas. The three main families of these fish are cyprinids (minnows, shiners, carp), centrarchids (sunfish, crappies, bass), and ictalurids (catfish). Juveniles of all species are most abundant in the shallows, often using submerged marsh vegetation for protection from predators. Predator species, the bluegill (*Lepomis macrochirus*), largemouth bass (*Micropterus salmoides*), sunfish (*Lepomis* spp.), warmouth (*Lepomis gulosus*), and black crappie (*Pomoxis nigromaculatus*), are all important for sport fishing. Gars (*Lepisosteus* spp.), pickerels (*Esox* spp.), and bowfin (*Amia calva*) are other common predators often found in coastal freshwater marshes.

Some oligohaline or estuarine fish and shellfish that complete their entire life cycle in the estuary extend their range to include the freshwater marshes. Killifish (*Fundulus* spp.), particularly the banded killifish (*F. diaphanus*) and the mummichog (*F. heteroclitus*), are abundant in schools in shallow freshwater marshes, where they feed opportunistically on whatever food is available. The bay anchovy (*Anchoa mitchilli*) and tidewater silverside (*Menidia beryllina*) are often abundant in freshwater areas also. The latter breeds in this habitat more than in saltwater areas. Juvenile hog chokers (*Trinectes maculatus*) and naked gobies (*Gobiosoma bosci*) use tidal freshwater areas as nursery grounds (W. E. Odum et al., 1984).

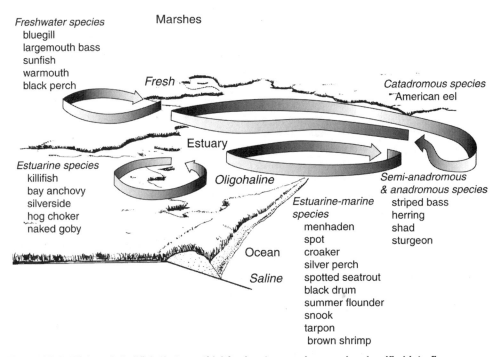

Figure 10-8 Fish and shellfish that use tidal freshwater marshes can be classified into five groups: freshwater, estuarine, anadromous, catadromous, and estuarine–marine.

Anadromous fish, which live as adults in the ocean, or semi-anadromous species, whose adults remain in the lower estuaries, pass through coastal freshwater marshes on their spawning runs to freshwater streams. For many of these species, the tidal freshwater areas are major nursery grounds for juveniles. Along the Atlantic Coast, herrings (*Alosa* spp.) and shads (*Dorosoma* spp.) fit into this category. The young of all of them, except the hickory shad (*A. mediocris*), are found in peak abundance in tidal fresh waters, where they feed on small invertebrates and, in turn, are an important forage fish for striped bass (*Morone saxatilis*), white perch (*Morone americana*), catfish (*Ictalurus* spp.), and others (W. E. Odum et al., 1984). As they mature late in the year, they migrate downstream to saline waters and offshore. Two species of sturgeon (*Acipenser brevirostrum* and *A. oxyrhynchus*) were formerly important commercially in East Coast estuaries, but were seriously overfished and presently are rare. Both species spawn in nontidal and tidal fresh waters, and juveniles may spend several years there before migrating to the ocean (Brundage and Meadows, 1982).

The striped bass is perhaps the most familiar semi-anadromous fish of the Mid-Atlantic Coast because of its importance in both commercial and sport fisheries. Approximately 90 percent of the striped bass on the East Coast are spawned in tributaries of the Chesapeake Bay system (Berggren and Lieberman, 1977). They spawn in spring in tidal fresh and oligohaline waters; juveniles remain in this habitat along marsh edges, moving gradually downstream to the lower estuary and nearshore zone as they mature. Because the critical period for survival of the young is the larval stage, conditions in the tidal fresh marsh area where these larvae congregate are important determinants of the strength of the year class (W. E. Odum et al., 1984).

The only catadromous fish species in Atlantic Coast estuaries is the American eel (*Anguilla rostrata*). It spends most of its life in fresh or brackish water, returning to the ocean to spawn in the region of the Sargasso Sea. Eels are common in tidal and nontidal coastal freshwater areas, in marsh creeks, and even in the marsh itself (Lippson et al., 1979).

The juveniles of a few species of fish that are marine spawners move into freshwater marshes, but most remain in the oligohaline reaches of the estuary. Species whose range extends into tidal freshwater marshes are menhaden (*Brevoortia tyrannus*), spot (*Leiostomus xanthurus*), croaker (*Micropogonias undulatus*), silver perch (*Bairdiella chrysoura*), spotted seatrout (*Cynoscion nebulosus*), black drum (*Pogonias cromis*), summer flounder (*Paralichthys dentatus*), snook (*Centropomus undecimalis*), and tarpon (*Megalops atlanticus*). Along the northern Gulf Coast, juvenile brown and white shrimp (*Penaeus* spp.) and male blue crabs (*Callinectes sapidus*) may also move into freshwater areas. These juveniles emigrate to deeper, more saline waters as temperatures drop in the fall.

Birds

Of all wetland habitats, coastal freshwater marshes may support the largest and most diverse populations of birds. W. E. Odum et al. (1984), working from a number of studies, compiled a list of 280 species of birds that have been reported from tidal freshwater marshes. They stated that, although it is probably true that this environment supports the greatest bird diversity of all marshes, the lack of comparative quantitative data makes it difficult to test this hypothesis (W. E. Odum, 1988). Bird species include waterfowl (44 species); wading birds (15 species); rails and shorebirds (35 species); birds of prey (23 species); gulls, terns, kingfishers, and crows (20 species); arboreal birds (90 species); and ground and shrub birds (53 species). A major reason for the intense use of these marshes is the structural diversity of the vegetation provided by broad-leaved plants, tall grasses, shrubs, and interspersed ponds.

Dabbling ducks (family Anatidae) and Canada Geese actively select tidal freshwater areas (R. E. Stewart, 1962) on their migratory flights from the north. They use the Atlantic Coast marshes in the late fall and early spring, flying farther south during the cold winter months. Most of these species winter in fresh coastal marshes of the northern Gulf, but some fly to South America. Their distribution in apparently similar marshes is variable; some marshes support dense populations, others few birds. For example, Fuller et al. (1988) found extensive use of new Atchafalaya River delta marshes by many species of ducks. Although the vegetation was dominated by arrowhead throughout the newly created islands, duck populations were twice as dense in the western islands of the delta compared to the east and central islands. On the central islands, ducks preferentially selected stands of three-square sedge over arrowhead; on the western islands where there was no three-square sedge, they frequented stands with mixed grass species over arrowhead. The reason for this selectivity is unclear. In the Atchafalaya Delta, it may be because the western islands remain fresh year-around, whereas the other islands sometimes experience saltwater encroachment (Holm, 1998). The birds feed in freshwater marshes on the abundant seeds of annual grasses and sedges and the rhizomes of perennial marsh plants and also in adjacent agricultural fields. They are opportunistic feeders, on the whole, ingesting from the available plant species. An analysis by Abernethy (1986) suggests that many species that frequent the fresh marsh early in the winter move seaward to salt marshes before beginning their northward migration in the spring. The reason for this behavior pattern is not known, but Abernethy speculated that the preferred foods of the freshwater marshes are depleted by early spring and the birds move into salt marshes that have not been previously grazed.

The Wood Duck (*Aix sponsa*) is the only duck species that nests regularly in coastal freshwater tidal marshes, although an occasional Black Duck (*Anas rubripes*) or Mallard (*A. platyrhynchos*) nest is found in Atlantic Coast marshes.

Wading birds are common residents of coastal freshwater marshes. They are present year-round in Gulf Coast marshes but only during the summer along the Atlantic Coast. An exception is the Great Blue Heron (*Ardea herodias*), which is seen throughout the winter in the northern Atlantic states. Nesting colonies are common throughout the southern marshes, and some species [Green-backed Herons (*Butorides striatus*) and Bitterns (*Ixobrychus exilis* and *Botaurus lentiginosus*)] nest along the Mid-Atlantic Coast. They feed on fish and benthic invertebrates, often flying long distances each day from their nesting areas to fish.

Rails (*Rallus* spp.) and shorebirds, including the Killdeer (*Charadrius vociferus*), sandpipers (Scolopacidae), and the American Woodcock (*Scolopax minor*), are common in coastal freshwater marshes. They feed on benthic macroinvertebrates and diverse seeds. Gulls (*Larus* spp.), terns (*Sterna* spp.), Belted Kingfisher (*Ceryle alcyon*), and crows (*Corvus* spp.) are also common. Some are migratory; some are not. A number of birds of prey are seen hovering over freshwater marshes, including the Northern Harrier (*Circus cyaneus*), the American Kestrel (*Falco sparverius*), falcons (*Falco* spp.), eagles , Ospreys (*Pandion haliaetus*), owls (Tytonidae), vultures (Cathartidae), and the Loggerhead Shrike (*Lanius ludovicianus*). The number of these beautiful birds has been declining in recent years. Two of them, the Bald Eagle (*Haliaeetus leucocephalus*) and the Peregrine Falcon (*Falco peregrinus*), are listed as endangered.

Arboreal birds use the coastal freshwater marshes intensively during short periods of time on their annual migrations. Flocks of tens of thousands of swallows (Hirundinidae) have been reported over the upper Chesapeake freshwater marshes (R. E. Stewart and Robbins, 1958). Flycatchers (Tyrannidae) are also numerous. They often perch on trees bordering the marsh, darting out into the marsh from time to time to capture insects (W. E. Odum et al., 1984). Although coastal marshes may be used for only short periods of time by a migrating species, they may be important temporary habitats. For example, the northern Gulf coastal marshes are the first landfall for birds on their spring migration from South America. Often, they reach this coast in an exhausted state and the availability of forested barrier islands for refuge and marshes for feeding is critical to their survival.

Sparrows, finches (Fringillidae), juncos (*Junco* spp.), blackbirds (Icteridae), wrens (Troglodytidae), and other ground and shrub birds are abundant residents of coastal freshwater marshes. W. E. Odum et al. (1984) indicated that 10 species breed in Mid-Atlantic Coast marshes, including the Ring-necked Pheasant (*Phasianus colchicus*), Red-winged Blackbird (*Agelaius phoeniceus*), American Goldfinch (*Carduelis tristis*), Rufous-sided Towhee (*Pipilo erythrophthalmus*), and a number of sparrows. The most abundant are the Red-winged Blackbirds, Dickcissels (*Spiza americana*), and Bobolinks (*Dolichonyx oryzivorus*), which can move into and strip a wild rice marsh in a few days.

Amphibians and Reptiles

Although W. E. Odum et al. (1984) compiled a list of 102 species of amphibians and reptiles that frequent coastal freshwater marshes along the Atlantic Coast, many are poorly understood ecologically, especially with respect to their dependence on this type of habitat. None is specifically adapted for life in tidal freshwater marshes. Instead, they are able to tolerate the special conditions of this environment. River turtles, the most conspicuous members of this group, are abundant throughout the southeastern United States. Three species of water snakes (*Nerodia*) are common. *Agkistrodon piscivorus* (the cottonmouth) is found south of the James River. In the South, especially along the Gulf Coast, the American alligator's preferred habitat is the tidal freshwater marsh. These large reptiles used to be listed as threatened or endangered, but they have come back so strongly in most areas that they are presently harvested legally (under strict control) in Louisiana and Florida. They nest along the banks of coastal freshwater marshes, and the animal, identified by its high forehead and long snout, is a common sight gliding along the surface of marsh streams.

Mammals

The mammals most closely associated with coastal freshwater marshes are all able to get their total food requirements from the marsh, have fur coats that are more or less impervious to water, and are able to nest (or hibernate in northern areas) in the marsh (W. E. Odum et al., 1984). These include the river otter (*Lutra canadensis*), muskrat (*Ondatra zibethicus*), nutria (*Myocastor coypus*), mink (*Mustela vison*), raccoon (*Procyon lotor*), marsh rabbit (*Silvilagus palustris*), and marsh rice rat (*Oryzomys palustris*). In addition, the opossum (*Didelphis virginiana*) and white-tailed deer (*Odocoileus virginianus*) are locally abundant. The nutria was introduced from South America some years ago and has spread throughout the Gulf Coast states and into Maryland, North Carolina, and Virginia. It is not likely to spread farther north because of its intolerance to cold, but the South Atlantic marshes would seem to provide an ideal habitat. The nutria is more vigorous than the muskrat and has displaced it from the freshwater marshes in many parts of the northern Gulf (O'Neil, 1949). As a result, muskrat density is highest in oligohaline marshes. The muskrat, for some reason, is not found in coastal Georgia and South Carolina, or in Florida, although it is abundant farther north along the Atlantic Coast. Muskrat, nutria, and beaver (*Castor canadensis*) can influence the development of a marsh. The first two species destroy large amounts of vegetation with their feeding habits (they prefer juicy rhizomes and uproot many plants when digging for them), their nest building, and their underground passages. Beavers have been observed in tidal freshwater marshes in Maryland and Virginia. Their influence on forested habitats is well known, but their impact on tidal freshwater marshes needs to be studied more closely (W. E. Odum et al., 1984).

Ecosystem Function

Primary Production

Many production estimates have been made for freshwater coastal marshes. Productivity is generally high, usually falling in the range of 1,000 to 3,000 g m^{-2} yr^{-1} (Table 10-2). The large variability reported from different studies stems, in part, from a lack of standardization of measurement techniques, but real differences can be attributed to several factors:

1. *Type of Plant and Its Growth Habit.* Fresh coastal marshes, in contrast to saline marshes, are floristically diverse, and productivity is determined, at least to some degree, by genetic factors that regulate the species' growth habits. Tall perennial grasses, for example, appear to be more productive than broad-leaved herbaceous species such as arrow arum and pickerelweed.

2. *Tidal Energy.* The stimulating effect of tides on production has been shown for salt marshes (see Chapter 9). Whigham and Simpson (1977) showed that the fresh marsh grass *Zizania aquatica* responded positively to tides, and the general trends in production shown by Brinson et al. (1981a) in nonforested freshwater wetlands support the idea that moving water generally stimulates production.

3. *Other Factors.* Soil nutrients (Reader, 1978), grazing, parasites, and toxins are other factors that can limit production (de la Cruz, 1978).

The elevation gradient across a fresh coastal marsh and the resulting differences in vegetation and flooding patterns account for three broad zones of primary production. The low marsh bordering tidal creeks, dominated by broad-leaved perennials, is characterized by apparently low production rates. Biomass peaks early in the growing season. Turnover rates, however, are high, especially along the northern Gulf Coast (Visser, 1989), suggesting that annual production may be much higher than can be determined from peak biomass. Much of the production is stored in below-ground biomass (root : shoot $\gg 1$) in mature marshes; this biomass is mostly rhizomes rather than fibrous roots. Decomposition is rapid, the litter is swept from the marsh almost as fast as it forms, the soil is bare in winter, and erosion rates are high. In new marshes, below-ground biomass is initially low, increasing to a below-ground : above-ground ratio of about 0.2 to 0.4 after the third growing season (White, 1993).

The parts of the high marsh dominated by perennial grasses and other erect, tall species are characterized by the highest production rates of freshwater species, and

Table 10-2 Peak standing crop and annual net primary production (NPP) estimates for tidal freshwater marsh associations in approximate order from highest to lowest productivity[a]

Vegetation Type[b]	Peak Standing Crop (g m^{-2})	Annual NPP (g m^{-2} yr^{-1})
Extremely high productivity		
Spartina cynosuroides (big cordgrass)	2,311	—
Lythrum salicaria (spiked loosestrife)	1,616	2,100
Zizaniopsis miliacea (giant cutgrass)	1,039	2,048
Panicum hemitomon (maidencane)	1,160	2,000
Phragmites communis (common reed)	1,850	1,872
Moderate productivity		
Zizania aquatica (wild rice)	1,218	1,578
Amaranthus cannabinus (water hemp)	960	1,547
Typha sp. (cattail)	1,215	1,420
Bidens spp. (bur marigold)	1,017	1,340
Polygonum sp./Leersia oryzoides (smartweed/rice cutgrass)	1,207	—
Ambrosia tirifida (giant ragweed)	1,205	1,205
Acorus calamus (sweet flag)	857	1,071
Sagittaria latifolia (duck potato)	432	1,071
Low productivity		
Peltandra virginica/Pontederia cordata (arrow arum/pickerelweed)	671	888
Hibiscus coccineus (rose mallow)	1,141	869
Nuphar adventa (spatterdock)	627	780
Rosa palustris (swamp rose)	699	—
Scirpus deltarum	—	523
Eleocharis baldwinii	130	—

[a]Values are means of one to eight studies.
[b]Designation indicates the dominant species in the association.
Source: W. E. Odum et al. (1984), Sasser and Gosselink (1984), Visser (1989), White (1993), and Sasser et al. (1995).

root : shoot ratios are approximately one. Because tidal energy is not as strong and the plant material is not so easily decomposed, litter accumulates on the soil surface, and little erosion occurs. The high-marsh mixed-annual association typically reaches a large peak biomass late in the growing season. Most of the production is above ground (root : shoot < 1), and litter accumulation is common.

Organic and Energy Fluxes

Energy Flow

W. E. Odum et al. (1984) identified three major sources of organic carbon to tidal freshwater marshes. The largest source is probably the vascular marsh vegetation, but organic material brought from upstream (terrestrial carbon) may be significant, especially on large rivers and where domestic sewage waters are present. Phytoplankton productivity is a largely unknown quantity. Most of the organic energy flows through the detrital pool and is distributed to benthic fauna and deposit-feeding omnivorous nekton. These groups feed fish, mammals, and birds at higher trophic levels. The magnitude of the herbivore food chain, in comparison to the detritus one, is poorly understood. Insects are more abundant in fresh marshes than in salt marshes, but most do not appear to be herbivorous. Marsh mammals apparently can "eat out" significant areas of vegetation (Evers et al., 1998), but direct herbivory is probably small in comparison to the flow of organic energy from destroyed vegetation into the detrital pool. Nevertheless, these rodents may exert strong control on species composition and on primary production (Evers et al., 1998). Herbivores also act in synergy with other stresses, for example, saltwater intrusion, flooding, and fire (Grace and Ford, 1996; Taylor et al., 1994).

The phytoplankton–zooplankton juvenile fish food chain in fresh marshes is of interest because of its importance to humans. Zooplankton is an important dietary component for a variety of larval, postlarval, and juvenile fish of commercial importance that are associated with tidal freshwater marshes (Van Engel and Joseph, 1968).

Birds are major seasonal or year-around consumers in all types of tidal freshwater marshes. In addition, they move materials out of the system, processing it into guano, which in some areas may be a significant source of nutrients. They also modify plant composition and production by "eat-outs" (T. J. Smith and Odum, 1981).

Litter Decomposition

As with other marshes, little plant production is consumed directly by grazers. Although nutrias and muskrats are common in freshwater tidal marshes, herbivory is thought to account for the consumption of less than 10 to 40 percent of plant production (W. E. Odum, 1988). The remaining 60 to 90 percent becomes available to consumers through the detrital food chain. The vegetation is attacked by bacteria and fungi, aided by the fragmenting action of small invertebrates, and the bacteria-enriched decomposed broth feeds benthic invertebrates, which, in turn, are prey to larger animals.

Temperature is the major factor that controls litter decomposition; the higher the temperature, the higher the decay rate. The combined availability of oxygen and water is a second factor; plants submerged in anaerobic environments decompose slowly, as do exposed plants in dry environments. Optimum conditions for decomposition are found in a moist aerobic environment such as a regularly flooded tidal marsh. An important third factor in decomposition rates is the kind of plant tissue involved. In freshwater tidal marshes, two groups of plants can be identified (W. E. Odum et al., 1984). The broad-leaved perennials (e.g., *Pontederia cordata, Sagittaria latifolia,* and *Nuphar luteum*) generally contain high leaf concentrations of nitrogen (2–4 percent). Their detritus has high nutritional quality, as indicated by a low C : N ratio and their selection by detritus consumers over *Spartina* detritus (Dunn, 1978; Smock and Harlowe, 1983). In contrast, the high-marsh grasses (e.g., *Zizania aquatica* and *Panicum hemitomon*) are low in nitrogen (the tissue concentration is usually less than 1 percent) and are composed primarily of long stems with much structural tissue that is resistant to decay. Their slow decay rates, combined with lower tidal energy on the high marsh, may explain why litter accumulates there and erosion rates are low.

Organic Import and Export

Figure 10-2 is a simplified representation of the energy–carbon flow in freshwater tidal marshes. In mature tidal freshwater marshes (Fig. 10-2a), most organic production is decomposed to litter and peat within the marsh system and nutrients are extracted and recycled. This is confirmed by the studies of Bowden et al. (1991) discussed later under Nutrient Budgets. Floating tidal freshwater marshes may have even more closed cycles (Fig. 10-2b). Because they float, no surface flows export or import organic material. This limits fluxes to subsurface dissolved materials (Sasser et al., 1991). The largest loss of organic energy in these mature marshes is probably to deep peats in the case of anchored marshes or to an organic sludge layer under the water column in floating marshes (Sasser et al., 1991). The magnitude of this loss was measured as 145 to 150 g C m^{-2} yr^{-1} in a Gulf Coast freshwater marsh (Hatton, 1981).

Other losses of organic carbon from marshes occur through flushing from the marsh surface, conversion to methane that escapes as a gas, and export as biomass in the bodies of consumers that feed on the marsh.

In highly reduced freshwater sediments, where, in contrast to salt marshes, little sulfur is available as an electron acceptor, it is expected that methanogenesis from carbon dioxide and fermentation should be the predominant pathways of respiratory energy flow (Delaune et al., 1983a; W. E. Odum, 1988). The annual loss of carbon as methane from Gulf Coast freshwater marshes has been estimated as 160 g CH_4 m^{-2} yr^{-1} (C. J. Smith et al., 1982). In comparison, Lipschultz (1981) estimated a loss of only 10.7 g CH_4 m^{-2} yr^{-1} from a *Hibiscus*-dominated freshwater marsh in the Chesapeake Bay. Crozier and Delaune (1996) reported that methane production is correlated with labile sediment organic carbon. Since, on an areal basis, there is no clear gradient of labile carbon from fresh to salt marshes, their data did not show a clear consistent increase in methane production in freshwater marshes over saline ones.

In marked contrast to mature tidal freshwater marshes, the new marshes of the Mississippi River delta are extremely open systems (Fig. 10-2c). They store almost no organic matter in the sediment (organic carbon $\approx$ 0.5 percent). Large areas of broad-leaved herbaceous vegetation die at the surface every fall and readily disappear so that the marsh surface is bare during the winter. No carbon or energy budgets are available for this system, but it appears that, instead of the cycle shown in Figure 10-2a and b, the organic production in new marshes is rapidly exported. Even the underground rhizomes are a renewable commodity that accumulates only seasonally.

Nutrient Budgets

Mature Marshes

In general, nutrient cycling and nutrient budgets in coastal freshwater wetlands appear to be similar to those of salt marshes; they are fairly open systems that have the capacity to act as long-term sinks, sources, or transformers of nutrients. Even though East Coast mature marshes generally are vigorously flooded by tides, they recycle a major portion of the nitrogen requirements of the vegetation (Fig. 10-2a). Figure 10-9 illustrates nitrogen cycling in a 22.8-ha tidal freshwater marsh near Boston, Massachusetts. In this budget, most nutrient inputs are inorganic from the North River, and the marsh and river nutrient cycles are mostly independent. Within the marsh itself, the major cycle is from peat to ammonium nitrogen to live plants. Some of the ammonium nitrogen is nitrified to nitrate nitrogen and denitrified. Yet overall nitrate loss always exceeded denitrification measurements by the acetylene block method, suggesting that other sinks of nitrate such as assimilatory nitrate reduction may be important (Bowden et al., 1991). Peat mineralization is sufficient to satisfy

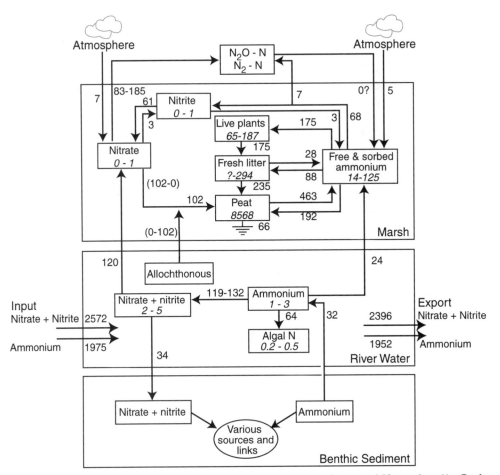

Figure 10-9 Nitrogen budget for a 22.8-ha tidal freshwater marsh in coastal Massachusetts. Pool sizes are in kilimoles N, fluxes in kilimoles N per year. (*After Bowden et al., 1991*)

the nitrogen demands of the vegetation, and nearly all of the nitrogen flowing across the marsh from the adjacent river is exported again. Mineralized litter and peat are conserved in the marsh by plant uptake and by microbial and litter immobilization. Despite this closed mineral cycle, Bowden et al. (1991) suggest that the small uptake of nitrogen from the river may be important in a cumulative sense over time.

New Marshes

In these marshes, plant biomass is unrelated to sediment stores of nutrients (Johnson et al., 1985), so it is likely that the enormous flux of nutrients carried over the marsh by river and tides supplies the nutrient requirements of the system. The nutrients trapped in live plant biomass are nearly all exported again at the end of the growing season.

Even though most nutrient requirements of mature marshes can be met by recycling, the system may still be seasonally importing, transforming, and releasing materials. Whether this is true of freshwater coastal marshes depends on the age and ecological maturity of the marsh, the magnitude of upland runoff, anthropogenic effects such as sewage loading, and the magnitude of tidal action (Stevenson et al., 1977). Unfortunately, most nutrient studies of freshwater coastal marshes have been carried out in areas heavily influenced by nearby urban communities, and so it is difficult to know how representative they are. From a study of the Woodbury Creek in New Jersey (Simpson et al., 1983), tidal freshwater marshes appear to be net importers of nitrogen and phosphorus during the spring, primarily because of the magnitude of upland runoff during this period. During spring and summer, nutrients were tied up in plant biomass. Since sediment concentrations do not seem to change much seasonally, the living biomass reflects seasonal storage. After the vegetation dies in the fall, there appears to be a rapid net export of nutrients associated with the high decomposition rate and tidal flushing. During the year as a whole, the net balance of nitrogen and phosphorus indicated a net export. This may reflect the fact that the marshes studied were subject to unusually high loading with domestic sewage and were saturated with respect to those nutrients.

Metal Accumulation

Because tidal freshwater marshes, almost more than any other type of wetland, are found where major cities and industries have developed (because the tidal rivers are sufficiently large for large oceangoing ships yet the river water is fresh and can be used for residential, commercial, and industrial purposes), they are often the first aquatic ecosystem to be impacted by pollution from all of this human development. Because, like all wetlands, freshwater tidal marshes are good sinks for metals, they are often places where trace metals accumulate. The accumulation of metals was measured in a high marsh in New Jersey and found to be variable (Dubinski et al., 1986). Vegetation and litter appeared to play only minor roles, but cadmium, copper, chromium, lead, and zinc accumulated in the top 5 cm of the soil at the end of the growing season. All but the cadmium and chromium had disappeared by the following year (Table 10-3).

Table 10-3 Percentage retention of heavy metals in the top 5 cm of sediment in a tidal freshwater marsh in a Delaware River, New Jersey, estuary. Sludge containing heavy metals in two loadings was applied in March. Sediments were sampled in October and the following March

	Treatment Level			
	October Sample		March Sample	
Element	Low	High	Low	High
Cadmium	43	47	46	15
Chromium	28	53	28	12
Copper	0	52	0	0
Lead	0	31	0	0
Nickel	0	0	0	0
Zinc	0	31	0	0

Source: Dubinski et al. (1986).

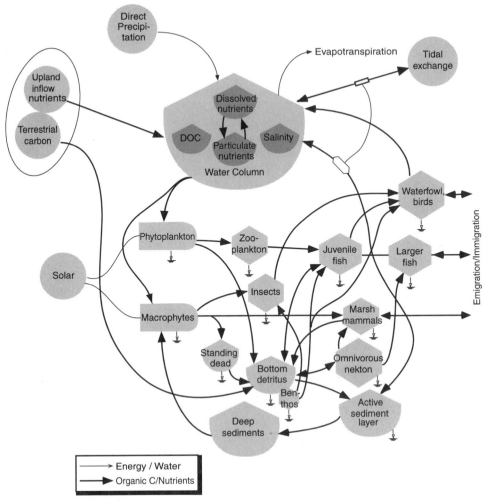

Figure 10-10 Conceptual model of a tidal freshwater marsh.

Ecosystem Models

Quantitative models of carbon and energy flow through the detrital food web in freshwater coastal wetlands are practically nonexistent. Conceptual models showing the principal nutrient and energy flows of three different types of tidal freshwater marshes were introduced in Figure 10-2. The more detailed conceptual model illustrated in Figure 10-10 is based on the work of W. E. Odum et al. (1984), who developed a simple energy consumption model for some consumers. The production of fish and waterfowl, both of which immigrate to and emigrate from the marsh, is based on complex food webs. This model, just as several of the marsh types discussed previously, shows little export of organic carbon but some tidal exchange of water and associated nutrients. There are few simulation models of tidal freshwater ecosystems. Bowden et al. (1991) calculated the nitrogen budget in Figure 10-9, aided by the insights and results from simulation models of the transport and sediment processes of the total nitrogen budget (Vorosmarty et al., 1982; Morris and Bowden, 1986).

Recommended Readings

Bowden, W. B., C. J. Vorosmarty, J. T. Morris, B. J. Peterson, J. E. Hobbie, P. A. Steudler, and B. Moore. 1991. Transport and processing of nitrogen in a tidal freshwater wetland. Water Resources Research 27:389–408.

Odum, W. E., T. J. Smith III, J. K. Hoover, and C. C. McIvor. 1984. The Ecology of Tidal Freshwater Marshes of the United States East Coast: A Community Profile. U.S. Fish and Wildlife Service, FWS/OBS-87/17, Washington, DC. 177 pp.

Simpson, R. L., R. E. Good, M. A. Leck, and D. F. Whigham. 1983. The ecology of freshwater tidal wetlands. BioScience 33:255–259.

Chapter 11

Mangrove Swamps

The mangrove wetland replaces the salt marsh as the dominant coastal ecosystem in subtropical and tropical regions. An estimated 240,000 km^2 of mangrove wetlands are found throughout the world. Mangrove wetlands are limited in the United States to the southern extremes of Florida (where there are approximately 2,700 km^2 of mangroves) and to Puerto Rico. Mangrove wetlands have been classified according to their hydrodynamics and topography as fringe mangroves, riverine mangroves, basin mangroves, and dwarf or scrub mangroves. The dominant mangrove plant species are known for several adaptations to the saline wetland environment, including prop roots, pneumatophores, salt exclusion, salt excretion, and the production of viviparous seedlings. There is generally little understory, and secondary productivity is dominated by several species of crabs. Mangrove wetlands have definite vegetation zonation patterns, although the importance of successional stages and physical conditions to this zonation has been much debated. The zonation may result from each species's optimal niche for productivity. Mangroves require a greater percentage of their energy for maintenance under high-salinity conditions. Massive diebacks of mangrove trees can occur when environmental change is rapid or during hurricanes. Damage can be mitigated if there is freshwater or tidal flushing and exacerbated if the system is already under stress. Highest productivity occurs in riverine forests, which are most open to both tidal action and inputs of nutrients from adjacent uplands. The least productive systems are dwarf mangroves, which occur under nutrient-poor conditions, in hypersaline soils, and at the northern extreme of the mangrove's range. Several complete energy budgets have been developed to describe mangrove wetlands, and organic export studies and comparisons with estuarine productivities have verified the importance of these ecosystems to the secondary productivity of adjacent estuaries.

The coastal salt marsh of temperate middle and high latitudes gives way to its analog, the mangrove swamp, in tropical and subtropical regions of the world. The mangrove swamp is an association of halophytic trees, shrubs, and other plants growing in brackish to saline tidal waters of tropical and subtropical coastlines. This coastal, forested wetland (the wetland is called a *mangal* by some researchers) is infamous for its impenetrable maze of woody vegetation, its unconsolidated peat that seems to have no bottom, and its many adaptations to the double stresses of flooding and salinity. The word *mangrove* comes from the Portuguese word *mangue* for "tree" and the English word *grove* for "a stand of trees" and refers to both the dominant trees and the entire plant community.

Many myths have surrounded the mangrove swamp. It was described at one time or another in history as a haven for wild animals, a producer of fatal "mangrove root gas," and a wasteland of little or no value (Lugo and Snedaker, 1974). Researchers, however, have established the importance of mangrove swamps in exporting organic matter to adjacent coastal food chains, in providing physical stability to certain shorelines to prevent erosion, in protecting inland areas from severe damage during hurricanes and tidal waves, and in serving as sinks for nutrients and carbon. There is an extensive literature on the mangrove swamp on a worldwide basis—more than 5,000 titles. This interest probably stems from the worldwide scope of these ecosystems and the many unique features that they possess. Much of the early literature on mangroves concerned floristic and structural topics. Since the early 1970s, the focus has been on the functional aspects of mangrove swamps. Since that time, a significant literature on ecophysiology, primary productivity, stressors, food chains, and the detritus dynamics of mangrove ecosystems has been produced, along with some work on nutrient cycling, mangrove restoration, valuation of mangrove resources, and responses of mangroves to sea-level changes (Lugo, 1990a). Good summaries of mangrove ecosystem function are provided by Lugo and Snedaker (1974), Chapman (1976b), W. E. Odum et al. (1982), Tomlinson (1986), Armentano (1990), Lugo et al. (1990a,b), W. E. Odum and McIvor (1990), Twilley et al. (1996), and Twilley (1998). Rützler and Feller (1996) give a general description of mangroves in the Caribbean, and Alongi (1998) provides an interesting contrast between mangroves and salt marshes (Chapter 9) in a book on coastal ecosystem processes.

Geographical Extent

Global Distribution

Mangrove swamps are found along tropical and subtropical coastlines throughout the world, usually between 25° N and 25° S latitude (Fig. 11-la). Their limit in the Northern Hemisphere generally ranges from 24° to 32° N latitude, depending on the local climate and the southern limits of freezing weather. There are an estimated 240,000 km² of mangrove swamps in the world, with more that half of those swamps found in the latitudinal belts between 0° and 10° (Fig. 11-1b). Mangroves are divided into two groups—the Old World mangrove swamps and the New World and West African mangrove swamps. An estimated 68 species of mangroves exist, and their distribution is thought to be related to continental drift in the long term and possibly to transport by primitive humans in the short term (Chapman, 1976b). The distribution of these species, however, is uneven. The swamps are particularly dominant in the Indo–West Pacific region (part of the Old World group), where they contain the greatest diversity of species. There are 36 species of mangroves in that region, whereas there are only about 10 mangrove species in the Americas. It has been argued,

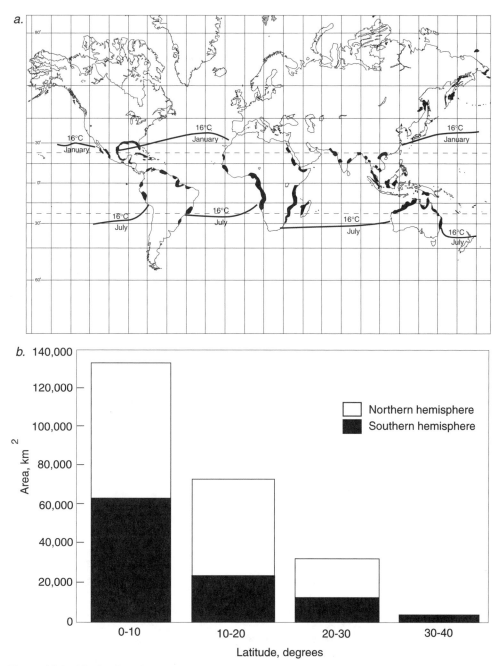

Figure 11-1 Distribution of mangrove wetlands: **a.** in the world and **b.** by latitude. (*After Chapman, 1977; Twilley et al., 1992*)

therefore, that the Indo-Malaysian region was the original center of distribution for the mangrove species (Chapman, 1976b). Certainly some of the most intact mangrove forests in the world are found in Malaysia and in Micronesia, the small islands east of the Philippines in the western Pacific. Studies are only now beginning to illustrate how important these mangrove swamps are to local economies in these regions (Ewel et al., 1998; Cole et al., 1999). Several mangrove species, not native to the Hawaiian Archipelago despite its appropriate climate and coastal geomorphology, invaded the islands in the early 20th century and are now permanent fixtures on coastlines there (Allen, 1998).

There is also a great deal of segregation between the mangrove vegetation found in the Old World region and that found in the New World of the Americas and West Africa. Two of the primary genera of mangrove trees, *Rhizophora* (red mangrove) and *Avicennia* (black mangrove), contain separate species in the Old and New Worlds, suggesting "that speciation is taking place independently in each region" (Chapman, 1976b).

Florida Mangroves

J. H. Davis (1940) described the Florida coast as supporting more than 2,600 km^2 of mangrove swamps, although Craighead (1971) revised the estimate down to 1,750 km^2. According to the National Wetlands Inventory, completed in 1982, there are 2,730 km^2 of mangroves in Florida (W. E. Odum and McIvor, 1990). The best development of mangroves in Florida is along the southwest coast, where the Everglades and the Big Cypress Swamp drain to the sea (see Fig. 11-2). Mangroves extend up to 30 km inland along water courses on this coast. The area includes Florida's Ten Thousand Islands, one of the largest mangrove swamps in the world at 600 km^2 (Smith et al., 1994). Because of development pressure, a significant fraction of the original mangroves on these islands has been lost or altered. Patterson (1986) reported that there was a loss of 24 percent of mangroves on one of the most developed islands in this region, Marco Island, from 1952 to 1984. Mangroves are now protected in Florida, and it is illegal to remove them from the shoreline.

Mangrove swamps are also common farther north along Florida's coasts, north of Cape Canaveral on the Atlantic and to Cedar Key on the Gulf of Mexico, where mixtures of mangrove and salt marsh vegetation appear. One species of mangrove (*Avicennia germinans*) is found as far north as Louisiana and in the Laguna Madre of Texas, although the trees are more like shrubs at these extreme locations (Chapman, 1976b, 1977). Extensive mangrove swamps are also found throughout the Caribbean Islands, including Puerto Rico. Lugo (1988) estimated that there were originally 120 km^2 of mangroves in Puerto Rico, although only half of those remained by 1975.

Geographic Limitation by Frost and Competition

The frequency and severity of frosts are the main factors that limit the extension of mangroves beyond tropical and subtropical climes (Twilley, 1998). For example, in the United States, mangrove wetlands are found primarily along the Atlantic and Gulf coasts of Florida up to 27° to 29° N latitude (Fig. 11-2), north of which they are replaced by salt marshes (Kangas and Lugo, 1990). The red mangrove can survive

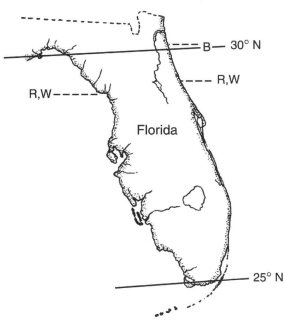

Figure 11-2 Distribution of mangrove wetlands along the Florida coastline. R, B, and W indicate the northern extent of black, red, and white mangroves, respectively, along the Florida coastline. (*After W. E. Odum et al., 1982, W. E. Odum and McIvor, 1990*)

temperatures as low as 2° to 4°C for 24 hours, whereas the black mangrove can withstand several days at this temperature, allowing black mangroves to extend farther north on Florida's east coast than red mangroves, as far north as 30° N (Fig. 11-2). Similarly, Schaeffer-Novelli et al. (1990) described mangroves as extending to 28° to 30° S latitude along the Brazilian coast. Chapman (1976b) suggested that three to four nights of a light frost are sufficient to kill even the hardiest mangrove species. Lugo and Patterson-Zucca (1977) showed that mangroves survived approximately five nonconsecutive days of frost in January 1977 in Sea Horse Key, Florida (latitude 29° N), but estimated that it would take 200 days for the forest to recover from frost damage. They also hypothesized that soil salinity stress could modify frost stress on mangroves, suggesting that the latitudinal limit of mangroves reflects a number of stresses rather than one factor.

Kangas and Lugo (1990) suggested that the boundary between tropical mangroves and temperate salt marshes can be attributed to a combination of frost stress on mangroves and, in the absence of stress, a competitive advantage by mangrove vegetation over salt marsh grasses. They also suggested that the replacement of salt marsh grasses by mangrove trees "may be a special case of the more general phenomena of tree vegetation replacing herbaceous vegetation in successional sequences and along environmental gradients" such as the replacement of tundra by boreal forests along a temperature gradient and the replacement of grasslands by deciduous forests along moisture gradients.

Geomorphology and Hydrology

Geomorphological Setting

There are several different types of mangrove wetlands, each having a unique set of topographic and hydrodynamic conditions. A classification scheme of five geomorphological settings where mangrove forests occur, as developed by Thom (1982), includes systems dominated by waves, tides, and rivers, or most often, by combinations of these three energy sources (Fig. 11-3).

Like the coastal salt marsh, the mangrove swamp can develop only where there is adequate protection from high-energy wave action. Several physiographic settings favor the protection of mangrove swamps, including (1) protected shallow bays, (2) protected estuaries, (3) lagoons, (4) the leeward sides of peninsulas and islands, (5) protected seaways, (6) behind spits, and (7) behind offshore shell or shingle islands. Unvegetated coastal and barrier dunes usually develop where this protection does not exist, and mangroves are also often found behind these dunes (Chapman, 1976b).

In addition to the required physical protection from wave action, the range and duration of the flooding of tides exert a significant influence over the extent and functioning of the mangrove swamp. Tides constitute an important subsidy for the mangrove swamp, importing nutrients, aerating the soil water, and stabilizing soil salinity. Salt water is important to the mangroves in eliminating competition from freshwater species. The tides provide a subsidy for the movement and distribution of the seeds of several mangrove species (see Mangrove Adaptations later in this chapter). They also circulate the organic sediments in some fringe mangroves for the benefit of filter-feeding organisms such as oysters, sponges, and barnacles and for deposit

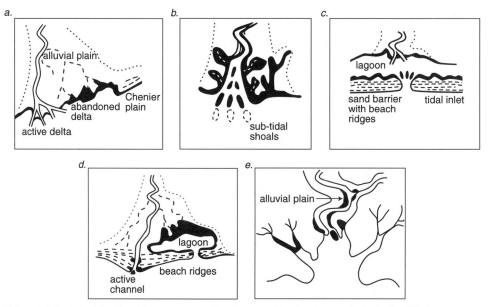

Figure 11-3 Classification of mangrove wetlands according to geomorphological settings: **a.** river dominated; **b.** tide dominated; **c.** wave-dominated barrier lagoon; **d.** composite river and wave dominated; **e.** drowned bedrock valley. (*After Thom, 1982*)

feeders such as snails and crabs. Like salt marshes, mangrove swamps are intertidal, although a large tidal range is not necessary. Most mangrove wetlands are found in tidal ranges of 0.5 to 3 m or more. Mangrove tree species can also tolerate a wide range of inundation frequencies. *Rhizophora* spp., the red mangrove, is often found growing in continually flooded coastal waters below normal low tide.

At the other extreme, mangroves can be found several kilometers inland along river banks where there is less tidal action. These mangroves depend on river discharge and are nourished by river flooding in addition to infrequent tidal inundation and the stability of groundwater and surface water levels near the coast. Lugo (1981) found that these inland mangroves depend on storm surges and "are not isolated from the sea but critically dependent on it as a source of fresh sea water."

Hydrodynamic Classification

The development of mangrove swamps is the result of topography, substrate, and freshwater hydrology, as well as tidal action. A classification of mangrove wetland ecosystems according to their physical hydrologic conditions was developed by Lugo and Snedaker (1974) and Lugo (in Wharton et al., 1976) and included six types. Cintrón et al. (1985) suggested a simplification of that classification system to four major types:

1. Fringe mangroves, including overwash islands
2. Riverine mangroves
3. Basin mangroves
4. Dwarf (or scrub) mangroves

The features of these types of mangrove wetlands are shown in Figure 11-4 and are discussed next.

Fringe Mangroves

Fringe mangrove wetlands are found along protected shorelines and along some canals, rivers, and lagoons (Fig. 11-4a). They are common along shorelines that are adjacent to land higher than mean high tide but are exposed to daily tides. In contrast to the overwash mangroves discussed in the next paragraph, fringe mangrove wetlands tend to accumulate organic debris because of the low-energy tides and the dense development of prop roots. Because the shoreline is open, these wetlands are often exposed to storms and strong winds that lead to the further accumulation of debris. Fringe mangroves are found on narrow berms along the coastline or in wide expanses along gently sloping beaches. If a berm is present, the mangroves may be isolated from freshwater runoff and would then have to depend completely on rainfall, the sea, and groundwater for their nutrient supply. These wetlands are found throughout south Florida along both coasts and in Puerto Rico.

A special case of fringe mangroves are small islands and narrow extensions of larger land masses (spits) that are "overwashed" on a daily basis during high tide. These are sometimes called overwash mangrove islands (Fig. 11-4b). The forests are dominated by the red mangrove (*Rhizophora mangle*) and a prop root system that obstructs the tidal flow and dissipates wave energy during periods of heavy seas. Tidal velocities are high enough to wash away most of the loose debris and leaf litter into the adjacent bay. The islands often develop as concentric rings of tall mangroves

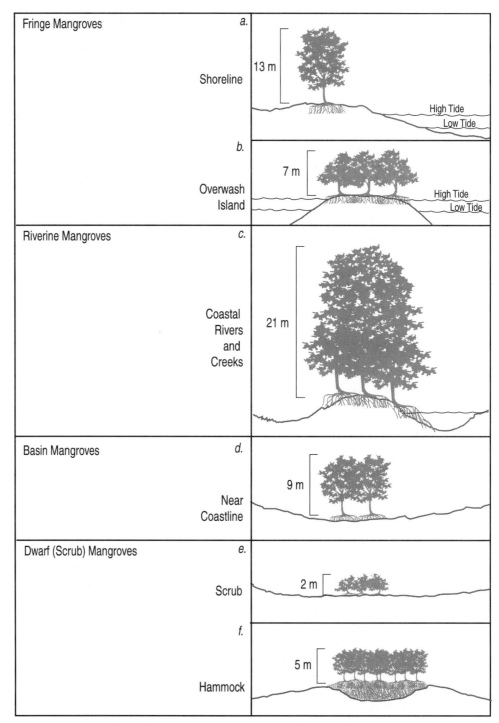

Figure 11-4 Classification of mangrove wetlands according to hydrodynamic conditions. General classification is **a.** and **b.** fringe mangroves; **c.** riverine mangroves; **d.** basin mangroves; **e.** and **f.** scrub (dwarf) mangroves. (*After Wharton et al., 1976; Lugo, 1980; Cintrón et al., 1985*)

around smaller mangroves and a permanent, usually hypersaline, pool of water. These wetlands are abundant in the Ten Thousand Islands region of Florida and along the southern coast of Puerto Rico. They are particularly sensitive to the effects of ocean pollution.

Riverine Mangroves

Tall, productive riverine mangrove forests are found along the edges of coastal rivers and creeks, often several miles inland from the coast (Fig. 11-4c). These wetlands may be dry for a considerable time, although the water table is generally just below the surface. In Florida, freshwater input is greatest during the wet summer season, causing the highest water levels and the lowest salinity in the soils during that time. Riverine mangrove wetlands export a significant amount of organic matter because of their high productivity. These wetlands are affected by freshwater runoff from adjacent uplands and from water, sediments, and nutrients delivered by the adjacent river, and, hence, they can be significantly affected by upstream activity or stream alteration. The combination of adequate fresh water and high inputs of nutrients from both upland and estuarine sources causes these systems to be generally very productive, supporting large (16–26 m) mangrove trees. Salinity varies but is usually lower than that of the other mangrove types described here. The flushing of fresh water during wet seasons causes salts to be leached from the sediments (Cintrón et al., 1985).

Basin Mangroves

Basin mangrove wetlands occur in inland depressions, or basins, often behind fringe mangrove wetlands, and in drainage depressions where water is stagnant or slowly flowing (Fig. 11-4d). These basins are often isolated from all but the highest tides and yet remain flooded for long periods once tide water does flood them. Because of the stagnant conditions and less frequent flushing by tides, soils have high salinities and low redox potentials. These wetlands are often dominated by black mangroves (*Avicennia* spp.) and white mangroves (*Laguncularia* spp.), and the ground surface is often covered by pneumatophores from these trees.

Dwarf Mangroves

There are several examples of isolated, low-productivity mangrove wetlands that are usually limited in productivity because of the lack of nutrients or freshwater inflows. Dwarf mangrove wetlands are dominated by scattered, small (often less than 2 m tall) mangrove trees growing in an environment that is probably nutrient poor (Fig. 11-4e). The nutrient-poor environment can be a sandy soil or limestone marl. Hypersaline conditions and cold at the northern extremes of the mangrove's range can also produce "scrub," or stressed mangrove trees, in riverine, fringe, or basin wetlands. True dwarf mangrove wetlands, however, are found in the coastal fringe of the Everglades and the Florida Keys and along the northeastern coast of Puerto Rico. Some of these wetlands in the Everglades are inundated by sea water only during spring tides or storm surges and are often flooded by freshwater runoff in the rainy season. These types of wetlands are actually an intermix of small red mangrove trees with marsh vegetation such as sawgrass (*Cladium jamaicense*) and rush (*Juncus roemarianus*) (Twilley, 1998).

Hammock mangrove wetlands also occur as isolated, slightly raised tree islands in the coastal fringe of the Florida Everglades and have characteristics of both basin and scrub mangroves. They are slightly raised as a result of the buildup of peat in what was once a slight depression in the landscape (Fig. 11-4f). The peat has accumulated from many years of mangrove productivity, actually raising the surface from 5 to 10 cm above the surrounding landscape. Because of this slightly raised level and the dominance by mangrove trees, these ecosystems look like the familiar tree islands or "hammocks" that are found throughout the Florida Everglades. They are different in that they are close enough to the coast to have saline soils and occasional tidal influences and, thus, can support only mangroves.

Chemistry

Salinity

Mangrove swamps are found under conditions that provide a wide range of salinity. J. H. Davis (1940) summarized several major points about salinity in mangrove wetlands from his studies in Florida:

1. There is a wide annual variation in salinity in mangrove wetlands.
2. Salt water is not necessary for the survival of any mangrove species but only gives mangroves a competitive advantage over salt-intolerant species.
3. Salinity is usually higher and fluctuates less in interstitial soil water than in the surface water of mangroves.
4. Saline conditions in the soil extend farther inland than normal high tide because of the slight relief, which prevents rapid leaching.

Figure 11-5 shows the wide spatial and temporal range of salinity, from the constant seawater salinity of outer-coast overwash and mangrove swamps to the brackish water in coastal rivers and canals, found in a region in Florida dominated by mangrove wetlands. Seasonal oscillations in salinity are a function of the height and duration of tides, the seasonality and intensity of rainfall, and the seasonality and amount of fresh water that enters the mangrove wetlands through rivers, creeks, and runoff. In Florida, as illustrated in Figure 11-5, summer wet season convective storms and associated freshwater flow in streams and rivers as well as an occasional hurricane in the late summer or early fall leads to the dilution of salt water and the lowest salinity concentrations. Salinity is generally the highest during the dry season, which occurs in the winter and early spring.

Soil salinity in mangrove ecosystems varies from season to season and with mangrove type (Table 11-1). In riverine mangrove systems, the soil salinity is less than that of normal sea water because of the influx of fresh water. In basin mangroves, on the other hand, salinity can be well above that of sea water because of evaporative losses (>50 ppt). Boto and Wellington (1984) found soil salinity in mangroves in North Queensland, Australia, to range from an average of 30 to 50 ppt, generally above that of the overlying waters. Highest salinities were found where tidal exchange was restricted.

Dissolved Oxygen

Reduced conditions exist in most mangrove soils when they are flooded. The degree of reduction depends on the duration of flooding and the openness of the wetland to freshwater and tidal flows. Some oxygen transport to the rhizosphere occurs through

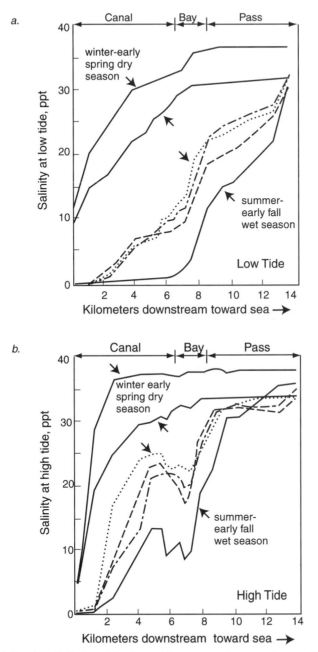

Figure 11-5 Variation in salinity from inland canal to open sea in a mangrove region of southwestern Florida during **a.** low tide and **b.** high tide. Canal is channelized stream that flows into the bay. (*After Carter et al., 1973*)

Table 11-1 Soil salinity ranges for major mangrove types

Hydrodynamic Type	Soil Salinity (ppt)
Fringe mangroves	
Avicennia zone	59
Rhizophora zone	39
Riverine mangroves	10–20[a]
Basin mangroves	
Avicennia zone	>50
Laguncularia zone	Low salinity
Mixed forest zone	30–40

[a]Higher in dry season when less freshwater streamflow is available.
Source: Cintrón et al. (1985).

the vegetation, and this contribution is locally significant (Thibodeau and Nickerson, 1986; McKee et al., 1988). When creeks and surface runoff pass water through mangrove wetlands, the reduced conditions are not as severe because of the increased drainage and the continual importing of oxygenated waters.

Mazda et al. (1990) reported on the variations in dissolved oxygen, salinity, and temperature in a Japanese mangrove that is connected by tides to the ocean through a coral reef but is occasionally isolated by a sandy sill. The oxygen content of the water that reaches the mangrove is the result of (1) the semidiurnal tidal cycle, which brings water to the mangrove at flood tide; (2) the diurnal pattern of dissolved oxygen from the productive coral reef; and (3) the presence or absence of the sandy sill that isolates the wetland (Fig. 11-6). The highest dissolved oxygen resulted when the flood tide occurred at the same time as peak dissolved oxygen in the coral reef; when

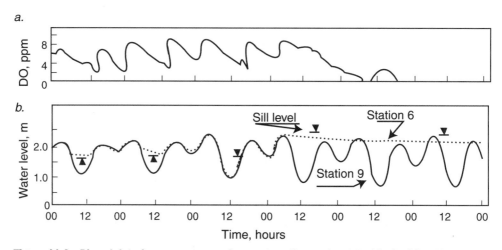

Figure 11-6 Diurnal data from mangrove embayment on the west coast of Amitori Bay, Japan, showing: **a.** dissolved oxygen (DO) in overlying water and **b.** water level. Station 6 refers to the water level in the mangrove area; station 9 refers to the water level in the open ocean. Sill level indicates relative height of sand barrier that separated mangrove from ocean between August 3 and 4, 1986. (*After Mazda et al., 1990*)

the flood tide occurred in the early morning, the dissolved oxygen in the mangrove waters was less. When the mangrove was isolated by an increase in the sandy sill caused by a nearby typhoon (between August 3 and 4, as shown in Fig. 11-6b), the dissolved oxygen decreased rapidly to anoxic conditions within two days. Anoxia leads to decreased activity and the death of benthic algae in the swamp, and continues until the barrier beach is breached. These data show the importance of physical flushing by ocean water and offshore productivity to mangrove oxygen supply.

Soil Acidity

Mangrove soils are often acidic, although, in the presence of carbonate as is often the case in south Florida, the soil pore water can be close to neutrality. The soils are often highly reduced, with redox potentials ranging from -100 to -400 mV. The highly reduced conditions and the subsequent accumulation of reduced sulfides in mangrove soils cause extremely acidic soils in many mangrove areas. Dent (1986, 1992) reported a measured accumulation of 10 kg S m^{-3} of sediment per 100 years in mangroves. When these soils are drained and aerated for conversion to agricultural land, the reduced sulfides, generally stored as pyrites, oxidize to sulfuric acid (see Chapter 6, Eq. 6-18), producing what are known as "cat clays." These highly acidic soils make traditional agriculture difficult (W. E. Odum and McIvor, 1990; Dent, 1992). Dent (1992) argued that the "dereclamation" of some previously "reclaimed" marginal coastal soils back to mangroves and salt marshes may be the best strategy for these acidic soils.

Ecosystem Structure

Canopy Vegetation

As is evident in coastal salt marshes, the stresses of waterlogged soils and salinity lead to a relatively simple flora in most mangrove wetlands, particularly when compared to their upland neighboring ecosystem, the tropical rain forest (Cintrón et al., 1985). There are more than 50 species of mangroves throughout the world (Stewart and Popp, 1987), representing 12 genera in 8 families. Fewer than 10 species of mangroves are found in the New World, and only 3 species are dominant in the south Florida mangrove swamps—the red mangrove (*Rhizophora mangle* L.), the black mangrove (*Avicennia germinans* L., also named *A. nitida* Jacq.), and the white mangrove (*Laguncularia racemosa* L. Gaertn.). Buttonwood (*Conocarpus erecta* L.), although strictly not a mangrove, is occasionally found growing in association with mangroves or in the transition zone between the mangrove wetlands and the drier uplands. Each of the hydrologic types of mangrove wetlands described previously is dominated by different associations of mangrove plants. Fringe mangrove wetlands are dominated by red mangroves (*Rhizophora*) that contain abundant and dense prop roots, particularly along the edges that face the open sea. Riverine mangrove wetlands are also numerically dominated by red mangroves, although they are straight trunked and have relatively few, short prop roots (Lugo and Snedaker, 1974). Black (*Avicennia* spp.) and white (*Laguncularia* spp.) mangroves also frequently grow in these wetlands. Basin wetlands support all three species of mangroves, although black mangroves are the most common in basin swamps and hammock wetlands are mostly composed of red mangroves. Scrub mangrove wetlands are typically dominated by widely spaced, short (less than 2 m tall) red or black mangroves.

A comparison of the structural characteristics of the major hydrodynamic types of mangrove wetlands is provided in Table 11-2. These data were compiled from more than 100 mangrove research sites throughout the New World. Fringe mangroves generally have a greater density of large trees ($>$ 10 cm dbh) compared to riverine and basin mangroves. Riverine wetlands, however, have the largest trees and, hence, a much greater basal area and tree height than do fringe or basin mangroves. The biomass of riverine mangroves is generally the highest, although data are difficult to compare because of the different methods and sample sizes used in various observations. Cintrón et al. (1985) reported a range of above-ground biomass for the Florida mangroves of 9 to 17 kg/m^2 for riverine mangroves and 0.8 to 15 kg/m^2 for fringe mangroves. Single measures of 0.8 kg/m^2 for a dwarf mangrove wetland and 9.8 kg/m^2 for a hammock mangrove (both in Florida) were also reported.

Understory Vegetation

One of the interesting aspects of certain mangrove swamps is the lack of conspicuous understory vegetation (Janzen, 1985; Corlett, 1986; Lugo, 1986), except for those in high-rainfall regions of the world or in ecotones of low soil salinity (Lugo, 1986). Janzen (1985), with the agreement of Lugo (1986), hypothesized that "plants with low light resources cannot accumulate enough metabolites fast enough to meet the metabolic demands of the drain of the machinery and morphology of salt tolerance," but Lugo (1986) suggested that energy sources other than light (e.g., tidal flushing and freshwater inflow) and stresses other than salt tolerance (e.g., hydrogen sulfide, low oxygen, frost) are also important factors. He suggested a more general hypothesis: "Understory plants grow in those mangrove ecosystems where combinations of nutrients, light energy, soil oxygen, and freshwater meet the metabolic demands of drains caused by all environmental stressors converging on the site."

One understory genus that is found in many mangrove wetlands is the mangrove fern (*Acrostichum* spp.). There are three species of this fern found in the world, usually in mangrove regions of high rainfall or low soil salinity (Medina et al., 1990). One species studied in Puerto Rico, *A. aureum* L., was found to have a wide range of light tolerance, growing in full sun in a disturbed mangrove forest and in the understory of a white mangrove forest. It was hypothesized that the fern tolerates higher salinity in the shade because of lower evaporative demands and, hence, less salt accumulation in its leaf tissues in shady conditions (Medina et al., 1990).

Plant Zonation and Succession

In trying to understand the vegetation of mangrove wetlands, most early researchers were concerned with describing plant zonation and successional patterns. Some attempts were made to equate the plant zonation found in mangrove wetlands with successional seres, but Lugo (1980) warned that "zonation does not necessarily recapitulate succession because a zone may be a climax maintained by a steady or recurrent environmental condition." J. H. Davis (1940) is generally credited with the best early description of plant zonation in Florida mangrove swamps, especially in fringe and basin mangrove wetlands (Fig. 11-7). He hypothesized that the entire ecosystem was accumulating sediments and was, therefore, migrating seaward. Elabo-

Table 11-2 Structural characteristics of canopy vegetation for major mangrove types[a]

Hydrodynamic Type	Number of Tree Species	Number of Trees (#/(ha))		Basal Area (m²/ha)		Stand Height (m)	Above-ground Biomass (kg/m²)
		>2.5 cm dbh	>10 cm dbh	>2.5 cm dbh	>10 cm dbh		
Fringe mangroves	1.7 ± 0.1 (33)	4005 ± 642 (33)	852 ± 115 (31)	22.2 ± 1.5 (33)	14.6 ± 1.9 (31)	13.3 ± 2.6 (32)	0.8–15.9 (8)
Riverine mangroves	1.9 ± 0.1 (36)	1979 ± 209 (28)	661 ± 71 (32)	30.4 ± 3.5 (5)	32.6 ± 4.7 (32)	21.2 ± 4.8 (26)	1.6–28.7 (8)
Basin mangroves	2.3 ± 0.1 (31)	3599 ± 400 (31)	573 ± 102 (21)	18.5 ± 1.6 (31)	10.6 ± 2.2 (21)	9.0 ± 0.7 (31)	—

[a]Data are based on mangrove sites in Florida, Mexico, Puerto Rico, Brazil, Costa Rica, Panama, and Ecuador. Values are the average ± standard error (number of observations) except for above-ground biomass, which is the range (number of observations).
Source: Cintrón et al. (1985).

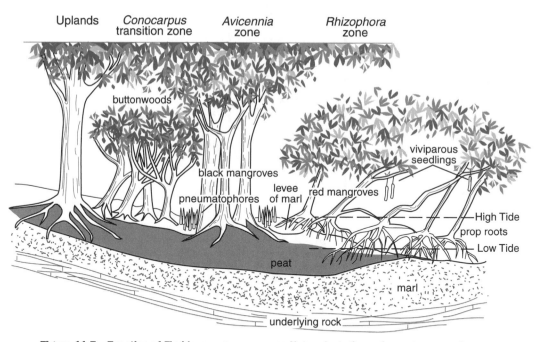

Figure 11-7 **Zonation of Florida mangrove swamp. Note adaptations of mangroves such as prop roots, viviparous seedlings, and pneumatophores (see Fig. 11-8). (*After J. H. Davis, 1940*)**

ration of the zonation pattern and theories for its occurrence were also provided by Egler (1952). Typically, *Rhizophora mangle* is found in the lowest zone, with seedlings and small trees sprouting even below the mean low tide in marl soils. Above the low-tide level but well within the intertidal zone, full-grown *Rhizophora* with well-developed prop roots predominate. There tree height is approximately 10 m. Behind these red mangrove zones and the natural levee that often forms in fringe mangrove wetlands, basin mangrove wetlands, dominated by black mangroves (*Avicennia*) with numerous pneumatophores, are found. Flooding occurs only during high tides. Buttonwood (*Conocarpus erecta*) often forms a transition between the mangrove zones and upland ecosystems. Flooding occurs there only during spring tides or during storm surges, and soils are often brackish to saline (Chapman, 1976b).

The zonation of plants in mangrove wetlands led some researchers (e.g., J. H. Davis, 1940) to speculate that each zone is a step in an autogenic successional process that leads to freshwater wetlands and, eventually, to tropical upland forests or pine forests. Other researchers, led by Egler (1952), considered each zone to be controlled by its physical environment to the point that it is in a steady state or at least a state of arrested succession (allogenic succession). For example, with a rising sea level, the mangrove zones migrate inland; during periods of decreasing sea level, the mangrove zones move seaward. Egler thought that the impact of fire and hurricanes made conventional succession impossible in the mangroves of Florida. Another theory, advanced by Chapman (1976b), is that mangrove succession may be a combination of both autogenic and allogenic strategies, or a "succession of successions." If that is the case, successional stages could be repeated a number of times before the next successional level is attained.

Lugo (1980) reviewed mangrove succession in light of E. P. Odum's criteria (1969; see Chapter 8) and found that, except for mangroves on accreting coastlines, traditional successional criteria do not apply. Succession in mangroves is primarily cyclic, and it exhibits patterns of stressed or "youthful" ecosystems, including slowed or arrested succession, low diversity, $P:R$ greater than one, and open material cycles, even in mature stages. Lugo (1980) concluded that mangroves are true steady-state systems in the sense that they are the optimal and self-maintaining ecosystems in low-energy tropical saline environments. In such a situation, high rates of mortality, dispersal, germination, and growth are the necessary tools of survival. Unfortunately, these attributes could lead many to the identification of mangroves as successional systems.

It is no longer accepted dogma that mangrove wetlands are "land builders" that are gradually encroaching on the sea, as was suggested by J. H. Davis (1940). In most cases, mangrove vegetation plays a passive role in the accumulation of sediments, and the vegetation usually follows, not leads, the land building that is caused by current and tidal energies. It is only after the substrate has been established that the vegetation contributes to land building by slowing erosion and by increasing the rate of sediment accretion (Lugo, 1980). The mangrove's successional dynamics appear to involve a combination of (1) peat accumulation balanced by tidal export, fire, and hurricanes over years and decades; and (2) advancement or retreat of zones according to the fall or rise of sea level over centuries.

Ball (1980) has suggested another allogenic succession model for mangroves, in which interspecific competition predominates. She found that red mangroves did not grow in dry upland locations because they did not have a competitive advantage there but did dominate where salinity and intertidal water levels gave them a competitive advantage. In the same type of argument, Thibodeau and Nickerson (1986) suggested that red mangroves have a much lower ability to tolerate high sulfides typical of extremely reduced conditions than do black mangroves, and so red mangroves occur in regions that are frequently flushed by tides, whereas black mangroves are found in isolated basin settings where strongly reduced substrates containing high sulfides are found and pneumatophores can be of the greatest use (see the following section).

Mangrove Adaptations

Mangrove vegetation, particularly the dominant trees, has several adaptations that allow it to survive in an environment of high salinity, occasional harsh weather, and anoxic soil conditions. These physiological and morphological adaptations have been of interest to researchers and are among some of the most distinguishing features that the layperson notices when first viewing these wetlands. Some of the adaptations, shown in Figures 11-7 and 11-8, are: (1) salinity control, (2) prop roots and pneumatophores, and (3) viviparous seedlings (see also Chapter 7).

Salinity Control

Mangroves are facultative halophytes; that is, they do not require salt water for growth but are able to tolerate high salinity and, thus, outcompete vascular plants that do not have this salt tolerance. The ability of mangroves to live in saline soils depends on their ability to control the concentration of salt in their tissues. In this respect, mangroves are similar to other halophytes (see Chapter 7). Mangroves have the ability both to prevent salt from entering the plant at the roots (*salt exclusion*) and to excrete

a.

Figure 11-8 Adaptations of mangroves, including **a.** prop and drop roots of red mangroves, **b.** pneumatophores of black mangroves, **c.** viviparous propagule hanging in red mangrove canopy, and **d.** red mangrove seedling germinated and rooted in sediments. (*Photograph a by M. T. Vogel, by permission; all others by W. J. Mitsch*)

salt from the leaves (*salt secretion*). Salt exclusion at the roots is thought to be a result of reverse osmosis, which causes the roots to absorb only fresh water from salt water. The root cell membranes of mangroves species of *Rhizophora, Avicennia,* and *Laguncularia,* among others, may act as ultrafilters that exclude salt ions. Water is drawn into the root through the filtering membrane by the negative pressure in the xylem developed through transpiration at the leaves; this action counteracts the osmotic pressure caused by the solutions in the external root medium (Scholander et al., 1965; Scholander, 1968). There are also a number of mangrove species (e.g., *Avicennia* and *Laguncularia*) that have salt-secreting glands on the leaves to rid the plant of excess salt. The solutions that are secreted often contain several percent NaCl, and salt crystals may form on the leaves. Another possible way, still questioned as to its importance, in which mangroves discharge salt is through leaf fall. This leaf fall may

b.

Figure 11-8 *(Continued).*

c.

Figure 11-8 *(Continued).*

be significant because mangroves produce essentially two crops of leaves per year (Chapman, 1976b).

Prop Roots and Pneumatophores

Some of the most notable features of mangrove wetlands are the *prop roots* and *drop roots* of the red mangrove (*Rhizophora;* Fig. 11-8a) and the numerous, small pneumatophores of the black mangrove (*Avicennia*) (reaching 20–30 cm above the sediments, although they can be up to 1 m tall; Fig. 11-8b). The drop roots are special cases of the prop roots that extend from branches and other upper parts of the stem directly down to the ground, rooting only a few centimeters into the sediments. Oxygen enters the plant through small pores, called *lenticels,* that are found on both pneumatophores and prop and drop roots. When lenticels are exposed to the atmosphere during low tide, oxygen is absorbed from the air and some of it is transported to and diffuses out of the roots through a system of aerenchyma tissue. This maintains an aerobic microlayer around the root system. When the prop roots or pneumatophores of mangroves are continuously flooded by stabilizing the water levels, those mangroves that have submerged pneumatophores or prop roots soon die (Macnae, 1963; J. H. Day, 1981).

d.

Figure 11-8 *(Continued)*.

In an interesting experiment to determine the importance of oxygen transport from the aerial organs to the sediments, Thibodeau and Nickerson (1986) "capped" with plastic tubing the pneumatophores of *Avicennia germinans* in a fringe mangrove forest in the Bahamas. They observed that the aerobic zone surrounding the roots was reduced in the area by capping, indicating that the pneumatophores help the plant produce an oxidized rhizosphere. They also found that the greater the number of pneumatophores present in a given area, the more oxidized the soil. They described the relationship as

$$E_{\mathrm{H}} = -307 + 1.1\mathrm{pd} \tag{11.1}$$

where E_{H} = redox potential (mV)

pd = pneumatophore density (number per 0.25 m^2)

Viviparous Seedlings

Red mangroves (and related genera in other parts of the world) have seeds that germinate while they are still in the parent tree; a long, cigar-shaped hypocotyl (viviparous seedling) develops while hanging from the tree (Fig. 11-8c). This is apparently an adaptation for seedling success where shallow anaerobic water and sediments would otherwise inhibit germination. The seedling (or propagule) eventually falls and often will root if it lands on sediments or will float and drift in currents and tides if it falls into the sea. After a time, if the floating seedling becomes stranded and the water is shallow enough, it will attach to the sediments and root (Fig. 11-8d). Often, the seedling becomes heavier with time, rights itself to a vertical position, and develops roots if the water is shallow. It is not well understood whether contact with the sediments stimulates root growth or if the soil contains some chemical compound that promotes root development (Chapman, 1976b). The value of the floating seedlings for mangrove dispersal and for invasion of newly exposed substrate is obvious. Rabinowitz (1978) reported that the obligate dispersal time (the time required during propagule dispersal for germination to be completed) was 40 days for the red mangrove and 14 days for the black mangrove propagules. She also estimated that red and black mangrove propagules could survive for 110 and 35 days, respectively. In contrast, J. H. Davis (1940) had found that red mangrove propagules could float for more that one year.

Consumers

W. E. Odum et al. (1982) found the following data from the literature describing faunal use of mangroves in Florida in terms of the number of species: 220 fish; 181 birds, including 18 wading birds, 29 water birds, 20 birds of prey, and 71 arboreal birds; 24 reptiles and amphibians; and 18 mammals. In general, a wide diversity of animals is found in mangrove wetlands; their distribution sometimes parallels the plant zonation described previously. Many of the animals that are found in mangrove wetlands are filter feeders or detritivores, and the wetlands are just as important as a shelter for most of the resident animals as they are a source of food. Some of the important filter feeders found in Florida mangroves include barnacles (*Balanus eburneus*), coon oysters (*Ostrea frons*), and the eastern oyster (*Crassostrea virginica*). These organisms often attach themselves to the stems and prop roots of the mangroves within the intertidal zone, filtering organic matter from the water during high tide.

Crabs are among the most important animal species in mangrove wetlands around the world, and they appear to play an important role in maintaining biodiversity in mangrove ecosystems (D. A. Jones, 1984; T. J. Smith et al., 1991; Camilleri, 1992; Twilley et al., 1996). They burrow in the sediments, prey on mangrove seedlings, facilitate litter decomposition, and are the key transfer organism for converting detrital energy to wading birds and fish in the mangrove forest itself and to offshore estuarine systems. Mangroves around the world are dominated by 6 of the 30 families of Brachyura that collectively make up about 127 species (D. A. Jones, 1984). *Uca* and *Sesarma* are the most abundant crab genera in mangrove wetlands. *Uca* spp. (fiddler crabs) are particularly abundant in mangrove wetlands in Florida, living on the prop roots and high ground during high water and burrowing in the sediments during low tide. *Sesarma* (sesarmids) is the most abundant genus of crabs in the world, with dozens of species in the Indo-Malaysian and East African mangroves (D. A. Jones, 1984) and many fewer in tropical America.

One of the most significant ways in which crabs may influence the distribution of mangroves is by selective predation of mangrove propagules (T. J. Smith, 1987; T. J. Smith et al., 1989; Osborne and Smith, 1990). Yet this effect is not common to all mangrove wetlands. Smith et al. (1989), in comparing the effect of this predation on mangroves around the world, found that crabs consumed up to 75 percent of the mangrove propagules in Australian mangrove swamps but very little of the litterfall or propagules in Panamanian and Florida mangrove swamps.

The role of crabs in leaf litter removal (burial and consumption) has been illustrated in a number of studies in Malaysia (Malley, 1978), Australia (Robertson, 1986; Robertson and Daniel, 1989; T. J. Smith et al., 1991), and East Africa (Micheli et al., 1991). Robertson and Daniel (1989) estimated that, in some mangrove forests, leaf processing by Sesarmid crabs alone was over 75 times the rate as microbial degradation and that crabs removed more that 70 percent of the litter of the high-intertidal *Ceriops* and *Bruguiera* mangroves in tropical Australia. Smith et al. (1991) developed an experiment where crabs were removed from experimental plots in *Rhizophora* forests. They found that the removal of crabs caused significantly higher concentrations of sulfide and ammonium in the mangrove soils, due primarily to the absence of burrowing, which oxygenates the soil. Taking into account the role that they have on seedling survival, carbon cycling, sediment microtopography, and soil chemistry, Smith et al. (1991) suggested that crabs are the keystone species of the mangrove ecosystem.

Many other invertebrates, including snails, sponges, flatworms, annelid worms, anemone, mussels, sea urchins, and tunicates, are found growing on roots and stems in and above the intertidal zone. Wading birds that are frequently found in Florida mangroves include the Wood Stork (*Mycteria americana*), White Ibis (*Eudocimus albus*), Roseate Spoonbill (*Ajaia ajaja*), cormorant (*Phalacrocorax* spp.), Brown Pelican (*Pelicanus occidentalis*), egrets, and herons (W. E. Odum and McIvor, 1990). Vertebrates that inhabit mangrove swamps include alligators, crocodiles, turtles, bears, wildcats, pumas, and rats.

Ecosystem Function

Certain functions of mangroves such as net primary productivity, litter production, organic export, and nutrient cycling have been studied extensively. A picture of the dynamics of mangrove wetlands has emerged from several key studies. These studies have demonstrated the importance of the physical conditions of tides, salinity, and nutrients to these wetlands and have shown where natural and human-induced stresses have caused the most effect.

Primary Productivity

A wide range of productivity has been measured in mangrove wetlands due to the wide variety of hydrodynamic and chemical conditions encountered. Table 11-3 presents a balance of carbon flow in several fringe and basin mangrove swamps in Florida and

Table 11-3 Mass balance of carbon flow (g C m^{-2} yr^{-1}) in mangrove forests in Florida and Puerto Rico

	Rookery Bay, Florida[a]		Puerto Rico Fringe[b]	Fahkahatchee Bay, Florida[c]		
	Fringe	Basin		Basin	Fringe	Fringe
Gross primary productivity (GPP)						
Canopy	2,055	3,292	3,004	3,760	4,307	5,074
Algae	402	26	276			
Total	2,457	3,318	3,280			
Respiration (plants)						
Leaves, stems	671	2,022	1,967	1,172	1,416	3,084
Roots, above ground	22	197	741	146	182	215
Roots, below ground	?	?	?	?	?	?
Total	693	2,219	2,708	1,318	1,598	3,299
Net primary production	1,764	1,099	572	2,442	2,709	1,775
Growth		186	153			
Litterfall		318	237			
Respiration (heterotrophs)		197	135			
Respiration (total)		2,416	2,843			
Export		64	500			
Net ecosystem production (NEP)		838	−63			
Burial		?	?			
Growth		186	153			

[a]Lugo et al. (1975), Twilley (1982, 1985), and Twilley et al. (1986).
[b]Golley et al. (1962).
[c]Carter et al. (1973).
Source: Twilley (1988).

Puerto Rico, and Table 11-4 summarizes daily productivity and litter production data for riverine, basin, and scrub mangrove wetlands from a number of field studies. Net primary productivity ranges from approximately 1,100 to 5,400 g m^{-2} yr^{-1} (assuming 1 g C = 2 g dry weight). Gross and net primary productivity was highest in riverine mangrove wetlands, lower in basin mangrove wetlands, and lowest in dwarf mangrove wetlands. In a Mexican mangrove forest (J. W. Day et al., 1987), net above-ground primary productivity ranged from 1,607 g m^{-2} yr^{-1} for a fringe mangrove to 2,458 g m^{-2} yr^{-1} for a riverine mangrove system. Day et al. attributed the higher productivity to the greater influence of nutrient loading and freshwater turnover at the riverine site. This range compares well with a study by S. Y. Lee (1990) of *Kandelia cancel* mangroves in Hong Kong, where mangrove plant net primary productivity averaged 1,950 to 2,440 g m^{-2} yr^{-1}, depending on the method used for the productivity determination. Total net primary productivity, including all plants at the Hong Kong site, was approximately 4,400 g m^{-2} yr^{-1}; *Phragmites communis* contributed about 46 percent of the site productivity and macroalgae and phytoplankton another 3.2 percent.

Table 11-4 Primary productivity, respiration, and litterfall measurements for three types of mangrove wetlands

	Mangrove Wetland		
	Riverine	Basin	Scrub
Gross primary productivity[a] (kcal m^{-2} day^{-1})	108	81	13
Total respiration[a] (kcal m^{-2} day^{-1})	51	56	18
Net primary productivity[a] (kcal m^{-2} day^{-1})	57	25	0
Litter production[b] (kcal m^{-2} day^{-1})	14 ± 2	9.0 ± 0.4	1.5

[a]From several sites in Florida; based on CO_2 gas exchange measurements; assumes 1 g organic matter = 4.5 kcal.
[b]Average ± standard error from several sites in Florida and Puerto Rico; measured with litter traps; assumes 1 g organic matter = 4.5 kcal.
Source: Brown and Lugo (1982).

In a subsequent 7-year study of net primary productivity in Mexico, J. W. Day et al. (1996) found low net primary productivity in basin mangroves (400–695 g m^{-2} yr^{-1}), as expected, given their hydrologically isolated position in the landscape and the buildup of soil salinity.

Litterfall

Organic production, as measured by litterfall, is summarized in Figure 11-9 for scrub, basin, fringe, and riverine mangrove systems, illustrating the same connection as described previously between hydrologic conditions and productivity. Litterfall ranges from approximately 200 g m^{-2} yr^{-1} in scrub wetlands to 1,200 g m^{-2} yr^{-1} in riverine swamps. The greater the hydrologic turnover is (riverine > fringe > basin > scrub), the greater the litter production. S. Y. Lee (1989) found litter productivity in his mangrove site in a highly managed tidal shrimp pond near Hong Kong to be about 1,100 g m^{-2} yr^{-1}, identical to the litterfall measured by Flores-Verdugo et al. (1987) at a mangrove site on the Pacific coast of Mexico. The Hong Kong wetland could be considered to be a fringe mangrove because water changes in the pond every spring tide but remains in the pond for 4 to 5 days during neap tide periods, whereas the Mexican site was located in a lagoon that contains an ephemeral inlet open to the coastline for 3 to 4 months per year.

Factors Affecting Productivity

The important factors that control mangrove function in general and primary productivity in particular are: (1) tides and storm surges; (2) freshwater discharge; (3) parent substrate; and (4) water–soil chemistry, including salinity, nutrients, and turbidity. These factors are not mutually exclusive, for tides influence water chemistry and, hence, productivity by transporting oxygen to the root system, by removing the buildup of toxic materials and salt from the soil water, by controlling the rate of sediment accumulation or erosion, and by indirectly regenerating nutrients lost from the root zone. The principal chemical conditions that affect primary productivity are soil water salinity and the concentration of major nutrients. Lugo and Snedaker (1974) have concluded that, compared to mangrove wetlands such as dwarf mangroves that are isolated from the influence of daily tides, "environments flushed adequately and frequently by seawater and exposed to high nutrient concentrations are more favorable

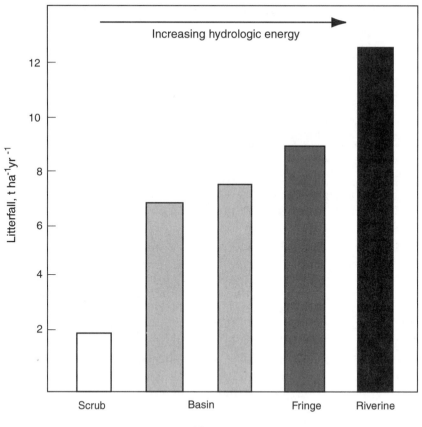

Figure 11-9 Litterfall rates in different hydrologic types of mangrove wetlands. The two values for basin mangroves are for monospecific and mixed forests. (*After Twilley, 1988*)

for mangrove ecosystem development; forests in these areas exhibit higher rates of net primary productivity."

The importance of chemical conditions to mangrove productivity is difficult to document because the chemical conditions include both stimulants (nutrients) and stressors (salinity).

High soil salinities, which, in turn, are a function of the local hydrology and geomorphology, appear to be the most important variable that influences the productivity of mangroves in a given region. For example, one study of mangroves in Puerto Rico (Cintrón et al., 1978) found that tree height of the mangroves, as a measure of productivity, was inversely related to soil salinity according to the following relationship:

$$h = 16.6 - 0.20C_s \qquad (11.2)$$

where h = tree height (m)
C_s = soil salinity concentration (ppt)

In a similar analysis, J. W. Day et al. (1996) were able to demonstrate a relationship between total litterfall in an *Avicennia*-dominated basin mangrove forest in Mexico and soil salinity as

$$L = 3.915 - 0.039C_s \qquad (11.3)$$

where L = litterfall (g m^{-2} yr^{-1})

Nevertheless, both salinity and nutrients appear to be important for mangrove growth. Kuenzler (1974) suggested that, even with low transpiration rates in mangroves as a result of high salinities, productivity can be high if nutrients are abundant.

Productivity and Zonation

A series of detailed investigations in the 1970s on the pattern of metabolic zonation of Florida mangroves led to a number of rules that partially explain the zonation of species as described earlier in this chapter. Carter et al. (1973) examined the productivity of mangrove canopy leaves through gas exchange measurements along a gradient of fresh to saline water in southwestern Florida. They found that both respiration and gross primary productivity increased with increased salinity (the salinity range remained well below the seawater concentration) but at different rates so that net primary productivity available for plant growth actually decreased at the highest salinity of 16 ppt (Fig. 11-10). Mangroves put more of their captured energy into growth rather than physiological maintenance when the water is low in salts. In saline water, more respiratory work is necessary as a metabolic cost of adapting to high salinities. Lugo and Snedaker (1974) summarized the functional zonation in these studies as follows:

1. Gross primary productivity of red mangroves decreases with increasing salinity.
2. Gross primary productivity of black and white mangroves increases with increasing salinity.
3. In areas of low salinities and under equal light conditions, gross primary productivity of red mangroves is four times that of black mangroves.
4. In areas of intermediate salinity, white mangroves have rates of gross primary productivity twice that of red mangroves.
5. In areas of high salinities, white mangroves exhibited gross primary productivity higher than that of black mangroves, which, in turn, was higher than red mangroves.

Thus, the zonation of mangroves, as described by J. H. Davis (1940), Egler (1952), and others, has a functional basis. Species that are found growing outside of their optimal zone have lower productivity than do those that are adapted to conditions in that zone, and competition will eventually eliminate them from that zone.

Decomposition

The decomposition process in mangroves has been studied with litter bag measurements for a number of different plants (mainly red mangroves and black mangroves) and in a number of different mangrove types (Table 11-5). Decay rates range from 0.30 yr^{-1} for red mangrove leaves in a basin mangrove wetland in Florida to 8.4 yr^{-1} for red mangrove leaves in a fringe mangrove wetland. Lugo and Snedaker (1974)

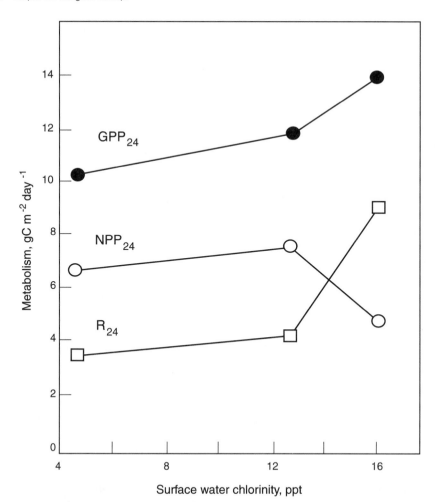

Figure 11-10 Changes in mangrove metabolism, as measured by carbon gas exchange with increased salinity (as measured by chlorine content). GPP_{24} indicates gross primary productivity over 24 hours; NPP_{24} indicates net primary productivity over 24 hours; R_{24} indicates respiration over 24 hours. (*After Lugo, 1990b*)

reported previously that decomposition is accelerated by moisture, with optimal decomposition occurring at about 50 percent moisture. Subsequent studies have shown that black mangrove leaves decay three times faster than do red mangrove leaves, a condition attributed to the higher C/N ratio in red mangrove leaves (Twilley, 1982). Mangrove leaves in a rapidly flushing environment also appear to decay more rapidly than those in a slowly flushing environment (Table 11-5). Because basin mangroves are generally dominated by black mangroves, it has been hypothesized that the faster decomposition of black mangrove leaves may lead to a greater export of dissolved organic material, rather than particulate organic matter, from basin mangroves when compared to riverine or fringe mangroves (Cintrón et al., 1985).

Crabs and other invertebrates, when they are abundant in mangrove wetlands, may play a significant role in the decomposition of mangrove litter due to their leaf

Table 11-5 Litter decay rates in mangrove wetlands

Mangrove Type	Location	Decay Rate (yr^{-1})	Half-Life (days)
Fringe			
Rhizophora mangle	Florida	8.39	30
	Florida	2.55	99
	Florida	1.46	173
	Florida	1.10	231
	Florida	0.85	346
	Puerto Rico	2.55	99
	Mexico	1.42	178
	Brazil	1.23	206
Laguncularia racemosa	Brazil	3.08	82
Basin			
Rhizophora mangle	Florida	0.30	—
	Florida	1.22	231
	Florida	2.01	139
	Florida	1.28	231
	Florida	1.63	173
(fast flushing)	Florida	1.90	133
(slow flushing)	Florida	0.97	260
Avicennia germinans			
(fast flushing)	Florida	6.04	60
(slow flushing)	Florida	3.89	65
Riverine	Mexico	2.33	99
	Colombia	2.81	87
Laguncularia racemosa	Mexico	1.71–4.7	54–147
Dwarf	Florida	1.46	173
	Florida	2.29	115
Hammock	Florida	2.29	115

Source: Cintrón et al. (1985) and Flores-Verdugo et al. (1990).

litter consumption and shredding of leaves into smaller particles (Robertson and Daniel, 1989; Camilleri, 1992; Twilley et al., 1997). Robertson and Daniel (1989) showed that Sesarmid crabs removed 580 to 803 g m^{-2} yr^{-1}, or an amazingly high 71 to 79 percent of total litterfall, in two intertidal mangrove forests in Australia. This compares to only 194 to 252 g m^{-2} yr^{-1} of energy that was exported as litter by tides. A similar study by Robertson (1986) showed that the crab *Sesarma messa* buried or consumed about 28 percent of the annual litterfall of a *Rhizophora* mangrove forest; he argued in this and in several subsequent papers that many of the early mangrove carbon cycling models from southern Florida did not recognize this important role of crabs.

The importance of the crab's biological control of decomposition via shredding was summarized by Camilleri (1992) as causing several significant outcomes in the wetland, including the following:

1. The shredded leaf particles are less likely to be exported from the mangrove ecosystem than are the larger leaves.

2. The shredded material makes particulate organic material available to a larger number of detritivores.

3. The shredded particles are more easily colonized by microfauna and microorganisms, leading to more rapid decomposition and better recycling of nutrients.

Mortality

Natural tree mortalities on large scales are frequently cited in the literature for mangrove wetlands (see the summary in Jiménez et al., 1985). These mortalities are due to both natural causes, for example, hurricanes, droughts, and frosts, and human-influenced changes, for example, alteration of hydrology, dredging, and increased sedimentation (Lugo et al., 1981; Jiménez et al., 1985; Lugo, 1990a). Changes in the hydrologic conditions "either through alteration of regional hydrology or modification of the geomorphology of the mangrove basin" (Lugo, 1990a) are particularly important stresses on mangrove systems, often serving to exacerbate natural stresses. Stresses and massive mortalities are particularly noticeable in mangrove wetlands before or during a senescence stage when tree growth naturally slows down and wide gaps in the canopy and lack of regeneration are typical (Jiménez et al., 1985). When massive diebacks of mangrove trees occur, a common denominator often appears to be rapid environmental change. The damage is mitigated if there is freshwater or tidal flushing, or is exacerbated if the system is already under stress from factors such as excessive salinity, low temperature, excessive siltation, or altered hydrology (Lugo et al., 1981; Jiménez et al., 1985). Lugo et al. (1981) and Jiménez et al. (1985) argued that even massive diebacks of mangroves reportedly due to biological causes—such as gall disease in the mangroves of Gambia (Teas and McEwan, 1982)—are manifestations of changes in riverine or tidal hydrology, not solely due to biological factors alone.

Hurricane Effects

Hurricanes have a particular effect on mangrove forests that, until recently, had not been quantified. Earlier simulations of models by Lugo et al. (1976) and Sell (1977) suggested that the time required for the attainment of a steady state in mangrove biomass in Florida is approximately the same as the average period between tropical hurricanes (approximately 20 to 24 years for Caribbean systems). This match suggests that mangroves may have adapted or evolved to go through one life cycle, on average, between major tropical storms. On average, a mangrove forest reaches maturity just as the next hurricane or typhoon hits.

When Hurricane Andrew passed over south Florida in the late summer of 1992, an opportunity existed for detailed studies of the immediate impact of hurricanes on mangroves. Major damage to mangroves in the vicinity of the Everglades occurred due to trunk snapping and uprooting rather than due to any storm surge (T. J. Smith et al., 1994). Mortality was greatest for red mangrove in the 15- to 30-cm DBH age class and over a wider band of 10 to 35 cm dbh for black and white mangrove (Fig. 11-11). In addition, two interesting observations were made as a result of interpreting the recovery of the mangroves from the hurricane (T. J. Smith et al., 1994):

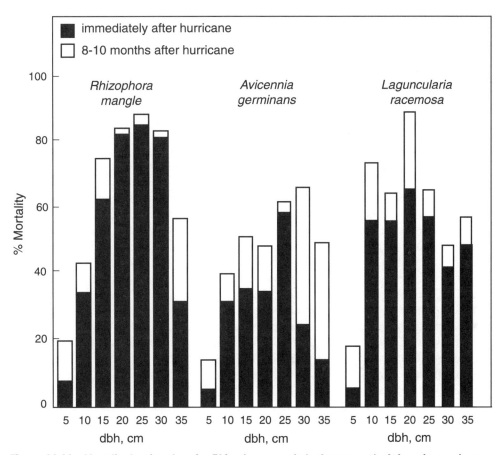

Figure 11-11 Mortality by size class for *Rhizophora mangle* (red mangrove), *Avicennia germinans* (black mangrove), and *Laguncularia racemosa* (white mangrove) as a result of Hurricane Andrew as measured immediately (September–October 1992) and 8 to 10 months after the hurricane struck south Florida mangroves in September 1992. (*After T. J. Smith et al., 1994*)

1. Gaps that developed in the mangrove forests due to lightning prior to the hurricane were, after the hurricane, small green patches amid the gray matrix of dead mangroves. Apparently, the small-sized mangrove trees that were spared in these patches would now serve as propagule regeneration sites for the much larger scale area of catastrophic disturbance. Small-scale disturbance nested in large-scale disturbance provides a positive feedback for more rapid mangrove recovery.

2. The loss of mangrove trees in a hurricane removes a major source of aeration of mangrove soils, the trees themselves (see, e.g., McKee et al., 1988). As a result of the loss of this aeration, soils in hurricane-impacted mangrove wetlands might become even more reduced, producing even more toxic hydrogen sulfide that could preclude mangrove regeneration for a number of years. Recovery of the mangrove forest to a system similar to

that prior to the disturbance is not assured because of this negative feedback.

Export of Organic Material

Mangrove swamps are important exporters of organic material to the adjacent estuary through a process referred to as *outwelling* (S. Y. Lee, 1995). In one of the first studies on outwelling from mangroves, Heald (1971) estimated that about 50 percent of the above-ground productivity of a mangrove swamp in southwestern Florida was exported to the adjacent estuary as particulate organic matter. From 33 to 60 percent of the total particulate organic matter in the estuary came from *Rhizophora* (red mangrove) material. The production of organic matter was greater in the summer (the wet season in Florida) than other seasons, although detrital levels in the swamp waters were greatest from November through February, which is the beginning of the dry season. Thirty percent of the yearly detrital export occurred during November. Heald also found that, as the debris decomposed, its protein content increased. The apparent cause of this enrichment, also noted in salt marsh studies, is the increase of bacterial and fungal populations.

Since those early studies, an abundance of studies have been undertaken on the outwelling from mangroves to adjacent estuaries. S. Y. Lee (1995) identified 21 such studies where export of particulate and dissolved organic carbon, nutrients, and even organisms themselves were investigated. Almost all of the studies agree that there is export of particulate organic carbon from mangroves to the adjacent estuary. One study (Rivera-Monroy et al., 1995) did illustrate, contrary to many other studies, that a mangrove forest in Mexico was a sink for inorganic nitrogen and a source of organic nitrogen. A summary of several studies on carbon export from mangrove wetlands gave a range of 2 to 400 g C m^{-2} yr^{-1} with an average of about 200 g C m^{-2} yr^{-1}, about double that exported from salt marshes (Twilley, 1998). In a comparison of leaf litter production and organic export for riverine, fringe, and basin mangrove systems (Fig. 11-12), riverine mangrove systems exported a majority of their organic litter (94 percent, or 470 g C m^{-2} yr^{-1}), whereas basin mangroves exported much less (21 percent, or 64 g C m^{-2} yr^{-1}), leaving the leaf litter to decompose or accumulate as peat. The proportion of litterfall production that is exported as well as the total amount of litter that is exported increase as the tidal influence increases.

Flores-Verdugo et al. (1987) estimated that "a minimum of 74 percent, but probably closer to 90 percent, of mangrove litter fall reached the lagoon waters" at their semiriverine Mexican mangrove site. By contrast, S. Y. Lee (1989) estimated that the export of litter from his mangrove site located above the mean water level was "unimportant" and that most of the litter production was consumed by crabs or microbial decomposition. In a subsequent paper, S. Y. Lee (1991) found that the grazing of the living biomass in this mangrove was also low, with an estimated 2.8 to 3.5 percent of the net above-ground primary productivity consumed by herbivores.

Effects on the Estuary

The role of mangrove wetlands as both a habitat and a source of food for estuarine fisheries is one of the most often cited functions of these ecosystems. The fact that organic carbon is exported from mangrove wetlands does not guarantee that it enters estuarine food chains. Yet, several independent studies have verified that mangrove wetlands are important nursery areas and sources of food for sport and commercial

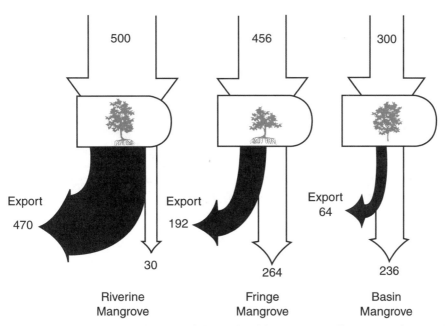

Figure 11-12 Summary of organic carbon inflows (litterfall), export to adjacent aquatic systems, and other losses (decomposition and peat production). Width of each pathway is proportional to flow (in g C m^{-2} yr^{-1}). (*After Twilley et al., 1986*)

fisheries (e.g., W. E. Odum and Heald, 1972; Yáñez-Arancibia et al., 1988; W. E. Odum and McIvor, 1990; Robertson and Duke, 1990; Fig. 11-13). Early studies by W. E. Odum (1970) and W. E. Odum and Heald (1972) established that detrital export is important to sport and commercial fisheries (Fig. 11-13a). Through the examination of the stomach contents of more than 80 estuarine animals, Odum found that mangrove detritus, particularly from *Rhizophora,* is the primary food source in the estuary. Important consumers include the spiny lobster (*Panulirus argus*), pink shrimp (*Penaeus duorarum*), mullet (*Mugil cephalus*), tarpon (*Megalops atlanticus*), snook (*Centropomus undecimalis*), and mangrove snapper (*Lutjanus apodus*). The primary consumers also used the mangrove estuarine waters during their early life stages as protection from predators and as a source of food. In a series of studies in Terminos Lagoon in Mexico (Yáñez-Arancibia et al., 1988, 1993), the use of mangrove forests by estuarine fish was clearly illustrated (Fig. 11-13b). There are clear connections among seasonal pulsing of mangrove detrital production, adjacent planktonic and seagrass productivity, and fish movement and secondary productivity. It is reasonable to extrapolate from this and similar studies that the removal of mangrove wetlands would cause a significant decline in sport and commercial fisheries in adjacent open waters.

Fleming et al. (1990) demonstrated for a site in southeast Florida that the amount of mangrove detritus relative to seagrass (*Thalassia testudinum*) detritus decreased rapidly from 80 percent mangrove detritus along the stream, where the mangroves were found, to 10 percent mangrove detritus about 90 m beyond the mouth of the stream. Using a technique involving the measurement of $^{13}C/^{12}C$ ratios of organic

a.

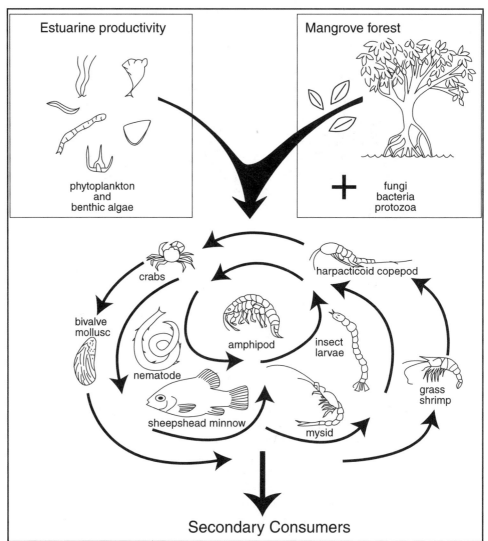

Figure 11-13   Illustrations of the support of fisheries by mangrove wetlands.   a. Detritus-based food web in south Florida estuary illustrating major contribution of mangrove detritus to the estuarine food chain. (*After W. E. Odum and Heald, 1972*)   b. Life histories and habitat utilization of six fish species including marine–estuarine spawners, estuarine spawners, and freshwater spawners in a lagoon on the Gulf of Mexico, Mexico. (*After Yáñez-Arancibia et al., 1988*)

matter, they found that mangroves were providing 37 percent and seagrasses 63 percent of the carbon to organisms in the bay, suggesting that mangrove detrital export is important to offshore water only in regions local to the wetlands and that it may be unimportant relative to other carbon sources farther offshore.

In a series of conceptual models that include the connection of mangrove ecosystems to their adjacent estuaries, Twilley (1988) illustrated that mangroves contributed an estimated 338 to 345 g C m^{-2} estuary yr^{-1} to the estuary, or from 83 to 86 percent

b.

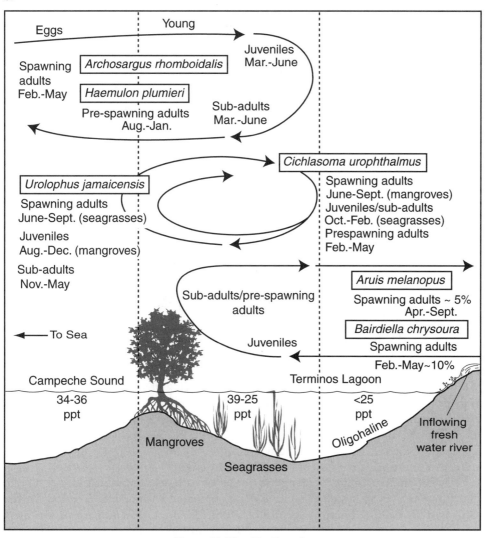

Figure 11-13 *(Continued).*

of the allochthonous inputs, and from 39 to 52 percent of the entire fixed carbon pool available for secondary productivity in the estuary. These numbers illustrate the importance of mangrove export to the overall secondary productivity of adjacent estuarine waters.

Organic carbon export from mangroves includes not only particulate organic matter (POM), which is most often measured, but also dissolved organic matter (DOM), which has not received as much attention as POM in export studies. Snedaker (1989) argued that future research needs to focus on the role of DOM as (1) an alternative food source for animals in the lower parts of the food chain, (2) an energy source for heterotrophic microorganisms, and (3) a source of chemical cues for estuarine species. Wafar et al. (1997) measured all of these fluxes as well as those of particulate and dissolved nitrogen and phosphorus in a *Rhizophora* swamp on the west coast of India. They made the following three conclusions:

1. Mangrove production is only important for the carbon budget of the adjacent estuary, not the nitrogen or phosphorus budget.
2. The energy flux coming from the mangroves is more important for sustaining microbial food chains than for sustaining particulate food chains.
3. The influence of the mangrove forest on the adjacent estuary relative to phytoplankton production in the estuary is, of course, a function the size of the adjacent estuary relative to the size of the mangrove forests.

Energy Flow

Golley et al. (1962) developed a synoptic energy budget of a Puerto Rican mangrove wetland for an average day in May. Of the total gross productivity of the canopy trees of 82 kcal m^{-2} day^{-1}, a major portion is used by the plants themselves in respiration. Mangrove leaves and stems accounted for 54 kcal m^{-2} day^{-1} of respiration and above-ground prop roots accounted for another 20 kcal m^{-2} day^{-1}. Export was 14 kcal m^{-2} day^{-1} and soil respiration was 4 kcal m^{-2} day^{-1}. Animal metabolism, estimated to be 0.8 kcal m^{-2} day^{-1}, made up a very minor part of the energy flow in this ecosystem. Lugo et al. (1975) found similar patterns of metabolism in mangrove swamps in southwest Florida, although the respiration of the red mangrove prop roots was much lower (0.6 kcal m^{-2} day^{-1}). They also found a higher production : respiration ratio than was found in the Puerto Rico study, convincing them of the importance of organic export from mangrove wetlands. A significant contribution to energy fixation by periphyton growing on red mangrove prop roots—a net productivity of 11 kcal m^{-2} day^{-1}—was also noted in the Florida study. Periphyton may have an important function of capturing and concentrating nutrients from incoming tidal waters for eventual use by the mangroves themselves.

Overall annual energy budgets (in terms of carbon) are summarized for basin and fringe mangrove systems in Florida and compared with annualized data for the Golley et al. (1962) Puerto Rico study (Table 11-3). Gross primary productivity values, which ranged from 25,000 to 50,000 kcal m^{-2} yr^{-1} (assuming 1 g C = 10 kcal), establishes these systems as among the most productive in the world. In some of the wetlands, productivity due to epiphytic algae was significant, contributing as much as 16 percent of the gross productivity. Heterotrophic respiration was a relatively minor energy flow at about 1,300 to 2,000 kcal m^{-2} yr^{-1}, well below the autotrophic respiration. Net ecosystem productivity (NEP), which is the amount of energy (or organic carbon) remaining after respiration and export, was calculated to be a deficit for the Puerto Rican fringe wetland (i.e., the system was not accumulating carbon in the peat) compared to a substantial energy accumulation of 8,000 kcal m^{-2} yr^{-1} in the basin mangrove in Florida's Rookery Bay. This comparison was summarized by Twilley (1988) as follows:

> These estimates of NEP between a fringe and a basin mangrove suggest that a large proportion of the NPP [net primary productivity] in the more inundated forests [fringe mangroves] is exported, while in the basin forests more of the net production is accumulated or utilized within the system. This supports the "open"

versus "closed" concept of fringe and basin mangroves, respectively, as proposed by Lugo and Snedaker (1974) in relation to hydrologic energy.

Ecosystem Models

Several qualitative and quantitative compartment models have been developed from research on the functional characteristics of mangroves in south Florida. In a quantitative model, Chen and Twilley (1998) described a simulation model that predicts the biomass and composition of mangrove forests through time (see also Chapter 8; Fig. 8-9). In a simple conceptual model of energy and material flows (Fig. 11-14), the mangrove ecosystem is shown to be affected principally by tidal exchange and upland inflow but also internally by the role of crabs capturing fresh litter and consuming and burying it, thus reducing the amount of outwelling. Occasional hurricane pulses

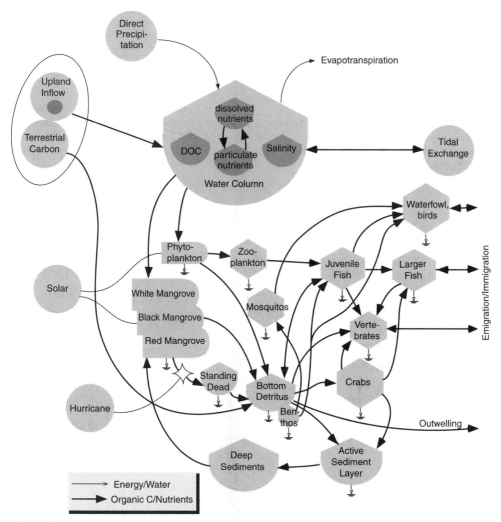

Figure 11-14   Conceptual model of mangrove wetlands that illustrates the importance of freshwater inflow, tidal exchange, crabs, hurricanes, salinity, and nutrients.

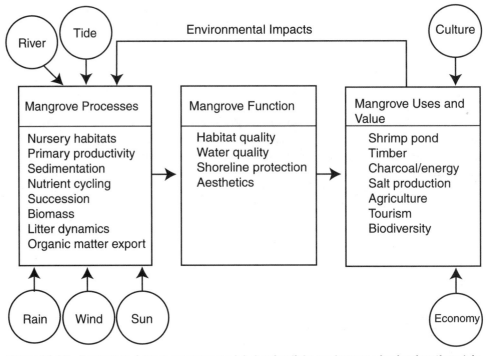

Figure 11-15 Ecosystem–human economy model showing linkages between forcing functions (circles), mangrove processes and functions, and mangrove uses by and values for humans. Note environmental impacts that provide a negative feedback to mangrove processes and functions, which, in turn, would lower human uses and values. (*After Twilley, 1995*)

turn live mangroves into standing dead wood, but the effects are temporary rather than long-lasting.

A model that describes the interactions between mangrove ecosystems and human economy (Fig. 11-15) illustrates the connections between mangrove processes and functions and the uses and values of mangroves by humans. An important aspect of this diagram is the illustration that these uses of mangrove ecosystems by humans, whether the use is heavy such as the conversion of mangrove forests to shrimp ponds or light such as tourism stimulated by productive fisheries or clean water, have negative feedbacks to the mangrove processes themselves and too much of any of these human uses will amplify the loss of these important ecosystems. Finding the balance between maintaining productive coastal mangrove ecosystems and developing human economies is both difficult yet essential for sustaining coastlines throughout the tropical world.

Recommended Readings

Alongi, D. M. 1998. Coastal Ecosystem Processes. CRC Press, Boca Raton, FL.

Chapman, V. J. 1976. Mangrove Vegetation. J. Cramer, Vaduz, Germany. 447 pp.

Lugo, A. E., and S. C. Snedaker. 1974. The ecology of mangroves. Annual Review of Ecology and Systematics 5:39–64.

Rützler, K., and I. C. Feller. 1996. Caribbean mangrove swamps. Scientific American (March 1996):94–99.

Smith, T. J., M. B. Robblee, H. R. Wanless, and T. W. Doyle. 1994. Mangroves, hurricanes, and lightning strikes. BioScience 44:256–262.

Twilley, R. R. 1998. Mangrove wetlands. In M. G. Messina and W. H. Conner, eds. Southern Forest Wetlands: Ecology and Management. Lewis Publishers, Boca Raton, FL, pp. 445–473.

Part **4**

Inland Wetland Ecosystems

Freshwater Marshes

Freshwater inland marshes are perhaps the most diverse of the marsh types discussed in this book. The pothole marshes of the north-central United States and south-central Canada are individually small, occurring in the moraines of the last glaciation. Playas of north Texas and western New Mexico are similar in size but have different geological origins and are found in an arid climate. Coastal nontidal freshwater marshes and the Florida Everglades are other freshwater marshes. Vegetation in freshwater marshes is characterized by graminoids such as the tall reeds Typha *and* Phragmites, *the grasses* Panicum *and* Cladium, *the sedges* Cyperus *and* Carex, *broad-leaved monocots such as* Sagittaria *spp., and floating aquatic plants such as* Nymphaea *and* Nelumbo. *Some inland marshes such as the prairie glacial marshes follow a 5- to 20-year cycle of (1) drought, when the marsh dries out, exposing large areas of mud flat on which dense seedling stands germinate; (2) reflooding, after rain drowns out the annual seedlings but allows the perennials to spread rapidly and vigorously; (3) deterioration, sometimes associated with concentrated muskrat activity; and (4) a lake stage, after most of the emergent vegetation has gone. This resets the cycle to stage (1). In contrast to bogs, inland marshes have high-pH substrates, high available soil calcium, medium or high loading rates for nutrients, high productivity, and high soil microbial activity that leads to rapid decomposition, rapid recycling, and nitrogen fixation. Peat may or may not accumulate. Most of the primary productivity is routed thorough detrital pathways, but herbivory can be important, particularly by muskrats and geese. Consumers can have significant yet indirect effects on the detrital pathways as well. Inland marshes are valuable as wildlife islands in the middle of agricultural land and have been tested extensively as sites that can assimilate nutrients.*

The wetlands discussed in this chapter are a diverse group. Nevertheless, they can be treated as a unit because of the fact that they are nontidal, freshwater systems dominated by grasses, sedges, and other freshwater emergent hydrophytes. Otherwise, they differ in their geological origins and in their driving hydrologic forces, and they vary in size from the small pothole marshes of less than a hectare in size to the immense sawgrass monocultures of the Florida Everglades. Terminology for wetlands, especially inland freshwater wetlands, is often confusing and contradictory. The major differences between European and North American words used to describe freshwater wetlands are illustrated in Table 3-5. In Europe, for example, the term *reedswamp* is often used to describe one type of freshwater marsh dominated by *Phragmites* spp., whereas, in the United States, the word *swamp* usually refers to a forested wetland. Although the use of classifying terms connotes clear boundaries between different wetland types, in reality they form a continuum. The extremes of freshwater marshes are clearly different, but, at the boundaries between two wetland types (e.g., marsh and bog), the distinction is not always clear.

As used in Europe, the term *fen*, one kind of marsh, refers to a peat-forming wetland that receives nutrients from sources other than precipitation, usually through groundwater movement. Its peats are not acidic. *Marshes* (and *reedswamps*, as the term is used in Europe) have mineral soils rather than peat soils. American terminology has developed without much regard to whether the system is peat forming. The fact is that most of the North American marshes, regardless of how far south they are and regardless of their geological origins, are peat forming. Based on the classification developed by the U.S. Fish and Wildlife Service (Cowardin et al., 1979), most of the marshes described in this chapter have been classified as palustrine, riverine, or lacustrine, persistent or nonpersistent, emergent wetlands.

For the purposes of this book, we have divided inland nonforested wetlands into two groups. Bogs and fens are communities characterized by deep accumulation of peat. These peatlands are a boreal or high-altitude phenomenon generally associated with low temperatures and short growing seasons and are discussed in Chapter 13. All other nonforested, inland freshwater wetlands have been included in this chapter as freshwater *marshes*.

A number of good references on inland freshwater marshes exist, with the most extensive coverage of many aspects of these systems by Weller (1994). The reader is also referred to Good et al. (1978), Dykyjova and Kvet (1978), Gore (1983a), Hofstetter (1983), Prince and D'Itri (1985), Denny (1985a), G. D. Adams (1988), van der Valk (1989), Kushlan (1990), and Gopal and Masing (1990) for additional reading on various aspects or types of freshwater marshes.

Geographical Extent and Geological Origins

There are few accurate measures of how many freshwater marshes there are in the world for several reasons. First, they are often ephemeral or convert to other types of wetlands such as unvegetated flats or forested wetlands over a relatively short period of time. Second, they can be confused with and miscounted with peatlands, particularly fens. Third, there is such a wide variety of dominant vegetation covers and water depths (from saturated soils to 1 m of water depth) in freshwater marshes as to make their classification and inventory very difficult. Rodin et al. (1975) and Matthews and Fung (1987) estimated, respectively, that there are 89 to 101 million ha of "nonforested swamps" that are also not peatlands in the world (Table 12-1). Most of these wetlands, by a process of elimination, would be freshwater marshes. This range probably represents a reasonable measure of the extent of inland marshes in the world and suggests, therefore, that inland marshes represent less than 20 percent of the total amount of wetlands in the world (see Chapter 3). Aselmann and Crutzen (1989, 1990) estimated that there are 27 million

Table 12-1 Estimated area of inland marshes in the world and North America

	Area (million ha)	Reference
World	89.4[a]	Rodin et al. (1975)
	100.8[a]	Matthews and Fung (1987)
United States (lower 48 states)	7.7	Hofstetter (1983); Table 12-2
	9.9	Hall et al. (1994); Table 4-3
United States (including Alaska)	26.9	Hall et al. (1994); Table 4-3
Canada	15.9[b]	Zoltai (1988)

[a] Referred to as "nonforested swamp."
[b] Total wetlands minus peatlands.

ha of marshes in the world, but this estimate appears to be quite low, given that Canada and the United States alone have an estimated 42 million ha (Table 12-1). Based on the estimates presented in Table 12-1, there appear to be 8 to 10 million ha of inland marshes in the lower 48 states, another 17 million ha of inland marshes in Alaska, and another 16 million ha of inland marshes in Canada. Table 12-2 summarizes an early estimate of the extent of freshwater marshes in the conterminous United States, classified essentially by depth (and duration) of flooding. This classification system reflects the study's primary interest in habitat value and waterfowl and illustrates that over 70 percent of these inland marshes were extremely shallow (meadows) or shallow freshwater marshes. Most are freshwater marshes, although a small percentage (7 percent) are saline inland marshes (Table 12-2). Countless other small marshes occur throughout the continent, many of which have not been inventoried. Several major groups of freshwater marshes, however, can be identified.

Prairie Potholes and Nebraska Sandhills

One of the dominant areas of freshwater marshes in the world is the prairie pothole region of North America (see Chapter 4 and Figs. 4-11 and 4-13). Individual pothole marshes are usually small; they originated in millions of depressions formed by glacial

Table 12-2 Area of inland marsh types in the United States (lower 48 states)

Description	Area ($\times 10^3$ ha)	Percentage of Total Marshes
Freshwater		
Meadows	3,041	39.6
Shallow marshes	2,502	32.6
Deep marshes	1,602	20.8
Total freshwater	7,145	93.0
Saline		
Flats	431	5.6
Marshes	110	1.4
Total saline	541	7.0
Total	7,686	100.0

Source: Hofstetter (1983) and Shaw and Fredine (1956).

action. They are found in greatest abundance in moraines of undulating glacial till, especially in a 780,000-km^2 region west of the Great Lakes in Minnesota, Iowa, and the Dakotas in the United States and in Alberta, Saskatchewan, and Manitoba in Canada (Fig. 4-11). It has been estimated that before the Europeans settled in this country there were 80,000 km^2 of these potholes in the region (Frayer et al., 1983; Leitch, 1989), so the prairie pothole region was approximately 10 percent wetland in presettlement times. They occur as far south as the southernmost advance of the glaciers. These mid-latitude marshes are among the richest in the world because of the fertile soils and warm summer climate. Similar habitats farther north of the pothole region are dominated by mosses and are properly considered peatlands. It has been estimated that much more than half of the prairie pothole marshes have been converted to other uses, particularly to agriculture (Leitch, 1989).

South of the prairie pothole region is the Nebraska sandhill region, a large (52,000 km^2) stabilized dune field that comprises one-fourth of the state of Nebraska. The region is now dominated by cattle ranches and irrigated crops. In lowlands between the dune formations, there are about 5,400 km^2 of wetlands, mostly wet meadows whose water table is a few decimeters below the land surface (Rundquist, 1983; Novacek, 1989). The area is hydrologically significant not only because of the wetlands, but also because the sandhills are important groundwater recharge and storage areas. The remaining wet meadows are thought to be threatened by declining water tables ostensibly caused by groundwater extraction for crop irrigation.

Near-Coast Marshes

Tidally influenced freshwater marshes were discussed in Chapter 10. Inland of tides and especially concentrated along the northern coast of the Gulf of Mexico and along the South Atlantic Coast of the United States, however, are large tracts of freshwater marshes that have vegetation and ecological functions similar in many ways to those of the prairie pothole marshes. One major difference is the relatively stable water level in the coastal systems that results from the influence of the adjacent ocean. Except for Mississippi River delta marshes, these marshes originated in the same way as coastal salt marshes did; that is, they were formed as the sea gradually rose and inundated river valleys and the shallow coastal shelf after the last glacial period. Sediment and peat deposition associated with the inundation kept the marsh surface in the intertidal zone, and marshes spread inland when more land was flooded. The extensive delta marshes of the Mississippi River were similarly influenced by rising sea level, but the immense sediment load of the river repeatedly built delta lobes into shallow coastal bays across the northern Gulf Coast. Thus, even after sea levels stabilized following the retreat of the last glaciers, river sediments continued to build marshes along the coast.

The Everglades

The largest single marsh system in the United States is in the Everglades of south Florida (see Chapter 4 and Fig. 4-12). It originally occupied an area of almost 10,000 km^2 extending in a strip up to 65 km wide and 170 km long from Lake Okeechobee to the brackish marshes and mangrove swamps of the southwest coast of Florida (Kushlan, 1991). Although some areas have been drained for agriculture, the remaining Everglades is still immense. An area of 4,200 km^2 has been preserved in a near-natural state, most of it as sawgrass (*Cladium jamaicense*) marshes. South Florida was built on a flat, low limestone formation, and so limestone outcrops appear throughout the Everglades. The average slope of the Everglades is only 2.8 cm/km so that fresh water from Lake Okeechobee flows slowly southward across the land as a broad sheet during the wet season. The freshwater head prevents salt intrusion from the Gulf of Mexico in the south. During the dry season, surface water dries up and is found only in the deepest sloughs and in some solution holes that serve as wildlife refuges (locally called *gator holes*). The Florida Everglades marshes are threatened by a combination of altered hydroperiods caused by human development, drainage for development, and polluted runoff from upstream agricultural activity. The water flow through this "river of grass" is heavily managed, particularly in the upper half of the original Everglades. A major $8 billion restoration of the Everglades is being undertaken by the U.S. Army Corps of Engineers and the state of Florida to try to restore some of the hydrologic integrity of the Everglades.

California Central Marshes

On the West Coast, in the valley between the Cascade Range and the Coast Range in Washington, Oregon, and northern California, there used to be large tracts of marshland. Only a few pockets of marsh remain. The rest has been drained for agriculture.

Vernal Pools

In the western United States, particularly in California west of the Sierra Nevada Mountains, in the Central Valley, and on coastal terraces in the southwestern part of the state, shallow, intermittently flooded wet meadows called *vernal pools* dot the landscape (P. H. Zedler, 1987). These wetlands range from 50 m^2 to about 0.5 ha in size and typically comprise less than 10 percent of a given landscape. They are wet in the cool winter and spring, when there is a bloom of plant life, invertebrates, and migratory waterfowl, but they are generally dry during the warm summer months (Zedler, 1987). The term "vernal pools," although now used by wetland managers as a description of any wetland in North America that dries up in the summer, is classically defined as seasonally flooded wetlands associated with a Mediterranean-type climate of wet winters and dry summers.

Great Lakes Marshes

Marshes occur along the shores of lakes, especially along the Laurentian Great Lakes of North America (Prince and D'Itri, 1985; Krieger et al., 1992; Mitsch and Bouchard, 1998; Keough et al., 1999). They were formed originally in the deltas of rivers that flow into lakes and in protected shallow areas, often behind natural barrier beaches or levees thrown up by wave action on the shore or behind ancient beach ridges (Herdendorf, 1987). Since shorelines have been stabilized along the Great Lakes, many of the remaining Great Lakes marshes are now managed and protected from water-level changes by artificial dikes (Herdendorf, 1987; Mitsch, 1989; Wilcox and Whillans, 1999). There are an estimated 1,200 km^2 of wetlands around the Great

Table 12-3 Extent of coastal freshwater marshes around the Laurentian Great Lakes within the United States

Lake	Shore Length (km)	Number of Wetlands	Area of Wetlands (km^2)
Lake Superior	1,598	348	267
Lake Michigan	2,179	417	490
Lake Huron	832	177	249
Lake St. Clair	256	20	36
Lake Erie	666	96	83
Lake Ontario	598	312	84
Total	6,129	1,370	1,209

Source: Herdendorf (1987).

Lakes in the United States (Table 12-3) and probably a larger area around the Great Lakes in Canada. Glooschenko and Grondin (1988) estimated that there are 9,000 km^2 of wetlands in southern Ontario alone, some of which are associated with Lakes Erie, Ontario, and St. Clair. Even though one wetland site on Lake Erie was designated as a National Estuarine Research Reserve, it is debated in the literature as to whether the wetlands and coastal rivers of the Laurentian Great Lakes should be called estuaries (Schubel and Pritchard, 1990).

Riverine Marshes

Riverine wetlands are common throughout the world and the North American continent. Wetlands are particularly extensive in the floodplain of the Mississippi River valley and the many smaller rivers that empty into the South Atlantic. Although these wetlands are mostly forested (Chapter 15), marshes often border the forests or occupy pockets within them, sometimes in abandoned *oxbows*. In Australia, these same backwater marshes are called *billabongs*. At the headwaters of rivers, beavers build wetlands by damming small streams. These small marshes often attract much wildlife, especially waterfowl.

Playas

Playas are an interesting group of marshes found on the high plain of northern Texas and eastern New Mexico. They are small basins that contain a clay or fine sandy loam hydric soil. Typically, a playa has a watershed area of about 55 ha and a wetland area of about 7 ha. Because the climate of the high plain is arid, the size of the wetland is closely related to the size of the watershed that drains into it. An estimated 25,000 such basins occur on the high plain, but no complete inventory has been made. These wetlands are particularly important waterfowl habitats, and most of the ecological information available about them is related to habitat value. Virtually all playa watersheds are farmed. The water draining into the playa is, therefore, rich in fertilizers; furthermore, playas are often a source of irrigation water. Both irrigation and eutrophication have had strong effects on the size and quality of playa marshes (Guthery et al., 1982; Bolin and Guthery, 1982; Bolin et al., 1989).

Inland Saline Marshes

In many parts of the world where arid climates persist, inland marshes can be saline rather than fresh water. These marshes, then, have characteristics of both coastal salt marshes (because of the salinity) and inland marshes (because they are not tidal). Good examples of these kinds of marshes are the fringe marshes around the Great Salt Lake in Utah or the Salton Sea in California. There are also inland "salt marshes" in places such as eastern Nebraska where these marshes, some as large as 15 to 20 ha, are still found northwest of Lincoln along Salt Creek and its tributaries. These marshes are the result of saline seeps from deep groundwater and the consistently high evapotranspiration/precipitation ratio of the region.

Hydrology

As with any other wetland, the flooding regime, or hydroperiod, of freshwater marshes determines their ecological character. The critical factors that determine the character of these wetlands are the presence of excess water and sources of water other than direct precipitation. Excess water occurs either when precipitation exceeds evapotranspiration or when the watershed draining into the marsh is large enough to provide adequate runoff or groundwater inflow. The hydroperiods of several freshwater marsh systems were illustrated in Figure 5-3. Along sea coasts, water levels tend to be stable over the long term because of the influence of the ocean (Fig. 5-3c). Water levels in inland marshes, in contrast, are much more controlled by the balance between precipitation and evapotranspiration, especially for marshes in small watersheds that are affected by restricted throughflow. Water levels of lacustrine marshes, such as those found along the Laurentian Great Lakes, are generally stable but are influenced by the year-to-year variability of lake levels and whether the wetland is diked or open to the lake (Fig. 5-3d). Many marshes such as wet meadows, sedge meadows, vernal pools, and even prairie potholes dry down seasonally (Fig. 5-3e and g), but the plant species found there reflect the hydric conditions that exist during most of the year. The seasonality of these marshes is due to the fact that they are primarily fed by runoff and precipitation.

Some marshes intercept groundwater supplies. Their water levels, therefore, reflect the local water table, and the hydroperiod is less erratic and seasonal (Fig. 5-3f). These types of marshes, such as those found in the prairie pothole region (LaBaugh et al., 1998), can be either recharge or discharge wetlands. Other marshes collect surface water and entrained nutrients from watersheds that are large enough to maintain hydric conditions most of the time. For example, overflowing lakes supply water and nutrients to adjacent marshes, and riverine marshes are supplied by the rise and fall of the adjacent river.

Climatic Variability

Because river and stream discharge, lake levels, and precipitation itself are often notoriously variable due to weather shifts from year to year, the water regime of most inland marshes also varies in a way that is predictable only in a statistical sense. Those marshes that are fed by groundwater in addition to one or more of the previous sources are less variable, but even groundwater levels vary somewhat with season and often dramatically with climate shifts or human use. The importance of groundwater and climatic shifts on freshwater marsh hydrology in the prairie pothole was illustrated by LaBaugh et al. (1996, 1998). Even in the same region, water levels responded differently to shifts in the balance between precipitation and evapotranspiration (Fig. 12-1). Stewart and Kantrud (1971) emphasized the duration of flooding in their classification of prairie pothole marshes. They classified wetlands as ephemeral, temporary, seasonally semipermanent, and permanent. In addition, they recognized that marshes may move through several of these classes over the span of a few years. Thus, a marsh that would ordinarily be considered permanent might be in a "drawdown" phase that gives the appearance of an ephemeral marsh.

Biogeochemistry

The water and soil chemistry of freshwater marshes is dominated by a combination of mineral rather than peat soils, overlain with autochthonous inputs of organic matter from the productivity of the vegetation. Given these conditions, there is still a wide range of chemical possibilities for the water and soil in freshwater marshes (Fig. 12-2). Using conductivity as a measure of general salinity, LaBaugh et al. (1996, 1998) found that inland marshes ranged from less than 100 μS cm^{-1} in soft-water systems dominated by rainfall to over 300,000 μS cm^{-1} in inland salt marshes dominated by saline seeps. The chemistry of inland marshes can be described in contrast to that of ombrotrophic bogs (Chapter 13) at one extreme and eutrophic tidal wetlands at the other (Chapter 10). Differences are related to the magnitude of nutrient and other chemical inputs and to the relative importance of groundwater and surface water inflow. Inland marshes are generally minerotrophic in contrast to bogs. That is, the inflowing water has higher amounts of dissolved materials, including nutrients, resulting from the presence of dissolved cations in streams, rivers, and groundwater compared to bogs that are fed simply by rainfall. The organic substrate of freshwater marshes, while shallow compared to bogs and fens, is saturated with bases, and the pH, as a result, is close to neutral. Because nutrients are plentiful, productivity is higher in freshwater marshes than it is in bogs, bacteria are active in nitrogen fixation and litter decomposition, and turnover rates are high. The accumulation of organic matter that does occur results from high production rates, not from the inhibition of decomposition by low pH (as occurs in bogs).

Although inland freshwater wetlands are minerotrophic, they generally lack the high nutrient loading associated with tidal inundation of freshwater tidal marshes (Chapter 10). Flooding in inland marshes tends to be controlled by seasonal changes in local rainfall, subsequent runoff, and evapotranspiration; inundation of tidal areas occurs regularly, once or twice a day. In the latter case, even if tidewater nutrient levels are low, the large volumes of flooding water result in high loading rates. Because of these differences in surface flooding between inland and freshwater tidal marshes, groundwater flow is usually more important as a source of nutrients in inland marshes.

Table 12-4 lists the nutrient concentrations reported for sediments in several inland freshwater marshes. The values vary widely, depending on the substrate, parent material, open or closed nature of the basin, connection with groundwater, and even nutrient uptake by plants. Ion concentrations of freshwater marshes are high, and water is generally in a pH range of 6 to 9. Organic matter can vary from a very high content (75 percent) as can be found in freshwater marshes in coastal Louisiana to a low content (10–30 percent) in marshes fed by inorganic sediments from agricultural watersheds or open to organic export (Table 12-4). Concentrations of total (as distin-

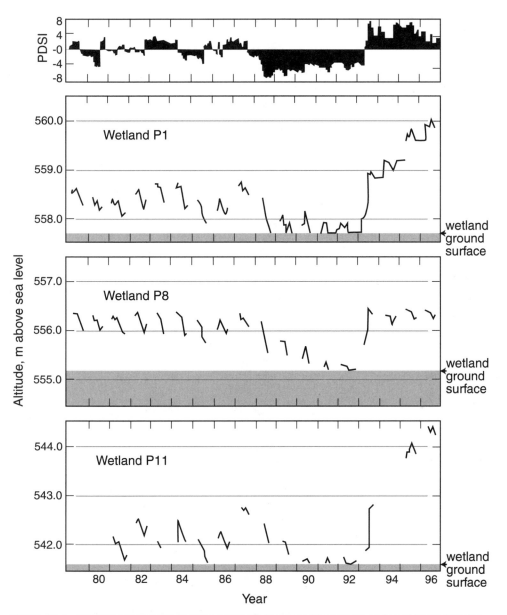

Figure 12-1 Water-level patterns in three wetlands in the prairie pothole region of North America, illustrating the uneven effect that climate has on supposedly similar wetlands in the same region, probably due to nonuniform groundwater effects on the wetlands. PDSI is the Palmer Drought Severity Index and is a relative measure of climatic "wetness." Its value decreases during drought conditions. (*After LaBaugh et al., 1996*)

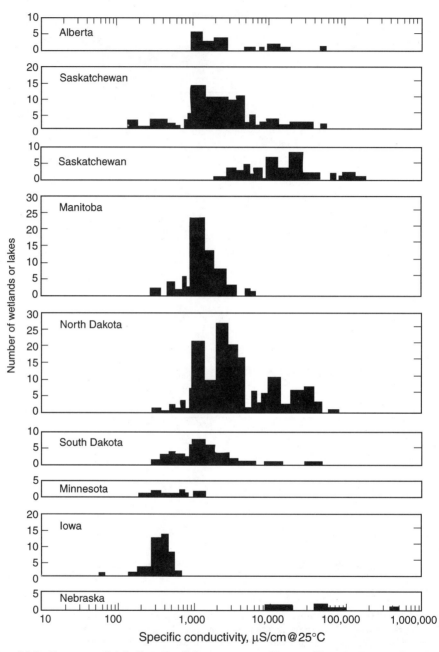

Figure 12-2 Frequency distribution of salinity, as measured by specific conductance, for prairie lakes and wetlands in the United States and Canada as measured by several investigators. Concentrations shown for Nebraska are typical of inland salt marshes. (*After LaBaugh, 1989*)

Table 12-4 Soil chemistry (top 20 cm) of selected inland freshwater marsh soils

		Marsh Type		
Parameter	Riverine, Wisconsin[a]	Coastal, Louisiana[b]	Lake Erie Marshes, Ohio[c]	Lacustrine, Czech Republic[d]
pH	6.4–6.5	6–7	6.8–8.6 (natural) 5.6–10.5 (diked)	5.1
Organic matter (%)	40.4–43.4	75	9–17 (natural) 11–31 (diked)	39
Total N (%)	1.36–1.94	1.8–2.4	—	2.4
Total P (%)	—	0.07–0.1	—	0.013
Available P (ppm)	50–203	0.3[e]	26 ± 19[f]	130
Available K (ppm)	98–230	—	662 ± 351[f]	550
Available Ca (ppm)	5,700–12,700 (exchangeable)	—	5,528 ± 2,501[f]	5,140
Available Mg (ppm)	1,219–2,770	—	268 ± 185[f]	570

[a] Klopatek (1978).
[b] Sasser and Gosselink (1984).
[c] Mitsch (1989, 1992b).
[d] Dykyjova and Kvet (1982), *Typha angustifolia* marsh.
[e] PO_4 in interstitial water.
[f] Average ± std. dev. for nine marshes.

guished from available) nutrients are reflections of the kinds of sediments in the marsh. Mineral sediments are often associated with high phosphorus content, for example, whereas total nitrogen is closely correlated to organic content. Dissolved inorganic nitrogen and phosphorus—the elements that most often limit plant growth—often vary seasonally from very low concentrations in the summer, when plants take them up as rapidly as they become available, to high concentrations in the winter, when plants are dormant but mineralization continues in the soil.

In Table 12-4, the high total soil nitrogen (1–2 percent) and phosphorus (0.01–0.1 percent) reflect an enormous reservoir of organic nutrients that can be mineralized and made available for plant use. Available inorganic nitrogen and phosphorus are one to three orders of magnitude lower. This available supply varies seasonally, depending on both the rate of formation (mineralization) and the uptake by marsh plants. In most freshwater marshes, the inflow probably contributes little to the concentration of available nutrients, but it does contribute significantly to the long-term pool of nutrients available to the vegetation.

Ecosystem Structure

Vegetation

The vegetation of fresh inland marshes has been detailed in many studies. The dominant species vary from place to place, but the number of genera common to all locations in the temperate zone is quite remarkable. Table 12-5 lists dominant emergent species from sites that represent a wide range of different inland marshes. Common species include the graminoids *Phragmites australis* (=*P. communis;* reed grass), *Typha* spp. (cattail), *Sparganium eurycarpum* (bur reed), *Zizania aquatica* (=*Z. palustris;* wild rice), *Panicum hemitomon, Cladium jamaicense;* and the sedges *Carex* spp., *Schoenoplectus tabernaemontani* (=*Scirpus validus;* bulrush), and *Eleocharis* spp. (spike rush). In addition, broad-leaved monocotyledons such as *Pontederia cordata* (pickerelweed) and *Sagittaria* spp. (arrowhead) are frequently found. Herbaceous dicotyledons are represented by a number of species, typical examples of which are *Ambrosia* spp. (ragweed) and *Polygonum* spp. (smartweed). Frequently represented also are such ferns as *Osmunda regalis* (royal fern) and *Thelypteris palustris* (marsh fern), and the horsetail, *Equisetum* spp. One of the most productive species in the world is the tropical sedge *Cyperus papyrus,* which flourishes in marshes and on floating mats in southern and eastern Africa.

Marsh Vegetation Zonation

These typical plant species do not occur randomly mixed together in marshes. Each has its preferred habitat. Different species often occur in rough zones on slight gradients, especially flooding gradients. Figure 12-3 illustrates the typical distribution of species along an elevation gradient in a midwestern North American freshwater marsh. Sedges (e.g., *Carex* spp., *Scirpus* spp.), rushes (*Juncus* spp.), and arrowheads (*Sagittaria* spp.) typically occupy the shallowly flooded edge of a pothole. Two species of cattail (*Typha latifolia* and *T. angustifolia*) are common. The narrow-leaved species (*T. angustifolia*) is more flood tolerant than the broad-leaved cattail (*T. latifolia*) and may grow in water up to 1 m deep. The deepest zone of emergent plants is typically vegetated with hardstem bulrushes (*Scirpus acutus*) and softstem bulrushes (*Schoenoplectus tabernaemontani*). Beyond these emergents, floating-leaved and submersed vegetation will

Table 12-5 Typical dominant emergent vegetation in different freshwater marshes

Marsh Type and Location	Dominant Species	Reference
Prairie glacial marsh, Iowa	*Typha latifolia* *Typha angustifolia* *Scirpus validus* *Scirpus fluviatilis* *Scirpus acutus* *Sparganium eurycarpum* *Carex* spp. *Sagittaria latifolia*	van der Valk and Davis (1978b); Weller (1994)
Riverine marsh, Wisconsin	*Typha latifolia* *Scirpus fluviatilis* *Carex lacustris* *Sparganium eurycarpum* *Phalaris arundinacea*	Klopatek (1978)
Lake Erie wetlands, Ohio	*Typha angustifolia*[a, b] *Cyperus erythrorhizos*[a, b] *Scirpus validus*[a] *Leersia oryzoides*[a] *Echinochloa walteri*[a] *Ludwigia palustris*[a] *Phragmites australis*[a] *Scirpus acutus*[a] *Scirpus fluviatilis*[b] *Sparganium eurycarpum*[b] *Amaranthus tuberculatus*[b] *Nuphar advena*[b] *Nelumbo lutea*[b]	Robb (1989)
"Tule" marshes, California and Oregon	*Scirpus acutus* *Scirpus californicus* *Scirpus olneyi* *Scirpus validus* *Phragmites australis* *Cyperus* spp. *Juncus patens* *Typha latifolia*	Hofstetter (1983)
Floating freshwater coastal marsh, Louisiana	*Panicum hemitomon* *Thelypteris palustris* *Osmunda regalis* *Vigna luteola* *Polygonum sagittatum* *Sagittaria latifolia* *Decodon verticillatus*	Sasser and Gosselink (1984)
Everglades, Florida	*Cladium jamaicense* *Panicum hemitomon* *Rhynchospora* spp. *Eleocharis* spp. *Sagittaria latifolia* *Pontederia lanceolata* *Crinum americanum* *Hymenocallis* spp.	Hofstetter (1983)

[a] Diked marshes (impoundments).
[b] Undiked marshes.

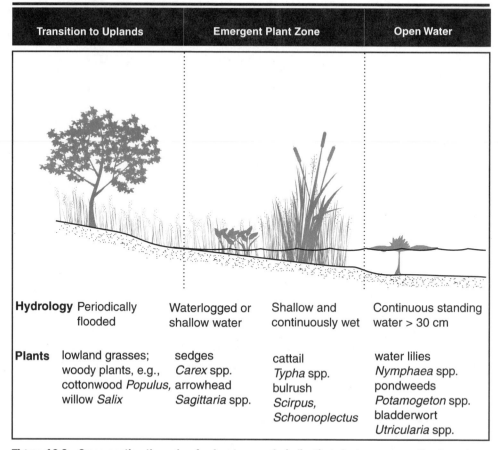

Transition to Uplands	Emergent Plant Zone		Open Water
Hydrology Periodically flooded	Waterlogged or shallow water	Shallow and continuously wet	Continuous standing water > 30 cm
Plants lowland grasses; woody plants, e.g., cottonwood *Populus,* willow *Salix*	sedges *Carex* spp. arrowhead *Sagittaria* spp.	cattail *Typha* spp. bulrush *Scirpus, Schoenoplectus*	water lilies *Nymphaea* spp. pondweeds *Potamogeton* spp. bladderwort *Utricularia* spp.

Figure 12-3 Cross section through a freshwater marsh, indicating plant zone according to water depth and typical plants found in each zone for temperate-zone midwestern North America.

grow, the latter to depths dictated by light penetration. Typical floating-leaved aquatic hydrophytes include rhizomatous plants such as water lilies (*Nymphaea tuberosa* or *N. odorata*), water lotus (*Nelumbo lutea*), and spatterdock (*Nuphar advena*) and stoloniferous plants such as water shield (*Brasenia schreberi*) and smartweed (*Polygonum* spp.). Submersed hydrophytes include coontail (*Ceratophyllum demersum*), water millfoil (*Myriophyllum* spp.), pondweed (*Potamogeton* spp.), wild celery (*Vallisneria americana*), naiad (*Najas* spp.), bladderwort (*Utricularia* spp.), and waterweed (*Elodea canadensis*).

Figure 12-4 shows the plant zonation of a freshwater marsh/littoral zone in sub-Saharan Africa. The shallow flooded emergent zone is dominated by *Typha, Phragmites,* and *Cyperus papyrus. Typha* taxonomy in Africa has been somewhat confused, but it now appears that there are two distinct species—*T. domingensis* Pers. *sensu lato* in tropical and warm-temperate climates and *T. capensis* Rohrb. in more temperate climates of northern and southern Africa (Denny, 1993). Although *Phragmites* is represented by the cosmopolitan *P. communis* (=*P. australis*), there are two other species in Africa as well. *Cyperus papyrus* grows in the emergent zone but also develops floating islands when it breaks free from the shoreline, similar in function

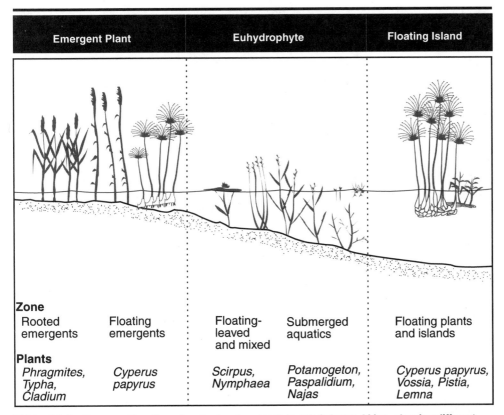

| Emergent Plant | | Euhydrophyte | | Floating Island |

Figure 12-4 Cross section through a freshwater marsh in sub-Saharan Africa, showing different vegetation zones, including emergent, euhydrophyte, and floating-island zones. (*After Denny, 1993*)

to the floating marshes described in Chapter 10. The euhydrophyte (true water plant) zone in Figure 12-4 was coined by Denny (1985b) to refer to rooted, floating-leaved, and submerged macrophytes. In Africa, examples of plants in this zone are *Chara, Fontinalis, Nymphaea, Ceratophyllum, Vallisneria, Potamogeton,* and *Paspalidium.* Thus, despite these differences and vastly different climates, freshwater marshes around the world share some common species, many common genera, and are functionally much the same.

It is difficult to generalize about vegetation zonation in and adjacent to freshwater marshes as there is some evidence that vegetation zonation depends on the size of the wetland. W. C. Johnson et al. (1987), using ordination techniques in some North Dakota prairie potholes, found that vegetation in small marshes formed essentially a continuum from upland edge to deep water, whereas vegetation had strong discontinuities (i.e., sharp boundaries) between marsh vegetation and upland meadows in large marshes. This difference was attributed to greater wave action and ice scour that occur in large wetlands.

Marsh Cycles

A unique structural feature of prairie pothole marshes is the 5- to 20-year cycle of dry marsh, regenerating marsh, degenerating marsh, and lake (Weller and Spatcher,

1965; van der Valk and Davis, 1978b) that is related to periodic droughts (Fig. 12-5). During drought years, standing water disappears. Buried seeds in the exposed mud flats germinate to grow a cover of annuals (*Bidens, Polygonum, Cyperus, Rumex*) and perennials (*Typha, Scirpus, Sparganium, Sagittaria*). When rainfall returns to normal, the mud flats are inundated. Annuals disappear, leaving only the perennial emergent species. Submersed species (*Potamogeton, Najas, Ceratophyllum, Myriophyllum, Chara*) also reappear. For the next year or more, during the regenerating stage, the emergent population increases in vigor and density. After a few years, however, these populations begin to decline. The reasons are poorly understood, but often muskrat populations explode in response to the vigorous vegetation growth. Their nest and trail building can decimate a marsh. Whatever the reason, in the final stage of the cycle, there is little emergent marsh; most of the area reverts to an open shallow lake or pond, setting the stage for the next drought cycle. The wildlife use of these wetlands follows the same cycle. The most intense use occurs when there is good interspersion of small ponds with submersed vegetation and emergent marshes with stands diverse in height, density, and potential food.

Saline Marsh Vegetation

Where evapotranspiration exceeds precipitation and/or saline groundwater seeps occur, inland salt marshes are often found. For example, the Nebraska salt marshes located in eastern Nebraska near Lincoln (Fig. 12-6) support many plant genera familiar to coastal

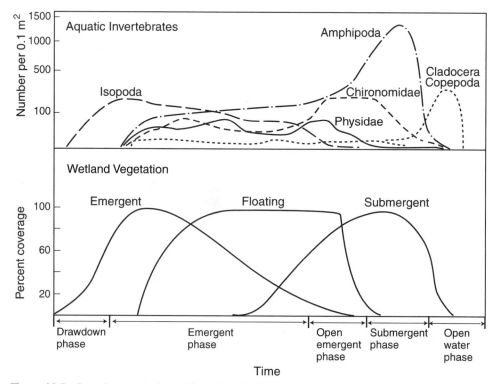

Figure 12-5 Drawdown cycle in prairie pothole freshwater marshes, showing changes in vegetation and dominant aquatic macroinvertebrates. Time horizon is 5 to 20 years with drawdown cycle caused by periodic droughts. (*After Voigts, 1976*)

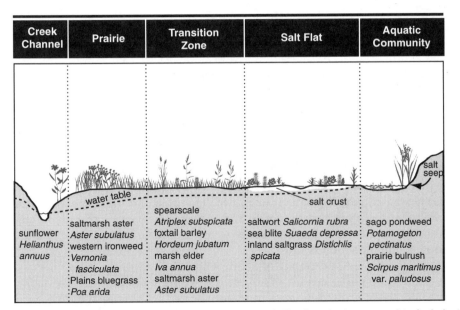

Creek Channel	Prairie	Transition Zone	Salt Flat	Aquatic Community

water table

salt crust

salt seep

		spearscale		
	saltmarsh aster	*Atriplex subspicata*	saltwort *Salicornia rubra*	sago pondweed
sunflower	*Aster subulatus*	foxtail barley	sea blite *Suaeda depressa*	*Potamogeton*
Helianthus	western ironweed	*Hordeum jubatum*	inland saltgrass *Distichlis*	*pectinatus*
annuus	*Vernonia*	marsh elder	*spicata*	prairie bulrush
	fasciculata	*Iva annua*		*Scirpus maritimus*
	Plains bluegrass	saltmarsh aster		var. *paludosus*
	Poa arida	*Aster subulatus*		

Figure 12-6   Cross section through an inland salt marsh, indicating plant zones and typical plants found in each zone. (*After Farrar and Gersib, 1991*)

salt marsh ecologists (Farrar and Gersib, 1991). The most saline parts of these marshes are dominated by salt-tolerant macrophytes such as saltwort (*Salicornia rubra*), sea blight (*Suaeda depressa*), and inland saltgrass (*Distichlis spicata*), whereas the open ponds and their fringes are dominated by plants such as sago pondweed (*Potamogeton pectinatus*), wigeon grass (*Ruppia maritima*), prairie bulrush (*Scirpus maritimus* var. *paludosus*), and even cattails (*Typha angustifolia* and *T. latifolia*). In California, the major species in brackish marshes include pickleweed (*Salicornia virginica*) and alkali bulrush (*Scirpus robustus*) (deSzalay and Resh, 1996, 1997).

Reed Die-Back

A marsh vegetation dynamic that has been affecting marshes in Europe and elsewhere is the phenomenon known as *reed die-back*. Since the early 1950s, scientists have noted a retreat of the common reed *Phragmites australis* around lakes throughout Europe (Ostendorp, 1989; Ostendorp et al., 1995). Several factors that cause the loss of reeds have been suggested (Ostendorp et al., 1995):

1. Direct destruction through human development
2. Mechanical damage due to waves, winds, boat traffic, etc.
3. Grazing by animals such as geese, swans, coots, muskrats, and cattle
4. Lake eutrophication and sedimentation
5. Lake level manipulation and bank erosion
6. Replacement of reeds by species such as *Typha* sp. and *Glyceria maxima*

In what may seem surprising to Eastern United States coastal marsh managers, who have been trying to reduce the advance of *Phragmites* into tidal freshwater and brackish marshes for years, the loss of *Phragmites* beds around lakes in Europe is considered to be a major environmental concern. Their loss has been cited as a main reason for the general deterioration of the lake littoral zone, including increased erosion and deposition of shorelines; loss of bird species such as warblers, bitterns, grebes, and herons; uprooting of shoreline trees and bushes; loss of submersed vegetation; and reduction in fish populations (Ostendorp et al., 1995).

Seed Banks and Germination

Seed banks and fluctuating water levels interact in complicated ways to produce vegetation communities in freshwater marshes (see Chapter 8). As a general rule, seed germination is maximized under shallow water or damp soil conditions, after which many perennials can reproduce vegetatively into deeper water. Keddy and Reznicek (1986) illustrated that fluctuating water levels along the Great Lakes allowed greater diversity of plant types and species in the coastal marshes and that these marshes sometimes have a density of buried seeds an order of magnitude greater than that of the more studied prairie marshes.

The importance of the timing of flooding and drying was illustrated in an experiment involving uniform seed banks subjected to several hydroperiods. Although plant density and above-ground biomass were not affected by the different hydroperiods, species composition, diversity, and richness were affected (Fig. 12-7). Highest richness and diversity occurred in continuously moist soils. Flooding followed by a drawdown to moist soils, as is a typical hydroperiod in the midwestern United States, encouraged obligate wetland species, whereas moist soil conditions followed by flooding encouraged the growth of fewer wetland species and more annual species.

The importance of other variables, particularly nutrients, for the germination and success of freshwater marsh plants is not as well understood. In one series of experiments, Gerritsen and Greening (1990) illustrated the importance of water levels and nutrient conditions for seed germination in seed banks from two marshes in the Okefenokee Swamp in Georgia. Although water level was the important variable in determining which plants germinated, nitrogen was shown to limit marsh plant growth in drawdown conditions, and phosphorus limited the growth of a few species while they were inundated. In a series of experiments using natural (many species) and synthetic seed banks (two species only) under different hydrologic and nutrient conditions, Willis and Mitsch (1995) found similar effects of the water level on germination but were unable to verify the importance of nutrient additions.

Other Factors

The particular species found in freshwater wetlands are also determined by many other environmental factors. Nutrient availability determines, to a large degree, whether a wetland site will support mosses or angiosperms (i.e., whether it is a bog or a marsh) and what the species diversity will be. It is not obvious for freshwater marshes that highly fertile wetlands are highly diverse. In fact, studies of freshwater marsh plant diversity published in the literature (e.g., Wheeler and Giller, 1982; Vermeer and Berendse, 1983; R. T. Day et al., 1988; D. R. J. Moore and Keddy, 1989; Moore et al., 1989) suggest the opposite conclusion. For example, Moore et al. (1989) contrasted several fertile and infertile sites (as measured by the plant standing crop) in eastern Ontario and found the greatest species richness between 60 and 400 g/m² standing crop and much less richness at higher plant standing crop (>600 g/m²) (Fig. 12-8). They also found rare species only at the infertile sites, suggesting that the conservation of infertile wetlands should be part of overall wetland management strategies.

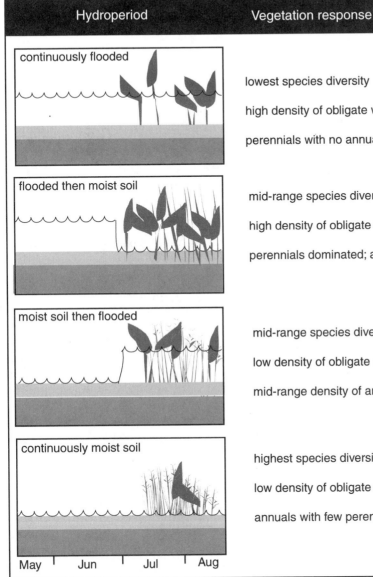

Hydroperiod	Vegetation response

continuously flooded

lowest species diversity

high density of obligate wetland plants

perennials with no annuals

flooded then moist soil

mid-range species diversity

high density of obligate wetland plants

perennials dominated; almost no annuals

moist soil then flooded

mid-range species diversity

low density of obligate wetland plants

mid-range density of annuals

continuously moist soil

highest species diversity

low density of obligate wetland plants

annuals with few perennials

May   Jun   Jul   Aug

Figure 12-7 Experimental results of four different growing season hydroperiods on a common seed bank of annual and perennial freshwater marsh plants in Ohio. Continuously deep water favored low density of obligate wetland perennials; flooded then moist, typical of natural hydroperiod for the midwestern United States, favored perennial-dominated and diverse wetland plants; moist then flooded, typical of some managed marshes around the Laurentian Great Lakes, favored fewer obligate wetland plants and higher density of annuals; continuously flooded soil had highest diversity but fewest obligate wetland plants and highest density of annuals. (*Unpublished data from S. A. Johnson, W. J. Mitsch and J. E. Christensen, Ohio State University, Columbus, OH.*)

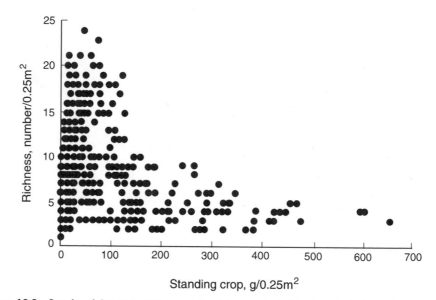

Figure 12-8   **Species richness versus vegetation biomass in 0.25-m² quadrats from three wetland areas in Ontario, Quebec, and Nova Scotia. (*After Moore et al., 1989*)**

Plant species also change with latitude, that is, as temperatures increase or decrease and the winters become more or less severe. Although the same genera may be found in the Tropics and in the Arctic, the species are usually different, reflecting different adaptations to cold or heat. Because of their long isolation from one another, the flora of the North American continent differ at the species level from the European flora. Finally, soil salts, even in low concentrations, determine the species found on a site (McKee and Mendelssohn, 1989). Because many inland marshes are potholes that collect water that leaves only by evaporation, salts may become concentrated during periods of low precipitation, adversely affecting the growth of salt-intolerant species.

Invasive Species

Nonnative plant species are often a part of the vegetation of freshwater marshes, particularly in areas that have been disturbed. Plants such as *Eichhornia crassipes* (water hyacinth), *Salvinia molesta* (salvinia), and *Alternanthera philoxeroides* (alligator weed) have invaded tropical and subtropical regions of the world. In the southeastern United States, *E. crassipes* is considered to be a nuisance plant because of its prolific growth rate. This free-floating aquatic plant can double the area that it covers in two weeks and has choked many waterways that have received high nutrient loads for almost a century (Penfound and Earle, 1948). On the other hand, the plant has been praised for its ability to sequester nutrients and other chemicals from the water and has often been proposed as part of natural wetlands to purify wastewater (Ma and Yan, 1989). Although there are many theories about alien aquatic plants, there is some validity to the concept that disturbed ecosystems are most susceptible to biological invasions (Mitchell and Gopal, 1991).

It has been hypothesized that tropical regions are more susceptible to invasion than temperate regions because invading plants grow much more rapidly and are more noticeable in the Tropics than in temperate latitudes. In the freshwater marshes of the St. Lawrence and Hudson River valleys and in the Great Lakes region of North America, however, *Lythrum salicaria* (purple loosestrife), a tall purple-flowered emergent hydrophyte, has spread at an alarming rate in this century, causing much concern to those who manage these marshes for wildlife (Stuckey, 1980; Balogh and Bookhout, 1989). The plant is aggressive in displacing native grasses, sedges, rushes, and even *Typha* spp. Many freshwater marsh managers have implemented programs designed to control purple loosestrife by chemical and mechanical means. Other aquatic aliens such as the submersed *Hydrilla verticillata,* a plant native to Africa, Asia, and Australia, and *Myriophyllum spicatum* have invaded open, shallow-water marshes in the United States (Steward, 1990; Galatowitsch et al., 1999) but rarely compete well with emergent vegetation.

Consumers

Perhaps one reason that small marshes of the prairie region and the western high plains harbor such a rich diversity of organisms and wildlife is that they are often natural islands in a sea of farmland. Cultivated land does not provide a diversity of either food or shelter, and many animals must retreat to the marshes, which have become their only natural habitats. In cases in which flow from watersheds is seasonal, freshwater marshes can serve as biological and hydrologic "oases" during low-flow and drought conditions.

Decomposers

Like other wetland systems, inland marshes are detrital ecosystems. Unfortunately, we know very little about the many small benthic organisms that are the primary consumers in wetlands (inland marshes are no exception). It is probable that small decomposers such as nematodes and enchytraeids are relatively more important than larger decomposers in marshes compared to terrestrial woodlands.

Invertebrates

Invertebrates, similar to amphibians, are the links between plants and their detritus, on the one hand, and animals such as fish, ducks and other birds, and even several mammals, on the other. In addition, they serve as plant material shredders and detritivores (Weller, 1994). The most conspicuous invertebrates are the true flies (Diptera), which often make one's life miserable in the marsh. These include midges, mosquitoes, and crane flies. Many, especially in adult stages, are herbivorous; Cragg (1961) attributed about one-third of consumer respiration in a *Juncus* moor to herbivores, chiefly Diptera. However, in the larval stages, many of the insects are benthic. Midge larvae, which are called *bloodworms* because of their rich red color, "are found submerged in bottom soils and organic debris, serving as food for fish, frogs, and diving birds. When pupae surface and emerge as adults, they are exploited as well by surface-feeding birds and fish" (Weller, 1994). Odonata, represented by dragonflies and damselflies, are a notable feature of freshwater marshes; their very presence generally indicates good water quality. Crustaceans such as crayfish and mollusks such as snails can be common in some freshwater marshes. The former are food for large fish and mammals alike, whereas the latter are often found grazing on mats of filamentous algae.

Temporal cycles and spatial patterns of invertebrate species and concentrations reflect the natural seasonal cycle of insect growth and emergence superimposed on the vegetation cycles described previously. Voigts (1976) described patterns of inverte-brates in prairie potholes as they changed with hydrologic and vegetation conditions (Fig. 12-5). McLaughlin and Harris (1990) investigated insect emergence from diked and undiked marshes along Lake Michigan and found more insects, more insect biomass, and a greater number of taxa in the diked marsh and the greatest numbers and biomass in the sparsely vegetated zones of the wetlands rather than in open water or dense vegetation.

Mammals

A number of mammals inhabit inland marshes. The most noticed is probably the muskrat (*Ondatra zibethicus*). This herbivore reproduces rapidly and can attain popula-tion densities that decimate the marsh, causing major changes in its character. Like plants, each mammalian species has preferred habitats. For example, Figure 12-9 shows the distribution of muskrats and other mammals on an elevation gradient in a Czech fishpond littoral marsh. Muskrats are found in the most aquatic areas, the water vole in overlapping but higher elevations, and other voles in the relatively terrestrial parts of the reed marsh. Most of the mammals are herbivorous.

Birds

Birds, particularly waterfowl, are also abundant in freshwater marshes. Many of these are herbivorous or omnivorous. Waterfowl are plentiful in almost all wetlands probably because of the food richness and the diversity of habitats for nesting and resting. Water-fowl nest in northern freshwater marshes, winter in southern marshes, and rest in other

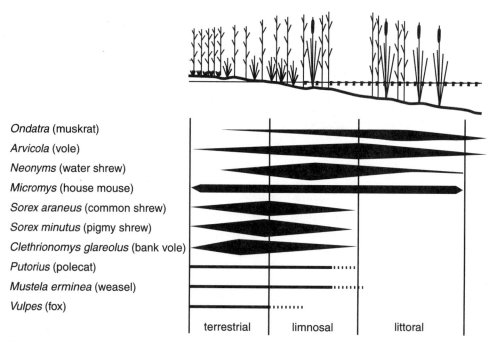

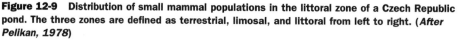

Figure 12-9 Distribution of small mammal populations in the littoral zone of a Czech Republic pond. The three zones are defined as terrestrial, limosal, and littoral from left to right. (*After Pelikan, 1978*)

marshes before resuming their migrations. Weller and Spatcher (1965) described how birds partition a typical marsh as they compete for food and nesting sites (Fig. 12-10). Different species distribute themselves along an elevation gradient according to how well they are adapted to water. In northern marshes, the Loon (*Gavia immer*) usually uses the deeper water of marsh ponds, which may hold fish populations. Grebes (*Podilymbus* sp. and *Podiceps* sp.) prefer marshy areas, especially during the nesting season. Some ducks (dabblers) such as Mallards (*Anas platyrhynchos*) nest in upland sites, feeding along the marsh–water interface and in shallow marsh ponds. Others (diving ducks) such as the Ruddy Duck (*Oxyura jamaicensis*) nest over water and fish by diving. For example, the Black Duck (*Anas rubripes*), one of the most popular ducks for naturalists and hunters alike, uses the emergent marsh as its preferred habitat (Frazer et al., 1990a, b). The Northern Shoveler (*A. clypeata*), the "whale of the waterfowl," uses its large bill and lateral lamellae to filter plankton. Geese (*Branta canadensis* and *Chen* sp.) and swans (*Cygnus* sp.), the "cattle of the waterfowl," along with Canvasback Ducks (*Aythya valisineria*) and the Wigeon (*Anas americana*), are major marsh herbivores. Wading birds such as the Great Blue Heron (*Ardea herodias*) and the Great Egret (*Casmerodius albus*) usually nest colonially in wetlands and fish along the shallow ponds and streams. The Least Bittern (*Ixobrychus exilis*) builds nests a meter or less above the water in *Typha*

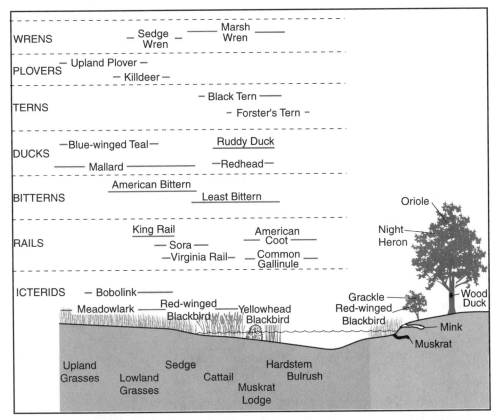

Figure 12-10 Typical distribution of birds across a freshwater marsh from open water edge across shallow water to upland grasses. Placement of muskrat and mink are also illustrated. (*After Weller and Spatcher, 1965*)

or *Scirpus/Schoenoplectus* stands. Rails live in the whole range of wetlands; many of them are solitary birds that are seldom seen. Marsh Wrens (*Cistothorus plaustris*), Virginia Rails (*Rallus limicola*), Soras (*Porzana carolina*), and Swamp sparrows (*Melospiza georgiana*) live amid the dense vegetation of freshwater marshes, often heard but not seen. Songbirds are also abundant in and around marshes. They often nest or perch in adjacent uplands and fly into the marsh to feed. Swallows (*Riparia riparia* and *Stelgidopteryx serripennis*) and swifts are common around freshwater marshes, flying above the marsh, with their mouths ever open to capture emerging insects, often in swarms of dozens or even hundreds of birds.

One of the most conspicuous wetland birds is the blackbird, represented by the Yellow-headed Blackbird (*Xanthocephalus xanthocephalus*) in parts of the midwestern United States and the Red-winged Blackbird (*Agelaius phoeniceus*) in other areas. The Red-winged Blackbird is a very social species and is quite territorial, especially during the nesting season (see Özesmi and Mitsch, 1997).

Fish

One of the most difficult questions about which to generalize is whether freshwater marshes support much fish life. As a general rule, the deeper the water in the marsh and the more open the system is to large rivers or lakes, the more variety and abundance of fish that can be supported. The positive aspect of freshwater marshes as habitats and nurseries for fish was investigated by Derksen (1990) for a large Manitoba marsh complex and by Stephenson (1990) for Great Lakes marshes. Derksen found extensive use of marshes by northern pike (*Esox lucius*) with emigration from the marsh occurring primarily in the autumn. Stephenson investigated fish utilization for spawning and rearing of young in five marshes connected to Lake Ontario in Ontario, Canada. A total of 36 species of fish were collected, 23 to 27 species per marsh. The spawning adults of 23 species and the young-of-the-year of 31 species were collected in the marshes, indicating the importance of these marshes for fish reproduction in Lake Ontario. Eighty-nine percent of the species encountered were using the marshes for reproduction (Stephenson, 1990).

Along sea coasts, where a large portion of the fish and shellfish species are estuarine, young larvae spawned at sea enter the estuary for food and shelter during their postlarval and juvenile stages. Many of these organisms, including the commercially important brown and white shrimp, menhaden, and male blue crab, often migrate up estuary into fresh marshes. Similarly, blue and channel catfish, threadfin, and gizzard shad, although usually found in fresh streams and lakes, migrate down estuary into the oligohaline zone in fall and winter (Gosselink et al., 1998).

Common carp (*Cyprinus carpio*) are able to withstand the dramatic fluctuations of water temperature and dissolved oxygen typical of shallow marshes and are, thus, abundant in many inland wetlands. They affect marsh vegetation by direct grazing, uprooting vegetation while searching for food (King and Hunt, 1967), and causing severe turbidity in the water column. For these reasons, carp are not considered desirable by many freshwater marsh managers.

Amphibians

Amphibians are an important group of organisms in freshwater marshes, often serving as the link between insect populations and wading birds, mink, raccoons, and some fish in complex food webs. Larval tadpoles, which can be quite abundant in some freshwater marshes, eat small plants and animals and are, in turn, eaten by large fish and wading birds. The adult frogs feast on emerging insects. Even terrestrial toads use freshwater marshes as mating and breeding grounds in the spring (Weller, 1994). There has been much concern about declining amphibian populations (Blaustein and Wake, 1990); one of the causes that has been suggested has been the loss of wetland habitat. Richter and Azous (1995) investigated the relationships between amphibian richness and variables such as wetland size, vegetation type, presence of competitors and predators, hydrologic characteristics, hydroperiod fluctuations, and land use. The variables that explained the highest correlation with amphibian richness were water-level fluctuations and percentage of the watershed that was urbanized. These data do not explain the exact cause of the loss of amphibians, but urban pollution and stream and hydrological modifications appear to be likely causes.

Ecosystem Function

Primary Production

The productivity of inland marshes has been reported in a number of studies (Table 12-6). Estimates are generally quite high, ranging upward from about 1,000 g m^{-2} yr^{-1}. Some of the best estimates, which take into account underground production as well as that above ground, come from studies of fishponds in former Czechoslovakia. (These are small artificial lakes and bordering marshes used for fish culture.) These estimates, some indicating values of over 6,000 g m^{-2} yr^{-1}, are high compared with most of the North American estimates. The productivity is higher than even the productivity of intensively cultivated farm crops.

The emergent monocotyledons *Phragmites* and *Typha*, two of the dominant plants in freshwater marshes, have high photosynthetic efficiency. For *Typha*, efficiency is highest early in the growing season, gradually decreasing as the season progresses. *Phragmites*, in contrast, has a fairly constant efficiency rate throughout most of the growing season. The efficiencies of conversion by these plants in optimum environments of 4 to 7 percent of photosynthetically active radiation are comparable to those calculated for intensively cultivated crops such as sugar beets, sugarcane, or corn.

Productivity variation is undoubtedly related to a number of factors. For example, Gorham (1974) established the close positive relationship between above-ground biomass and summer temperatures (Fig. 12-11). Innate genetic differences among species accounts for part of the variability. For example, in one study that used the same techniques of measurement (Kvet and Husak, 1978), *Typha angustifolia* production was determined to be about double that of *T. latifolia*. *T. angustifolia*, however, is typically found in deeper water than that of the habitats of the other species, and so environmental factors were not identical for purposes of comparison.

The dynamics of underground growth are much less studied than those of above-ground growth (Bradbury and Grace, 1983; Hogg and Wein, 1987). In freshwater tidal wetlands (Chapter 10), relative root growth was shown to be related to the plant's life history. Annuals generally use small amounts of photosynthate to support root growth, whereas species with perennial roots and rhizomes often have root : shoot ratios well in excess of one. This relationship appears to hold true for inland marshes also. Perennial species in freshwater marshes generally have more below-ground than above-ground biomass (Fig. 12-12). It is interesting to note, however, that, whereas biomass root : shoot ratios are usually greater than one, root-production-to-shoot biomass ratios are generally less than one. Since above-ground production is approxi-

Table 12-6 Selected primary production estimates for inland freshwater marshes

Dominant Species	Location	Net Primary Productivity (g m^{-2} yr^{-1})	Reference
REEDS AND GRASSES			
Glyceria maxima	Lake, Czech Republic	900–4,300[a]	Kvet and Husak (1978)
Phragmites communis	Lake, Czech Republic	1,000–6,000[a]	Kvet and Husak (1978)
P. communis	Denmark	1,400[a]	Anderson (1976)
Panicum hemitomon	Floating coastal marsh, Louisiana	1,700[b]	Sasser et al. (1982)
Schoenoplectus lacustris	Lake, Czech Republic	1,600–5,500[a]	Kvet and Husak (1978)
Sparganium eurycarpum	Prairie pothole, Iowa	1,066[b]	van der Valk and Davis (1978a)
Typha glauca	Prairie pothole, Iowa	2,297[b]	van der Valk and Davis (1978a)
T. latifolia	Oregon	2,040–2,210[a]	McNaughton (1966)
Typha sp.	Lakeside, Wisconsin	3,450[a]	Klopatek (1974)
SEDGES AND RUSHES			
Carex atheroides	Prairie pothole, Iowa	2,858[b]	van der Valk and Davis (1978a)
larex lacustris	Sedge meadow, New York	1,078–1,741[a]	Bernard and Solsky (1977)
Juncus effusus	South Carolina	1,860[a]	Boyd (1971)
Scirpus fluviatilis	Prairie pothole, Iowa	943[a]	van der Valk and Davis (1978a)
BROAD-LEAVED MONOCOTS			
Acorus calamus	Lake, Czech Republic	500–1,100[a]	Kvet and Husak (1978)

[a] Above- and below-ground vegetation.
[b] Above-ground vegetation.

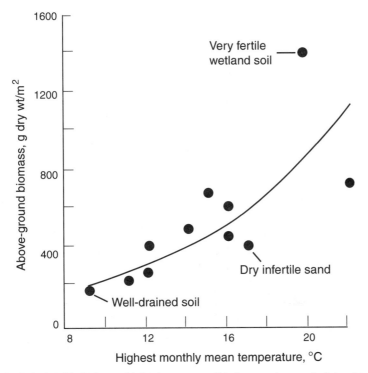

Figure 12-11 Relationship between highest mean monthly temperature and above-ground stand-ing crop of various sedges in freshwater wetlands and uplands. Data points are for wetlands ex-cept where noted otherwise. (*After Gorham, 1974*)

mated by above-ground biomass, this latter ratio is an index of the allocation of resources by the plant, and it indicates that less than one-half of the photosynthate is translocated to the roots. The coexistence of large root biomass and relatively small root production suggests that the root system is generally longer lived (i.e., it renews itself more slowly) than the shoot.

Decomposition

With some notable exceptions such as muskrat and geese grazing, herbivory is consid-ered fairly minor in inland marshes where most of the organic production decomposes before entering the detrital food chain. The decomposition process is much the same for all wetlands. Variations stem from the quality and resistance of the decomposing plant material, the temperature, the availability of inorganic nutrients to the microbial decomposers, and the flooding regime of the marsh. A conceptual diagram of decom-position in freshwater marshes that shows the complexity and interaction of the products of decomposition (Fig. 12-13) illustrates that the action of microbial organ-isms is not undertaken simply to incorporate plant organic matter into microbial cells. In addition, in the process, organic material is dissolved, nutrients are released to the substrate, other nutrients are absorbed by the microflora or adsorbed to fine organic particles, respiration oxidizes and releases organic carbon as carbon dioxide, ingested organic materials are released as repackaged fecal material, and dissolved organic

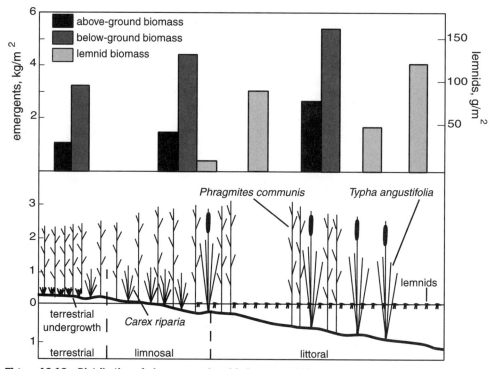

Figure 12-12 Distribution of above-ground and below-ground biomass of emergent vegetation and lemnids across a reed bed (*Phragmites*) transect, showing relation to elevation and flooding. Three zones are as in Figure 12-9. (*After Kvet and Husak, 1978*)

matter may aggregate and flocculate into fine particles. The dynamics of these reactions in freshwater marshes are as poorly understood as are those in other wetlands. It is probably true, as it is in other detrital ecosystems, that the microbial decomposers are preyed on heavily by microscopic meiobenthic organisms, chiefly nematodes; the meiobenthic organisms are, in turn, a food source for larger macrobenthic organisms. Thus, several links in the food chain may precede those that provide the commonly visible birds and other carnivores with their dinners.

Consumers play a significant role in detrital cycles. Most litter decomposition studies in freshwater marshes were done with senesced plant material during the winter, and low ($k = 0.002–0.007$ day^{-1}) rates were generally measured (Table 12-7). However, in a comparison of the decay of fresh biomass and senesced wetland plant leaves, Nelson et al. (1990a, b) found that samples of freshly harvested wetland plant material (*Typha glauca*) decomposed more than twice as fast ($k = 0.024$ day^{-1}) as naturally senesced material did ($k = 0.011$ day^{-1}). This comparison illustrates the more rapid decomposition that results when animals such as muskrats harvest live plant material.

Muskrats also may play a positive role in the energy flow of a marsh as they harvest aquatic plants and standing detritus for their muskrat mounds. Wainscott et al. (1990) found in culturing experiments that litter from muskrat mounds supports substantially higher densities of microbes than litter from the marsh floor does. They suggested that these muskrat mounds may act like "compost piles" as they accelerate the decomposition and microbial growth that have become familiar to organic gardeners.

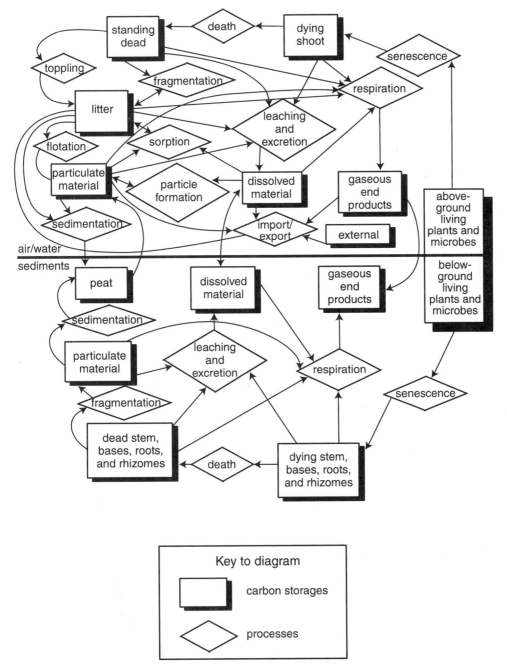

Figure 12-13 Conceptual model of decomposition in freshwater marshes. Boxes indicate storages of organic carbon; diamonds indicate processes. (*After Gallagher, 1978*)

Table 12-7 Decomposition rates of *Typha* spp. in various freshwater marshes

Species	Start Date	Duration (day)	Decay Rate (k, day⁻¹)	Reference
Typha latifolia	November	180	0.0035	Boyd (1970)
T. angustifolia	October	626	0.0019	Mason and Bryant (1975)
T. glauca	Late autumn	525	0.0012	Davis and van der Valk (1978a)
T. glauca	Late autumn	330	0.0020	Davis and van der Valk (1978b)
T. latifolia	Late autumn	300	0.0104	Webster and Simmons (1978)
T. latifolia	September	348	0.0019	Puriveth (1980)
T. angustifolia	Early autumn	154	0.0047	Hill (1985)
T. glauca (green)	April	138	0.0240	Nelson et al. (1990a)
T. glauca (senesced)	April	138	0.0110	Nelson et al. (1990a)

Source: From Nelson et al. (1990a).

Food Webs

Even though food chains begin in the detrital material of freshwater marshes, they develop into detailed webs that are still poorly understood. Benthic communities that feed on detritus form the basis of food for fish and waterfowl in the marshes. DeRoia and Bookhout (1989) found that chironomids made up 89 percent of the diet of Blue-winged Teal (*Anas discors*) and 99 percent of the diet of Green-winged Teal (*A. crecca*) in a Great Lakes marsh. The direct grazing of freshwater marsh vegetation has occasionally been reported in the literature. Crayfish are often important consumers of macrophytes, particularly of submersed aquatic plants, in freshwater marshes. The red swamp crayfish (*Procambarus clarkii*) was shown to have effectively grazed on *Potamogeton pectinatus* in a freshwater marsh in California, where the plant decreased from 70 to 0 percent cover of the marsh while the crayfish population almost doubled (Feminella and Resh, 1989). The direct consumption of marsh plants by geese, muskrats, and other herbivores is common in some parts of the world. "Eat-outs" causing large expanses of open water are the result of the inability of plants to survive after being clipped below the water surface by animals (Middleton, 1999).

In one study that disputed the low-herbivory assumption of inland marshes, deSzalay and Resh (1996) found herbivores in the benthic community of brackish inland marshes in California to represent about 27 percent of the total number of invertebrates collected and to be seasonally dominant. These herbivores fed primarily on filamentous algae and diatoms. Yet herbivory on marsh macrophytes generally remains low except for large-animal grazing from time to time.

Organic Export

Very little information is available about the export of organic energy from freshwater marshes. If other ecosystems are any guide, export is heavily governed by the flow of water across the marsh. Thus, pothole marshes, which have small outflows, must export very little. Some dissolved organic materials may flow out in groundwater, but otherwise the primary loss is through living organisms that feed in the marshes and then move away. In contrast, lakeside and riverine marshes may export considerable organic material when periodically flushed.

Energy Flow

It is not possible to calculate a tight organic energy budget for inland marshes from the information that is available. Several components, however, were estimated for the littoral fishpond system in Czechoslovakia (Table 12-8). This allows at least some perspective to be developed about the major fluxes of energy. Net organic energy fixed by emergent plants ranges from 1,600 to 16,000 kcal m^{-2} yr^{-1}. (For the purpose of comparison, for a site producing 1,000 g m^{-2} yr^{-1} of biomass per year, the energetic equivalent is about 4,500 kcal m^{-2} yr^{-1}.) Most of this net production is lost through consumer respiration. An early study by Cragg (1961) suggested that microbial respiration in peat was about 1,760 kcal m^{-2} yr^{-1} in a *Juncus* moor. No other estimates appear to be available.

Invertebrates, especially the microinvertebrates, play an important role in sediments in the flow of organic energy through the ecosystem. Without doubt, they are important in fragmenting the litter so that it can be more readily attacked by bacteria

Table 12-8 **Partial energy budget for a fishpond littoral marsh in Czech Republic**

	kcal m^{-2} yr^{-1}
PRODUCERS	
Gross production	4,500–27,000
Respiration	2,900–11,000
Net production	1,600–16,000
CONSUMERS	
Decomposers (bacteria and fungi)	1760[a]
Decomposers (small invertebrates)	300[a]
Invertebrate macrofauna	—
Mammal consumption	232
Bird consumption	20[b]
Mammal production	5
Bird production	1

[a] Cragg (1961).
[b] Assuming production = 5% of consumption.
Source: Based on data from Dykyjova and Kvet (1978).

and fungi (Fig. 12-13). These benthic organisms are also important intermediates in the transfer of energy to higher trophic levels. In Czech fishponds, Dvorak (1978) calculated the average benthic macrofaunal biomass, composed mostly of mollusks and oligochaetes, to be 4,266 g/m^2. They were selectively fed on by fish. In quite a different marsh, a *Juncus* moor, Cragg (1961) estimated the respiration rate of the small soil invertebrates to be about 300 kcal m^{-2} yr^{-1}.

Pelikan (1978) calculated the energy flow through the mammals of a reedswamp ecosystem. The total energy consumption was 235 kcal m^{-2} yr^{-1}, mostly by herbivores that ingested 220 kcal m^{-2} yr^{-1}. Insectivores ingested 10 kcal m^{-2} yr^{-1}, and carnivores ingested 1 kcal m^{-2} yr^{-1}. This amounted to about 0.55 percent of above-ground and 0.18 percent of below-ground plant production. Most of the assimilated energy was respired. The total mammal production was only 4.84 kcal m^{-2} yr^{-1}, which amounts to less than 1 g m^{-2} yr^{-1}. It seems evident that the indirect control of plant production by the muskrat is much more significant than the direct flow of energy through this group.

The same reedswamp ecosystem supported an estimated 83 nesting pairs of gulls (*Larus ridibundus*) and 20 pairs of other birds per hectare—about 13 passerines, 3 grebes, and the rest rails and ducks (Hudec and Stastny, 1978). The mean biomass was 44.4 kg/ha (fresh weight) for the gulls and 11.2 kg/ha for the remaining species. The production of eggs and young amounted to about 6,088 kcal/ha for gulls and 3,096 kcal/ha for the remaining species. If this production were considered to be 5 percent of the total consumption, the total annual flow of organic energy through birds would be about 20 kcal/m^2, or roughly 10 percent of the mammal contribution.

Collectively, these estimates of energy flow through invertebrates, mammals, and birds account for less that 10 percent of net primary production. Most of the rest of the energy used for organic production must be dissipated by microbial respiration, but some organic production is stored as peat, reduced to methane, or exported to adjacent waters. Export has been particularly difficult to measure because so many of these marshes intercept the water table and may lose organic materials through groundwater flows.

Nutrient Budgets

A number of attempts have been made to calculate nutrient budgets for wetlands, but the results form no consistent picture because freshwater wetlands vary so widely in so many different ways. In a number of studies, freshwater wetlands have been evaluated as nutrient traps. These studies often emphasize input–output budgets.

Inflow–Outflow Budgets

One of the first studies that identified freshwater wetlands for their role as nutrient sinks was of Tinicum Marsh near Philadelphia (Grant and Patrick, 1970). The study, entitled "Tinicum Marsh as a Water Purifier," had among its goals to determine "the role of Tinicum Marsh wetlands in the reduction of nitrates and phosphates. . . ." This study found decreases in phosphorus (as PO_4^{3-}), nitrogen (as NO_3 and NH_3), and organic materials (as biochemical oxygen demand, BOD) in water flowing from the adjacent river over the marsh. G. F. Lee et al. (1975) summarized the results of two research projects on the effects of freshwater marshes on water quality in Wisconsin and concluded that there were both beneficial (i.e., marsh acts as sink) and detrimental (i.e., marsh acts as source) effects of the marshes on water quality, but that "from an overall point of view, it appears that the beneficial effects outweigh the detrimental effects." Mitsch (1977) found that 49 percent of the total nitrogen and 11 percent of the total phosphorus were removed by a floating water hyacinth (*Eichhornia crassipes*) marsh receiving wastewater in north-central Florida, whereas Vega and Ewel (1981) found a 16 percent retention of phosphorus for the same system a few years later. The former study found a seasonal change in the loss of nitrate nitrogen, with greater uptake in the summer months. A study by Klopatek (1978) of a Wisconsin riverine marsh showed the capacity of a marsh to be at least a seasonal sink for inorganic forms of nitrogen and phosphorus. A 2-year study of the potential of a managed marsh wetland in upper New York State to remove nutrients from agricultural drainage gave inconsistent results, with the wetland acting as a source of nitrogen and phosphorus in the first year and a net sink in the second year (Peverly, 1982). Studies of a freshwater marsh along Lake Erie's shoreline have shown that the wetland is effective in ameliorating nutrient loading from an agricultural watershed to the lake and that the effectiveness depends on the amount of annual runoff and the level of the lake (Klarer and Millie, 1989; Mitsch and Reeder, 1991, 1992). Several studies [e.g., by Fetter et al. (1978) in Wisconsin and Dolan et al. (1981) in Florida] have demonstrated the capacity of natural and created freshwater marshes to be sinks for nutrients when wastewaters are added to wetlands (see Chapter 20).

Reflecting interest in controlling nonpoint source pollution, studies have consistently shown that freshwater marshes can be sinks of nutrients when receiving point and nonpoint sources from both urban and rural areas (see Chapter 20). In a major experiment simulating the retention of nutrients and sediments from nonpoint source pollution in created marshes, the Des Plaines River wetlands in northeastern Illinois showed effective retention of phosphorus and nitrogen over a 3-year experiment, with retention rates averaging 0.4 to 3 g P m^{-2} yr^{-1} (Mitsch, 1992a; Mitsch et al., 1995) and 3 to 38 g N m^{-2} yr^{-1} (Phipps and Crumpton, 1994). In similar studies of artificial marshes in mid-Ohio, phosphorus retention was 7 g P m^{-2} yr^{-1} and nitrate–nitrogen retention 63 to 65 g N m^{-2} yr^{-1} (Nairn and Mitsch, 2000; Spieles and Mitsch, 2000). In the warmer climates of Florida, surprisingly, phosphorus retention rates in both the Everglades itself and a constructed wetland north of the Everglades were about 0.4 to 1.0 g P m^{-2} yr^{-1}

(Richardson et al., 1997; Moustafa et al., 1996; Moustafa, 1999), as low or even lower than rates seen in wetlands farther north. This lower retention rate may be because sedimentation as a phosphorus removal process may not be as important in sandy Florida as in the north where more phosphorus-rich clay is present.

Seasonal Patterns of Nutrient Storage

Vegetation traps nutrients in biomass, but the storage of these nutrients is seasonally partitioned in above-ground and below-ground stocks. For example, the seasonal dynamics of phosphorus in *Typha latifolia* is described in Figure 12-14. Phosphorus stocks in the roots and rhizomes of *Typha* are mobilized into the shoots early in the growing season. Total stocks increase to more than 4 g P m^{-2} during the summer. In the fall, some phosphorus in the shoot is remobilized into the below-ground organs before the shoots die, but most of it is lost by leaching and in the litter. The calculated below-ground deficit is an indication of the magnitude of the phosphorus demand by the plant. It cannot be met by shifting internal supplies; it is largest during the period of active growth in the summer.

In another set of measurements of nutrient conditions in below-ground organs of freshwater marsh macrophytes, C. S. Smith et al. (1988) measured the seasonal changes in several elements of above-ground and below-ground parts of *Typha latifolia* from a Wisconsin marsh. The ratio of the elements varied generally between 1 : 1 and 2 : 1 in above-ground : below-ground ($A : B$) stores, compared with a 2.2 : 1 ratio for $A : B$ biomass, because nutrients are more concentrated in root than in shoot tissues. They found that below-ground biomass concentrations of nitrogen, phosphorus, and potassium decreased significantly during the spring, supposedly because of shoot growth, whereas calcium, magnesium, manganese, sodium, and strontium showed little decrease in spring, suggesting that they are not limiting mineral reserves for spring growth.

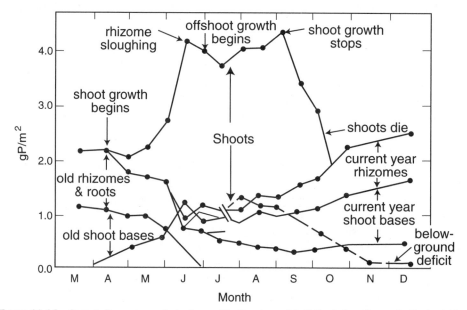

Figure 12-14 Cumulative seasonal stocks and below-ground deficit of phosphorus in *Typha latifolia* plant parts in a freshwater marsh on Lake Mendota, Wisconsin. (*After Prentki et al., 1978*)

Studies like those described previously lead to several generalizations about nutrient cycling in freshwater marshes:

1. *The size of the plant stock of nutrients in freshwater marshes varies widely in contrast to the much more abundant storage of nutrients in marsh soils.* More nitrogen and phosphorus are usually retained in above-ground plant parts in mineral substrate wetlands (freshwater marshes) than in peat wetlands due to the higher productivity and higher concentrations of nutrients. The above-ground stock of nitrogen ranges from as low as 3 g/m² to as high as 29 g/m² in freshwater marshes. The nitrogen and phosphorus budgets in Figure 12-15 show a peak biomass storage of 20.7 g N/m² and 5.3 g P/m² in a *Scirpus* wetland in Wisconsin. This plant storage is very small compared to the nutrients that are stored within the root zone of the peat and mineral soils, shown in Figure 12-15 to be 1,700 g N/m² and 12 g P/m² for total nitrogen and available phosphorus, respectively.

2. *The biologically inactivated stock of nutrients in plants is only a temporary storage that is released to flooding waters and sediments when the plant shoots die in autumn.* Where this occurs, the marsh may retain nutrients during the summer and release them in the winter.

3. *Nutrients retained in biomass often account for a small portion of nutrients that flow into marshes, and that percentage decreases with increased nutrient input.* Shaver and Melillo (1984) illustrated the importance of nutrient availability for the

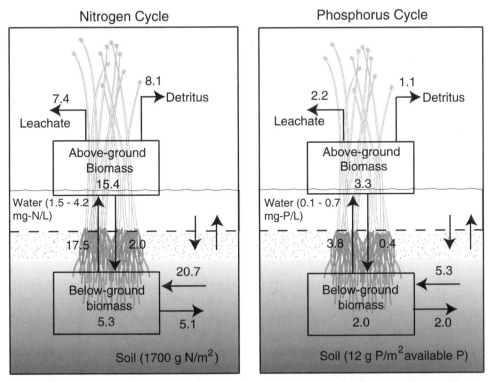

Figure 12-15   Flow of nitrogen and phosphorus through a river bulrush (*Scirpus fluviatilis*) stand in Wisconsin. Flows are in g m⁻² y⁻¹ and storages are in g/m² of nitrogen and phosphorus, respectively. (*From Klopatek, 1978*)

efficiency of nutrient uptake of three freshwater marsh plants, *Carex lacustris, Calamagrostis canadensis,* and *Typha latifolia*. The efficiency of nutrient uptake, defined as the increase in plant nitrogen or phosphorus mass divided by the nitrogen or phosphorus mass available, decreased with increasing nutrient availability, suggesting that plant uptake becomes less important with higher nutrient inputs even though the nutrient content of the plants may increase. The higher concentrations of nutrients in the plants also cause higher concentrations in the litter, which, in turn, can stimulate microbial release. Thus, as more nutrients become available to a freshwater marsh, the marsh becomes more "leaky." Nutrients are lost from the system, and nutrient turnover in the vegetation increases. Even if the uptake rate of nutrients is high in wetlands, much of those nutrients are returned via detrital decomposition to the nutrient pool in the sediments and overlying waters. If wetlands are used for nutrient removal (e.g., Mitsch et al., 2000a), then estimates of only 10 to 20 percent of the nutrient inflow going into plant biomass are common.

4. *Marsh vegetation often acts as a nutrient pump, taking up nutrients from the soil, translocating them to the shoots, and releasing them on the marsh surface during senescence*. The effect of this pumping mechanism may be to mobilize nutrients that have been sequestered in the soil. This has been demonstrated in several nutrient budgets (e.g., Figs. 12-15 and 12-16). In some cases (e.g., Fig. 12-16b), the uptake of nutrients by macrophytes from the sediments is considerably higher than the inflow. Most of this uptake is translocated back to the roots or lost through leaching and shoot senescence, so biomass storage of nutrients is generally low compared to annual inflow.

5. *In general, precipitation and dry fall account for less than 10 percent of plant nutrient demands in freshwater marshes*. Similarly, groundwater flows are usually small sources of phosphorus, but, in agricultural settings with artificial drainage, nitrate–nitrogen inflow can be high. Surface inflow is usually a major source of phosphorus because of its ability to sorb onto sediments, particularly clay. Considering all of these variables, it is not surprising that each marsh seems to have its own unique nutrient budget.

Nutrient Limitations

Both nitrogen and phosphorus limitations are generally factors in freshwater marsh productivity, and their uptake rates are not independent of each other. Shaver and Melillo (1984) illustrated that their freshwater marsh plants accumulated nutrients at a relatively narrow range and that all plants had a similar "optimum" N : P ratio of approximately 8 : 1 by mass. Higher nitrogen concentrations in available solutions did increase the N : P ratios of the plants, and low N : P ratios in available solutions decreased the N : P ratios of the plants. Thus, nitrogen and phosphorus uptakes were not independent of each other.

McJannet et al. (1995) investigated the nitrogen and phosphorus content of 41 freshwater marsh plants after they were grown in excess fertilizer for one growing season. There was a wide range of nitrogen (0.25–2.1 percent N) and phosphorus (0.13–1.1 percent P) that was not related to where the plants came from. However, plants that were from ruderal life histories (i.e., annuals or functional annuals) did have significantly lower nitrogen and phosphorus tissue concentrations than did perennials.

Koerselman and Meuleman (1996), in a study of several wetlands in Europe, found that the N : P ratios in wetland plant tissues were correlated with the N : P supply ratio and that any N : P ratio less than 14 : 1 suggests nitrogen limitation.

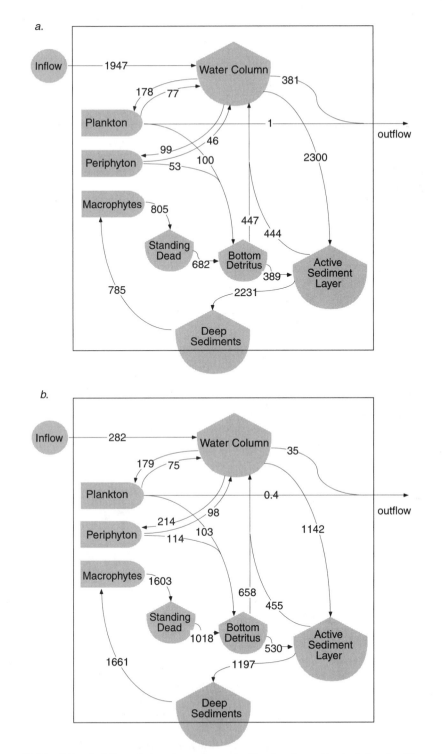

Figure 12-16 Estimated flow of phosphorus in created freshwater wetlands determined by a simulation model calibrated with several years' field data for **a.** high-flow and **b.** low-flow conditions. Site is in northeastern Illinois and inflow was pumped from adjacent Des Plaines River. Flows are in mg m^{-2} yr^{-1}. (*After Wang and Mitsch, in press*)

This is twice the often-used Redfield ratio ($N/P = 7.2$) that is used in planktonic systems to indicate relative nutrient limitation.

For a Manitoba *Scirpus acutus* marsh, Neill (1990c) found that neither nitrogen nor phosphorus increased net productivity when applied alone but that above-ground biomass nearly doubled when nitrogen and phosphorus were applied together. On the other hand, similar studies of a nearby marsh showed nitrogen limitation, indicating that differences in limiting factors are possible even in the same region (Neill, 1990c). Under conditions in which water levels are more stable such as Louisiana's Gulf Coast, the addition of nitrogen fertilizer at a rate of 10 g NH_4^+–N/m^2 caused approximately a 100 percent increase in the growth of *Sagittaria lancifolia* (Delaune and Lindau, 1990).

Experiments by Svengsouk and Mitsch (in review) support the multiple-nutrient limitation of some freshwater marsh plants. Their study investigated the relative limitations of nitrogen and phosphorus in mesocosms planted with both bulrush (*Schoenoplectus tabernaemontani*) and cattail (*Typha* sp.) together. Results suggested that, when both nitrogen and phosphorus are available, *Typha* competed well with *Schoenoplectus*. When only one of the nutrients was in abundance, *Schoenoplectus* did much better than *Typha* (Fig. 12-17).

In contrast to this study, enrichment studies by Craft et al. (1995) on the low-nutrient sawgrass (*Cladium jamaicense*) and other macrophyte communities illustrated that the most important limiting factor in the Florida Everglades appears to be phosphorus. Nitrogen additions had no effect on biomass production, nutrient uptake rates, or nitrogen enrichment of peat.

The role that nutrients play in plant productivity and species composition in freshwater marshes is also influenced by hydrologic conditions. Fluctuating water levels may affect nutrient limitations; when nitrogen and phosphorus fertilizers were added to prairie lacustrine marshes over 2 years, nitrogen additions stimulated the productivity of emergent macrophytes in the first year by increasing the *Scolochloa festucacea* biomass more when the soil was flooded and the *Typha latifolia* biomass more in intermediate depths (0–20 cm) than in deep water (20–40 cm) or dry conditions (Neill, 1990a). This response to added nitrogen indicated nitrogen limitation in these marshes. In the second year of the experiment, nitrogen was not as limiting as it was in the first year, and some phosphorus limitation was noted. The species composition in these experiments changed when nitrogen was added, causing a decrease in the *Scolochloa festucacea* biomass and an increase in moist-soil annuals. The phosphorus additions, however, had little effect on species composition (Neill, 1990b).

Ecosystem Models

An ecosystem model of a freshwater marsh is illustrated in Figure 12-18. Nutrient cycling and food webs are complex because of the important roles of algae (periphyton and sometimes phytoplankton) as well as macrophytes such as the major plants and a wide variety of consumers, including benthos, amphibians, muskrats, fish, and waterfowl and other birds. This model shows the importance of the detrital pathways, although some grazing of live plants occurs with zooplankton, fish, and muskrats.

Several other models have investigated the dynamics of freshwater marshes, whether for predicting wildlife use or nutrient dynamics. Oliver and Legovic (1988) used a simulation model to evaluate the importance of nutrient enrichment in a freshwater marsh near a wading-bird rookery in the Okefenokee Swamp in Georgia.

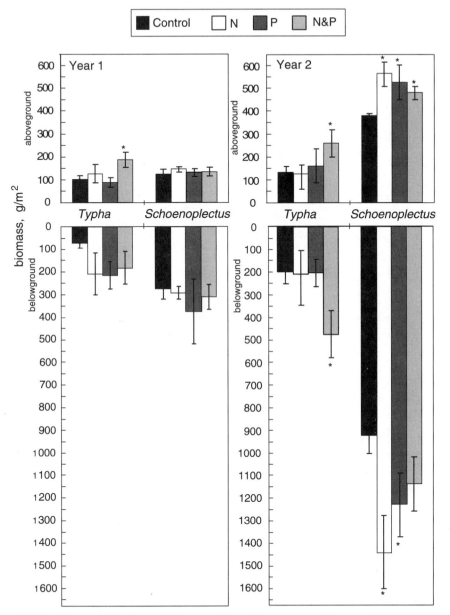

Figure 12-17 Biomass of *Typha* and *Schoenoplectus tabernaemontani* in mesocosms where they were planted together and fertilized with nitrogen (N), phosphorus (P), and nitrogen plus phosphorus (N&P) after one and two growing seasons in experimental mesocosms in Ohio (mean ± std. error; * = significant difference from control at $\alpha = 0.05$).

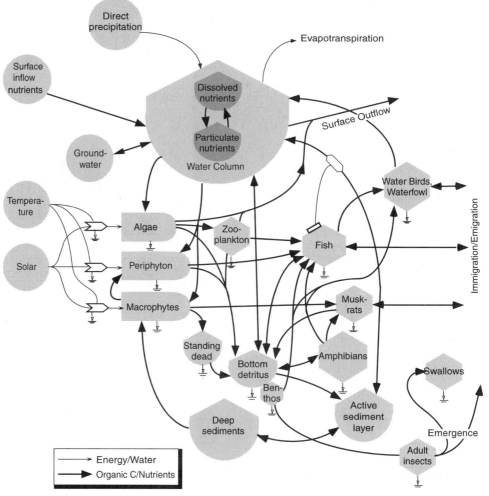

Figure 12-18   A conceptual ecosystem model for freshwater marshes.

The trophic-level model illustrated the influence of the avian fauna on several aspects of the ecosystem, including detritus, macrophytes, and aquatic fauna. Benthic detritus increased 8.9 times, macrophytes 4.5 times, and fish 1.4 times the background levels as a result of the simulated influx of 8,000 wading birds into the marsh.

Özesmi and Mitsch (1997) developed a habitat model that predicted the spatial distribution of breeding sites for red-winged blackbirds (*Agelaius phoeniceus*) in marshes along Lake Erie in northern Ohio. The model concluded that nest site locations were most closely related to water depth and vegetation durability. Metzker and Mitsch (1997) tested the concept of self-design in another model of the food chain in a newly constructed wetland in central Ohio. The model predicted that the general development of the fish community in constructed wetlands with and without river inflows was very sensitive to certain environmental variables that could lead the aquatic community to a less desirable high-biomass domination by carp (*Cyprinus carpio*) or an alternative steady state dominated by bass (*Micropterus salmoides*). The model illustrated the importance of the inflow of propagules to self-design in an ecosystem—depending on whether other fish were introduced by flowing river water or by importation of fish fry, the system behaved differently than if there were no outside importation of fish by inflows or stocking.

Spatial models have been applied to understanding patterns of landscape fragmentation, invasion of undesirable vegetation, and ecosystem dynamics. Wu et al. (1996) developed a spatial model to understand the dynamics of fire in the Florida Everglades. They found that with *Typha* invasion, the area burned and the frequency of fire decreased by 23 and 21 percent, respectively. Deeper water in the Everglades has decreased the frequency of fire by 63 percent, making less frequent fires during drought seasons more severe. Cattail (*Typha*) expansion into the sawgrass (*Cladium jamaicense*) of the Florida Everglades was investigated with another spatial model that predicted that cattails would invade 50 percent of a major (43,000 ha) water conservation area in the Everglades in 6 to 10 years if nutrient loading remained the same.

Modeling has been applied to estimating the fate and effects of nutrients in freshwater marshes. A simulation model was developed for a freshwater wetland along the coast of Lake Erie to investigate the roles of primary productivity, sedimentation, resuspension, and hydrology in the phosphorus retention capability of that wetland (Mitsch and Reeder, 1991). Open-water plankton and aquatic beds dominated by *Nelumbo lutea* were the major producers in this wetland, and a barrier beach restricted outflow to certain seasons. In general, allochthonous and autochthonous productivity resulted in a gross sedimentation of 13.3 mg P m^{-2} day^{-1} in the model but a resuspension of 10.4 mg P m^{-2} day^{-1}. The model was calibrated using data from a drought year (1988), for which a 17 percent retention of phosphorus was calculated. Predicted retention of phosphorus from 27 to 52 percent resulted from simulations of normal to high inflow (typical of normal and wet years), suggesting that the highest retention occurs during periods of high Lake Erie levels (Mitsch and Reeder, 1991). This model illustrated the importance of wetlands as buffers between agricultural watersheds and downstream aquatic ecosystems, but it suggests the importance of hydrologic conditions, both upstream and, in this case, downstream, for controlling the efficiency of this buffering capacity.

In an updated version of the preceding freshwater marsh model, Wang and Mitsch (in press) developed a generalized, yet detailed, wetland ecosystem model that was calibrated and validated with 3 years of data on three constructed wetlands in northeastern Illinois. Total phosphorus retained with sedimentation is simulated at a rate of 1.08 to 2.47 g P m^{-2} yr^{-1}, in the range reported for natural wetlands. The model also showed where the phosphorus was going on an annual rate (see the simulation results in Fig. 12-16). The simulation results indicated that macrophytes pumped about 0.31 to 1.66 g P m^{-2} yr^{-1} out of deep sediments and increased total phosphorus in the water column mostly during the nongrowing season. In an interesting calculation, the model showed that total phosphorus retention increased 5.1 percent if macrophytes were removed from the wetland.

Recommended Readings

Dykyjova, D., and J. Kvet, eds. 1978. Pond Littoral Ecosystems. Springer-Verlag, Berlin. 464 pp.

Weller, M. W. 1994. Freshwater Marshes, 3rd ed. University of Minnesota Press, Minneapolis, MN. 192 pp.

Chapter **13**

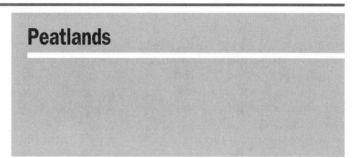

Peatlands

Bogs and fens are peatlands distributed primarily in the cool boreal zones of the world where excess moisture is abundant. Bogs and fens can be formed in several ways, originating either from aquatic systems, as in flowthrough succession or quaking bogs, or from terrestrial systems, as with blanket bogs. Although many types of peatlands are identifiable, classification according to chemical conditions usually defines three types: (1) minerotrophic (true fens), (2) ombrotrophic (raised bogs), and (3) transition (poor fens). When classified according to hydrology, peatlands are (1) geogenous (subject to external flows) or (2) ombrogenous (inflow from precipitation only). Ombrotrophic or ombrogenous peatlands are isolated from mineral-bearing groundwater and, thus, display lower pH, lower minerals, and more dominance by mosses than do minerotrophic or geogenous peatlands. Bog acidity is caused by cation exchange with mosses, oxidation of sulfur compounds, and organic acids. In bogs, the low nutrients and low pH lead to low primary productivity, slow decomposition, adaptive nutrient-cycling pathways, and peat accumulation. Several energy and nutrient budgets have been developed for peatlands, with the 1942 energy budget by Lindeman one of the most well known. Peatlands have been shown to be nutrient sinks in very long term studies as long as sufficient area is used in the analysis, and they are effective sinks for excess atmospheric carbon in undisturbed watersheds.

As described in the introduction to Chapter 12, we have divided inland non-forested wetlands, for the purposes of this book, into mineral soil marshes and peatlands. Bogs and fens belong to a major class of wetlands called *peatlands, moors,* or *mires* that occur as freshwater wetlands throughout much of the boreal zone of the world. *Bogs* are acid peat deposits with no significant inflow or outflow of surface water or groundwater, and support *acidophilic* (acid-loving) vegetation, particularly mosses. *Fens,* on the other hand, are open peatland systems that generally receive some drainage from surrounding mineral soils and are often covered by grasses, sedges, or reeds. They are in many respects transitional between marshes and bogs (see Table 3-5). As

419

a successional stage in the development of bogs, fens are important and will be considered in that context here.

Bogs and fens have been studied and described on a worldwide basis more extensively than any other type of freshwater wetland. The older European peatland literature is particularly rich, establishing most of our basic understanding of these wetlands. Hundreds of North American studies, however, have since appeared in print. Peatlands have been studied because of their vast area in temperate climates, their unique biota and successional patterns, their economic importance of peat as a fuel and soil conditioner, and, recently, their importance in the global atmospheric carbon balance. Bogs have intrigued and mystified many cultures for centuries because of such discoveries as the Iron Age "bog people" of Scandinavia, who were preserved intact for up to 2,000 years in the nondecomposing peat (see, e.g., Glob, 1969).

Because bogs and other peatlands are so ubiquitous in northern Europe and North America, many definitions and words, some unfortunate, that now describe wetlands in general originated from bog terminology; there is also considerable confusion in the use of terms such as *bog, fen, swamp, moor, muskeg, heath, mire, marsh, highmoor, lowmoor,* and *peatland* to describe these ecosystems. The use of the words *peatlands* in general and *bogs* and *fens* in particular will be limited in this chapter to deep peat deposits, mostly of the cold, northern, forested regions of North America and Eurasia. Peat deposits also occur in warm temperate, subtropical, or tropical regions, and we refer briefly to a major example of these, specifically the pocosins of the southeastern Coastal Plain.

Good references on peatlands are Heinselman (1963, 1970), Moore and Bellamy (1974), Radforth and Brawner (1977), Clymo (1983), Gore (1983a), Moore (1984), C. W. Johnson (1985), Damman and French (1987), Glaser (1987), Crum (1988), Verhoeven (1992), Bridgham et al. (1996), Post (1996), and Warner and Rubec (1997). Godwin (1981) provides an interesting personal history of the study of bogs in the United Kingdom from a historical, archeological viewpoint.

Geographical Extent

As described by Curtis (1959): "The bog . . . is a common feature of the glaciated landscapes of the entire northern hemisphere and has a remarkably uniform structure and composition throughout the circumboreal regions." Bogs and fens are distributed in cold temperate climates of high humidity, mostly in the Northern Hemisphere (Fig. 13-1), where precipitation exceeds evapotranspiration, leading to moisture accumulation. Extensive areas of bogs and fens occur in Scandinavia, eastern Europe, western Siberia, Alaska, and Canada. Major areas where a very large percentage of the landscape is peatland include the Hudson Bay lowlands in Canada, the Fennoscandian Shield in northern Europe, and the western Siberian lowland around the Ob and Irtysh Rivers.

Estimates range from 2.4 to 4.1 million km^2 for northern boreal and subarctic wetlands that are mostly peatlands (Kivinen and Pakarinen, 1981; Matthews and Fung, 1987; Aselmann and Crutzen, 1989; Gorham, 1991). There is some convergence with recent estimates to suggest that there are about 3.5 million km^2 of peatlands in the world (Bridgham et al., in press). Specific country or regional estimates of peatland extent include 1.6 million km^2 for the former Soviet Union (Botch et al., 1995), 1.1 to 1.2 million km^2 for Canada (see Chapter 4), 220,000 km^2 for Fennoscania (Gorham, 1991), and 130,000 km^2 for Alaska (Bridgham et al., in press).

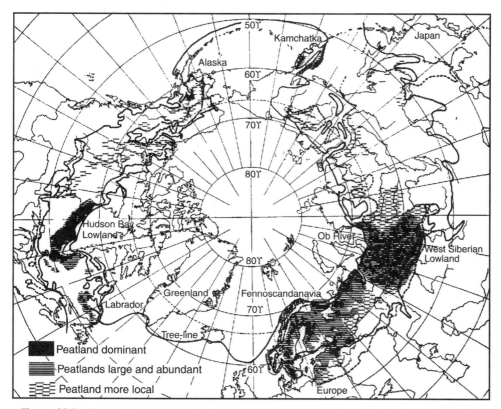

Figure 13-1 Area of abundant peatlands in the boreal zone or taiga of the Northern Hemisphere. Peatlands are associated with boreal regions and their subalpine equivalents in mountainous regions. South of the tree line, woodland tundra or subalpine areas extend to the northern broken line. South of the boreal zone, transition zones are shown to the broken line. (*After Sjörs, 1961b*)

Some major peatlands, not illustrated in Fig. 13-1, are found in the Southern Hemisphere in southern South America and in New Zealand, but the size of these peatlands collectively is small compared to those in the Northern Hemisphere. For example, the New Zealand Land Resource Inventory (Cromarty and Scott, 1996) lists 3,113 km^2 of wetlands in the entire country, many of which are peatlands. There are 43,900 ha of *pakihi* (shallow-peat heathland) and another 35,600 ha of forest–pakiha associations, some of which support sphagnum moss (Buxton et al., 1996). Sanders and Winterbourn (1993) estimate that there may be as much as 8,000 ha on South Island where *Sphagnum* moss is being harvested for horticultural purposes. Otherwise, raised bogs in New Zealand are not characterized by *Sphagnum* moss or ericaceous species common in the Northern Hemisphere but by rushlike plants co-dominated by restiad (coming from the family Restionaceae) bog species. In the Hamilton region of North Island, there were originally 50,000 ha of these peatlands, but less than 140 ha remain; another 9,200-ha bog is further to the north (Clarkson et al., 1999).

The distribution of mires in North America is shown in Figure 13-2. Canada has approximately 1.10 million km^2 of peatlands (Zoltai, 1988), giving it the largest peat resources in the world. Their distribution approximates the boreal forest region

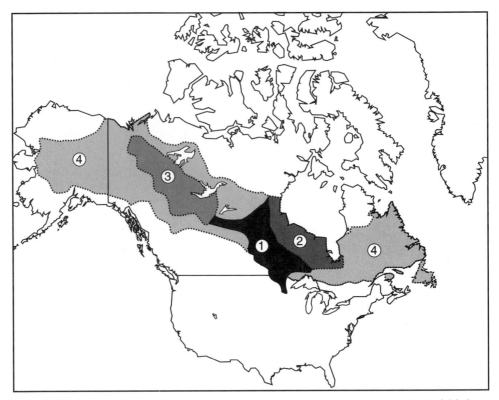

Figure 13-2 Major boreal peatlands of North America. Major peatland areas are (1) glacial Lake Agassiz region, (2) Hudson Bay lowlands, (3) Great Bear/Great Slave Lake region, and (4) other boreal peatlands including the interior of Alaska. (*After Glaser, 1987, as reported by Viereck and Little, 1972; Rowe, 1972; Zoltai and Pollet, 1983*)

delineated by the occurrence of black spruce. Combining this with an estimated 0.55 million km^2 of peatlands in the United States, including Alaska (Tables 4-1 and 4-3), the 1.66 million km^2 of northern peatlands in North America represent at least one-third of the world's peatlands.

These peatlands dip into the conterminous United States from northern Minnesota to northern Maine. Minnesota peatland distribution, depicted in Figure 13-3, was estimated by Soper (1919) in Glaser (1987) to be about 2.7 million ha. In the northeast United States, the zone of ombrogenous bogs dips into Maine, New York, and Vermont. South of these peatlands, there tend to be small depressional systems that receive at least some nutrient inflow (Damman and French, 1987). In the United States, bogs usually develop in basins scoured out by the Pleistocene glaciers. One of the most studied bog areas in the United States is the Lake Agassiz region in northern Minnesota, which extends well into Canada (Fig. 13-2).

Although predominantly a northern phenomenon, peats can accumulate wherever drainage is impeded and anoxic conditions predominate regardless of temperature. Thus, bogs are found as far south as northern Illinois, Indiana, and Ohio in the north-central United States and are also common in the unglaciated Appalachian Mountains in West Virginia. The Middle Atlantic Coastal Plain supports an expansive area of

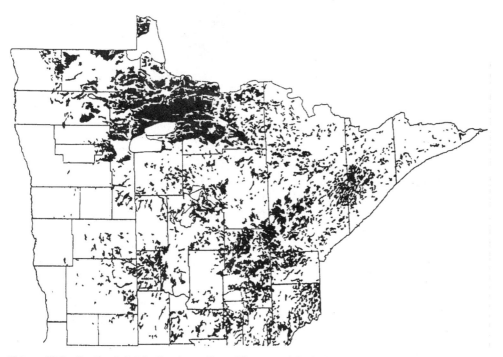

Figure 13-3   Peatland distribution in northern Minnesota (shaded area). (*After Glaser, 1987, based on Minnesota Department of Natural Resources peatlands map, 1978*)

poorly drained peatlands, called *pocosins,* that are similar to more northern peatlands in that they are nutrient poor and dominated by evergreen woody plants such as bilberry, whortleberry, cranberry, heather, and Labrador tea belonging to the Ericaceae or heath family. Pocosins once covered 1.2 million ha, 70 percent of which were in North Carolina. Only about 370,000 ha of the original 1.2 million ha remained in 1980 (Richardson, 1983).

Hydrology and Peatland Development

Two primary processes necessary for peatland development are a positive water balance and peat accumulation. First, a positive water balance, meaning that precipitation is greater than evapotranspiration, is essential for peatland development and survival. Water budgets for a fen, bog, and pocosin, illustrated in Figure 5-6f, h, and i, show that evapotranspiration is generally only 50 to 70 percent of precipitation. Just as important, the seasonal distribution of precipitation and excess water is important because peatlands require a humid environment year-around. In seasonally wet climates with cold winters, such as in the midwestern United States south of Minnesota, Wisconsin, and Michigan, peatlands will not be common if hot, dry summers persist. The southern limit to bog species and, hence, to the bog wetland is thought to be determined by the intensity of solar radiation in the summer months when precipitation and humidity are otherwise adequate to support bogs farther south (Larsen, 1982, after Transeau, 1903).

A second requirement for peatland development is a surplus of peat production over decomposition, or accumulation greater than decomposition ($A > D$). Although primary production is generally low in northern peatlands compared to other ecosystems, decomposition is even more depressed and so peat accumulates. This is a necessary condition for the development of ombrotrophic bogs (see description later in this section). The continued development of the ecosystem is directly related to the amount of surplus water and peat. For example, in a cool, moist maritime climate, peatlands can develop over almost any substrate, even on hill slopes. In contrast, in warm climates where both evapotranspiration and decomposition are elevated, ombrotrophic peatlands seldom develop even when a precipitation surplus occurs.

Once formed, a bog is remarkably resistant to conditions that alter the water balance and peat accumulation. The perched water table, the water-holding capacity of the peat, and its low pH create a microclimate that is stable under fairly wide environmental fluctuations.

Ecosystem Development

Given the conditions of water surplus and peat accumulation, peatlands develop through variations of two processes: *terrestrialization* (the infilling of shallow lakes) or *paludification* (the blanketing of terrestrial ecosystems by overgrowth of peatland vegetation). Three major bog formation processes are commonly seen:

1. *Quaking Peatland Succession:* This is the classical process of terrestrialization, as described in most introductory botany or limnology courses (Weber, 1908; Kratz and DeWitt, 1986). Bog development in some lake basins involves the filling in of the basin from the surface, creating a *quaking bog* (or *Schwingmoor* in German; Fig. 13-4). Plant cover, only partially rooted in the basin bottom or floating like a raft, gradually develops from the edges toward the middle of the lake (Ruttner, 1963). A mat of reeds, sedges, grasses, and other herbaceous plants develops along the leading

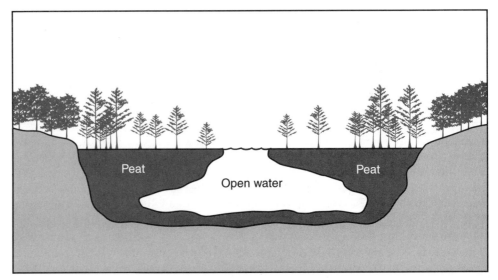

Figure 13-4 Typical profile of a quaking bog.

edge of a floating mat of peat that is soon consolidated and dominated by *Sphagnum* and other bog flora. The mat has all of the characteristics of a raised bog except hydrologic isolation. The older peat is often colonized by shrubs and then forest trees such as pine, tamarack, and spruce, which form uniform concentric rings around the advancing floating mat. These peatlands develop only in small lakes that have little wave action; they receive their name from the quaking of the entire surface that can be caused by walking on the floating mat. After peat accumulates above the water table, isolating the *Sphagnum*-dominated flora from their nutrient supply, the bog becomes increasingly nutrient poor. The development of a perched water table also isolates the peatland from groundwater and nutrient renewal. The result is a classic concentric, or excentric, raised ombrotrophic bog.

2. *Paludification:* A second pattern of bog evolution occurs when blanket bogs exceed basin boundaries and encroach on formerly dry land. This process of paludification can be brought about by climatic change, geomorphological change, beaver dams, logging of forests, or the natural advancement of a peatland. Often, the lower layers of peat compress and become impermeable, causing a perched water table near the surface of what was formerly mineral soil. This causes wet and acid conditions that kill or stunt trees and allow only ombrotrophic bog species to exist. In some situations, the progression from forest to bog can take place in only a few generations of trees (Heilman, 1968).

3. *Flowthrough Succession:* Intermediate between terrestrialization and paludification is flowthrough succession (also termed *topogenous development*), in which the development of peatland modifies the pattern of surface water flow. The development of a bog from a lake basin that originally had continuous inflow and outflow of surface and groundwater is shown in Figure 13-5 in five stages. In the first stage, the inflow of sediments and the production of excess organic matter in the lake begin the buildup of material on the bottom of the lake. The growth of marsh vegetation continues the buildup of peat (stage 2), until the bottom rises above the water level and the flow of water is channelized around the peat. As the peat continues to build, the major inflow of water may be diverted (stage 3) and areas may develop that become inundated only during high rainfall (stage 4). In the final stage (stage 5), the bog remains above the groundwater level and becomes a true ombrotrophic bog. Figure 13-6 illustrates how a sedge peat bog can build up over 5,000 years to become a poor fen and a raised ombrotrophic bog.

Landscape Development

The developmental processes described previously also determine large-scale patterns of landform development.

1. *Raised Bogs:* These are peat deposits that fill entire basins, are raised above groundwater levels, and receive their major inputs of nutrients from precipitation (Fig. 13-7a and b). These bogs are primarily found in the boreal and northern deciduous biomes. Pine trees sometimes grow in these bogs in areas that have drier climates, although many are treeless. When a concentric pattern of pools and peat communities forms around the most elevated part of the bog, the bog is called a *concentric domed bog* (Fig. 13-7a). Bogs that form from previously separate basins on sloping land and form elongated hummocks and pools aligned perpendicular to the

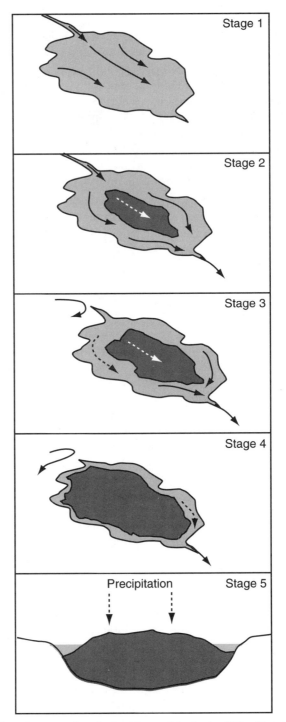

Figure 13-5 Flowthrough succession of a bog from a lake basin. (*After Moore and Bellamy, 1974*)

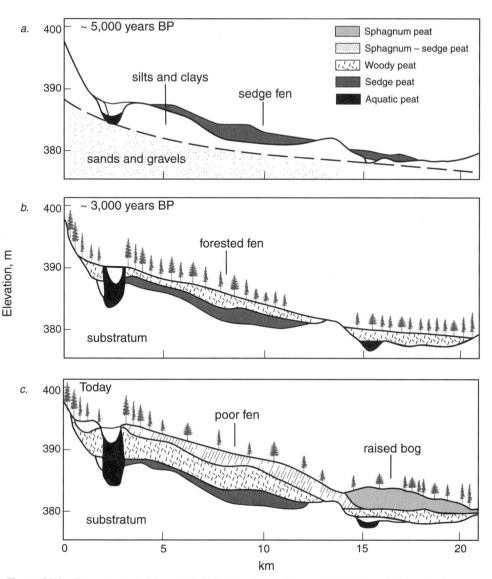

Figure 13-6 Peat accumulation over 5,000 years in a built-up peatland in the Lake Agassiz peat-lands of northern Minnesota from **a.** a sedge fen to **b.** a forested fen to **c.** a landscape of poor fen and raised bog. Peat description is based on plant remains. (*After Boelter and Verry, 1977, based on data from Heinselman, 1963, 1970*)

slope are called *excentric raised bogs* (Fig. 13-7b). The former are found near the Baltic Sea, and the latter are found primarily in the North Karelian region of Finland.

The hydrology of raised bogs has been investigated and found to be more compli-cated than originally thought, particularly for bogs that are on the edge of the boreal zone (Siegel et al., 1995; Glaser et al., 1997a, b). Studies in the Lake Agassiz region of Minnesota showed that bogs and fens are really part of a regional hydrology, with fens receiving groundwater and raised bogs generally recharging groundwater (Fig. 13-8a). Raised bogs are normally assumed to be disconnected from groundwater and

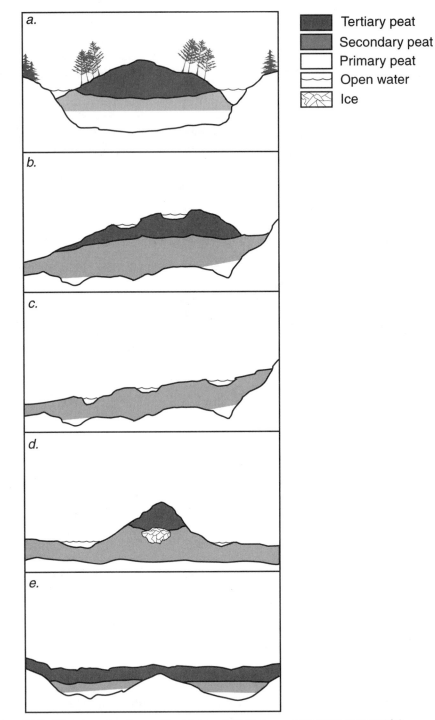

Tertiary peat
Secondary peat
Primary peat
Open water
Ice

Figure 13-7 Diagrams of major peatland types, including **a.** raised bog (concentric), **b.** excentric raised bog, **c.** aapa peatland (string bogs and patterned fens), **d.** paalsa bog, and **e.** blanket bog. (*After Moore and Bellamy, 1974*)

a.

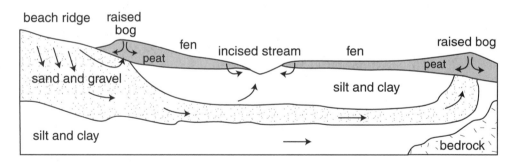

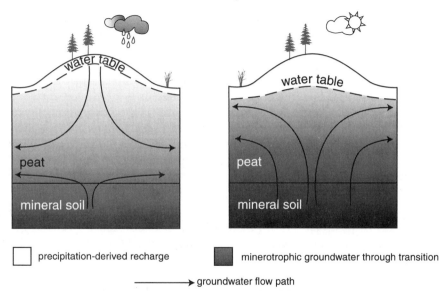

b. Wet climate   *c.* Dry climate

☐ precipitation-derived recharge   ■ minerotrophic groundwater through transition

→ groundwater flow path

Figure 13-8   a. Regional linkages between groundwater and raised bogs in the Lake Agassiz region of Minnesota (area is approximately 10 km long and 30 m thick). Detailed patterns of subsurface hydrology in the raised bogs are illustrated for   **b.** wet climate and   **c.** dry climate. During wet periods, precipitation-derived recharge maintains a head that flushes mineral-rich groundwater from the peat. During droughts, the water mound drops and mineral-rich groundwater can move upward into the raised-bog peat. (*After Siegel et al., 1995; Glaser et al., 1997a*)

fed only by precipitation. In a normal wet climate, this pattern of bog hydrology is true as a downward flow of excess precipitation deflects upwardly moving groundwater from mineral soil well below the surface (Fig. 13-8b). This accelerates peat accumulation, which, in turn, maintains the peat and, hence, hydrologic mound in the landscape. During droughts, which can be frequent events in peatlands on the edge of the boreal region, groundwater can move upward to within 1 to 2 m of the peat surface (Fig. 13-8c) and dramatically influence the peatland chemistry (see Biogeochemistry later in this chapter).

2. *Aapa Peatlands:* These wetlands (Figs. 13-7c and 13-9), also called *string bogs* and *patterned fens,* are found throughout the boreal region, often north of the raised-bog region. The dominant feature of these wetlands is the long, narrow alignment of the higher peat hummocks (strings) that form ridges perpendicular to the slope of the peatland and are separated by deep pools (*flarks* in Swedish). In appearance, they resemble a hillside of terraced rice fields.

Glaser (1983a, b, 1987), Siegel (1983), and Foster et al. (1983) described the formation of a patterned landscape in northern Minnesota that is similar to the aapa peatlands or string bogs shown in Figure 13-9. It consists of bogs and water tracks. The latter have been further differentiated into *tree islands, strings,* and *flarks* (Fig. 13-9a). Water flowing across the peatlands is channeled by topographic features into definite zones, usually along the path of minerotrophic runoff. These water tracks are invaded by nutrient-demanding sedges such as *Carex lasiocarpa.* The porous peat of the sedges further channels the flow of water. Raised bogs develop wherever topographic obstructions divert the path of runoff. In these areas, sphagnum grows and spreads outward until it is blocked by the main flows of the water tracks. The boundary between fen and bog is sharpened by the accumulation of dense, relatively impermeable decomposing sphagnum peat in the bog.

In the water track, a network of parallel, linear peat ridges (strings) and water-filled depressions (flarks) develops perpendicular to the direction of the water flow (Glaser et al., 1981; Glaser, 1983c). The pattern begins as a series of scattered pools on the down slope, wetter edge of the water track. These pools gradually coalesce into linear flarks. The peat accumulation in the adjacent strings and the increasing impermeability of decomposing peat in the flarks accentuate the pattern. Within the large water tracks, tree islands appear to be remnants of continuous swamp forests that were replaced by sedge lawns in the expanding water tracks (Glaser and Wheeler, 1980).

3. *Paalsa Bogs:* These bogs, found in the southern limit of the tundra biome, are large plateaus of peat (20–100 m in breadth and length and 3 m high) generally underlain by frozen peat and silt (Fig. 13-7d). The peat acts like an insulating blanket, actually keeping the ground ice from thawing and allowing the southernmost appearance of the discontinuous permafrost. In Canada, as much as 40 percent of the land area is influenced by cyrogenic factors. When peat overlies frozen sediments, it influences the pattern of the landscape. Many distinctive forms are similar to European aapa and paalsa peatlands but are embedded in a continuous peat-covered landscape (Gore, 1983a).

4. *Blanket Bogs:* These wetlands (Fig. 13-7e) are common along the northwestern coast of Europe and throughout the British Isles. The favorable humid Atlantic climate allows the peat literally to "blanket" very large areas far from the site of the original peat accumulation. Peat in these areas can generally advance on slopes of up to 18 percent; extremes of 25 percent have been noted on slopes covered by blanket bogs in western Ireland (Moore and Bellamy, 1974). Thompson (1983) reported the distribution of patterned peatlands farther south in North America than had been reported previously.

Stratigraphy

Accumulated deep peat deposits in northern peatlands have left a record of deposition thousands of years old, documenting the pathway of bog formation. This wealth of

Figure 13-9 Two oblique aerial images of string bogs in North America: **a.** Cedarburg Bog in southwestern Wisconsin, showing a pattern of parallel peat ridges (strings) alternating with water-filled depressions (flarks) running diagonally across the lower half of the photograph and **b.** a string fen in Labrador, Canada. The strings stand out because they are vegetated with ericaceous shrubs and scrub trees over sphagnum moss, whereas flarks are dominated by mosses and herbs or, in the case of the Canadian site, extensive standing water. (*Photograph a by G. Guntenspergen; photograph b by D. Wells, reprinted by permission of C. Rubec and reprinted from Mitsch et al., 1994a, p. 30, Fig. 30, with permission from Elsevier Science*)

information has been exploited to study the history of climatic and associated geological and ecological change during the Quarternary epoch, that is, the geological period covering the waxing and waning of the great glaciations (Godwin, 1981; A. M. Davis, 1984; Solem, 1986). As an example, Figure 13-10 shows the terrestrialization of Cedar Bog Lake in south-central Minnesota, which has been reconstructed from the peat record (Glaser, 1987, from data of Lindeman, 1941, and Cushing, 1963). Beginning about 8000 B.P. (Fig. 13-10a) when the water surface was about 4 m below present levels, a sedge-dominated wetland spread over exposed mud flats on the

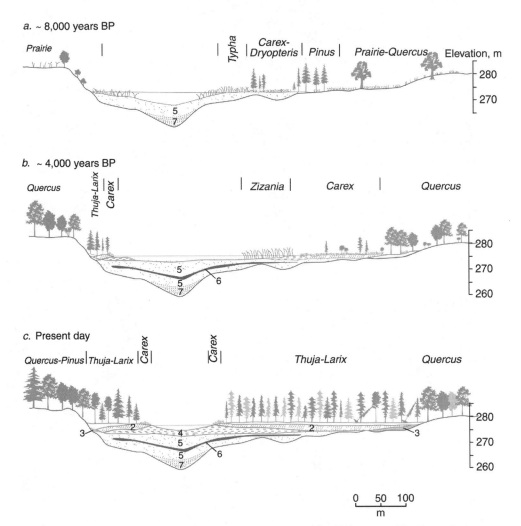

Figure 13-10 Terrestrialization at Cedar Bog Lake, south-central Minnesota. The three stages represent **a.** the inferred vegetation and stratigraphy during the Early Holocene approximately 8,000 years ago, **b.** the inferred vegetation and lake level 4,000 years ago during the Mid-Holocene, and **c.** the present-day vegetation and lake level. The lake was 4 m lower during the Early Holocene and the subsequent rise in the water table was accompanied by growth of the surrounding peatland. Sediment layers, defined by Lindeman (1941) are (1) woody peat, (2) sedge peat, (3) sapropsammite, (4) coarse-detritus copropel, (5) marly copropel, (6) dark copropel, (7) sideritic copropel. Elevations are height above sea level. (*After Glaser, 1987; Cushing, 1963*)

sheltered southeast end of the lake basin (Fig. 13-10b). About 4000 B.P., the warmer, drier period of the Mid-Holocene ended and a cooler, moister period followed, during which most of the great peatlands of the Lake Agassiz plain formed. During that period, a cedar swamp grew across Cedar Bog Lake (Fig. 13-10c) over the sedge marsh. Apparently, that growth restricted the penetration of minerotrophic runoff from surrounding uplands, leading to the present condition (Fig. 13-10c).

Classification of Peatlands

Peatlands develop within a complex interaction of climate, hydrology, topography, chemistry, and vegetation development (succession). Because the physical and biotic processes that form peatlands are complex and differ somewhat from region to region, many different classification systems have been proposed over the past century (Table 13-1). Moore (1984) described seven features on which classification schemes have been based: (1) floristics, (2) vegetation structure, (3) geomorphology (succession or development), (4) hydrology, (5) chemistry, (6) stratigraphy, and (7) peat characteristics. The last is used primarily for economic exploitation purposes. The other six

Table 13-1   Historical classification schemes for peatlands

Principal Basis for Classification	Mineral-influenced Peatlands	Transition Peatlands	Precipitation-dominated Peatlands	Reference
Topography	Fen		Bog or rasied bog	General use
	Niedermoore (low moor)	Ubergangsmoore	Hochmoore (high moor)	Weber (1907)
Hydrology	Geogenous Limnogenous Topogenous Soligenous		Ombrogenous	von Post and Granlund (1926), Sjörs (1948), Du Rietz (1949), Damman (1986)
	Rheophilous	Transition	Ombrophilous	Kulczynski (1949)
	Soligenous		Ombrogenous	Walter (1973)
	Minerogenous		Ombrogenous	Warner and Rubec (1997)
Water chemistry	Rich fen	Poor fen	Bog	General use; Sjörs (1948)
	Minerotrophic Rheotrophic	Mesotrophic	Ombrotrophic Ombrotrophic	Moore and Bellamy (1974) Moore and Bellamy (1974)
Nutrition	Nährstoffreichere Eutrophic	Mittelreiche Mesotrophic	Nährstoffearme Oligotrophic	Weber (1907) Weber (1907), Pjavchenko (1982)
Vegetation	Emergent or forested fen	Transitional	Moss–lichen or forested bog	Cowardin et al. (1979), Gorham and Janssens (1992)

Source: Bridgham et al. (1996).

are closely interrelated, leading to classification schemes that combine several natural features. Bridgham et al. (1996) simplified the list to (1) topography (geomorphology), (2) ontogeny (genesis and succession), (3) hydrology, (4) water chemistry, and (5) vegetation species, and summarized the bases of various peatland classifications (Table 13-1).

Chemistry-based Classification

The developmental processes described previously lead to increasing isolation of bogs from surface and subsurface flows of both water and mineral nutrients. The degree of hydrologic isolation of mires leads to a simple classification that is probably the most frequently used today and is based on the degree to which the peatland receives groundwater inflow as compared to only precipitation (Du Rietz, 1954; Sjörs, 1961a; Moore and Bellamy, 1974):

1. *Minerotrophic Peatlands:* These are true fens that receive water that has passed through mineral soil (Gorham, 1967). These peatlands generally have a high groundwater level and occupy a low point of relief in a basin. They are also referred to as *rheotrophic peatlands* by Moore and Bellamy (1974) and *rich fens* in general use.

2. *Mesotrophic Peatlands:* These peatlands are intermediate between mineral-nourished (minerotrophic) and precipitation-dominated (ombrotrophic) peatlands. Another term used frequently for this class is *transitional peatlands* or *poor fens.*

3. *Ombrotrophic Peatlands:* These are the true raised bogs that have developed peat layers higher than their surroundings and that receive nutrients and other minerals exclusively by precipitation.

Another "trophic" classification of peatlands, found in older European literature (Weber, 1907) and originally developed to classify peatlands and not lakes (Hutchinson, 1973), is the three-level trophic classification now familiar to limnologists:

1. *Eutrophic Peatlands:* Nutrient-rich peatlands; described by Weber (1907) as *Nährstoffreichere* (eutrophe)

2. *Mesotrophic Peatlands:* Same as before; described by Weber (1907) as *Mittelreiche* (mesotrophe)

3. *Oligotrophic Peatlands:* Nutrient-poor peatlands; described by Weber (1907) as *Nährstoffearme* (oligotrophe)

Hutchinson (1973) suggests that the process of peatland development could be called *oligotrophication.* The terms *eutrophic* and *oligotrophic* were only applied to lakes and their current limnological use a decade later by Naumann (1919). Russian scientists such as Pjavchenko (1982) and Bazilevich and Tishkov (1982) continued to use this nomenclature well into the 1980s.

Bridgham et al. (1996) argue for caution in the use of the "-trophic" suffix for classifying peatlands because the classic peatland gradient from minerotrophic to ombrotrophic, characterized by surface water chemistry such as pH, conductivity, and alkalinity, does not necessarily correlate with the eutrophic to oligotrophic gradient, which is defined in terms of nutrient. (e.g., nitrogen, phosphorus, and potassium) availability (see Limiting Nutrients later in this chapter). In fact, one study (Bridgham et al., 1998) found evidence to suggest that there was higher phosphorus availability in bogs and higher nitrogen availability in fens. In other words, a strict correlation between measures of dissolved minerals and available nutrients has never been established. They suggest resurrecting the terms eutrophic and oligotrophic, which are rarely seen in peatland literature today, because they clearly refer to nutrients and not to other minerals.

Hydrology-based Classification

Heinselman (1970) and Moore and Bellamy (1974), arguing for simplicity in the description of peatlands, stated that terms such as *soligenous* and *ombrogenous* actually refer to the hydrological and topographic origins of the peatlands and not to the mineral conditions of the inflowing water. Bridgham et al. (1996) called for the return of a true hydrologic classification of peatlands based on the following categories (von Post and Granlund, 1926; Sjörs, 1948; Du Rietz, 1949; Damman, 1986; Fig. 13-11):

1. *Ombrogenous Peatlands:* Open only to precipitation
2. *Geogenous Peatlands:* Open to outside hydrologic flows other than precipitation:
   a. *Limnogenous Peatlands:* Develop along slow-flowing streams or lakes
   b. *Topogenous Peatlands:* Develop in topographic depressions with at least some regional groundwater flow
   c. *Soligenous Peatlands:* Develop with regional interflow and surface runoff

One of the more complete classifications developed for wetlands in general and peatlands in particular is the new Canadian Wetland Classification System (Warner and Rubec, 1997). This classification uses *minerotrophic* and *ombrotrophic* in its water chemistry classification and *minerogenous* and *ombrogenous* in its hydrological classification. This simple classification of peatlands is:

1. *Bog:* Peatland receiving water exclusively from precipitation and not influenced by groundwater; sphagnum-dominated vegetation
2. *Fen:* Peatland receiving water rich in dissolved minerals; vegetation cover composed dominantly of graminoid species and brown mosses
3. *Swamp:* Peatland dominated by trees, shrubs, and forbs; waters rich in dissolved minerals

A nonpeatland class (mineral wetland) also, of course, includes another type of swamp. These are discussed in Chapter 14). General wetland classifications are discussed in more detail in Chapter 21.

Biogeochemistry

Soil and water chemistry are among the most important factors in the development and structure of the bog ecosystem (Heinselman, 1970). Factors such as pH, mineral concentration, available nutrients, and cation exchange capacity influence the vegetation types and their productivity. Conversely, the plant communities influence the chemical properties of the soil water (Gorham, 1967). In few wetland types is this interdependence so apparent as in northern peatlands. Table 13-2 gives typical pH values and cation concentrations for different Northern Hemisphere peatlands.

1. ombrogenous

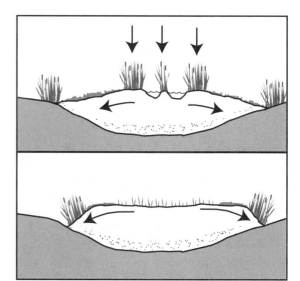

2. geogenous

    a. limnogenous

    b. topogenous

    c. soligenous

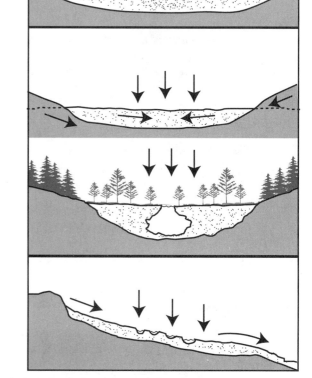

Figure 13-11 Classification of peatlands based on hydrology. Two major categories are geogenous peatlands, which are open to surface and groundwater flow, and ombrogenous peatlands, which only receive precipitation. (*After Damman, 1986*)

Table 13-2 pH and cations of water in selected peatlands

Location/Peatland Type	Cations (mg/L)					Reference
	pH	Ca^{2+}	Mg^{2+}	Na^+	K^+	
General Classification						
Ombrotrophic bog	3.8					Moore and Bellamy
Transitional fen	4.1–4.8					(1974)
Minerotrophic fen	5.6–7.5					
Oligotrophic	2.6–3.3					Pjavchenko (1982)
Mesotrophic	4.1–4.8					
Eutrophic	>4.8					
Minnesota Peatlands						
Bog	<4.0					Bridgham et al.
Acidic fen	4.0–4.1					(1996)
Intermediate fen	4.9–5.6					
Swamp forest	4.3–6.6					
Northwest Canada						Nicholson et al.
Low-boreal bog	4.2 ± 0.05	1.7	0.4	5.1	—	(1996)
Poor fen	4.9 ± 0.9	8.7	2.5	4.1	—	
Moderately rich fen	6.4 ± 0.9	19.6	4.6	7.3	—	
Extremely rich fen	6.5 ± 1.0	57	11.7	7.5	—	
Hudson Bay lowland, Canada						Reeve et al. (1996)
distance from mineral soil						
<1 m	6.2	34	3.3	8.9	1.7	
1–2 m	5.4	9.1	0.9	3.1	1.9	
>2 m	4.5	2.9	0.5	1.1	1.1	
Canadian peatlands, undisturbed						Wind-Mulder et al.
New Brunswick	3.9	0.4	0.5	4.1	0.3	(1996)
Eastern Quebec	3.7	4.6	3.1	16.5	<0.4	
Central Quebec	3.7	0.6	0.3	0.3	0.2	
Alberta	3.7	3.0	1.1	1.0	0.2	
Canadian peatlands, harvested						Wind-Mulder et al.
New Brunswick	3.7	0.3	1.3	7.2	0.9	(1996)
Eastern Quebec	5.3	5.3	2.8	11	2.2	
Central Quebec	4.0	0.9	0.3	0.5	0.4	
Alberta	5.6	12.6	4.1	6.1	2.6	

Hemond (1980) summarized biogeochemical processes in a small floating-mat *Sphagnum* bog (Thoreau's Bog) in New England (Fig. 13-12). The bog is ombrotrophic and its chemistry is primarily determined by the chemistry of bulk precipitation. Nitrates are taken up by vegetation or quickly denitrified. Metal ions are accumulated from precipitation primarily by ion exchange or through active uptake by *Sphagnum* in the case of potassium. Their retention depends on the exchange affinity of different metals. For example, lead, principally an anthropogenic input, is almost quantitatively retained, whereas most potassium is leached. The largest source of acidity for this bog is precipitation and the release of organic acids from mosses, other vegetation, and detritus. Acidity is counteracted by biological processes in the bog such as sulfate reduction, which decreases sulfuric acid, and photosynthesis, which can reduce carbonic acid in the water.

The major features of peatland biogeochemistry are discussed next.

Acidity and Exchangeable Cations

The pH of peatlands generally decreases as the organic content increases with the development from a minerotrophic fen to an ombrotrophic bog (Fig. 13-13). Fens are dominated by minerals from surrounding soils, whereas bogs rely on a sparse supply of minerals from precipitation. Therefore, as a fen develops into a bog, the supply of metallic cations (Ca^{2+}, Mg^{2+}, Na^+, K^+) drops sharply (compare the concentration of cations in precipitation with the concentrations in bogs and fens in Table 13-2). At the same time, as the organic content of the peat increases because of the slowing of the decomposition rate, the capacity of the soil to adsorb and exchange cations increases (see Chapter 6). These changes lead to the domination by hydrogen ions, and the pH falls sharply. Gorham (1967) found that bogs in the English Lake

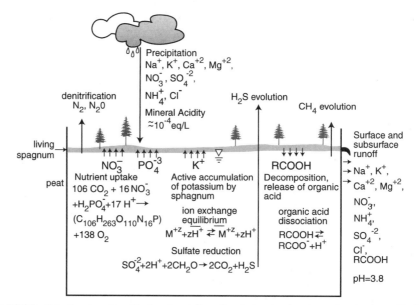

Figure 13-12 A bog ecosystem biogeochemistry, showing major inputs and outputs and indicating chemical reactions within the bog ecosystem that are biogeochemically significant. The water budget for this bog is shown in Figure 5-6h. (*After Hemond, 1980*)

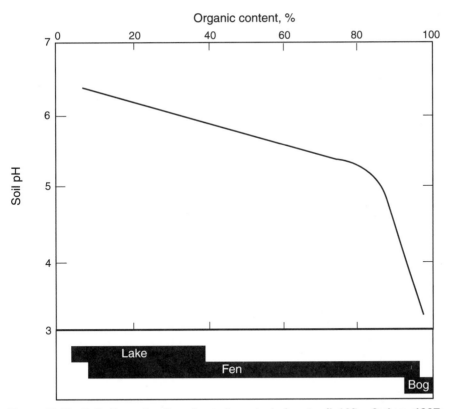

Figure 13-13   **Soil pH as a function of organic content of peat soil. (*After Gorham, 1967*)**

District had a pH range of 3.8 to 4.4 compared to noncalcareous fens, which had a pH range of 4.8 to 6.0. Pjavchenko (1982) assigned a pH range of 2.6 to 3.3 to oligotrophic bogs and a range of 4.1 to 4.8 to mesotrophic bogs; a pH greater than 4.8 defined a eutrophic (minerotrophic) fen (Table 13-2).

Over the gradient from ombrotrophic bogs to minerotrophic fens, pH and Ca^{2+} ion concentrations are reliably interrelated (Fig. 13-14). The rise in pH is related to the input of cations from a mineral source, either surface runoff or groundwater discharge. In addition, peatlands may receive a significant input of Ca^{2+} from atmospheric deposition that is transported as dust from adjacent prairies (Glaser, 1987). The airborne source is apparently insufficient to account for calcium concentrations in rich fens, however. The theoretical curve relating pH and calcium (solid line in Fig. 13-14a) pertains to solutions in equilibrium with the atmosphere with a Ca-balancing charge (Glaser, 1987). In rich-fen samples, the pH is much lower than would have been predicted from the Ca^{2+} concentration. These samples have apparently been oversaturated with calcite because of the complexing of Ca^{2+} to organic compounds or the release of organic anions (Glaser, 1987). Although the data in Figure 13-14 indicate a continuum of pH and Ca^{2+} values over the range of peatlands sampled, bogs are clearly different chemically from rich and intermediate fens.

Glaser et al. (1990) emphasized the importance of soligenous and groundwater sources of minerals. As little as 10 percent of the water supply from groundwater may change the pH of a bog from 3.6 to 6.8, that is, from an ombrotrophic bog to a

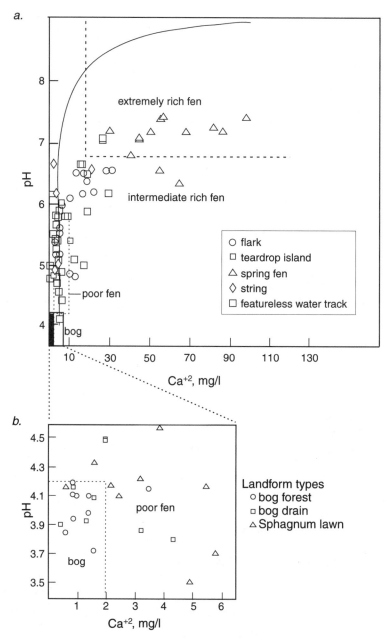

Figure 13-14   a. The relationship between calcium and pH for bog and fen samples in northern Minnesota.   **b.** is an expansion of the bog section of **a.** (*After Glaser, 1987*)

minerotrophic rich fen. In the upper peat of a Minnesota raised bog [the study described previously on the reversal of flow patterns (Siegel et al., 1995; Glaser et al., 1997a,b)], pH and conductivity, both indicators of mineral groundwater, increased dramatically in a drought year compared to a wet year (Fig. 13-15).

The causes of bog acidity are not entirely clear (Clymo and Hayward, 1982; Kilham, 1982; Sikora and Keeney, 1983; Gorham et al., 1984b; Urban et al., 1985). Glaser (1987) cited five causes of low pH:

1. *Cation Exchange by Sphagnum:* Clymo (1963) and Clymo and Hayward (1982) concluded that this is the most important mechanism for the generation of acidity in peatlands. Figure 13-16 shows the direct relationship between pH and exchangeable hydrogen on peat, presumably the result of the metabolic activity of the plants. Note that *Sphagnum* peats have a higher exchangeable hydrogen and, consequently, a lower pH than sedge-dominated peats.

2. *Oxidation of Sulfur Compounds to Sulfuric Acid:* Organic sulfur reserves in peat may be oxidized to acidic compounds.

3. *Acid Atmospheric Deposition:* Sulfur deposition is a significant source of acidity, depending on the oxidation state of the sulfur and the location of the bog. Acid sources in precipitation and dry deposition are usually small except close to sources of atmospheric pollution (Gorham, 1967; Brackke, 1976). Novák and Wieder (1992) found, in studies of peatlands in the United States and the Czech Republic, that there were no trends of increasing sulfate concentrations in peatland soils with increasing sources of atmospheric sulfate deposition, due both to the rapid turnover of sulfate and to the much larger pools of organic and other inorganic sulfur.

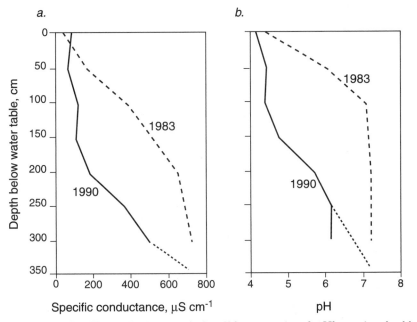

Figure 13-15   a. Specific conductance and   b. pH in pore water of a Minnesota raised bog after a period of drought (1983) and after seven wet years (1990). (*After Siegel et al., 1995*)

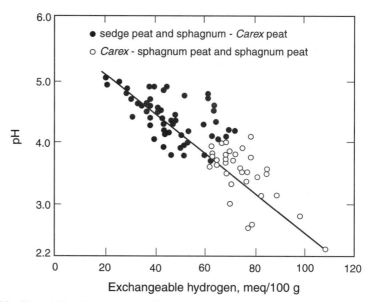

Figure 13-16 The relationship between pH and exchangeable hydrogen. (*After Puustjarvi, 1957; Walmsley, 1977*)

4. *Biological Uptake of Nutrient Cations by Plants:* Ions in the peat water are concentrated by evaporation and are differentially absorbed by the mosses. This affects acidity, for example, by the uptake of cations that are exchanged with plant hydrogen ions to maintain the charge balance.

5. *Buildup of Organic Acids by Decomposition:* Gorham et al. (1984b) presented evidence supporting this source of bog acidity. Organic acids help buffer the system against the alkalinity of metallic cations brought in by rainfall and local runoff.

A detailed hydrogen budget constructed for a Minnesota bog complex implicated nutrient uptake as the major source of acidity (Urban et al., 1985; Table 13-3). About 15 percent of this represents ion exchange on the cell walls of *Sphagnum*. Most of this acidity is neutralized by the release of cations during decomposition. Most of the rest of the acidity is generated by organic acid production from fulvic and other acids that result from the incomplete oxidation of organic matter (McKnight et al., 1985) and that buffer the pH of bogs throughout the world at a value of about 4. In addition to decomposition, the major source of alkalinity to neutralize the acids, the weathering of iron and aluminum and runoff are major processes.

Limiting Nutrients

Bogs are exceedingly deficient in available plant nutrients; fens that contain groundwater and surface water sources generally have considerably more nutrients. The paucity of nutrients in bogs leads to two significant results, which are discussed in more detail later in this chapter: (1) The productivity of nutrient-poor bogs is lower than that of nutrient-rich fens; and (2) the characteristic plants, animals, and microbes have many special adaptations to the low-nutrient conditions.

Table 13-3 Acidity balance for a Minnesota bog complex

Sources	Acidity (meq m^{-2} yr^{-1})
Wet and dry deposition	-0.20 ± 10.7
Upland runoff	-44.3 ± 18.6
Nutrient uptake	827 ± 248
Organic acid production	263 ± 50
Total	1,044
Sinks	
Denitrification	12.2
Decomposition	784
Weathering	76
Outflow	142 ± 50
Total	1,044

Source: Urban et al. (1985).

Many studies have attempted to find the ultimate limiting factor for bog primary productivity; this may be a complex and academic question because all available nutrients are in short supply and the growing season is short and cool. Bridgham et al. (1996) found 33 peatland limiting factor studies in the literature. Calcium, one of the major components of alkalinity, has rarely been found to be limiting to plant growth and, in fact, it could cause an opposite effect by being toxic (Boatman and Lark, 1971) or by binding available phosphorus and making it unavailable to plants (Bridgham et al., 1996). Goodman and Perkins (1968) found that potassium was the major limiting factor for the growth of *Eriophorum vaginatum* (cotton grass) in a bog in Wales, and Heilman (1968) found that levels of phosphorus and, to a lesser extent, potassium were deficient in black spruce (*Picea mariana*) foliage in a *Sphagnum* bog in Alaska. In almost all of the other studies, nitrogen and phosphorus have been shown to be the major limiting factors in bog and fen productivity. When these nutrients are added in significant amounts to peatlands, major vegetation shifts occur (Wassen et al., 1990; Kadlec and Knight, 1996); with management such as mowing, the limiting factor can change from nitrogen to phosphorus (Verhoeven and Schmitz, 1991). Bog formation in its latter stages is essentially limited to nutrients brought in by precipitation. The effects on peatlands of increased atmospheric sources of nitrogen throughout the developed world due to fossil fuel burning has yet to be adequately assessed.

The nutrient status of peatlands is controlled by a number of processes. Historically, the major focus of research has been on surface and subsurface solute inflows, but the rate of peat accumulation and atmospheric inputs is also important (Hemond et al., 1987; Damman, 1990).

Figure 13-17a shows the decrease in nitrogen content of peat soils as the organic content increases. The nitrogen content is above 4 percent in fens but decreases to less than 2 percent in bogs. The increased dominance of sphagnum moss in bogs, generally with a nitrogen content of less than 1 percent (Gorham, 1967), contributes to this drop. Figure 13-17b shows the pattern of phosphorus with depth in ombrotrophic bogs and presents a comparison of that pattern with those in mineral soils.

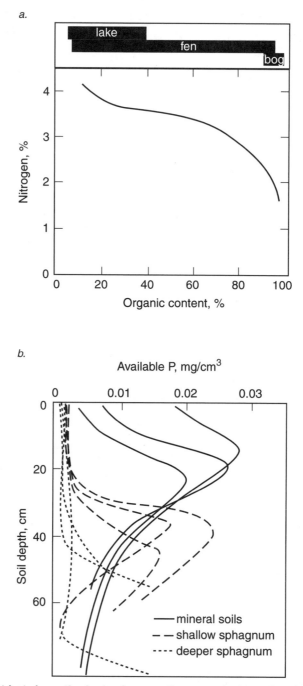

Figure 13-17 Nutrients in peatlands showing **a.** nitrogen content as a function of organic content and **b.** available phosphorus as a function of depth for mineral soils and sphagnum soil. (*After Gorham, 1967; Heilman, 1968*)

Nutrients are concentrated deeper in bogs than in fens because of peat compression, the influence of past minerotrophic conditions, and a complex process of solute partitioning by freezing (Kadlec and Li, 1990). Most of the nutrients, sequestered in organic compounds, appear to be resistant to decomposition. The surface peat has an "insulating" effect, isolating surface plant life from the mineral water below (Gorham, 1967).

Ecosystem Structure

Northern peatlands, particularly ombrotrophic bogs, support plants, animals, and microbes that have many adaptations to physical and chemical stresses. The organisms must deal with waterlogged conditions, acid waters, low nutrients, and extreme temperatures. The result is that specialized and unique flora and fauna have evolved in this wetland habitat.

Vegetation

Bogs can be simple sphagnum moss peatlands, sphagnum–sedge peatlands, sphagnum–shrub peatlands, bog forests, or any number or combination of acidophilic plants. Mosses, primarily those of the genus *Sphagnum,* are the most important peat-building plants in bogs throughout their geographical range. Mosses grow in cushionlike, spongy mats; water content is high, with water sometimes held higher than it normally would be held by capillary action. *Sphagnum* grows shoots actively only in the surface layers (at a rate of about 1–10 cm annually); the lower layers die off and convert to peat.

    Sphagnum often grows in association with cotton grass (*Eriophorum vaginatum*), various sedges (*Carex* spp.), and certain ericaceous shrubs such as heather (*Calluna vulgaris*), leatherleaf (*Chamaedaphne calyculata*), cranberry and blueberry (*Vaccinium* spp.), and Labrador tea (*Ledum palustre*). Trees such as pine (*Pinus sylvestris*), crowberry (*Empetrum* spp.), spruce (*Picea* spp.), and tamarack (*Larix* spp.) are often found in bogs as stunted individuals that may be scarcely 1 m high yet several hundred years old (Ruttner, 1963; Malmer, 1975).

Minnesota Peatland

Heinselman (1970) described seven vegetation associations in the Lake Agassiz peatlands of northern Minnesota that are typical of many of those in North America. These occur in an intricate mosaic across the landscape, reflecting the topography, chemistry, and previous history of the site that slopes from southeast to northwest, as shown in the cross section in Figure 13-6. The vegetation zones correspond closely to the underlying peat and to the present nutrient status of the site. The major zones are:

1. *Rich Swamp Forest:* These forested wetlands form narrow bands in very wet sites around the perimeter of peatlands. The canopy is dominated by northern red cedar (*Thuja occidentalis*); there are also some species of ash (*Fraxinus* spp.), tamarack, and spruce. A shrub layer of alder, *Alnus rugosa,* is often present, as are hummocks of *Sphagnum* moss.

2. *Poor Swamp Forest:* These swamps, occurring downslope of the rich swamp forests, are nutrient-poor ecosystems and are the most common peatland type in the Lake Agassiz region. Tamarack is usually the dominant canopy

tree, with bog birch (*Betula pumila*) and leatherleaf in the understory and *Sphagnum* forming 0.3- to 0.6-m-high hummocks.

3. *Cedar String Bog and Fen Complex:* This is similar to zone 2 except that trees'-edge fens alternate with cedar (*Thuja occidentalis*) on the bog ridges (strings) and treeless sedge (mostly *Carex*) in hollows (flarks) between the ridges.

4. *Larch String Bog and Fen:* In this type of string bog, similar to zones 2 and 3, tamarack (*Larix*) dominates the bog ridges.

5. *Black Spruce–Feathermoss Forest:* This type is a mature black spruce (*Picea marina*) forest that also contains a carpet of feathermoss (*Pleurozium*) and other mosses. The trees are tall, dense, and even aged. This peatland occurs near the margins of ombrotrophic bogs and generally does not have standing water.

6. Sphagnum–*Black Spruce–Leatherleaf Bog Forest:* This is a widespread wetland type in northern North America. Stunted black spruce is the only tree, and there is a heavy shrub layer of leatherleaf, laurel (*Kalmia* spp.), and Labrador tea growing in large ''pillows'' of *Sphagnum* moss between spruce patches. This association is found in convex relief and is isolated from mineral-bearing water.

7. Sphagnum–*Leatherleaf*–Kalmia–*Spruce Heath:* A continuous blanket of *Sphagnum* moss is the most conspicuous feature; a low shrub layer and stunted trees (usually black spruce) are present in 5 to 10 percent of the area. Zones 6 and 7, as shown in the lower end of the Figure 13-6 transect, occur on a raised bog.

In the water chemistry classification presented earlier in the chapter, zones 1 through 4 would be classified as minerotrophic, zone 5 as transitional, zone 6 as semi-ombrotrophic, and zone 7 as ombrotrophic.

Vegetation Gradients

Although *Sphagnum* species are the characteristic peat-forming ground cover of bogs, as sedges are of poor fens, there is a considerable overlap of species along the chemical gradient from mineral poor to mineral rich and from low pH to high pH. In a direct gradient analysis of vascular plants found in both bogs and fens in northern Minnesota (Fig. 13-18), the sedges *Carex oligosperma* and *Eriophorum spissum* decrease in cover abundance with mineral enrichment of the peat, whereas tamarack (*Larix laricina*) increases in abundance. Black spruce and the ericaceous shrubs Labrador tea and leatherleaf, however, show dual peaks, indicating that their distribution is not controlled by mineral water chemistry but by another gradient such as water level or possibly nitrogen or phosphorus availability.

Nicholson et al. (1996) investigated climatic and ecological gradients and how they affected bryophyte distribution in the Mackenzie River basin in northwestern Canada. They found that the most important variables that explained bryophyte species distributions were water chemistry (Mg^{2+}, Ca^{2+}, H^+), height above the water table, precipitation, and annual temperature. As a result of examining these gradients, seven peatland groups were clustered from the original 82 sites in the basin: (1) poor fens, (2) peat plateaus with thermokarst pools, (3) low-boreal bogs, (4) bogs and peat plateaus without thermokarst pools, (5) low-boreal dry poor fens, (6) wet

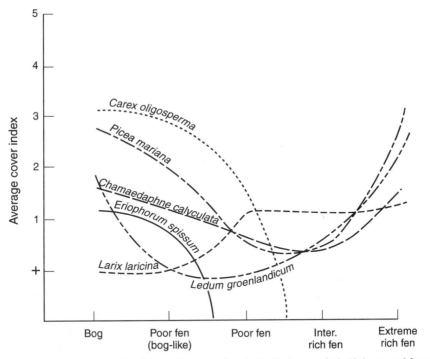

Figure 13-18 Gradient analysis of the major vascular plants that occur in both bogs and fens in Minnesota. Ordinate from Braun–Blanquet scale. Ranges in water chemistry are bog (pH<4.2; Ca^{2+}<2 mg/L), poor fen (boglike) (pH = 4.1–4.6; Ca^{2+} = 1.5–5.5), poor fen (pH = 4.1–5.8, Ca^{2+}<10 mg/L), intermediate rich fen (pH = 5.8–6.7; Ca^{2+} = 10–32 mg/L), extremely rich fen (pH>6.7; Ca^{2+}>30 mg/L). (*After Glaser, 1987*)

moderate-rich fens, and (7) wet extremely rich fens. Thermokarst pools are features of a permafrost landscape where permafrost thawing and subsequent ice melting creates an uneven topography of mounds, sinkholes, caverns, and lake basins.

In another study that attempted to relate vegetation directly to water chemistry, Gorham and Janssens (1992) investigated two families of mosses (Sphagnaceae and Amblystegiaceae) at 440 sites across northern North America (Fig. 13-19). They found a clear bimodal split in their occurrence, with Sphagnaceae found most often in low-pH peatlands (mode pH, 4.0–4.25) and Amblystegiaceae found in high-pH peatlands (mode pH, 6.76–7.0).

Black Spruce Peatlands

One of the dominant forested wetlands in the world is the black spruce peatland of the taiga of Canada and Alaska. Black spruce (*Picea mariana*), often growing in association with tamarack (*Larix laricina*), is the tree species most associated with forested peatlands in the boreal regions of North America (Post, 1996). These wetlands are estimated to encompass about half of the palustrine shrub–scrub wetlands in Alaska and cover an estimated 14 million ha in the state. Black spruce is mostly associated with ombrotrophic (bog) rather than minerotrophic (fen) communities. In bogs, it is found in associations with leatherleaf (*Chamaedaphne calyculata*), Labrador tea (*Ledum* spp.), laurel (*Kalmia latifolia*), blueberry (*Vaccinium* spp.), and bog

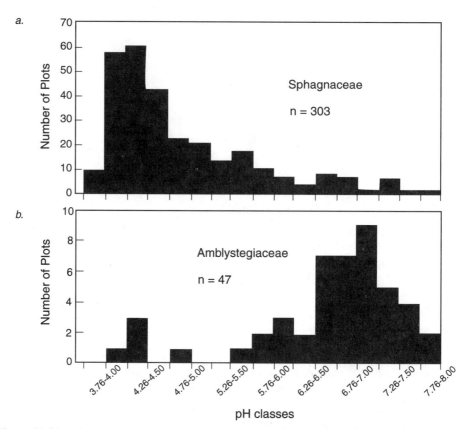

Figure 13-19 Distribution of two bryophyte families (Sphagnaceae and Amblystegiaceae) versus surface water pH for 440 peatland plots across North America. Plots were counted if they had at least one species of the family covering more than 25 percent of the total area. The bimodal pattern suggests a classification of peatlands based on moss vegetation. (*After Gorham and Janssens, 1992*)

rosemary (*Andromeda polifolia*). *Sphagnum* spp., of course, is found as ground cover in these bogs. In Alaska, common associations include *P. mariana* with *Vaccinium uliginonsum, Ledum groenlandicum,* and feathermoss (*Pleurozium schreberi*) and *P. mariana* with *Sphagnum* spp. and *Cladina* spp. (Post, 1996). In regions where permafrost is prevalent, black spruce wetlands often occur in paalsa hummocks.

Pocosin Vegetation

In contrast to the more northern peatlands, the woody vegetation of pocosins is dominated by evergreen trees and shrubs. Two broad community classes have been identified, and their presence was related to fire frequency, soil type, and hydroperiod. A *Pinus–Ericalean* (pine and heath shrub) community develops on deep organic soils with long hydroperiods and frequent fire. Three associations within this community are (1) pond pine (*Pinus serotina*) canopy with titi (*Cyrilla racemiflora*) and zenobia (*Zenobia pulverulenta*) shrubs, (2) pond pine and loblolly bay (*Gordonia lasianthus*) canopy with fetterbush (*Lyonia lucida*), and (3) pond pine canopy with titi and fetterbush shrubs.

A *conifer–hardwood* community type is found on shallow organic soils with slightly shorter hydroperiods. Two associations in this group are (1) pond pine canopy with titi, fetterbush, red maple (*Acer rubrum*), and black gum (*Nyssa sylvatica*) shrubs; and (2) pond pine and pond cypress (*Taxodium distictium* var. *nutans*) canopy with red maple, titi, fetterbush, and black gum shrubs.

Peat buildup in pocosins has a major effect on hydrology and nutrients. Fire is a recurring influence that has been proposed as a major control of plant succession. Pocosins are found on the deepest peats that are always saturated at depth. Roots do not penetrate these deep peats into the underlying mineral soils. As a result, nutrients are limited and growth is stunted (Otte, 1981; Sharitz and Gibbons, 1982). Shallow peat burns allow the regeneration of stunted pocosins, whereas deeper burns would lead to a nonpocosin community.

Vegetation Adaptations

The vegetation in bogs and peatlands both controls and is controlled by its physical and chemical environment. Some of the conditions for which adaptations are necessary in peatlands are discussed next.

Waterlogging

Many bog plants, in common with wetland vegetation in general, have anatomical and morphological adaptation to waterlogged anaerobic environments. These include (1) the development of large intercellular spaces (aerenchyma or lacunae) for oxygen supply, (2) reduced oxygen consumption, and (3) oxygen leakage from the roots to produce a locally aerobic root environment. These adaptations were discussed in Chapter 7. *Sphagnum,* conversely, is morphologically adapted to maintain waterlogging. The compact growth habit, overlapping leaves, and rolled branch leaves form a wick that draws up water and holds it by capillarity. These adaptations enable *Sphagnum* to hold water up to 15 to 23 times its dry weight (Vitt et al., 1975b).

Acidification of the External Interstitial Water

Sphagnum has the unique ability to acidify its environment, probably through the production of organic acids, especially polygalacturonic acids located on the cell walls (Clymo and Hayward, 1982). The galacturonic acid residues in the cell walls increase the cation exchange capacity to double that of other bryophytes (Vitt et al., 1975a). The adaptive significance of this peculiarity of *Sphagnum* is unclear. The acid environment retards bacterial action and, hence, decomposition, enabling peat accumulation despite low primary production rates. It has been suggested that the high cation exchange capacity also enables the plant to maintain a higher and more stable pH and cation concentration in the living cells than in the surrounding water (Glaser, 1987).

Nutrient Deficiency

Many bog plants have adaptations to the low nutrient supply that enable them to conserve and accumulate nutrients. Adaptations seen in bog plants include evergreenness; sclerophylly, or the thickening of the plant epidermis to minimize grazing; uptake of amino acids; and high root biomass (Bridgham et al., 1996). Some bog plants, notably cotton grass (*Eriophorum* spp.), translocate nutrients back to perennating organs prior to litterfall in the autumn. These nutrient reserves are available for the following year's growth and seedling establishment. The roots of other bog plants penetrate deep into peat zones to bring nutrients to the surface. Bog litter has been demonstrated to release potassium and phosphorus, often the most limiting nutrients, more rapidly than other nutrients, an adaptation that keeps these nutrients in the upper layers of peat (Moore and Bellamy, 1974). Many ericaceous plants have adapted to low concentrations of nitrogen by effectively utilizing ammonium nitrogen in place of limited nitrate nitrogen under low-pH conditions, by efficiently using nitrogen and even by utilizing organic nitrogen sources (Bridgham et al., 1996).

Another well-known adaptation to nutrient deficiency in bogs is the ability of carnivorous plants to trap and digest insects. This special feature is seen in several unique insectivorous bog plants, including the pitcher plant (*Sarracenia purpurea;* Fig. 13-20) and sundew (*Drosera* spp.). A nutrient limitation study developed for *Sarracenia* in Minnesota showed, that although nutrient and insect additions did not increase biomass, there were respective nutrient increases in the leaves of the plant. A nutrient budget was developed and it was estimated that insect capture accounts for approximately 10 percent of the plant's nitrogen and phosphorus needs (Chapin and Pastor, (1995).

Some bog plants also carry out symbiotic nitrogen fixation. The bog myrtle (*Myrica gale*) and the alder develop root nodules characteristic of nitrogen fixers and have been shown to fix atmospheric nitrogen in bog environments.

Overgrowth by Peat Mosses

Many flowering plants are faced with the additional problem of being overgrown by peat mosses as the mosses grow in depth and in area covered. Adapting plants must

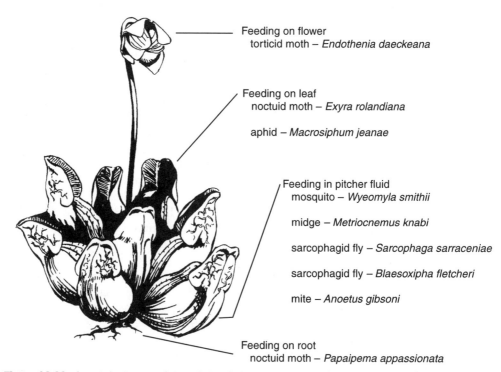

Figure 13-20 Invertebrate associates of the pitcher plant (*Sarracenia purpurea*). (*After Damman and French, 1987*)

raise their shoot bases by elongating their rhizomes or by developing adventitious roots. Trees such as pine, birch, and spruce are often severely stunted because of the moss growth and poor substrate; they grow better on bogs where the vertical growth of moss has ceased.

Consumers

Mammals

The populations of animals in bogs are generally low because of the low productivity and the unpalatability of bog vegetation. Animal density is closely related to the structural diversity of the peatland vegetation. For example, forested peatlands tend to support the greatest number of small-mammal species, especially close to upland habitats (Stockwell and Hunter, 1985). Table 13-4 lists small mammals associated with peatlands in northern Minnesota. Large mammals tend to roam over larger landscapes and are, thus, not specific to individual peatland types. In northern Minnesota and New England, moose (*Alces alces*) are frequently found in small peatlands. White-tailed deer (*Odocoileus virginianus*) browse heavily in white cedar bogs in winter. Black bear (*Ursus americanus*) use peatlands for escape cover and for food. The woodland caribou (*Rangifer tarandus*) was the largest mammal that was largely restricted to peatlands, but it disappeared from Minnesota in 1936 (Nelson, 1947) probably as a result of hunting pressure. Smaller mammals closely associated with peatlands are beaver (*Castor canadensis*), lynx (*Lynx canadensis*), fishers (*Martes pennant*), and snowshoe hares (*Lepus americanus*). The beaver is a fairly recent import into Minnesota peatlands. It moved in along drainage ditches, seldom penetrating deep into large peatlands, but it has had a significant effect on peatland flooding in northern Minnesota (Naiman et al., 1991). Wet forests are becoming the only habitats where wide-ranging mammals such as the black bear, otter, and mink are found (Sharitz and Gibbons, 1982; L. D. Harris, 1989). This is not so much because peatlands are obligate habitats but because the clearing of upland forests has forced the remaining population into the remaining large tracts of forested wetlands.

Amphibians and Reptiles

Glaser (1987) reported only seven species of amphibians and four species of reptiles in northern Minnesota peatlands. Acid waters below pH 5 appear to be the major limiting factor in their ability to colonize bogs.

Birds

Many bird species are seen in peatlands during different times of the year. For example, Warner and Wells (1980) reported 70 species during the breeding season. Most of these are also common on upland sites, but a few depend on peatlands for survival (Glaser, 1987). These include the Sandhill Crane (*Grus canadensis*), Great Gray Owl (*Strix nebulosa*), Short-eared Owl (*Asio flammeus*), Sora (*Porzana carolina*), and Sharp-tailed Sparrow (*Ammospiza caudacuta*). In New England, Damman and French (1987) described the distribution of typical bird species in different habitat types of a lake-border bog (Fig. 13-21). As one moves from the Canadian border south, the species change, but the new species have analogous positions along the gradient.

Invertebrates in Pitcher Plants

Pitcher plants are interesting for the unique fauna associated with different parts of the plant. Pitcher plants (*Sarracenia* spp.) are obligate host to more invertebrate

Table 13-4 **Presence of small mammals in general peatland types in northern Minnesota**

Species	Fen	Swamp Thicket	Swamp Forest	Forested Bog	Open Bog	Adjacent Upland
Masked shrew (*Sorex cinereus*)	4	4	4	4	4	4
Water shrew (*Sorex palustris*)	2				2	
Arctic shrew (*Sorex articus*)	4	4	1–4		1	1
Pygmy shrew (*Sorex hoyi*)	2–4	3	2–3	3	2	3
Short-tailed shrew (*Blarina brevicauda*)	2–4	4	3–4	2	1	4
Star-nosed mole (*Condylura cristata*)		2	0–4			
Eastern chipmunk (*Tamias striatus*)	0–1		0–1			4
Least chipmunk (*Eutamias minimus*)	0–1		0–2			3
Franklin ground squirrel (*Spermophilus franklinii*)	0–1				1	1
Red squirrel (*Tamiasciurus hudsonicus*)	0–1	1	4	4		4
Northern flying squirrel (*Glaucomys sabrinus*)			0–2			3
Deer mouse (*Peromyscus maniculatus*)			3–4	1	1	4
White-footed mouse (*Peromyscus leucopus*)	1	2	2–3			4
Southern red-backed vole (*Clethrionomys gapperi*)	4	4	4	4	4	4
Heather vole (*Phenacomys intermedius*)[a]						
Meadow vole (*Microtus pennsylvanicus*)	2–4	4	1–3	1	4	1
Southern bog lemming (*Synaptomys cooperi*)			0–4	0–4	2	
Northern bog lemming (*Synaptomys borealis*)	0–1				2	
Meadow jumping mouse (*Zapus hudsonius*)	2	3	0–3		1	3
Least weasel (*Mustela nivalis*)[a]						

Key:
4—characteristic   2—occasional   0 or blank—not found
3—frequent   1—occurred   [a]—reported to occur in peatlands
Source: Glaser (1987), Nordquist and Birney (1980), and Minnesota Department of Natural Resources (1984).

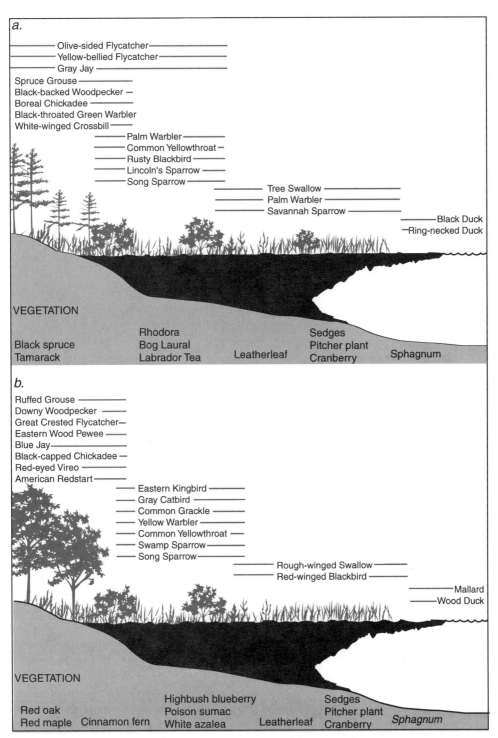

Figure 13-21 Comparison of bird distribution, typical of a lake-border bog in the northern and southern parts of the northeastern United States. (*After Damman and French, 1987*)

species than any other bog plant (Rymal and Folkerts, 1982; Fig. 13-20). In the water-filled pitcher are found a mosquito, a midge, two sarcophagid flies, and a mite. An aphid and three moths feed exclusively on the tissue. Other insects are associated with other parts of the plant.

Ecosystem Function

The dynamics of peatlands reflect the realities of the harsh physical environment and the scarcity of mineral nutrients. These conditions result in several major features:

1. Bogs are systems of low primary productivity; fens are generally more productive; *Sphagnum* mosses often dominate bogs and other vegetation is stunted in growth.

2. Bogs and fens are peat producers whose rates of accumulation are controlled by a combination of complex hydrologic, chemical, and topographic factors. This peat contains a great store of nutrients, most of it below the rooting zone and, thus, unavailable to plants.

3. Low-nutrient peatlands in cold climates have developed several unique pathways to obtain, conserve, and recycle nutrients. The amount of nutrients in living biomass is small. Cycling is slow because of the low temperatures, the nutrient deficiency of the litter, and the waterlogging of the substrate. It is more active when peat production stagnates and when bogs receive increased nutrient inputs.

Primary Productivity

Major organic inputs to bog systems come from the primary production of the vascular plants, liverworts, mosses, and lichens. Little is known about algal productivity. Allochthonous organic carbon inputs can be significant in small bogs, but their importance decreases with the size of the peatland. Among vascular plants, ericaceous shrubs and sedges are the most important primary producers, and much of this production is below ground (Dennis and Johnson, 1970; Svensson and Rosswall, 1980). Mosses, especially sphagnum, account for one-third to one-half of the total production (Forrest and Smith 1975; Grigal, 1985).

Bogs and fens are usually less productive than most other wetland types and are generally less productive than the climatic terrestrial ecosystems in their region. Moore and Bellamy (1974) described the productivity of a forested *Sphagnum* bog as about half that of a coniferous forest and a little more than a third that of a deciduous forest. Table 13-5 summarizes many measurements and ranges of peatland biomass and productivity. According to Pjavchenko (1982), forested peatlands produce a range of 260 to 400 g organic matter m^{-2} yr^{-1}, with the low value that of an ombrotrophic bog and the high value that of a minerotrophic fen. Malmer (1975) cited a typical range of 400 to 500 g m^{-2} yr^{-1} for nonforested, raised (ombrotrophic) bogs in western Europe. In contrast, Lieth (1975) estimated the net primary productivity in the boreal forest to average 500 g m^{-2} yr^{-1} and that in the temperate forest to average 1,000 g m^{-2} yr^{-1}. The estimate for boreal forests probably includes bog forests as well as upland forests.

Thormann and Bayley (1997) investigated the above-ground net primary productivity of a bog, fens, and marshes in the southern boreal region of western Canada using similar field techniques (Fig. 13-22). They found that, when all above-ground strata were combined, the productivity of the bog, which was dominated by the moss *Sphagnum fuscum*, was 390 g m^{-2} yr^{-1}, about the same as the productivity measured in three fen sites (277–409 g m^{-2} yr^{-1}), a riverine marsh (323 g m^{-2} yr^{-1}), and a lacustrine marsh (757 g m^{-2} yr^{-1}). The authors concluded that, particularly for the fens and marshes, the absence of below-ground productivity measurements probably greatly underestimated the overall productivity of these systems much more than that of the bog dominated by mosses.

Moss Productivity

The measurement of the growth or primary productivity of *Sphagnum* mosses presents special problems not encountered in productivity measurements of other plants (Clymo, 1970; Moore and Bellamy, 1974). The upper stems of the plant elongate, and the lower portions gradually die off, become litter, and eventually form peat. It is difficult to measure the sloughing off of dead material to litter. It is equally hard to measure the biomass of the plant at any one time because it is difficult to separate the living and dead material of the peat. The following two methods for measuring *Sphagnum* growth give comparable results: (1) the use of "innate" time markers such as certain anatomical or morphological features of the moss and (2) the direct measurement of changes in weight. Growth rates for *Sphagnum* determined by these two techniques generally fall in the range of 300 to 800 g m^{-2} yr^{-1} (Moore and Bellamy, 1974).

Table 13-6 summarizes production data from a number of sources on three species of *Sphagnum* (Rochefort et al., 1990), arranged by decreasing latitude. Although Damman (1979) and Wieder and Lang (1983) suggested that annual production should increase with decreasing latitude, only *S. magellanicum* showed such a trend. Evidently, local and regional factors are more important than latitude. For example, production varied from year to year by a factor of 3 in one 4-year study (Rochefort et al., 1990).

Decomposition

The accumulation of peat in bogs is determined by the production of litter (from primary production) and the destruction of organic matter (decomposition). As with primary production, the rate of decomposition in peat bogs is generally low because of (1) waterlogged conditions, (2) low temperatures, and (3) acid conditions. In fact, the accumulation of peat in peatlands is due more to slow decomposition processes than to net community productivity. Besides leading to peat accumulation, slow decomposition leads to slower nutrient recycling in an already nutrient-limited system.

The pattern of *Sphagnum* decomposition with depth in a bog in southern England is shown in Figure 13-23. The decomposition rate is highest near the surface, where aerobic conditions exist. By 20 cm depth, the rate is about one-fifth of that at the surface. This is caused by anaerobic conditions, as illustrated by the sulfide production curve, also shown in Figure 13-23. Clymo (1965) attributed the bulk of the organic decomposition that does occur in peat bogs to microorganisms, although the total numbers of bacteria in these wetland soils are much fewer than in aerated soils. As pH decreases, the fungal component of the decomposer food web becomes more important relative to bacterial populations.

Table 13-5 Biomass and net primary productivity of selected northern bogs and other peatlands

Location	Type of Peatland	Living Biomass (g dry wt/m^2)	Net Primary Productivity (g dry wt m^{-2} yr^{-1})	Reference
EUROPE				
Western Europe	General nonwooded raised bog	1,200	400–500	Malmer (1975)
Western Europe	Forested raised bog	3,700	340	Moore and Bellamy (1974)
Russia	Eutrophic forest bog	9,700–11,000	400	Pjavchenko (1982)
	Mesotrophic forest bog	4,500–8,900	350	Pjavchenko (1982)
	Oligotrophic forest bog	2,200—3,600	260	Pjavchenko (1982)
Russia	Mesotrophic (transition) *Pinus–Sphagnum* bog	8,500	393	Bazilevich and Tishkov (1982)
England	Blanket bog (*Calluna–Eriophorum–Sphagnum*)		659 ± 53[a]	Forrest and Smith (1975)
England	Blanket bog		635	Heal et al. (1975)
Ireland	Blanket bog		316	Doyle (1973)
NORTH AMERICA				
Michigan	Rich fen (*Chamaedaphne–Betula*)		341[b]	Richardson et al. (1976)
Minnesota	Forested peatland	15,941	1,014[b]	Reiners (1972)
	Fen forest	9,808	651[b]	Reiners (1972)
	Forested parched bog	10,070 (above-ground)	360	Grigal et al. (1985)
	Forested raised bog	3,100 (above-ground)	300	Grigal et al. (1985)

Location	Wetland type		Value	Reference
Manitoba	Peatland bog		1,943	Reader and Stewart (1972)
			—	
Alberta	Bog		280[b]	Szumigalski and Bayley (1996a)
	Poor fen		310[b]	
	Moderately rich fen		360[b]	
	Lacustrine sedge fen		214[b]	
	Extremely rich fen		245[b]	
Alberta	Bog		390[b]	Thormann and Bayley (1997)
	Floating sedge fen		356[b]	
	Lacustrine sedge fen		277[b]	
	Riverine sedge fen		409[b]	
Quebec	Poor fen		114[b]	Bartsch and Moore (1985)
	Rich fen		335[b]	
	Transitional fen		176[b]	
GENERAL				
	Northern bog marshes (Does not include bog forests or ombrotrophc bogs)	Above-ground	101—1,026[c]	Reader (1978)
		Below-ground	141—513[d]	Reader (1978)

[a] Mean ± standard deviation for seven sites.
[b] Above-ground only.
[c] Range for nine bog marshes.
[d] Range for five bog marshes.

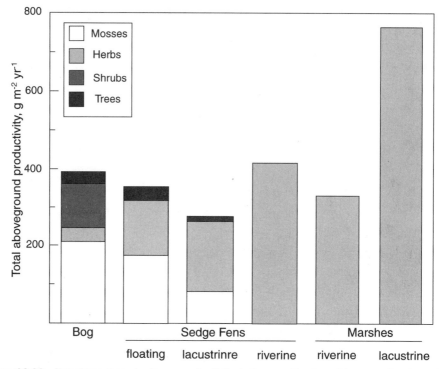

Figure 13-22 Net above-ground primary productivity in four peatlands and two marshes in southern boreal Alberta, Canada. (*After Thormann and Bayley, 1997*)

Chamie and Richardson (1978) described rates of decomposition of plant material from a rich fen in central Michigan. Weight losses for leaves of sedge, willow, and birch were about 36 to 37 percent after 1 year ($k = 0.45–0.46$ yr^{-1}); the ericaceous leatherleaf (*Chamaedaphne* sp.) leaves and stems decomposed more slowly, showing a 16 percent loss in 1 year ($k = 0.17$ yr^{-1}). This decomposition rate is one-half to one-quarter the rates found either in more aerobic environments or at lower latitudes. Bridgham et al. (1991) compared the decay rate in pocosins in North Carolina using the cotton-strip decay technique of Latter and Howson (1977), Hill et al. (1985), and Harrison et al. (1988). They found that decay was 2 to 10 times faster in disturbed peatlands than in natural ones and that the fastest decomposition occurred in a high-nutrient, neutral-pH, and low-soil-moisture site. The removal of waterlogged conditions and subsequent anaerobic conditions was the dominant factor that explained faster decomposition rates. In a comparison of peatlands and mineral soil swamps in the United States and the Netherlands, Verhoeven et al. (1994) used the same cotton-strip method and found substantially lower decay rates in ombrogenous bogs in both countries compared to other peatlands and mineral soil wetlands. Total phosphorus (positive correlation) and soil organic matter (negative correlation) explained 75 percent of the decay rates. Thus, low nutrients and high organic matter (which keeps the soils reduced) were argued as having contributed to the low decay rates in the bogs. Szumigalski and Bayley (1996b) found the following progression of rates of decay of litter from peatlands in central Alberta: *Carex* > *Betula* > mosses. The highest decomposition rates were with plant material with the highest nitrogen

Table 13-6 Comparison of selected data on production of *Sphagnum* species in order of decreasing latitude

Species[a]	Growth (mm/yr)	Production (g m⁻² yr⁻¹)	Latitude (N)	Location	Mean Annual Precipitation (mm)	Mean Annual Temperature (°C)	Source
fus	1.4–3.2	70	68°22'	N Sweden	600	2.9	Rosswall and Heal (1975)
fus	—	250	63°09'	S Finland	532	3.5	Silvola and Hanski (1979)
fus	7–16	220–290	63°09'	S Finland	532	3.5	K. Tolonen, in Rochefort et al. (1990)
fus	9.5	195	60°62'	S Finland	632	4–4.8	Pakarinen (1978)
mag	14.7	70	59°50'	S Norway	1,250	5.9	Pedersen (1975)
ang	9.8	500	—	—	—	—	
fus	7.8	90	56°05'	S Sweden	800	7.9	Damman (1978)
mag	10–18	100	—	—	—	—	
mag	28–34	50–100	55°09'	England	1,270	9.3	S. B. Chapman (1965)
ang	14–15	110–240	54°46'	England	1,980	7.4	Clymo and Reddaway (1971)
mag	—	230	54°46'	England	1,980	7.4	Forrest and Smith (1975)
ang		240–330	—	—	—	—	
ang	38–43	110–440	54°46'	England	1,980	7.4	Clymo (1970)
fus	6–7	75–83	54°43'	Quebec	791	4.9	Bartsch and Moore (1985)
ang	4–17	19–127	—	—	—	—	T. R. Moore (1989)
fus	30	270	54°28'	England	1,375	7.4	Bellamy and Rieley (1967)
fus	35–51	424–801	54°20'	N Germany	714	8.4	Overbeck and Happach (1957)
mag	120–160	252–794	—	—	—	—	
ang	—	488–1,656	—	—	—	—	
fus	17–24	50	49°53'	S Manitoba	517	2.5	Reader and Stewart (1971)
fus	7–31	240	49°52'	NE Ontario	858	0.8	Pakarinen and Gorham (1983)
fus	11–34	69–303	49°40'	NW Ontario	714	2.6	Rochefort et al. (1990)
mag	20–39	52–240	—	—	—	—	
ang	62	97–198	—	—	—	—	
mag		540	39°07'	West Virginia	1,330	7.9	Wieder and Lang (1983)

[a] fus, *Sphagnum fuscum*; mag, *S. magellanicum*; ang, *S. angustifolium*.
Source: Rochefort et al. (1990).

459

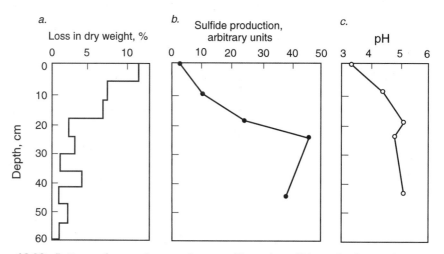

Figure 13-23 Patterns of **a.** sphagnum decomposition, **b.** sulfide production, and **c.** pH with depth in an English bog. Decomposition is during a 103-day period. (*After Clymo, 1965*)

content. Using a standard litter material of *Carex lasiocarpa*, litter losses were in the following order: poor fen > wooded rich fen > bog > open rich fen > sedge fen.

There has been considerable speculation about factors that give rise to patterned peatlands. The pattern of strings and flarks or hummocks and hollows, for example, appears to be related to differential rates of peat accumulation. Rochefort et al. (1990) determined that differential accumulation in a poor-fen system in northwestern Ontario, Canada, was caused more by differences in peat decomposition rates than by differences in primary production rates. They found that, even though the production rates of *Sphagnum* in hummocks were generally about equal to the rates in hollows or even lower than the rates in minerotrophic hollows, hummock species had slower decomposition rates than those of hollow species. As a result, peat accumulated faster on hummocks than in hollows, and hummocks may be expanding at the expense of hollows.

Mineralization

Closely related to the process of decomposition is the mineralization of organic material to its basic elements (e.g., inorganic carbon, phosphorus, and nitrogen). Verhoeven et al. (1994), in a comparison of nitrogen mineralization and phosphorus release rates in Dutch and American wetlands, found much lower rates in bogs compared to other peatlands and mineral soil wetlands. Yet several studies have investigated the relative availability of nutrients in peatland and have found some bogs or bog species such as *Sphagnum* to have higher nutrient availability than some fens or fen species such as *Carex* (Waughman, 1980; Verhoeven et al., 1990; Koerselman et al., 1993). Waughman (1980) found concentrations of NH_4^+ and soluble PO_4^{3-} to be greater in bogs than in fens in Germany. Bridgham et al. (1998) found high rates of carbon, nitrogen, and phosphorus turnover in ombrotrophic bogs compared to other wetlands and concluded that nutrient turnover and availability are not necessarily lower in bogs and higher in fens and other wetlands across the ombrotrophic–minerotrophic gradient. Ombrotrophic bogs have adapted to low nutrient availability and low soil bulk density (the plant roots encounter fewer nutrients because there is not as much soil per unit volume) by turning over what organic nutrients they do have more rapidly.

Peat Formation

The vertical accumulation rate of peat in bogs and fens is generally thought to be between 20 and 80 cm/1,000 yr in European bogs (Moore and Bellamy, 1974), although Cameron (1970) gave a range of 100 to 200 cm/1,000 yr for North American bogs and Nichols (1983) reported an accumulation rate for peat of 150 to 200 cm/1,000 yr in warm, highly productive sites. Malmer (1975) described a vertical growth rate of 50 to 100 cm/1,000 yr as typical for western Europe. Assuming an average density of peat of 50 mg/mL, this rate is equivalent to a peat accumulation rate of 25 to 50 g m^{-2} yr^{-1}. This range compares reasonably well with the accumulation rate of approximately 100 g m^{-2} yr^{-1} (Fig. 13-24b) measured for a transition forested bog in the former Soviet Union by Bazilevich and Tishkov (1982) (see Energy Flow next section). In comparison, Hemond (1980) estimated a rapid accumulation rate of 430 cm/1,000 yr, or 180 g m^{-2} yr^{-1}, for Thoreau's Bog, Massachusetts, close to the 170 g m^{-2} yr^{-1} rate determined from Lindeman's model (Fig. 13-24a) in that same section.

Durno (1961) compared rates of peat growth in vertical sections of peat in England and related them to climatic periods (Table 13-7). The growth of peat was rapid (110 cm/1,000 yr) during the Boreal period, but it slowed down considerably in the wetter Atlantic climate (14–36 cm/1,000 yr). This was not expected, except that, at the time it happened, the peatland was transforming itself from a minerotrophic fen to an ombrotrophic bog; the slower peat accumulation was, thus, probably based on low productivity during the bog stage. The surface layer, produced in the cool, moist Sub-Atlantic period, had a higher rate of peat accumulation (48–96 cm/1,000 yr), but Durno (1961) suggested that this layer might subsequently be subjected to compression.

Energy Flow

Bog ecosystems have energy flow patterns similar to those in many other wetlands although the magnitudes of the flows are reduced. Low temperatures, flooding, and the chemical conditions of anaerobic sediments limit primary production (input) and slow decomposition (output). The net result is that inputs, though low, generally exceed outputs and there is a buildup of peat.

Although few detailed studies of energy flow have been developed for bog ecosystems, one of the earliest energy budgets for any ecosystem was determined in a now classical study by Lindeman (1942) of Cedar Bog Lake, a small bog in northern Minnesota (Fig. 13-24a). Although this energy budget is crude, the main features have stood the test of time. Very little of the incoming radiation (< 0.1 percent) is captured in photosynthesis. The two largest flows of organic energy are to respiration (26 percent) and to storage as peat (70 percent). Energy flow in the simplified food web is primarily to herbivores (13 percent), and about 3.5 percent goes to decomposers. As the following two more recent budget measurements show, the peat storage term is exceedingly high, and decomposition losses are probably underestimated.

Heal et al. (1975) described the energy flows and storage of an English blanket bog. Net primary productivity (635 g m^{-2} yr^{-1}) primarily stemmed from *Sphagnum*

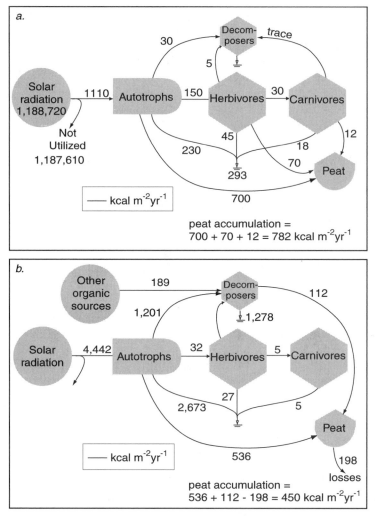

Figure 13-24 Diagrams of the energy flow in **a.** Cedar Bog Lake, Minnesota, and **b.** a Russian transition peatland. Flows in kcal m^{-2} yr^{-1}. Flows in (a) were originally published in calories whereas flows in (b) were published in gram dry weight and converted to energy as 4.5 kcal g^{-1}. (*a after Lindeman, 1942; b after Bazilevich and Tishkov, 1982; Alexandrov et. al., 1994*)

Table 13-7 Rates of peat growth (depth) for four bogs in southern Pennines region, England, for different climatic periods

Climatic Period	Age of Peat (years)	Growth Rate (cm/1,000 yr)
Sub-Atlantic	2,500	48–96
Sub-Boreal	5,000	12–48
Atlantic	7,200	14–36
Boreal	8,900	111

Source: Durno (1961).

mosses (300 g m^{-2} yr^{-1}). In contrast to Lindeman's results, only 1 percent of the productivity was consumed in the grazing food web, primarily by psyllid and tipulid flies. The primary energy flow was through the detrital food web, where about 85 percent was decomposed by microflora and 10 percent was accumulated as peat beneath the water table. This system showed a relatively low amount of peat buildup during the period of measurement.

Bazilevich and Tishkov (1982) and Alexandrov et al. (1994) presented a detailed breakdown of energy flow through a mesotrophic (transition) bog in the European region of Russia (Fig. 13-24b and Table 13-8). The bog is a sphagnum–pine (*Sphagnum girgensohnii–Pinus sylvestris*) community containing shrubs such as bilberry (*Vaccinium myrtillus*). The total energy stored in the bog was estimated to be in excess of 137 kg dry organic matter/m^2, with dead organic matter (to a depth of 0.6 m of peat) accounting for 94 percent of the storage. Living biomass was 8.5 kg/m^2, or about 6 percent of the organic storage. Gross primary productivity was 987 g m^{-2} yr^{-1}, or about 4,400 kcal m^{-2} yr^{-1} (assuming 1 g organic matter = 4.5 kcal), with about 60 percent consumed by plant respiration. The distribution of the net primary production came from trees (39 percent), algae (28 percent), shrubs (21 percent), mosses and lichens (9 percent), and grasses (3 percent). The net annual primary production, combined with other abiotic and biotic energy flows, was primarily

Table 13-8   Balance of energy inputs and outputs of mesotrophic bog in Russia

Description of Energy Flow	Energy Flow (g m^{-2} yr^{-1})	Percentage of Total Input
INPUTS		
Gross primary productivity	987	95.9
Precipitation and runoff	34	3.3
Subsurface flow	6	0.6
Biotic input from other ecosystems	2	0.2
Total inputs	1,029	100.0
OUTPUTS		
Plant respiration	594	57.7
Phytophage respiration	6	0.6
Carnivore respiration	1	0.1
Decomposer respiration	284	27.6
Consumption from other ecosystems	4	0.4
Abiotic oxidation	15	1.5
Subsurface outflows	5.5	0.5
Surface outflows	19.5	1.9
Total outputs	929	90.3
ACCUMULATION		
Peat formation	86	8.3
Growth retention	14	1.4
Total accumulation	100	9.7

Source: Bazilevich and Tishkov (1982) and Alexandrov et al. (1994).

consumed by decomposers. Much less was consumed in grazing food webs. Net accumulation of peat was 100 g m^{-2} yr^{-1} (or about 450 kcal m^{-2} yr^{-1}). Losses other than biotic decomposition, which accounted for most of the loss of organic matter, were chemical oxidation and surface and subsurface flows.

Comparison of these two energy budgets from Russia and the United States (Fig. 13-24), carried out on separate continents and several decades apart, illustrates several points. First, Lindeman's Cedar Bog was approximately one-fourth as productive as Bazilevich and Tishkov's Russian peatland, a possibility, given that the Russian site was described as transitional between a bog and a fen. Second, the American bog accumulated more peat than did the Russian peatland, results that could reflect either the sophistication of the measuring techniques for the time or what happens to peatlands as they transition from fens to true bogs. Lindeman shows a high 71 percent of the gross primary productivity stored permanently as peat, whereas Bazilevich and Tishkov found only 10 percent of the gross primary productivity stored as peat. The former high number is unlikely. Finally, assuming 50 g/L as the density of peat, Lindeman's peat accumulation results in a very high 350 cm/$1,000$ yr, whereas the Russian study has a more reasonable 200 cm/$1,000$ yr. North American and European studies generally bracket a range of 50 to 200 cm/$1,000$ yr (see Peat Formation earlier in this chapter).

Nutrient Budgets

There have been a number of studies on the nitrogen retention capacity of natural bogs and fens because they generally have either no outflows or simple outflows and rely, to a great extent, on inputs from precipitation (Johnston, 1991). Although peatlands are usually anaerobic, denitrification has not generally been considered a major pathway for nitrogen loss in these systems, at least in studies in Alaska, Massachusetts, and Minnesota (respectively, Barsdate and Alexander, 1975; Hemond, 1980; Urban and Eisenreich, 1988). A multiple-year study by Kadlec (1983) demonstrated that a peatland in Michigan that received wastewater was consistently a sink for nitrogen (75–81 percent sequestered) but began to export phosphorus after several years of phosphorus retention. In one of the longest continuously operating projects with wetlands receiving nutrients in the world, a fen-type peatland in Michigan dominated by sedge–willow and leatherleaf–bog birch communities has been subjected to treated wastewater application since 1975 and as a full-scale system since 1978 (Kadlec and Knight, 1996). The area affected by the nutrient influx in this large peatland stabilized at approximately 80 ha after about 20 years (see Chapter 20). Beyond a zone of about $1,000$ m downstream, phosphorus and nitrogen concentrations in the water were reduced by about 97 percent to background levels (see Chapter 20).

Crisp (1966) developed an overall nutrient budget for nitrogen, phosphorus, sodium, potassium, and calcium for a blanket bog–dominated watershed in the Pennines, England (Table 13-9). The only input considered was precipitation, which has relatively high concentrations of dissolved materials because of the proximity of the bog to the sea. The author found that the outputs of nutrients greatly exceeded the inputs from precipitation. This finding was partially the result of the omission of any estimations of input caused by the weathering of the parent rock in the watershed. The budget does illustrate that the erosion of peat results in a major loss of nitrogen, exceeding the input of nitrogen by precipitation. The erosional losses of phosphorus, calcium, and potassium are 50 percent or more of the input by precipitation. Major

Table 13-9   **Nutrient budget for blanket bog watershed in Pennines region of England**

	Nutrient Flow (g m^{-2} yr^{-1})				
	Na$^+$	K$^+$	Ca^{2+}	P	N
Inflow					
Precipitation	2.5	0.3	0.9	0.07	0.8
Output					
Sale of sheep	0.02	0.005	0.02	0.01	0.005
Dissolved material in stream	4.5	0.9	5.4	0.04	0.3
Erosion of peat in stream	0.03	0.2	0.5	0.04	1.5
Net loss	2.0	0.8	5.0	0.02	1.0

Source: Crisp (1966).

outputs of calcium, sodium, and potassium in dissolved form illustrate that this budget is incomplete without the inclusion of weathering inputs. Sheep grazing and harvesting and the stream drift of organisms were insignificant losses from the system.

Richardson et al. (1978) presented some nutrient budgets for a fen in central Michigan. The peat layer, measured to only 20 cm, represented more than 97 percent of nutrient storage in the fen. The uptake of nutrients by plants and litterfall was generally very low, lower, for example, than in the blanket bog described previously. Although the results of this Michigan study are not representative of a typical fen and do not represent a complete budget, a comparison of available and total storages shows a general peatland phenomenon: Available nutrients are a small percentage of the total nutrients stored in peat.

Nitrogen budgets for Thoreau's Bog (Hemond, 1983) and for a perched raised-bog complex (Urban and Eisenreich, 1988) make an interesting comparison (Fig. 13-25). Although total nitrogen input is comparable in both systems, the Minnesota perched-bog system catches some runoff from surrounding uplands, whereas nitrogen fixation is the largest source of biologically active nitrogen in Thoreau's Bog. Otherwise, the budgets are remarkably similar despite the differences in bog type and location. Both accumulate nitrogen in peat and lose a significant portion through runoff. Denitrification is an uncertain term. Peatlands appear to have the capacity for denitrification (Hemond, 1983), but the magnitude *in vivo* is uncertain.

Carbon budgets for peatlands have drawn a great deal of interest, given the importance of these ecosystems in global carbon dynamics. It is accepted that boreal peatlands were once carbon sinks (Gorham, 1991; Mitsch and Wu, 1995), but there is little consensus that they are contemporary sinks (Rivers et al., 1998). Carbon budgets have been developed for small peatlands (Carroll and Crill, 1997; Waddington and Roulet, 1997) and for substantially sized peatland-dominated watersheds (Rivers et al., 1998). The latter, a 1,500-km^2 watershed in the Lake Agassiz peatlands in Minnesota, illustrated that the peat watershed had a net carbon storage of 12.7 g C m^{-2} yr^{-1} but that there was a tenuous balance between the watershed being a source and a sink of carbon. Inflows of carbon are groundwater, precipitation, and net community productivity, whereas outflows are groundwater and surface flow and outgassing of methane (Fig. 13-26). It was estimated from a companion study (Glaser et al., 1997b) that peat is accumulating at a rate of 1 mm/yr (100 cm/1,000 yr).

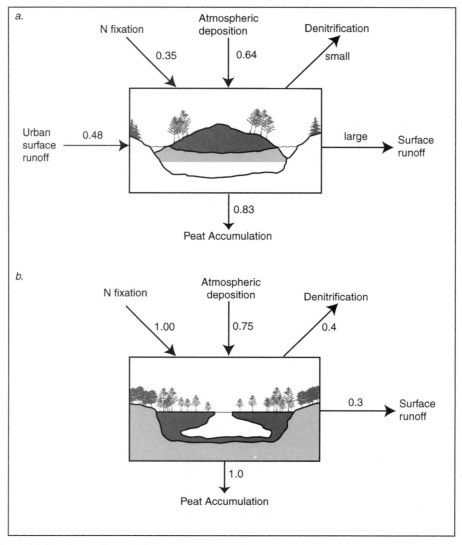

Figure 13-25   Comparison of nitrogen budgets for two northern ombrotrophic bogs: **a.** perched raised-bog complex in northern Minnesota and   **b.** small floating-mat sphagnum bog (Thoreau's Bog) in Massachusetts. Values are in g N m^{-2} yr^{-1}. (*Data from Urban and Eisenreic, 1988; Hemond, 1983*)

This budget illustrates the importance of accurate hydrologic measurements as well as biological productivity measurements in determining accurate nutrient budgets.

Ecosystem Models

Relatively few ecosystem models have been developed for northern peatlands, despite the abundance of research on those ecosystems. Conceptual models such as those by Heal et al. (1975) for an English blanket bog, Lindeman (1942) for Cedar Bog in Minnesota, and Bazilevich and Tishkov (1982) and Alexandrov et al. (1994) for a

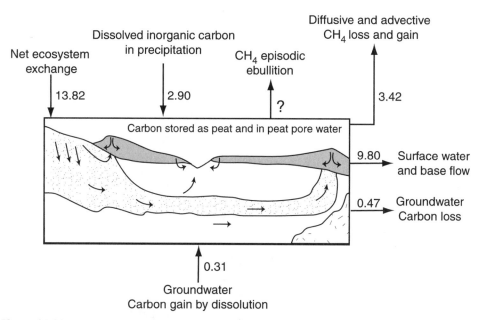

Figure 13-26 Carbon dynamics of the 1,500-km^2 Rapid River watershed in the Lake Agassiz peatland basin of northern Minnesota. Fluxes are in g m^{-2} yr^{-1}. (*After Rivers et al., 1998*)

transition bog in Russia (see Fig. 13-24) have been useful for understanding peatland dynamics. Logofet and Alexandrov (1984, 1988) simulated carbon and nitrogen flux through a simple multicompartment model of a dwarf shrub–*Sphagnum* mesotrophic bog in the former Soviet Union. Their simulation suggested that the assimilated organic matter is almost entirely lost through plant (62 percent) and microorganism (26 percent) respiration. About 7.4 percent of assimilated organic matter and 20 percent of allochthonous material were deposited as peat. Organic material had a total residence time of about 26 years in the system (8.5 years in plant material and 17 in other compartments). These slow turnover rates are in sharp contrast with rates in most temperate wetlands.

Recommended Readings

Bridgham, S. D., J. Pastor, J. A. Janssens, C. Chapin, and T. J. Malterer. 1996. Multiple limiting gradients in peatlands: A call for a new paradigm. Wetlands 16:45–65.

Glaser, P. H., D. I. Siegel, E. A. Romanowicz, and Y. P. Shen. 1997. Regional linkages between raised bogs and the climate, groundwater, and landscape of north-western Minnesota. Journal of Ecology 85:3–16.

Moore, P. D., ed. 1984. European Mires. Academic Press, London. 367 pp.

Moore, P. D., and D. J. Bellamy. 1974. Peatlands. Springer-Verlag, New York. 221 pp.

Freshwater Swamps

This chapter covers mineral soil–dominated, freshwater forested wetlands. Deepwater swamps, dominated by bald cypress–tupelo (Taxodium distichum–Nyssa aquatica*) and pond cypress–black gum* (Taxodium distichum *var.* nutans–Nyssa sylvatica *var.* biflora*) associations and found throughout the Coastal Plain and Mississippi embayment of the southeastern United States, can occur under nutrient-poor conditions (cypress domes and dwarf cypress swamps) or under nutrient-rich conditions (lake-edge swamps, cypress strands, and alluvial river swamps). Two other dominant forested wetland types in the eastern United States are Atlantic white cedar* (Chamaecyparis thyoides*) swamps found along the eastern seaboard of the United States and red maple* (Acer rubrum*) swamps found throughout New England and the Mid-Atlantic states. Trees in forested wetlands have developed several unique adaptations to the wetland environment, including knees, wide buttresses, adventitious roots, and gas transport to the rhizosphere. Swamp primary productivity is closely tied to hydrologic conditions. The highest productivity is in a pulsing hydroperiod wetland that receives high inputs of nutrients; lower productivity occurs in either drained or continuously flooded swamps. Consumption is primarily through detrital pathways, and decomposition rates depend on flooding, type of material, and average annual temperature. Swamps have been shown to be nutrient sinks, particularly in studies of phosphorus budgets, and they have been investigated for their value as nutrient sinks when wastewater is applied. Modeling has been used to investigate deepwater swamp regions, particularly for the effects of logging, hydrologic modifications, fire, and wastewater additions.*

In the nomenclature of this book, swamps are forested wetlands. We have discussed tidal saltwater mangrove swamps in Chapter 11 and northern peatlands that include swamp forests in Chapter 13. This chapter discusses forested wetlands that are neither tropical saltwater nor northern peatland; these can be simply described as nontidal, non-peat-accumulating, nonriparian palustrine forested wetlands or, more simply, as freshwater swamps.

The freshwater deep swamp was defined by Penfound (1952) as having "fresh water, woody communities with water throughout most or all of the growing season." In the southeastern United States, the cypress (*Taxodium* sp.) and tupelo/gum (*Nyssa* sp.) swamps are the major deepwater forested wetlands and are characterized by bald cypress–water tupelo and pond cypress–black gum communities with permanent or near-permanent standing water. These include cypress domes and alluvial cypress swamps along rivers. Along the middle eastern seaboard and along the panhandle of Florida, the cypress swamp partially gives way to another forested wetland, the Atlantic white cedar (*Chamaecyparis thyoides*) swamp. Farther northeast through New England and well into the Midwest, other types of freshwater forested wetlands occur, although they are not as wet as the cypress–tupelo swamps, nor are the tree species coniferous as are cypress and Atlantic white cedar. These broad-leaved deciduous forested wetlands include forests found along river floodplains and a multitude of forested wetlands that are found in isolated upland depressions. Forested wetlands on floodplains that receive only seasonal pulses of flooding are discussed in Chapter 15.

There are several good publications on the wetlands covered in this chapter. The reader is referred to an excellent tome called *Cypress Swamps* (Ewel and Odum, 1984) for more details about southern deepwater cypress swamps. The book *Forested Wetlands* (Lugo et al., 1990), although not specifically limited to systems covered in this chapter, has much information on the functional aspects of southern deepwater swamps. Laderman (1998) presents "coastally restricted" forests from around the world, many of which are true swamps such as cypress swamps and Atlantic white cedar swamps. This complements very well her previous publications on the ecology of Atlantic white cedar wetlands (Laderman, 1987, 1989). Messina and Conner (1998) present both the ecology and the management of "southern forested wetlands," which includes cypress–tupelo swamps, mangrove swamps, (covered in Chapter 11), and southern peatlands such as the pocosins (covered in Chapter 13). Golet et al. (1993) present a community profile of the red maple swamps of northeastern United States.

Geographical Extent

Cypress–Tupelo Swamps

Bald cypress (*Taxodium distichum* [L.] Rich.) swamps are found as far north as southern Illinois and western Kentucky in the Mississippi River embayment and southern New Jersey along the Atlantic Coastal Plain (Fig. 14-1a). Pond cypress (*Taxodium distichum* var. *nutans* [Ait.] Sweet), described variously as either a different species or a subspecies of bald cypress, has a more limited range than bald cypress and is found primarily in Florida and southern Georgia; it is not present along the Mississippi River floodplain except in southeastern Louisiana. Another species indicative of the deepwater swamp is the water tupelo (*Nyssa aquatica* L.), which has a range similar to that of bald cypress along the Atlantic Coastal Plain and the Mississippi River, although it is generally absent from Florida except for the western peninsula. Water tupelo occurs in pure stands or is mixed with bald cypress in floodplain swamps.

Cypress–tupelo swamps were once common throughout what is now the southeastern United States. Certainly only a small portion of the original cypress–tupelo swamps found in the New World by early European explorers still exists because logging of these swamps was extensive in Florida, Louisiana, the Mississippi embay-

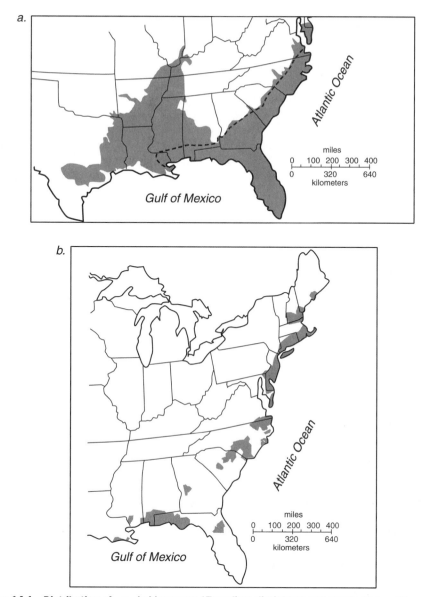

Figure 14-1 Distribution of **a.** bald cypress (*Taxodium distichum*) and pond cypress (*Taxodium distichum* var. *nutans*) (with dotted line indicating northern extent of pond cypress) and **b.** white cedar (*Chamaecyparis thyoides*) in the eastern United States. (*After Little, 1971; Laderman, 1989*)

ment, and along the big rivers of the Atlantic Coastal Plain. Cypress (pond and bald), known by the Seminoles in Florida as "hatch-in-e-haw," which means everlasting (Conner and Buford, 1998), and later by the timber industry as the "wood eternal," was recognized for its rot-resistant characteristics and strength, which led it to be used for boxes for transportation of fruit and vegetables and for river pilings as well as support beams for antebellum southern mansions along the lower Mississippi. It was heavily harvested around the turn of 20th century, particularly from 1880 to

1925 in Louisiana (Conner and Toliver, 1990). Brandt and Ewel (1989) estimated that only about 10 percent of cypress swamps found in presettlement times still remain in the United States.

Shepard et al. (1998) estimated that there are 11.4 million ha of "oak–gum–cypress," or about 14.5 percent of all of the timberland (78.5 million ha) in the southeastern United States. By comparison, the 1992 National Resources Inventory of all nonfederal land in the United States found about 20 million ha of palustrine wooded wetlands in southeastern United States; this probably includes some of the riparian wetlands covered in Chapter 15. Both of these estimates omit outlier cypress swamps found in "non-southeastern" states such as Illinois, Missouri, and Oklahoma. A much lower estimate of the full extent of cypress–tupelo swamps in the United States is the estimate of 1.2 to 2 million ha of second-growth cypress-tupelo swamps provided by Williston et al. (1980) and Kennedy (1982). The greatest stocks of bald cypress, pond cypress, and water tupelo, when measured in timber volume, are found respectively, in Louisiana, Florida, and Georgia (Fig. 14-2).

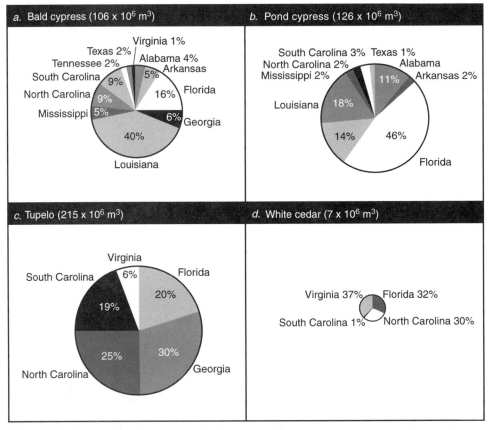

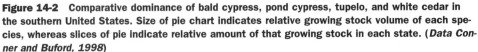

Figure 14-2 Comparative dominance of bald cypress, pond cypress, tupelo, and white cedar in the southern United States. Size of pie chart indicates relative growing stock volume of each species, whereas slices of pie indicate relative amount of that growing stock in each state. (*Data Conner and Buford, 1998*)

White Cedar Swamps

White cedar swamps, dominated by Atlantic white cedar (*Chamaecyparis thyoides* [L.] BSP), are found along the Atlantic and Gulf of Mexico coastlines of the United States as far north as southeastern Maine (Fig. 14-1b). These wetlands are not nearly as plentiful as are cypress–tupelo swamps (Fig. 14-2). White cedar occurs in about 215,000 ha of forestland (Table 14-1), but the species accounts for a majority of the trees in only about 44,200 ha (Sheffield et al., 1998). Only 5,300 ha of Atlantic white cedar swamps remain in the glaciated northeastern United States (Motzkin, 1991), with red maple swamps much more prevalent there. Virginia has about 37 percent of the timber volume of Atlantic white cedar in southeastern United States (Fig. 14-2), yet the three states that have the most area of timberland with Atlantic white cedar are North Carolina, Florida, and New Jersey (Table 14-1). The regions with the highest concentrations of Atlantic white cedar are the Pinelands of southeastern New Jersey; the Dismal Swamp of Virginia and North Carolina; and the floodplains of the Escambia, Appalachicola, and Blackwater Rivers in Florida (Sheffield et al., 1998).

Red Maple Swamps

One of the most common of the broad-leaved deciduous forested wetlands in the northeastern United States is the red maple (*Acer rubrum* L.) swamp. Toward the

Table 14-1 Estimated extent of Atlantic white cedar (*Chamaecyparis thyoides*) and red maple (*Acer rubrum*) swamps in the United States

State	Atlantic White Cedar Swamps[a] (ha)	Red Maple Swamps[b] (ha)
North Atlantic		
Connecticut	2,000	37,000
Delaware	2,000	—
Maryland	6,000	—
Massachusetts	3,000	121,000
New Jersey	44,000	29,000
New York	2,000	124,000
Rhode Island	—	18,000
Vermont	—	24,000
South Atlantic		
North Carolina	77,000	—
South Carolina	1,000	—
Virginia	5,000	—
Gulf		
Alabama	5,000	—
Florida	64,000	—
Mississippi	4,000	—
Total	215,000	353,000

[a] Forests in the United States with Atlantic white cedar trees but not necessarily dominated by that species.
[b] Assumes all broad-leaved deciduous forested wetlands in these states, as classified by U.S. Fish & Wildlife's National Wetland Inventory, are red maple swamps.
Source: Golet et al. (1993) and Sheffield et al. (1998).

west into Pennsylvania and Ohio, red maple swamps are replaced by swamps dominated by trees such as ash (*Fraxinus* spp.), American elm (*Ulmus americana*), swamp white oak (*Quercus bicolor*), and a number of other species, but in the northeastern United States, the red maple swamp is the most common swamp. Using an approximation that all broad-leaved deciduous forested wetlands in several of the coastal states of the northeastern United States are red maple swamps (this approximation would not apply west or south of New York), Golet et al. (1993) estimated that there are 353,000 ha of red maple swamps in these six states. Red maple forests also occur in the Upper Peninsula of Michigan and northeastern Wisconsin. The range of the species *Acer rubrum* extends westward to the Mississippi River and northward through much of Ontario and parts of Manitoba and Newfoundland, but the tree can grow in both wetlands and dry, sandy or rocky uplands. Thus, the presence of red maple does not always indicate wetlands as would the presence of cypress, tupelo, or white cedar.

Geomorphology and Hydrology

Cypress Swamps

Southern cypress–tupelo swamps occur under a variety of geologic and hydrologic conditions, ranging from the extremely nutrient-poor dwarf cypress communities of southern Florida to the rich floodplain swamps along many tributaries of the lower Mississippi River basin. A useful classification of deepwater swamps in terms of their geological and hydrological conditions includes the following types:

1. Stillwater cypress domes
2. Dwarf cypress swamps
3. Lake-edge swamps
4. Slow-flowing cypress strands
5. Alluvial river swamps

The physical features and flow conditions of these wetlands are summarized in Figure 14-3 and are described next.

Cypress Domes

Cypress domes (sometimes called cypress ponds or cypress heads) are poorly drained to permanently wet depressions dominated by pond cypress. They are generally small in size, usually 1 to 10 ha, and are numerous in the upland pine flatwoods of Florida and southern Georgia. Cypress domes are found in both sandy and clay soils and usually have several centimeters of organic matter that has accumulated in the wetland depression. These wetlands are called *domes* because of their appearance when viewed from the side: The larger trees are in the middle, and smaller trees are toward the edges (Fig. 14-3a). Ewel and Wickenheiser (1988) confirmed that trees grow slowest at the edges and fastest near the center of the domes but found no significant differences in tree growth among small, medium, and large cypress domes. This "dome" phenomenon, it has been suggested, is caused by deeper peat deposits in the middle of the dome, fire that is more frequent around the edges of the dome, or a gradual increase in the water level that causes the dome to "grow" from the center outward (Vernon, 1947; Kurz and Wagner, 1953). A definite reason for this profile has not been determined, nor do all domes display the characteristic shape.

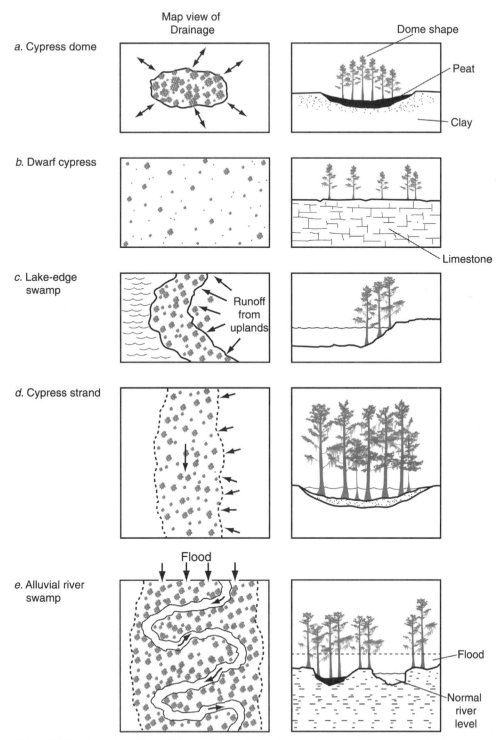

Figure 14-3 General profile and flow pattern of major types of deepwater swamps, showing
a. cypress dome, **b.** dwarf cypress, **c.** lake-edge swamp, **d.** cypress strand, and **e.** alluvial
river swamp. (*After H. T. Odum, 1982*)

A typical hydroperiod for a Florida cypress dome is shown in Figure 5-3h. The wet season is in the summer, and dry periods occur in the fall and spring. An example of a water budget for a Florida cypress dome is given in Figure 14-4a. The standing water in cypress domes is often dominated by rainfall and surface inflow, and there is little or no groundwater inflow. The cypress domes are sometimes underlain by an impermeable clay layer and sometimes by a *hardpan,* a layer of consolidated and relatively impermeable material. Both layers impede downward drainage, although there is often some loss of groundwater to the surrounding upland as it radiates outward from the dome rather than vertically. The major loss of water from the cypress dome is evapotranspiration. Radial groundwater loss is rapid during the dry season but relatively slow during the wet summers, when water levels surrounding the cypress dome are also high.

Dwarf Cypress Swamps

There are major areas in southwestern Florida, primarily in the Big Cypress Swamp and the Everglades, where pond cypress is the dominant tree but it grows stunted and scattered in a herbaceous understory marsh (Fig. 14-3b). The trees generally do not grow more than 6 or 7 m high and are more typically 3 m in height. These wetlands are called dwarf cypress or pigmy cypress swamps. The poor growing conditions are primarily caused by the lack of suitable substrate overlying the bedrock limestone that is found in outcrops throughout the region. The hydroperiod includes a relatively short period of flooding as compared with other deepwater swamps, and fire often occurs. The cypress, however, are rarely killed by fire because of the lack of litter accumulation and buildup of fuel. Although distinct from dwarf cypress swamps, individual small cypress trees are also found in scattered locations throughout the Florida Everglades. These trees are often grouped in clusters that have the appearance of small cypress domes.

Lake-Edge Swamps

Bald cypress swamps are also found as margins around many lakes and isolated sloughs in the Southeast, ranging from Florida to southern Illinois (Fig. 14-3c). Tupelo and water-tolerant hardwoods such as ash (*Fraxinus* spp.) often grow in association with the bald cypress. A seasonally fluctuating water level is characteristic of these systems and is necessary for seedling survival. The trees in these systems receive nutrients from the lake as well as from upland runoff. The lake-edge swamp has been described as a filter that receives overland flow from the uplands and allows sediments to settle out and chemicals to adsorb onto the sediments before the water discharges into the open lake. The importance of this filtering function, however, has not been adequately investigated.

Cypress Strands

A cypress strand is "a diffuse freshwater stream flowing through a shallow forested depression on a gently sloping plain" (Wharton et al., 1976). Cypress strands (Fig. 14-3d) are found primarily in south Florida, where the topography is slight, and rivers are replaced by slow-flowing strands with little erosive power. The substrate is primarily sand, and there is some mixture of limestone and remnants of shell beds. Peat deposits are shallow on higher ground and deeper in the depressions. The hydroperiod has a seasonal wet-and-dry cycle. The deeper peat deposits usually retain moisture even in extremely dry conditions. Much is known about south Florida strands from studies

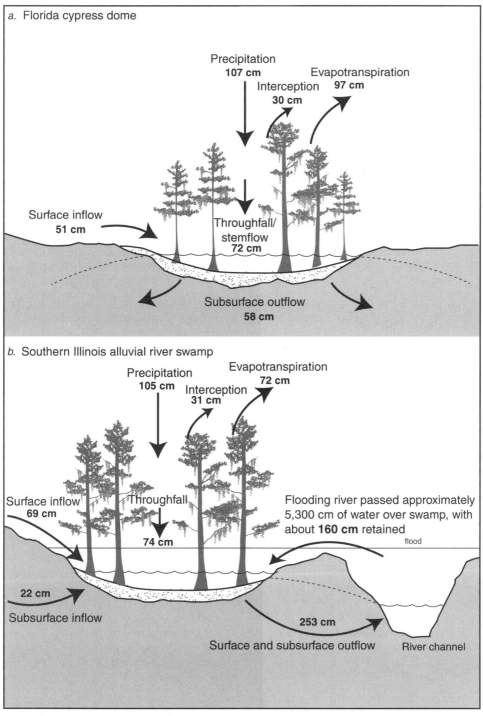

Figure 14-4 Annual water budget for deepwater swamps for **a.** Florida cypress dome and **b.** southern Illinois cypress–tupelo alluvial swamp. (*a after Heimburg, 1984; b after Mitsch et al., 1979a*)

of the Fakahatchee Strand by Carter et al. (1973) and the Corkscrew Swamp by Duever et al. (1984).

Alluvial River Swamps

The broad alluvial floodplains of southeastern rivers and creeks support a vast array of forested wetlands, some of which are permanently flooded deepwater swamps (Fig. 14-3e). (Those forested wetlands that are temporarily flooded by streams and rivers are described in Chapter 15.) Alluvial river swamps, usually dominated by bald cypress or water tupelo or both, are confined to permanently flooded depressions on floodplains such as abandoned river channels (*oxbows*) or elongated swamps that usually parallel the river (*sloughs*). These alluvial wetlands sometimes are called *backswamps,* a name that distinguishes them from the drier surrounding bottomlands and indicates their hydrologic isolation from the river except during the flooding season. The water budget for an alluvial bald cypress–water tupelo swamp is shown in Figure 14-4b. The backswamps are noted for a seasonal pulse of flooding that brings in water and nutrient-rich sediments. Alluvial river swamps are continuously or almost continuously flooded. The hydrologic inflows are dominated by runoff from the surrounding uplands and by overflow from the flooding rivers.

White Cedar Swamps

White cedar swamps occupy a narrow hydrologic niche generally between deepwater cypress–tupelo swamps and moist-soil red maple swamps. The hydrologic regime of cedar swamps can be classified as seasonally flooded (see Chapter 5), with flooding for an extended period during the growing season. Golet and Lowry (1987) found that a group of swamps in Rhode Island had a wide variability in annual water-level fluctuations, ranging from 17 to 75 cm in amplitude and averaging 42 cm over a 7-year period. The percentage of wetland flooded during the growing season ranged from 18 to 76 percent.

Red Maple Swamps

Red maple swamps and mineral soil forested wetlands occur, in general, in several different hydrogeomorphic regimes, the most common being isolated basins in glacial till or glaciofluvial deposits left behind by glaciation. Golet et al. (1993) suggest that red maple swamps occur in the four types of hydrologic settings originally described by Novitzki (1979, 1982):

1. Surface water depression swamps
2. Surface water slope swamps
3. Groundwater depression swamps
4. Groundwater slope swamps

These four types of swamps are described next and illustrated in Figure 14-5.

Surface Water Depression Swamps

This type of swamp is dominated by surface runoff and precipitation, with little groundwater outflow due to a layer of low-permeability soils (Fig. 14-5a). These swamps belong to the perched wetland type described in Chapter 5 (Fig. 5-15e),

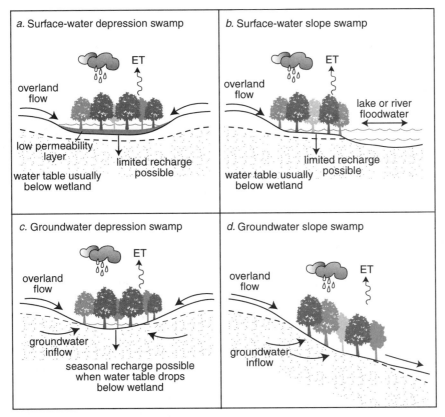

Figure 14-5   **Groundwater flow patterns for red maple swamps: a. surface water depression swamp,   b. surface water slope swamp,   c. groundwater depression swamp, and d. groundwater slope swamp. (*After Golet et al., 1993*)**

which are separated from the water table by an unsaturated zone. Water levels in these swamps will be 20 to 30 cm above the ground surface during the nongrowing season but drop well below the surface due to evapotranspiration during the growing season.

Surface Water Slope Swamps

This type of swamp is generally found in alluvial soil adjacent to a lake or stream and is fed, to some degree, by precipitation and surface runoff but, more important, by overbank flooding from the adjacent stream, river, or lake (Fig. 14-5b). The hydroperiods of these wetlands match the seasonal patterns of the adjacent body of water, with relatively rapid wetting and drying. Some groundwater recharge is possible, but that groundwater shortly discharges back to the stream, river, or lake. When a lake is adjacent to the swamp, water fluctuations are generally less dramatic.

Groundwater Depression Swamps

These are the groundwater discharge wetlands described in Chapter 5, where the swamp is in a depression low enough to intercept the local groundwater table (Fig. 14-5c). These kinds of red maple swamps occur in coarse-textured glaciofluvial deposits, where the interchange between groundwater and surface water is enhanced by relatively coarse soil material. Water-level fluctuations in these types of swamps are less dramatic than fluctuations in surface flow wetlands because of the relative stability of the groundwater levels.

Groundwater Slope Swamps

Forested wetlands often develop on slopes or hillsides where groundwater discharges to the surface as springs and seeps (Fig. 14-5d). The red maple swamp found in the headwaters of a stream is representative of this type of swamp (Golet et al., 1993). Groundwater flow into these swamps can be continuous or seasonal, depending on the local geohydrology and on the evapotranspiration rates of the swamp and adjacent uplands.

Surface water and even saturated soil levels in red maple swamps are seasonal with wet and dry conditions mainly driven by seasonal evapotranspiration in the swamp (Fig. 14-6). Golet et al. (1993) describe the three levels of saturation found in red maple swamps:

1. *Seasonally Flooded:* Surface water is present for extended periods, especially early in the growing season, but is generally absent by the end of the season in most years. Most of these wetlands are influenced by groundwater and are groundwater depression swamps.

2. *Temporarily Flooded:* Surface water is present for short periods during the growing season but is otherwise well below the surface. This condition occurs primarily in wetlands dependent on surface flow such as the surface water depression swamps and surface water slope swamps described previously. If the seasonal flooding is too temporary, red maple will be replaced by other less water-tolerant trees.

3. *Seasonally Saturated:* Soil is saturated during the early part of the growing season, after which unsaturated conditions prevail due to evapotranspiration. There may not be much surface standing water in these

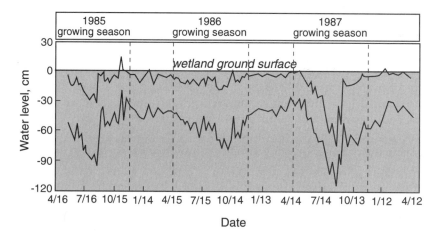

Figure 14-6   Relative water levels in two seasonally saturated red maple swamps in Rhode Island for 1985–1987. Growing season precipitation amounts for 1985, 1986, and 1987 were 104, 76, and 59 cm, respectively. (*After Golet et al., 1993*)

wetlands as depicted in Figure 14-6. These hydroperiods are characteristic of groundwater slope wetlands, which are more saturated than those areas flooded by groundwater, and yet have high evapotranspiration rates during the growing season.

Biogeochemistry

Lockaby and Walbridge (1998) describe the biogeochemistry of forested wetlands as "the most complex and difficult to study with any forest ecosystem type." Forested wetlands have soil and water chemistry that varies from the rich sediments of alluvial cypress swamps to the extremely low mineral and acidic waters of surface water depression red maple swamps and cypress domes. Wide ranges of pH, dissolved substances, and nutrients are found in the soils and waters of these swamps (Table 14-2). Several facts should be noted from this wide range of soil and water chemistry:

1. Swamps are generally acidic to circumneutral, depending on the accumulation of peat and the degree to which precipitation dominates the hydrology.
2. Nutrient conditions vary from nutrient- and mineral-poor conditions in rainwater-fed swamps to nutrient- and mineral-rich conditions in alluvial river swamps and groundwater discharge swamps.
3. An alluvial river swamp often has water quality very different from that of the adjacent river. Swamps in alluvial settings are generally fed by both groundwater discharge and flooding rivers and can have water chemistry quite different from either source.

Many freshwater swamps, particularly alluvial river swamps, are "open" to river flooding and other inputs of neutral and generally well-mineralized waters. The pH of many alluvial swamps in the southeastern United States is 6 to 7, and there are high concentrations of dissolved ions. Cypress domes and perched-basin swamps, on the other hand, are fed primarily by rainwater and have acidic waters, usually in the pH range of 3.5 to 5.0, caused by humic acids produced within the swamp. Colloidal humic substances contribute to both the low pH and the tea-colored or "blackwater" appearance of the standing water in many forested wetlands.

The buffering capacity of the water in cypress domes is low. Little or no alkalinity and low concentrations of dissolved ions and nutrients are characteristic of precipitation-dominated swamps such as cypress domes and some Atlantic white cedar and red maple swamps. These wetlands have much in common with the oligotrophic or ombrotrophic peatlands described in the previous chapter. On the other hand, swamps open to major surface water and groundwater inputs are generally rich in alkalinity, dissolved ions, and nutrients. For example, conductivity ranges from only 60 μS/cm in cypress domes in Florida to 200 to 400 μS/cm in alluvial cypress swamps in Kentucky and Illinois (Table 14-2). Cypress domes and dwarf cypress swamps are low in nutrients because of their relative hydrologic isolation. For example, average phosphorus levels of only 50 to 240 μg P/L in cypress domes and 10 μg P/L in a scrub cypress swamp were observed in Florida by S. L. Brown (1981). Mitsch (1984) noted the range of total phosphorus to be 50 to 160 μg P/L in the central pond of a similar cypress dome, with most of the phosphorus in inorganic form, whereas Dierberg and Brezonik (1984) found an average of 180 μg P/L for

Table 14-2 Water chemistry in selected deepwater swamps[a]

	North-Central Florida Cypress Dome[b]	Lousiana Bayou Swamp[c]	Louisiana Atchafalaya Basin[d]	Southern Illinois Cypress–Tupelo Swamp[e]	Western Kentucky Cypress Slough[f]
pH	4.51 ± 0.36 (51)	—	6.4–8.4	5.8–6.5 (4)	6.6–7.2 (8)
Conductivity (µS/cm)	60 ± 17 (41)	—	—	51–240 (9)	360–550 (8)
Alkalinity (mg CaCo$_3$/L)	1.8 ± 222.9 (13)	—	38–179	12–84 (9)	—
Na$^+$ (mg/L)	4.95 ± 1.60 (38)	—	—	0.7–7.8 (7)	26.5–43.6 (4)
K$^+$ (mg/L)	0.34 ± 0.24 (38)	—	—	1.0–7.0 (7)	3.1–4.0 (4)
Mg^{2+} (mg/L)	1.37 ± 0.59 (39)	—	—	1.0–4.3 (7)	7–52 (4)
Ca^{2+} (mg/L)	2.87 ± 0.99 (39)	—	—	2.3–10.6 (7)	—
SO$_4^=$ (mg/L)	2.6 ± 2.7 (25)	—	—	0.5–4 (4)	9.2–38.6 (9)
Dissolved oxygen (mg/L)	2.0 ± 1.8 (21)	—	1.8–9.9	0.9–4.0 (5)	5.8–10.9 (8)
Turbidity (NTU)	2.8 ± 8.7 (34)	—	—	23–690 (8)	0.9–29 (9)
Color (mg Pt/L)	456 ± 162 (43)	—	—	—	—
Total organic C (mg/L)	40 ± 13 (32)	—	3.3–31.6	—	—
NO$_3$–N (mg N/L)	0.08 ± 0.19 (63)	0.01–0.13 (24)	0.03–0.19	<.01 (6)	—
NH$_3$–N (mg N/L)	0.14 ± 0.19 (63)	0.01–0.62 (24)	0.02–1.71	0.10–4.1 (6)	—
Total N (mg N/L)	1.6 ± 1.3 (62)	0.58–1.82 (23)	0.47–9.70	0.60–4.7 (6)	—
ortho-P (mg P/L)	0.07 ± 0.11 (61)	0.05–0.44 (24)	0.01–0.24	0.06–0.28 (9)	0.02–0.03 (33)
Total P (mg P/L)	0.18 ± 0.38 (63)	0.15–0.66 (24)	0.08–0.56	0.17–0.47 (9)	0.08–0.20 (33)

[a] Numbers given as range (number of samples) except as otherwise noted.
[b] Dierberg and Brezonik (1984); average ± standard deviation.
[c] Kemp and Day (1984); range for three swamp sites.
[d] Hern et al. (1980).
[e] Mitsch et al. (1977).
[f] Hill (1983) and Mitsch et al. (1991); range for average of three sites.

5 years of sampling in the same cypress dome (Table 14-2). On the other hand, phosphorus levels are often considerably higher in alluvial river swamps, particularly during flooding from the adjacent river. Kemp and Day (1984) reported phosphorus concentrations as high as 660 μg P/L in Louisiana swamps, and Mitsch et al. (1977) reported concentrations as high as 470 μg P/L in an alluvial cypress swamp in southern Illinois.

In a comparison of an Atlantic white cedar swamp with adjacent forested wetlands and nonforested peatlands, Whigham and Richardson (1988) found the cedar swamp soils to be significantly higher in pH, calcium, and magnesium than the other sites (Table 14-3), suggesting a groundwater or brackish-water source might be important for Atlantic white cedar to compete with other swamp trees. Phosphorus was lowest in the white cedar swamp, suggesting this was the most significant limiting nutrient. The high pH measured in this study suggests that Atlantic white cedar may do best in sites with high pH, although these swamps have been reported to occur under low-pH (3.2–4.4) conditions in the Great Dismal Swamp (F. Day, 1984).

The isolation of alluvial river swamps from their nearby streams and rivers for most of the year often leads to differences in the water chemistry of the swamps and rivers. Denitrification and sulfate reduction, which cause loss of nitrates and sulfates, are dominant in the stagnant swamp but are less prevalent in the flowing river. Furthermore, dissolved ions in the backswamp are often lower in concentration than the same ions in the river. Dissolved ion concentrations in the river are low when it is flooding the backswamp, because of dilution due to precipitation and possibly because of nutrient uptake by vegetation. This temporary physicochemical isolation of alluvial river swamps was noted in studies in Louisiana and southern Illinois. Water chemistry in Louisiana backswamps in the Atchafalaya Basin is distinct from that of the adjacent rivers and streams except during the flooding season, when the waters of the entire region are well mixed (Beck, 1977; Hern et al., 1980). In southern Illinois, several dissolved substances were significantly lower in the standing water of a riparian cypress–tupelo swamp than in the adjacent river. The swamp had lower values of calcium, magnesium, sodium, potassium, sulfate, and nitrate than the river

Table 14-3 Soil chemistry of *Chamaecyparis thyoides* and *Acer rubrum* swamps in Maryland compared to nonforested peatlands

Soil Parameters (top 50 cm)	White Cedar Swamp	Red Maple Swamp	Nonforested Peatlands
pH	5.34	4.23	4.54
Organic matter (%)	59 ± 5	67 ± 3	68 ± 2
Nitrogen (%)	1.6 ± 0.1	1.5 ± 0.1	1.7 ± 0.1
Phosphorus (%)	0.07 ± 0.01	0.24 ± 0.03	0.10 ± 0.01
NO_3–N (μg/g)	0.8 ± 0.1	0.3 ± 0.1	0.5 ± 0.1
NH_4–N (μg/g)	67 ± 4	72 ± 19	76 ± 10
Ca^{2+} (μg/g)	1,810	339	710
Mg^{2+} (μg/g)	1,420	493	477
K^+ (μg/g)	1,054	1,622	857
Na^+ (μg/g)	841	134	383
Fe (mg/g)	6.3	5.9	5.4
Al (mg/g)	8.0	5.4	7.6

Source: Whigham and Richardson (1988).

did, despite the fact that the swamp was flooded annually (Mitsch, 1979; Dorge et al., 1984).

Ecosystem Structure

Vegetation

Cypress Swamps

Southern deepwater swamps, particularly cypress wetlands, have plant communities that either depend on or adapt to the almost continuously wet environment. There are several distinctions between bald cypress and pond cypress swamps (Table 14-4). The dominant canopy vegetation found in alluvial river swamps of the southeastern United States includes bald cypress (*Taxodium distichum*) and water tupelo (*Nyssa aquatica*). The trees are often found growing in association in the same swamp, although pure stands of either bald cypress or water tupelo are also frequent in the southeastern United States. Many of the pure tupelo stands are thought to be the result of the selective logging of bald cypress (Penfound, 1952). The pond cypress– black gum (*Taxodium distichum* var. *nutans*–*Nyssa sylvatica* var. *biflora* [Walt.] Sarg.) swamp is more commonly found on the uplands of the southeastern Coastal Plain, usually in areas of poor sandy soils without alluvial flooding. These same conditions are usually found in cypress domes.

Table 14-4   Distinction between bald cypress and pond cypress swamps

Characteristic	Bald Cypress Swamp	Pond Cypress Swamp
Dominant cypress	*Taxodium distichum*	*Taxodium distichum* var. *nutans*
Dominant tupelo or gum (when present)	*Nyssa aquatica* (water tupelo)	*Nyssa sylvatica* var. *biflora* (black gum)
Tree physiology	Large, old trees, high growth rate, usually abundance of knees and spreading buttresses	Smaller, younger trees, low growth rate, some knees and buttresses but not as pronounced
Location	Alluvial floodplains of Coastal Plain, particularly along Atlantic seaboard, Gulf seaboard, and Mississippi embayment	"Uplands" of Coastal Plain, particularly in Florida and southern Georgia
Chemical status	Neutral of slightly acid, high in dissolved ions, usually high in suspended sediments and rich in nutrients	Low pH, poorly buffered, low in dissolved ions, poor in nutrients
Annual flooding from river	Yes	No
Types of deepwater swamps	Alluvial river swamp, cypress strand, lake-edge swamp	Cypress dome, dwarf cypress swamp

One of the main features that distinguishes bald cypress trees from pond cypress trees is the leaf structure (Fig. 14-7). Bald cypress has needles that spread from the twig in a flat plane, whereas pond cypress needles are appressed to the twig. Both species are intolerant of salt and are found only in freshwater areas. Pond cypress is limited to sites that are poor in nutrients and are relatively isolated from the effects of river flooding or large inflows of nutrients.

Bald cypress
Taxodium distichum

Pond cypress
Taxodium distichum var. *nutans*

Figure 14-7 Distinction of leaves between   **a.** bald cypress (*Taxodium distichum*) and
b. pond cypress (*Taxodium distichum* var. *nutans*).

A listing of some of the dominant canopy and understory vegetation in various cypress swamps in the southeastern United States is given in Table 14-5. When deepwater swamps are drained or when their dry period is extended dramatically, they can be invaded by pine (e.g., *Pinus elliottii*) or hardwood species. In north-central Florida, a cypress–pine association indicates a drained cypress dome (Mitsch and Ewel, 1979). Hardwoods that characteristically are found in cypress domes include swamp red bay (*Persea palustris*) and sweet bay (*Magnolia virginiana*). In lake-edge and alluvial river swamps, several species of ash (*Fraxinus* sp.) and maple (*Acer* sp.) often grow as subdominants with the cypress or tupelo or both. In the Deep South, Spanish moss (*Tillandsia usneoides*) is found in abundance as an epiphyte on the stems and branches of the canopy trees. Schlesinger (1978) found that Spanish moss has a relatively high biomass and productivity compared to epiphytic communities in many other temperate forests.

The abundance of understory vegetation in cypress–tupelo swamps depends on the amount of light that penetrates the tree canopy. Many mature swamps appear as quiet, dark cathedrals of tree trunks devoid of any understory vegetation. Even when enough light is available for understory vegetation, it is difficult to generalize about its composition (Table 14-5). There can be a dominance of woody shrubs, of herbaceous vegetation, or of both. Fetterbush (*Lyonia lucida*), wax myrtle (*Myrica cerifera*), and Virginia willow (*Itea virginica*) are common as shrubs and small trees in nutrient-poor cypress domes. Understory species in higher nutrient river swamps include buttonbush (*Cephalanthus occidentalis*) and Virginia willow. Some continually flooded cypress swamps that have high concentrations of dissolved nutrients in the water develop dense mats of duckweed (e.g., *Lemna* spp., *Spirodela* spp., or *Azolla* spp.) on the water surface during most of the year. An experiment in enriching cypress domes with high-nutrient wastewater caused a thick mat of duckweed to develop in what was otherwise a nutrient-poor environment (H. T. Odum et al., 1977a).

Floating logs and old tree stumps often provide substrate for understory vegetation to attach and to flourish (Dennis and Batson, 1974). In some swamps, free or attached floating mats develop in much the same way that they develop in freshwater marshes (see Chapter 10). Huffman and Lonard (1983) described floating mats from 0.5 m² to 0.5 ha in a 1,200-ha swamp in Arkansas. The mats were dominated by water willow (*Decodon* spp.) in association with southern wild rice (*Zizania miliacea*) and water pennywort (*Hydrocotyle verticillata*). The floating mats had the following successional pattern: (1) a pioneer stage as particles accumulate on partially decaying logs and water willow takes root in the mat; (2) a water willow–herbaceous stage as the mat, now floating, is invaded by a variety of marsh plants; (3) a water willow–herb stage when additional woody plants, including bald cypress, become established; and (4) a tree stage after the submergence of the floating mat when only bald cypress can survive. Thus, the mats are an important mechanism for *Taxodium* seed germination and survival, especially when drawdowns are infrequent (Huffman and Lonard, 1983).

White Cedar Swamps

Cedar swamps occur within a wide climatic range along the East Coast of the United States and in an intermediate hydrology between deepwater cypress swamps in the South and forested swamps such as red maple swamps in the North. Often thought of as monospecific, even-aged stands with tightly spaced *Chamaecyparis thyoides* trees, cedar swamps in these cases would have no subcanopy, few shrubs, and little in the herbaceous layer. However, the tree is often found in mixed stands, with co-dominants

Table 14-5 Dominant or abundant vegetation of deepwater swamps of the southeastern United States

	Cypress Dome[a]	Alluvial River Swamp[b]
Location	North-central Florida	Louisiana
Dominant canopy trees	*Taxodium distichum* var. *nutans* (pond cypress) *Nyssa sylvatica* var. *biflora* (swamp black gum)	*Taxodium distichum* (bald cypress) *Nyssa aquatica* (water tupelo)
Subdominant trees	*Pinus elliottii* (slash pine) *Persea palustris* (swamp red bay) *Magnolia virginiana* (sweet bay)	*Acer rubrum* var. *drummondii* (Drummond red maple) *Fraxinus tomentosa* (pumpkin ash)
Shrubs	*Lyonia lucida* (fetterbush) *Myrica cerifera* (was myrtle) *Acer rubrum* (red maple) *Cephalanthus occidentalis* (buttonbush) *Itea virginica* (Virginia willow)	*Cephalanthus occidentalis* (buttonbush) *Celtis laevigata* (hackberry) *Salix nigra* (black willow)
Herbs and aquatic vegetation	*Woodwardia virginica* (Virginia chain fern) *Saururus cernuus* (lizard's tail) *Lachnanthes carolinina* (red root)	*Lemna minor* (duckweed) *Spirodela polyrihza* (duckweed) *Riccia* sp. *Limnobium spongia* (common frog's bit)

	Alluvial River Swamp[c]	Scrub Cypress[d]
Location	Southern Illinois	Southern Florida
Dominant canopy trees	*Taxodium distichum* (bald cypress) *Nyssa aquatica* (water tupelo)	*Taxodium distichum* var. *nutans* (pond cypress)
Subdominant trees	—	*Pinus elliottii* (slash pine)
Shrubs	*Cephalanthus occidentalis* (buttonbush) *Itea virginica* (Virginia willow) *Rosa palustris* (swamp rose)	*Myrica cerifera* (wax myrtle) *Ilex cassine* *Stylingia sylvatica*
Herbs and aquatic vegetation	*Azolla mexicana* (water fern) *Spirodela polyrhiza* (duckweed)	*Panicum hermitomon*

	Stillwater Cypress Swamp[e]	Cypress Strand[f]
Location	Okefenokee Swamp, Georgia	Southwestern Florida
Dominant canopy trees	*Taxodium distichum* var. *nutans* (pond cypress)	*Taxodium distichum* (bald cypress)

Table 14-5 *(Continued).*

	Stillwater Cypress Swamp[e] (cont.)	Cypress Strand[f] (cont.)
Subdominant trees	*Ilex cassine* *Nyssa sylvatica* var. *biflora* (swamp black gum)	*Taxodium distichum* var. *nutans* (pond cypress) *Salix caroliniana* (willow) *Annona glabra* (pond apple) *Acer rubrum* (red maple) *Sabal palmetto* (cabbage palm)
Shrubs	*Lyonia lucida* (fetterbush) *Itea virginica* (Virginia willow) *Leucothoe racemosa* *Clethra alnifolia*	*Myrica cerifera* (wax myrtle) *Persea borbonia* (red bay) *Hippocratea volubilis* (*liana*) *Toxicodendron radicans* (poison ivy)
Herbs and aquatic vegetation	*Eriocaulon compressum* (pipewort)	*Blechnum serrulatum* *Nephrolepsis exaltata* (Boston fern) *Thelypteris kunthii* (swamp fern) *Chloris neglecta* (finger grass) *Andropogon virginicus* (broom sedge) *Ludwigia repens*

[a] After Monk and Brown (1965), S. L. Brown (1981), and Marois and Ewel (1983).
[b] After Conner and Day (1976).
[c] After Anderson and White (1970) and Mitsch et al. (1979a).
[d] After S. L. Brown (1981).
[e] After Schlesinger (1978).
[f] After Carter et al. (1973).

such as *Betula populifolia* (gray birch), *Picea mariana* (black spruce), *Pinus strobus* (Eastern white pine), and *Tsuga canadensis* (Eastern hemlock) (Laderman, 1989). In the South, co-dominant trees include *Gordonia lasianthus* (loblolly bay), *Persea borbonia* (red bay), *P. palustris* (swamp red bay), and *Taxodium distichum* (bald cypress).

The shrub layer in cedar swamps with relatively open canopies includes many ericaceous shrubs such as *Aronia arbutifolia* (red chokeberry), *Clethra alnifolia* (sweet pepperbush), *Ilex glabra* (gallberry), *Leucothoe racemosa* (fetterbush), and *Vaccinium corymbosum* (highbush blueberry) (Laderman, 1989).

Red Maple Swamps

The canopy of red maple swamps is obviously dominated by *Acer rubrum* L. Canopy cover generally exceeds 80 percent, although trees in these northern swamps tend to be shorter with less biomass than those in southern swamps. General characteristics of the tree layer are summarized in Table 14-6. Although up to 50 tree species have been found in a red maple swamp, the red maple can account for up to 90 percent of the stem density and basal area (Golet et al., 1993). In general, a specific site will have about four species of trees in the canopy/subcanopy, depending on which region of the glaciated Northeast these red maple swamps occur (Table 14-7).

Shrubs include *Ilex verticillata* (winterberry), *Vaccinium corymbosum* (highbush blueberry), *Lindera benzoin* (spicebush), *Viburnum* spp. (arrowwood), *Alnus rugosa*

Table 14-6 General vegetation characteristics of red maple swamps in the northeastern United States

Attribute	General Range of Averages
Trees	
Stand height (m)	13–15
Tree density (stems/ha)	470–1,960
Basal area (m²/ha)	16.6–37.8
Shrubs	
Cover (%)	6–87
Shrub density (stems/ha)	250–91,000
Basal area (m²/ha)	6.7–8.2
Herbs	
Cover (%)	19–43

Source: Golet et al. (1993).

(speckled alder), *Cephalanthus occidentalis* (buttonbush), *Corylus cornuta* (hazelnut), and *Rhododendron viscosum* (swamp azalea), with dominance depending on the region in which the swamps are found (Table 14-7). Shrub cover is generally greater than 50 percent, although some red maple swamps have shrub cover as low as 6 percent (Table 14-7).

One of the most interesting features of many red maple swamps is the predominance of a great variety of ferns in the herbaceous layer, including *Osmunda cinnamomea* (cinnamon fern), *Onoclea sensibilis* (sensitive fern), *Osmunda regalis* (royal fern), *Thelypteris thelypteroides* (marsh fern), *Matteuccia struthiopteris* (ostrich fern), *Osmunda claytoniana* (interrupted fern), and various *Dryopteris* spp. (wood ferns). Other common herbaceous plants include *Symplocarpus foetidus* (skunk cabbage), *Caltha palustris* (marsh marigold), several species of *Glyceria* (manna grass), and several of more than 32 species of *Carex*.

Other Swamps of Glaciated Regions

Forested swamps occur throughout the glaciated midwestern United States, and, in fact, most of the wetlands remaining in states such as Ohio, Indiana, and Illinois are forested wetlands that occur in isolated basins or floodplains amid agricultural fields (Table 14-8). They were the fields that were too wet to plant and gradually were invaded by tree species. They often are a remnant of a gradual process of ponds of glacial origin slowly infilling and becoming forested (a true hydrarch succession; see Chapter 8). However, they may also occur in wet basins on mineral hydric soils rather than peat deposits. As with red maple swamps, the trees generally have replaced herbaceous marshes that once occupied those sites, because of natural succession or because of artificial drainage (Fig. 14-8). However, the succession of these systems is poorly understood.

Fire

Fire is generally infrequent in swamps because of the standing water, but it can be a significant ecological factor during droughts or in swamps that have been artificially

Table 14-7 Vegetation growing in association with *Acer rubrum* in red maple swamps in the northeastern United States

Subdominant Trees	Shrubs	Ferns and Other Herbs

ZONE 1. SOUTHERN NEW ENGLAND, SEABOARD LOWLAND, COASTAL PLAIN

Subdominant Trees	Shrubs	Ferns and Other Herbs
Betula alleghaniensis	*Rhododendron viscosum*	*Osmunda cinnamomea*
Nyssa sylvatica	*Vaccinium corymbosum*	*Onclea sensibilis*
Fraxinus americana	*Ilex verticillata*	*Osmunda regalis*
Ulmus americana	*Clethra alnifolia*	*Thelypteris thelypteroides*
Tsuga canadensis	*Lindera benzoin*	*Woodwardia virginica*
Quercus alba	*Viburnum* spp.	*Carex* spp.
	Toxicodendron vernix	*Calamagrostis canadensis*
		Glyceria spp.
		Symplocarpus foetidus
		Veratrum viride
		Caltha palustris
		Impatiens capensis
		Aralia nudicaulis
		Iris versicolor
		Mosses, including *Sphagnum,* *Dicranum*

ZONE 2. GREAT LAKES AND ALLEGHENY PLATEAU

Subdominant Trees	Shrubs	Ferns and Other Herbs
Conifer–hardwood swamps		
Betula alleghaniensis	*Vaccinium corymbosum*	*Osmunda cinnamomea*
Pinus strobus	*Ilex verticillata*	*Onclea sensibilis*
Tsuga canadensis	*Lindera benzoin*	*Osmunda regalis*
Ulmus spp.	*Viburnum* spp.	*Thelypteris thelypteroides*
Hardwood swamps	*Aronia melanocarpa*	*Matteuccia struthiopteris*
Fraxinus pennsylvanica	*Alnus rugosa*	*Osmunda claytoniana*
Fraxinus nigra	*Hamamelis virginiana*	*Dryopteris* spp.
Quercus bicolor	*Toxicodendron vernix*	*Carex* spp.
Tilia americana		*Symplocarpus foetidus*
Juglans cinerea		*Impatiens capensis*
Ulmus americana		*Aralia nudicaulis*
		Arisaema triphyllum
		Saururus cernuus
		Polygnum spp.
		Coptis trifolia
		Clintonia spp.

ZONE 3. ST. LAWRENCE VALLEY/LAKE CHAMPLAIN BASIN

Subdominant Trees	Shrubs	Ferns and Other Herbs
Lake floodplain swamps		
Acer saccharinum	*Nemopanthus mucronata*	*Onclea sensibilis*
Quercus bicolor	*Cephalanthus occidentalis*	*Osmunda claytoniana*
Populus deltoides		*Osmunda cinnamomea*
Platanus occidentalis		
Juglands cinerea		
Non-floodplain swamps		
Juglands cinerea	*Ilex verticillata*	*Osmunda cinnamomea*
Fraxinus nigra	*Carpinus caroliniana*	*Matteuccia struthiopteris*
Pinus strobus	*Vaccinium corymbosum*	*Osmunda regalis*
Betula populifolia	*Alnus rugosa*	*Onclea sensibilis*
Betula papyrifera	*Corylus cornuta*	*Osmunda claytoniana*
Fraxinus pennsylvanica	*Viburnum* spp.	*Thalictrum pubescens*
Betula alleghaniensis	*Nemopanthus mucronata*	*Aralia nudicaulis*

Table 14-7 *(Continued).*

Subdominant Trees	Shrubs	Ferns and Other Herbs
Tsuga canadensis	*Cornus stolonifera*	*Solidago* spp.
Thuja occidentalis	*Spiraea latifolia*	*Impatiens capensis*
Populus tremuloides		*Glyceria* spp.
Larix laricina		*Rumex verticillatus*
Abies balsamea		

ZONE 4. NORTHEASTERN MOUNTAINS

Abies balsamea	*Alnus rugosa*	*Osmunda cinnamomea*
Betula populifolia	*Ilex verticillata*	*Onclea sensibilis*
Betula papyrifera	*Viburnum* spp.	*Glyceria* spp.
Betula alleghaniensis	*Salix* spp.	*Carex* spp.
Ulmus americana	*Abies balsamea*	*Aster* spp.
Populus tremuloides	*Spiraea latifolia*	*Solidago* spp.
Fraxinus spp.		

Source: Golet et al. (1993).

drained. In general, fire is more frequent in the forested swamps of Florida than anywhere else, because of the more frequent lightning storms and because of a predictable dry season. For example, from 1970 to 1977, there were four fires in the Big Cypress National Preserve in southern Florida. An average of 500 ha burned per fire (Duever et al., 1986, as cited in Ewel, 1990). It appears that fire is rare in most alluvial river swamps but can be more frequent in cypress domes or dwarf cypress swamps—as frequent as several times per century (Ewel, 1990; Fig. 14-9). Ewel and Mitsch (1978) investigated the effects of fire on a cypress dome in northern Florida and found that fire had a "cleansing" effect on the dome, selectively killing almost all of the pines and hardwoods but relatively few pond cypress. This suggests a possible advantage of fire to some shallow cypress ecosystems in eliminating competition that is less water tolerant.

Fire in white cedar swamps can have much the same effect. If water is low, fire can be quite destructive, killing cedar trees and burning the peat deeply. If water levels are high, light fire can have a cleansing effect, eliminating shrubs and brush and favoring cedar seedling germination (Laderman, 1989). In a study of *C. thyoides* swamps of the Atlantic Coast, Motzkin et al. (1993) found that the highly flammable cedar foliage burned frequently (five fires per each 100- to 200-year interval) during pre-European settlement time. When fires became more rare after European settlement, stands of cedar became the familiar dense monospecific systems that are common today (Motzkin et al., 1993).

Tree Adaptations

Vascular plants, particularly trees, have a difficult time surviving under continuously flooded conditions. Only a handful of species of trees in North America can stay viable in continuous flooding and, even then, their growth is generally slowed; trees that are found in freshwater swamps are stressed with the wet conditions but have found ways to adapt. Many of these adaptations were discussed in Chapter 7. The most conspicuous adaptations specific to the major tree species in forested swamps are discussed here.

Table 14-8 Typical vegetation in a hardwood swamp forest in Ohio[a]

Trees	Wetland Indicator Status[b]
TREES	
Quercus palustrus (pin oak)	FACW
Quercus bicolor (swamp white oak)	FACW
Acer saccharinum (silver maple)	FACW
Quercus rubrum (red maple)	FAC
Ulmus americana (American elm)	FACW
Fraxinus pennsylvanica (green ash)	FACW
SHRUBS/UNDERSTORY	
Lindera benzoin (spicebush)	FACW
Cephalanthus occidentalis (buttonbush)	OBL
Rosa multiflora[c] (multiflora rose)	FACU
Carpinus caroliniana (hornbeam, ironwood)	FAC
HERBS	
Polygonum spp. (smartweed)	FAC → OBL
Symplocarpus foetidus (skunk cabbage)	OBL
Lemna spp. (duckweed)	OBL
Alisma plantago-aquatica (water plantain)	OBL
Aster spp. (asters)	FAC → FACW (mostly)
Carex spp. (sedges)	FACW → OBL
Ranunculus septentrionalis (swamp buttercup)	OBL
Saxifraga pennsylvanica (swamp saxifrage)	OBL
Onoclea sensibilis (sensitive fern)	FACW
Bidens comosa (leafy-bracted beggar-ticks)	FACW
Bidens frondosa (devil's beggar-ticks)	FACW
Scirpus atrovirens (green bulrush)	OBL
Scirpus cyperinus (wool grass)	FACW

[a] Wetland species at Gahanna Woods Nature Preserve, Franklin County, Ohio.
[b] Use of wetland indicator status for the northeastern United States. In order of wet to dry: OBL, obligate wetland plant; FACW, facultative wet plant; FAC, facultative plant; FACU, facultative upland plant (see Chapter 21).
[c] Nonnative species.

Knees and Pneumatophores

Cypress (bald and pond), water tupelo, and black gum are among a number of wetland plants that produce pneumatophores. In deepwater swamps, these organs extend from the root system to well above the average water level (Fig. 14-10a). On cypress, these "knees" are conical and typically less than 1 m in height, although some cypress knees are as tall as 3 to 4 m. Knees are much more prominent on cypress than on tupelo. Hall and Penfound (1939) discovered that cypress had more knees (three per tree) than tupelo did (one per tree) in a cypress–tupelo swamp in Louisiana. Pneumatophores on black gum in cypress domes are actually arching or "kinked" roots that approximate the appearance of cypress knees.

The functions of the knees have been speculated about for the last century. It was thought that the knee might be an adaptation for anchoring the tree because of the appearance of a secondary root system beneath the knee that is similar to and smaller than the main root system of the tree (Mattoon, 1915; C. A. Brown, 1984).

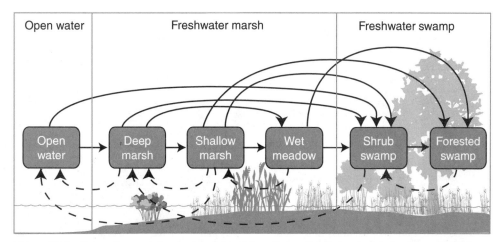

Figure 14-8 General succession to mineral soil forested wetlands in glaciated regions of North America. (*After Golet et al., 1993*)

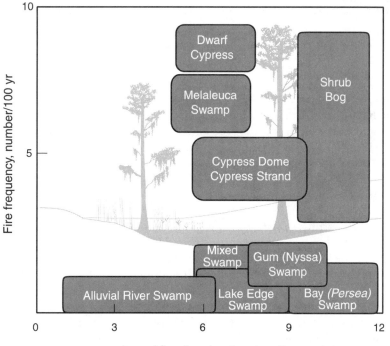

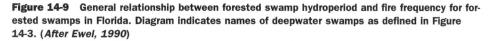

Figure 14-9 General relationship between forested swamp hydroperiod and fire frequency for forested swamps in Florida. Diagram indicates names of deepwater swamps as defined in Figure 14-3. (*After Ewel, 1990*)

a.

Figure 14-10 Among the several features of vegetation in cypress swamps are **a. cypress knees and b. buttresses and large size of cypress trees.**

Observations of swamp and upland damage in South Carolina following Hurricane Hugo in 1990 showed that cypress trees often remained standing while hardwoods and pines did not, supporting the tree-anchoring theory for cypress root, knee, and buttress systems (K. Ewel, personal communication).

Other discussions of cypress knee function have centered on their possible use as sites of gas exchange for the root systems. Penfound (1952) argued that cypress knees are often absent where they are most needed—in deep water—and that the wood of the cypress knee is not aerenchymous; that is, there are no intercellular gas spaces capable of transporting oxygen to the root system. Kramer et al. (1952) concluded that the knees did not provide aeration to the rest of the trees. However, gas exchange does occur at the knees. Carbon dioxide evolved at rates of 4 to 86 g C/m^2 of tissue per day from cypress knees in Florida cypress domes (Cowles, 1975). S. L. Brown (1981) estimated that gas evolution from knees accounted for 0.04 to 0.12 g C m^{-2} yr^{-1} of the respiration in a cypress dome, and 0.23 g C m^{-2} yr^{-1} in an alluvial river swamp. This accounted for 0.3 to 0.9 percent of the total tree respiration, but 5 to 15 percent of the estimated woody tissue (stems and knees) respiration. The fact that CO_2 is exchanged at the knee, however, does not prove that oxygen transport is taking place there or that the CO_2 evolved was the result of oxidation of anaerobically produced organic compounds in the root system.

Buttresses

Taxodium and *Nyssa* species and, to a lesser degree, *Chamaecyparis thyoides* often produce swollen bases or buttresses (stem hypertrophy) when they grow in flooded

b.

Figure 14-10 *(Continued).*

conditions (Fig. 14-10b). The basal swelling can extend from less than 1 m above the soil to several meters, depending on the hydroperiod of the wetland. Swelling generally occurs along the part of the tree that is flooded at least seasonally, although the duration and frequency of the flooding necessary to cause the swelling are unknown. One theory described the height of the buttress as a response to aeration: The greatest swelling occurs where there is a continual wetting and soaking of the tree trunk but where the trunk is also above the normal water level (Kurz and Demaree, 1934). The value of the buttress swelling to ecosystem survivability is unknown; it may simply be a relict response that is of little use to the plant.

Seed Germination and Dispersal

The seeds of swamp trees require oxygen for germination. For example, it has been demonstrated that even cypress seeds and seedlings require moist but not flooded soil for germination and survival (Mattoon, 1915; DuBarry, 1963). Occasional drawdowns, if only at relatively infrequent intervals, are therefore necessary for the survival of trees in these swamps unless floating mats develop. Otherwise, continuous flooding will ultimately lead to an open-water pond.

The dispersal and survival of the seeds of many swamp trees depend on hydrologic conditions. Schneider and Sharitz (1986) found a relatively low number of viable seeds in a seed bank study of a cypress–tupelo swamp in South Carolina: An average of 127 seeds/m² were found for woody species (88 percent as cypress or tupelo) in the swamp compared to a seed density of 233 seeds/m² from an adjacent bottomland hardwood forest. They speculated that the continual flooding in the cypress–tupelo swamp leads to reduced seed viability. Huenneke and Sharitz (1986) and Schneider and Sharitz (1988) elaborated further on the importance of *hydrochory* (seed dispersal by water) in these swamps. Hydrologic conditions, particularly scouring by flooding waters, are important factors in determining the composition, dispersal, and survival of seeds in riverine settings. Seeds are transported relatively long distances; Schneider and Sharitz (1988) reported that a distance of 1.8 km was necessary in their study to trap 90 percent of seeds from a given tree. They found that flowing water distributes seeds nonrandomly: The highest seed densities accumulate near obstructions such as logs, tree stumps, cypress knees, and tree stems, and the lowest seed densities in open-water areas. They also found higher densities of seeds near the edge of the swamps (415 seeds/m²) than near the center (175 seeds/m²). Bald cypress and tupelo seeds are produced mainly in the fall and winter, between the periods of both lowest (October) and highest (March) streamflows, giving the fallen seeds the widest possible range of hydrologic conditions. Schneider and Sharitz (1988) concluded that "seed dispersal processes of many wetland species are sensitively linked to the timing and magnitude of hydrologic events."

Longevity

Some swamp trees may live for centuries and achieve great sizes. One individual cypress in Corkscrew Swamp in southwestern Florida was determined to be about 700 years old; Laderman (1998) reports that the maximum age of *Taxodium* is 1,000 years. By contrast, *Chamaecyparis thyoides* lives to a maximum of 300 years (Clewell and Ward, 1987). Mature bald cypress are typically 30 to 40 m in height and 1 to 1.5 m in diameter. Anderson and White (1970) reported a very large cypress tree in a cypress–tupelo swamp in southern Illinois that measured 2.1 m in diameter. C. A.

Brown (1984) summarized several reports that documented bald cypress as large as 3.6 to 5.1 m in diameter.

Shallow or Adventitious Roots

Some species such as *Acer rubrum* develop very shallow root systems in response to flooding, in all likelihood because the surface soil is closest to the atmospheric source of oxygen. In aerated soils, the same species will develop deep roots. Other swamp species such as willows (*Salix* sp.), green ash (*Fraxinus pennsylvanica*), and cottonwoods (*Populus deltoides*) develop adventitious roots above ground from the stem in response to flooding.

Gaseous Diffusion

Woody trees have a particular problem getting oxygen to their rhizosphere when they are flooded, and few species do it well enough to survive continual flooding. The swamp trees, including *Taxodium, Nyssa, Alnus, Fraxinus,* among others, have the ability to supply oxygen to their root systems in amounts adequate for rhizospheric demands. Solar radiation, which heats up the tree stems by a couple of degrees, causes a light-induced gas flow that can be considerably greater in selected swamp seedlings than in the same trees in the dark; this thermally induced flow of air through vascular plants is called *thermo-osmosis* by some (Grosse et al., 1998), and is enhanced by the development of aerenchymatous stem and root tissue (see also Chapter 7).

Consumers

Invertebrates

Invertebrate communities, particularly benthic macroinvertebrates, have been analyzed in several cypress–tupelo swamps. A wide diversity and high number of invertebrates have been found in permanently flooded swamps. Species include crayfish, clams, oligochaete worms, snails, freshwater shrimp, midges, amphipods, and various immature insects. The organisms are highly dependent, either directly or indirectly, on the abundant detritus found in these systems. Beck (1977) found that a cypress–tupelo swamp in the Louisiana Atchafalaya Basin had a higher number of organisms (3,768 individuals/m^2) than the bayous (3,292/m^2), lakes (1,840/m^2), canals (1,593/m^2), and rivers (327/m^2). These high densities were attributed to the abundance of detritus and to the pulse of spring flooding. Sklar and Conner (1979), working in contiguous environments in Louisiana, found a higher number of benthic organisms in a cypress–tupelo swamp (7,500 individuals/m^2) than was found in a nearby impounded swamp (3,000/m^2) or in a swamp managed as a crayfish farm. Their study suggests that the natural swamp hydroperiod results in the highest number of benthic invertebrates.

Ziser (1978) and Bryan et al. (1976) surveyed the benthic fauna of alluvial river swamps. Oligochaetes and midges (Chironomidae), both of which can tolerate low-dissolved-oxygen conditions, and amphipods such as *Hyalella azteca,* which occur in abundance amid aquatic plants such as duckweed, usually dominate. In nutrient-poor cypress domes, the benthic fauna are dominated by Chironomidae, although crayfish, isopods, and other Diptera are also found there. Stresses stemming from low dissolved oxygen and periodic drawdowns account for the low diversity and number in these domes.

The production of wood in deepwater swamps results in an abundance of substrate for invertebrates to colonize, although few studies have documented the importance of this substrate in swamps for invertebrates. Thorp et al. (1985) found that suspended *Nyssa* logs had three times as many invertebrates and twice as many taxa when they were placed in a swamp-influent stream than in the swamp itself or by its outflow stream. The swamp inflow had the highest number of mayflies (Ephemeroptera), stoneflys (Plecoptera), midges (Chironomids), and caddie flies (Trichoptera), whereas Oligochaetes were greatest in the swamp itself, supposedly because of anoxic, stagnant conditions.

Fish

Fish are both temporary and permanent residents of alluvial river swamps (Wharton et al., 1976, 1981). Several studies have noted the value of sloughs and backswamps for fish and shellfish spawning and feeding during the flooding season (Lambou, 1965, 1990; R. J. Patrick et al., 1967; Bryan et al., 1976; Wharton et al., 1981). The deepwater swamp often serves as a reservoir for fish when flooding ceases, although the backwaters are less than optimum for aquatic life because of fluctuating water levels and occasional low-dissolved-oxygen levels. Some fish such as bowfin (*Amia calva*), gar (*Lepisosteus* sp.), and certain top minnows (e.g., *Fundulus* spp. and *Gambusia affinis*) are better adapted to periodic anoxia through their ability to utilize atmospheric oxygen. Several species of forage minnows often dominate alluvial river swamps, where most larger fish are temporary residents of the wetlands (Clark, 1979). Fish are sparse to nonexistent in the shallow cypress domes, white cedar swamps, and red maple swamps because of the lack of continuous standing water.

Reptiles and Amphibians

Reptiles and amphibians are prevalent in swamps because of their ability to adapt to fluctuating water levels. Nine or ten species of frogs are common in many southeastern cypress–gum swamps (Clark, 1979). Two of the most interesting reptiles in southeastern deepwater swamps are the American alligator and the cottonmouth. The alligator ranges from North Carolina through Louisiana, where alluvial cypress swamps and cypress strands often serve as suitable habitats. The cottonmouth, or water moccasin (*Agkistrodon piscivorus*), a poisonous water snake that has a white inner mouth, is found throughout much of the range of cypress wetlands and is the topic of many a "snake story" of those who have been in these swamps. Other water snakes, particularly several species of *Nerodia*, however, are often more important in terms of number and biomass and are often mistakenly identified as cottonmouth. The snakes feed primarily on frogs, small fish, salamanders, and crayfish.

Red maple swamps are important areas in the forested northeastern United States for the breeding and feeding of reptiles and amphibians (Golet et al., 1993). DeGraaf and Rudis (1986) found that there were 45 species of reptiles and amphibians that required forest cover sometime during the year in New England and that, of the 11 types of forests studied, red maple swamps were actually the preferred habitat of 12 of those 45 species. In a later study, DeGraaf and Rudis (1990) found that red maple swamps with streams supported twice as many individuals of reptiles and amphibians as did red maple swamps without streams, with wood frog (*Rana sylvatica*), redback salamander (*Plethodon cinereus*), and American toad (*Bufo americanus*) accounting for 90 percent of the abundance.

Ecosystem Function

Several generalizations about the ecosystem function of swamps will be discussed in this section:

1. Swamp productivity is closely tied to its hydrologic regime.
2. Nutrient inflows, often coupled with hydrologic conditions, are major sources of influence on swamp productivity.
3. Swamps can be nutrient sinks whether the nutrients are a natural source or are artificially applied.
4. Decomposition of woody and nonwoody material in swamps is affected by the water regime and the subsequent degree of anaerobiosis.

Primary Productivity

There is a wide range of productivity reported for forested swamps, with almost all of the studies carried out in the southeastern United States. Brinson et al. (1981a), Conner and Day (1982), and Lugo et al. (1990a) compiled summaries of much of the data on deepwater swamp productivity. The studies show that primary productivity depends on hydrologic and nutrient conditions (Table 14-9) in a diverse number of deepwater swamps: Florida cypress strands (Carter et al., 1973; Burns, 1978), Louisiana cypress–tupelo swamps (Conner and Day, 1976; Day et al., 1977; Conner et al., 1981), Florida cypress swamps (Mitsch and Ewel, 1979; S. L. Brown, 1981; Marois and Ewel, 1983), an Illinois alluvial river swamp (Mitsch, 1979; Mitsch et al., 1979a; Dorge et al., 1984), a Georgia pond cypress swamp (Schlesinger, 1978), a North Carolina alluvial swamp (Brinson et al., 1980), the Great Dismal Swamp of Virginia (Megonigal and Day, 1988; Powell and Day, 1991), and western Kentucky swamps (Mitsch et al., 1991). In almost all of these studies, only above-ground productivity was estimated.

Powell and Day (1991) made direct measurements of below-ground productivity and found that it was highest in a mixed hardwood swamp (989 g m^{-2} yr^{-1}) and much lower in a more frequently flooded cedar swamp (366 g m^{-2} yr^{-1}), a cypress swamp (308 g m^{-2} yr^{-1}), and a maple–gum swamp (59 g m^{-2} yr^{-1}). These results suggest that the allocation of carbon to the root system decreases with increased flooding.

Figure 14-11 shows three similar relationships that have been suggested to explain the importance of hydrology on swamp productivity. In a study of cypress tree productivity in Florida, Mitsch and Ewel (1979) concluded that growth was low in pure stands of cypress (characterized by deep standing water) and in cypress–pine associations (characterized by dry conditions). The productivity of cypress was high in cypress–tupelo associations that were characterized by moderately wet conditions and in cypress–hardwood associations that have fluctuating water levels characteristic of alluvial river swamps (Fig. 14-11a). Conner and Day (1982) suggested a similar relationship between swamp productivity and hydrologic conditions and produced a similar curve, including some data points (Fig. 14-11b). Golet et al. (1993) developed a similar curve for radial growth of *Acer rubrum* as a function of hydrologic conditions for six swamps in Rhode Island (Fig. 14-11c). They showed that productivity was greatest when the water level was average and that productivity, as measured by radial growth, was lower when conditions were either wetter or drier than normal. All of

Table 14-9 Biomass and net primary productivity of deepwater swamps in the southeastern United States

Location/Forest Type	Tree Standing Biomass (kg/m²)	Litterfall (g m⁻² yr⁻¹)	Stem Growth (g m⁻² yr⁻¹)	Above-Ground NPP[a] (g m⁻² yr⁻¹)	Reference
LOUISIANA					
Bottomland hardwood	16.5[b]	574	800	1,374	Conner and Day (1976)
Cypress–tupelo	37.5[b]	620	500	1,120	Ibid.
Impounded managed swamp	32.8[b,c]	550	1,230	1,780	Conner et al. (1981)
Impounded stagnant swamp	15.9[b,c]	330	560	890	Ibid.
Tupelo stand	36.2[b]	379	—	—	Conner and Day (1982)
Cypress stand	27.8[b]	562	—	—	Ibid.
NORTH CAROLINA					
Tupelo swamp	—	609–677	—	—	Brinson (1977)
Floodplain swamp	26.7[d]	523	585	1,108	Mulholland (1979)
VIRGINIA					
Cedar swamp	22.0[b]	758	441	1,097[j]	Dabel and Day (1977), Gomez and Day (1982), Megonigal and Day (1988)
Maple gum swamp	19.6[b]	659	450	1,050[j]	Ibid.
Cypress swamp	34.5[b]	678	557	1,176[j]	Ibid.
Mixed hardwood swamp	19.5[b]	652	249	831[j]	Ibid.
GEORGIA					
Nutrient-poor cypress swamp	30.7[e]	328	353	681	Schlesinger (1978)
ILLINOIS					
Cypress tupelo	45[d]	348	330	678	Mitsch (1979), Dorge et al. (1984)

Cypress–ash slough	31.2	136	498	634	Taylor (1985), Mitsch et al. (1991)
Cypress swamp	10.2	253	271	524	Ibid.
Stagnant cypress swamp	9.4	63	142	205	Ibid.
FLORIDA					
Cypress–tupelo (6 sites)	19 ± 4.7[f]	—	289 ± 58[f]	760[g]	Mitsch and Ewel (1979)
Cypress–hardwood (4 sites)	15.4 ± 2.9[f]	—	336 ± 76[f]	950[g]	Ibid.
Pure cypress stand (4 sites)	9.5 ± 2.6[f]	—	154 ± 55[f]	—	Ibid.
Cypress–pine (7 sites)	10.1 ± 2.1[f]	—	117 ± 27[f]	—	Ibid.
Floodplain swamp	32.5	521	1,086	1,607	S. L. Brown (1978)
Natural dome[h]	21.2	518	451	969	Ibid.
Sewage dome[i]	13.3	546	716	1,262	Ibid.
Scrub cypress	7.4	250	—	—	S. L. Brown and Lugo (1982)
Drained strand	8.9	120	267	387	Carter et al. (1973)
Undrained strand	17.1	485	373	858	Ibid.
Larger cypress strand	60.8	700	196	896	Duever et al. (1984)
Small tree cypress strand	24.0	724	818	1,542	Ibid.
Sewage enriched cypress strand	28.6	650	640	1,290	Nessel (1978)

[a] NPP = net primary productivity = litterfall + stem growth.
[b] Trees defined as >2.54 cm DBH (diameter at breast height).
[c] Cypress, tupelo, ash only.
[d] Trees defined as >10 cm DBH.
[e] Trees defined as >4 cm DBH.
[f] Average ± std error for cypress only.
[g] Estimated.
[h] Average of five natural domes.
[i] Average of three domes; domes were receiving high nutrient wastewater.
[j] Above-ground NPP is less than sum of litterfall plus stem growth because some stem growth is measured as litterfall.

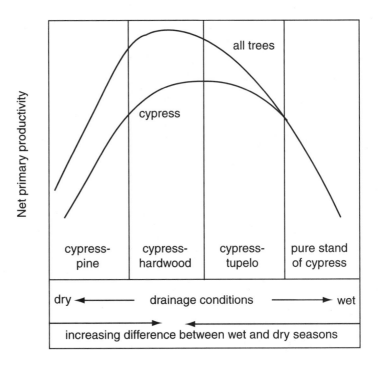

a.

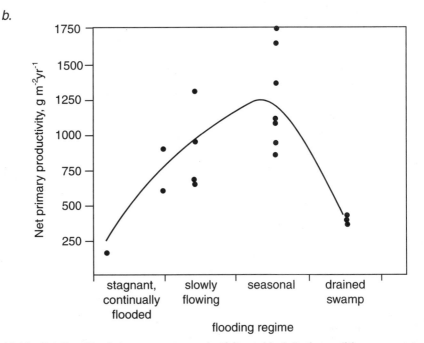

b.

Figure 14-11 Relationships between swamp productivity and hydrologic conditions: **a.** general relationship for cypress swamps in north-central Florida, **b.** relationship between flooding regime and net primary productivity of Louisiana swamps, and **c.** relationship between radial growth of red maple (*Acer rubrum*) and annual water level for six Rhode Island red maple swamps over 6 years. (*a after Mitsch and Ewel, 1979; b after Conner and Day, 1982; c after Golet et al., 1993*)

c.

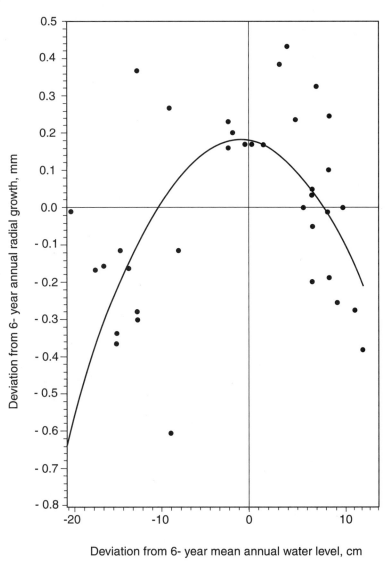

Deviation from 6- year mean annual water level, cm

Figure 14-11 *(Continued).*

the curves in Figure 14-11 suggest that the highest productivity occurs in systems that are neither very wet nor too dry but that have either average hydrologic conditions or seasonal hydrologic pulsing.

The importance of flooding to the productivity of alluvial river swamps is further illustrated in Figure 14-12a, where the basal-area growth of bald cypress in an alluvial river swamp in southern Illinois was strongly correlated with the annual discharge of the adjacent river. This graph suggests that higher tree productivity in this wetland occurred in years when the swamp was flooded more frequently than average or for longer durations by the nutrient-rich river. Similar correlations were also obtained when other independent variables that indicate degree of flooding were used. Data

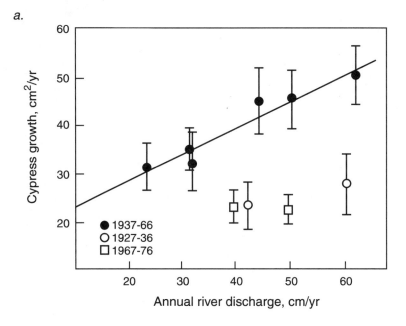

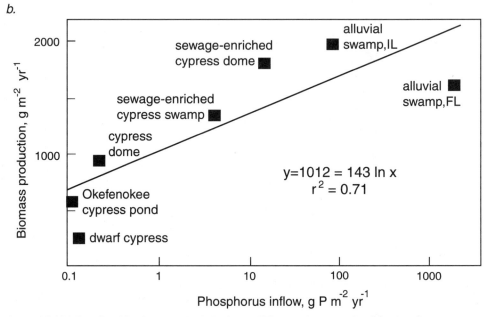

Figure 14-12 Relationships between hydrologic conditions and tree productivity in cypress swamps: **a.** increase in basal area of bald cypress trees in southern Illinois alluvial swamp as a function of river discharge for 5-year periods and **b.** cypress swamp biomass production as a function of phosphorus inflow. Data points in **a.** indicate mean; bar indicates one standard error. (*a after Mitsch et al., 1979a; b after S. L Brown, 1981*)

points for tree growth and flooding for 1927–1936 and 1967–1976 were not included in the regression analysis due to extensive logging activity and the invasion of the pond by beavers during those periods (Mitsch et al., 1979a).

S. L. Brown (1981) and Brinson et al. (1981a) found a similar relationship between productivity and hydrology. Brown emphasized the importance of nutrient inflows as well as hydrologic conditions to productivity in cypress swamps (Fig. 14-12b). She pointed out that a stillwater cypress dome that received a low rate of nutrient inflow increased in productivity when it was enriched with high-nutrient wastewater. Nessel et al. (1982) and Lemlich and Ewel (1984) also demonstrated an increased growth of individual *Taxodium* trees in cypress domes when high-nutrient wastewater was applied. Hydrologic inflows and nutrient inflows are coupled in most swamps, and both charts in Figure 14-12 reflect the same phenomenon.

The influence of the alteration of the hydroperiod on the productivity of cypress swamps is apparent from the preceding studies. When flooding of an alluvial river swamp is decreased, lower productivity may result. When stillwater swamps are drained, productivity of the swamp species will eventually decrease as water-intolerant species invade. Impounding wetlands, whether by artificial levees or beaver dams, leads to deeper and more continuous flooding and decreases productivity, sometimes to the point of killing the canopy trees.

Decomposition

Biological utilization of organic matter is primarily through detrital pathways in deep-water swamps, although decomposition is impeded by the anaerobic conditions usually found in the sediments. The first-order decay coefficients for litter in a variety of swamps are summarized in Table 14-10. Brinson (1977) found that decomposition of *Nyssa aquatica* leaves and twigs in a North Carolina riparian tupelo swamp was greatest in the wettest site, whereas Duever et al. (1984) found that decomposition in a southern Florida cypress strand region was slowest in an area where no flooding occurred and more rapid in areas that were flooded from 50 to 61 percent of the time. Deghi et al. (1980) found no difference between decomposition of pond cypress needles in deep and shallow sites in Florida cypress domes, although decomposition was generally more rapid in wet sites than in dry sites.

F. P. Day (1982) and Yates and Day (1983) investigated the decomposition of several types of forested wetland leaves in forested wetland communities in the Great Dismal Swamp in Virginia (Table 14-10). Decomposition was generally slower there than in the Florida and North Carolina alluvial swamps. Variations in decomposition were seen from site to site. There was a faster breakdown of litter in seasonally flooded sites than in a mixed hardwood site that was not inundated. The data suggested, however, that the type of litter was often as important as flooding in differentiating among sites. Water tupelo leaves had the highest decomposition rates, maple leaves showed a middle range of decomposition, and cedar and bald cypress had the slowest rates (F. P. Day, 1982). There was also quite a difference in decomposition rates measured between 1978 and 1979 (F. P. Day, 1982) and in 1980 (Yates and Day, 1983). The slower decomposition in the cypress and maple–gum communities in 1980 was attributed to a drought during the summer and autumn and an abbreviated period of inundation in the spring (Yates and Day, 1983).

The decomposition of tree roots in forested wetlands has also been investigated for cypress swamps. F. P. Day et al. (1989) studied *Taxodium* root decomposition

Table 14-10 Decay coefficients of litter in deepwater swamps

	Material	Site description	Decay Coefficient (k, yr^{-1})[a]	Reference
FLORIDA				
Cypress strand	Site leaves	On forest floor	0.86–1.39[b]	Burns (1978)
	Site leaves	On debris pile	0.69–0.75[b]	Ibid.
	Site leaves	Dry site	0.47	Duever et al. (1984)
	Site leaves	Flooded 50% of time	0.23	Ibid.
	Site leaves	Flooded 61% of time	0.30	Ibid.
Cypress dome	Pond cypress leaves	Flooded 100% of time	1.21–1.40	Deghi et al. (1980)
	Pond cypress leaves	Dry site	0.50–0.69	Ibid.
Scrub cypress	Site leaves		0.25	S. L. Brown et al. (1984)
NORTH CAROLINA				
	Water tupelo leaves		1.89	Brinson (1977)
	Water tupelo twigs		0.28	Ibid.
VIRGINIA				
Cypress community	Bald cypress leaves		0.33	F. P. Day (1982)
	Mixed litter (1978–1979)		0.51–0.59	Ibid.
	Mixed litter (1980)		0.28	Yates and Day (1983)
Maple gum communities	Tupelo leaves		0.65	F. P. Day (1982)
	Maple leaves		0.47	Ibid.
	Mixed litter (1978–1979)		0.51–0.67	Ibid.
	Mixed litter (1980)		0.29	Yates and Day (1983)
Cedar community	Cedar leaves		0.34	F. P. Day (1982)
	Mixed litter (1978–1979)		0.35–0.43	Ibid.
	Mixed litter (1980)		0.49	Yates and Day (1983)

[a] Decay coefficinet (k) based on exponential decay: $y = y_0 e^{-kt}$

[b] Range due to different mesh size used in litterbags.

where y = final biomass

y_0 = initial biomass

t = time in years

Source: Brinson et al. (1981).

in experimental mesocosms and found little difference in mass loss between a continuously flooded and a periodically flooded mesocosm, although the periodically flooded roots generally had higher concentrations of nutrients (nitrogen and phosphorus) than the continuously flooded mesocosm did. The decomposition of root material was extremely slow after the initial rapid leaching losses in the first 6 months, and inhibited decay was correlated with abiotic variables such as redox potential, low oxygen concentrations, and pH. Tupacz and Day (1990) estimated the rate of root decomposition with litter bags and sediment cores in a continuation of the previously mentioned Great Dismal Swamp studies, and found slowest decay on sites with the longest soil saturation and greatest soil depths. Decay rates, expressed as linear rather than exponential quantities, ranged from 0.48 to 1.00 mg g^{-1} day^{-1} in the litter bags. Roots of *Quercus* rather than *Taxodium* or *Nyssa* were most resistant to decay.

Generally, decomposition of both leaves and roots seems to be maximum in wet but not permanently flooded sites. The rates also increase with increases in average ground temperature and depend strongly on the quality (species, type of litter or roots) of the decomposing material.

Organic Export

Little organic matter is exported from stillwater and slow-flowing swamps such as cypress domes and strands and most red maple swamps and cedar swamps. Export from lake-edge swamps and alluvial river swamps, however, can be significant. Organic export appears to be higher from watersheds that contain significant swamps than from those that do not (Mulholland and Kuenzler, 1979; see Fig. 6-23). J. W. Day et al. (1977) found a high rate of 10.4 g C m^{-2} yr^{-1} exported as total organic carbon from a swamp forest in Louisiana. Further discussions of organic export from swamp-dominated watersheds are contained in Chapters 6 and 15.

Energy Flow

The energy flow of deepwater swamps is dominated by primary productivity of the canopy trees. Energy consumption is primarily accomplished by detrital decomposition. Significant differences exist, however, between the energy flow patterns in low-nutrient swamps such as dwarf cypress swamps and cypress domes and high-nutrient swamps such as alluvial cypress swamps (Table 14-11). All of the cypress wetlands

Table 4-11 Estimated energy flow (kcal m^{-2} day^{-1}) in selected Florida cypress swamps[a]

Parameter	Dwarf Cypress Swamp	Cypress Dome	Alluvial River Swamp
Gross primary productivity[b]	27	115	233
Plant respiration[c]	18	98	205
Net primary productivity	9	17	28
Soil or water respiration	7	13	18
Net ecosystem productivity	2	4	10

[a] Assume 1 g C = 10 kcal.
[b] Assumes GPP = net daytime photosynthesis + nighttime leaf respiration.
[c] Plant respiration = 2 × (nighttime leaf respiration) + stem respiration + knee respiration.
Source: S. L. Brown (1981).

are autotrophic—productivity exceeds respiration. Gross primary productivity, net primary productivity, and net ecosystem productivity are highest in the alluvial river swamp that receives high-nutrient inflows. Buildup and/or export of organic matter are characteristic of all of these deepwater swamps but are most characteristic of the alluvial swamp. There are few allochthonous inputs of energy to the low-nutrient wetlands, and energy flow at the primary producer level is relatively low. The alluvial cypress–tupelo swamp depends more on allochthonous inputs of nutrients and energy, particularly from runoff and river flooding. In alluvial deepwater swamps, productivity of aquatic plants is often high, whereas aquatic productivity in cypress domes is usually low.

Nutrient Budgets

The functioning of forested wetlands as nutrient sinks was suggested by Kitchens et al. (1975) in a preliminary winter–spring survey of a swamp forest alluvial river swamp complex in South Carolina. They found a significant reduction in phosphorus as the waters passed over the swamp. They assumed this to be the result of biological uptake by aquatic plant communities. In a similar study in Louisiana, J. W. Day et al. (1977) found that nitrogen was reduced by 48 percent and phosphorus decreased by 45 percent as water passed through a lake–swamp complex of Barataria Bay to the lower estuary. They attributed this decrease in nutrients to sediment interactions, including nitrate storage/denitrification and phosphorus adsorption to the clay sediments. Several studies in Florida were initiated to investigate the value of wetlands as nutrient sinks. These studies, from the Center for Wetlands at the University of Florida, included the purposeful disposal of high-nutrient wastewater in cypress domes (H. T. Odum et al., 1977a; Ewel and Odum, 1984; Dierberg and Brezonik, 1984, 1985) and the long-term inadvertent disposal of wastewater by small communities in forested wetlands (Boyt et al., 1977; Nessel and Bayley, 1984). In all of these studies, the wetlands acted as sinks for nitrogen and phosphorus. In the cypress dome experiments, the nutrients were essentially retained in the water, sediments, and vegetation with little surface outflow. Boyt et al. (1977) described nutrient uptake that occurred in a mixed hardwood swamp that had received domestic sewage effluent for 20 years. They found that the total phosphorus and the total nitrogen in the outflow were reduced by 98 and 90 percent, respectively, compared with the inflow. In at least one contrasting study, a net export of phosphorus was seen in 6 out of 7 years in a multiyear study of a forested wetland in Florida that received treated wastewater (Knight et al., 1987). The wetland, then, was serving as a source of phosphorus, probably the result of the entrainment of phosphorus that had accumulated in the forested wetland litter and soils.

Nutrient budgets of deepwater swamps vary from "open" alluvial river swamps that receive and export large quantities of materials to "closed" cypress domes that are mostly isolated from their surroundings (Table 14-12). Mitsch et al. (1979a) developed a nutrient budget for an alluvial river swamp in southern Illinois and found that 10 times more phosphorus was deposited with sediments during river flooding (3.6 g P m^{-2} yr^{-1}) than was returned from the swamp to the river during the rest of the year (Fig. 14-13). Thus, the swamp was a sink for a significant amount of phosphorus and sediments during that particular year of flooding, although the percentage of retention was low (3–4.5 percent) because a very large volume of water passed over the swamp during flooding conditions.

Table 14-12 Phosphorus inputs to deepwater swamps (g P m^{-2} yr^{-1})

Swamp	Rainfall	Surface Inflow	Sediments from River Flooding	Reference
FLORIDA				
Dwarf cypress	0.11	—[a]	0	S. L. Brown (1981)
Cypress dome	0.09	0.12	0	Ibid.
Alluvial river swamp	—[a]	—[a]	3.1	Ibid.
SOUTHERN ILLINOIS				
Alluvial river swamp	0.11	0.1	3.6	Mitsch et al. (1979a)
NORTH CAROLINA				
Alluvial tupelo swamp	0.02–0.04	0.01–1.2	0.2	Yarbro (1983)

[a] Not measured.

A study by Kuenzler et al. (1980) typifies several chemical budget studies that have been developed for the Coastal Plain floodplain swamps of North Carolina. They found that 94 percent of the phosphorus transported to these wetlands was carried by surface water. They also found that there was a significant retention of phosphorus by the swamp, resulting in low concentrations of phosphorus downstream of the wetland. Yarbro (1983) found that an alluvial swamp in North Carolina retained 0.3 to 0.7 g P m^{-2} yr^{-1} over 2 years of measurement. Intrasystem cycling through the vegetation accounted for 0.5 g P m^{-2} yr^{-1}, but only a fraction was retained in the woody biomass.

Ecosystem Models

A conceptual model of a Florida cypress swamp is shown in Figure 14-14. This model illustrates the importance of water level, fire, nutrients, and tree harvesting on ecosystem functioning. Several simulation models have also been developed to study the ecosystem dynamics of deepwater swamps. A review of some of these models is given in Mitsch et al. (1982), Mitsch (1983, 1988), and H. T. Odum (1983, 1984). One such simulation model of a Florida cypress swamp, similar to the model shown in Figure 14-14, was used to investigate the management of cypress domes in Florida. The model predicted long-term (100 years) effects of drainage, harvesting, fire, and nutrient disposal. Simulations showed that, when water levels were lowered and trees were logged at the same time, the cypress swamp did not recover and was replaced by shrub vegetation. The model also suggested that if tree harvesting occurred without drainage, the cypress would recover because of the absence of fire.

In a model that investigated understory productivity in cypress domes, Mitsch (1984) demonstrated annual patterns of aquatic productivity that peaked in the spring when maximum solar radiation is available through the deciduous cypress canopy. M. T. Brown (1988) developed a model that demonstrated the annual patterns of hydrology and nutrients in forested swamps typical of central and northern Florida. The model illustrated the impact of surface water diversion, lowering groundwater,

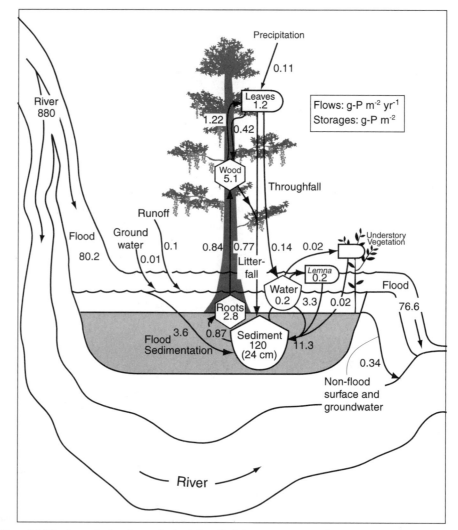

Figure 14-13 Annual phosphorus budget for alluvial cypress swamp in southern Illinois. (*After Mitsch et al., 1979a*)

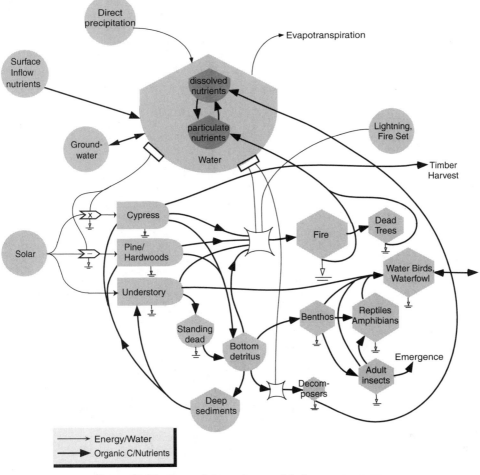

Figure 14-14   General ecosystem model of cypress swamp.

and additions of advanced wastewater treatment effluents on the hydroperiod and nutrient dynamics of the swamps.

Recommended Readings

Ewel, K. C., and H. T. Odum, eds. 1984. Cypress Swamps. University Presses of Florida, Gainesville, FL. 472 pp.

Laderman, A. D., ed. 1998. Coastally Restricted Forests. Oxford University Press, Oxford, UK. 334 pp.

Lugo, A. E., M. M. Brinson, and S. L. Brown, eds. 1990. Forested Wetlands: Ecosystems of the World 15. Elsevier, Amsterdam. 527 pp.

Messina, M. G., and W. H. Conner, eds. 1998. Southern Forested Wetlands. Lewis Publishers, Boca Raton, FL. 616 pp.

Chapter **15**

Riparian Ecosystems

Riparian wetlands, ecosystems in which soils and soil moisture are influenced by the adjacent stream or river, are unique because of their linear form along rivers and streams and because they process large fluxes of energy and materials from upstream systems. Major expanses of riparian ecosystems, called bottomland hardwood forests, are found in the eastern United States, although many have been drained and cleared for agriculture. Less extensive but ecologically critical riparian systems in the western United States, especially in the arid Southwest, have nearly all been greatly modified as a result of water management practices, logging, and use by livestock. Riparian zones respond to structures and processes operating at continental scales (climate, geology); river system (intrariparian) scales (elevation, timing, magnitude, duration of flooding, stream valley landform dynamics); and local transriparian scales (slope and moisture gradients, sediment sorting, biotic processes). Riparian zones span a wide range of environments and processes, the common thread being the linkage between riparian zone, river, and adjacent upland. Riparian systems of the American West often occur on low-order streams and are characterized by extreme and variable fluvial conditions and geomorphic responses. Southeastern riparian systems, which are the most extensive wetlands in the United States, in contrast, are generally low-lying, flat, extensive floodplains, with strong seasonal hydrologic pulses and well-developed soils.

Flooding of the riparian zone affects soil chemistry by producing anaerobic conditions, importing and removing organic matter, and replenishing mineral nutrients. The plant communities of riparian ecosystems, which form in response to continental and watershed gradients, transriparian moisture, valley form, and soil gradients, are generally productive and diverse. The riparian zone is valuable for many animals that seek its refuge, diversity of habitat, and abundant water or that use it as a corridor for migration. This is particularly true in arid regions, where riparian zones may support the only vegetation within many kilometers. The functions of these ecosystems are poorly understood, except that they are open to large fluxes of energy and nutrients. The riparian ecosystem acts as a nutrient sink for lateral runoff from uplands but as a nutrient transformer for upstream–downstream flows. Few energy–nutrient models have been developed to describe these systems, although tree growth models have been applied to bottomland hardwood forests with some success.

The riparian zone of a river, stream, or other body of water is the land adjacent to that body of water that is, at least periodically, influenced by flooding. E. P. Odum (1981) described the riparian zone as "an interface between man's most vital resource, namely, water, and his living space, the land." A National Symposium on Strategies for the Protection and Management of Floodplain Wetlands and Other Riparian Ecosystems (R. R. Johnson and McCormick, 1979) developed a definition of riparian ecosystems:

> Riparian ecosystems are ecosystems with a high water table because of proximity to an aquatic ecosystem or subsurface water. Riparian ecosystems usually occur as an ecotone between aquatic and upland ecosystems but have distinct vegetation and soil characteristics. Aridity, topographic relief, and presence of depositional soils most strongly influence the extent of high water tables and associated riparian ecosystems. These ecosystems are most commonly recognized as bottomland hardwood and floodplain forests in the eastern and central U.S. and as bosque or streambank vegetation in the west. Riparian ecosystems are uniquely characterized by the combination of high species diversity, high species densities and high productivity. Continuous interactions occur between riparian, aquatic, and upland terrestrial ecosystems through exchanges of energy, nutrients, and species.

Minshall et al. (1989) incorporated the U.S. Fish and Wildlife Service wetland definition (Cowardin et al., 1979) into their definition of riparian zones of the western states' Great Basin region: "Land inclusive of hydrophytes and/or with soil that is saturated by ground water for at least part of the growing season within the rooting depth of potential native vegetation."

This definition includes both "wetlands" as defined by Cowardin et al. (1979) and more mesic adjacent lands called *transitional* ecosystems by Kovalchik (1987). The latter are subirrigated sites such as inactive floodplains, terraces, toe-slopes, and meadows that have seasonally high water tables that recede to below the rooting zone in late summer. Thus, wetlands are a subset of riparian habitats, based on the more restrictive wetland definition of water dependency (Fig. 15-1). We make no attempt to distinguish between wetland and nonwetland parts of the riparian ecosystem in this chapter.

Gregory et al. (1991) emphasized landscape pattern in their characterization of the riparian zone:

> Riparian zones are the interfaces between terrestrial and aquatic ecosystems. As ecotones, they encompass sharp gradients of environmental factors, ecological processes and plant communities. Riparian zones are not easily delineated but are composed of mosaics of landforms, communities, and environments within the larger landscape.

In general terms, riparian ecosystems are found wherever streams or rivers at least occasionally cause flooding beyond their channel confines or where new sites for vegetation establishment and growth are created by channel meandering (e.g., point

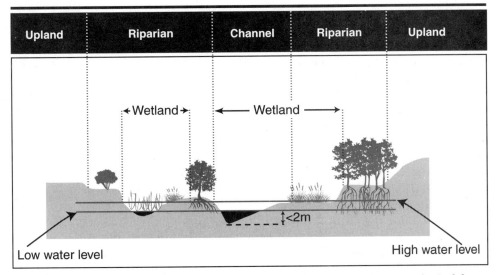

Figure 15-1 The relationship of riparian ecosystems to wetlands. Wetlands are a subset of riparian ecosystems. (*After Minshall et al., 1989*)

bars). In arid regions, riparian vegetation may be found along or in ephemeral streams as well as on the floodplains of perennial streams. In most nonarid regions, floodplains and, hence, riparian zones tend to appear first along a stream "where the flow in the channel changes from ephemeral to perennial—that is, where groundwater enters the channel in sufficient quantity to sustain flow through nonstorm periods" (Leopold et al., 1964).

Riparian ecosystems can be broad alluvial valleys several tens of kilometers wide in the southern United States or narrow strips of streambank vegetation in the arid western United States. Brinson et al. (1981b) described the "abundance of water and rich alluvial soils" as factors that make riparian ecosystems different from upland ecosystems. They listed three major features that separate riparian ecosystems from other ecosystem types:

1. Riparian ecosystems generally have a linear form as a consequence of their proximity to rivers and streams.

2. Energy and material from the surrounding landscape converge and pass through riparian ecosystems in much greater amounts than those of any other wetland ecosystem; that is, riparian systems are open systems.

3. Riparian ecosystems are functionally connected to upstream and downstream ecosystems and are laterally connected to upslope (upland) and downslope (aquatic) ecosystems.

In the United States, the most extensive riparian ecosystems are the mesic bottomland hardwood forest ecosystems of the Southeast, including the broad (often tens of kilometers wide), flat floodplain of the Mississippi River as well as the floodplains of numerous smaller rivers emptying into the Gulf of Mexico and the Atlantic Ocean. They are part of a flooding continuum that includes deepwater swamps (Chapter 14). These systems are quite different from the steep, narrow, continuously changing riparian zones in the Northwest mountains and from the floodplains of ephemeral rivers of the arid Southwest.

Conference proceedings, reports, and review articles about riparian wetlands include the proceedings of national symposia on riparian ecosystems (R. R. Johnson and McCormick, 1979; Abell, 1984; R. R. Johnson et al., 1985; Hook and Lea, 1989), the proceedings of a series of workshops on southern bottomland hardwood forests (Clark and Benforado, 1981; Gosselink et al., 1990b), and reviews on riparian ecosystems (Brinson et al., 1981b; Wharton et al., 1982; Minshall et al., 1989; Faber et al., 1989; Sharitz and Mitsch, 1993; Naiman and Decamps, 1997). In addition, 16 papers from the Symposium on Semiarid Riparian Ecosystems were published in a special issue of the journal *Wetlands* (Friedman et al., 1998a). Another special issue of *Wetlands* (Kleiss, 1996a) was devoted to research on the bottomland ecosystem of the Cache River basin in Arkansas.

Types and Geographical Distribution

Table 15-1 gives the distribution of land in riparian forests in the contiguous United States based on riparian forest vegetation type groups from U.S. Forest Service surveys. Abernethy and Turner (1987) estimated a total 1985 riparian forest area of about 22.9 million ha, excluding California, Arizona, New Mexico, Hawaii, and Alaska. Forested wetland area in Arizona, California, and New Mexico is about 360,000 ha (Brinson et al., 1981b), and 12 million ha are estimated in Alaska. Between 1940 and 1980, the national loss rate was 0.27 percent per year, or about 2.8 million ha. Most of this loss occurred in the south-central and southeastern United States, which has 58 percent of the total U.S. forested wetlands. The Forest Service surveys do not include many western U.S. riparian strips because the minimum surveyed size was 38 m wide, or 4 ha in area. Thus, the Pacific Coast and Rocky Mountain estimates are probably low. These figures differ slightly from those given in Chapter 4 because they were compiled by a different methodology. The trends, however, are the same from both sources.

Table 15-1 Wetland forest area in the United States, 1980, and rates of change 1940–1980

Region	Area, 1940 (× 1,000 ha)	Area, 1980 (× 1,000 ha)	Cumulative change 1940–1980 (× 1,000 ha/yr)	Percentage Change per Year
North central	6,900	5,000	−47.6	−0.69
Northeast	1,650	3,000	+33.8	+2.0
Pacific Coast (excluding California)	730	1,250	+12.9	+1.8
Rocky Mountain (excluding Arizona and New Mexico)	730	375	−9.0	−1.2
South central	9,000	7,100	−47	−0.52
Southeast	6,600	6,100	−11.7	−0.18
Total	25,600	22,800	−68.6	−0.27

Source: Abernethy and Turner (1987).

Mesic Riparian Ecosystems

Mesic riparian ecosystems, commonly called bottomland hardwood forests or "bottomland hardwoods," are one of the dominant types of riparian ecosystems in the United States. Historically, the term "bottomland hardwood forest" has been used to describe the vast forests that occur on river floodplains of the eastern and central United States, especially in the Southeast. Bottomland hardwood forests have been characterized as follows:

1. The habitat is inundated or saturated by surface water or groundwater periodically during the growing season.
2. The soils within the root zone become saturated periodically during the growing season.
3. The prevalent woody plant species associated with a given habitat have demonstrated the ability, because of morphological and/or physiological adaptation(s), to survive, achieve maturity, and reproduce in a habitat where the soils within the root zone may become anaerobic for various periods during the growing season.

Bottomland hardwood forests are particularly notable wetlands because of the large areas that they cover in the southeastern United States (Fig. 15-2) and because

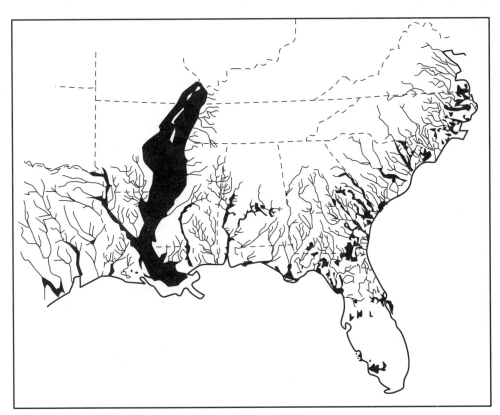

Figure 15-2 Extent of bottomland hardwood forests of the southeastern United States. (*After Putnam et al., 1960*)

of the rapid rate at which they are being converted to other uses such as agriculture and human settlements (see also Chapter 4). This ecosystem is particularly prevalent in the lower Mississippi River alluvial valley as far north as southern Illinois and western Kentucky (Taylor et al., 1990) and along many streams that drain into the Atlantic Ocean on the South Atlantic Coastal Plain. The Nature Conservancy (1992) estimated that before European settlement the Mississippi River alluvial plain supported about 8.5 million ha of riparian forests; about 2.0 million ha remained in 1991 (Fig. 4-9). The loss of riparian wetlands has been accompanied by fragmentation: Whereas single forest blocks used to cover hundreds of thousands of hectares, there now remain isolated fragments, most less than 100 ha in size, surrounded by agricultural fields (Gosselink et al., 1990c). The Atlantic Coastal Plain from Maryland to Florida is another area of dense riparian forests lining the many rivers that flow into the ocean. These forests have been logged, as have most in this country, but many are otherwise fairly intact.

Western U.S. Riparian Wetlands

Abernethy and Turner (1987) estimated a total 1980 forested riparian area of 375,000 ha in the Rocky Mountain states (excluding Arizona and New Mexico) and 1.25 million ha in the Pacific Coast states of Washington and Oregon. Arizona was reported to contain about 113,000 ha of riparian ecosystems (Babcock, 1968), New Mexico probably less. California has been reported to have lost more than 91 percent of its original wetland area (Dahl, 1990), including most of its riparian wetlands (Faber et al., 1989).

In contrast to the broad, flat, expansive southeastern riparian forests, riparian zones in the western United States tend to be narrow, linear features of the landscape, often lining streams with steep gradients and narrow floodplains. Along southeastern high-order rivers, the contrast in elevation and vegetation between bottomland and upland is often subtle and the gradient gradual, whereas in the West gradients are usually sharp and the visual distinctions are usually clear (Figs. 15-3 and 15-4).

Western riparian zones have been extensively modified by human activity. Conversion to housing or agriculture is widespread. Damage from grazing animals is ubiquitous. In an area where vegetation is generally limited by the lack of water, riparian vegetation and the availability of water inevitably draw and concentrate cattle. Grazing along these primarily low-order streams results in increased erosion and channel downcutting (Kauffman and Kreuger, 1984). Higher order streams have been modified for water use. They have been dammed; water withdrawals have depleted instream flows; the timing and size of floods and low flows have been modified; vegetation has been removed to reduce transpiration losses. In some areas, essentially all water has been withdrawn and perennial streams have become ephemeral below a dam. In other cases, return flows from agricultural use enhance streamflow and have led to expanded and more stable riparian zones (W. C. Johnson et al., 1976; Williams and Wolman, 1984; Bradley and Smith, 1986; Faber et al., 1989; Szaro, 1989; Knopf and Scott, 1990; Rood and Mahoney, 1990).

In the Northwest, the major impact on streams and riparian zones has occurred because of logging. Beginning in the early 1800s, streams were used to float logs from the forest to estuarine holding pens. Because many streams were too small to move logs efficiently, they were dammed, banks were cleared, and were otherwise modified to enhance logging (Sedell and Duval, 1985).

Figure 15-3 A view of Red Meadow Creek, Montana, about 2,200 m elevation. The photo shows typical linked series of riparian wetlands in western U.S. mountains from high-elevation seeps fed by melting snow, through shrub-dominated steep avalanche wetland chutes, to extensive wet meadows dominated by *Carex* and *Juncus*. (*Photograph by L. C. Lee, reprinted by permission*)

Figure 15-4 A riparian zone in the southwestern United States, showing a distinct pattern of trees in an otherwise almost treeless landscape. (*Photograph by L. C. Lee, reprinted by permission*)

Geomorphology and Hydrology

One way to describe these different systems is in terms of landform and process gradients that result in the changing continua of riparian zones. Throughout this chapter, we refer to processes on three major gradients that are nested in both space and time (Kovalchik, 1987; Minshall et al., 1989):

1. A *continental* gradient that includes the effects of east–west and latitudinal climatic gradients acting at the hydrologic basin level

2. An *intrariparian* longitudinal continuum (R. R. Johnson and Lowe, 1985), reflecting changes in elevation, stream gradient, fluvial processes, and sediments along the length of a stream system (Frissell et al., 1986; Baker, 1989; Gregory et al., 1991)

3. A lateral *transriparian* gradient (R. R. Johnson and Lowe, 1985) across the riparian zone; this is a local topographic gradient that reflects stream valley cross-sectional form and determines the local moisture regime and soil development

Continental and Regional Scales

Climate

Climate—related to latitude, elevation, and continental weather patterns—determines precipitation patterns and the presence or absence of a significant spring thaw. In the eastern and south-central United States, the climate is mesic. Precipitation exceeds evapotranspiration on an annual basis, and even during late summer, moisture deficits are not usually severe. Mountains are nonexistent or not as high as in the West. As a result, streams are usually perennial from source to mouth (Table 15-2). In a westward direction, moisture decreases; in the American West, evapotranspiration exceeds precipitation. In addition, along a latitudinal gradient in the Mid- and Southwest, precipitation declines so that in the southwestern United States in general evapotranspiration is much greater than precipitation (precipitation is less than 20 to 50 cm, and evapotranspiration is greater than 100 cm). The picture is complicated by the rainshadow effect of the Rocky Mountain range, yielding major climatic contrasts between the moist, maritime northwestern Coast Range and the arid eastern slope of the Rocky Mountains.

Evapotranspiration, which is governed primarily by air temperature, can control water levels even in mesic regions (Sharitz and Gibbons, 1982). In arid regions, it may be the single most important influence on riparian zones. Hence, western riparian zones are much more variable than the mesic eastern part of the United States. As examples of the foregoing, the delta of the Colorado River on the Gulf of California to its confluence with the Gila River is (or was before dam construction) perennial with hydroriparian vegetation; upstream, one encounters the mesoriparian vegetation of the intermittent San Pedro River. The intermittent headwater tributaries of the San Pedro are dry, desert, xeroriparian habitats (R. R. Johnson and Lowe, 1985). In the montane northwestern states, where high altitudes moderate evapotranspiration and increase precipitation, a mesic to semiarid climate gives rise to lush mesoriparian vegetation.

Watershed

In addition to continental factors—precipitation and temperature—the character of a watershed exerts a strong influence on river flooding. The local flooding regime of a riparian zone is determined by the elevation of the watershed (which, in turn, is related to precipitation and evapotranspiration), by the size of the watershed, and by its slope.

Drainage Area: Discharge volume, channel width, flow variability, and flooding duration are all related to drainage area of the stream basin upstream of the floodplain. For unconfined, mature streams, channel width is directly proportional to discharge volume, which, in turn, depends on upstream drainage basin area (Fig. 15-5). This relationship does not hold for immature, steep streams, especially in the arid West. In streams in the midwestern and southern United States, the channel is sized so that overbank flooding, on average, occurs about once every 1.5 years (Fig. 5-13). Similarly, for streams of the Great Plains, channel width is directly related to mean annual discharge. Flow variability is damped by increasing watershed size and discharge; the larger the annual discharge, the smaller the variation in flow. For example, streams with a low annual discharge of less than 1 m^3 s^{-1} tend to experience 10-year floods at least 100 times the mean discharge, whereas streams with an annual discharge of 100 m^3 s^{-1} have 10-year floods only 10 times the mean (Fig. 15-5b). This apparently occurs because in large watersheds the runoff volume is buffered by unequally distributed precipitation.

Flood duration is also affected by watershed area. For two basins in Arkansas, Bedinger (1981) found that sites that had larger upstream drainage areas had correspondingly longer times in standing water because small watersheds have rapid runoff

Table 15-2 A comparison of physical, geomorphic, and ecological features of western and eastern United States riparian ecosystems

Attribute	Southwest Upstream	Southwest Downstream	Northwest Upstream	Northwest Downstream	Southeast Upstream	Southeast Downstream
Example	San Pedro River	Colorado River		North N Fork Flathead River		Mississippi River alluvial valley
Climate	Arid	Arid	Semiarid	Semiarid	Mesic	Mesic
Net Precipitation	ET≥P	ET>P	ET≥P	ET>P	P>ET	P>ET
Landscape pattern	Linear	Braided	Linear	Linear	Linear	Broad floodplain
Stream gradient	Steep	Flat	Steep	Moderate/flat	Steep/moderate	Flat, meandering
Floodplain width	Narrow	Broad	Narrow	Moderate	Moderate	Very broad
Valley shape	V	Alluvial fan	V	Alluvial fan	U	Sinuous depositional flats
Fluvial type	Very flashy	Flashy	Very flashy	Flashy	Flashy	Seasonally pulsed
Fluvial stability (annual peak flood to mean flow)		10–100		10–100	5–10	2–5
Geomorphic system stability	Very low disequilibrium	Low	Very low disequilibrium	Fairly stable	Moderately stable	Stable (1,000+ yr)
Landscape contrast	Very high	High	Moderate	High	Fairly low	Low
Vegetation	Alder, poplar	Cottonwood, willow, ash, walnut, screw bean, salt cedar	Willow, poplar, birch, hemlock	Willow, sycamore, walnut, cottonwood	Sycamore, oak, hackberry, silver maple, river birch	Cypress, tupelo, many oak spp., willow, hickory, loblolly pine
Human impacts	Water diversion, dams, grazing, clearing for water retention		Logging, water diversion, dams		Logging, farming, flood control levees and draining	

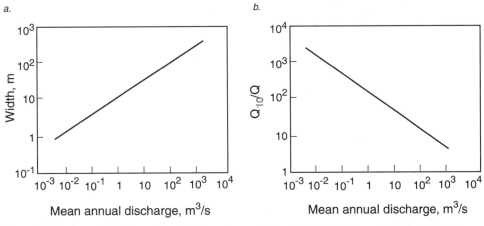

Figure 15-5 Characteristics of Great Plains streams in the United States. **a.** Active channel width as a function of mean annual discharge. The line separates streams characterized as narrow from those characterized as wide in relation to the average width for a given discharge. **b.** Flow variability (10-year flood divided by mean annual discharge) as a function of mean annual discharge. (*After Friedman et al., 1998*)

and sharp flood peaks, whereas large watersheds have flood peaks that are less sudden and longer lasting (Table 15-3). Typically, midwestern and eastern U.S. riparian ecosystems on small watersheds or low-order streams are flooded for several days to several weeks during the spring thaw, whereas the flooding of bottomland forests on the high-order Mississippi River lasts for long periods of up to several months. In northern latitudes, snow stores precipitation over large areas of the watershed. When it is released during the short spring thaws, great floods may occur that are outliers in the relationship between discharge and flow variability. For example, in Figure 15-5b, points above the regression line would be instances where the 10-year flood is greater than expected from the mean annual discharge. These discrepancies occur more frequently in streams with low mean discharges (small watersheds) than in high-discharge streams (large watersheds).

Channel Slope: The relationship between flood peak and mean flow is also influenced by the stream gradient. A stream that has a steep slope or gradient floods less frequently but with sharper flood peaks than a stream that contains a gentle slope

Table 15-3 Relationship of drainage basin size to duration of flooding of bottomland forests in Arkansas

Drainage Basin Area (mi^2)	Flooding Duration (% of time)
300	5–7
500–700	10–18
Several tens of thousands	18–40

Source: Bedinger (1981).

(Fig. 15-6). In the steeply sloped rivers of California, the flow of water tends to be extreme and to be associated with storms. For example, a flood 50 times the mean annual flood can be expected every 20 years on the Santa Ana River, and a 50-year flood on the Gila River is 280 times the mean annual discharge (Graf, 1988). In comparison, a 20-year flood on eastern U.S. rivers is only about 2 times the average annual flood.

Intrariparian Scale

Within the constraints of the continental or regional environment, the riparian zone changes along the course of a river system from its headwaters to its mouth. Vannote et al. (1980) described the structure and function of aquatic communities along a river system. The *intrariparian continuum* (R. R. Johnson and Lowe, 1985) refers to the same continuum in the riparian corridor along a river's course—the characteristics of communities along the river are determined by the geologic setting and by the geomorphic, hydrologic, and other physical processes that provide the physical environment within which a riparian community develops.

Most rivers have three major geomorphic zones (Fig. 15-7): erosion, storage and transport, and sediment deposition. These zones have been variously characterized (Brakenridge, 1988; Graf, 1988; R. R. Harris, 1988; Minshall et al., 1989; Faber et al., 1989).

1. The *zone of erosion* is in the headwaters and upper reaches of low-order streams. This zone is at high altitudes, at least in many areas of western North America. The stream course tends to be steep and straight, and the valleys are often V-shaped because they are scoured. The steep banks have narrow riparian zones. Flood frequency and duration vary widely, depending on precipitation. Contrary to the preceding generalization, if

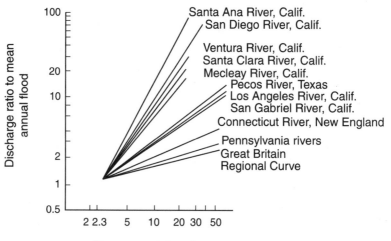

Figure 15-6 Ratio of flood to mean annual flood (which has a recurrence interval of 2.33 years) and recurrence intervals for several rivers in England and the United States. (*After Faber et al., 1989*)

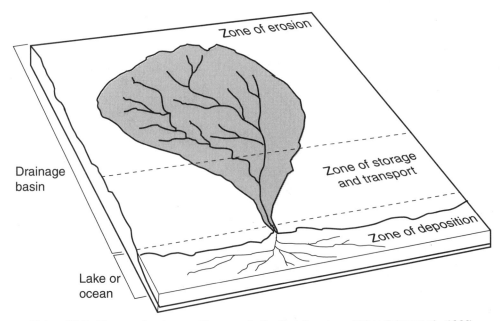

Figure 15-7 Three major geomorphic zones in the fluvial system. (*After Faber et al., 1989*)

local geology permits, some headwater areas may contain fairly extensive, flat meadows (fens) that may accrete considerable organic peat at high altitudes.

2. The *zone of storage and transport* occurs below the zone of erosion. These mid-order streams are primarily conduits for sediment, nutrients, and water. They tend to be fairly steep, and their straight V- or U-shaped channels, with some coarse-sediment deposition, form a narrow floodplain. These sediments are often scoured during high-energy floods. Flooding is variable and depends on the size of the watershed, the gradient, and the local precipitation.

3. The *zone of deposition* is characteristic of high-order, low-gradient streams. Sediment deposition is much greater than erosion and transport, and valley slopes are gentle. These two factors lead to the development of broad floodplains and sinuous and meandering stream channels. Sediments grade from coarse at the channel to fine at the periphery of the floodplain. The flooding of the riparian zone tends to be seasonal, characterized by one or a few long-duration spring floods. At their downstream ends, rivers typically debouch into flat, broad valleys, where the channels become braided and the flow is often unconfined. These depositional rivers are characteristic of the southeastern U.S. Coastal Plain, running into coastal estuaries. In the American West, evapotranspiration is the dominant process that determines the volume of the flow of these braided streams. As water evaporates and flow decreases, soil salts may accumulate.

A stream system can be broken down into eight reaches (Fig. 15-8). Headwater streams are typically erosional features, whereas the transition zones are zones of

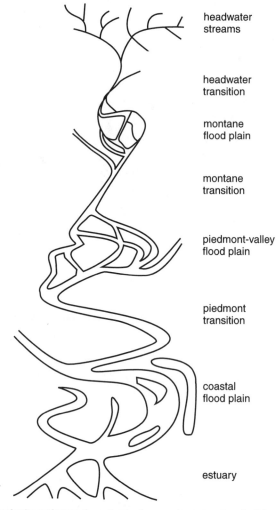

headwater
streams

headwater
transition

montane
flood plain

montane
transition

piedmont-valley
flood plain

piedmont
transition

coastal
flood plain

estuary

Figure 15-8 **Biophysical continuum for a large river system, progressing from headwater streams of an alpine area, through the montane, piedmont valley, and coastal floodplains, to an estuary. (*After Standford et al., 1996*)**

storage or transport. The three floodplain environments, montane, piedmont valley, and coastal floodplain, typically occur below valley constrictions or sites where tributaries flow into the main channel. They are typically depositional and the streams are braided or meandering. These areas form complex habitat mosaics with rich biodiversity.

Arid U.S. Riparian Systems

In the eastern Sierra Nevadas, glaciated bedrock valleys at elevations greater than 2,500 m tend to have incised stream channels and narrow floodplains, moderate gradients, and limited sediment deposits (R. R. Harris, 1988; Fig. 15-9). U-shaped valleys in montane floodplains in glacial till occur at slightly lower elevations (2,000–2,500 m). The floodplain gradient is moderate, and the floodplain width is the widest of all types classified. V-shaped valleys incised in glacial till span the widest elevation

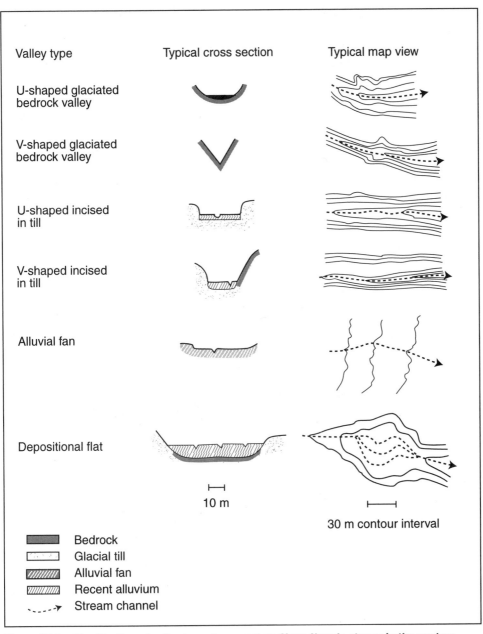

Valley type	Typical cross section	Typical map view
U-shaped glaciated bedrock valley		
V-shaped glaciated bedrock valley		
U-shaped incised in till		
V-shaped incised in till		
Alluvial fan		
Depositional flat		

10 m

30 m contour interval

Bedrock
Glacial till
Alluvial fan
Recent alluvium
Stream channel

Figure 15-9 **Classification of valley types in an eastern Sierra Nevada stream in the western United States.** (*After R. R. Harris, 1988*)

range and are almost always associated with diversions that reduce the mean annual streamflow below the diversion to less than 18 percent of the natural flow. These three channel types probably occur predominantly in erosional and transport sections of the stream (Fig. 15-7).

Depositional stream reaches bordered by alluvial fans occur at elevations of less than 2,000 m, all containing braided channels. The floodplains are characterized by gentle gradients, broad floodplains, meandering channels, and high groundwater tables.

Structurally, the temporal and spatial stability of these arid-zone rivers is fundamentally different from those of rivers in mesic climates. Because of the dramatic differences in peak floods compared to mean flows, the temporally unpredictable nature of flooding, and the coarseness of most sedimentary material in this region, arid-region channels are time-dependent systems that seldom reach any kind of equilibrium (Graf, 1988). The two primary factors governing channel morphology are the sediment supply and flow variability.

Sediment Supply: The variability of channel morphology due to sediment type and supply is illustrated by experience in Great Plains riparian systems. Here, glacial deposits found in the northeastern Great Plains include considerable silt and clay, which form cohesive stream banks. Meandering channels should predominate in these areas. Aeolian deposits of unconsolidated dunes and loess hills produce large bedloads that lead to braided channels. Ancient fluvial deposits of Pliocene age are found in some areas. The sand and gravel released from these formations promote braiding. Locally derived sediments may be rich in coarse or fine sediments and yield braided or meandering channels, respectively (Friedman et al., 1998b).

Temporal Variation in Flow. The response to flow variation depends on geologic, climatic, and biotic factors. Dams constructed on western streams serve to illustrate the effects of reduced flows. In a braided channel, the response is typically narrowing. This results in an expansion of the riparian zone as the stream bed infills and is invaded by pioneer tree species. With a meandering stream, flow reduction typically results in reduction in the meandering rate. Point bar development results in even-aged stands arranged in narrow arcs on the inside of the meander, with bank cutting on the outside. There is little change in overall riparian area. In contrast to reduced flows, a large flood may deposit sediments in the floodplain and promote tree establishment high enough above the channel bed to minimize future flow-related disturbances. For example, along a relatively undisturbed reach of the Missouri River between 1880 and the present, 70 percent of the dominant cottonwood trees (*Populus deltoides*) were established in the year of a flow greater than 1,400 $m^3 s^{-1}$ or in the two following years (Fig. 15-10). Nearly all of the trees that survived the most recent flood (1978) were established at more than 1.2 m above the lower limit of perennial growth. Younger trees (established since 1978) are at lower elevations, but unlikely to survive in the long term (Scott et al., 1997).

Mesic Riparian Systems

Most of the riparian ecosystems in mesic climates, such as the extensive bottomland forests of the Mississippi River are characterized as zones of deposition. These river systems are dominated by spring floods and late summer flow minima. Typical water budgets show that throughflow is far more important than other water sources and sinks in the riparian zone (Table 15-4). A typical broad floodplain in mesic climates such as eastern North America (Fig. 15-11) contains several major features:

1. The river channel meanders through the area, transporting, eroding, and depositing alluvial sediments.

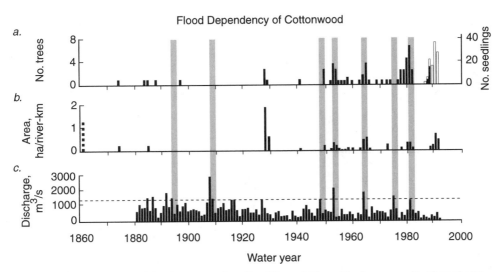

Figure 15-10 Dynamics of riparian trees along the Missouri River, Montana. **a.** Number of sampled cottonwood trees and seedlings established in each year. Solid bars represent tree, sapling, and pole size classes. Open bars represent seedling size class, with a different scale. **b.** Area established each year per kilometer of river. **c.** Maximum daily discharge for each water-year. The dotted line indicates a discharge of 1,400 m³ s⁻¹. Shaded vertical sections are years with maximum discharge greater than 1,400 m³ s⁻¹ and the two following years. (*After Scott et al., 1997*)

2. Natural levees adjacent to the channel are composed of coarse materials that are deposited when floods flow over the channel banks. Natural levees, sloping sharply toward the river and more gently away from the floodplain, are often the highest elevation on the floodplain.

3. Point bars are areas of sedimentation on the convex sides of river curves. As sediments are deposited on the point bar, the meander curve of the river tends to increase in radius and migrate downstream. Eventually, the point bar begins to support vegetation that stabilizes it as part of the floodplain.

4. Meander scrolls are depressions and ridges on the convex side of bends in the river. They are formed from point bars as the stream migrates laterally across the floodplain. This type of terrain is often referred to as ridge-and-swale topography.

Table 15-4   Annual water budget on the Black Swamp, Arkansas

Variable	Annual Flow (m yr⁻¹)
Inflow	14
Outflow	16
Rainfall	1
Evapotranspiration	1
Groundwater discharge	<1
Infiltration	<1

Source: Walton et al. (1996).

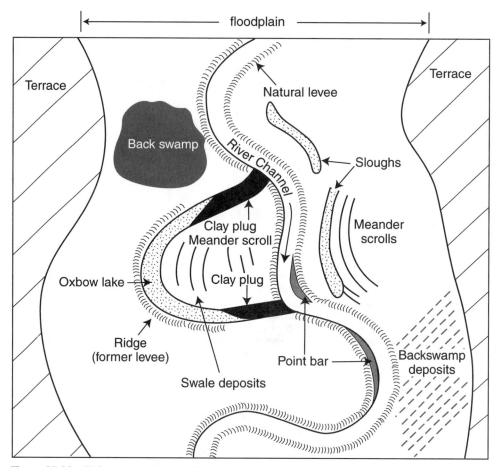

Figure 15-11 Major geomorphologic features of a southeastern U.S. floodplain. (*After Leopold et al., 1964; Brinson et al., 1981b*)

5. Oxbows, oxbow lakes, or billabongs (in Australia) are bodies of permanently standing water that result from the cutoff of meanders. Deepwater swamps (Chapter 14) or freshwater marshes (Chapter 12) often develop in oxbows in the southern United States.

6. Sloughs are areas of dead water that form in meander scrolls and along valley walls. Deepwater swamps can also form in the permanently flooded sloughs.

7. Backswamps are deposits of fine sediments that occur between the natural levee and the valley wall or terrace.

8. Terraces are "abandoned floodplains" that may have been formed by the river's alluvial deposits but are not hydrologically connected to the present river.

As is true of arid riparian systems, the importance of the river to the floodplain and the floodplain to the river in the mesic systems cannot be overemphasized. If either is altered, the other will change over time because floodplains and their rivers are in a continuous dynamic balance between the building and the removal of structure. In the long term, floodplains result from a combination of the deposition of alluvial materials (aggradation) and the downcutting of surface geology (degradation).

Two major aggradation processes are thought to be responsible for the formation of most floodplains: deposition on the inside curves of rivers (point bars) and deposition from overbank flooding. "As a river moves laterally, sediment is deposited within or below the level of the bankfull stage on the point bar, while at overflow stages the sediment is deposited on both the point bar and over the adjacent flood plain" (Leopold et al., 1964). The resulting floodplain is made up of alluvial sediments (or alluvium) that can range from 10 to 80 m thick. For example, in the lower reaches of the Mississippi River, the alluvium, derived from the river over many thousands of years, generally progresses from gravel or coarse sand at the bottom to fine-grained material at the surface (Bedinger, 1981).

Degradation (downcutting) of floodplains occurs when the supply of sediments is less than the outflow of sediments, a condition that could be caused naturally with a shift in climate or synthetically with the construction of an upstream dam. There are few long-term data to substantiate the first cause, but a considerable number of "before-and-after" studies have verified stream degradation downstream of dams attributed to the trapping of sediments (Meade and Parker, 1985). In the absence of geologic uplifting, rivers tend to degrade slowly, and "downcutting is slow enough that lateral swinging of the channel can usually make the valley wider than the channel itself" (Leopold et al., 1964). The process is, thus, difficult to observe over short periods; both aggradation and degradation can only be inferred from the study of floodplain stratigraphy.

The formation of a riparian floodplain and terrace is shown in sequences A to B and C to E in Figure 15-12. When degradation occurs but some of the original floodplain is not downcut, that "abandoned" floodplain is called a terrace. Although it may be composed of alluvial fill, it is part of the active floodplain only during peak floods. Aggradation and degradation can alternate over time, as shown in the sequence C to E in Figure 15-12. A third case, a dynamic steady state, can exist if aggradation resulting from the input of sediments from upstream is balanced by the degradation or downcutting of the stream (Brinson et al., 1981b). Figure 15-12 demonstrates that the same surface geometry can result from two dissimilar sequences of aggradation and degradation.

Transriparian Scale

At a local site on a stream, the riparian vegetation is determined by the cross-sectional morphology, including braiding of the stream, width of the floodplain, soil type, and elevation and moisture gradients. These are all determined, in part, by larger scale (continental, basin, stream system) processes that are modified by local biotic and physical processes.

Typical stream geomorphic valley types of the western United States (Fig. 15-9) are predictably associated with distinctive vegetation patterns (Table 15-5; see also Ecosystem Structure later in this chapter). The riparian soil moisture regime, in large part, explains the plant associations. The relationship, however, is seldom that simple. Soil moisture and depth to groundwater are not the only factors governing plant establishment (Fig. 15-13). Low floodplain elevations are often swept clean of plants by floods, so that seedlings do not survive, and the vegetation is limited to annuals

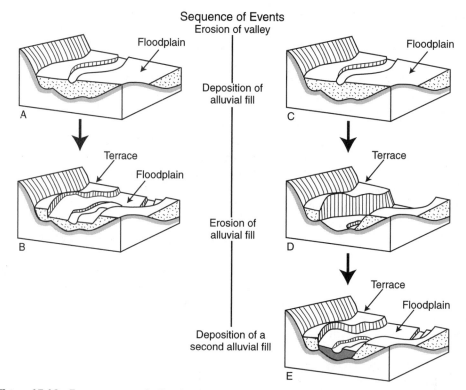

Figure 15-12 Two sequences in the development of floodplains and river terraces. (*After Brinson et al., 1981b; Leopold et al., 1964*)

and perennials that survive until the next flood. Trees mature only at elevations above moderate floods, where they can become well enough established to withstand the severe floods.

The extensive bottomland hardwood forests of the southeastern United States show clear zonation patterns related to elevation and flood frequency (Fig. 15-14). In these broad floodplains, elevations are not monotonic gradients because the floodplain level does not necessarily rise uniformly from the river. Although a description of zonation across these extensive floodplains is useful in a descriptive sense, it has dubious value as a management tool because it ignores the strong interactions among the zones and of the whole riparian area with rivers and uplands. The zones occur because a gradient exists across the floodplain. The gradient itself presupposes flows of materials and energy to maintain it. The coupling is functionally tight for it extends to large landscapes in which river, floodplain, and uplands form both longitudinal and transverse continua that are important to the integrity of the whole river system (Vannote et al., 1980; Forman and Godron, 1986; M. G. Turner, 1989; Gosselink et al., 1990c; Naiman and Decamp, 1997).

Biogeochemistry

Soil Structure

The physiochemical characteristics of the soils of riparian ecosystems are different from those of either upland ecosystems or permanently flooded swamps. In erosional

Table 15-5 Vegetation characteristics of different valley types in the eastern Sierra Nevada

Valley type[a]	Mean Riparian Zone Width[b] (m)	Mean Species Richness[c]	Associated Vegetation Types	Vegetation Type Associations		
				Substrate	Cross-sectional Type	Diversion Status
U-shaped bedrock	7.0	8.0	*Pinus contorta*–meadow	Sand, cobble	Incised	Undiverted
			Salix spp.–*Glyceria striata*	Gravel	Braided	Undiverted
U-shaped till	27.0	13.0	*Salix* spp.–*Cornus stolonifera*	Cobble	One-sided	NS[d]
			Populus tremuloides–*Salix* spp.	NS[d]	One- or two-sided	Undiverted
V-shaped till	21.0	9.0	*Betula occidentalis*–*Salix* spp.	NS	NS	NS
Alluvial fan	27.0	9.0	*Chrysothamnus nauseosus*–*Artemisa tridutula*	Gravel	Braided	Diverted
			Populus trichocarpa–*Rosa woodsii*	Gravel	NS	Diverted
All			*Populus tremuloides*–*Populus trichocarpa*	Boulders	NS	NS

[a] See Figure 15-9 for description of valley types.
[b] Average width of vegetation by valley type.
[c] Average number of species per plot by valley type.
[d] NS = not stated.
Source: R. R. Harris (1988).

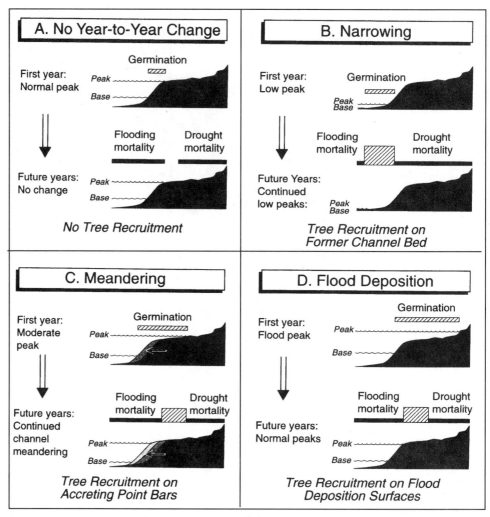

Figure 15-13 **Hydrogeomorphic control of cottonwood recruitment. Diagrammatic representations of cottonwood seed germination, early seedling mortality, and tree recruitment in relation to annual high- and low-flow lines along a bottomland elevation gradient. Four idealized situations depicted, including a. little to no recruitment because of no year-to-year change; b. channel narrowing; c. recruitment on point bars of a meandering river; d. tree recruitment at high elevations caused by infrequent floods and no channel movement. (*Auble and Scott, 1998; reprinted by permission, Society of Wetland Scientists*)**

and transport systems, sediments are coarse and soils poorly developed. In general, riparian soils on floodplains and low river terraces of the American Southwest are youthful and poorly developed. Outcrops of bedrock occur frequently. Because of the coarse texture and immaturity of these soils, they may not have an aquic moisture regime, even though the deep-rooted trees are in contact with shallow groundwater. Some remnant plant communities are established on nonaquic soils that used to be wetter than they are now (Padgett et al., 1989). Aquic soils must be saturated with

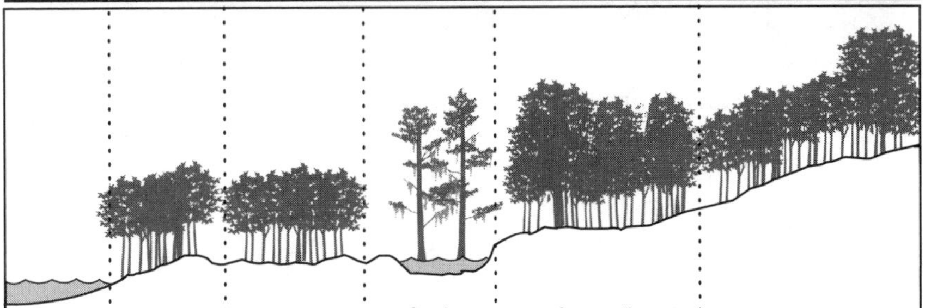

	River channel	Natural levee	First Terrace	Swamp	Second Terrace	Upland Forest
Flooding extent	Continuous flooding	Seasonally temporarily flooded	Intermittently flooded	Semi-permanently flooded	Seasonally temporarily flooded	Seldom flooded
Flooding frequency, % of years	100	11 - 50	51 - 100	~100	11 - 50	1 - 10
Flooding duration, % of growing season	100	2 - 25	>25	~100	2 - 25	<2
Tree species	Submerged aquatics; open water	Sycamore *Platanus occidentalis* Sweetgum *Liquidambar styraciflua* American elm *Ulmus americana*	Overcup oak *Quercus lyrata* Water hickory *Carya aquatica* Green ash *Fraxinus pennsylvanica* Sugarberry *Celtis laevigata* American elm *Ulmus americana*	Bald cypress *Taxodium distichum* Water tupelo *Nyssa aquatica* Red maple *Acer rubrum* Green ash *Fraxinus pennsylvanica*	Sweetgum *Liquidambar styraciflua* Willow oak *Quercus phellos* Water oak *Quercus nigra* Cherrybark oak *Q. falcata var. pagodifolia* Swamp chestnut oak *Q. michauxii*	Upland forest Species

Figure 15-14 General relationship between vegetation associations and floodplain topography, flood frequency, and flood duration of a southeastern U. S. bottomland hardwood forest. (*After Wharton et al., 1982; Clark and Benforado, 1981*)

water long enough to result in anaerobic conditions at least a few days a year, causing distinct mottles or gleyed soils.

In erosional and transport segments of streams, riparian soils are characterized by large fragments, for example, 10 to 23 percent fragments greater than 2 mm, 65 to 71 percent sand and remaining silt and clay, in aquic soils of the upper Gila River basin (Brock, 1985). Riverwash consists of 63 percent coarse material, 35 percent sand, 2 percent silt, and 2 percent clay. Most soils in the mountainous region of Utah developed on alluvium, but colluvial and residual soils occur where underground flows are shallow, where there is late snowmelt, or where groundwater surfaces (Padgett et al., 1989). Organic soils may occur, primarily at high elevations where temperatures are cold and water is abundant.

In these dynamic systems, the substrate can change rapidly and dramatically. For example, at one location on the San Francisco River, major areas of well-drained sandy soil 1.5 to 2 m deep were replaced by riverwash from a single severe flood (Brock, 1985). Depositional systems in xeric areas, such as the Great Basin region of Nevada and western Utah riparian soils on alluvial fans, can be desert pavement that contains a smooth wind-formed surface of tightly fitting pebbles. Lime- and silica-cemented hardpans are common. The lower basin of these types of rivers typically has deep loam or silt clay loam soils that are saline or sodic (Minshall et al., 1989).

Depositional southeastern U.S. riparian forests, especially on high-order rivers, develop deep alluvial fine-grained soils. Some of the most important soil properties are aeration, organic and clay content, and nutrient content. All of these characteristics are influenced by the flooding and subsequent dewatering of these ecosystems; the characteristics of the soil, in turn, greatly affect the structure and function of the plant communities that are found in the riparian ecosystem. Table 15-6 presents the typical physicochemical characteristics of southeastern U.S. floodplain soils, including oxygenation, organic matter content, and mineral nutrients.

Soil Oxygen

Soil oxygen is one of the most important (yet changeable) characteristics of bottomland soils. Anaerobic conditions are created rapidly when the floodplain is flooded, sometimes in a period as short as a few days (see Fig. 6-6). When the floodplain is dewatered, aerobic conditions quickly return. Most riparian plants are unable to function normally under extended periods of anoxia, although some plants have special adaptations that enable them to survive extended periods of little soil aeration (see Chapter 7). Soil aeration, defined as the capacity of a soil to transmit oxygen from the atmosphere to the root zone, is dramatically curtailed by flooding water, but it is also affected by several other characteristics of the soil, including texture, amount of organic matter, permeability to water flow, elevation of groundwater, and degree of compaction. In southeastern riparian systems, soils whose clay content is high and whose pore size is small impede drainage more than sandy or loamy soils do, thereby increasing the likelihood that they will be poorly aerated. The high organic matter content of these bottomland soils can both increase and deplete soil oxygen. Organic matter usually improves the soil structure and, thus, increases the aeration of clay soils. On the other hand, decomposing organic matter creates an oxygen demand of its own. Soil aeration also depends on how close the groundwater level is to the soil surface. Finally, when bottomland soils are compacted, the air-filled pores may decrease dramatically, thereby decreasing soil aeration (W. H. Patrick, 1981).

In the American West, the coarse-textured immature riparian soils generally drain more rapidly and are better aerated. They often show no mottling or other signs of prolonged saturation (Padgett et al., 1989). Conversely, saline or sodic soils may have sharp cemented horizons that prevent drainage.

Organic Matter

The organic content of southeastern bottomland soils is usually intermediate (2–5 percent) compared with the highly organic peats (20-60 percent) at one extreme and upland soils (0.4–1.5 percent) at the other (Wharton et al., 1982). Alternating aerobic and anaerobic conditions apparently slow down but do not eliminate decomposition. In addition, the high clay content provides a degree of protection against decomposition of litter and other organic matter on the floodplain. Furthermore, bottomland forests are generally more productive than upland forests, and, therefore, more organic litter is produced. Wharton et al. (1982) suggested that a 5 percent organic content of soils is a good indicator of periodically flooded areas as compared to more permanently flooded swamps.

Organic content is low in the immature soils of western riparian systems where coarse-grained substrates and good aeration exist. The input of large woody organic matter and other allochthonous organic carbon, and the senescence of roots may increase organic content (Fetherston et al., 1995; Hedman et al., 1996). In high-altitude zones that experience poor drainage, generally on more or less level terrain, organic peats may develop.

Nutrients

The bottomland hardwood wetlands of the southeastern United States generally have ample available nutrients because of several processes. The high clay content of the soils and their continual replenishment during flooding result in high concentrations and availability of nutrients such as phosphorus, which has a high affinity for fine soil particles. The high organic content results in higher concentrations of nitrogen than would be found in upland soils of low organic content (W. H. Patrick, 1981). Mitsch et al. (1979a), for example, found that clay-rich sediments that flooded an alluvial swamp in southern Illinois had phosphorus contents of 8.0 to 9.8 mg/g dry weight.

Anoxic conditions during flooding have several other effects on nutrient availability. Flooding causes soils to be in a highly reduced oxidation state and often causes a shift in pH, thereby increasing mobilization of certain minerals such as phosphorus, nitrogen, magnesium, sulfur, iron, manganese, boron, copper, and zinc (see Chapter 6). This can lead to both greater availability of certain nutrients and also to an accumulation of potentially toxic compounds in the soil. The low oxygen levels also cause a shift in the redox state of several nutrients to more reduced states, changing the availability of these nutrients to plants (see Chapter 6). Denitrification may be prevalent in soils that have adjacent oxidized and reduced zones, as is the case in periodically flooded riparian soils. Phosphorus is more soluble in flooded soils than in dry soils, but whether a shift from reduced to oxidized conditions and back again makes phosphorus more or less available has not been determined. Nevertheless, the periodic wetting and drying of riparian soils are important in the release of nutrients from leaf litter. The generally high concentrations of nutrients and the relatively good soil texture during dry periods suggest that the major limiting factor in riparian ecosystems is the physical stress of inadequate root oxygen during flooding rather than the inadequate supply of any mineral.

In riparian systems of the American West, nutrient levels are probably more closely related to the flux of nutrients than to soil stocks, except in high-altitude organic fens. The alder (*Alnus tenuifolia*) is a common nitrogen-fixing tree of the western riparian zones. In the Pacific Northwest, concentrations of available nitrogen are higher in alder-dominated stands than in coniferous riparian zones, reflecting this capability (Goldman, 1961). In arid zones, nutrients and salts accumulate through evapotranspiration, sometimes to toxic levels, where streams carry water and dissolved salts to the riparian zone.

Ecosystem Structure

Hierarchical Organization and Landscape Pattern

Throughout this chapter, we have emphasized the role that scale plays in the organization and functioning of riparian ecosystems. Specifically, riparian structures and pro-

Table 15-6 Physiochemical characteristics of floodplain and terrace soils

Characteristic	Swamp	Bottomland	First Terrace	Upland Transition
Soil texture	Dominated by dense clays or sands	Dominated by dense clays at surface; some coarser fractions underneath	Clay and sandy loams dominate; sandy soils frequent	Sands to clay
Soil composition (%sand : silt : clay)				
Blackwater	69 : 20 : 12	74 : 14 : 12	—	—
Alluvial	29 : 23 : 48	34 : 21 : 45	71 : 16 : 14	—
Organic matter (%)				
Blackwater	18	8	—	—
Alluvial	4.5	2.8–3.4	3.8	—
Oxygenation	Moving water aerobic; stagnant water anaerobic	Alternating seasonally anaerobic and aerobic	Alternating; mostly aerobic; occasionally anaerobic	Aerobic year-around
Soil color	Gray to olive, gray with greenish-gray, bluish-gray, and grayish-green mottles	Mostly gray on backwater floodplains and reddish on alluvial with brownish-gray and grayish-brown mottles	Mostly gray or grayish brown with brown, yellowish-brown, and reddish-brown mottles	Mostly red, brown, reddish brown, yellow, yellowish red, and yellowish brown, with a wide range of mottle colors

pH			
Blackwater	5.0	5.1	—
Alluvial	5.0	5.3–5.5	5.6
Phosphorus (ppm)			
Blackwater	11.2	9.8	—
Alluvial	9.1	6.3–8.1	4.8
Calcium (ppm)			
Blackwater	607	346	—
Alluvial	1,079	669–752	186
Magnesium (ppm)			
Blackwater	98	36	—
Alluvial	154	140–145	23
Sodium (ppm)			
Blackwater	46	31	—
Alluvial	94	28–31	23
Potassium (ppm)			
Blackwater	48	29	—
Alluvial	51	28–32	20

Source: Wharton et al. (1982).

539

cesses are determined by continental/regional climatic and geologic processes, intrariparian processes along the length of a stream system, and transriparian phenomena across a stream and its riparian zone. These scales are closely interrelated. The plants growing at a point on a western stream may be confined geographically to the west by continental barriers to diffusion; they are adapted to the elevation of the site (intrariparian scale); and they respond to the moisture content of the soil, which is related to the regional climate, the transriparian elevation, and local flooding phenomena. The spatial scales of structures are related to the time scales of events. Thus, in southeastern bottomland systems, sediment deposition and erosion occurring over thousands of years produce major geomorphic features—soil type, point bars, ridge-and-swale topography, floodplains. Vegetation communities develop over hundreds of years as a function of soil type, topography, and hydrology. River flow varies in irregular wet-and-dry cycles of tens of years. Seedling establishment is often closely tied to these cycles.

The annual cycle is the dominant time scale, reflecting the seasonality of temperature and hydrology, driving plant production, decomposition, and secondary biotic production. At shorter time scales, chemical processes and biota respond to short-term flooding and drying at spatial microscales and minute-to-hour time scales (Gosselink et al., 1990c).

These types of hierarchies were clearly described for aquatic systems by Vannote et al. (1980), Frissell et al. (1986), and others, and applied to riparian zones by Baker (1989), Gosselink et al. (1990c), Gregory et al. (1991), Naiman and Decamp (1997), and Patten (1998). Figure 15-15 shows a representation of this spatial and temporal hierarchy in the organization of stream–valley forms and riparian biota.

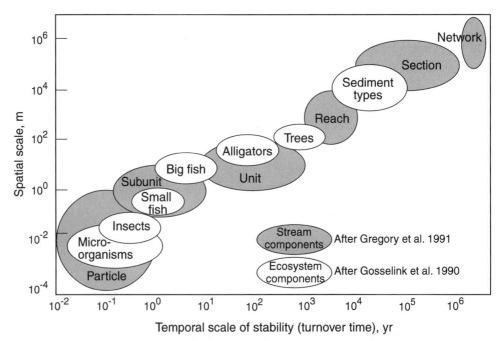

Figure 15-15 Temporal and spatial scales of hierarchical organization of valley landforms and biota. The turnover rate is a measure of the longevity or persistence of the features. (*After Gregory et al., 1991; Gosselink et al., 1990c*).

The vegetation of riparian ecosystems develops in response to processes at all three scales. Although local flooding or the moisture regime often appears to be the immediate cause of specific vegetation associations, that regime itself reflects both micro- and macroscale processes.

Gregory et al. (1991) described these landscape scale interactions in riparian systems:

> The narrow, ribbon-like networks of streams and rivers intricately dissect the landscape, accentuating the interaction between aquatic and surrounding terrestrial ecosystems . . . . The linear nature of lotic ecosystems enhances the importance of riparian zones in landscape ecology. River valleys connect montane headwaters with lowland terrains, providing avenues for the transfer of water, nutrients, sediments, particulate organic matter, and organisms . . . . Nutrients, sediments, and organic matter move laterally and are deposited onto floodplains, as well as being transported off the land into the stream. River valleys are important routes for the dispersal of plants and animals . . . and provide corridors for migratory species.

Vegetation

Southeastern Bottomland Forests

The vegetation of high-order southeastern riparian ecosystems is dominated by diverse trees that are adapted to the wide variety of environmental conditions on the floodplain. The most important local environmental condition is the hydroperiod, which determines the "moisture gradient," or—as Wharton et al. (1982) prefer—the "anaerobic gradient," which varies in time and space across the floodplain. The plant species found along this gradient respond to elevation relative to the river's flooding regime (Fig. 15-14). The lowest parts of the bottomland, nearly always flooded, form the cypress–tupelo gum swamps discussed in Chapter 14. At slightly higher bottomland elevations than the deep swamps, the soils are semipermanently inundated or saturated and support an association of black willow (*Salix nigra*), silver maple (*Acer saccharinum*), and sometimes cottonwood (*Populus deltoides*) in the pioneer stage. A more common association in this zone includes overcup oak (*Quercus lyrata*) and water hickory (*Carya aquatica*), which often occur in relatively small depressions on the floodplains. Also found in this zone are green ash (*Fraxinus pennsylvanica*), red maple (*Acer rubrum*), and river birch (*Betula nigra*). New point bars that form in the river channel often are colonized by monospecific stands of willow, silver maple, river birch, or cottonwood.

Higher still on the bottomland floodplain, in areas flooded or saturated one to two months during the growing season, are found an even wider array of hardwood trees. Common species include laurel oak (*Quercus laurifolia*), green ash (*Fraxinus pennsylvanica*), American elm (*Ulmus americana*), and sweetgum (*Liquidambar styraciflua*) as well as sugarberry (*Celtis laevigata*), red maple (*Acer rubra*), willow oak (*Quercus phellos*), and sycamore (*Platanus occidentalis*). Pioneer successional communities in this zone can consist of monotypic stands of river birch or cottonwood.

Temporarily or infrequently flooded terraces at the highest elevations of the floodplain (second terrace in Fig. 15-14) are flooded for less than a week to about a month during each growing season. Several oaks, tolerant of occasionally wet soils, appear here. These include swamp chestnut oak (*Quercus michauxii*), cherrybark oak (*Quercus falcata* var. *pagodifolia*), and water oak (*Quercus nigra*). Hickories (*Carya* spp.) are often present in this zone in associations with the oaks. Two pines, spruce pine (*Pinus glabra*) and loblolly pine (*Pinus taeda*), occur at the edges of this zone in many bottomlands.

Plant zonation is not linear topographically, nor is it vegetationally discrete. Figure 15-14 is a theoretical cross section of the microtopography of an alluvial floodplain in the southeastern United States. The actual complex microrelief does not show this smooth transition from one zone to the next. The natural levee next to the stream (Fig. 15-14), in fact, is often one of the most diverse parts of the floodplain because of fluctuations in its elevation.

Similarly, plant associations have diffuse edges. For example, Gemborys and Hodgkins (1971) showed the overlap of dominant plants species along a moisture gradient in a southwestern Alabama riparian zone (Fig. 15-16), and Robertson (1987) used ordination techniques to show that the distribution of plant species formed a continuum along a complex flooding–soil texture (drainage–aeration) gradient in a southern Illinois upland swamp.

Low-order streams in the southeastern United States, as might be expected, have smaller watersheds and narrower riparian zones. Headwater reaches are dominated by *Liquidambar styraciflua* (sweetgum), *Nyssa sylvatica* var *biflora* (black gum), and *Acer rubrum* (red maple). Midreaches are typically dominated by *Taxodium distichum* (bald cypress) and tupelo gum (Rheinhardt et al., 1998).

Arid and Semiarid Riparian Forests

The riparian forests of the semiarid grasslands and arid western United States differ from those found in the humid eastern and southern United States. The natural upland ecosystems of this region are grasslands, deserts, or other nonforested ecosystems, and so the riparian zone is a conspicuous feature of the landscape (Brinson et al., 1981b; R. R. Johnson and Lowe, 1985). The riparian zone in arid regions is also narrow and sharply defined in contrast with the wide alluvial valleys of the southeastern United States. As stated by R. R. Johnson (1979), "when compared to the drier surrounding uplands, these riparian wetlands with their lush vegetation are attractive oases to wildlife and humans alike."

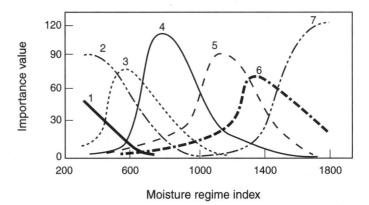

Figure 15-16 Importance value curves for the seven dominant species in a small stream bottomland in the Alabama coastal plain. 1—*Cornus florida*; 2—*Pinus palustris*; 3—*Quercus nigra*; 4—*Liquidambar styraciflua*; 5—*Nyssa sylvatica* var. *sylvatica*; 6—*Magnolia virginiana*; 7—*Nyssa sylvatica* var. *biflora*. (*After Gemborys and Hodgkins, 1971*)

Western U.S. riparian ecosystem tree species are *phreatophytes;* that is, they are plants that obtain their water from *phreatic* sources, such as groundwater or the capillary fringe of the groundwater table (Fig. 15-17). Many species use surface water supplies when seedlings (hence, the general germination requirement of bare, moist soil), but put down long, deep roots that later supply water requirements from groundwater. By comparing the deuterium content of groundwater and surface water to plant sap water, Smith et al. (1998) determined that cottonwood (*Populus tricho-carpa*) used soil water (derived from surface flows) in the spring, both soil water and groundwater during the summer, and groundwater exclusively in the fall. At another site where both willow (*Salix goodingii*) and cottonwood (*P. fremontii*) used ground-water exclusively in the summer, the invasive alien *Tamarix ramisissima* (salt cedar) was able to obtain water from both sources.

Phreatophytes are also termed drought avoiders, in contrast to many arid-zone plants such as cactus that are drought tolerators. Their deep roots intercept the groundwater supply and the trees continue to transpire, even when the soil surface is extremely dry. It is for this reason that for decades riparian vegetation has been removed along managed streams to conserve water for human use. Table 15-7 summa-rizes the ecophysiological traits of three dominant riparian plant species in the south-western United States. Cottonwood was considered an obligate phreatophyte, whereas both *Prosopis pubescens* (screwbean mesquite) and salt cedar are facultative. Mesquite is found in the riparian zone at the upper (driest) edge in the arid Southwest. Salt cedar is an introduced species that is rapidly replacing cottonwood in many areas. All three species maintain high transpiration rates, especially on an areal basis, but cottonwood has a low water-use efficiency (photosynthesis rate, or carbon dioxide uptake, per unit water transpired). This difference helps explain the competitive edge salt cedar seems to have over cottonwood.

There are significant differences between the flora of riparian ecosystems in the eastern and western United States. One notable feature is the general absence of oak

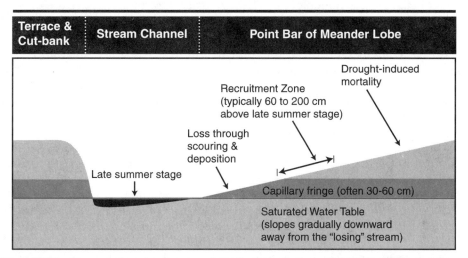

Figure 15-17 Cross section of a stream and riparian zone in the western United States at the point bar of a meander lobe, showing the riparian water table, the capillary fringe, and the suitable band for cottonwood establishment and survival. (*After Mahoney and Rood, 1998*)

Table 15-7 Summary of ecophysiological attributes of the dominant taxa of southwestern United States riparian ecosystems

Attribute	Cottonwood/Willow (*Populus/Salix*)	Mesquite (*Prosopis*)	Salt Cedar (*Tamarix*)
Phreatophytic status	Obligate	Facultative (spatial)	Facultative (temporal)
Physiognomy	Arborescent (some shrub sp.)	Varies (spatial)	Shrubs (thickets)
Leaf size	Large (some narrow)	Small (compound)	Small (cladophylls)
Stress tolerance (water/salinity)	Low	Moderate	High
Peak transpiration rate			
Leaf area basis	High	Moderate	Moderate
Stand basis	High	High	Very high
Water use efficiency	Low	Moderate	High

Source: Smith et al. (1998), based on studies by Busch and Smith (1993, 1995), Busch et al. (1992), Sala et al. (1996), Cleverly et al. (1997), and Devitt et al. (1997).

(*Quercus* spp.) in the West (Brinson et al., 1981b). Also, few species are common to both regions because of the differences in climate, although species of cottonwood (*Populus* spp.) and willow (*Salix* spp.) are found across the United States. Diagrams of the major riparian plant associations in the arid Southwest (Fig. 15-18a) and the montane Northwest (Fig. 15-18b) show the transriparian gradient (channel to upland) across the bottom and the altitudinal gradient from base to apex of the triangle. As altitude increases toward the stream headwaters, channel and riparian zone width decreases, and the stream tends to be confined in deeply cut valleys with little riparian relief. As a result, species differences across the flooding zone disappear. In the arid Southwest, cottonwood and willow communities dominate the most hydric part of the riparian zone, with mesquite and hackberry at higher locations relative to the stream bed. At higher elevations, alder (*Alnus* spp.) grows next to the stream channel, replaced by sycamore and walnut (*Juglans* sp.) communities on slightly higher terraces. In the headwater areas, Engelmann spruce (*Picea engelmannii*) grows down to the lowest perennial plant zone, with grass meadows along the channel where floodplains exist.

In the montane Northwest, the most hydric sites across the riparian gradient support the same communities as in the Southwest at low and intermediate elevations, and are replaced by spruce and alpine tundra at the highest elevations. Cottonwood (*Populus* spp.) replaces mesquite at low elevations and aspen replaces sycamore and walnut at intermediate elevations.

Plant communities change in the arid western United States on a broader latitudinal and elevation scale (Fig. 15-19). At high elevations, temperatures are lower than at low elevations and the moisture stress is reduced. As a result, plants tend to move down the elevation gradient as latitude increases. For example, conifers found only above about 2,000 m at a latitude of 30° can be found 1,000 m lower at a latitude of 50°. Cottonwood (*Populus* spp.) communities appear to be an exception to this rule for the following reasons: (1) Cottonwood communities occur at sea level in the Southwest, but at high latitudes the lowest locations are at 300 to 500 m; and (2) the longer growing season in southern latitudes combined with hydrologic factors may give rise to a cottonwood, willow, sycamore, maple mixed deciduous forest, which has no comparable community in higher latitudes (Patten, 1998).

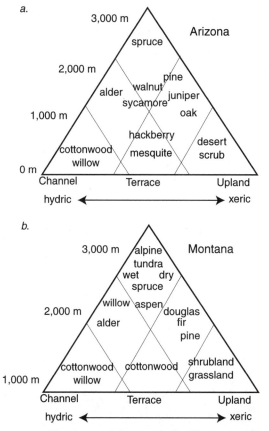

Figure 15-18 Riparian communities in **a.** Arizona and **b.** Montana relative to valley location (channel, terrace, and upland) and elevation. The hydric/xeric gradient from channel to upland becomes less distinct with elevation, resulting in little difference between channel and upland vegetation at high elevations. (*After Patten, 1998*)

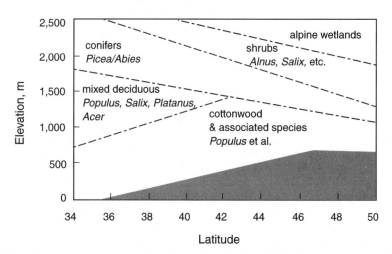

Figure 15-19 Riparian plant communities along elevational and latitudinal gradients in the arid West. (*After Patten, 1998*)

545

Keammerer et al. (1975) and W. C. Johnson et al. (1976) described the riparian forest in the Great Plains along the Missouri River in North Dakota, where common species included cottonwood, willow, and green ash in the lower terraces and American elm, box elder (*Acer nugundo*), and bur oak (*Quercus macrocarpa*) at the higher elevations (Fig. 15-20).

Several studies have related riparian vegetation associations in the western United States to physical site factors. R. R. Harris (1988) found a close relationship between eight vegetation types (identified by ordination analysis) and six valley forms in the eastern Sierra Nevada (Table 15-5). Similarly, Szaro (1989) found a strong relationship between site factors and plant associations for riparian forests in Arizona and New Mexico. He classified 24 forest and scrub community types based on cluster analysis and found that elevation was the most important determinant of vegetation at a site. Elevation, stream bearing, and stream gradient together explained 84 percent of the variation in the data. Baker (1989) found elevation and valley width to be closely correlated with species composition. Finally, L. C. Lee (1983) determined that, for riparian vegetation community types in a Pacific Northwest river system, the floristic gradient determined by discriminant analysis was significantly related to a similar gradient of soil development. He interpreted the vegetation data to represent a canopy structure gradient from nonforested wetlands to a nonwetland forested association.

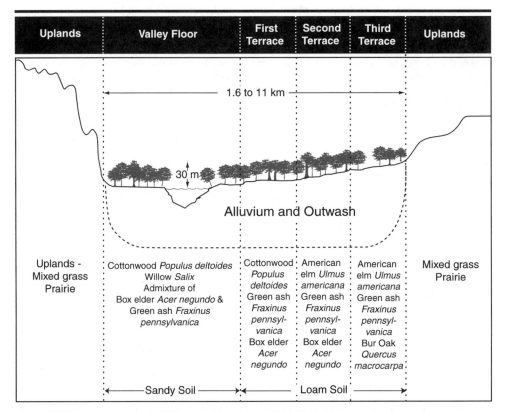

Figure 15-20 Cross section of Missouri River in semiarid North Dakota, showing major riparian tree species on various terraces. (*After Keammerer et al., 1975*)

Recruitment for woody plant seedlings is generally related to the availability of bare moist soils and to safety from physical disturbance, flooding, and drought (Fig. 15-13). A model of recruitment and survival of cottonwood depends on previous flooding. When no year-to-year changes occur, no seedlings survive; after channel narrowing from reduced flows, tree recruitment occurs on the former channel bed; in meandering streams, trees recruit on point bars; and after major floods, they recruit on flood depositional surfaces well above the stream bed.

Riparian systems in arid areas support many plant species unique to these environments. With the rapid modification of almost every stream for water management and the invasion of a number of exotic plant species, some native plants are becoming increasingly rare. For example, Faber et al. (1989) list nine rare and endangered riparian plant species in southern California.

Consumers

The riparian ecosystem provides a valuable and diverse habitat for many animal species. Some studies (e.g., Blem and Blem, 1975) have documented that forested floodplains are generally more populated with wildlife than are adjacent uplands. Because riparian ecosystems are at the interface between aquatic and terrestrial systems, they are a classic example of the ecological principle of the "edge effect" (E. P. Odum, 1979a). The diversity and abundance of species tend to be greatest at the ecotone, or "edge" between two distinct ecosystems such as a river and uplands. Brinson et al. (1981b) described four ecological attributes that are important to the animals of the riparian ecosystem:

1. *Predominance of Woody Plant Communities.* This is particularly important in regions where the forested riparian zone is the only wooded region remaining, as in heavily farmed regions, or where the riparian forest was the only wooded area to begin with, as is the case in the arid West. Trees and shrubs not only provide protection, roosting areas, and favorable microclimates for many species, but they also provide standing dead trees and "snags" in streams that represent habitat value for both terrestrial and aquatic animals. The riparian vegetation also shades the stream, stabilizes the stream bank with tree roots, and produces leaf litter, all of which support a greater variety of aquatic life in the stream.

2. *Presence of Surface Water and Abundant Soil Moisture.* The stream or river that passes through a riparian ecosystem is an important source of water, especially in the Far West, and of food for consumers such as waterfowl and fish-eating birds; an area of protection and travel for beavers and muskrats; and a reproduction haven for amphibians. The floodplain is also important for many aquatic species, particularly as a breeding and feeding area for fish during the flood season.

3. *Diversity and Interspersion of Habitat Features.* The riparian zones in both eastern and western North America form an array of diverse habitats, from permanently flooded swamps to organic fens and infrequently flooded forests. This diversity, coupled with the edge effect discussed previously, provides for a great abundance of wildlife.

4. *Corridors for Dispersal and Migration.* The linear nature of bottomland and other riparian ecosystems along rivers provides protective pathways for

animals such as birds, deer, elk, and small mammals to migrate among habitats. Fish migration and dispersal may also depend on maintaining riparian ecosystems along the stream or river.

Eastern Riparian Systems

A comprehensive description of the fauna that inhabit floodplain wetlands is not possible here because of the great number of birds, reptiles, amphibians, fish, mammals, and invertebrates that use the floodplain environment. Several good reviews on the fauna of bottomland and riparian wetlands are given by Fredrickson (1979), Wharton et al. (1981, 1982), and Brinson et al. (1981b). The surveys that have been made in these areas show a rich diversity of animals. The animals, like the plants, are accustomed to certain zones in the floodplain, although animal, plant, and geomorphic zones do not always coincide (Fig. 15-21). Several animals such as the beaver, river otter, snakes of the genus *Nerodia,* the cottonmouth, and several frogs are present near the water. Deer, foxes, squirrels, certain species of mice, the copperhead, and the canebrake rattlesnake exist at the upland edge. Most of the food chains are detrital, based on the organic production of the vegetation. Other characteristics of the fauna, associated with the sequence of flooding and dewatering, are high mobility, arboreal (tree-inhabiting) abilities, swimming ability, and the ability to survive inundation (Wharton et al., 1981). Most of the species in Figure 15-21 have one or more of these traits.

Several studies have reported a strong relationship between fishery yields and floodplains of lowland rivers because fish spawn and feed in floodplains during flood stages on the river (Welcomme, 1979; Risotto and Turner, 1985; Kwak, 1988; R. E. Turner, 1988a, b; Lambou, 1990; Hall and Lambou, 1990; Killgore and Baker, 1996) and because the productivity of large, lowland rivers depends on the exchange of nutrients with the floodplains (Junk et al., 1989).

Western Riparian Systems

Arid western riparian areas are particularly diverse and under enormous pressure from human development. Faber et al. (1989) summarized the fauna of southern California riparian systems as follows: Insect fauna are estimated at 27,000 to 28,000 species that play an important role in the riparian community as both predators and prey. Fish populations are limited since most of the streams are intermittent. They are disappearing rapidly because of habitat destruction. Amphibians are present around undisturbed mountain streams. Eighty-eight species of breeding birds are strictly riparian and another 23 use riparian habitats extensively. Many species of nonbreeding birds also use riparian zones for food and rest during migration. The group most affected by riparian habitat loss consists of 76 species of passerine birds, of which 59 nest in the riparian zone. Forty-five mammal species in southern California are associated with the riparian habitat.

In the northwestern United States and western Canada, salmon success is closely related to streams that have intact riparian zones (Sedell et al., 1980, 1982; Sedell and Luchessa, 1982). Snags and other large, woody material from the riparian zone historically created shoals, dammed sloughs, and caused log jams, enhancing the salmon habitat. Many rivers in this region have been isolated from their floodplains by clearing these snags and improving the channel for navigation (Sedell and Froggatt, 1984).

Animals and Riparian Landscapes

Large animals, by their activities, modify and reshape the structure and processes of riparian ecosystems (Naiman and Rogers, 1997). Human impacts tend to simplify

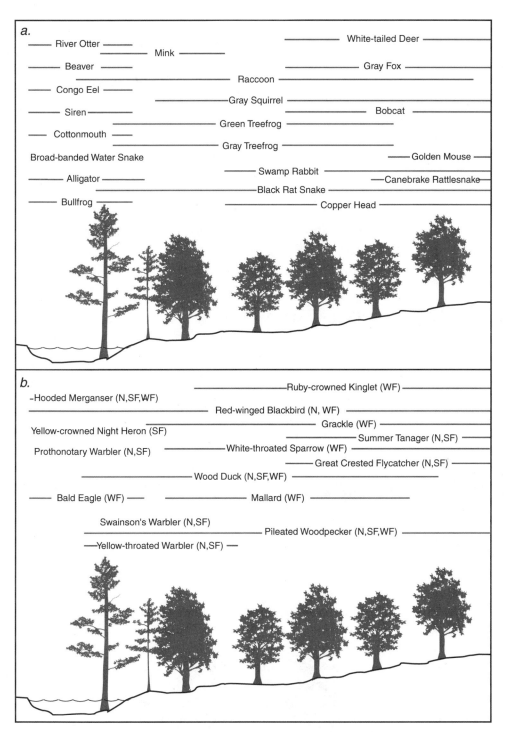

Figure 15-21 Distribution of animals in southern bottomland hardwood forests: **a.** amphibians, reptiles, and mammals; and **b.** birds in relation to nesting (N), summer foraging (SF), and winter foraging (WF). (*After Fredrickson, 1979*)

stream corridors, whereas large animals tend to increase patch heterogeneity and connectivity (Fig. 15-22). These animals selectively eat vegetation, burrow, wallow in mud holes, make trails, and build dams, all of which have long-term ecosystem-level consequences. For example, many large mammals modify habitat. In southern Africa, hippopotamus making nightly feeding forays in the river from surrounding riparian areas create a maze of trails and canals that become corridors for other animals and that modify river flows and riparian flooding. Elephants and cape buffalo also create trails and otherwise modify habitat. They enlarge natural depressions with their wallowing, creating additional water storage in the dry season. Warthogs dig

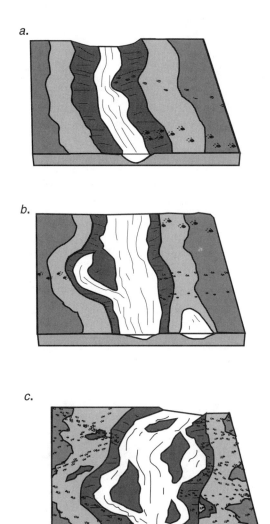

a.

b.

c.

Figure 15-22 The impact of humans and large animals on the vegetative patch structure of river corridors: **a.** Channel processes are constrained, creating a simple linear zonation; **b.** under the influence of natural hydrology and channel processes, heterogeneity of zones is improved; **c.** patch size and heterogeneity of riparian vegetation is optimized with the feeding and movement of large animals. (*After Naiman and Rogers, 1997*)

extensively, uprooting vegetation and plowing the soil. In North America, beavers dam streams and harvest riparian vegetation, creating ponds that attract ducks and riparian birds (Brown et al., 1996) and opening the forest canopy. Moose, elk, and brown bear trample paths to the water. All of these activities tend to increase the spatial and temporal heterogeneity of the riparian zone.

Geese and other birds, prairie dogs, elk, moose, crocodile, and reedbuck, by their selective feeding habits and their destruction of vegetation, modify plant species competition, disperse seed, and, thus, change resource distribution and processing rates. All of these activities lead to high diversity of habitats and species, modify and accelerate energy and nutrient cycles, improve patch connectivity, and expand resilience and resistance to disturbance (Naiman and Rogers, 1997).

Ecosystem Function

Riparian ecosystems have many functional characteristics that result from the unique physical environment. It is generally recognized that they are highly productive because of the convergence of energy and materials that pass through riparian wetlands in great amounts.

Throughout this chapter, we have emphasized the close interrelationship of stream and riparian zone. Figure 15-23 illustrates schematically the role of riparian vegetation in the stream community. The interaction is much broader than that shown. At the drainage basin level, the frequency and duration of flooding of the riparian zone are functions of drainage density (length of channel/area of watershed). The flooding regime, in turn, controls sediment aeration and reduction, nutrient availability and the mineralization rate, and nutrient quality and flux to the adjacent stream.

The river–floodplain system, while an integrated whole, has been described differently by ecologists, depending on whether they have done most of their research in the stream itself or in the riparian systems of the floodplain.

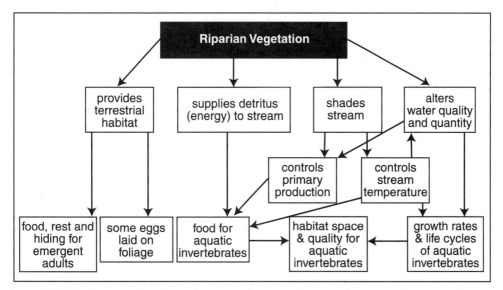

Figure 15-23 Some relationships between riparian vegetation and stream aquatic communities. (*After Knight and Bottorff, 1984*)

The River

The River Continuum Concept

The *river continuum concept* (RCC) is a theory developed in the early 1980s to describe the longitudinal patterns of biota found in streams and rivers (Vannote et al., 1980; Minshall et al., 1983, 1985). According to the RCC, most organic matter is introduced to streams from terrestrial sources in headwater areas (Fig. 15-24). The production/respiration ($P : R$) ratio is less than one (i.e., the stream is heterotrophic), and invertebrate shredders and collectors dominate the fauna. Biodiversity is limited by low temperatures, low light, and low nutrients (Fig. 15-25a). Organic matter is reduced in size as it travels downstream. In river midreaches, more light is available, phytoplankton prosper, and biodiversity is highest. The $P : R$ ratio is greater than one. Organic matter input from upstream is fine; filter feeders dominate the flora. In braided reaches or where the floodplain is broad, however, the bank habitat is a major source of snags and logs that lead to debris dams that slow water flow and increase stream habitat diversity. The increased input of riparian coarse debris increases food diversity and increases heterotrophy. The $P : R$ ratio is less than one. Terrestrial inputs are probably more nutritious than the fine reworked material from upstream. Even bacteria respond to the concentrations and sources of organic matter. The allelic frequency of metabolic enzymes is correlated with the habitat (J. V. McArthur, 1989). This suggests that bacterial flora and genetic selection pressures in a specific habitat (i.e., a floodplain–stream reach) are functions of linkages and interactions between stream and floodplain, including timing, quantity, quality, and source of organic material inputs. Finally, in the highest order streams, riparian litter inputs are minor and turbidity reduces primary productivity. Hence, the system is heterotrophic again ($P : R < 1$) and diversity is often low (Figs. 15-24 and 15-25a).

There are two major corollaries to the RCC (Johnson et al., 1995). *Nutrient spiraling,* or *resource spiraling,* refers to the process whereby resources (organic carbon, nutrients, etc.) are temporarily stored and then released as they "spiral" downstream from organic to inorganic form and back again. In headwater streams, nutrient spirals are long. In large high-order rivers, nutrient spirals tend to be tight and short (nutrients are rapidly processed). The *serial discontinuity concept,* developed by Ward and Stanford (1983, 1995), involves the effects that floodplains, dams, and the lateral dimension in general have on the functioning of a river system. The effect on biodiversity of three types of river reaches is illustrated in Figure 15-25b, where braided streams are shown to have the lowest diversity because of shifting sediments typical of these systems, whereas meandering streams have the highest diversity because of the frequent lateral movement of organisms from bodies of water on the wide floodplain and the spatial heterogeneity of the river–floodplain system.

Flood Pulse Concept

The RCC initially considered the importance of the riparian zone only in an indirect way by noting that small low-order streams are seen to be influenced by shading and abundant contributions of allochthonous organic matter. The importance of backwaters, oxbows, and floodplains to river ecosystem function was virtually ignored. Junk et al. (1989) developed a flood pulse concept (FPC) for floodplain–large river systems based on their experience in both temperate and tropical regions of the world (Fig. 15-26). They dispute the river continuum concept as a generalizable theory because (1) most of the theory was developed from experience on low-order temperate

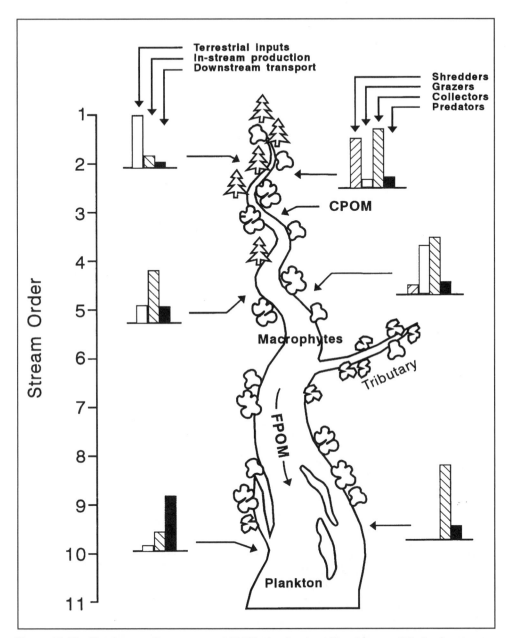

Figure 15-24 The river continuum concept (RCC) showing transition from small first-order stream to very large eleventh-order river. Charts on the left indicate relative importance of terrestrial, instream, or upstream energy sources to the aquatic food chain, while the charts on the right indicate the relative importance of different feeding groups of invertebrates. CPOM, coarse particulate organic matter; FPOM, fine particulate organic matter. (*Johnson et al., © 1995 American Institute of Biological Sciences; reprinted by permission*)

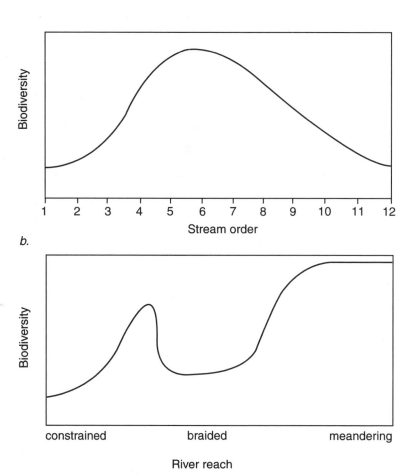

Figure 15-25 Species richness of benthic invertebrates as a function of **a.** stream order according to the river continuum concept, and **b.** three different reaches of a river. (*After Ward, 1998*)

streams, and (2) the concept is mostly restricted to habitats that are permanent and lotic. In the FPC, the pulsing of the river discharge is the major force controlling biota in river floodplains, and lateral exchange between the floodplain and river channel and nutrient cycling within the floodplain "have more direct impact on biota than nutrient spiraling discussed in the RCC" (Junk et al., 1989). The FPC thus considers the river-floodplain exchange to be of enormous importance in determining the productivity of both the river and the adjacent riparian zone (Fig. 15-26). The alternating dry-and-wet cycles optimize productivity of the littoral zone and the adjacent forest, decomposition of all that is produced, and fish spawning and feeding. Bayley (1995) argues that biological productivity in general and multispecies fish yield in particular are higher in river–floodplain systems such as shown in Figure 15-26 over equivalent stable bodies of water.

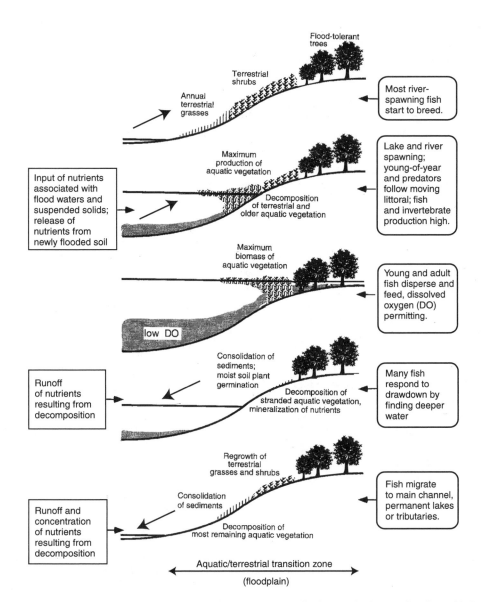

Figure 15-26 The flood pulse concept (FPC) for a river and its floodplain, illustrating five periods over the wet and dry seasons of a river. (*From Bayley, 1995, after Junk et al., 1989; © 1995 American Institute of Biological Sciences, reprinted by permission*)

The Riparian Ecosystem

Primary Productivity

Riparian wetlands are generally more productive than adjacent upland ecosystems because of their unique hydrologic conditions. Periodic flooding (flood pulses) usually contributes to higher productivity in at least three ways (Brinson et al., 1981b):

1. Flooding provides an adequate water supply for the vegetation, particularly in arid riparian wetlands.

2. Nutrients are supplied and favorable alteration of soil chemistry results from periodic overbank flooding. These alterations include nitrification, sulfate reduction, and nutrient mineralization.

3. The flowing water of overbank flooding offers a more oxygenated root zone than if the water were stagnant. The periodic "flushing" also carries away many waste products of soil and root metabolism such as carbon dioxide and methane.

Yet with all of these positive features, flooding can be a physiological stress as well (see Chapters 7 and 14) to the vegetation of these systems that is often facultative in its ability to deal with it. Table 15-8 presents measurements of biomass and net primary productivity for several mideastern U.S. forested riparian ecosystems. In general, floodplain wetlands that have an annual unaltered cycle of wet and dry periods have an above-ground net biomass production (litterfall + stem growth) greater than $1,000 \text{ g m}^{-2} \text{ yr}^{-1}$. This level of productivity is generally higher than that of forested wetlands that are permanently flooded or have sluggish flow, a point described in detail in Chapter 14. It is usually the case that both permanently flooded zones and rarely flooded zones are less productive than those zones that alternate between wet and dry conditions more frequently, although exceptions have been reported (S. L. Brown and Peterson, 1983; Mitsch and Rust, 1984).

The subsidy–stress model of H. T. Odum (1971) and E. P. Odum (1979a), later refined as the *pulse stability* concept by all three Odums (W. E. Odum et al., 1995), includes concepts that potentially apply well to floodplain forests. It is logical to assume that the seasonal pulsing of flooding can be both a subsidy and a stress in much the same way that twice-per-day flooding is a subsidy to salt marshes and mangroves. As discussed in Chapter 8, pulsing is frequent in nature and systems such as bottomland forests appear to be well adapted to taking advantage of this subsidy. Despite this clear theoretical basis for riparian ecosystem productivity, it has been difficult to confirm or deny these theories in practice. Megonigal et al. (1997) investigated productivity in floodplain swamps throughout the southeastern United States and concluded that, while permanently flooded floodplain swamps did have lower productivity, there was no evidence that sites that were seasonally pulsed were any more productive than sites that were clearly upland (Fig. 15-27). They suggest that the model developed by Mitsch and Rust (1984; Fig. 15-28) may be a more appropriate description of what goes on with productivity of bottomland forests, where "subsidies and stresses may occur simultaneously and cancel one another" (Megonigal et al., 1997). In the Mitsch–Rust model, flood intensity and duration affect moisture, available nutrients, anaerobiosis, and even length of growing season in a complex and nonlinear "push–pull" arrangement.

There have been few studies of the primary productivity of western U.S. riparian wetlands. Because these systems span a broader range of fluvial energy than southeast-

Table 15-8 Biomass and net primary productivity (NPP) of riparian forested wetlands, including comparisons with upland and wet forests

Location/Forest Type	Above-Ground Biomass (kg m⁻²)	Litterfall (g m⁻² yr⁻¹)	Stem Growth (g m⁻² yr⁻¹)	Above-Ground NPP (g m⁻² yr⁻¹)	Reference
LOUISIANA					
Bottomland hardwood	16.5	574	800	1,374	Conner and Day (1976)
Seasonally flooded forest	—			1,306±410	Gosselink et al. (1981b)
Upland forest				1,002±271	
Wet forest (cypress) (n=7)	—	425±164	269±117	765±245	Megonigal et al. (1997)
Bottomland hardwood (n=7)	—	670±109	440±135	1,110±177	
Upland forest (n=2)	—	690±26	329±288	1,019±262	
SOUTH CAROLINA					
Wet forest (cypress) (n=2)	—	385±98	242±64	625±35	Megonigal et al. (1997)
Bottomland hardwood (n=10)	—	720±98	488±122	1,208±198	
Upland forest (n=4)	—	714±91	415±82	1,130±84	
ILLINOIS					
Floodplain forest	29.0	—	—	1,250	F. L. Johnson and Bell (1976)
Transition forest	14.2	—	—	800	
Floodplain forest	—	491	177	668	S. L. Brown and Peterson (1983)
KENTUCKY					
Bottomland forest	30.3	420	914	1,334	Mitsch et al. (1991)
Bottomland forest	18.4	468	812	1,280	
Wet forest (cypress/ash)	31.2	136	498	634	

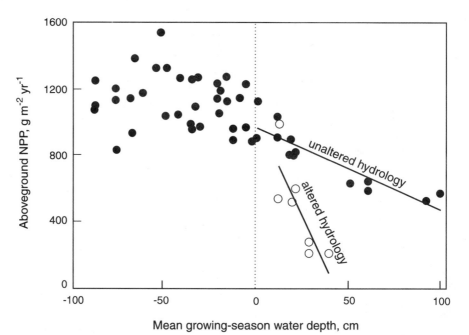

Figure 15-27 The relationship between net primary productivity of floodplain forests and mean growing season water depth in bottomland hardwood forests of the southeastern United States. (*After Megonigal et al., 1997*)

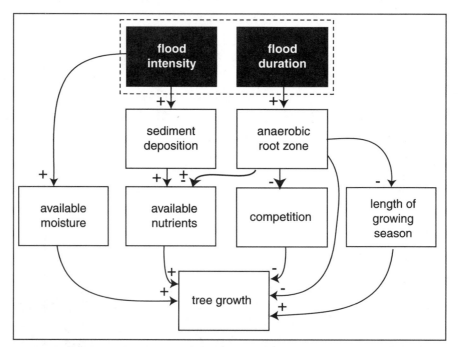

Figure 15-28 Causal model that describes the major causes for increases and decreases in individual tree growth in riparian floodplain forests. Plus (+) sign indicates a positive effect; minus sign (−) indicates a negative effect. (*After Mitsch and Rust, 1984*)

ern floodplain forests (for which the most productivity data are available), it would be interesting to determine whether the hydrologic energy–productivity relationship discussed previously holds up in western U.S. riparian ecosystems, or whether the energies of many montane streams are beyond subsidy levels, acting as stresses and reducing primary production.

Decomposition

The decomposition of organic matter in riparian ecosystems is undoubtedly related to the intensity and duration of flooding, although consistent relationships have not been found. The most rapid decomposition has been reported in the wettest sites and the slowest on dry levee sites (Brinson, 1977); that is, decomposition is faster on dry sites compared to sites flooded 16 to 50 percent of the time (Duever et al., 1984). McArthur and Marzolf (1987) reported that the dissolved nutrient concentration of decomposing leaves depended on both the species and their location on the floodplain. The relationship was not straightforward between leachate concentrations and precipitation or dissolved organic carbon. The rate of decomposition in riparian wetlands is probably the greatest in areas that are generally aerobic but are supplied with adequate moisture, slightly slower at sites that are continually dry and lacking in moisture, and slowest in permanently anaerobic wetlands (Brinson et al., 1981a). The research of J. V. McArthur (1989) and his associates on the correlation of bacterial allele frequency with specific habitats illustrates the extreme complexity of the decomposition process.

Organic Export

There is considerable evidence that watersheds that drain areas of wetlands export more organic carbon than watersheds that do not have wetlands. The organic carbon is in both dissolved and particulate forms, although the particulate fraction is generally a small percentage of the total carbon in most rivers (Brinson et al., 1981b). The large organic carbon export from watersheds dominated by riparian ecosystems is attributed to the following factors:

1. A large surface area of litter, detritus, and organic soil that is exposed to river water during flooding
2. Rapid leaching of soluble organic carbon from some riparian soils and litter exposed to flooding
3. The long time during which water is in contact with the floodplain, allowing for significant passive leaching
4. River movement over floodplains during flooding that can physically erode and transport particulate organic carbon from the floodplain

Both particulate and dissolved carbon have been shown to be valuable for downstream ecosystems, particularly for lacustrine and marine ecosystems. The particulate fraction is important as a source of energy for shredders and for filter-feeding organisms (Vannote et al., 1980). The dissolved organic carbon probably is most valuable as a source of carbon for microorganisms, which, in turn, convert it to particulate form (Brinson et al., 1981a, b).

In addition to export through the river system, some riparian systems lose considerable carbon as methane. Most of this is apparently produced in organic-rich riparian and river-bottom sediments. In the Amazon River floodplain, for example, methane is found in supersaturated concentrations in all water. Fluxes to the atmosphere varied from 0.3 to 3,700 mg m^{-2} day^{-1} methane, and the Amazon floodplain was estimated to account for between 5 and 10 percent of global methane emissions (Devol et al., 1994).

Sedell and his co-workers (Sedell et al., 1980; Sedell and Froggatt, 1984) have documented the contribution of large woody material from the riparian zone not only as an energy source to the adjacent stream but as an important source of structure that modifies fluvial processes and enhances the interaction of the stream with its riparian zone.

Energy Flow

Studies of the overall energy flow of riparian wetlands are generally lacking, although there are several studies related to primary productivity (see earlier discussion). It is, therefore, possible to make only broad generalizations about riparian ecosystem energy flow. The food webs that develop in these ecosystems begin with the production of detritus, although a great complexity and diversity of animals develop in the food webs. Large animals may ingest considerable live vegetation, but their primary influence is probably in reshaping structure and resource use. A unique feature of riparian wetlands is that the detrital production supports both aquatic and terrestrial communities. The wetlands also receive and transport to downstream ecosystems a large amount of detrital energy.

Nutrient Cycling

Several important points should be made about nutrient cycling in riparian ecosystems:

1. Riparian ecosystems have "open" nutrient cycles that are dominated by the flooding stream or river, runoff from upslope forests, or both, depending on stream order and season.

2. Riparian forests exert significant biotic control over the intrasystem cycling of nutrients, and seasonal patterns of growth and decay often match available nutrients.

3. Contact of water with sediments on the forest floor leads to several important nutrient transformations. Riparian wetlands can serve as effective sinks for nutrients that enter as lateral (transriparian) runoff and groundwater flow.

4. When the entire river system that flows through a watershed dominated by riparian wetlands is investigated, the riparian wetlands often appear to be nutrient transformers that change a net import of inorganic forms of nutrients to a net export of organic forms.

The nutrient cycles of riparian wetlands are open cycles characterized by large inputs and outputs caused by flooding. Cycling is also significantly affected by the many chemical transformations that occur when the soil is saturated or under water. Brinson et al. (1981b) described the cycling of nitrogen in a cypress–tupelo stream–floodplain complex. In winter, flooding contributes dissolved and particulate nitrogen that is not taken up by the canopy trees because of their winter dormancy. The nitrogen is retained, however, by filamentous algae on the forest floor and through immobilization by detritivores. In spring, the nitrogen is released by decomposition as the waters warm and as the filamentous algae are shaded by the developing canopy.

Much of the nitrate is probably immobilized in the decaying litter and slowly made available for plant uptake. The vegetation canopy begins to develop, increasing plant nitrogen uptake and lowering water levels through evapotranspiration. When the sediments are exposed to the atmosphere as summer progresses, ammonification and nitrification are stimulated further, making the nitrogen more available for plant uptake. Nitrification, in turn, produces nitrates that are lost through denitrification when they are exposed to anaerobic conditions caused by subsequent flooding. Fail et al. (1989) measured the stocks and the annual accretion of tissue nutrients in a riparian forest in south Georgia (Table 15-9). In this forest, most of the tissue nutrients were tied up in above-ground live biomass. The soil stores (as distinguished from below-ground tissue stores) were not measured. In some western riparian systems dominated by alder, soil nitrogen levels are enhanced by nitrogen fixation, and these enhanced concentrations influence the adjacent stream (Goldman, 1961).

Longitudinal Nutrient Exchange

Inflows and outflows of nutrients in riparian systems occur along two gradients, longitudinal, that is, along the stream course, and transverse from uplands across the riparian zone to the stream. At the stream system level, Elder and Mattraw (1982) and Elder (1985) investigated the flux and speciation of nitrogen and phosphorus in the Apalachicola River in northern Florida, a river that is dominated by forested riparian ecosystems along much of its length. The authors of this study concluded that the floodplain forests were nutrient transformers rather then nutrient sinks, when the total export of nutrients by the watershed was compared to the upstream inputs; that is, the inputs and outputs were similar for both total nitrogen and total phosphorus but the forms of the nutrients were different. Compared to the influx, the stream discharge contained net increases in particulate organic nitrogen, dissolved organic nitrogen, particulate phosphorus, and dissolved phosphorus, and net decreases in dissolved inorganic phosphorus and soluble reactive phosphorus. This means that the riparian forests were net importers of inorganic forms of nutrients and net exporters of organic forms. These transformations, the authors argued, are important to secondary productivity in downstream estuaries.

Table 15-9 Above- and below-ground vegetation, nutrient pools, and accretion rates for nitrogen, phosphorus, potassium, calcium, and magnesium in a riparian zone along a southeastern United States stream (pools are in g m^{-2}; accretion in g m^{-2} yr^{-1})

| | Above-Ground | | Below-Ground | | |
| | | | | Accretion | |
Nutrient	Pool	Accretion	Pool	Gross	Net
Biomass	16,500	583	3,160	220	83
N	114	4	16	3.8	1.5
P	6.8	0.2	1.2	0.3	0.1
K	31.5	1.1	7.0	2.0	0.1
Ca	97.2	3.5	4.4	0.3	0.2
Mg	14.1	0.4	2.5	0.15	0.1

Source: Fail et al. (1989).

Despite this study, it is clear that depositional riparian systems do trap considerable amounts of nutrients from upstream. The inflow of upstream nutrients is a major pathway in the local nutrient cycle. Hupp and Morris (1990) and Hupp and Bazemore (1993) developed a dendrogeomorphic approach to determining sedimentation rates in forested wetlands of the Black Swamp on the Cache River in eastern Arkansas and two river sites in western Tennessee. They found that sedimentation rates averaged 0.25 cm/yr with a maximum of 0.60 cm/yr in cypress–tupelo swamps, in contrast to a sedimentation rate of 0.10 cm/yr when cypress and tupelo trees were absent. A distinct decrease in sedimentation correlated with an increase in land elevation in both studies (Fig. 15-29), suggesting that there is a tendency of low-elevation alluvial swamps to "fill in" with sediments. The study also showed that sedimentation rates have increased significantly at the Arkansas site from 0.01 cm/yr prior to 1945 to an average of 0.28 cm/yr over the past 19 years.

Aust et al. (1991) described an experiment in which sedimentation was measured in a southwestern Alabama riverine swamp that had been subjected to three types of land disturbances. They found that areas logged by helicopter and skidder trapped more sediments (0.12–0.22 cm/flood season) than the natural control area (0.11 cm/flood season), leading to an interesting speculation that logging improved the wetland's ability to retain sediments from the flooding river. Their study was apparently done in a season of relatively low flood conditions, as ^{137}Cs analysis indicated a mean sedimentation rate of 0.76 cm/yr during the preceding 35 years.

Kleiss (1996b), in a study of a reach of the Cache River in Arkansas, found that 14 percent of the river's suspended sediment load was retained. The greatest sediment deposition occurred in the lowest cypress–tupelo wetland areas and in low-lying backwater areas, rather than on the natural levees. Overall, the removal efficiency of the riparian system was estimated for total nitrogen at 21 percent and total phosphorus at 3 percent (Dortch, 1996). Phillips (1989) calculated that in 10 study basins repre-

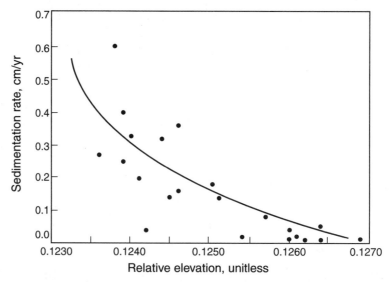

Figure 15-29 Sedimentation rate, primarily from flooding river, in Black Swamp, Arkansas, versus relative elevation of sampling site. Data suggest relatively higher sedimentation rates in lower floodplain elevations. (*After Hupp and Morris, 1990*)

sentative of most of the different types of mesic riparian systems, 23 to 93 percent of upland sediment that reached the stream was subsequently stored in riparian wetlands. Table 15-10 illustrates the range of measurements of phosphorus input to riverine forests. Although these data are only approximations of the phosphorus contribution stemming from overbank and transriparian nutrient fluxes, it is worth noting that they are usually several orders of magnitude greater than phosphorus inflows from precipitation and are at least on the same order of magnitude as intrasystem cycling flows. The relative importance of these inflows for alluvial river swamps was shown in Chapter 14.

Transverse Nutrient Exchange

Other studies have emphasized the role of riparian wetlands in transriparian nutrient fluxes. Thus, riparian ecosystems have been measured as effective "filters" for nutrient materials that enter through lateral runoff and groundwater. Peterjohn and Correll (1984) found that a 50-m-wide riparian forest in an agricultural watershed near the Chesapeake Bay in Maryland removed an estimated 89 percent of the nitrogen and 80 percent of the phosphorus that entered it from upland runoff, groundwater, and bulk precipitation. The study estimated that there was a net removal of 1.1 g N m^{-2} yr^{-1} of particulate organic nitrogen, 4.7 g N m^{-2} yr^{-1} of nitrate nitrogen, and 0.3 g P m^{-2} yr^{-1} of particulate phosphorus. Typically, nitrate–nitrogen losses through riparian forests are between 2 and 6 g N m^{-2} yr^{-1} (Table 15-11).

The Maryland study is only one of many (e.g., Schlosser and Karr, 1981a, b; Johnston et al., 1984; Lowrance et al., 1984, 1985, 1997; Jacobs and Gilliam, 1985; Cooper et al., 1986, 1987; Kuenzler and Craig, 1986; Jordan et al., 1986; Whigham et al., 1986; Cooper and Gilliam, 1987; Fail et al., 1989; Groffman et al., 1991; Hanson et al., 1994a, b; Groffman and Hanson, 1997) that have demonstrated the way in which riparian ecosystems can be effective in removing as well as modifying nutrients and sediments from agricultural runoff before it reaches a stream or river. For example, Jacobs and Gilliam (1985) suggested that riparian buffer strips less than 16 m in width in North Carolina were effective in denitrifying a substantial portion

Table 15-10   **Rates of phosphorus retention in riparian forested wetlands**

Location	Sedimentation Rate	Reference
Cache River, southern Illinois	3.6 g P m^{-2} yr^{-1} contributed by flood as sedimentation for flood of 1.13-yr recurrence interval	Mitsch et al. (1979a)
Prairie Creek, north-central Florida	3.25 g P m^{-2} yr^{-1} as sedimentation from river overflow	S. L. Brown (1978)
Creeping Swamp, Coastal Plain, North Carolina	0.17 g P m^{-2} yr^{-1} sedimentation on floodplain floor from river overflow	Yarbro (1979)
Creeping Swamp, Coastal Plain, North Carolina	0.315–0.730 g P m^{-2} yr^{-1} based input–output budget of floodplain (most was filterable reactive phosphorus)	Yarbro (1979)
Kankakee River, northwestern Illinois	1.36 g P m^{-2} yr^{-1} contributed by unusually large spring flood lasting 62–80 days	Mitsch et al. (1979b)

Source: Brinson et al. (1981b).

Table 15-11   Nitrogen loss rates as reported in the literature for riparian zone studies

Rates		Conditions	Reference
g N m^{-2} day^{-1}	g N m^{-2} yr^{-1}		
	4.5–6.0	Riparian forest, Chesapeake Bay, Maryland	Peterjohn and Correll (1984)
0.22		NO$_3$ + glucose; buffer zones	Groffman et al. (1991)
1.58		NO$_3$ + glucose; grass strips	
	0.5–1.6	Riparian maple swamp (unenriched)	Hanson et al. (1994a, b)
	2.0–3.6	Riparian maple swamp (enriched)	
	6.9	Restored riparian wetland	Lowrance et al. (1985)
	4.3	Young hardwood riparian forest	
0.87–1.3[a]		Moderately well drained soil	Groffman and Hanson (1997)
2.6–24.4[a]		Very poorly drained soils	
	1.5–15.5[a]	Alluvial soil	Groffman and Hanson (1997)
	1.0–2.0[a]	Light till	

[a] Range of means (14–16 samples/yr).
Source: Partially from Groffman (1994).

of nitrates in agricultural subsurface salvage water before it reached the adjacent stream. Care must be taken in the assumption that riparian forests are unlimited sinks for nitrate nitrogen. Hanson et al. (1994b) found that, although a riparian forest receiving enriched nitrate nitrogen from a housing development remained a sink of nitrogen, there were clear signs of nitrogen saturation in the soil, litter, and groundwater.

Ecosystem Models

Few energy–nutrient models have been developed for bottomland hardwood forests in particular or riparian wetlands in general. A general energy–nutrient model for a typical bottomland hardwood forest is shown in Figure 15-30. The flux of a river past and sometimes into the floodplain (longitudinal exchange) and the important inflow of groundwater and surface water from adjacent uplands (transverse exchange) are two features that typify riparian ecosystems. The primarily detrital food chain includes a major export of detrital material into the river, particularly during flooding. The water, nutrients, and sediments that are transported into the bottomland forest during flooding represent the energy subsidy that makes these systems among the most productive forested systems in a given region.

The scarcity of specific energy–nutrient simulation models can be attributed to the difficulty in quantifying the relationships between stream flooding and ecosystem productivity as much as any cause. The plant composition or vegetation change in a riparian site has been modeled in several ways: (1) direct specification, (2) environmental suitability, and (3) transition simulation (Auble, 1991). Direct specification involves a description of the current vegetation and a set of relations that specify the vegetation resulting from various actions. This kind of model is most useful for management decisions involving changes that are abrupt, for example, clearing for cropland and reservoir filling. This approach may not involve any mathematical characterization

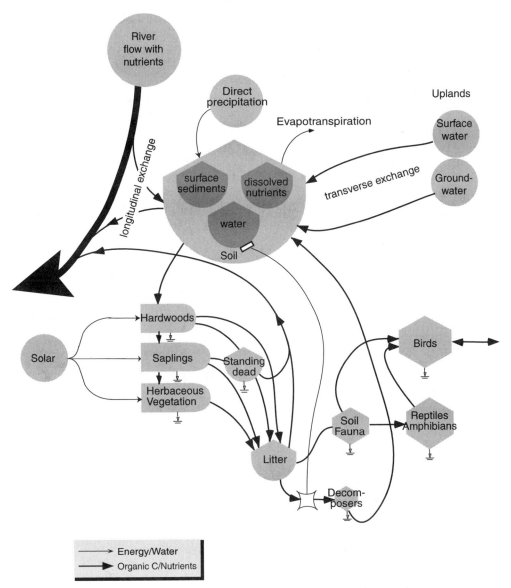

Figure 15-30 Conceptual model of a bottomland hardwood ecosystem.

at all; it may simply represent a conceptual approach to vegetation change (Auble, 1991).

Environmental suitability models add environmental relations to the description of vegetation in a site. When the characterization of vegetation in various sites is linked to local environmental conditions, changes in the environment can be used to predict vegetation change. Environmental conditions need not be instantaneous; instead, they may represent long-term averages, extreme events, or other similar parameters. For example, Stromberg and Patten (1990) used site-specific regressives between discharge and stem radial growth increment of cottonwood to establish a flow standard based on average tree growth. W. C. Johnson (1998) described a graphical model comparing the responses of a braided river (Platte River) and a meandering river (Missouri River) to regulation by dams and diversions. Preregulation, the riparian area shifts according to weather shifts and climatic change. Initially after regulation, the braided river narrows and the riparian zone expands, whereas the meandering river ceases to meander. After major geomorphic adjustments are complete, both rivers show sharp declines in pioneer woodland in favor of late-succession woodlands or grasslands. Most simulation models include some aspects of the environmental suitability approach since the relationship between individual species or species groups and the environment is a key aspect of most ecological models.

Transition simulation applies the relationship of vegetation and the environment to a set of incremental changes; that is, an incremental change in one or more environmental variables leads to a change in vegetation. The new vegetation condition then becomes the starting point for the next incremental change (Auble, 1991). This is the most flexible and commonly used modeling approach in ecology and has been applied to a large class of riparian systems.

Transition models based on individual species growth and their responses to the environment include a suite of models based on SWAMP, a tree growth simulation model of southern wetland forests. This model was developed by Phipps (1979) and applied to a bottomland forest in Arkansas and later to a forested wetland in Virginia (Phipps and Applegate, 1983). Unlike energy–nutrient models, this model simulates the growth of individual trees in the forest, summing the growth of all trees in a plot to determine plot dynamics. The model contains several subroutines, including GROW, which "grows" trees on the plot according to a parabolic growth form; KILL, which determines the survival probabilities of trees and occasionally "kills" trees; and CUT, which enables the modeler to remove trees, as in lumbering or insect damage. Another important subroutine, WATER, describes the influence of water-level fluctuations on tree growth. This subroutine assumes that tree growth will be suboptimal during the peak May–June growth period if water is either too high or too low. The effect of water levels on tree productivity is hypothesized to be a parabolic function as follows:

$$H = 10.05511 \, (T-W)^2 \tag{15.1}$$

where H = growth factor related to the water table
T = water table depth of sample plot
W = optimum water table depth

The model begins with all tree species greater than 3 cm in diameter on a 20-m × 20-m plot and "grows" them on a year-by-year basis, generating results that depend on hydrologic conditions such as flooding frequency, depth to the water table, and other factors, including shading and simulated lumbering. Typical results from runs of the model based on data from the White River in Arkansas indicate the importance of altered hydroperiod and lumbering for the structure of the bottomland forest.

FORFLO, a model based on SWAMP, was applied to evaluate the impact of an altered hydrologic regime on succession in a bottomland forest in South Carolina (Pearlstein et al., 1985; Brody and Pendleton, 1987). Mitsch (1988) developed a simple simulation model that illustrated the simultaneous effects of river flooding and nutrient dynamics on forested wetland productivity. Model simulations illustrated that productivity (net biomass accumulation) ranged from highest to lowest in the following order of hydrologic conditions in agreement with the hydrologic theory of Brinson et al. (1981a):

$$Pulsing > Flowing > Stagnant$$

Mitsch (1988) summarized these findings by stating that "The pulsing system had the highest productivity because of high nutrients and because the pulsing hydrology allowed the water level to decrease between floods to levels closer to the optimum requirements of the vegetation." This model also illustrated the flood pulse concept (FPC) of floodplain systems as proposed by Junk et al. (1989).

Recommended Readings

Friedman, J. M., M. L. Scott, and D. T. Patten, eds. 1998. Semiarid Riparian Ecosystems. Special Issue of Wetlands 18:497–696.

Gosselink, J. G., L. C. Lee, and T. A. Muir, eds. 1990. Ecological Processes and Cumulative Impacts: Illustrated by Bottomland Hardwood Wetland Ecosystems. Lewis Publishers, Chelsea, MI. 708 pp.

Kleiss, B. A., ed. 1996. Bottomland Hardwoods of the Cache River, Arkansas. Special Issue of Wetlands 16:255–396.

Naiman, R. J., and H. Decamps. 1997. The ecology of interfaces: Riparian zones. Annual Review of Ecology Systematics 28:621–658.

Sharitz, R. R., and W. J. Mitsch. 1993. Southern floodplain forests, in W. H. Martin, S. G. Boyce, and A. C. E. Echternacht, eds. Biodiversity of the Southeastern United States: Lowland Terrestrial Communities. John Wiley & Sons, New York. pp. 311–372.

Part **5**

Wetland Management

Values and Valuation of Wetlands

Wetlands provide many services and commodities to humanity. At the population level, wetland-dependent fish, shellfish, fur animals, waterfowl, and timber provide important and valuable harvests and millions of days of recreational fishing and hunting. At the ecosystem level, wetlands moderate the effects of floods, improve water quality, and have aesthetic and heritage value. They also contribute to the stability of global levels of available nitrogen, atmospheric sulfur, carbon dioxide, and methane.

The valuation of these services and commodities for purposes of wetland management is complicated by the difficulty of comparing by some common denominator the various values of wetlands against human economic systems, by the conflict between a private owner's interest in the wetlands and the values that accrue to the public at large, and by the need to consider the value of a wetland as a part of an integrated landscape. Valuation techniques include nonmonetized scaling and weighting approaches for comparing different wetlands or different management options for the same wetland, and common-denominator approaches that reduce the various values to some common term such as dollars or embodied energy. These common-denominator methodologies can include willingness to pay, opportunity costs, replacement value, and energy analysis. None of these approaches is without problems, and no universal agreement about their use has been reached.

The term *value* imposes an anthropocentric orientation on a discussion of wetlands. The term is often used in an ecological sense to refer to functional processes, as, for example, when we speak of the "value" of primary production in providing the food energy that drives the ecosystem. However, in ordinary parlance, the word connotes something worthy, desirable, or useful to humans. The reasons that wetlands are often legally protected have to do with their value to society, not with the abstruse ecological processes that occur in wetlands; this is the sense in which the word *value* is used in this chapter. Perceived values arise from the functional ecological processes described in previous chapters but are determined also by human perceptions, the location of a particular wetland, the human population pressures on it, and the extent of the resource. Regional wetlands are integral parts of larger landscapes—drainage basins, estuaries. Their functions and their values to people in these landscapes depend on both their extent and their location. Thus, the value of a forested wetland varies. If it lies along a river, it probably plays a greater functional role in stream water quality and downstream flooding than if it were isolated from the stream. If situated at the headwaters of a stream, a wetland functions differently from a wetland located near the stream's mouth. The fauna it supports depend on the size of the wetland relative to the home range of the animal. Thus, to some extent, each wetland is ecologically unique. This complicates the measurement of its "value."

Wetland Values

Wetland values can conveniently be considered from the perspective of three hierarchical levels—population, ecosystem, and global.

Population Values

The easiest wetland values to identify are the populations that depend on wetland habitats for their survival.

Animals Harvested for Pelts

Fur-bearing mammals and the alligator are harvested for their pelts (Fig. 16-1). In contrast to most other commercially important wetland species, these animals typically have a limited range and spend their lives within a short distance of their birthplaces. The most abundant are the muskrats (*Ondatra zibethicus*), with about 10 million pelts harvested each year (Fig. 16-1b) out of a total of 12 million for all species (Fig. 16-1a). Muskrats (Fig. 16-2a) are found in wetlands throughout the United States except, strangely, the South Atlantic Coast. They prefer fresh inland marshes, but along the northern Gulf Coast are more abundant in brackish marshes. About 50 percent of the nation's harvest is from the Midwest and 25 percent from along the northern Gulf of Mexico, mostly Louisiana (Chabreck, 1979).

The nutria (*Myocastor coypus*) is the next most abundant species. It is very much like a muskrat but is larger and more vigorous (Figs. 16-1c and 16-2b; Kinler et al., 1987). This species was imported from South America to Louisiana and escaped from captivity in 1938, spreading rapidly through the state's coastal marshes. It is now abundant in freshwater swamps and in coastal freshwater marshes from which it may have displaced muskrats to more brackish locations. Ninety-seven percent of the U.S. nutria harvest occurs in Louisiana. In order of decreasing abundance (but not dollar value), other harvested fur animals are beaver (Fig. 16-1d), mink (Fig. 16-1e), and otter. Raccoons are also taken commercially and in the north-central states are second only to muskrats. Beavers are associated with forested wetlands, especially in the Midwest. Minnesota harvests 27 percent of the nation's catch.

The alligator represents a dramatic success story of the return from the edge of extinction to a healthy population (at least in the central Gulf states). It is now harvested under close regulation in Florida and Louisiana. The species was threatened by severe hunting pressure, not by habitat loss. When that pressure was removed, its numbers increased rapidly. Alligators are abundant in fresh and slightly brackish lakes and streams and build nests in adjacent marshes and swamps. In Louisiana, the indus-

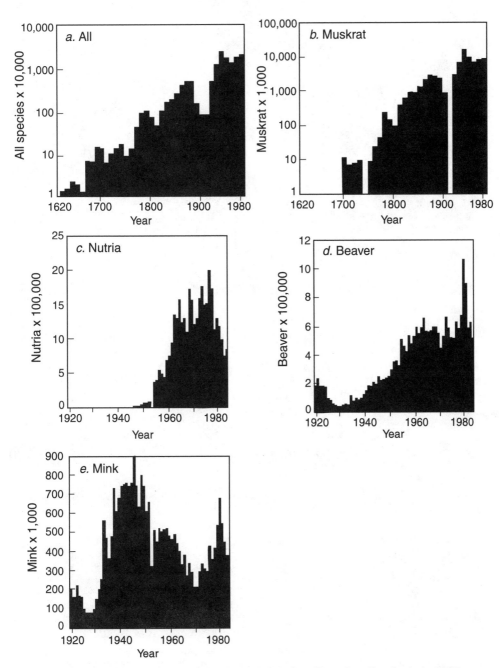

Figure 16-1 Harvests of fur species in North America. (*Data from Novak et al., 1987*)

Figure 16-2 Two animals that are frequently harvested for their pelts include   **a.** the muskrat (*Ondatra zibethicus*) and   **b.** the nutria (*Myocastor coypus*). (*Photograph of muskrat by Ohio Department of Natural Resources, reprinted by permission; photograph of nutria by Greg Linscombe, Louisiana Department of Wildlife and Fisheries*)

try was worth over $16 million in 1992, for both wild and farm-raised animals (G. Linscombe, Louisiana Department of Wildlife and Fisheries, personal communication).

Waterfowl and Other Birds

Birds, as our only remaining evolutionary link to the dinosaurs that once roamed the Earth, may have actually survived the dinosaur die-off precisely because of wetlands (Gibbons, 1997; Weller, 1999). Although not all present bird species require wetlands as their primary habitat, a great many do (Figs. 16-3 and 16-4). Eighty percent of America's breeding bird population and more than 50 percent of the 800 species of protected migratory birds rely on wetlands (Wharton et al., 1982). Wetlands, which are probably known best for their waterfowl abundance, also support a large and valuable recreational hunting industry. We use the term "industry" because hunters spend large sums of money in the local economy (estimated at $58 million for the Mississippi flyway alone) for guns, ammunition, hunting clothes, travel to hunting spots, food, and lodging. Most of the birds hunted are hatched in marshes in the far North, sometimes above the Arctic Circle, but are shot during their winter migrations to the southern United States and Central America. There are exceptions—the Wood Duck (*Aix sponsa*) breeds locally throughout the continent—but the generalization holds for most species. Different groups of geese and ducks have different habitat preferences, and these preferences change with the maturity of the duck and the season.

A broad diversity of wetland habitat types is important for waterfowl success. The freshwater prairie potholes of North America are the primary breeding place for waterfowl in North America. There, an estimated 50 to 80 percent of the continent's main game species are produced (Batt et al., 1989). Wood Ducks prefer forested wetlands. During the winter, diving ducks (*Aythya* spp. and *Oxyura* spp.) are found in brackish marshes, preferably adjacent to fairly deep ponds and lakes. Dabbling ducks (*Anas* spp.) prefer freshwater marshes and often graze heavily in adjacent rice fields and in very shallow marsh ponds. Gadwalls (*Anas strepera*) like shallow ponds with submerged vegetation.

The waterfowl value of wetlands such as the prairie pothole region of North America (see Chapters 4 and 12) is unmistakable. The loss of wetlands in this region is estimated to be much more than half the original pothole area. In one study, Batt et al. (1989) examined waterfowl census data for the prairie pothole region over the 30-year period 1955 to 1985 and found that there was an average of almost 22 million waterfowl (dabbling and diving ducks) in the region, dominated by the Mallard (average 3.7 million). Population size fluctuated from year to year, however. When these fluctuations were compared to the number of potholes flooded in May of each year, there was a clear positive correlation (Fig. 16-5), indicating the importance of wetland hydrology in the breeding success of waterfowl.

Generally, the duck population of North America has shown a 10- to 20-year cycle of increase and decline, with low points in the early 1960s and 1990s and highs in the mid-1950s, mid-1970s, and late 1990s (Table 16-1 and Fig. 16-5). Populations of nine of the ten duck species listed in Table 16-1 were lower than historical averages after the dry years 1987–1991, while populations of seven of the same ten duck species were higher than historical averages after the wet years 1995–1998 (see Fig. 16-5). Over that period, from dry period to wet period, the total number of ducks increased by 60 percent. The trends of below-average populations during dry periods and above-average populations during wet periods are particularly apparent for Mal-

Figure 16-3 Two wetland waterfowl known around the world: **a.** Mallard (*Anas platyrhynchos*) and **b.** Canada Goose (*Branta canadensis*).

Figure 16-4 Herons are consummate symbols of wetlands throughout the world. Different species that dominate this wading niche in parts of the world include: **a.** Great Blue Heron (*Ardea herodias;* North America); **b.** White-necked Heron (*Ardea cocoi;* South America); **c.** Black-headed Heron (*Ardea melanocephala; eastern Africa*); **d.** White-faced Heron (*Ardea novaehollandiae;* Australia/New Zealand); **e.** Gray Heron (*Ardea cinerea;* Europe and Africa). (*Photographs a by T. Daniel, Ohio Department of Natural Resources; b, c by W. J. Mitsch; d by B. Harcourt, courtesy of New Zealand Department of Conservation; e by P. Marion; reprinted by permission.*)

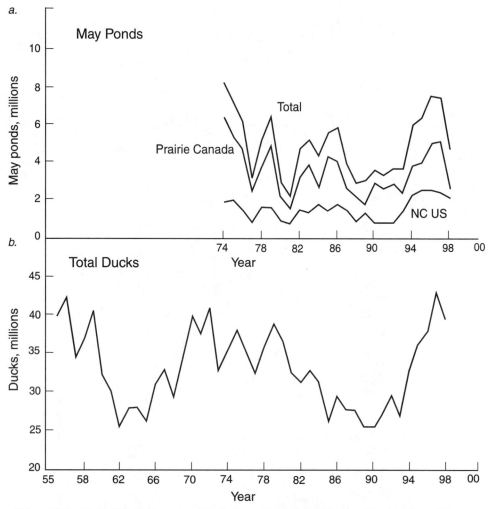

Figure 16-5 Illustration of the connection between duck populations and how wet conditions are in the prairie pothole region: a. estimated number of ponds in May in the prairie pothole region; b. duck breeding populations in North America, 1955–1998, excluding scooters, eiders, mergansers, and oldsquaws. (*From U.S. Fish and Wildlife Service, Washington, DC*)

lards, Green-winged and Blue-winged Teals, Northern Shovelers, and Canvasback (Table 16-1).

Habitat degradation and loss, both in the northern breeding grounds and in the wintering areas, are certainly factors in population declines; Northern Pintail and Scaup have showed a steady decline. However, climatic changes that influence the number of ponds from year to year in the breeding grounds appear to be the major cause of year-to-year fluctuations.

Hunting is closely regulated and tailored to the local region. The Mallard makes up about one-third of the U.S. total of harvested ducks. About 50 percent are shot in wetlands; most of the rest are shot in agricultural fields. In Louisiana, one-third of the Mallard population is killed each year. The percentage is lower for other species—about 8 to 13 percent. The vast flocks of geese that used to be so abundant

Table 16-1 Population estimates of the 10 most common species of breeding ducks and 4 species of goose in North America for a dry year (1991) and a wet year (1998) in the prairie pothole region, with percentage change in 1991 and 1998 compared to 1955–1990 and 1955–1997 averages, respectively

Species	Population (×1,000)		Percentage Change	
	1991 (Dry Year)	1998 (Wet Year)	1991[a]	1998[b]
All species	24,200	39,100		+20
Mallard (*Anas platyrhynchos*)	5,353 ± 188	9,640 ± 302	−27	+32
Gadwall (*Anas strepera*)	1,573 ± 94	3,742 ± 206	+22	+149
American Wigeon (*Anas americana*)	2,328 ± 135	2,858 ± 145	−14	−5
Green-winged Teal (*Anas crecca*)	1,601 ± 88	2,087 ± 139	−4	+16
Blue-winged Teal (*Anas discors*)	3,779 ± 245	6,399 ± 332	−10	+36
Northern Shoveler (*Anas clypeata*)	1,663 ± 84	4,120 ± 194	−8	+106
Northern Pintail (*Anas acuta*)	1,794 ± 199	3,558 ± 194	−62	−36
Redhead (*Aythya americana*)	437 ± 37	918 ± 77	−26	+48
Canvasback (*Atythya valisneria*)	463 ± 57	689 ± 57	−16	+28
Scaup (*Aythya* spp.)	5,247 ± 333	4,112 ± 234	−7	−35
Average of 10 duck species			−15	+34
Canada Goose (*Branta canadensis*)	3,750	4,683		
Snow Goose (*Chen caerulsecens*)	2,440	3,776		
White-fronted Goose (*Anser albifrons*)	492	941		
Brant (*Branta bernicla*)	275	276		

[a] Compared to average for 1955–1990

[b] Compared to average for 1955–1997

Source: U.S. Fish and Wildlife Service. Duck surveys on summer breeding grounds; goose surveys during summer, fall, and winter.

along the eastern seaboard and the Gulf Coast are smaller now but are still abundant and are considered to be important as hunted species in some areas.

Fish and Shellfish

Over 95 percent of the fish and shellfish species that are harvested commercially in the United States are wetland dependent (Feierabend and Zelazny, 1987). The fishing industry contributed $1.9 billion to the U.S. gross national product in 1998. The degree of dependence on wetlands varies widely with species and with the type of wetland. Some important species are permanent residents; others are merely transients that feed in wetlands when the opportunity arises. Some shallow wetlands, which may exhibit several other wetland "values," may be virtually devoid of fish, whereas other types of deepwater and coastal wetlands may serve as important nursery and feeding areas. Table 16-2 lists some of the major nektonic species associated with wetlands. Table 9-5 shows the dollar value of coastal fisheries. Virtually all of the freshwater species are dependent, to some degree, on wetlands, often spawning in marshes bordering lakes or in riparian forests during spring flooding. These species are primarily

Table 16-2   **Dominant commercial and recreational fish and shellfish associated with wetlands**

Species Common Name	Scientific Name	Commercial Harvest (Metric Tons)[a]
FRESHWATER		
Catfish and bullhead	*Ictalurus* sp.	16,800
Carp	*Cyprinus carpio*	11,800
Buffalo	*Ictiobus* sp.	11,300
Crayfish	*Procambarus clarkii*	11,300
Perch	*Perca* sp., *Istizostedion* sp.	—
Pickerel	*Esox* sp.	—
Sunfish	*Lepomis* sp., *Micropterus* sp., *Pomoxis* sp.	—
Trout	*Salmo* sp., *Salvelinus* sp.	—
ANADROMOUS		
Salmon	*Oncorhynchus* sp.	396,600
Shad and alewife	*Alosa* sp.	4,400
Striped bass	*Morone saxatilis*	3,600
SALTWATER		
Menhaden	*Brevoortia* sp.	901,300
Shrimp	*Penaeus* sp.	135,200
Blue crab	*Callinectes sapidus*	102,100
Oyster	*Crassostrea* sp.	17,200
Mullet	*Mugil* sp.	11,000
Atlantic croaker	*Micropogonias undulatus*	7,700
Hard clam	*Mercenaria* sp.	5,800
Bluefish	*Pomatomus saltatrix*	4,300
Seatrout	*Cynoscion* sp.	4,000
Spot	*Leiostomus xanthurus*	3,300
Drum	*Pogonias cromis, Sciaenops ocellatus*	2,200
Soft clam	*Mya arenaria*	1,300

[a] Landings are 1993–1997 averages, except 1971–1975 landings for freshwater fish.
Source: National Marine Fisheries Service, U.S. Commercial Landings.

recreational, although some small local commercial fisheries exploit them. The saltwater species tend to spawn offshore, move into the coastal marsh "nursery" during their juvenile stages, and then emigrate offshore as they mature. They are often important for both commercial and recreational fisheries. The menhaden is caught only commercially, but competition between commercial and sport fishermen for shrimp, blue crab, oyster, catfish, seatrout, and striped bass can be intensive and acrimonious. Anadromous fish probably use wetlands less than the other two groups. However, young anadromous fish fry sometimes linger in estuaries and adjacent marshes on their migrations to the ocean from the freshwater streams in which they were spawned.

The relationship between the area of marsh or the marsh–water interface of coastal wetlands and the production of commercial fisheries has been illustrated for shrimp and fish harvests in Louisiana and worldwide (R. E. Turner, 1982; Gosselink, 1984; Costanza et al., 1989; Browder et al., 1989), blue crab production in western Florida (Lynn et al., 1981), and oyster production in Virginia (Batie and Wilson, 1978). In one example, R. E. Turner (1977, 1982) showed a direct relationship between shrimp and fish harvests and wetland area for a number of fisheries around the world, including marine, freshwater, and pond raised (Fig. 16-6).

Analyses of fishery harvests from wetlands show the importance of the recreational catch. Although the commercial harvest is usually much better documented, several

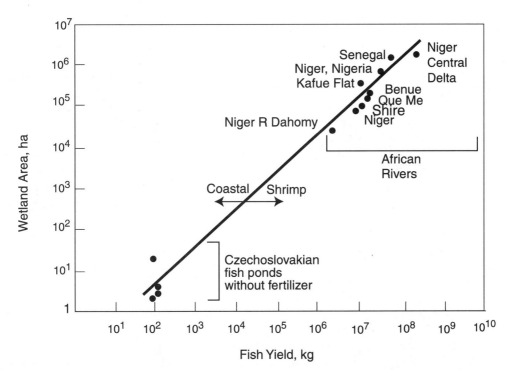

Figure 16-6 Relationships between wetland area and fish harvests. The linear slope describes the line of about 60 kg/ha yield. Pond fisheries are unfertilized, managed ponds in Czechoslovakia. African floodplain river fisheries after Welcomme (1976). Wetland-dependent coastal fisheries yields per ha (adapted from Turner, 1977) are generally 10 times higher than inland ecosystems. (*After Turner, 1982*)

studies have shown that the recreational catch far outweighs the commercial catch for certain species. Furthermore, the value to the economy of recreational fishing is usually far greater than the value of the commercial catch because sports fishermen spend more money per fish caught (they are less efficient) than their commercial counterparts. For example, DeSylva (1969) estimated that in California it costs an angler $18.11 to catch one salmon; its value from the angler's standpoint is five times what it is to the commercial fisherman. These excess dollars feed the California economy.

Timber and Other Vegetation Harvest

The area of wetland timber in the United States is about 22 million ha. Two-thirds of that area is east of the Rocky Mountains. The Mississippi River alluvial floodplain and the floodplains of rivers entering the South Atlantic are mostly deciduous wetlands, whereas the forested wetlands along the northern tier of states are primarily evergreen. The former are more extensive and potentially more valuable commercially because of the much faster growth rates in the South. R. L. Johnson (1979) estimated that the 13 million ha of bottomland hardwood and cypress swamps in the southeastern part of the nation contained about 112 m^3 of merchantable timber per hectare, worth about $620/ha, or about $8 billion total. As timber prices climb and as the land becomes more and more valuable for agriculture, these wetlands are being clear-cut and drained (see Chapter 17), although by using sound silviculture practices, the timber industry can coexist with productive wetlands.

In addition to the timber harvest, the production of herbaceous vegetation in marshes is a potential source of energy, fiber, and other commodities. These prospects have not been explored widely in North America but are viable options elsewhere. For example, Chung (1989), Mitsch (1991a, b), and Yan (1992) described several commercial products that are harvested from restored and natural salt marshes and freshwater marshes in China. The productivity of many wetland species (e.g., *Spartina alterniflora*, *Phragmites communis*, *Typha angustifolia*, *Eichhornia crassipes*, *Cyperus papyrus*) is as great as our most vigorous agricultural crops. Ryther et al. (1979) estimated that energy put into growing wetland crops can be returned five- to ninefold in their harvest. (Only about half as much energy, however, can be recovered if the crop is fermented to produce methane, one use for harvested wetland vegetation.) The economics of commercial production has not been completely worked out, but Ryther et al. (1979) stated that a 1,000-ha water hyacinth (*Eichhornia crassipes*) energy farm in the southern United States could produce on the order of 10^{12} Btu of methane per year and, at the same time, remove all of the nitrogen from the wastewater of a city of about 700,000 people. It should be understood that the use of wetlands for any purpose involving the harvesting of the vegetation is bound to have a significant effect on the way the system functions. Therefore, the benefits of the harvest should be weighed against any functional values lost through the harvesting operation.

In addition to the annual production of living vegetation in wetlands, great reservoirs of buried peat exist around the world. Kivinen and Pakarinen (1981) estimated world peat resources (30 cm or more in thickness) at 420 million ha, mostly in the former Soviet Union and Canada. The United States has only about 21 million ha, most of it in Alaska, Minnesota, and Michigan. This buried peat is, of course, a nonrenewable energy source that destroys the wetland habitat when it is mined. In the United States and Canada, peat is mined primarily for horticultural peat production, but in other parts of the world—for example, several republics of the former

Soviet Union—it has been used as a fuel source for hundreds of years. It is used to generate electricity, formed into briquettes for home use, and gasified or liquified to produce methanol and industrial fuels (see Chapter 17, p. 622).

Endangered and Threatened Species

Wetland habitats are necessary for the survival of a disproportionately high percentage of endangered and threatened species. Table 16-3 summarizes the statistics but imparts no information about the particular species involved, their location, wetland habitat requirements, degree of wetland dependence, and factors contributing to their demise. Although wetlands occupy only about 3.5 percent of the land area of the United States, of the 209 animal species listed as endangered in 1986, about 50 percent depend on wetlands for survival and viability. Almost one-third of native North American freshwater fish species are endangered, threatened, or of special concern. Almost all of these were adversely affected by habitat loss (J. D. Williams and Dodd, 1979). Ernst and Brown (1989) listed 63 species of plants and 34 species of animals that are considered endangered, threatened, or candidates for listing and that occupy southern U.S. forested wetlands. Of these, amphibians and many reptiles are especially linked to wetlands. In Florida, where the number of amphibian and reptile species is about equal to the number of mammal and breeding bird species, 18 percent of all amphibians and 35 percent of all reptiles are considered threatened or endangered or their status is unknown (Harris and Gosselink, 1990).

The fate of several endangered species is discussed here to illustrate the ecological complexity of species endangerment. Whooping Cranes nest in wetlands in the Northwest Territories of Canada, in water 0.3 to 0.6 m deep, during the spring and summer. In the fall, they migrate to the Aransas National Wildlife Refuge, Texas, stopping off in riverine marshes along the migration route. In Texas, they winter in tidal marshes. All three types of wetlands are important for their survival (Williams and Dodd, 1979).

Table 16-3 Threatened and endangered species associated with wetlands

Taxon	Number of Species Endangered	Number of Species Threatened	Percentage of U.S. Total Threatened or Endangered
Plants	17	12	28
Mammals	7	—	20
Birds	16	1	68
Reptiles	6	1	63
Amphibians	5	1	75
Mussels	20	—	66
Fish	26	6	48
Insects	1	4	38
Total	98	25	

Source: Niering (1988).

The decline in the once-abundant species has been attributed both to hunting and to habitat loss. The last Whooping Crane nest in the United States was seen in 1889 (R. S. Miller et al., 1974). In 1941, the flock consisted of 13 adults and 2 young. Since then, the flock has been gradually built up to about 75 birds.

American alligator populations were reduced by hunters and poachers to such low levels that the species was declared endangered in the 1970s. Under close control, the population has subsequently been rebuilt. Alligators have a reciprocal relationship with wetlands—they depend on them, but, in return, the character of the wetland is shaped by the alligator, at least in the south Florida Everglades. As the annual dry season approaches, alligators dig "gator holes." The material thrown out around the holes forms a berm high enough to support trees and shrubs in an otherwise treeless prairie. The trees provide cover and breeding grounds for insects, birds, turtles, and snakes. The hole itself is a place where the alligator can wait out the dry period until the winter rains. It also provides a refuge for dense populations of fish and shellfish (up to $1,600/m^2$). These organisms, in turn, attract top carnivores, and so the gator holes are sites of concentrated biological activity that may be important for the survival of many species.

The slackwater darter (*Etheostoma boschungi*) is an example of the specificity sometimes required by wetland species. This small fish, found in small and moderate creeks in Alabama and Tennessee, migrates 1 to 5 km upstream at certain times of the year to spawn in small marshy areas associated with water seeps. It requires shallow (2–8 cm) water and deposits its eggs on a single species of rush (*Juncus acuminatus*) in spite of the presence of other species. The larvae remain in the vicinity of these marshes for 4 to 6 weeks before returning to their home streams (Williams and Dodd, 1979).

Ecosystem Values

At the level of the whole ecosystem, wetlands have value to the public for flood mitigation, storm abatement, aquifer recharge, water quality improvement, aesthetics, and general subsistence. Some of the ecosystem values of wetlands vary from year to year or from season to season. For example, Figure 16-7 illustrates several of the potential ecosystem values of riparian forested wetlands during flooding (spring) and dry (summer) seasons.

Flood Mitigation

Chapter 5 dealt with the importance of hydrology in determining the character of wetlands. Conversely, wetlands influence regional water flow regimes. One way they do this is to intercept storm runoff and to store storm waters, thereby changing sharp runoff peaks to slower discharges over longer periods of time (Fig. 16-8). Because it is usually the peak flows that produce flood damage, the effect of the wetland area is to reduce the danger of flooding (Novitzki, 1979; Verry and Boelter, 1979). Riverine wetlands are especially valuable in this regard. On the Charles River in Massachusetts, the floodplain wetlands were deemed so effective for flood control by the U.S. Army Corps of Engineers that it purchased them rather than build expensive flood control structures to protect Boston (U.S. Army Corps of Engineers, 1972). The study on which the Corps' decision was made is a classic in wetland hydrologic values. It demonstrated that if the 3,400 ha of wetlands in the Charles River basin were drained and leveed off from the river, flood damages would increase by $17 million per year.

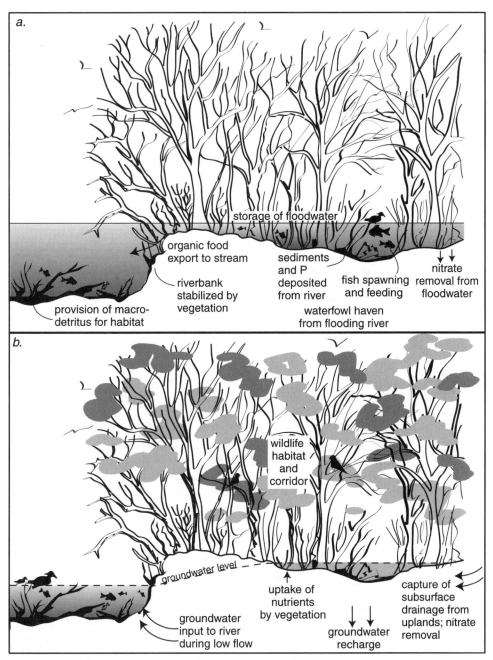

a.

storage of floodwater

organic food
export to stream

sediments
and P
deposited
from river

fish spawning
and feeding

nitrate
removal from
floodwater

riverbank
stabilized by
vegetation

provision of macro-
detritus for habitat

waterfowl haven
from flooding river

b.

wildlife
habitat
and
corridor

groundwater level

groundwater
input to river
during low flow

uptake of
nutrients
by vegetation

groundwater
recharge

capture of
subsurface
drainage from
uplands; nitrate
removal

Figure 16-7 Illustration of several of the potential wetland values for riparian wetlands during
a. flood season and   **b.** dry season.

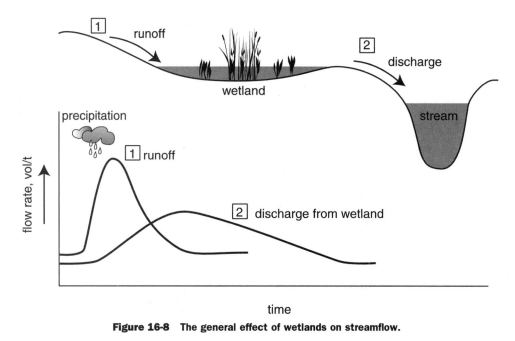

Figure 16-8 **The general effect of wetlands on streamflow.**

Bottomland hardwood forests along the Mississippi River before European settlement stored floodwater equivalent to about 60 days of river discharge. Storage capacity was reduced to only about 12 days (Gosselink et al., 1981a) as a result of leveeing the river and draining the floodplain. The consequences—the confinement of the river to a narrow channel and the loss of storage capacity—are major reasons that flooding is increasing along the lower Mississippi River (Belt, 1975).

Novitzki (1985) analyzed the relationship between flood peaks and the percentage of basin area in lakes and wetlands. In the Chesapeake Bay drainage basin, where the wetland area was 4 percent, flood flow was only about 50 percent of that in basins containing no wetland storage. However, in Wisconsin river basins that contained 40 percent lakes and wetlands, spring streamflow was as much as 140 percent of that in basins that do not contain storage. This apparent anomaly is probably related to a reduction in the proportion of precipitation that can infiltrate the soil and to a lack of additional storage capacity in lakes and wetlands that are already at full capacity during spring floods. Thus, the location of wetlands in the river basin can complicate the response downstream. For example, detained water in a downstream wetland of one tributary can combine with flows from another tributary to increase the flood peak rather than to desynchronize flows.

A quantitative approach to the flood mitigation potential of wetlands was undertaken by Ogawa and Male (1983, 1986), who used a hydrologic simulation model to investigate the relationship between upstream wetland removal and downstream flooding. The study found that for rare floods—that is, those predicted to occur only once in 100 or more years—the increase in peak streamflow was significant for all sizes of streams when wetlands were removed (Fig. 16-9). The authors concluded that the usefulness of wetlands in reducing downstream flooding increases with (1) an increase in wetland area, (2) the distance the wetland is downstream, (3) the

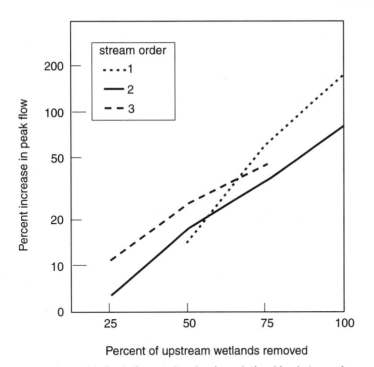

Percent of upstream wetlands removed

Figure 16-9 Hydrologic model simulation results showing relationships between increase in peak flood streamflow and percentage wetland removal for Massachusetts watershed. Results are for various stream orders for 130-year flood. (*After Ogawa and Male, 1983*)

size of the flood, (4) the closeness to an upstream wetland, and (5) the lack of other upstream storage areas such as reservoirs.

Storm Abatement

Coastal wetlands absorb the first fury of ocean storms as they come ashore. Salt marshes and mangrove wetlands act as giant storm buffers. This value can be seen in the context of marsh conservation versus development. Natural marshes, which sustain little permanent damage from these storms, can shelter inland developed areas. Buildings and other structures on the coast are vulnerable to storms, and damage has often been high. Inevitably, the public pays much of the cost of this damage through taxes for public assistance, rebuilding public services such as roads and utilities, and federally guaranteed insurance (Farber, 1987). Two hurricanes that struck the southeastern United States in the early 1990s (Hugo in 1989; Andrew in 1992) passed through major inland wetland areas in South Carolina, Florida, and Louisiana. The damage caused to human development increases as these remaining coastal wetlands (salt marshes and mangroves) are drained and developed.

Aquifer Recharge

Another value of wetlands related to hydrology is groundwater recharge. This function has received too little attention, and the magnitude of the phenomenon has not been well documented. Some hydrologists believe that, although some wetlands recharge groundwater systems, most wetlands do not (Carter et al., 1979; Novitzki, 1979;

Carter, 1986; Carter and Novitzki, 1988). The reason for the absence of recharge is that soils under most wetlands are impermeable (Larson, 1982). In the few studies available, recharge occurred primarily around the edges of wetlands and was related to the edge : volume ratio of the wetland. Thus, recharge appears to be relatively more important in small wetlands such as prairie potholes than in large ones. These small wetlands can contribute significantly to recharge of regional groundwater (Weller, 1981). Heimburg (1984) found significant radial infiltration from cypress domes in Florida and concluded that the rate of infiltration was proportional to the area of the wetland and the depth of the surficial water table. He also concluded that these wetlands represent hydrologic "highs" in the surface water table and are, therefore, "closely coupled to groundwater." There did not appear to be any direct percolation to deep aquifers, however.

Water Quality

Wetlands, under favorable conditions, have been shown to remove organic and inorganic nutrients and toxic materials from water that flows across them. The concept of wetlands as sinks for chemicals was discussed in detail in Chapter 6; the practice of using wetlands for wastewater treatment is discussed in Chapter 20. Wetlands have several attributes that influence the chemicals that flow through them, whether the chemicals are naturally added or artificially applied (Sather and Smith, 1984):

1. A reduction in water velocity as streams enter wetlands, causing sediments and chemicals sorbed to sediments to drop out of the water column

2. A variety of anaerobic and aerobic processes in close proximity, promoting denitrification, chemical precipitation, and other chemical reactions that remove certain chemicals from the water

3. A high rate of productivity in many wetlands that can lead to high rates of mineral uptake by vegetation and subsequent burial in sediments when the plants die

4. A diversity of decomposers and decomposition processes in wetland sediments

5. A large contact surface of water with sediments because of the shallow water, leading to significant sediment–water exchange

6. An accumulation of organic peat in many wetlands that causes the permanent burial of chemicals

Aesthetics

A real but difficult aspect of a wetland to capture is its aesthetic value, often hidden under the dry term "nonconsumptive use values," which simply means that people enjoy being out in wetlands. There are many aspects of this kind of wetland use. Wetlands are excellent "biological laboratories" where students in elementary, secondary, and higher education can learn natural history first hand. They are visually and educationally rich environments because of their ecological diversity. Their complexity makes them excellent sites for research. Many visitors to wetlands use hunting and fishing as excuses to experience their wildness and solitude, expressing that frontier pioneering instinct that may lurk in all of us. In addition, wetlands are a rich source of information about our cultural heritage. The remains of prehistoric Native American villages and mounds of shells or middens have contributed to our understanding of

Native American cultures and of the history of the use of our wetlands. Many artists—the Georgia poet Sidney Lanier, the painters John Constable and John Singer Sargent, and many others who paint and photograph wetlands—have been drawn to them.

Subsistence Use

In many regions of the world, including Alaska, Canada, and several developing nations, the subsistence use of wetlands is extensive. There, wetlands provide the primary resources on which village economies are based. These societies have adapted to the local ecosystems over many generations and are integrated into them. Few alternative lifestyles exist for these cultures (Ellanna and Wheeler, 1986; see also Chapter 1).

Regional and Global Values

The wetlands function of maintaining water and air quality influences a much broader scale than that of the wetland ecosystem itself. Wetlands may be significant factors in the global cycles of nitrogen, sulfur, methane, and carbon dioxide.

Nitrogen Cycle

The natural supply of ecologically useful nitrogen comes from the fixation of atmospheric nitrogen gas (N_2) by a small group of plants and microorganisms that can convert it into organic form. Currently, ammonia is manufactured from N_2 for fertilizers, at more than double the rate of all natural fixation. Wetlands may be important in returning a part of this "excess" nitrogen to the atmosphere through denitrification. Denitrification requires the proximity of an aerobic and a reducing environment such as the surface of a marsh as well as a source of organic carbon, something abundant in most wetlands. Because most temperate wetlands are the receivers of fertilizer-enriched agricultural runoff and are ideal environments for denitrification, it is likely that they are important to the world's available nitrogen balance. The phenomenon of nitrogen enrichment of coastal waters causing "dead zones" or hypoxia (dissolved oxygen < 2.0 mg/L) in the hypolimnion has already begun to occur worldwide (Kessler and Jansson, 1994; Rabalais et al., 1996, 1998). Wetlands have been recommended as a key ecosystem in providing a solution to this eutrophication (Mitsch, 1999).

Sulfur Cycle

Sulfur is another element whose cycle has been modified by humans. The atmospheric load of sulfate has been greatly increased by fossil fuel burning. It is almost equally split between anthropogenic sources (104×10^{12} g/yr), chiefly caused by fossil fuel burning; and natural biogenic sources (103×10^{12} g/yr) of which salt marshes account for about 25 percent (Cullis and Hirschler, 1980; Andreae and Raembonck, 1983; Gosselink and Maltby, 1990). When sulfates are washed out of the atmosphere by rain, they acidify oligotrophic lakes and streams. When sulfates are washed into marshes, however, the intensely reducing environment of the sediment reduces them to sulfides. Some of the reduced sulfide is recycled to the atmosphere as hydrogen, methyl, and dimethyl sulfides, but most forms insoluble complexes with phosphate and metal ions (see Chapter 6). These complexes can be more or less permanently removed from circulation in the sulfur cycle.

Carbon Cycle

The carbon cycle may also be significantly affected in several ways on a global scale by wetlands. Carbon dioxide concentrations in the atmosphere are steadily increasing because of the burning of fossil fuels and because the rapid clear-cutting of tropical forests results in the oxidation of organic matter in trees and soils.

The huge volume of peat deposits in the world's wetlands, particularly in Canada and Russia, has the potential to serve as a sink for carbon dioxide and provide major relief to increasing carbon in the atmosphere (Mitsch and Wu, 1995). These peat deposits, if disturbed, however, could contribute significantly to worldwide atmospheric carbon dioxide levels, depending on the balance between draining and oxidation of the peat deposits and their formation in active wetlands. Figure 16-10 illustrates the global carbon budget with wetland fluxes superimposed. An estimated 0.08 Pg C (Pg = 10^{15} g) of carbon is currently sequestered annually in wetland peats (Gorham, 1991), a small percentage of the estimated global net primary productivity of wetlands of 4 to 9 Pg C/yr. The carbon sequestration is also a small fraction of the total storage of carbon in the Earth's soils of 1,400 Pg C, one-third to one-half of which is in wetlands.

Gorham (1991) provided two other estimates of the role of wetlands in global carbon cycling. First, combustion of peat and oxidation of peat provide an estimated 0.026 Pg/yr of carbon back to the atmosphere. Second, drainage of wetlands is estimated to contribute 0.008 to 0.042 Pg C/yr back to the atmosphere (the low number is for long term; the high number for short term). Thus, wetland drainage and peat burning could be releasing back to the atmosphere from 45 to 89 percent of the carbon being sequestered.

Another aspect of wetlands and climate change is the emission of methane. Aselmann and Crutzen (1989) estimated a release of 0.03 to 0.12 Pg C/yr of methane

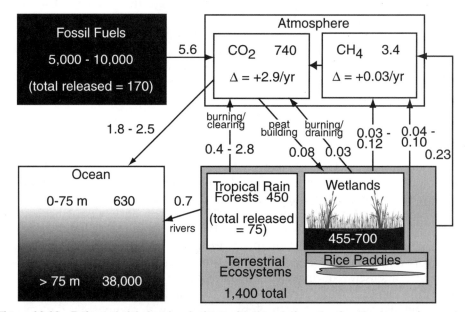

Figure 16-10 Estimated global carbon budget, with the relative role of wetlands superimposed. Sources for numbers shown are given in Mitsch and Wu (1995). Storages are in Pg carbon; fluxes are in Pg carbon/yr.

from natural wetlands and another 0.04 to 0.10 Pg C/yr from rice paddies (Fig. 16-10). This is thought by some to be a serious problem with wetlands because methane is estimated to be 20 times more effective as a greenhouse gas than is carbon dioxide. However, the factor of 20 is probably not justified because of the rapid degradation of methane in the atmosphere compared to carbon dioxide (Lashof and Ahuja, 1990; Gorham, 1991). Thus, wetlands, like tropical rain forests, may be shifting from being a net sink to a net source of carbon to the atmosphere. Their protection should thus be encouraged to prevent the release of yet more carbon to the atmosphere. To put these numbers in perspective, the burning of fossil fuel contributes an estimated 5.6 Pg C/yr, a number far in excess of sequestrations or releases from wetlands (Fig. 16-10). Furthermore, methane, which is released from anaerobic organic soils, may also function as a sort of homeostatic regulator for the ozone layer that protects us from the deleterious effects of ultraviolet radiation (Sze, 1977).

Quantifying Wetland Values

A number of efforts have been made to quantify the "free services" and amenities that wetlands provide to society. Ascribing economic value to wetlands has been an exercise frequently practiced in the past 25 years. Starting with the publication of "The Southern River Swamp—A Multiple-Use Environment" (Wharton, 1970) and "The Value of the Tidal Marsh" (Gosselink et al., 1974), a significant number of papers have discussed the values attributed to wetlands for the services that they provide (e.g., Mitsch, 1977b; Farber and Costanza, 1987; Costanza et al., 1989; Turner, 1991; Barbier, 1994; Gren, 1994; Gren et al., 1994; Bell, 1997; Mitsch and Gosselink, 2000). Costanza et al. (1997) took these types of calculations one step further by estimating the public service functions of all the Earth's ecosystems, including wetlands. References for this subject are also found in several literature reviews and planning guides (e.g., Leitch and Ekstrom, 1989; Folke, 1991; Whitehead, 1992; Barbier et al., 1997; Economic Research Service, 1998; Söderquist et al., 2000).

As described previously, the values of wetlands occur at three levels of ecological hierarchy—population, ecosystem, and biosphere (Fig. 16-11). This "value" also accrues to different segments of the economy. Ecological populations, generally harvested for food or fiber, are the easiest values to estimate and agree on, and they generally accrue to the landowners and local population. At the ecosystem scale, wetlands provide flood control, drought prevention, and water quality protection. These ecosystem values are real, but their quantification is difficult and the benefits are generally regional and less specific to individual landowners. At the highest level of ecosystem worth, the biosphere, we know the least about how to estimate the values, and benefits accrue to the entire world.

A number of approaches to the valuation of wetlands have been advanced. Because of the complexities described previously, there is no universal agreement about which is preferable. In part, the choice depends on the circumstances. Valuations fall broadly into two classes, ecological (or functional) evaluation and economic (or monetary) evaluation. The former are generally necessary before attempting the latter, for it is the valued ecological functions that determine the monetary value.

Ecological Valuation: Scaling and Weighting Approaches

The ecological-valuation approach has been widely used as a means of forming a rational basis for deciding on different management options. Probably the best devel-

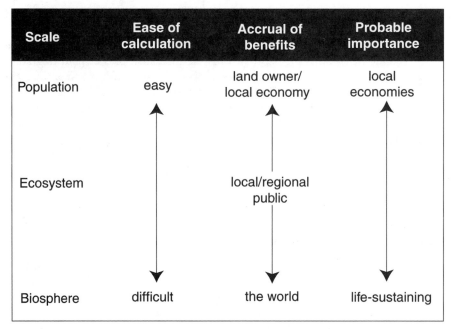

Scale	Ease of calculation	Accrual of benefits	Probable importance
Population	easy	land owner/ local economy	local economies
Ecosystem		local/regional public	
Biosphere	difficult	the world	life-sustaining

Figure 16-11 The ease of quantification, accrual, and probable importance of values for population, ecosystem, and biosphere. (*After Mitsch and Gosselink, 2000*)

oped procedures assess the relative value of wildlife habitats. E. P. Odum (1979b) described the general procedure, as follows:

a. Make a list of all the values that a knowledgeable person or panel can apply to the situation in question, and assign a numerical value of "1" to each.

b. Scale each factor in terms of a maximum level; for example, if 200 ducks per acre could be supported by a first-class marsh but only 100 are supported by the marsh in question, then the scaled factor is 0.5, or 50 percent of the maximum value for that item.

c. Weigh each scaled factor in proportion to its relative importance; for example, if the value 2 is considered 10 times more important to the region than the value 1, then multiply the scale value of 2 by 10.

d. Add the scaled and weighted values to obtain a value index. Because the numbers are only arbitrary and comparative, the index is most useful in comparing different wetlands or the same wetland under different management plans. It is desirable that each value judgment reflect the consensus of several "experts," for example, determined by the "Delphi method" (Dalkey, 1972).

Habitat Evaluation Procedures

Table 16-4 shows an example from the Habitat Evaluation Procedure (HEP) of the U.S. Fish and Wildlife Service of the application of this technique to evaluate different development plans for a cypress–gum swamp ecosystem. The present value of the swamp for a representative group of terrestrial and aquatic animals was evaluated (baseline condition) using a habitat suitability index (HSI) based on a range of 0 to

Table 16-4 Comparison of the impact of two management plans and a no-management control in a cypress–gum swamp[a]

Species	Baseline Condition	Future with Project Plan A[b]		Future with Project Plan B[c]		Future Without Project	
		50 Years	100 Years	50 Years	100 Years	50 Years	100 Years
TERRESTRIAL							
Raccoon	0.7	0.5	0.6	0.8	0.8	0.7	0.9
Beaver	0.7	0.2	0.2	0.4	0.3	0.6	0.4
Swamp rabbit	0.7	0.2	0.2	0.8	0.8	0.7	0.4
Green heron	0.9	0.2	0.1	0.8	0.9	0.9	1.0
Mallard	0.8	0.3	0.2	1.0	0.9	0.9	1.0
Wood duck	0.8	0.3	0.2	0.9	1.0	1.0	1.0
Prothonotary warbler	0.8	0.3	0.1	0.6	0.7	0.8	0.9
Snapping turtle	0.8	0.4	0.3	0.8	0.7	0.8	0.9
Bullfrog	0.9	0.3	0.2	0.8	0.9	1.0	1.0
Total terrestrial HSI	7.1	2.7	2.1	6.9	7.0	7.4	7.5
Mean terrestrial HSI	0.8	0.3	0.2	0.8	0.8	0.8	0.8
AQUATIC							
Channel catfish	0.3	0.3	0.4	0.4	0.4	0.4	0.4
Largemouth bass	0.4	0.2	0.3	0.7	0.8	0.4	0.4
Total aquatic HSI	0.7	0.5	0.7	1.1	1.2	0.8	0.8
Mean aquatic HSI	0.4	0.3	0.4	0.6	0.6	0.4	0.4

[a] Numbers in the tables are habitat suitability index (HSI) values, which have a maximum value of 1 for an optimal habitat.
[b] Channelization of water and clearing of swamp for agricultural development with a loss of 324 ha of wetland.
[c] Construction of levees around swamp for flood control with no loss of wetland area.
Source: Schamberger et al. (1979).

1 for the optimum habitat for the species in question. The evaluation resulted in a mean terrestrial HSI of 0.8 and a mean aquatic HSI of 0.4. This baseline condition was compared with the projected habitat condition in 50 and 100 years under three projected scenarios—Plan A, Plan B, and a no-project projection. The results suggest that Plan A would be detrimental to the environment, whereas Plan B would have no effect on terrestrial habitat values and would improve aquatic ones. Whether to proceed with either of these plans is a decision that requires weighing the projected environmental effects against the projected economic benefits of the project.

One often neglected feature of the analysis is the effect of aggregating HSIs for different species. Although, overall, Plan B appears to be about equivalent environmentally to the no-project option, scrutiny of Table 16-4 shows that Plan B is expected to improve the habitat for swamp rabbits and large-mouthed bass but decrease its value for warblers and turtles. This kind of detailed scrutiny may be important because it indicates a change in the quality of the environment, but it is often neglected when the "apples and oranges" are combined into "fruit."

Wetland Evaluation Techniques

Several wetland evaluation procedures have been developed that attempt to deal with two shortcomings of the habitat evaluation discussed previously by (1) evaluating all relevant goods and services (not just biotic ones) derived from the site and (2) incorporating a landscape focus. The Wetland Evaluation Technique (WET) (Adamus et al., 1987), once in favor with the U.S. Army Corps of Engineers but now hardly used, rates a broad range of functional attributes on a scale of high, moderate, and low. The result is a list of functions, each involving a quality rating for three attributes: (1) *Social significance* assesses the value of a wetland to society in terms of its economic value, strategic location (e.g., upstream from an urban area that requires flood protection), or any special designations it carries (e.g., habitat for an endangered species); (2) *effectiveness* is the site's capacity to carry out a function because of its physical, chemical, or biological characteristics (e.g., to store floodwaters); (3) *opportunity* refers to the opportunity of a wetland to perform a function to its level of capability (e.g., whether the upstream watershed is capable of producing floodwaters). The evaluator is charged with the task of weighing each function to get an integrated evaluation.

Hydrogeomorphic Analysis

A more recent wetland evaluation technique, the hydrogeomorphic (HGM) classification (Brinson, 1993a), also allows a quantification of the functions of wetlands. Its uniqueness lies in its valuation of natural wetland functions without regard to their significance to society. This is done by comparing the wetland of interest to a reference site that is characteristic of the same hydrogeomorphic class. Brinson et al. (1994) summarized the assessment procedure:

1. Group wetlands into HGM classes with shared properties (the classification is discussed in Chapter 21).
2. Define the relationship between hydrogeomorphic properties and the functions of wetlands. The goal is to select functions that are linked clearly and logically to wetland HGM properties and that have hydrologic, geomorphic, and ecological significance. This step represents the scientific basis for the presence of the function.

3. Develop functional profiles for each wetland class. These can range from descriptive narratives to multivariate data sets covering numerous sites.

4. Develop a scale for expressing functions within each wetland class, by using indicators and profiles from the reference wetlands of that class. These scales serve as benchmarks for each wetland class. Reference wetlands should include the full range of natural and human-induced variations due to stress and disturbance.

5. Develop the assessment methodology itself. The assessment relies on indicators to reveal the likelihood that the functions being evaluated are present in the wetland and depends on reference populations to scale the assessment. The reference wetlands are also used to set goals for compensatory mitigation.

In an illustration of the method to estimate the impact of a project or restoration on wetland functions, Rheinhardt et al. (1997) apply the HGM method to evaluate mitigation strategies in mineral soil forested pine (*Pinus palustris*) flats in North Carolina. Fourteen variables were used to estimate the function of both study and reference wetlands (Table 16-5). Absolute values of some of the variables (e.g., tree density) are then translated into indices on a scale of 0.0 to 1.0 by comparing those functions to a reference wetland site. Such indices, in turn, are applied to model

Table 16-5 Field parameters used to estimate ecosystem function in a hydrogeomorphic assessment of forested wetlands in southeastern North Carolina

Variable	Description
HYDROLOGY/TOPOGRAPHY	
V_{DITC}	Lack of ditches nearby (< 50 m)
V_{MICR}	Microtopographic complexity
HERBACEOUS VEGETATION	
V_{GRAM}	Percentage cover of graminoids
V_{FORB}	Percentage cover of forbs
CANOPY VEGETATION	
V_{TREE}	Total basal area for trees (m²/ha; >10 cm DBH)
V_{TDEN}	Density of canopy trees (stems/ha; >10 cm DBH)
V_{TDIA}	Average tree diameter (m)
V_{CVEG}	Sorensen similarity index of canopy importance value
SUBCANOPY VEGETATION	
V_{SUBC}	Density of subcanopy (stems/ha)
V_{SDLG}	Percentage cover of trees and shrubs <1 m tall
V_{SVEG}	Sorensen similarity index of subcanopy importance value
LITTER/STANDING DEAD	
V_{LTR}	Litter depth (cm)
V_{SNAG}	Density of standing dead stems (stems/ha)
V_{CWD}	Volume of coarse woody debris (cm³/ha)

Source; Rheinhardt et al. (1997).

functions such as "maintain hydrologic regime" as in Table 16-6, and comparisons of human impact on wetlands can be assessed. Table 16-6 shows a hypothetical case in which an airport is destroying a wetland (with an overall loss index of 0.71 when compared to a nearby reference, which, by definition, has an index of 1.0) and two restoration alternatives are being considered. The analysis shows that restoration of a cropland back to a wetland (Restoration Alternative 1) would be a good alternative as the cropland has 0.0 value in maintaining hydrologic regime presently. Thus, the restoration is estimated to require only 1 ha of that cropland (gain = +0.71) for every hectare of wetland lost due to the airport (loss = −0.71). This is a 1 : 1 mitigation ratio (the ratio of area of wetland restored to wetland lost). Restoration of an existing pine plantation to a natural pine wetland, on the other hand, would probably be easier but, functionally, the plantation already has some of the desired values of wetlands (it rates a functional index of 0.64 before any restoration takes place and would rate an index of 0.91 after restoration, a net change of +0.27). So that restoration strategy would require 2.6 ha (0.71/0.27) of pine plantation to be restored for every hectare of wetland lost for the airport (mitigation ratio = 2.6 : 1).

Economic Evaluation: Common-Denominator Approaches

Evaluation systems that seek to compare natural wetlands to human economic systems usually reduce all values to monetary terms (thus losing sight of the apples and oranges). Conventional economic theory assumes that in a free economy the economic benefit of a commodity is the dollar amount that the public is willing to pay for the good or service rather than be without it. This measure is formally expressed as *willingness-to-pay*, or, more accurately, *net* willingness-to-pay, "the amount society would be willing to pay to produce and/or use a good beyond that which it actually does pay" (Scodari, 1990). The principle is illustrated as follows (Scodari, 1990): Suppose a fisherman were willing to pay $30 a day to use a particular fishing site but had to spend only $20 per day in travel and associated costs. The net benefit, or economic value, to the fisherman of a fishing day at the site is not the $20 expenditure but the $10 difference between what he was willing to spend and what he had to spend. If the fishing opportunity at the site were eliminated, the fisherman would lose $10 worth of satisfaction fishing; the $20 cost that he would have incurred would be available to spend elsewhere. In the case of commercial goods such as harvested fish, the total value of a wetland is the sum of the net benefit to the consumer plus the net benefit to the producer (the fisherman).

Economists estimate consumer benefits through a *demand curve* and producer benefits through a *supply curve* (Fig. 16-12). When a good is sold in an open, competitive market, its market price rises as the supply volume decreases. The price of the last unit of the good purchased measures the consumer's marginal willingness to pay. For all other units of the good, however, the consumer is willing to pay more than the marginal price; that is, the scarcer an item, the more one is willing to pay for it. The marginal price also approaches the producer's marginal cost because the price must represent at a minimum the cost of production. The area under the curve bounded by the marginal price—the *consumer surplus*—represents the net benefit of the good to the consumer. Similarly, a supply curve describes the relationship of supply to price; the excess of what producers earn over production costs—the area over the good's supply curve bounded by price—is the producer surplus, or *economic rent*. Although not an exact measure of social welfare, the sum of consumer surplus

Table 16-6 Predicted changes in hydrologic regime function resulting from a hypothetical airport construction on one wetland site and the comparison of the mitigation required for two different wetland restoration alternatives (variables are defined in Table 16-5)

Variable	Reference Wetland		Wetland Being Destroyed				Restoration Alternative 1[a]				Restoration Alternative 2[b]			
			Now		After Airport		Now		After Restoration		Now		After Restoration	
	Raw	Index	Raw	Index	Raw	Index	Raw	Index	Raw	Index	Raw	Index	Raw	Index
V_{TREE}	14.7	1.0	14.6	1.0	—	0.0	0.0	0.0	0.0	0.0	15.3	1.0	10.0	0.7
V_{SUBC}	12,550	1.0	13,314	1.0	—	0.0	0.0	0.0	6,963	0.5	18,402	0.5	9,800	0.8
V_{MICR}	2.5	1.0	2.5	1.0	—	0.0	0.0	0.0	2	1.0	4.2	1.0	4.2	1.0
V_{DITC}	1.0	1.0	0.5	0.5	—	0.0	0.0	0.0	1.0	1.0	0.5	0.5	1.0	1.0
Functional index[c]		1.0		0.71		0.0		0.0		0.71		0.64		0.91
Relative impact						−0.71				+0.71				+0.27
Mitigation ratio[d]								0.71/0.71 = 1 : 1				0.71/0.27 = 2.6 : 1		

[a] Restoration of an agricultural field (former wetland) to a forested wetland.
[b] Restoration of a pine plantation to a forested wetland.
[c] Hydrologic functional index = $[((V_{TREE} + V_{SUBC} + V_{MICR})/3) \times V_{DITC}]^{1/2}$.
[d] Ratio of wetland that must be restored to area of wetland destroyed to achieve functional equivalent hydrologic regime.
Source: Rheinhardt et al. (1997).

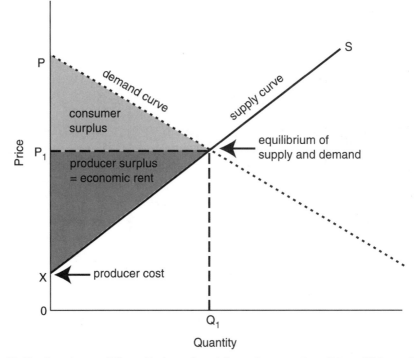

Figure 16-12 Superimposed theoretical supply-and-demand curves show the equilibrium price and quantity, the consumer surplus, and the economic rent.

and economic rent provides a useful approximation of the net benefit of a good or service (Scodari, 1990).

Although this characterization of the value of a commodity is reasonable under most conventional economic conditions, it leads to real problems in monetizing nonmarket commodities such as pure water and air and in pricing wetlands whose value in the marketplace is determined by their value as real estate, not by their "free services" to society. Consequently, attempts to monetize wetland values have generally emphasized the commercial crops from wetlands: fish, shellfish, furs, and recreational fishing and hunting for which pricing methodologies are available. As E. P. Odum (1979b) pointed out, this kind of pricing ignores ecosystem- and global-level values related to clean air and water and other "life-support" functions. Even in the cases of market commodities from wetlands, available data are seldom adequate to develop reliable demand curves (Bardecki, 1987).

Economists recognize four more or less independent aspects of "value" that contribute to the total. These aspects are (1) use value—the most tangible portion of total value derived from identifiable direct benefits to the individual; hunting, harvesting fish, and nature study are examples; (2) social value—those amenities that accrue to a societal group rather than an individual; examples are improved water quality, flood protection, and the maintenance of the global sulfur balance; (3) option value—the value that exists for the conservation of perceived benefits for future use; (4) existence value—the benefits deriving from the simple knowledge that the valued resource exists—irrespective of whether it is ever used. For example, the capacity of an extant wetland to conserve biological diversity is an existence value. As we have seen, use value is the easiest to estimate. The other three values, which are more difficult to quantify and also generally reflect longer term viewpoints, have been addressed by economists using alternative methods, as illustrated next.

Willingness-to-Pay Methods

In the absence of a well-developed free-market alternative, pricing methodologies have been applied. One of these, "willingness-to-pay," establishes a more or less hypothetical (contingency) market for nonmarket goods or services (Mitchell and Carson, 1989). The evaluation of willingness-to-pay has been carried out in a number of ways (Bardecki, 1987), including:

1. *Gross Expenditure.* This approach evaluates the total expenditures for a specific activity (e.g., the willingness of hunters to spend dollars to travel to a site, buy equipment, and rent hunting easements) as a measure of the value of the wetland site. Aside from the fact that this method measures only one aspect of the total value of a wetland, it is beset with serious methodological difficulties (Carey, 1965; Knetsch and Davis, 1966).

2. *Travel Cost.* In this method, the costs of travel from different locations to a common property resource such as a wetland site are used to create a demand curve for the goods and services of that site. The total net value of the site is represented by the integral of the demand function (Hammack and Brown, 1974; Gum and Martin, 1975). This approach has been used most effectively for recreational activity. It is not effective in estimating the demand for off-site services such as downstream water quality enhancement or flood mediation.

3. *Imputed Willingness-to-Pay.* In circumstances in which goods or services that depend on wetlands are produced and are recognized in the marketplace (e.g., commercial fish harvests), the values of these goods can be interpreted as measures of society's willingness-to-pay for the productivity of the wetlands and, hence, for the wetland itself (Farber and Costanza, 1987). This method is limited to specific products or services and does not cover the entire range of wetland values.

4. *Direct Willingness-to-Pay (Contingent Value).* This method uses direct surveys of consumers to generate willingness-to-pay values for the entire range of potential wetland benefits. This method can separate public from private benefits, derive marginal values rather than average ones, and deal with individual sites or the entire resource base (Bardecki, 1987). On the other hand, an individual's willingness-to-pay for a wetland "service" is probably strongly influenced by his or her understanding of the ecology and functions of wetlands. Thus, the contingent-value method, like the other approaches, has both strengths and weaknesses.

Opportunity Costs

A second approach to resource evaluation in the absence of a free-market model is the *opportunity cost* approach. In general terms, the opportunity cost associated with a resource is the net worth of that resource in its best alternative use. For example, "the opportunity cost of conserving a wetland area is the net benefit which might have been derived from the best alternate use of the area which must be foregone in order to preserve it in its natural state" (Bardecki, 1987). Because determining the opportunity cost associated with wetland conservation would require the evaluation of each wetland service as well as the identification and valuation of the best alternative use, in practice, a comprehensive evaluation of the opportunity cost of wetland conservation is far from possible. Nevertheless, it represents a useful approach to the valuation of specific wetland functions (Shabman, 1986).

Replacement Value

If one could calculate the cheapest way of replacing various services performed by a wetland and could make the case that those services would have to be replaced if the wetland were destroyed, then the figure arrived at would be the "replacement value." Some of the replacement technologies that might be necessary to replace services provided by wetland processes are listed in Table 16-7. A sample calculation of the replacement cost method is shown in Table 16-8. In this example, a fish hatchery is used to calculate fishery production, a flood reservoir to calculate flood and drought control, sediment dredging to estimate sediment retention, and wastewater treatment to estimate water quality enhancement.

This approach has the merit of being accepted in the world of conventional economists. For certain functions, it gives very high values compared with those of other valuation approaches discussed in this section. For example, the tertiary treatment of wastewater is extremely expensive, as is the cost of replacing the nursery function of marshes for juvenile fish and shellfish. Serious questions, however, have been raised about whether these functions would be replaced by treatment plants and fish nurseries if the wetlands were destroyed. Some ecologists and economists argue that in the long run either the services of wetlands would have to be replaced or the quality of human life would deteriorate. Other individuals argue that this assertion cannot be supported in any convincing manner.

Energy Analysis

A completely different approach uses the idea of energy flow through an ecosystem or the similar concept of "embodied energy." The concept of embodied energy (Costanza, 1980), or *emergy* (energy memory; H. T. Odum, 1988, 1989, 1996), the total energy required to produce a commodity, is assumed to be a valid index of the totality of ecosystem functions and is applicable to human systems as well. In this way, both natural and human systems can be evaluated on the basis of one common currency—energy. Because there is a linear relationship between embodied energy and money, energy flow can be translated to the more familiar currency of dollars. A simple calculation using the annual energy flow of a bottomland forested wetland in Illinois is illustrated in Table 16-8. Here, an estimated ecosystem energy flow of 20,000 kcal m^{-2} yr^{-1} yielded an estimated value of $714 ha^{-1} yr^{-1}. A conversion of 0.05 units fossil fuel energy/unit GPP energy (20 kcal GPP/kcal fossil fuel) was used as well as an estimated $14,000 kcal fossil fuel/$U.S. (1974 $).

Costanza (1984) and Costanza et al. (1989) showed that the economist's willingness-to-pay approach and energy analysis converge to a surprising degree for coastal marshes in Louisiana, although both methods result in a great deal of uncertainty (Table 16-9). The energy analysis approach yielded higher wetland values, but the ranges overlap. The sensitivity of both conventional and energy analysis methods to the choice of a discount rate, which has been vital for decades in the outcome of

Table 16-7 Some replacement technologies for societal support values provided by wetlands

Societal Support	Replacement Technologies
PEAT ACCUMULATION	
Accumulating and storing organic matter (peat)	Artificial fertilizers Artificial flooding
HYDROLOGIC FUNCTIONS	
Maintaining drinking water quality	Water transport Pipeline to distant source
Maintaining groundwater level	Well-drilling Saltwater filtering
Maintaining surface water level	Dams for irrigation Pumping water to dam Irrigation pipes and machines Water transport for domestic animals
Moderation of water flows	Regulating gate Pumping water to stream
BIOGEOCHEMICAL fUNCTIONS	
Processing sewage; cleansing nutrients and chemicals	Mechanical sewage treatment Sewage transport Sewage treatment plant Clear-cutting ditches and stream
Maintaining drinking water quality	Water quality inspections Water purification plant Silos for manure from domestic animals Nitrogen filtering Water transport
Filter to coastal waters	Nitrogen reduction in sewage treatment plants
FOOD CHAIN FUNCTIONS	
Providing food for humans and domestic animals	Agriculture production Import of food
Providing cover	Roofing materials
Sustaining anadromous trout populations	Releases of hatchery-raised trout Farmed salmon
Sustaining other fish species and wetland-dependent flora and fauna	Work by nonprofit organizations
Species diversity; storehouse for genetic material	Replacement not possible
Bird watching, sport fishing, boating, and other recreational values	Replacement not possible
Aesthetic and spiritual values	Replacement not possible

Source: Folke (1991).

cost–benefit studies, is also demonstrated in this comparison. The energy analysis method is based on using the total amount of energy captured by natural ecosystems as a measure of their ability to do useful work (for nature and, hence, for society). The gross primary productivity (GPP) of representative coastal marsh systems, which ranges from 48,000 to 70,000 kcal m^{-2} yr^{-1}, is converted to monetary units by multiplying by a conversion factor of 0.05 units fossil fuel energy/unit GPP energy

Table 16-8 Estimated value of 770-ha riparian wetlands along the Kankakee River, northeastern Illinois, estimated by replacement value approach and by energy analysis

REPLACEMENT COST APPROACH

Ecosystem Function (Replacement Technology)	$/Year	Total Value ($ ha^{-1} yr^{-1})
Fish productivity (fish hatchery)	$91,000	
Flood control/drought prevention (flood control reservoir)	$691,000	
Sediment control (sediment dredging)	$100,000	
Water quality enhancement (wastewater treatment)	$57,000	
Total replacement cost	$939,000	
Value/area $939,000 yr^{-1}/770 ha =		1,219

ENERGY FLOW APPROACH

Energy Flow Parameter	Number	Total Value
Ecosystem gross primary productivity (kcal m^{-2} yr^{-1})	20,000	
Energy conversion, (kcal GPP/kcal fossil fuel)	20	
Energy conversion in U.S. economy (kcal fossil fuel /U.S.$)	14,000	
Value/area $\dfrac{20{,}000 \text{ kcal GPP m}^{-2}\text{ yr}^{-1} \times 10{,}000 \text{ m}^2 \text{ ha}^{-1}}{20 \text{ kcal GPP/kcal ff} \times 14{,}000 \text{ kcal ff/\$}} =$		714

Source: Mitsch et al. (1979).

Table 16-9 Estimates of wetland values in $/ha of Louisiana coastal marshes based on willingness-to-pay and energy analysis at two discount rates

	Discount Rate	
Method	3 percent	8 percent
Willingness-to-pay		
Commercial fishery	$2,090	$783
Fur trapping (muskrat and nutria)	991	373
Recreation	447	114
Storm Protection	18,653	4,732
Total willingness-to-pay value	$22,181	$6,002
Energy analysis	$42,000–$70,000	$16,000–$26,000
Best estimate	$22,000–$42,000	$6,000–$16,000

Source: Costanza et al. (1989).

and dividing by the energy/money ratio for the economy (15,000 kcal fossil fuel/ 1983 $). These calculations resulted in an estimate of annual coastal wetland value of about $1,560 ha^{-1} yr^{-1}, which, when converted to present value for an infinite series of payments, yields the range of capitalized values of $16,000 to $70,000 ha^{-1} for the discount rates used in Table 16-9.

In comparison, the willingness-to-pay estimates reflect the assessment that a reasonable range of wetland value for coastal Louisiana is between $6,000 and $22,000 ha^{-1}, depending on the discount rate applied to determine the present value. Costanza et al. (1989) used this range from the willingness-to-pay and energy analysis approaches to suggest that the annual loss of Louisiana coastal wetlands is costing society from $77 million to $544 million per year.

The energy analysis method, although imprecise because of the several estimates used, is more satisfying to many wetland scientists than conventional cost-accounting methods because it is based on the inherent productivity of the ecosystem, not on perceived values that may change from generation to generation and from location to location.

In analyzing other comparisons of valuation methods, Mitsch et al. (1979c) and Folke (1991) compared the use of a replacement cost method with the use of an energy analysis method for case studies in Illinois and Sweden, respectively, and found them to be in a similar range. In the Illinois example, the energy analysis method gave a number about 60 percent of the replacement value (Table 16-8). In the example in Sweden, where a 2.5-km^2 peatland lake on the island of Gotland was threatened, the replacement cost was estimated to be about $1,600 ha^{-1} yr^{-1}, with most of the cost involved in replacing the biogeochemical processes of the wetland (52–82 percent of the cost) and less involved in replacing the hydrologic processes (7–40 percent) and food chain functions (8–11 percent). When the energy cost of the economic replacements was compared with the energy lost when the wetland was lost, the results were remarkably similar. If the 2.5-km^2 wetland was lost, the economic replacement cost in energy terms would range from 3.5 to 12×10^9 kcal yr^{-1}; the ecosystem loss calculation ranged from 13 to 18×10^9 kcal yr^{-1}. In this case, the energy analysis method gave a slightly higher estimate of the energy (and, hence, the money) cost of wetland loss than the replacement cost method did.

Multiple-Function Approach

Values placed on wetlands using these evaluation methods have ranged from very high to low. Although few people dispute that wetlands have many and varied values, the lack of consistent, accepted methodologies for comparing them with conventional economic goods and services limits the usefulness of the estimates that have been made. Perhaps the most comprehensive attempt to evaluate wetlands for management purposes is a joint project of Wildlife Habitat Canada and Environment Canada (Manning et al., 1990). Their process of valuing a wetland development project begins with applying a multiple-function screening that incorporates a series of specific standards or benchmarks reflecting applicable societal goals associated with wetlands. Most of the standards are identified from existing legislation, stated government goals and objectives, and scientific principles. A project that fails to satisfy one or more of these goals can often be eliminated, avoiding further, detailed analysis.

Major projects passing the initial screening must subsequently be evaluated in terms of their impact on identified social values. This process involves a cost accounting of the benefits of the project compared to the costs of natural wetland goods and services lost. In this process, both the willingness-to-pay and the opportunity cost methodologies are used. The results of four pilot studies showed limitations in each of the methods but also the possibility for developing a more useful process for evaluating major development projects. One key element of the evaluation process is the need to recognize and conceptually separate the relationships among the ecological functions of wetlands, the recognizable benefits to society, and the socioeconomic values that can be placed on those benefits.

Problems and Paradoxes of Quantifying Wetland Values

Regardless of which kind of ecosystem evaluation is used, several generic problems and paradoxes to quantifying wetland values need to be recognized and addressed.

1. *The term "value" is anthropocentric; hence, assigning values to different natural processes usually reflects human perceptions and needs rather than intrinsic ecological processes.* Since wetlands are multiple-value systems, the evaluator is often faced with the problem of comparing and weighing different commodities. For example, a fresh marsh area is more valuable for waterfowl than a salt marsh area is, but the salt marsh may be much more valuable as a fish habitat. Which is rated higher depends on the value judgment made by the evaluator, a judgment that has nothing to do with the intrinsic ecological viability of either area. This is the old "apples versus oranges" problem. The decision, to some extent, reflects a matter of preference. Furthermore, in most wetland evaluations, evaluators are not concerned with single commodities. Instead, they wish to approximate the overall value of an area, that is, the value of the whole fruit basket, rather than the apples, oranges, and pears. Complexity is added when the concern is to compare the value of a natural wetland with the same piece of real estate proposed to be used for economic development—a dammed lake, a parking lot, or an oil well. In that case, the comparison is not between apples and oranges but between apples and computer chips or a fruit basket and electrical energy. Conventional economics solves the comparison problem by reducing all commodities to a single index of value—dollars. This is difficult when some of the commodities are natural products of wetlands that do not compete in the marketplace.

2. *The most valuable products of wetlands are public amenities that have no commercial value for the private wetland owner.* The wetland owner, for example, has no control over the harvest of marsh-dependent fish that are caught in the adjacent estuary or even offshore in the ocean. The owner does not control and usually cannot capitalize on the ability of the wetland to purify wastewater, and certainly cannot control its value for the global nitrogen balance. Thus, there is often a strong conflict between what a private wetland owner perceives as his or her best interests and the best interests of the public. In coastal Louisiana, a marsh owner can earn revenues of perhaps $25 per acre annually from the renewable resources of his or her marsh by leasing it for hunting and fur trapping. In contrast, depending on where it is situated, a wetland may be worth hundreds of thousands of dollars per acre as a housing development or as a site for an oil well. Riparian wetlands in the Midwest and lower Mississippi River basin yield little economic benefit to the owner for the flood mitigation and water quality maintenance that they provide for downstream populations. Yet if cleared and planted in corn or soybeans, the land will provide economic benefits to the owner. Many of the current regulations that govern wetlands were initiated to protect the public's interest in privately owned wetlands (see Chapter 18).

3. *The ecological value, but not necessarily the economic value of a wetland depends on its context in the landscape.* This applies to both scale issues and location within the landscape. Wetland values are different, accrue to different "stakeholders," and probably have different importance, depending on the spatial scale on which we base our estimations, and their location in the watershed. Small wetlands are often important locally for their support of populations of ducks, geese, migratory birds, and fish and shellfish. This is true of prairie pothole wetlands, for example. Linear riparian wetlands that lie along rivers and streams, although narrow, may be extremely important because they remove nutrients and soil from upland runoff, thus helping maintain stream water quality. Large areas of wetland, for example, the great expanses of peatlands in northern latitudes, may have global value as regulators of global climate.

4. *The relationships among wetland area, surrounding human population, and marginal value are complex.* Conventional economic theory states that the less there is of a commodity, the more valuable the remaining stocks. This generalization is complicated in nature because different natural processes operate on different scales. This is an important consideration in parklands and wildlife preserves, for example. Large mammals require large ranges in which to live. Small plots below a certain size will not support *any* large mammals. Thus, the marginal-value generalization falls apart because the marginal value ceases to increase below a certain plot size. In fact, it becomes zero. In wetlands, the situation is even more complex because they are open ecosystems that maintain strong ties to adjacent ecosystems. Therefore, two factors that govern the ecological value (and, hence, the value to society) of the wetland are its *interspersion* with other ecosystems, that is, its place in the total regional landscape, and the *degree of linkage* with other ecosystems. A small wetland area that supports few endemic organisms may be extremely important during critical times of the day or during certain seasons for migratory species that spend only a day or a week in the area. A narrow strip of riparian wetland along a stream that amounts to very little acreage may efficiently filter nutrients out of runoff from adjacent farmland, protecting the quality of the stream water. Its value is related to its interspersion in the landscape, not to its size. A typical curve of marginal wetland value versus surrounding human population (Fig. 16-13) shows that, as demand increases for the goods and services of a wetland area, or supply decreases as the wetland is encroached upon, the marginal value at first increases. However, as wetland functions are compromised, either by overuse (e.g., overfishing or hunting, tree harvesting, increased fertilizer runoff, water-level stabilization) or by reduction in wetland area (which may, for example, eliminate wide-ranging predators or reduce water treatment capacity), the marginal value declines because the wetland can no longer produce the services that make it useful ecologically and economically. These kinds of considerations have only recently begun to be addressed in a quantitative way, and the methodology is not well developed.

5. *Commercial values are finite, whereas wetlands provide values in perpetuity. Wetland development is often irreversible.* The time frame for most human projects is 10 to 30 years. Private entrepreneurs expect to recoup their investments and profits in projects within this time frame and seldom consider longer term implications. Even large public-works projects such as dams for energy generation seldom are seen in terms longer than 50 to 100 years. The destruction of natural areas, on the other hand, removes their public services forever. Often, especially for wetlands, the decision to develop is irreversible. If an upland field is abandoned, it will gradually revert to

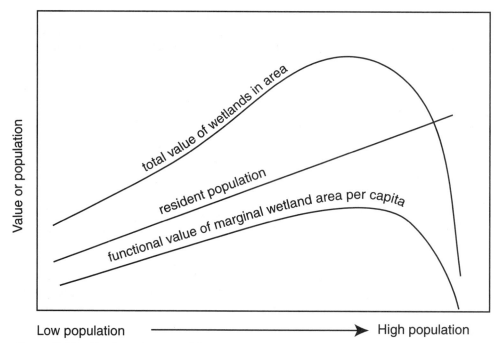

Figure 16-13 Marginal value of additional wetland area to a given region as a function of nearby human population (from Mitsch and Gosselink, 2000, based on King and Herbert, 1997). Total value is a product of population times functional value per capita. Functional marginal value of additional wetland per capita does initially increase as population increases as wetlands are becoming rare. At some point of population density, however, these functions become taxed with pollution, lost corridors, etc., and value drops precipitously for additional population increase.

a forest; but once a wetland is drained and developed, it is usually lost forever because of associated changes in the hydrologic regime of the area. For example, in Louisiana and elsewhere, marsh areas were diked and drained for agriculture early in the 20th century. Many of these developments have subsequently been abandoned. They did not revert to wetlands, but are now large, shallow lakes, distinguishable by their straight edges (Fig. 16-14).

6. *A comparison of economic short-term gains with wetland value in the long term is often not appropriate.* Wetland values, even when multiple functions are quantified, often cannot "compete" with short-term economic calculations of high-economic-yield projects such as commercial development or intensive agriculture. This is especially true because economic analyses typically discount the value of future amenities. The issue of wetland conservation versus development has an intergenerational component. Future generations do not compete in the marketplace, and decisions that will affect the public resources they inherit are often made without regard to their interests.

7. *Estimates of values, by their nature, are colored by the biases of individuals and society and by the economic system.* There are good reasons why we wish to protect nature; developed countries, having taken care of the basic needs of their citizens, are particularly involved in protecting ecosystems, including wetlands, for their aesthetic as well as more functional attributes, not all of which translate into direct economic

Figure 16-14 Sugar cane fields "reclaimed" from fresh marsh in coastal Louisiana. Natural marsh is in foreground. Such development projects are usually irreversible since land elevations and flooding patterns are permanently changed. (*Photograph by C. Sasser, reprinted by permission*)

benefit. Other cultures, where the basic needs of food and shelter cannot be taken for granted, have different views of the economics of wetlands. Many cultures do live in and among wetlands and use them for daily subsistence—the production of food and fiber. Yet they generally leave the normal wetland functions intact. The values that we ascribe to wetlands are not separate from the institutions and cultures from which we come.

8. *A landscape view of wetlands is required to make intelligent decisions about the values of created and managed wetlands.* A paradox of assigning values to ecosystems is that, unless we take a landscape view, it can be argued that we should replace a less valuable system (e.g., a grassland) with another more valuable one (e.g., a wetland). Costanza et al. (1997), when estimating the value of the world's ecosystem services, estimated that wetlands are 75 percent more valuable than lakes and rivers, 15 times more valuable than forests, and 64 times more valuable than grasslands and rangelands (Table 16-10). A straightforward economic analysis would thus argue for the replacement of forests and prairies with wetlands. While this physical substitution is, of course, not possible in most instances because climatic and hydrologic variables determine what ecosystem occurs in a particular landscape, on a microscale it is not only possible to substitute wetlands for grasslands and upland forests, but it is frequently done to meet regulatory requirements for wetland mitigation in the United States. Many question whether the created wetland can achieve the same functional and, hence, "economic" value as did the original ecosystem at that site. Some argue that these created ecosystems are doomed to failure (Roberts, 1993; Malakoff, 1998), whereas others are more optimistic that these systems do, indeed, provide real measurable value that might even exceed what was at the site previously (Young, 1996).

Table 16-10   Estimated unit values of ecosystems

Ecosystem	Unit Value (U.S.$ ha^{-1} yr^{-1})
Estuaries	22,832
Wetlands	14,785
Lakes/rivers	8,498
Forests	969
Grasslands	232

Source: Costanza et al. (1997).

If one ignores the technical problems of functional ecosystem substitution, the idea attracts many because of the common perception among economists that any commodity can be replaced. As scarcity of one product drives the price up, the creativity of the free market will surely result in the development of a cheaper substitute. This is not true of ecosystems, however. Much of the value of an ecosystem, especially an open system such as a wetland, depends on its landscape context and the strong interactions among the parts of the landscape. Thus, the value of a riparian forest depends on its ecological links to the adjacent stream on one side and the upland fields or forest on the other. Similarly, a coastal wetland derives much of its value from the movement of upstream and coastal ocean organisms, sediments, and chemicals into and out of the wetland. In this context, a landscape is like a tile mosaic: The individual tiles have little value in themselves. Rather, their value lies in their precise location and color within the entire mosaic. In the same way, the value of a wetland or a pond or a stream is in its location and ecological properties as they interact with other ecosystems to form a functional landscape.

Figure 16-15   This flock of white pelicans on a coastal marsh symbolizes the difficulty of placing a dollar value on wetlands. They are not hunted and so have no "sport-hunting value." Nevertheless, they are one species supported by wetlands. Have they any value? If so, how is it defined, and how quantified, in a social system in which "market values" often take precedence over any others? (*Photograph by R. Abernethy, used by permission*)

A Faustian Bargain

Because of the many problems documented in this chapter related to valuation of natural ecosystems, many ecologists oppose their economic valuation, since it implies that natural systems can be equated in the marketplace to other market products (Fig. 16-15). On the other hand, attempts to place dollar values on natural ecosystems, such as those cited in this chapter, have raised public awareness of the high value of the goods and services of nature, and in this way helped in efforts to conserve natural resources. Thus, ecologists are caught in a "Faustian bargain with the devil" trying to make the case for natural resource values in the common currency of our civilization, while, at the same time, clearly documenting the reasons why natural ecosystem conservation should not depend on the operation of free-market forces. There is no easy answer to this dilemma.

Recommended Readings

Barbier, E. B., J. C. Burgess, and C. Folke. 1994. Paradise Lost? The Ecological Economics of Biodiversity. Earthscan Publications, London. 267 pp.

Costanza, R., R. d'Arge, R. de Groot, S. Farber, M. Grasso, B. Hannon, K. Limburg, S. Naeem, R. V. O'Neill, J. Paruelo, R. G. Raskin, P. Sutton, and M. van den Belt. 1997. The value of the world's ecosystem services and natural capital. Nature 387:253–260.

Folke, C., and T. Kaberger, eds. 1991. Linking the Natural Environment and the Economy. Kluwar Academic, Dordrecht. 305 pp.

Söderquist, T., W. J. Mitsch, and R. K. Turner, eds. 2000. The Values of Wetlands: Landscape and Institutional Perspectives. Special Issue of Ecological Economics. 36 (in press).

Turner, K. 1991. Economics and wetland management. Ambio 20:59–63.

Chapter 17

Human Impacts and Management of Wetlands

Wetland impacts have included both wetland alteration and wetland destruction. In earlier times, wetland drainage was considered the only policy for managing wetlands. Significant wetland alteration continues, particularly by dredging, filling, drainage, hydrologic modification, peat mining, removal for mineral extraction, and water pollution. There is increasing recognition of the economic importance of the world's peat resources, which are estimated to be 1.9 trillion tons. Wetlands can also be managed in their more or less natural state for certain objectives such as fish and wildlife enhancement, agricultural and aquaculture production, water quality improvement, and flood control. A more recent concern that will possibly impact both coastal and inland wetlands is global climate change. There are few management alternatives that are available to either protect wetlands or make them emit fewer greenhouse gases if climate does change and/or sea level rises.

The concept of wetland management has had different meanings at different times to different disciplines and in different parts of the world. Until the middle of the 20th century, wetland management usually meant wetland drainage to many policymakers, except for a few resource managers who maintained wetlands for hunting, fishing, and waterfowl/wildlife protection. Landowners were encouraged through government programs to tile and drain wetlands to make the land suitable for agriculture and other uses. Countless coastal wetlands were destroyed by dredging for navigation and filling for land development. There was little understanding of and concern for the inherent values of wetlands. The value of wetlands as wildlife habitats, particularly for waterfowl, was recognized in the first half of this century by some fish and game managers to whom wetland management often meant the maintenance of hydrologic conditions to optimize fish or waterfowl populations. Only relatively recently have other values such as flood control and water quality enhancement been recognized.

Today, the management of wetlands means setting several objectives, depending on the priorities of the wetland manager. In some cases, objectives such as preventing pollution from reaching wetlands or using wetlands as sites of wastewater treatment or disposal can be conflicting. Many floodplain wetlands are now managed and zoned to minimize human encroachment and maximize floodwater retention. Coastal wetlands are now included in coastal zone protection programs for storm protection and as sanctuaries and subsidies for estuarine fauna. In the meantime, wetlands continue to be altered or destroyed through drainage, filling, conversion to agriculture, water pollution, and mineral extraction.

An Early History of Wetland Management

The early history of wetland management, a history that still influences many people today, was driven by the misconception that wetlands were wastelands that should be avoided or, if possible, drained and filled. Throughout the world, as long as there have been humans, there has been hydrologic alteration of the landscape. As described by Larson and Kusler (1979), "For most of recorded history, wetlands were regarded as wastelands if not bogs of treachery, mires of despair, homes of pests, and refuges for outlaw and rebel. A good wetland was a drained wetland free of this mixture of dubious social factors." In the United States, this opinion of wetlands and shallow-water environments led to the destruction of more than half of the total wetlands in the lower 48 states over about 200 years. In New Zealand, settlement by Europeans that began in earnest in the mid-1800s, contributed significantly to a 90 percent loss of wetlands in a relatively short time. In China, Europe, and regions of the Middle East, civilizations have both lived with but also significantly altered the natural hydrologic landscape for thousands of years. Very preliminary estimates suggest that, over human history, about half of the world's wetlands have been lost (Dugan, 1993; see Chapters 3 and 4).

Wetland Drainage History in the United States

Had not politics intervened, George Washington may have succeeded in draining the Great Dismal Swamp in Virginia in the mid-18th century (see Chapter 4). The practice of draining swamps and other wetlands was an acceptable and even desired practice from the time Europeans first settled in North America. In the United States, some public laws actually encouraged wetland drainage. Congress passed the Swamp Land Act of 1849, which granted to Louisiana the control of all swamplands and overflow lands in the state for the general purpose of controlling floods in the Mississippi River basin. In the following year, the act was extended to the states of Alabama, Arkansas, California, Florida, Illinois, Indiana, Iowa, Michigan, Mississippi, Missouri, Ohio, and Wisconsin. Minnesota and Oregon were added in 1860. The act was designed to decrease federal involvement in flood control and drainage by transferring federally owned wetlands to the states, leaving to them the initiative of "reclaiming" wetlands through activities such as levee construction and drainage.

By 1954, an estimated 26 million ha of land had been ceded to those 15 states for reclamation. Ironically, although the federal government passed the Swamp Land Act to get out of the flood control business, the states sold those lands to individuals for pennies per acre and the private owners subsequently successfully lobbied both national and state governments to protect these lands from floods. Further, governments are now paying enormous sums to buy the same lands back for conservation purposes. Although current government policies are generally in direct opposition to the Swamp Land Act and it is now disregarded, the act cast the initial wetland policy of the U.S. government in the direction of wetland elimination.

Other actions led to the rapid decline of the nation's wetlands. An estimated 23 million ha of wet farmland, including some wetlands, were drained under the U.S. Department of Agriculture's Agricultural Conservation Program between 1940 and 1977 (Office of Technology Assessment, 1984). An estimated 18.6 million ha of land, much former wetlands, were drained in seven states in the upper Mississippi River basin alone (Table 17-1). Some of the wetland drainage activity was hastened by projects of groups such as the Depression-era Works Progress Administration (WPA), the Soil Conservation Service, and other federal agencies. Coastal marshes were eliminated or drained and ditched for intercoastal transportation, residential developments, mosquito control, and even for salt marsh hay production. Interior wetlands were converted primarily to provide land for urban development, road construction, and agriculture.

Typical of the prevalent attitude toward wetlands is the following quote by Norgress (1947) discussing the "value" of Louisiana cypress swamps:

> With 1,628,915 acres of cutover cypress swamp lands in Louisiana at the present time, what use to make of these lands so that the ideal cypress areas will make a return on the investment for the landowner is a serious problem of the future. . . .
>
> The lumbermen are rapidly awakening to the fact that in cutting the timber from their land they have taken the first step toward putting it in position to perform its true function—agriculture. . . .
>
> It requires only a visit into this swamp territory to overcome such prejudices that reclamation is impracticable. Millions of dollars are being put into good roads. Everywhere one sees dredge boats eating their way through the soil, making channels for drainage.
>
> After harvesting the cypress timber crop, the Louisiana lumbermen are at last realizing that in reaping the crop sown by Nature ages ago, they have left a heritage to posterity of an asset of permanent value and service—land, the true basis for wealth.
>
> The day of the pioneer cypress lumberman is gone, but we need today in Louisiana another type of pioneer—the pioneer who can help bring under cultivation the enormous areas of cypress cutover lands suitable for agriculture. It

Table 17-1 Drainage statistics of selected states in the upper reaches of the Mississippi River basin

State	Total Area Drained (× 1,000 ha)	Percentage of All Land That Is Drained	Percentage of Cropland That Is Drained
Illinois	3,965	30	35
Indiana	3,273	30	50
Iowa	3,154	20	25
Ohio	3,000	20	50
Minnesota	2,580	15	20
Missouri	1,720	10	25
Wisconsin	910	6	10
Total	18,602		

Source: USDA (1987), as cited in Zucker and Brown (1998).

is important to Louisiana, to the South, and the Nation as a whole, that this be done. Would that there were some latter-day Horace Greeleys to cry, in clarion tones, to the young farmers of today, "Go South, young man; go South!"

As an example of state action leading to wetland drainage, Illinois passed the Illinois Drainage Levee Act and the Farm Drainage Act in 1879, which allowed counties to organize into drainage districts to consolidate financial resources. This action accelerated draining to the point that 30 percent of Illinois is now under some form of drainage (Table 17-1) and almost all of the original wetlands in the state (85 percent) have been destroyed.

Wetland Alteration

In a sense, wetland alteration or destruction is an extreme form of wetland management. One model of wetland alteration (Fig. 17-1) assumes that three main factors influence wetland ecosystems: water level, nutrient status, and natural disturbances. Through human activity, the modification of any one of these factors can lead to wetland alteration, either directly or indirectly. For example, a wetland can be disturbed through decreased water levels, as in draining and filling, or through increased water levels, as in downstream drainage impediments. Nutrient status can be affected through

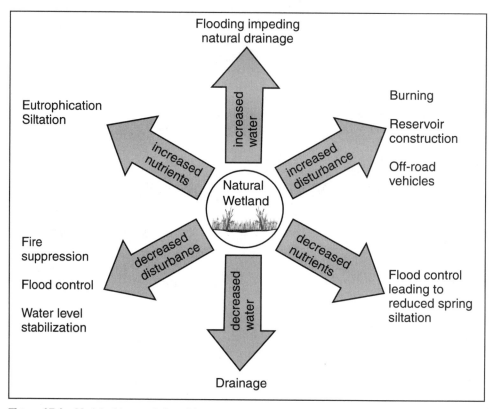

Figure 17-1 Model of human-induced impacts on wetlands, including effects on water level, nutrient status, and natural disturbance. By either increasing or decreasing any one of these factors, wetlands can be altered. (*After Keddy, 1983*)

upstream flood control that decreases the frequency of nutrient inputs or through increased nutrient loading from agricultural areas.

The most common alterations of wetlands have been (1) draining, dredging, and filling of wetlands; (2) modification of the hydrologic regime; (3) highway construction; (4) mining and mineral extraction; and (5) water pollution. These wetland modifications are described in more detail next.

Wetland Conversion: Draining, Dredging, and Filling

The major cause of wetland loss around the world continues to be conversion to agricultural use. Drainage for farms in the United States has progressed at a rate of 490,000 ha/yr since the early 1900s (slope of line in Fig. 17-2a) (Gosselink and Maltby, 1990). Little drainage occurred during the Depression and war years, and since 1985 little additional drainage has occurred, except in Minnesota and Ohio (Zucker and Brown, 1998). About 65 percent of this land was wetland (Office of Technology Assessment, 1984). This conversion was particularly significant in the vast midwestern U.S. "breadbasket," which has provided the bulk of the grain produced on the continent (Fig. 17-2b and Table 17-1). Some of the world's richest farming is in the former wetlands of Illinois, Iowa, and southern Minnesota. When drained and cultivated, the fertile soils of the prairie pothole marshes and east Texas playas produce excellent crops. With ditching and modern farm equipment, it has been possible to farm these small marshes. The modern farm equipment of today and mass-produced reels of plastic drainage pipe also make it possible to drain much more area per day than was ever possible with earlier equipment and the use of clay tiles (Fig. 17-3).

Since the mid-20th century, however, the most rapid changes have occurred in the bottomland hardwood forests of the Mississippi River alluvial floodplain. As populations increased along the river, the floodplain was channeled and leveed so that it could be drained and inhabited. Since colonial times, the floodplain has provided excellent cropland, especially for cotton and sugarcane. Cultivation, however, was restricted to the relatively high elevation of the natural river levees, which flooded regularly after spring rains and upstream snowmelts but drained rapidly enough to enable farmers to plant their crops. Because the river levees were naturally fertilized by spring floods, they required no additional fertilizers to grow productive crops. One of the results of drainage and flood protection is the additional cost of fertilization. The lower parts of the floodplain, which are too wet to cultivate, were left as forests but harvested for timber. As pressure for additional cropland increased, these agriculturally marginal forests were clear-cut at an unprecedented rate (Fig. 17-4). This was feasible, in part, because of the development of soybean varieties that mature rapidly enough to be planted in June or even early July, after severe flooding has passed. Often, the land thus "reclaimed" was subsequently incorporated behind flood control levees where it was kept dry by pumps (Stavins, 1987). Clear-cutting of bottomland forests is still proceeding from north to south. Most of the available wetland has been converted in Arkansas and Tennessee; Mississippi and Louisiana are experiencing large losses.

Along the nation's coasts, especially the East and West coasts, the major cause of wetland loss is draining and filling for urban and industrial development. Compared to land converted to agricultural use, the area involved is rather small. Nevertheless, in some coastal states, notably California, almost all coastal wetlands have been lost. The rate of coastal wetland loss from 1954 to 1974 was closely tied to population density (Fig. 17-5). This finding underscores two facts: (1) Two-thirds of the world's

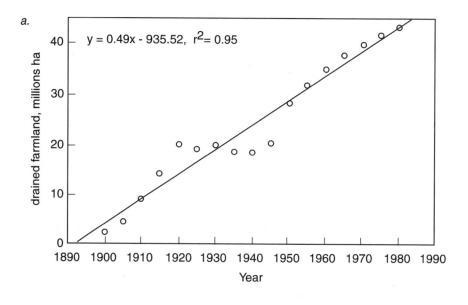

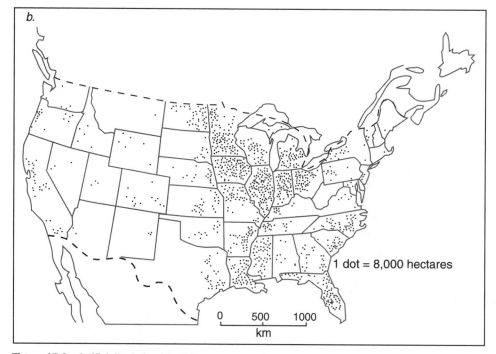

Figure 17-2 Artificially drained land in the United States: **a.** Trend from 1900–1980. (*After Gosselink and Maltby, 1990, based on data from Office of Technology Assessment, 1984*) **b.** Extent and location as of 1985. Each dot represents 8,000 ha (20,000 acres) and total area drained is 43,500,000 ha. (*After Dahl, 1990*)

a.

b.

Figure 17-3 Modern drainage machinery such as that illustrated in these photos is able to drain dozens of hectares per day: **a.** detail of the drainage machinery; **b.** results of about 1 minute of drainage, showing new ditch and plastic pipe installed. (*Photographs by W. J. Mitsch*)

Figure 17-4 Oblique aerial photograph of bottomland in the Tensas River basin, Louisiana, formerly a 1-million-ha forest. In the foreground, trees have been sheared off and felled with a bulldozer blade. Above, a line of standing trees remains along a slightly wet depression. Above the standing trees, felled trees have been pushed into a line to be burned. In the distance, the cleared land has been harrowed for planting, probably to soybeans. (*Larry Harper, U.S. Army Corps of Engineers, 1981*)

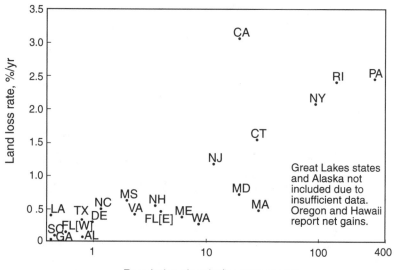

Figure 17-5 Relationship between coastal wetland loss (1954–1974) and population density for coastal counties. (*After Gosselink and Baumann, 1980*)

618

population lives along coasts and (2) population density puts great pressure on coastal wetlands as sites for expansion. The most rapid development of coastal wetlands occurred after World War II. In particular, several large airports were built in coastal marshes. Since the passage of federal legislation controlling wetland development, the rate of conversion has slowed.

Hydrologic Modifications

Ditching, draining, and levee building are hydrologic modifications of wetlands specifically designed to dry them out. Other hydrologic modifications destroy or change the character of thousands of hectares of wetlands annually. Usually these hydrologic changes were made for some purpose that had nothing to do with wetlands; wetland destruction is an inadvertent result. Canals, ditches, and levees are created for three primary purposes:

1. *Flood Control.* Most of the canals and levees associated with wetlands are for flood control. The canals have been designed to carry floodwaters off the adjacent uplands as rapidly as possible. Normal drainage through wetlands is slow surface sheet flow; straight, deep canals are more efficient. Ditching marshes and swamps to drain them for mosquito control or biomass harvesting is a special case designed to lower water levels in the wetlands themselves. Along most of the nation's major rivers are systems of levees constructed to prevent overbank flooding of the adjacent floodplain. Most of these levees were built by the U.S. Army Corps of Engineers after Congress passed flood control legislation following the disastrous floods of the 1920s and 1930s. (For a fascinating account of the great flood of 1927, the disruption it caused, and the social and political reverberations that led to flood control legislation, see Barry, 1997). These levees, by separating the river from its floodplain, isolated wetlands so that they could be drained expeditiously. For example, along the lower Mississippi River the construction of levees created a demand from farmers for additional floodplain drainage. The sequence of response and demand was so predictable that farmers bought and cleared floodplain forests in anticipation of the next round of flood control projects.

2. *Navigation and Transportation.* Navigation canals tend to be larger than drainage canals. They traverse wetlands primarily to provide water transportation access to ports and to improve transport among ports. For example, the Intracoastal Waterway was dredged through hundreds of miles of wetlands in the northern Gulf Coast. In addition, when highways were built across wetlands, fill material for the roadbed was often obtained by dredging soil from along the right-of-way, thus forming a canal parallel to the highway.

3. *Industrial Activity.* Many canals are dredged to obtain access to sites within a wetland for the purpose of sinking an oil well, building a surface mine, or other kinds of development. Usually pipelines that traverse wetlands are laid in canals that are not backfilled.

The result of all of these activities can be a wetland crisscrossed with canals, especially in the immense coastal wetlands of the northern Gulf Coast (Fig. 17-6). These canals modify wetlands in a number of ecological ways by changing normal hydrologic patterns. Straight, deep canals in shallow bays, lakes, and marshes capture flow, depriving the natural channels of water. Canals are hydrologically efficient,

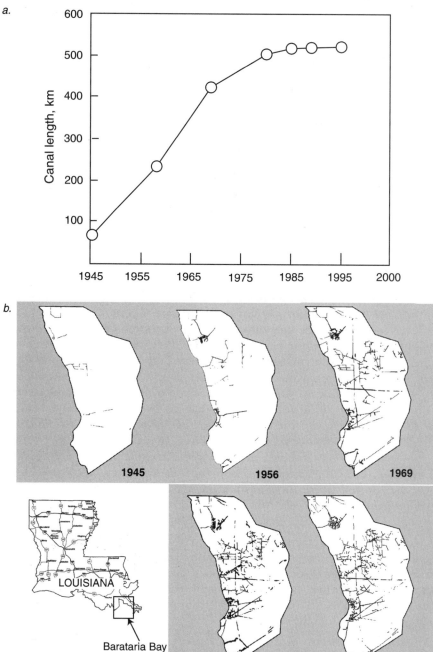

Figure 17-6 Navigational canals constructed in wetlands of the north-central coast of the Gulf of Mexico (Barataria Bay, Louisiana): **a.** growth in the length of the canals from 1945–1995; **b.** computer images of area for 1945–1985. The concentrated nodes of canals are sites of oil fields. Each short canal segment provides access to an oil well. Canal density has not increased since 1980 because of declining economically extractable oil reserves in the area, the low price of crude oil, and much stronger permit restrictions for wetland activities. (*After Sasser et al., 1986*)

allowing the more rapid runoff of fresh water than the normal shallow, sinuous channels do. As a result, water levels fluctuate more rapidly than they do in unmodified marshes, and minimum levels are lowered, drying the marshes. The sheet flow of water across the marsh surface is reduced by the spoil banks that almost always line a canal and by road embankments that block sheet flow. Consequently, the sediment supply to the marsh is reduced, and the water on the marsh is more likely to stagnate than when freely flooded. In addition, when deep, straight channels connect low-salinity areas to high-salinity zones, as with many large navigation channels, tidal water, with its salt, intrudes farther upstream, changing freshwater wetlands to brackish. In extreme cases, salt-intolerant vegetation is killed and is not replaced before the marsh erodes into a shallow lake. On the Louisiana coast, the natural subsidence rate is high; wetlands go through a natural cycle of growth followed by decay to open bodies of water. There, canals accelerate the subsidence rate by depriving wetlands of natural sediment and nutrient subsidies.

Highway Construction

Highway construction can have a major effect on the hydrologic conditions of wetlands (see Fig. 17-7). Although few definitive studies have been able to document the

Figure 17-7 This sign is typical of the conflict between wetland protection and highway construction. The sign was located near a proposed highway in central Florida. (*Photograph by W. J. Mitsch*)

extent of wetland damage caused by highways, the major effects of highways are alteration of the hydrologic regime, sediment loading, and direct wetland removal. In general, wetlands are more sensitive to highway construction than uplands are, particularly through the disruption of hydrologic conditions. Clewell et al. (1976) and Evink (1980) found that highway construction in Florida led to negative effects on coastal wetlands through hydrologic isolation. The authors of the former study discovered that isolated tidal marshes became less saline and began to fill with vegetation because of the construction of a filled roadway. Evink (1980) found that the decreased circulation that resulted from a causeway increased nutrient retention in the wetland and led to subsequent symptoms of eutrophication. Adamus (1983) concluded that the "best location for a highway that must cross a wetland is one which minimizes interference with the wetland ecosystem's most important driving forces." Other than solar energy and wind, the most important driving forces for wetlands are hydrologic, including tides, gradient currents (e.g., streamflow), runoff, and groundwater flow. The importance of protecting the hydrologic regime during highway construction is based on the contention presented in Chapter 5 that the hydrology of wetlands is the most important determinant of a wetland's structure and function.

Peat Mining

World resources of peat, principally in peatlands in the Northern Hemisphere (see Chapter 13), are estimated to be 1.9 trillion tons, of which countries that comprise the former Soviet Union have about 770 billion tons and Canada about 510 billion tons. In the United States, deposits of peat occur in all 50 states, with estimated resources of about 310 billion tons, or about 16 percent of the world total. Surface peat mining has been a common activity in several European countries, particularly Ireland and countries in eastern and northeastern Europe, since the 18th century. These countries account for over 19 million metric tons, or almost 75 percent of peat mining in the world (Table 17-2); most of this peat is used as a fuel for electric power production. Since its inception, peat mining in the United States and Canada has been primarily undertaken for horticultural and agricultural applications. The fibrous structure and porosity of peat promote a combination of water retention and drainage, which makes it useful for applications such as potting soils, lawn and garden soil amendments, and turf maintenance on golf courses. Peat is also used as a filtering medium to remove toxic materials and pathogens from wastewater, sewage effluent, and stormwater (Jasinski, 1999).

On an international basis, Canada ranks fifth in the global production of peat, after Finland, Ireland, Russia, and Germany (Table 17-2) and probably second behind Germany in the production of horticultural peat. In 1990, 749,000 metric tons of peat were sold by Canadian producers; by 1998, that production had increased significantly to 2,980,000 metric tons. Canadian peat, regarded as among the best quality peats in the world, is sold to markets in the United States and Japan as well as Canada.

Peat produced in the United States was about 676,000 metric tons in 1998, ranking ninth in total production in the world. It is generally classified as reed–sedge peat, whereas the imports from Canada typically are a weakly decomposed *Sphagnum* peat, which has a higher market value per ton. The estimated value of marketable

Table 17-2 World peat production by country for 1998

Country[a]	1998 Peat Production ($\times 10^3$ tons/yr)		
	Fuel	Horticulture	Total
Finland	7,000	400	7,400
Ireland	4,500	300	4,800
Russia[b]	3,000		3,000
Germany	180	2,800	2,980
Canada		1,127	1,127
Sweden	800	250	1,050
Ukraine[b]	1,000		1,000
Estonia			1,000
United States		676	676
United Kingdom		500	500
Latvia			450
Belarus[b]	300		300
Netherlands		300	300
Denmark		205	205
France		200	200
Poland			200
Lithuania			195
Spain			60
Hungary		45	45
Norway	1	30	31
Australia		15	15
Argentina		5	5
Burundi		5	5
Grand total	16,800	6,900	25,500[c]

[a] In addition to the countries listed, Austria, Iceland, and Italy produced negligible amounts of peat.
[b] Production appears to be for fuel use.
[c] Includes production for which use is not known.
Source: Jasinski (1999).

peat in the contiguous United States was about $16 million in 1995, with Alaskan peat output valued at $450,000. Geographically, about 85 percent of U.S. peat production was from the Great Lakes and Southeast regions, led by Florida, Michigan, and Minnesota, in order of importance. The remainder was produced in the Midwest, Northeast, and West. Approximately 95 percent of domestic peat is sold for horticulture/agriculture usage, including, in order of importance, general soil improvement, potting soils, earthworm culture, the nursery business, and golf course maintenance and construction. Mining for energy production is often proposed on a large scale for the peatlands in Minnesota and the pocosins in North Carolina. It has been estimated that Minnesota has enough peat reserves to supply its energy needs for 32 years (Williams, 1990).

Many believe that peat production for horticultural purposes will continue to grow while its use as a fuel will continue to decrease, as it has since the political restructuring of Eastern Europe in the early 1990s. Thus, there is concern that peat usage will decrease in Europe and Russia and increase in North America, especially Canada. The implications of this change in peat harvesting on both regional hydrology and climate change remain uncertain.

Mineral and Water Extraction

Surface mining activity for materials other than peat often affects major wetlands regions. Phosphate mining in central Florida has had a significant impact on wetlands in the region (M. T. Brown et al., 1992). Thousands of hectares of wetlands may have been lost in central Florida because of this activity alone, although the reclamation of phosphate-mined sites for wetlands is now a common practice. H. T. Odum et al. (1981) argued that "managed ecological succession" on mined sites could be an economical alternative to current expensive reclamation techniques involving massive earth moving and reclamation planting.

Surface mining of coal has also affected wetlands in some parts of the United States (Brooks et al., 1985). Forty-six thousand hectares of wetlands in western Kentucky in the early 1980s, mostly bottomland hardwood forests, were or could have been affected by surface coal mining (Mitsch et al., 1983b, c). The recognition of the potential benefits of including wetlands as part of the reclamation of coal mines has not been as widespread as one would have expected (Fig. 17-8) because of the strict interpretation of measures regulating the return of the land to its original contours and because of liability questions. This is in contrast to the widespread acceptance of the reclamation of wetlands on phosphorus mine sites in Florida.

In some parts of the country, the withdrawal of water from aquifers or minerals from deep mines has resulted in accelerated subsidence rates that are lowering the elevations of marshes and built-up areas alike, sometimes dramatically. For example, groundwater and mineral extraction has led to as much as 2.5 m of subsidence in northern Galveston Bay (Kreitler, 1977). Land subsidence, which can also result in the creation of lakes and wetlands, is a geologically common phenomenon in Florida. Often, in karst deposits, when excessive amounts of water are removed from the ground, underground cave-ins occur, causing surface slumpage. Some believe that

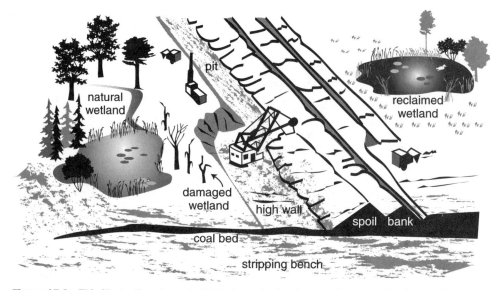

Figure 17-8 This illustration shows both the impact of surface mining on wetlands and the possible use of wetlands in reclamation of coal surface mines for wildlife enhancement and control of mine drainage.

the cypress domes in north-central Florida are an indirect result of a similar natural process whereby fissures and dissolutions of underground limestone cause slight surface slumpage and subsequent wetland development.

Water Pollution

Wetlands are altered by pollutants from upstream or local runoff and, in turn, change the quality of the water flowing out of them. The ability of wetlands to cleanse water has received much attention in research and development and is discussed elsewhere in this book (see Chapter 20). The effects of polluted water on wetlands has received less attention, although water quality standards for wetlands have been established in several regions of United States. Many coastal wetlands are nitrogen limited; one response to nitrogen as one of the pollutants is increased productivity of the vegetation and increased standing stocks of vegetation followed by increased rates of decay of the vegetation, at least initially, and higher community respiration rates.

Species composition may also change with eutrophication of wetlands. For example, increased agricultural runoff, laden with phosphorus, is believed to have caused a spread of *Typha domingensis* in conservation areas that are part of the original Everglades in Florida (Fig. 17-9; Koch and Reddy, 1992; Gunderson and Loftus, 1993; Craft and Richardson, 1993, 1997; Craft et al., 1995; Qualls and Richardson, 1995; Vaithiyanathan and Richardson, 1997). This, in turn, has increased fears that the phosphorus will eventually lead to invasion of *Typha* in the Everglades National Park itself replacing the natural sawgrass (*Cladium jamaicense*).

When metals or toxic organic compounds are pollutants, effects on the wetland can be dramatic. In severe cases of water pollution, wetland vegetation can be killed, as occurred when oil was spilled on a coastal marsh (J. M. Baker, 1973) or sulfates were discharged into a forested wetland (J. Richardson et al., 1983). Acid drainage from active and abandoned coal mines has been shown to affect wetlands seriously. In a study of wetlands adjacent to coal surface mining in western Kentucky, Mitsch et al. (1983a–c) described the extensive ecological damage that could occur where waters with low pH and high iron and sulfur were discharged from the mines into or through wetlands.

In one of the most publicized and dramatic cases of water pollution of a wetland, selenium from farm runoff contaminated marshes in Kesterson National Wildlife Refuge in California's San Joaquin Valley (Ohlendorf et al., 1986, 1990; Presser and Ohlendorf, 1987; T. Harris, 1991). The selenium contamination led to excessive death and deformities of wildlife and to eventual "closing" of the contaminated marsh in the mid-1980s amid much controversy.

Wetland Management by Objective

Wetlands are managed for environmental protection, for recreation and aesthetics, and for the production of renewable resources. Stearns (1978) lists 12 specific goals of wetland management: that are applicable today:

1. Maintain water quality
2. Reduce erosion
3. Protect from floods

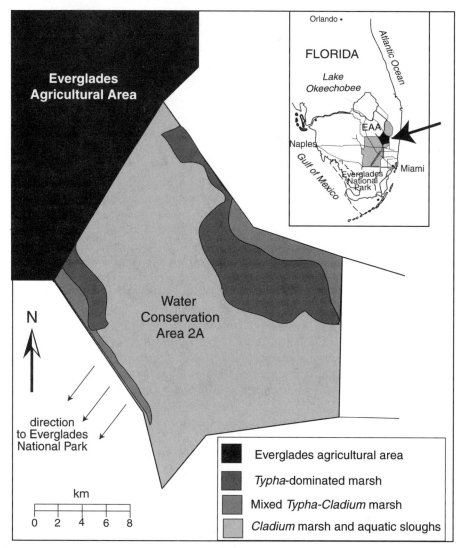

Figure 17-9 Everglades Water Conservation Area 2A (44,700 ha) in south Florida, showing the area that has received high-nutrient surface overflow from agricultural land drainage since the 1960s. Excess nutrients from the Everglades Agricultural Area to the northwest have caused the spread of *Typha domingensis* and the loss of *Cladium jamaicense* over the 8,000-ha area shaded. (*After Koch and Reddy, 1992*)

4. Provide a natural system to process airborne pollutants

5. Provide a buffer between urban residential and industrial segments to ameliorate climate and physical impact such as noise

6. Maintain a gene pool of marsh plants and provide examples of complete natural communities

7. Provide aesthetic and psychological support for human beings

8. Produce wildlife

9. Control insect populations

10. Provide habitats for fish spawning and other food organisms

11. Produce food, fiber, and fodder; for example, timber, cranberries, cattails for fiber

12. Expedite scientific inquiry

One excellent management decision is to fence in a wetland to preserve it. Although simple, this is an act of conservation of a valuable natural ecosystem involving no substantive changes in management practices. Often, however, management has one or more specific objectives that require positive manipulation of the environment. Efforts to maximize one objective may be incompatible with the attainment of others, although in recent years most management objectives have been broadly stated to enhance multiple objectives. Multipurpose management generally focuses on system-level support rather than individual species. This has often been achieved indirectly through plant species manipulation because plants provide food and cover for the animals. In the management of many small wetland areas in close proximity, the use of different practices or staggered management cycles so that the different areas are not all treated the same way at the same time, not only increases the diversity of the larger landscape but also attracts wildlife.

Wildlife Enhancement

The best wetland management practices are those that enhance the natural processes of the wetland ecosystem involved. One way to accomplish this is to maintain conditions as close as possible to the natural hydrology of the wetland, including hydrologic connections with adjacent rivers, lakes, and estuaries. Unfortunately, this cannot easily be accomplished in wetlands managed for wildlife; the vagaries of nature, especially in hydrologic conditions, make planning difficult. Hence, marsh management for wildlife, particularly waterfowl, has often meant water-level manipulation. Water-level control is achieved by dikes (impoundments), weirs (solid structures in marsh outflows that maintain a minimum water level), control gates, and pumps. In general, the results of the management activity depend on how well the water-level control is maintained, and control depends on the local rainfall and on the sophistication of the control structures. For example, weirs provide the poorest control; all they do is maintain a minimum water level. Pumps provide positive control of drainage or flooding depth at the desired time; and the management objectives can usually be met (Wicker et al., 1983), although the cost is much higher than fixed weirs.

To illustrate the trade-offs in wetland management for wildlife enhancement, some generalizations about water-level manipulation of Lake Erie (Ohio) coastal marshes are shown in Figure 17-10. Maximum migratory wildlife use of the marshes occurs in moist soil conditions, but these conditions are also the best for the invasion of potentially undesirable plants and are generally least favorable for the overall abundance and diversity of resident plant and animal populations. Shallow-water (hemi) conditions (around 15 cm depth in summer) usually result in the highest plant species diversity and greatest fish and resident wildlife use but less migratory wildlife. Deepwater conditions (>30 cm) offer the least potential for both annual emergent plants and invading, undesirable plants, and desirable migratory waterfowl use is only fair in deep water. Kroll et al. (1997) and Gottgens et al. (1998) point out that because

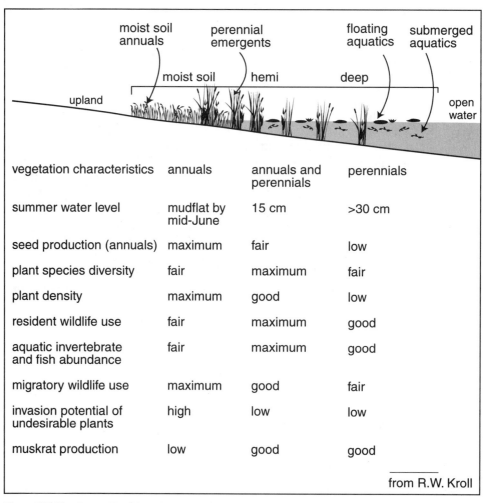

Figure 17-10 **Generalizations of water-level management for vegetation, wildlife use, and other characteristics as practiced on impounded marshes near Lake Erie in northern Ohio. (*From Roy Kroll, Winous Point Shooting Club, Port Clinton, Ohio*)**

the landward advance of marshes during high lake water times is restricted by human development, and because carp (*Cyprinus carpio*) are present in the lake, long-term above-average water levels probably mean that removal of dikes along the Great Lakes would lead to an irreversible loss of wetland vegetation fringing the lakes. Mitsch et al. (2000b) found that only 25 percent of the existing marshes encompassed by dikes would have had the right conditions to be emergent marshes more than 50 percent of the time during the last century.

Marsh Management

The set of management recommendations by Weller (1978) for prairie pothole marshes in the north-central United States and south-central Canada is another example of multipurpose wildlife enhancement. Those recommendations mimic the natural cycle of marshes in the middle of North America (see Chapter 12). Although they may seem drastic, they are entirely natural in their consequences. In sequence, the practices are as follows:

1. When a pothole is in the open stage and there is little emergent vegetation, the cycle should be initiated by a spring drawdown. This stimulates the germination of seedlings on the exposed mud surfaces.

2. A slow increase in water level after the drawdown maintains the growth of flood-tolerant seedlings without shading them out in turbid water. Shallowly flooded areas attract dabbling ducks during the winter.

3. The drawdown cycle should be repeated for a second year to establish a good stand of emergents.

4. Low water levels should be maintained for several more seasons to encourage the growth of perennial emergents such as *Typha*.

5. Maintaining stable, moderate water depths for several years promotes the growth of rooted submerged perennial aquatic plants and associated benthic fauna that make excellent food for waterfowl. During that period, the emergent vegetation will gradually die out and will be replaced by shallow ponds. When that occurs, the cycle can be initiated again, as described previously in step 1.

6. Different wetland areas maintained in staggered cycles provide all stages of the marsh cycle at once, maximizing habitat diversity.

Weller (1994) makes the distinction between a complete drawdown of water levels, as described before, a management option when vegetation is completely lost because of high water levels, herbivory, winter kill, or plant disease; and a partial drawdown, which can be implemented when vegetation is reduced but not eliminated or when wildife use has declined but not disappeared.

Wildlife management in coastal salt marshes such as those found in Louisiana uses a similar strategy, although the short-term cycle is not as pronounced there. Drawdowns to encourage the growth of seedlings and perennials preferred by ducks is a common practice, as is fall and winter flooding to attract dabbling ducks. As it happens, there is general agreement that stabilizing water levels is not good management, even though our society seems to feel intuitively that stability is a good thing. Wetlands thrive on cycles, especially flooding cycles, and practices that dampen these cycles also reduce wildlife productivity. Although the management practices described previously enhance waterfowl production, they are generally deleterious for wetland-dependent fisheries in coastal wetlands since free access between the wetlands and the adjacent estuary is restricted; the wetlands' role in regulating water quality is also often underutilized.

Managing Natural Ecosystem Managers

There is a tendency to want to control all external variables when we manage wetlands by objectives. This management tendency, although understandable, is particularly strong when herbivores such as geese, nutria, beavers, or muskrats "invade" a managed wetland. Of course, these animals can be discouraged and/or trapped to keep their influence on vegetation at a minimum. However, one has to remember that these animals are not invaders at all but are simply coming to a habitat that is generally well suited to their needs. In the ecosytem context, these animals are often nature's "ecosystem managers" (see Chapter 5) and provide many functions that, in the long term, may enhance marshes. Beavers cause water-level manipulations just as humans do. Muskrats and geese remove large areas of vegetation, but open up the system to allow for other vegetation to come into the wetland.

Whether the management of these ecosystem managers is a wise strategy is a complex issue. In coastal Louisiana, for example, muskrats and especially nutria (a South American immigrant) can "eat out" extensive marsh areas, which do not recover because of rising sea level and high marsh subsidence rates induced, in part, by human activities. Trapping used to keep the rodent populations in check, but the worldwide slump in fur salaes no longer makes trapping profitable, and rodent populations are rapidly escalating.

Agriculture and Aquaculture

When wetlands are drained for agricultural use, they no longer function as wetlands. They are, as the local farmer says, "fast lands" removed from the effects of periodic flooding and they grow terrestrial, flood-intolerant crops. Some use is made of more or less undisturbed wetlands for agriculture, but it is minor. In New England, high salt marshes were harvested for "salt marsh hay," which was considered an excellent bedding and fodder for cattle. In fact, Russell (1976) stated that the proximity of fresh and salt hay marshes was a major factor in selecting the sites for the emergence of many towns in New England before 1650. Subsequently, marshes were ditched to allow the intrusion of tides to promote the growth of salt marsh hay (*Spartina patens*), but the extent of this practice has not been well documented. On parts of the coast of the Gulf of Mexico where marshes are firm underfoot, they are still used extensively for cattle grazing. To improve access, small embankments or raised earthen paths are constructed in these marshes.

The ancient Mexican practice of *marceno* is unique. In the freshwater wetlands of the northern coast of Mexico, small areas were cleared and planted in corn during the dry season. These native varieties were tolerant enough to withstand considerable flooding. After harvest (or apparently sometimes before harvest), the marshes were naturally reflooded, and the native grasses were reestablished until the next dry season. This practice is no longer followed, but there has been some interest in reviving it (Orozco-Segovia, 1980).

On a global scale, the production of rice in managed wetlands contributes a major proportion of the world's food supply (Fig. 17-11). Aselmann and Crutzen (1989) estimated that there are approximately 1.3 million km^2 of rice paddies in the world, of which almost 90 percent are in Asia. In North America, especially in Minnesota, there are several commercial operations in the production of wild rice (*Zizania aquatica*) in wetlands and several other locations where Native American tribes have harvested wild rice in natural marshes for centuries.

Aquaculture, the farming of fish and shellfish, produces 21 million metric tons (mt), or about 20 percent of the annual worldwide total fish and shellfish harvest of 112 million mt (1995 production; World Resources Institute, Washington, DC). Most of this production occurs in Asia, with China by far the largest producer; however, in India, the second largest producer, shrimp farming is rapidly growing (Table 17-3). The United States is the major consumer of aquaculture products but accounts for only about 2 percent of worldwide production, mostly salmon and crayfish.

Figure 17-11 Rice cultivation is an important agricultural management practice throughout the world. This photo, taken in the Yangtze River Valley of China near Nanjing, shows typical rice plants being tilled early in the growing season while water is being added to the fields. (*Photograph by W. J. Mitsch*)

Fish farming practices vary. The most environmentally benign approach, similar to the Mexican *marceno* described previously, intercrops shellfish with a grain crop, usually rice. Typical is crayfish farming in the United States and Indian shrimp aquaculture in rotation with rice. The practice is described for crayfish in the southern United States. Crayfish are an edible delicacy in the southern United States and in many foreign countries. They live in burrows in shallow flooded areas such as swamp forests and rice fields, emerging with their young early in the year to forage for food. The young grow to edible size within a few weeks and are harvested in the spring. When

Table 17-3   Distribution of global aquaculture production

Country	Global Production (%)
China	57
India	9
Japan	4
Indonesia	4
Thailand	3
United States	2
Philippines	2
Korea, Republic of	2
Other countries	17

Source: Food and Agriculture Organization of the United Nations (1997).

floodwaters retreat, the crayfish construct burrows where they remain until the next winter flood. In crayfish farms, this natural cycle is enhanced by controlling water levels. An area of swamp forest is impounded; it is flooded deep during the winter and spring and drained during the summer. This cycle is ideal for crayfish, which thrive. Fish predators are controlled within the impoundments to improve the harvest. The hydrologic cycle is also favorable for forest trees. It simulates the hydrologic cycle of a bottomland hardwood forest; forest tree productivity is high, and seedling recruitment is good because of the summer drawdown. Species composition tends toward species typical of bottomland hardwoods.

Some rice farmers have also found that they can take advantage of the annual flooding cycle typically used to grow rice to combine rice and crayfish production. Rice fields are drained during the summer and fall when the rice crop matures and is harvested. Then the fields are reflooded, allowing crayfish to emerge from their burrows in the rice field embankments and forage on the vegetation remaining after the rice harvest. The crayfish harvest ends when the fields are replanted with rice. When this rotation is practiced, extreme care has to be exercised in the use of pesticides.

Today, most aquaculture practices involve much more intensive aquatic farming than the intercropping method. In India, intercropped shrimp yields are about 200 to 500 kg/ha per crop (James, 1999). A slightly more intensive farming technique that encloses natural ponds with mesh and uses weirs to control water levels and the influx and efflux of organisms yields as much as 1 mt/ha per crop, with one crop per year. Typically, in this type of culture, recruitment of postlarval juveniles to the aquaculture site occurs naturally, after which the area is sealed off and the shrimp are allowed to grow. They are harvested as they emigrate over the weirs or by seining or trawling within the enclosure.

The most intensive aquacultural techniques control all aspects of production. Wetlands, salt flats, mangrove forests, and even high-quality farmland are dredged to form ponds in which water levels are controlled by pumps. "Seed" organisms, the young postlarvae, are raised in separate hatcheries. The young organisms are fed in the ponds on synthetic diets, often composed of fish bycatch from commercial fisheries. Water quality is monitored, the ponds are aerated, and, in the most sophisticated operations, wastes are treated. Yields from this kind of operation can be several metric tons per hectare per crop and in tropical areas two crops per year are expected.

Whereas aquaculture farms in Asia have historically been small operations managed by local farmers, the worldwide boom in aquaculture, fueled by the high demand for fishery products, has led countries such as India to offer large incentives to initiate new fish farms and has drawn large corporations to invest in the industry. Aquaculture on this scale is a sophisticated enterprise. For the grower, one serious problem is that of diseases that infect the concentrated fish or shellfish populations. In India, for example, a viral disease nearly wiped out the shrimp crop in 1994 and 1995. More broadly, the concentration of fishponds in the coastal zone has serious environmental repercussions. Worldwide, 50,000 shrimp farms cover 9,800 km² of coastal lands. This has resulted in a serious loss of wetlands (coastal wetlands are required habitats for most commercial marine fish) and mangrove forests. These fish farms not only disrupt natural ecosystems but also bring in diseases, create enormous waste problems, deplete oxygen in shallow coastal waters, and reduce water quality. These disruptions have been cited as one reason for the decline in commercial fisheries in the areas where shrimp culture is concentrated.

Water Quality Enhancement

A number of studies have shown natural wetlands to be sinks for certain chemicals, particularly sediments and nutrients. The nutrient retention capacity of the wetland types covered in this book are discussed in the nutrient cycling sections of Chapters 9 through 15. It is now common to cite the water quality role of natural wetlands in the landscape as one of the most important reasons for their protection (Fig. 17-12). The idea of applying domestic, industrial, and agricultural wastewaters, sludges, and even urban and rural runoff to wetlands to take advantage of this nutrient sink capacity has also been explored in countless studies. The basic principles and practices of these so-called "treatment wetlands" are covered in detail in Chapter 20.

Flood Control and Groundwater Recharge

Wetlands can be managed, often passively, for their role in the hydrologic cycle. Hydrologic values of wetlands include streamflow augmentation, groundwater recharge, water supply potential, and flood protection (see Chapter 16). It is not altogether clear how well wetlands carry out these functions, nor do all wetlands perform these functions equally well. It is known, for example, that wetlands do not necessarily always contribute to low flows or recharge groundwater. Some wetlands,

Figure 17-12 Natural wetlands, when left as major parts of the landscape, often provide water quality roles in their natural condition without much human management. In this photo, two shades of water are noted, with the dividing line approximately at the canoe. The water in the foreground is highly polluted with sediments from a watershed artificially drained by a large ditch. The clearer water in the background is from a flooding river that is passing through a natural riparian wetland. The forested wetland, with water among the trees, can be seen in the background. The picture was taken during flooding conditions on the Kankakee River near Momence, Illinois, during a typical spring flood in this river. (*Photograph by W. J. Mitsch*)

however, should be and often are protected for their ability to hold water and slowly return it to surface water and groundwater systems during periods of low water. If wetlands are impounded to retain even more water from flooding downstream areas, considerable changes in vegetation will result as the systems adapt to the new hydrologic conditions.

Managing Wetlands in Changing Climate

Wetlands may be key ecosystems for mitigating the effects of fossil fuel emissions on climate. This was discussed in Chapter 16. Conversely, climate change and resulting sea-level change may have significant impacts on coastal and inland wetlands, as discussed by Mitsch and Wu (1995).

Coastal Wetlands

One of the major impacts of possible climate changes on wetlands is the effect that sea-level rise will have on coastal wetlands. Estimates of sea-level rise over the next century range from 50 to 200 cm (National Research Council, 1985). If the rise in sea level is not accompanied by equivalent vertical accretion of marsh sediments, then there will be a gradual disintegration of coastal marshes due to increased inundation, erosion, and saltwater intrusion. Since much of the coastline of the world is developed, efforts to protect dry upland from inundation by the construction of bulkheads or dikes will exacerbate the problem. In essence, the wetlands will be trapped between the rising sea and the protected dry land, a situation that has already occurred over the centuries in the Netherlands and China (Titus, 1991). Even in the absence of bulkheads in most of our regions where coastal wetlands exist, "the slope above the wetland is steeper than that of the wetlands; so a rise in sea level causes a net loss of wetland acreage" (Titus, 1991).

Estimates on the loss of coastal wetlands in the United States vary, with much of the variability dependent on the assumed sea-level rise and the degree to which dry land is protected at all cost (Table 17-4). If there is no shoreline protection, a sea-level rise of 1 m could reduce coastal wetlands by 26 to 66 percent. If the policy were to protect all dry land, the estimated loss of wetlands increases dramatically to 50 to 82 percent. How well these figures can be extrapolated to the rest of the world is unclear. In long-developed coastlines such as those of Europe and the Far East, the losses would probably be less.

A number of studies have evaluated regional changes in coastal wetlands that would result in the event of dramatic sea-level rise. A spatial cell-based simulation

Table 17-4 Estimated percentage coastal wetland loss in the United States with sea-level rise

	Sea-Level Rise		
	0.5 m	1 m	2 m
If no shores are protected	17–43%	26–66%	29–76%
If densely developed dry land is protected	20–45%	29–69%	33–80%
If all dry land is protected	38–61%	50–82%	66–90%

Source: Titus (1991).

model called SLAMM (Sea Level Affecting Marshes Model) has been used by Richard Park and colleagues (Park et al., 1991; Lee et al., 1991) to illustrate the effects of sea-level rise on coastal region. For example, Lee et al. (1991) predicted a 32 to 40 percent loss of wetlands in northeastern Florida with a rise of 1 to 1.25 m. Most of that loss is the low intertidal salt marsh.

Day and Templet (1989) and McKee and Mendelssohn (1989) discussed the Mississippi River delta in Louisiana as a model for global sea-level rise. Here, the apparent sea-level rise is already 1.2 m/100 yr, primarily because of sediment subsidence rather than actual sea rise (see Chapter 9). In this delta marsh, vertical accretion is not keeping up with subsidence, in part because the Mississippi River is carrying only about 20 percent of the sediment load it did in 1850 (Kesel and Reed, 1995), and its flow is contained within levees so riverborne sediments no longer reach the wetlands during spring floods. As a result, this region has the highest rate of wetland loss in the United States.

Patrick and Delaune (1990) measured the accretion and subsidence of sediments in the salt marshes of San Francisco Bay, California, and found that, because of an adequate supply of sediments, the salt marshes of south San Francisco Bay would probably have a net accretion rate despite sea-level rise. Lefeuvre (1990) presented a summary of the ecological effects of sea-level rise in coastal France near Mont-St.-Michel Bay. The salt marshes in this region are currently expanding at a rate of about 30 ha/yr due to a positive sediment influx of 2 cm/yr. He hypothesized that fragmentation would occur to the salt marshes and polders in the event of a maximum rise in sea level (1–2 m/100 yr) but that this fragmentation would enhance marine ecology by increasing marsh–tidal flat interfaces.

Management

The management possibilities for coastal wetlands in the face of sea-level rise offer some possibilities to save coastal wetlands. Figure 17-13 shows two future conditions. In Future 1, the house is protected with a bulkhead in the face of rising sea level and the salt marsh in lost. In Future 2, the house is moved upland to accommodate the wetland, which would begin to form if a gentle slope and adequate sediment sources were available. This models the wetlands of the Great Lakes, which, for centuries, were "wetlands on skateboards," moving inland and lakeward with frequent (over periods of decades) water-level changes in the Great Lakes (Mitsch, 1992b). With stabilization of the coastline in the past century, diking the remaining wetlands along the Great Lakes has been necessary for their survival.

Day and Templet (1989) concluded, after extensive investigation of the apparent sea-level rise in coastal Louisiana, that we can manage coastal wetlands in periods of rising sea level through comprehensive, long-range planning and through the application of the principles of ecological engineering by using nature's energies such as upstream riverine sediments and fresh water, vegetation productivity, winds, currents, and tides as much as possible. They cite the example of using vegetation to enhance sediment accretion such as with brush fences developed by the Dutch.

Inland Wetlands

In addition to the effects of climate change on coastal wetlands through sea-level rise, the change in climate will probably affect the health and distribution of inland wetlands. In the tundra, any melting of the permafrost would result in the loss of wetlands. In

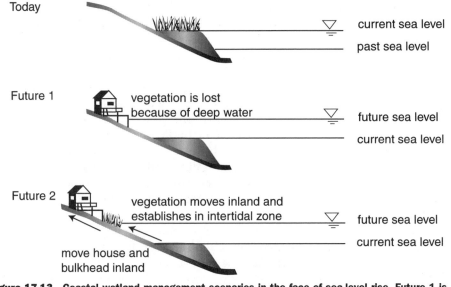

Figure 17-13 Coastal wetland management scenarios in the face of sea-level rise. Future 1 is without moving human habitation inland. Future 2 involves moving human activity inland to allow room for wetland to move inland. (*From Titus, 1991*)

boreal and temperate areas, climate change would result in changing rainfall patterns, thus affecting runoff and groundwater inflows to wetlands. In general, a decrease in precipitation or an increase in evapotranspiration will result in less frequent flooding of existing wetlands, although the types of wetlands may not change. Greater precipitation patterns, of course, would increase the length and depth of flooding of inland wetlands. Most susceptible to these effects are depressional wetlands that have very small watersheds and that are in regions between arid and mesic climates such as the prairie potholes of North America.

Feedbacks

The effects of climate change on inland wetlands, particularly peatlands, can lead to both positive and negative feedbacks on climate change. A possible positive feedback is increased temperatures expected from global change, leading to increased emissions of greenhouse gases (CO_2 and CH_4) as a result of increased metabolism (Table 17-5). Comparing these figures with the present fluxes shown in Figure 16-10 suggests that this feedback—climate change increasing the release of greenhouse gases from

Table 17-5 Estimated changes in CO_2 and CH_4 emissions (10^{15} g/yr = Pg/yr) from northern wetlands with change in climatic conditions. (assumes 5°C temperature rise)

	Warm/Wet		Warm/Dry	
	CO_2	CH_4	CO_2	CH_4
Tundra	+1.3	+0.1	+1.6	+0.1
Boreal peatland	—	+0.12	+0.83	+0.12

Source: Post (1990).

wetlands—may be significant. Christensen (1991) predicted that, as a result of a 5 percent global warming, the tundra would change from a net sink of CO_2 to a net source of up to 1.25 Pg/yr carbon due to a combination of thermokarst erosion, deepening of the active layer in permafrost areas, lowering of the water table, and higher temperatures.

In general, both the increase in temperature and the changes in water levels are important variables in the production of methane and carbon dioxide from wetlands, but their relative importance for methane generation is poorly understood (Moore and Knowles, 1989; Whalen and Reeburgh, 1990; Roulet et al., 1992b; Moore and Roulet, 1995). Using a model with inputs of a 3°C rise in temperature and a decrease in the water table between 14 and 22 cm for a subarctic fen, Roulet et al. (1992b) estimated that the increased temperature raised the methane flux between 5 and 40 percent, but the lowered water table decreased the methane flux by 74 to 81 percent. This decrease in methane flux in drier conditions was due to a decrease in the zone of active methanogenesis and to an increase in methane oxidation in the aerobic layer. Thus, the influence of global temperature rise would depend locally on the temperature increase relative to the induced change in the moisture regime.

Management

Limited experimentation, especially in rice paddies, suggests some management alternatives that might be appropriate for inland wetlands, especially to reduce methane emissions. Sass et al. (1992) measured the effects on methane emissions of four different water management methods in some rice fields in Texas and found that temporary drainage (midseason drainage and multiple aeration) decreased methane emission due to both increased CH_4 consumption in the aerobic layer and decreased CH_4 production. Such management may only be practical in flat systems with sufficient control of water levels.

Nutrient and compost management may also offer opportunities for reducing methane emissions. There may be a relationship between the C:N ratio of the organic matter in wetlands and CH_4 emissions, although the trends are not clear. Yagi and Minami (1990) found, in rice paddies in Japan, that compost with a low C:N ratio (enriched in nitrogen) causes lower emissions of methane than uncomposted rice straw with a high C:N ratio. Schutz et al. (1989), on the other hand, found high emissions in fields applied with compost. Because of its competition with methanogenesis, enhancing sulfate reduction is often suggested as a management alternative to reduce methane emissions (see Chapter 6). This has long been known as one of the primary reasons that methanogenesis is lower in saltwater wetlands than in freshwater wetlands.

We cannot estimate, at present, with much certainty whether wetlands are significant global carbon sources or sinks. Nevertheless, the opportunities cited previously for managing carbon dioxide and methane emissions in wetlands do not appear to be possible on a scale large enough to make much difference to the global carbon balance.

Recommended Readings

Harris, T. 1991. Death in the Marsh. Island Press, Washington, DC. 245 pp.

Mitsch, W. J., and X. Wu. 1995. Wetlands and global change. In R. Lal, J.
      Kimble, E. Levine, and B. A. Stewart, eds. Advances in Soil Science: Soil

Management and Greenhouse Effect. CRC Press, Boca Raton, FL. pp. 205–230.

Weller, M. W. 1994. Freshwater Marshes, 3rd ed. University of Minnesota Press, Minneapolis, MN. 192 pp.

Williams, M., ed. 1990. Wetlands: A Threatened Landscape. Basil Blackwell, Oxford. 419 pp.

Wetland Laws and Protection

In earlier times, wetland drainage was considered the only policy for managing wetlands throughout the world. With the recognition of wetland values, wetland protection has been emphasized by many laws and international agreements. The U.S. federal government has relied on executive orders, a "no net loss" policy, and the Section 404 dredge-and-fill permit program of the Clean Water Act for wetland protection, augmented by wetland protection programs in agriculture and aided by the development of wetland delineation procedures. Some states have wetland protection laws, although many more have relied on federal regulations. International cooperation in wetland protection, particularly through the Ramsar Convention and the North American Waterfowl Management Plan, has been emphasized in recent years as policymakers have come to realize that the functions of local wetlands cross political boundaries.

Wetlands are now the focus of legal protection efforts throughout the world, but, because of this focus, they are beginning to be defined by legal fiat as much as by the application of ecological principles. Chapter 2 reviewed the major definitions of wetlands that have developed both nationally and internationally. Some definitions are scientific, whereas others are principally to allow legal definition and thus protection of wetlands. Protection has been implemented through a variety of policies, laws, and regulations, ranging from animal and plant protection to land use and zoning restrictions to enforcement of dredge-and-fill laws. In the United States, wetland protection has historically been a national initiative, often with assistance and implementation provided by individual states. In the international arena, agreements to protect ecologically important wetlands throughout the world have been negotiated and ratified.

Legal Protection of Wetlands in the United States

As pointed out in Chapter 17, the policy of the United States for over 120 years was to drain wetlands. The Swamp Land Acts of 1849, 1850, and 1860 were precursors to one of the most rapid and dramatic changes in the landscape that has ever occurred in history, even though the acts themselves were deemed to be largely ineffective in their intended purpose (Hibbard, 1965; National Research Council, 1995). By the mid-1970s, about half of the wetlands in the lower 48 states were drained (see Chapter 4). In the early 1970s, interest in wetland protection increased significantly as scientists began to identify and quantify the many values of these ecosystems. This interest in wetland protection began to be translated at the federal level in the United States into laws, regulations, and public policies. Prior to this time, federal policy on wetlands was vague and often contradictory. Policies in agencies such as the U.S. Army Corps of Engineers, the Soil Conservation Service (now the Natural Resources Conservation Service), and the Bureau of Reclamation encouraged the destruction of wetlands, whereas policies in the Department of the Interior, particularly in the U.S. Fish and Wildlife Service, encouraged their protection. Some states have also developed inland and coastal wetland laws and policies, and activity in that area appears to be increasing.

The primary wetland protection mechanisms used by the U.S. federal government are summarized in Table 18-1. Some of the more significant activities of the federal government that led to a more consistent wetland protection policy have included presidential orders on wetland protection and floodplain management, implementation of a dredge-and-fill permit system to protect wetlands, coastal zone management policies, and initiatives and regulations issued by various agencies. Despite all of this activity related to federal wetland management, two major points should be emphasized:

1. *There is no specific national wetland law.* Wetland management and protection result from the application of many laws intended for other purposes. Jurisdiction over wetlands has also been spread over several agencies, and, overall, federal policy continually changes and requires considerable interagency coordination.

2. *Wetlands have been managed under regulations related to both land use and water quality.* Neither of these approaches, taken separately, can lead to a comprehensive wetland policy. The regulatory split mirrors the scientific split noted by many wetland ecologists, a split that is personified by people who have developed expertise in either aquatic or terrestrial systems. Rarely do individuals possess expertise in both areas.

Early Presidential Orders

President Jimmy Carter issued two executive orders in May 1977 that established the protection of wetlands and riparian systems as the official policy of the federal government. Executive Order 11990, Protection of Wetlands, required all federal agencies to consider wetland protection as an important part of their policies:

> Each agency shall provide leadership and shall take action to minimize the destruction, loss or degradation of wetlands, and to preserve and enhance the natural and beneficial values of wetlands in carrying out the agency's responsibilities for (1) acquiring, managing, and disposing of Federal lands and facilities; and (2) providing federally undertaken, financed, or assisted construction and improvement; and (3) conducting Federal activities and programs affecting land use, including but not limited to water and related land resources planning, regulating, and licensing activities.

Table 18-1 Major federal laws, directives, and regulations in the United States used for the management and protection of wetlands

	Date	Responsible Federal Agency
DIRECTIVE OR STATUTE		
Rivers and Harbors Act, Section 10	1899	U.S. Army Corps of Engineers
Fish and Wildlife Coordination Act	1967	U.S. Fish and Wildlife Service
Land and Water Conservation Fund Act	1968	U.S. Fish and Wildlife Service, Bureau of Land Management, Forest Service, National Park Service
National Environmental Policy Act	1969	Council on Environmental Quality
Federal Water Pollution Control Act (PL 92-500) as amended (Clean Water Act)	1972, 1977, 1982	
Section 404—Dredge-and-Fill Permit Program		U.S. Army Corps of Engineers with assistance from Environmental Protection Agency and U.S. Fish and Wildlife Service
Nationwide Permits		
State and Tribal 404 assumption		
Section 208—Areawide Water Quality Planning		Environmental Protection Agency
Section 303—Water Quality Standards		Environmental Protection Agency
Section 401—Water Quality Certification		Environmental Protection Agency (with state agencies)
Section 402—National Pollutant Discharge Elimination System		Environmental Protection Agency (or state agencies)
Coastal Zone Management Act	1972	Office of Coastal Zone Management Department of Commerce
Flood Disaster Protection Act	1973, 1977	Federal Emergency Management Agency
Federal Aid to Wildlife Restoration Act	1974	U.S. Fish and Wildlife Service
Water Resources Development Act	1976, 1990	U.S. Army Corps of Engineers
Executive Order 11990—Protection of Wetlands	May 1977	All agencies
Executive Order 11988—Floodplain Management	May 1977	All agencies
Food Security Act, swampbuster provisions	1985	Department of Agriculture, Natural Resources Conservation Service
Emergency Wetland Resources Act	1986	U.S. Fish and Wildlife Service
Executive Order 12630—Constitutionally Protected Property Rights	1988	All agencies
Wetlands Delineation Manual (various revisions)	1987, 1989, 1991	All agencies
"No Net Loss" Policy	1988	All agencies
North American Wetlands Conservation Act	1989	U.S. Fish and Wildlife Service
Coastal Wetlands Planning, Protection and Restoration Act	1990	U.S. Army Corps of Engineers
Wetlands Reserve Program	1991	Department of Agriculture, Natural Resources Conservation Service
Executive Order 12962—Conservation of Aquatic Systems for Recreational Fisheries	1995	All agencies
Federal Agriculture Improvement and Reform Act	1996	Department of Agriculture, Natural Resources Conservation Service
POLICY AND TECHNICAL GUIDANCE		
Water Quality Standards Guidance	1990	Environmental Protection Agency
Non-Point Source Guidance	1990	Environmental Protection Agency
Mitigation/Mitigation Banking	1990, 1995	U.S. Army Corps of Engineers
Wetlands on Agricultural Lands, memo of agreement	1990, 1994	U.S. Army Corps of Engineers, Department of Agriculture
Wetlands and Forestry Guidance	1995	U.S. Army Corps of Engineers, Department of Agriculture

Executive Order 11988, Floodplain Management, established a similar federal policy for the protection of floodplains, requiring agencies to avoid activity in the floodplain wherever practicable. Furthermore, agencies were directed to revise their procedures to consider the impact that their activities might have on flooding and to avoid direct or indirect support of floodplain development when other alternatives are available.

Both of these executive orders were significant because they set in motion a review of wetland and floodplain policies by almost every federal agency. Several agencies such as the U.S. Environmental Protection Agency and the Soil Conservation Service established policies of wetland protection prior to the issuance of these executive orders, but many other agencies such as the Bureau of Land Management were compelled to review or establish wetland and floodplain policies.

No Net Loss

A more significant initiative in developing a national wetlands policy was undertaken in 1987, when a National Wetlands Policy Forum was convened by the Conservation Foundation at the request of the U.S. Environmental Protection Agency to investigate the issue of wetland management in the United States (National Wetlands Policy Forum, 1988; Davis, 1989). The 20 distinguished members of this forum (which included three governors, a state legislator, state and local agency heads, the chief executive officers of environmental groups and businesses, farmers, ranchers, and academic experts) published a report that set significant goals for the nation's remaining wetlands. The forum formulated one overall objective:

> to achieve no overall net loss of the nation's remaining wetlands base and to create and restore wetlands, where feasible, to increase the quantity and quality of the nation's wetland resource base (National Wetlands Policy Forum, 1988).

The group recommended as an interim goal that the holdings of wetlands in the United States should decrease no further, and as a long-term goal that the number and quality of the wetlands should increase. In his 1988 presidential campaign and in his 1990 budget address to Congress, President George Bush echoed the "no net loss" concept as a national goal, shifting the activities of many agencies such as the Department of the Interior, the U.S. Environmental Protection Agency, the U.S. Army Corps of Engineers, and the Department of Agriculture toward achieving a unified and seemingly simple goal. Nevertheless, it was not anticipated that there would be a complete halt of wetland loss in the United States when economic or political reasons dictated otherwise. Consequently, implied in the no net loss concept is wetland construction and restoration to replace destroyed wetlands. The no net loss concept became a cornerstone of wetland conservation in the United States in the early 1990s.

The Clean Water Act Section 404 Program

The primary vehicle for wetland protection and regulation in the United States now is Section 404 of the Federal Water Pollution Control Act (FWPCA) amendments of 1972 (PL 92-500) and subsequent amendments (also known as the Clean Water Act). Section 404 required that anyone dredging or filling in "waters of the United States" must request a permit from the U.S. Army Corps of Engineers. This requirement was an extension of the 1899 Rivers and Harbors Act in which the Corps had responsibility for regulating the dredging and filling of navigable waters.

The use of Section 404 for wetland protection has been controversial and has been the subject of several court actions and revisions of regulations. The surprising point about the importance of the Clean Water Act in wetland protection is that wetlands are not directly mentioned in Section 404 and at first this directive was interpreted narrowly by the Corps to apply only to navigable waters. The definition of waters of the United States was expanded to include wetlands in two 1974–1975 court decisions, *United States v. Holland* and *Natural Resources Defense Council v. Calloway.* These decisions, along with Executive Order 11990, Protection of Wetlands, put the Army Corps of Engineers squarely in the center of wetland protection in the United States. On July 25, 1975, the Corps issued revised regulations for the Section 404 program that enunciated the policy of the United States on wetlands:

> As environmentally vital areas, [wetlands] constitute a productive and valuable
> public resource, the unnecessary alteration or destruction of which should be
> discouraged as contrary to the public interest (*Federal Register,* July 25, 1975).

Wetlands were defined in these regulations to encompass coastal wetlands ("marshes and shallows and . . . those areas periodically inundated by saline or brackish waters and that are normally characterized by the prevalence of salt or brackish water vegetation capable of growth and reproduction") and freshwater wetlands ("areas that are periodically inundated and that are normally characterized by the prevalence of vegetation that requires saturated soil conditions for growth and reproduction") (*Federal Register,* July 25, 1975). By these actions, the jurisdiction of the Corps has been extended to include 60 million ha of wetlands, 45 percent of which are in Alaska. Several times since 1975, the Corps has issued revised regulations for the dredge-and-fill permit program, and in 1985 the U.S. Supreme Court, in *United States v. Riverside Bayview Homes,* rejected the contention that Congress did not intend to include wetland protection as part of the Clean Water Act.

The procedure for obtaining a "404 permit" for dredge-and-fill activity in wetlands is complex (Fig. 18-1). As a starting point, no discharge of dredged or fill material can be permitted in wetlands if a practicable alternative exists. So in the initial screening of a project that involves potential effects on wetlands, the following three approaches are evaluated in sequence:

1. Avoidance—taking steps to avoid wetland impacts where practicable
2. Minimization—minimizing potential impacts to wetlands
3. Mitigation—providing compensation for any remaining, unavoidable
   impacts through the restoration or creation of wetlands (see Chapter 19)

An individual Section 404 permit is usually required for potentially significant impacts but for many activities that have minimal adverse effects, the Army Corps of Engineers has issued general permits. The best known are called nationwide, regional, or state permits for particular categories of activities (e.g., minor road crossings, utility line backfill, and bedding) as a means to expedite the permit process.

The decision to issue a permit rests with the Corps' district engineer, and it must be based on a number of considerations, including conservation, economics, aesthetics, and several other factors listed in Figure 18-1. Assistance to the Corps on the dredge-and-fill permit process in wetland cases is provided by the U.S. Environmental Protection Agency (EPA), the U.S. Fish and Wildlife Service, the National Marine Fisheries Service, and state agencies. The EPA has statutory authority to designate wetlands

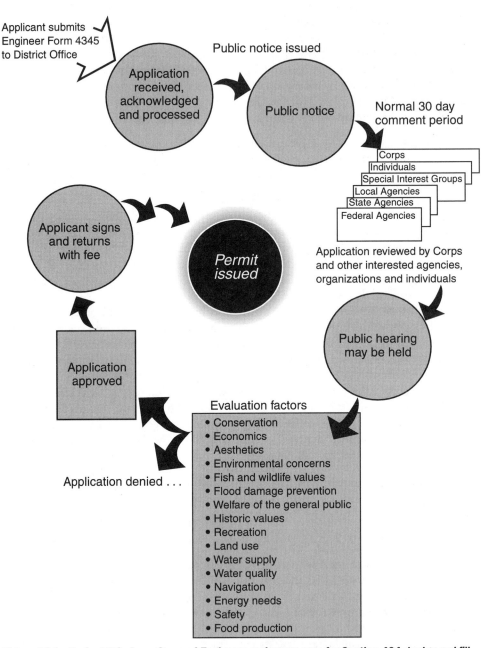

Applicant submits
Engineer Form 4345
to District Office

Application
received,
acknowledged
and processed

Public notice issued

Public notice

Normal 30 day
comment period

Corps
Individuals
Special Interest Groups
Local Agencies
State Agencies
Federal Agencies

Application reviewed by Corps
and other interested agencies,
organizations and individuals

Applicant signs
and returns
with fee

Permit
issued

Public hearing
may be held

Application
approved

Evaluation factors
• Conservation
• Economics
• Aesthetics
• Environmental concerns
• Fish and wildlife values
• Flood damage prevention
• Welfare of the general public
• Historic values
• Recreation
• Land use
• Water supply
• Water quality
• Navigation
• Energy needs
• Safety
• Food production

Application denied . . .

Figure 18-1 Typical U.S. Army Corps of Engineers review process for Section 404 dredge-and-fill permit request. (*From Kulser, 1983*)

644

subject to permits, and also has veto power on the Corps' decisions. Some states require state permits as well as Corps permits for wetland development. The district engineer, according to Corps regulations, should not grant a permit if a wetland is identified as performing important functions for the public such as biological support, wildlife sanctuary, storm protection, flood storage, groundwater recharge, or water purification. An exception is allowed when the district engineer determines "that the benefits of the proposed alteration outweigh the damage to the wetlands resource and the proposed alteration is necessary to realize those benefits" (*Federal Register,* July 19, 1977). The effectiveness of the Section 404 program has varied since the program began and has also varied from district to district.

Swampbuster

Normal agricultural and silvicultural activities were exempted from the Section 404 permit requirements, thereby still allowing wetland drainage on farms and in commercial forests. Allowing such exemptions created conflict within the federal government: The Army Corps of Engineers and the Environmental Protection Agency were encouraging wetland conservation through the Clean Water Act, and the Department of Agriculture was encouraging wetland drainage by providing federal subsidies for drainage projects. The conflict ended when Congress passed, as part of the 1985 Food Security Act, "swampbuster" provisions that denied federal subsidies to any farm owner who knowingly converted wetlands to farmland after the act became effective. The "swampbuster" provisions of the act drew the U.S. Soil Conservation Service (now the Natural Resources Conservation Service, or NRCS) into federal wetland management, primarily as an advisory agency helping farmers identify wetlands on their farms. The NRCS also administers a Wetlands Reserve Program that was set up in 1990 to acquire federal easements on up to 400,000 ha (1 million acres) of agricultural land that was formerly wetland.

In August 1993, the Clinton administration released a document entitled "Protecting America's Wetlands: A Fair, Flexible, and Effective Approach." The document reaffirmed no net loss, established that 21.5 million ha (53 million acres) of previously converted wetlands would not be subject to regulations, and established the NRCS as the lead agency for identifying wetlands on agricultural land under both the Clean Water Act and the Food Security Act "swampbuster" provisions. The policy was agreed to in a January 6, 1994, memorandum of agreement among the four principal federal agencies involved in wetland policy in the United States (U.S. Fish and Wildlife Service, Natural Resources Conservation Service, U.S. Army Corps of Engineers, and U.S. Environmental Protection Agency).

Wetland Delineation

To be able to determine whether a particular piece of land was a wetland and, therefore, if it was necessary to obtain a Section 404 permit to dredge or fill that wetland, federal agencies, beginning with the Army Corps of Engineers, began to develop guidelines for the demarcation of wetland boundaries in a process that came to be called *wetland delineation*. In 1987, the Army Corps of Engineers published a technical manual for wetland delineation (1987 Wetlands Delineation Manual). This manual specified three mandatory technical criteria—hydrology, soils, and vegetation—for a parcel of land to be declared a wetland. Subsequent to that, the Environmental Protection Agency, the Soil Conservation Service, and the U.S. Fish and Wildlife Service developed separate documents for their respective roles in wetland protection. Finally, in January 1989, after several months of negotiation, a single *Federal Manual for Identifying and Delineating Jurisdictional Wetlands* was published by the four federal agencies to unify the government's approach to wetlands. This 1989 Wetlands Delineation Manual, while also requiring the three mandatory technical criteria for a parcel of land to be declared a wetland, allowed one criterion to infer the presence of another (e.g., the presence of hydric soils to infer hydrology). The manual also provided some guidance about how to use field indicators such as water marks on trees or stains on leaves to determine recent flooding, wetland vegetation (from published lists), and hydric soil indicators such as mottling. The manual was used by developers and agencies alike to prove or disprove the presence of wetlands in the Section 404 permit process. Consulting firms specializing in wetland delineation sprung up overnight, and short courses on the methodology became very popular.

Beginning in early 1991, modifications of the manual were proposed in response to heavy lobbying by developers, agriculturalists, and industrialists for a relaxing of the wetland definitions, in order to lessen the regulatory burden on the private sector. A new manual was published for public comment in August 1991 (the 1991 Wetlands Delineation Manual) but was quickly and heavily criticized for its lack of scientific credibility and unworkability (Environmental Defense Fund and World Wildlife Fund, 1992); it was eventually abandoned in 1992. At present, the 1987 Corps technical manual, which was generally agreed to be a version between the "liberal" (politically speaking) 1989 manual and the "conservative" 1991 manual, has been used since 1992 as the official way in which wetlands are determined, and there is no reason to believe that this practice will change in the near future.

National Academy of Science Study

The development of a wetland delineation manual that all could agree on led to a contentious period in wetland policy between 1989 and 1991 when the 1989 and proposed 1991 manuals were introduced. About that time, many scientists began to call for the National Academy of Sciences, a nongovernmental agency set up in the 19th century to provide scientific reviews of subjects chosen and paid for by the federal government, to take up the issue of "what is a wetland?" In April 1993, the U.S. Environmental Protection Agency, at the request of the U.S. Congress, asked the National Research Council, the operating arm of the National Academy of Sciences, to appoint a committee to undertake a scientific review of scientific aspects of wetland characterization. The 17-member committee was selected in the summer of 1993 and met over a 2-year period. The committee was charged with considering (1) definition of wetlands; (2) adequacy of science for evaluating hydrologic, biological, and other ways that wetlands function; and (3) regional variation. The report from that committee (National Research Council, 1995) presented a new definition of wetlands (see Chapter 2) and gave 80 recommendations on topics such as fine-tuning the delineation procedure, dealing with especially controversial wetlands, regionalization, mapping, modeling, administrative issues, and functional assessment of wetlands. The report, in essence, suggested that use of the 1987 manual was appropriate with a few minor modifications. The report was released in early 1995, just as the U.S. Congress was considering two bills on wetlands that would have drastically changed the definitions and management of wetlands in the United States (Fig.

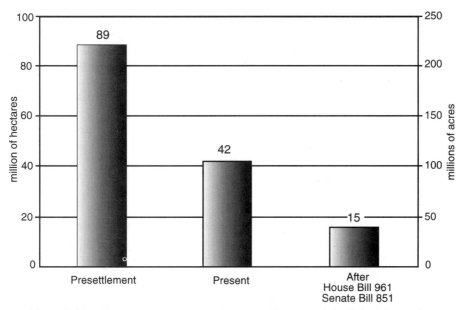

Figure 18-2 Estimated extent of wetlands in the lower 48 states of the United States for presettlement times (1780s) and present day. The numbers in the first two bars, already presented in Chapter 4, are compared with an estimate of the extent of wetlands that would have remained legally protected if House Bill 961 or Senate Bill 851 in the U.S. Congress had been passed in 1995. Each proposed law contained formal definitions of wetlands. These proposed laws would have protected from 11 million to 15 million ha of "legal" wetlands in the United States. Neither law passed, but this illustrates that wetlands can be lost either by drainage or by legal fiat that redefines wetlands. (*Source: Mike Davis, U.S. Army Corps of Engineers, March 7, 1996*)

18-2). It may have been a result of the release of the NRC report or just by coincidence, but neither bill became law.

Other Federal Activity

Several other federal laws and activities have led to wetland protection since the 1970s. The Coastal Zone Management Program, established by the Coastal Zone Management Act of 1972, has provided up to 80 percent of matching-funds grants to states to develop plans for coastal management based on establishing a high priority to protecting wetlands. The National Flood Insurance Program offers some protection to riparian and coastal wetlands by offering federally subsidized flood insurance to state and local governments that enact local regulations against development in flood-prone areas. The Clean Water Act, in addition to supporting the Section 404 program, authorized $6 million to the U.S. Fish and Wildlife Service to complete its inventory of wetlands of the United States (see Chapter 21). The Emergency Wetlands Resource Act passed by Congress in 1986 required the U.S. Fish and Wildlife Service to update its report on the status of and trends in wetlands every 10 years. The first report was issued in 1982 and published one year later (Frayer et al., 1983). The first update was published in 1991 as Dahl and Johnson (1991) (see Chapter 4 for the conclusions of these reports).

The North American Wetlands Conservation Act's purpose was to encourage voluntary, public–private partnerships to conserve North American wetland ecosystems. This law, passed in 1989, provides grants, primarily to state agencies and private and public organizations, to manage, restore, or enhance wetland ecosystems to benefit wildlife. From 1991 through mid-1999, almost 650 projects in Canada, Mexico, and the United States have been approved for funding and approximately 3.5 million ha (8.6 million acres) of wetlands and associated uplands have been acquired, restored, or enhanced in the United States and Canada. The act also paid for a significant amount of wetland conservation education and management plan projects in Mexico.

Wetland mitigation and wetland mitigation banking became an interest of the U.S. Congress and, although regulations were promulgated for wetland banks in the mid-1990s, no specific laws were enacted at the federal level.

The "Takings" Issue

One of the dilemmas of valuing and protecting wetlands is that the values accrue to the public at large but rarely to individual landowners who happen to have a wetland on their property. If government laws that protect wetlands or other natural resources lead to a loss of the use of that land by the private landowner, the restriction on that use has been referred to as a "taking" (denial of an individual's right to use his or her property). Many legal scholars believed that wetland and other land use laws could result in "takings" and thus be against the Fifth Amendment of the U.S. Constitution. In a major ruling in June 1992 (*Lucas v. South Carolina Coastal Council*), the U.S. Supreme Court ruled that regulations denying "economically viable use of land" require compensation to the landowner, no matter how great the public interest served by the regulations (Runyon, 1993). This case was referred back to the state of South Carolina to determine if the developer, David Lucas, was denied all economically viable use of his land (beach-front property that was rezoned by South Carolina in response to the 1980 Coastal Zone Management Act). The ultimate result of this Supreme Court decision on wetland legal protection is inconclusive and wetland protection regulations remain in effect in the United States.

State Management of Wetlands

Many individual states have issued wetland protection statutes or regulations. State wetland programs may become more important as the federal government attempts to delegate much of its authority to local and state governments. Kusler (1983) has suggested that, although local communities may also have wetland protection programs, states are much more probable governmental units for wetland protection for the following reasons:

1. Wetlands cross local government boundaries, making local control difficult.
2. Wetlands in one part of a watershed affect other parts that may be in different jurisdictions.
3. There is usually a lack of expertise and resources at the local level to study wetland values and hazards.
4. Many of the traditional functions of states such as fish and wildlife protection are related to wetland protection.

Many states that contain coastlines initially paid more attention to managing their coastal wetlands than to managing their inland wetlands as a result of an earlier interest in coastal wetland protection at the federal level and of the development of coastal zone management programs. In general, state programs can be divided into those that are based on specific coastal wetland laws and those that are designed as a part of broader regulatory programs such as coastal zone management. Several coastal states have coastal dredge-and-fill permit programs, whereas other states have specific wetland regulations administered by a state agency.

State programs for inland wetlands, although in an earlier stage of development, are more diverse, ranging from comprehensive laws to a lack of concern for inland wetlands. Comprehensive laws have been enacted in at least 16 states. Other states have few regulations governing inland wetlands. Between these two extremes, there are many states that rely on federal–state cooperation programs or on state laws that indirectly protect wetlands. Michigan has assumed responsibility from the federal government to issue Section 404 permits, although the Army Corps of Engineers retains control of the permit program for navigable waters (Meeks and Runyon, 1990) and the EPA retains federal oversight of the program. Floodplain protection laws or scenic and wild river programs are being implemented in more than half of the states and are often effective in slowing the destruction of riparian wetlands. States are also involved in wetland protection through wetland acquisition programs, conservation easement programs, preferential tax treatment for landowners who protect wetlands, and enforcement of state water quality standards as required by the Clean Water Act (Meeks and Runyon, 1990).

International Wetland Conservation

The Ramsar Convention

Intergovernmental cooperation on wetland conservation has been spearheaded by the Convention on Wetlands of International Importance, more commonly referred to as the *Ramsar Convention* because it was initially adopted at an international conference held in Ramsar, Iran, in 1971. The global treaty provides the framework for the international protection of wetlands as habitats for migratory fauna that do not observe international borders and for the benefit of human populations dependent on wetlands. The convention's mission is the conservation and wise use of wetlands by national action and international cooperation, as a means of achieving sustainable development throughout the world. A permanent secretariat headquartered at the International Union of Conservation of Nature and Natural Resources (IUCN) in Switzerland was established in 1987 to administer the convention, and a budget based on the United Nations scale of contributions was adopted.

The specific obligations of countries that have ratified the Ramsar Convention are the following:

1. Member countries shall formulate and implement their planning so as to promote the "wise use" of all wetlands in their territory, and develop national wetland policies.
2. Member countries shall designate at least one wetland in their territory for the "List of Wetlands of International Importance." The so-called Ramsar sites should be developed based on their "international significance in terms of ecology, botany, zoology, limnology or hydrology" (Navid, 1989).
3. Member countries shall establish nature reserves at wetlands.

Table 18-2 Ramsar Convention criteria for identifying wetlands of international importance, 1999

GROUP A. SITES CONTAINING REPRESENTATIVE, RARE, OR UNIQUE WETLAND TYPES

> Criterion 1: A wetland should be considered internationally important if it contains a representative, rare, or unique example of a natural or near-natural wetland type found within the appropriate biogeographic region.

GROUP B. SITES OF INTERNATIONAL IMPORTANCE FOR CONSERVING BIOLOGICAL DIVERSITY

Criteria based on species and ecological communities

> Criterion 2: A wetland should be considered internationally important if it supports vulnerable, endangered, or critically endangered species or threatened ecological communities.
>
> Criterion 3: A wetland should be considered internationally important if it supports populations of plant and/or animal species important for maintaining the biological diversity of a particular biogeographic region.
>
> Criterion 4: A wetland should be considered internationally important if it supports plant and/or animal species at a critical stage in their life cycles, or provides refuge during adverse conditions.

Specific criteria based on waterbirds

> Criterion 5: A wetland should be considered internationally important if it regularly supports 20,000 or more waterbirds.
>
> Criterion 6: A wetland should be considered internationally important if it regularly supports 1 percent of the individuals in a population of one species or subspecies of waterbird.

Specific criteria based on fish

> Criterion 7: A wetland should be considered internationally important if it supports a significant proportion of indigenous fish subspecies, species or families, life-history stages, species interactions, and/or populations that are representative of wetland benefits and/or values and thereby contributes to global biological diversity.
>
> Criterion 8: A wetland should be considered internationally important if it is an important source of food for fishes, spawning ground, nursery, and/or migration path on which fish stocks, either within the wetland or elsewhere, depend.

4. Member countries shall cooperate over shared species and development assistance affecting wetlands.

Early in the Ramsar process, the emphasis was on the protection of migratory fauna, particularly waterfowl. The importance of wetlands for many other biological functions have more recently been recognized, and currently eight criteria are used to evaluate potential wetland sites for formal designation as "wetlands of international importance" (Table 18-2). Group A sites must meet criterion 1 that they contain representative, rare, or unique wetland types. Group B sites, internationally important for conserving biological diversity, are judged on seven criteria involving questions of rare and endangered communities, biodiversity, habitat for waterfowl, or habitat or food source for indigenous fish species.

Currently, 117 countries have joined the Ramsar Convention. In 1993, there were 582 Ramsar wetland sites, comprising almost 37 million ha in the world. By early 2000, these numbers had almost doubled to 1,021 Ramsar wetland sites, totaling 74.8 million ha worldwide (Fig. 18-3 and Table 18-3). At that time, there were 17

Figure 18-3 Location of wetlands in the world identified as having international importance by the Ramsar Convention by 2000. (*Ramsar Convention Bureau, Gland, Switzerland, and Wetlands International, Wageningen, The Netherlands*)

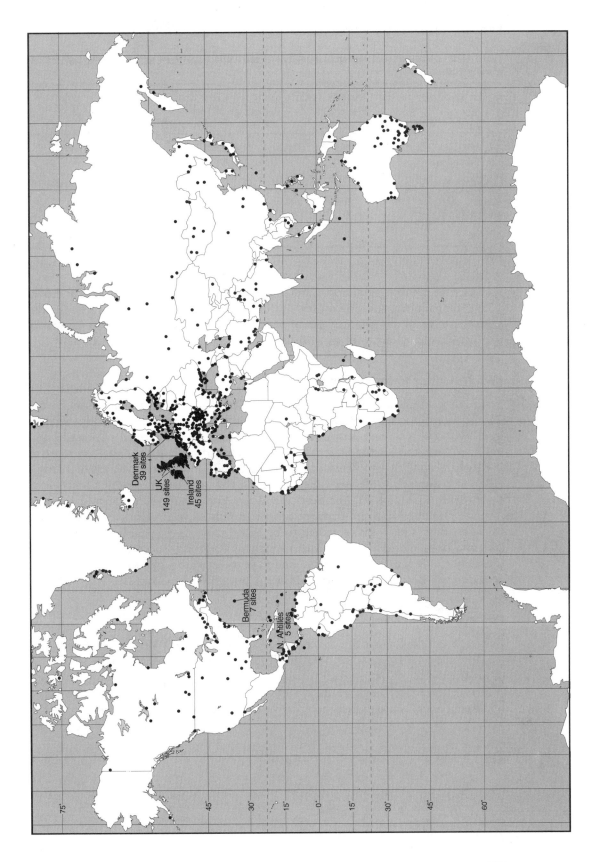

Denmark
39 sites

UK
149 sites

Ireland
45 sites

Bermuda
7 sites

N. Antilles
5 sites

Table 18-3 Distribution of wetlands in regions of the world subscribed to the Ramsar Convention in early 2000

Region of the World	Area of Wetlands (× 10⁶ ha)	Percentage of Total Ramsar Site Area
Europe	19.2	25.7
North America	15.3	20.5
Africa	14.5	19.4
Neotropics	14.3	19.1
Oceania	5.9	7.9
Asia	5.6	10.4
Total	74.8	100.0

Ramsar sites comprising almost 2 million ha in the United States, 36 sites and over 13 million ha in Canada, and 6 sites in Mexico. By contrast, the UK has 149 Ramsar sites, the most of any country. Europe and North America make up almost half of the Ramsar sites in terms of area (Table 18-3).

North American Waterfowl Management Plan

The United States and Canada, partially as a result of collaboration begun by the Ramsar Convention, established the *North American Waterfowl Management Plan* in 1986 to conserve and restore about 2.4 million ha of waterfowl wetland habitat in Canada and the United States. This treaty was formulated as a partial response to the steep decline in waterfowl in Canada and the United States that had become apparent in the early 1980s (see Chapter 16). This bilateral treaty is jointly administered by the U.S. Fish and Wildlife Service and the Canadian Wildlife Service but also involves public and private participation by groups such as Ducks Unlimited. The total cost of this plan was estimated to be $1.5 billion, to be paid by the two countries and private organizations. Major emphasis has been placed on sites that cross international borders, including the prairie pothole region, the lower Great Lakes–St. Lawrence River basin, and the Middle–Upper Atlantic Coastline.

Recommended Reading

Frazier, S. 1999. Ramsar Site Overview: A Synopsis of the World's Wetlands of International Importance. Wetland International, Wageningen, The Netherlands. 48 pp.

National Research Council (NRC). 1995. Wetlands: Characteristics and Boundaries. National Academy Press, Washington, DC. 306 pp.

Wetland Creation and Restoration

Loss rates of wetlands around the world and the subsequent recognition of wetland values have stimulated restoration and creation of these systems. Policies such as "no net loss" of wetlands in the United States have made wetland creation and restoration a veritable industry in that country. Wetland restoration involves returning a wetland to its original or previous wetland state, whereas wetland creation involves conversion of uplands or shallow open-water systems to vegetated wetlands. Wetland restoration and creation can occur for replacement of habitat, for coastal restoration, and for restoration of mined peatlands. Generally, wetland restoration and creation first involve establishment or reestablishment of appropriate natural hydrologic conditions, followed by establishment of appropriate vegetation communities. Although many of these created and restored wetlands have become functional, there have been some cases of "failure" of constructed or restored wetlands generally caused by a lack of proper hydrology. Creating and restoring wetlands should be based on the concept of self-design whereby the ecosystem adapts and changes according to its physical constraints. Giving these systems sufficient time to carry out their self-design is another factor that is generally overlooked. Wetland mitigation banks may be an approach that will overcome many of the limitations of current approaches to replacing lost wetlands.

Knowledge of the principles and techniques for wetland creation and restoration is one of the qualifications required of modern landscape managers and wetland scientists. The creation of wetlands in previously dry and/or nonvegetated areas and the restoration of wetlands where they once were found are exciting opportunities for reversing the trend of decreasing wetland resources and for providing aesthetic and functional units to the landscape. Wetland creation and restoration can range from the relatively simple building of farmland freshwater marshes by plugging existing drainage systems to the construction of more extensive wetlands for coastal protection or estuarine enhancement. Two basic principles should be the starting point for anyone involved in wetland restoration and creation:

1. Understanding wetland ecology and its principles (e.g., hydrology, biogeochemistry, adaptations, succession; see Chapters 5–8) is essential to successfully create and restore wetlands as part of a natural landscape.

2. In creating and restoring wetlands, managers must resist the ever-present temptation to overengineer by attempting either to channel natural energies that cannot be channeled or to introduce species that the landscape or climate do not support.

In all situations of wetland creation and restoration, human contribution to the design of wetlands should be kept simple and should strive to stay within the bounds established by the natural landscape, without reliance on dams, dikes, weirs, pumps, and other technological approaches that invite failure. Boule (1988) states that "simple systems tend to be self-regulating and self-maintaining."

A growing literature reviews creating and restoring wetlands. Starting with some early publications on the subject in the early 1990s (Kusler and Kentula, 1990; Marble, 1992; Kentula et al., 1992a; National Research Council, 1992), there are now several reviews on the creation and restoration of specific types of wetlands. These include special issues and review papers on tidal wetlands (Broome et al., 1988; Zedler, 1988; Broome, 1990; Shisler, 1990; Thayer, 1992; Zedler, 1996; Weinstein and Kreeger, 2000), mangroves (Lewis, 1990b, c), freshwater marshes (Erwin, 1990; Galatowitsch and van der Valk, 1994; Hammer, 1997; Mitsch and Bouchard, 1998), and forested wetlands (Clewell and Lea, 1990). Mitsch et al. (1998) discuss the general concept of self-design as a fundamental principle for the creation and restoration of wetlands, while Middleton (1999) has written a book that describes wetland restoration in the context of the importance of flood pulses and disturbance ecology.

Definitions

Several terms are frequently used in connection with the creation and restoration of wetlands. Lewis (1990a) points out that precise definitions are important and confusion about the exact meaning of wetland creation, restoration, and related terms is common. Bradshaw (1996) concurs that "we must be clear in what is being discussed" in ecological restoration in general. *Wetland restoration* refers to the return of a wetland from a disturbed or altered condition by human activity to a previously existing condition. The wetland may have been degraded or hydrologically altered and restoration then may involve reestablishing hydrologic conditions to reestablish previous vegetation communities. *Wetland creation* refers to the conversion of a persistent upland or shallow-water area into a wetland by human activity. *Wetland enhancement* refers to a human activity that increases one or more functions of an existing wetland. One type of created wetland, a *constructed wetland,* refers to a wetland that has been developed on a former upland environment to create poorly drained soils and wetland flora and fauna for the primary purpose of contaminant or pollution removal from wastewater or runoff (Hammer, 1997). This kind of wetland is also referred to as a *treatment wetland.* Constructed and treatment wetlands are discussed in Chapter 20. An additional term is *artificial wetland,* which is generally synonymous with constructed wetland. That term is no longer frequently used.

Reasons for Creating and Restoring Wetlands

Although the net trend around the world has been toward the destruction of wetlands (see Chapters 3 and 4), significant efforts are now being focused on the voluntary restoration and creation of these habitats. Part of the interest in wetland creation and restoration stems from the fact that we have lost a major part of our wetland resources and their societal values (see Chapter 16). Other interest is less voluntary and in response to government policies such as "no net loss" that require the replacement of wetlands for those unavoidably lost (see Chapter 18). There has also been great interest in constructing wetlands for habitat restoration or replacement, water quality enhancement, and coastal protection and restoration. New Zealand, which has lost 90 percent of its wetlands, has begun to restore marshes and other wetlands in the Waikato River basin on North Island and in the vicinity of Christchurch on South Island. In southeastern Australia, restoration of the Murray–Darling watersheds, particularly the riverine billabongs, has become a major undertaking, while coastal plain wetland restoration and creation are occurring in southwestern Australia. There are concerted efforts to restore mangrove forests in the Mekong Delta of Vietnam and along South American coastlines where shrimp farming has destroyed thousands of hectares of mangroves. Tidal marshes have been created along much of China's eastern coastline. In the United States, laws dictate that wetland loss must be mitigated through wetland creation and restoration. Now wetland restoration and creation are being proposed or implemented on very large scales to prevent more deterioration of existing wetlands (Everglades in Florida), to mitigate the loss of fisheries (Delaware Bay on East Coast), to reduce wetland loss (Mississippi Delta in Louisiana), and to solve serious cases of overenrichment of coastal waters (Baltic Sea in Scandinavia; Gulf of Mexico on southern coast of the United States).

Mitigating Wetland Habitat Loss

Enforcement of wetland and natural resource protection laws in the United States and elsewhere has led to the common practice of requiring that wetlands be created, restored, or enhanced to replace wetlands lost in development such as highway construction, coastal drainage and filling, or commercial development. These constructed wetlands, often called *mitigation wetlands* (Fig. 19-1), are built with the intent of replacing the wetland "function" lost by the development, usually in the same or an adjacent watershed. (To mitigate means to make less harsh or harmful. The term "mitigation wetland" or "wetland mitigation" is therefore poor use of English. We should rather refer to "mitigating the loss of a wetland." Perhaps a more appropriate term should be *replacement wetland*.) Replacement wetlands are designed to be at least the same size as the lost wetlands and often a *mitigation ratio* is applied so that more wetlands are created and/or restored than are lost. For example, a mitigation ratio of 2 : 1 means that 2 ha of wetlands will be restored or created for every hectare of wetland lost to development. Calculation of mitigation ratios for factors other than area have been attempted (see, e.g., the example in Chapter 16; Table 16-6), but this approach has not been widely accepted. There is considerable controversy as to whether wetland loss can be mitigated successfully or if it is essentially impossible (Roberts, 1993; Malakoff, 1998).

On paper, the U.S. Army Corps of Engineers' implementation of "no net loss" appears to be working in the United States (Fig. 19-2). An estimated 50,000 ha of wetland and associated uplands in the United States were gained from October 1993 through September 1999 due to enforcement of Section 404 of the Clean Water Act through wetland mitigation. This number is the result of the issuance of permits for the destruction of 68,900 ha of wetlands and the creation, restoration, enhancement, or preservation of 118,800 ha of wetlands and associated uplands. However, there

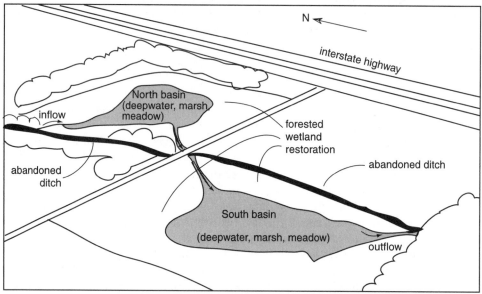

Figure 19-1 On-site mitigation for small wetlands in urban or suburban areas is common in the United States. This 6.1-ha wetland was constructed in Columbus, Ohio, in 1992 to mitigate the loss of about 3 ha of degraded wetland nearby. The photo is from 1996, four years after wetland restoration. The wetland was created by redirecting a channelized ditch shown in the lower sketch to adjacent excavated basins that had hydric soils. This wetland is generally considered to be one of the more successful freshwater marsh mitigation projects in this region. Long-term goals call for the development of forested wetlands surrounding the marshes; small planted samplings can be seen as small dots in the photo around both basins. Details of this wetland are presented in Niswander and Mitsch (1995).

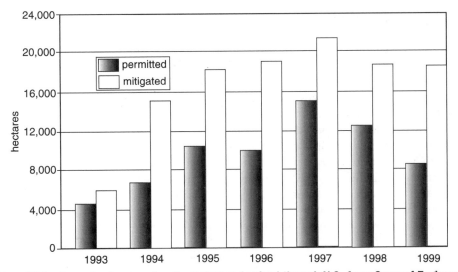

Figure 19-2 Approximate area of wetlands lost and gained through U.S. Army Corps of Engineers' Section 404 dredge-and-fill permit program, fiscal (October–September) years 1993–1999 in the United States. "Permitted" refers to approximate area of wetlands lost through the permit program. "Mitigated" refers to area of restored, created, enhanced, and preserved wetlands and associated uplands required for the permit (*J. Studt, U.S. Army Corps of Engineers, Washington, DC, January 19, 2000*)

are two reasons why one should not be so euphoric about this "net gain" of wetlands. First, it is impossible to tell from these general numbers just how "successful" this wetland trading has been as few statistics exist on what functions were lost versus what functions were gained. (Measuring success of replacement wetlands is discussed further at the end of this chapter.) Second, the estimated gain of 50,000 ha over 7 years does not make much of an impact on the loss of 47,300,000 ha of wetlands that occurred from presettlement time to the 1980s in the conterminous United States (see Table 4-6). At this rate of net gain, it would take over 600 years to recover 10 percent of our lost wetland resources. By way of contrast, the National Research Council (1992) called for the restoration of almost 10 percent of our lost wetlands (4,000,000 ha) by the year 2010.

Agricultural Land Restoration

For the past several decades in the United States, farm pond creation has been encouraged as a way of providing drinking water for domestic animals and other functions on the farm. Although individually quite small (about 0.2 ha), the total number of constructed ponds is large. Several years ago, ponds were being constructed at a rate of about 50,000 ponds per year. Marshes have developed around the perimeter of many of these ponds while others have converted to marshes. Many of these ponds were built with large, shallow areas to attract waterfowl, and these shallow zones have become typical pothole marshes. Dahl and Johnson (1991) determined that between the mid-1970s and the mid-1980s wetland pond areas increased by 320,000 ha in the United States. Most of this gain (170,000 ha) resulted from building ponds on nonagricultural uplands, but 91,000 ha were constructed farm ponds.

Conservation programs are now in place to encourage individual farmers in the United States to restore wetlands on their land. Both the Conservation Reserve Program (CRP) and the Wetlands Reserve Program (WRP) under the U.S. Department of Agriculture (USDA) have led to significant areas of wetlands being restored or protected. The CRP guidelines, announced in 1997, give increased emphasis to the enrollment and restoration of *cropped wetlands,* that is, wetlands that produce crops but serve wetland functions when crops are not being grown. The CRP also encourages wetland restoration, particularly through hydrologic restoration. In the CRP, participants voluntarily enter into contracts with the USDA to enroll erodible and other environmentally sensitive land in long-term contracts for 10 to 15 years. In exchange, participants receive annual rental payments and a payment of up to 50 percent of the cost of establishing conservation practices. The U.S. Fish and Wildlife Service estimated that as a result of the CRP about 36,000 ha of wetlands were restored in the United States from 1987 to 1990 (Josephson, 1992). The north-central office of the U.S. Fish and Wildlife service estimated that approximately 28,000 ha in the upper midwestern United States have been restored through 1997 (K. Kroonmeyer, personal communication, 1998), some the result of the CRP.

The Wetlands Reserve Program (WRP) is a newer voluntary program offering landowners the opportunity to protect, restore, and enhance wetlands on their property. The USDA Natural Resources Conservation Service (NRCS) provides technical and financial support to help landowners. The WRP options to protect, restore, and enhance wetlands and associated uplands include permanent easements, 30-year easements, or 10-year restoration cost-share agreements. As of mid-1999, approximately 318,000 ha of wetlands and adjacent uplands have been enrolled in the WRP in the United States, with the most intense activity in the lower Mississippi River basin, Iowa, and Florida (Fig. 19-3).

Forested and Riparian Wetland Restoration

There is less experience at forested wetland restoration and creation compared to herbaceous marshes, despite that fact that these wetlands have been lost at alarming rates, particularly in the southeastern United States (see Chapter 4). Forested wetland creation and restoration are different from marsh creation and restoration because forest regeneration takes decades rather than years to complete and there is more uncertainty about the results. For example, the aerial photograph in Figure 19-1 shows a wetland mitigation site 4 years after construction; one of the ultimate goals of that project was replacement of a forested wetland. While the marshes shown in the picture are well developed after 4 years, the area that is to be forested shows only tiny dots of small seedlings but hardly compares to the forest canopy of adjacent uplands. Furthermore, computer simulations suggested that the basal area of this created forest will approximate that of natural wet forests after 50 years (Niswander and Mitsch, 1995).

Most forested wetland restoration is carried out on sites where hydrology and soils are largely intact, and primarily involves planting appropriate vegetation (Clewell and Lea, 1990; Clewell, 1999). Forested wetland creation that involves the engineering of an entire wetland setting has been attempted in the phosphate mining region of central Florida (M. T. Brown et al., 1992; Clewell, 1999). One of the largest forested wetland creation projects of this type consisted of building 61 ha of marsh and forested

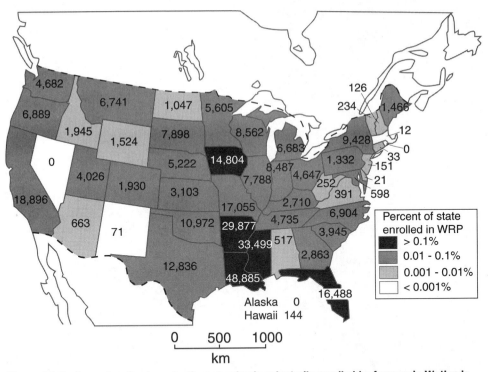

Figure 19-3 Area of wetlands and adjacent uplands voluntarily enrolled by farmers in Wetlands Reserve Program in the conterminous United States as of September 1999. Numbers are in hectares for each state; shading represents percentage of state area enrolled in the WRP. National total is 785,000 ha. (*U.S. Department of Agriculture, Natural Resources Conservation Service, November 1999*)

wetlands at a phosphate mine reclamation site. Fifty-five thousand trees, representing 12 wetland species, were planted (Erwin et al., 1984). The survival rate of the seedlings for the first year was 77 percent. In one of the longest documented histories of a forested wetland restoration, Clewell (1999) describes a 1.5-ha forested wetland planted in Florida in 1985 after extensive hydrologic and soil restoration of the phosphate mine site (Fig. 19-4). The state of Florida specified that this wetland have a minimum density of 988 trees/ha; 11 years after planting, there were 697 trees/ha over 10 cm in diameter, with domination by planted cypress *Taxodium* spp. and volunteer willows *Salix caroliniana*. By comparison, 10 reference forested wetlands in the region had a density of 469 trees/ha, considerably fewer than the stated goal and less than the density of the restored wetland. The basal area of the restored wetland after 11 years was 8.3 m^2/ha, about 38 percent of that of the reference forests.

Much riparian forest restoration in the United States has centered on the lower Mississippi River alluvial valley, where more than 78,000 ha were reforested by federal agencies over the 10-year period 1988–1997 (King and Keeland, 1999; Fig. 19-5), primarily with bottomland hardwood species (see Chapter 15) and, to a lesser extent, deepwater swamp species (see Chapter 14). This is a small contribution to the restora-

Figure 19-4 Cypress–bottomland hardwood forest restoration, including undergrowth, along a replacement headwater stream on phosphate-mined land in Florida that was restored 12 years prior to this photograph. The restoration was designed and installed by A. F. Clewell, Inc., for American Cyanamid Company, Hillsborough County, Florida. (*Photograph by A. F. Clewell, reprinted by permission*)

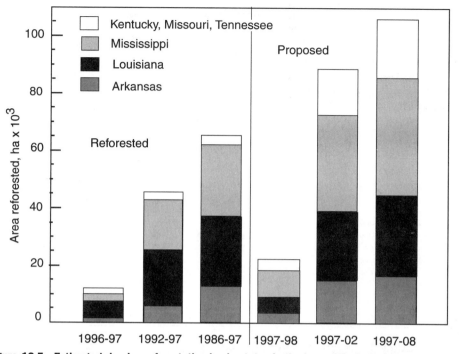

Figure 19-5   **Estimated riparian reforestation in six states in the lower Mississippi River valley from 1986–1997 and the amount of land projected for reforestation through 2008. Proposed numbers are considered underestimates as only 10 of 27 respondents provided estimates for proposed reforestation. Also, an estimated 12,100 ha of Conservation Reserve Program reforestation in Louisiana is not included in the illustration. (*After King and Keeland, 1999*)**

tion of this alluvial floodplain where 7,200,000 ha of bottomland hardwood forest have been lost (Hefner and Brown, 1985).

Hydrologic and Water Quality Restoration

Details of wetlands that are purposefully created to improve water quality from wastewater and nonpoint source pollution are described in the next chapter. However, lines are starting to blur between wetlands created and restored for habitat restoration and those restored for water quality improvement or flood control. In fact, most wetlands that are restored or created are done so for both habitat and water quality values. For example, a large-scale wetland and riparian forest restoration and creation, on the order of millions of hectares, is being proposed to help solve a major coastal pollution problem in the Gulf of Mexico (Mitsch et al., 1999; see Case Study 2 later in this chapter). Restoration done for water quality improvement would also have major advantages of habitat restoration and flood mitigation in addition to water quality improvement. Hey and Philippi (1995) estimated that restoration of 3 percent of the upper Mississippi River basin in wetlands (5.3 million ha) would have ameliorated the disastrous 1993 flood in that basin, which that resulted in billions of dollars of damage (Table 19-1).

Table 19-1 Indicators of early wetland and beaver pond area and restored wetland flood storage potential in the upper Mississippi and Missouri River basins

	Water Surface Area (× 10⁶ ha)	Percentage of Watershed[a]
Indicators of original wetlands in basin		
Hydric soils	17	9
Presettlement wetlands, 1780	18	10
Beaver ponds, 1600	21	11
Wetlands		
Existing, 1980	7.7	4
Restorable[b]	5.3	3
Total	13.0	7

[a] Based on watershed of 185 million ha above Thebes, Illinois.
[b] Estimated wetland area that would have stored the 1993 Upper Mississippi River flood to a depth of 1 m.
Source: Hey and Philippi (1995).

Case Study 1: Everglades Restoration

The restoration of the Florida Everglades, the largest wetland area in the United States, is actually several separate initiatives being carried out in the 4.6-million-ha Kissimmee–Okeechobee–Everglades (KOE) region in the southern third of Florida (see Fig. 4-12b). Overall, the Everglades restoration, as now planned by the U.S. Army Corps of Engineers, will cost almost $8 billion and will be carried out over 20 years or more. Problems in the Everglades have developed because of (1) excessive nutrient loading to Lake Okeechobee (Reddy and Flaig, 1995) and to the Everglades itself, primarily from agricultural runoff (Wu et al., 1997); (2) loss and fragmentation of habitat caused by urban and agricultural development; (3) spread of *Typha* and other invasives and exotics to the Everglades, replacing native vegetation (see Figure 17-9); and (4) hydrologic alteration due to an extensive canal and straightened river system built by the U.S. Army Corps of Engineers and others for flood protection, and maintained by several water management districts.

One major restoration project in the KOE region that has received a lot of attention is the restoration of the Kissimmee River (Dahm, 1995). As a result of the channelization of the river in the 1960s, a 166-km-long river was transformed into a 90-km-long, 100-m-wide canal, and the extent of wetlands along the river decreased by 65 percent (Table 19-2; Toth et al., 1995, 1998). The restoration of the Kissimmee River will be a major undertaking to reintroduce the sinuosity to the artificially straightened river. The river restoration work, expected to be completed in stages over the next several decades, will return some portion of lost wetland habitat to the riparian zone and will also provide sinks for nutrients that are otherwise causing increased eutrophication in downstream Lake Okeechobee.

Everglades restoration also involves halting the spread of high-nutrient cattail (*Typha domingensis*) throughout the low-nutrient sawgrass (*Cladium*

Table 19-2 Wetland changes due to channelization of the Kissimmee River in south Florida. Channelization took place between 1962 and 1971 and transformed a 166-km meandering river into a 90-km-long, 10-m-deep, 100-m-wide canal

Wetland Type	Prechannelization (ha)	Postchannelization (ha)	Percentage Change
Marsh	8,892	1,238	−86
Wet prairie	4,126	2,128	−48
Scrub–shrub wetland	2,068	1,003	−51
Forested wetland	150	243	+62
Other	533	919	+72
Total	15,769	5,531	−65

Source: Toth et al. (1995).

jamaicense) communities that presently dominate the Everglades (see Chapter 17). Since the main causes of the spread of cattails are nutrients and especially phosphorus emanating from agricultural areas in the basin, 16,000 ha of created wetlands, called Stormwater Treatment Areas (STAs), are planned for phosphorus control from the agricultural area. A prototype of the STAs, a 1,500-ha site called the Everglades Nutrient Removal (ENR) Project, has been in operation since mid-1994. Early results from this test site of the concept of creating and restoring wetlands to protect downstream wetlands from nutrient enrichment are encouraging. Experimental wetland basins are consistently decreasing phosphorus to levels below 10 μg P/L (Moustafa, 1999).

Case Study 2: Solving the Gulf of Mexico Hypoxia

A hypoxic zone has developed off the shore of Louisiana in the Gulf of Mexico where hypolimnetic waters with dissolved oxygen less than 2 mg/L O_2 now extend over an area of 1.6 to 2.0 million ha (Fig. 19-6; Rabalais et al., 1996, 1998, personal communication). Nitrogen, particularly nitrate nitrogen, is the most probable cause; 80 percent of the nitrogen input is from the 3-million-km^2 Mississippi River basin, which in turn comprises 41 percent of the coterminous United States. The control of this hypoxia is important in the Gulf of Mexico because the continental shelf fishery in the Gulf is approximately 25 percent of the U.S. total.

A number of approaches are being considered for controlling nitrogen flow into the Gulf; many of them involve large-scale modifications of land use practices in the Midwest. Among the options are modifying agricultural practices (e.g., reduced fertilizer use or alternate cropping techniques), tertiary treatment (biological, chemical, physical) of point sources, landscape restoration (e.g., riparian buffers and wetland creation) to control nonpoint source pollution from farmland, stream and delta restoration, and atmospheric controls of NO_x. The approach that appears to have the highest probability of success with a minimum impact on farming in the midwestern United States is landscape restoration. It has been suggested that 2 million ha of restored

(*continued on next page*)

Figure 19-6 Location of Gulf of Mexico hypoxia (low-dissolved-oxygen zone) and entire Mississippi River watershed that contributes to the problem. Restoration and creation of almost 10 million ha of wetlands and riparian zones have been proposed as a solution to this coastal pollution. (*Mitsch et al., 1999; Mitsch, 1999*)

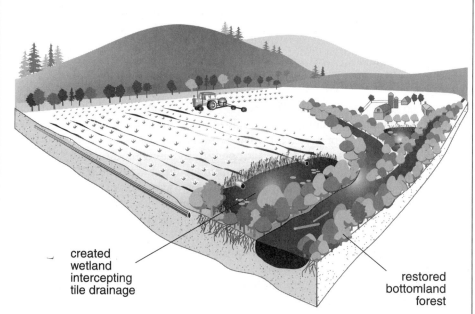

Figure 19-7 Wetland restoration in an agricultural setting for water quality improvement, particularly for nitrate nitrogen, should involve both wetlands and riparian buffers that intercept subsurface drainage.

and created wetlands and about 7.7 million ha of restored riparian buffers would be necessary to provide enough denitrification to substantially reduce the nitrogen entering the Gulf of Mexico (Fig. 19-7; Mitsch, 1999; Mitsch et al., 1999). Restoring the Gulf of Mexico requires restoring 3 percent of the Mississippi River basin. Interestingly, Hey and Philippi (1995) found that a similar wetland restoration effort would be required in the upper Mississippi River basin to mitigate the effects of very large and costly floods such as the one that occurred in the summer of 1993 (Table 19-1).

Coastal Protection and Restoration

A smaller source of new wetlands has been the creation and restoration of coastal marshes and mangroves, mostly in areas where considerable dredging occurs along navigable waterways. The U.S. Army Corps of Engineers annually dredges about 275 million m³ of material from the nation's waterways to keep them open. During the 1970s, this agency initiated a large study to determine the feasibility of creating marshes out of the spoil, thus turning a liability into an asset. In addition, there has recently been much interest in creating wetlands along the Gulf Coast by diverting the sediment-rich water of the Mississippi River into shallow bays, where the sediments settle out to intertidal elevations. Both processes appear to be feasible. Some of the early pioneering work on salt marsh restoration was done in Europe (Lambert, 1964; Ranwell, 1967) and China (Chung, 1982, 1989). Much of the early work on reclaiming coastal marshes in the United States was carried out on the North Carolina coastline (Woodhouse, 1979; Broome et al., 1988), in the Chesapeake Bay area (Garbisch et al., 1975; Garbisch, 1977), in the salt marshes and mangroves of Florida and Puerto Rico (Lewis, 1990a, b), and along the California coastline (Zedler, 1988; Josselyn et al., 1990). Much of this coastal wetland restoration has been undertaken as mitigation for coastal development projects (Zedler, 1988, 1996).

Case Study 3: Delaware Estuary Enhancement

A large coastal wetland restoration project in the eastern United States involves the restoration, enhancement, and preservation of 5,800 ha of coastal salt marshes on Delaware Bay in New Jersey and Delaware (Fig. 19-8). This estuary enhancement, being carried out by Public Service Electric and Gas (PSE&G) with advice from a team of scientists and consultants, was undertaken as mitigation for the potential impacts of once-through cooling from a nuclear power plant operated by PSE&G on the bay. The reasoning was that the impact of once-through cooling on finfish, through entrainment and impingement, could be offset by increased fisheries production from restored salt marshes. Because of the uncertainties involved in this kind of ecological trading, the area of restoration was estimated as the salt marshes that would be necessary to compensate for the impacts of the power plant on finfish, times a safety factor of 4. There are three distinct approaches being utilized in this project to restore the Delaware Bay coastline:

1. *Reintroduce Flooding:* The most important type of restoration involves the reintroduction of tidal inundation to about 1,800 ha of

(continued on next page)

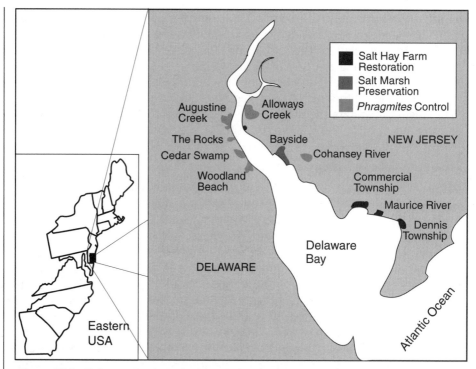

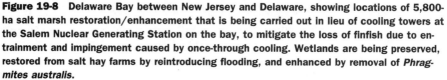

Figure 19-8   Delaware Bay between New Jersey and Delaware, showing locations of 5,800-ha salt marsh restoration/enhancement that is being carried out in lieu of cooling towers at the Salem Nuclear Generating Station on the bay, to mitigate the loss of finfish due to entrainment and impingement caused by once-through cooling. Wetlands are being preserved, restored from salt hay farms by reintroducing flooding, and enhanced by removal of *Phragmites australis*.

diked salt hay farms. Many marshes along Delaware Bay have been isolated from the bay by dikes, sometimes for centuries, and put into the commercial production of "salt hay" (*Spartina patens*). Restoration is being accomplished by excavating breaches in the dikes and, in most cases, connecting these new inlets to a system of recreated tidal creeks and existing canal systems.

2. *Reexcavate Tidal Marshes:* Additional restoration involves enhancing drainage by reexcavating higher order tidal creeks in salt marshes, thereby increasing tidal circulation in existing marshes. This is particularly important in marshes that were formerly diked, as the isolation from the sea has led to the filling in of former tidal creeks.

3. *Reduce* Phragmites *Domination:* In yet another set of restoration projects in Delaware and New Jersey, restoration involves the reduction in cover of the aggressive and invasive *Phragmites australis* in 2,100 ha of nonimpounded coastal wetlands. Alternatives that are being investigated include hydrological modifications such as channel excavation, breaching remnant dikes, and microtopographic changes, mowing, planting, and herbicide

application. The most common method that is currently being used is herbicide spraying followed by controlled burning and selected hydromodification (Weinstein et al., 1997).

Typical goals for the marsh restoration along Delaware Bay include a high percentage cover of desirable vegetation such as *Spartina alterniflora,* a relatively low percentage of open water, and the absence of the invasive reed grass *Phragmites australis.* Success of this coastal restoration project, subject to a combination of legal, hydrologic, as well as ecological constraints, is being estimated through comparison of restored sites to natural reference marshes.

Early results of this project are generally encouraging. In those marshes where tidal exchange was restored to formerly diked salt hay farms (the Dennis Township, Maurice River, and Commercial Township sites in Fig. 19-8), the reestablishment of *Spartina alterniflora* and other favorable vegetation has been rapid and extensive. At Dennis Township, approximately 64 percent of the site was already dominated by *Spartina alterniflora* after only two growing seasons. Tidal restoration was completed at the Maurice River site in early 1998 and major revegetation by *Spartina alterniflora* and some *Salicornia* has already occurred. At the third and largest salt hay farm restoration site, Commercial Township, revegetation is occurring rapidly from the bayside at this tidally restored site.

For the restoration of 1,600 ha of marshes being carried out through spraying and burning of *Phragmites* at the Cohansey River and Alloways Creek sites in New Jersey (Fig. 19-8), results are not as promising as those at the hydrologic restoration sites. Burning followed by spraying has not eliminated *Phragmites* at any given site, so repeated treatment is necessary. On these sites, alternative methods to spraying are being investigated for the control of *Phragmites,* including mowing and rhizome cutting, grading of dikes where *Phragmites* tends to dominate, selective planting, and microtopographic modifications of the marsh surface.

Case Study 4: Louisiana Delta Restoration

Louisiana is one of the most wetland-rich regions of the world with 36,000 km^2 of marshes, swamps, and shallow lakes. Yet Louisiana is suffering a rate of coastal wetland loss of 6,600 to 10,000 ha/yr due to natural (land subsidence) and human causes such as river levee construction, oil and gas exploration, urban development, sediment diversion, and possibly climate change (see Chapter 9). The passage by the U.S. Congress of the Coastal Wetland Planning, Protection and Restoration Act (CWPPRA) of 1990 has initiated comprehensive planning in Louisiana aimed at the protection, restoration, and creation of millions of hectares of coastal wetlands. The plans call for diverting the water and sediment of the Mississippi River to build new deltas and smaller *splays* to mimic spring floods and restore subsiding marshes; restoring barrier islands; and instituting measures to protect many smaller wetlands with dikes, plantings, and disposing of dredge spoil (Turner and Boyer, 1997). Several wetland enhancement and creation projects have simply created *crevasses* in the

(continued on next page)

natural levee of the lower Mississippi River. These crevasses allow river water and sediment to flow into shallow estuaries and create a crevasse splay or mini-delta that rapidly becomes a naturally vegetated marsh. Their extent and life span can seldom be predicted and they function in a natural manner because they mimic the natural geomorphic processes of the river.

The CWPPRA led to the Louisiana Coastal Wetlands Conservation and Restoration (LCWCR) Plan, which was completed in 1993, and the subsequent Coast 2050 plan in 1997. About $40 million is being spent annually on individual projects, with a total estimated price tag of about $14 billion. It has been estimated that if the restoration is not completed the cost in public use value over the next 50 years will be $37 billion (LCWCR Task Force, 1998).

Peatland Restoration

Peatland restoration has been a relatively new type of wetland restoration. Early attempts with peatlands occurred in Europe, specifically in Finland (Salonen, 1987, 1990), Germany (Eggelsmann and Schwaar, 1979), the United Kingdom (Meade, 1992), and The Netherlands (Schouwenaars, 1993). Increased peat mining in Canada and elsewhere has led to increased interest in understanding if and how mined peatlands can be restored. When peat surface mines are abandoned without restoration, the area rarely returns through secondary succession to the original moss-dominated system (Quinty and Rochefort, 1997). There is promise that restoration can be successful (Lavoie and Rochefort, 1996; Wind-Mulder et al., 1996; Quinty and Rochefort, 1997; Price et al., 1998), but because (1) surface mining causes major changes in local hydrology and (2) peat accumulates at an exceedingly slow rate, restoration progress will be measured in decades rather than years. In the 1960s and 1970s, block harvesting of peat was replaced by vacuum harvesting in southern Quebec and in New Brunswick, necessitating the development of different restoration techniques. While traditional block cutting of peat left a variable landscape of high ground and trenches, vacuum harvesting leaves relatively flat surfaces bordered by drainage ditches. Abandoned block-cut sites appear to revegetate with peatland species more easily than do vacuum-harvested sites and the latter can remain bare for a decade or more after mining (Rochefort and Campeau, 1997).

How to Create and Restore Wetlands

General Principles

Some general principles of ecotechnology that apply to the creation and restoration of wetlands include the following:

1. Design the system for minimum maintenance. The system of plants, animals, microbes, substrate, and water flows should be developed for self-maintenance and self-design.
2. Design a system that utilizes natural energies, such as the potential energy of streams, as natural subsidies to the system. Flooding rivers and tidal circulation transport great quantities of water and nutrients in relatively short time periods, subsidizing wetlands open to these flows.

3. Design the system with the hydrologic and ecological landscape and climate. Floods, droughts, muskrats, geese, and storms are expected disturbances and should not be feared. Natural ecosystems generally recover rapidly from natural disturbances to which they are adapted.

4. Design the system to fulfill multiple goals, but identify at least one major objective and several secondary objectives. If a wetland is being created or restored to replace a lost wetland, replacement of function should be an important consideration.

5. Design the system as an ecotone. This may require a buffer strip around the wetland site, but it also means that the wetland site itself will be a buffer system between upland and aquatic systems.

6. Give the system time. Wetlands do not become functional overnight. Several years may pass before plant establishment, nutrient retention, and wildlife enhancement can become optimal, and mature soils systems may take decades. Strategies that try to short-circuit ecological succession or overmanage it are doomed to failure.

7. Design the system for function, not form. If initial plantings and animal introductions fail but the overall function of the wetland, based on the fulfillment of initial objectives, is being carried out, then the wetland has not failed. The outbreak of plant diseases and the invasion of alien species are often symptomatic of other stresses and may indicate false expectations rather than ecosystem failure.

8. Do not overengineer wetland design with rectangular basins, rigid structures and channels, and regular morphology. Natural systems should be mimicked to accommodate biological systems.

Defining Goals

The design of an appropriate wetland or series of wetlands, whether for the control of nonpoint source pollution, for a wildlife habitat, or for wastewater treatment, should start with the formulation of the overall objectives of the wetland. One view is that wetlands should be designed to maximize ecosystem longevity and efficiency and minimize cost. The most important aspect of designing a wetland is to define the goal of the wetland project. Among the possible goals for wetland construction are the following:

1. Flood control
2. Wastewater treatment (e.g., domestic wastewater or acid mine drainage)
3. Stormwater or nonpoint source pollution control
4. Ambient water quality improvement (e.g., instream system)
5. Coastal restoration (including storm surge protection)
6. Wildlife enhancement
7. Fisheries enhancement
8. Replacement of similar habitat (wetland loss mitigation)
9. Research wetland

The goal, or a series of goals, should be determined before a specific site is chosen or a wetland is designed. If several goals are identified, one must be chosen as primary.

Placing Wetlands in a Riverine Landscape

In some cases, particularly when sites are being chosen for habitat replacement, many sites are available in which to locate a restored or created wetland. The restoration of entire rivers has been shown to be an elusive goal in many parts of the world because of years of drainage and channel "improvement," floodplain development, and significant loads of sediments and other nonpoint pollutants. The natural design for a riparian wetland fed primarily by a flooding stream or river (Fig. 19-9a) allows for flood events of a river to deposit sediments and chemicals on a seasonal basis in the wetland, and for excess water to drain back to the stream or river. Because there are natural and also often constructed levees along major sections of streams, it is often possible to create such a wetland with minimal construction. The wetland could be designed to capture flooding water and sediments and slowly release the water back to the river after the flood passes, or to receive flooding water and retain it through the use of flapgates.

Wetlands can be designed as instream systems by adding control structures to the streams themselves or by impounding a distributary of the stream (Fig. 19-9b). Blocking an entire stream is a reasonable alternative only in headwater streams, and it is not generally cost effective or ecologically advisable. This design is particularly vulnerable during flooding and its stability might be unpredictable, but it has the advantage of potentially "treating" a significant portion of the water that passes that point in the stream. The maintenance of the control structure and the distributary might mean making significant management commitments to this design.

A riparian wetland fed by a pump (Fig. 19-9c) creates the most predictable hydrologic conditions for the wetland but at an obvious extensive cost for equipment and maintenance. If it is anticipated that the primary objective of a constructed wetland is the development of a research program to determine design parameters for future wetland construction in the basin, then wetlands fed by pumps is a good design. If other objectives are more important, then the use of large pumps is usually not appropriate. Small pumps may be necessary to carry a riparian wetland through drought periods. Two examples of wetlands of this type constructed primarily for research and education are the Des Plaines River wetlands in northeastern Illinois (Hey et al., 1989; Sanville and Mitsch, 1994) and the Olentangy River wetlands in central Ohio (Mitsch et al., 1998).

The advantages of locating several small wetlands on small streams or intercepting ditches in the upper reaches of a watershed (but not in the streams themselves), rather than creating fewer larger wetlands in the lower reaches should be considered (Fig. 19-9d). Loucks (1990) argued that a better strategy for enabling wetlands to survive

Figure 19-9 Some landscape locations of created and restored wetlands in a riverine setting: **a.** riparian wetland that both intercepts groundwater from uplands but also receives annual flood pulse from adjacent river; **b.** riparian wetlands with natural flooding; **c.** riparian wetland with pump; **d.** multiple upstream wetlands versus single downstream wetland; **e.** lateral wetland intercepting groundwater carried by tile drains. (*After Mitsch, 1992a; Mitsch and Cronk, 1992; Kovacic et al., 1996*)

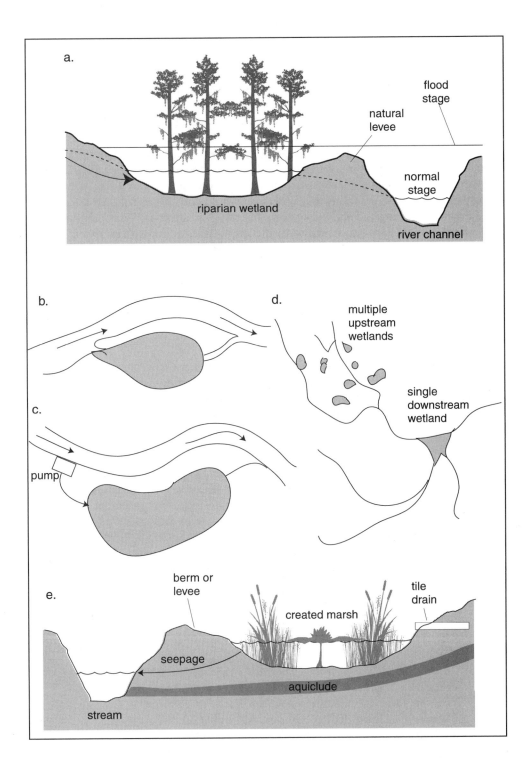

extreme events is to locate a greater number of low-cost wetlands in the upper reaches of a watershed rather than to build fewer high-cost wetlands in the lower reaches. A modeling effort on flood control by Ogawa and Male (1983, 1986) suggested the opposite: The usefulness of wetlands in decreasing flooding increases with the distance the wetland is downstream.

Figure 19-9e shows a design involving the creation of a wetland along a stream to intercept tile drains from agricultural fields. The stream itself is not diverted, but the wetlands receive their water, sediments, and nutrients from small tributaries, swales, and especially tile drains that otherwise would empty straight into the stream. If tile drains can be located and broken or blocked upstream to prevent their discharge into tributaries, they can be rerouted to make effective conduits to supply adequate water to constructed wetlands. Because tile drains are often the sources of the highest concentrations of chemicals such as nitrates from agricultural fields, the lateral wetlands would be an efficient means of controlling certain types of nonpoint source pollution while creating a needed habitat in an agricultural setting (Fig. 19-7).

Site Selection

Several important factors ultimately determine site selection. When the objective is defined, the appropriate site should allow for the maximum probability that the objective can be met, that construction can be done at a reasonable cost, that the system will perform in a generally predictable way, and that the long-term maintenance costs of the system are not excessive. These factors are elaborated below (Willard et al., 1989):

1. *Wetland restoration is generally more feasible than wetland creation:* Find a site where wetlands previously existed or where nearby wetlands still exist. In an area such as this, the proper substrate may be present, seed sources may be on site or nearby, and the appropriate hydrologic conditions may exist. A historical meander of a stream that has been abandoned or channelized makes an excellent potential site for restoring a wetland. Similarly, a tidal marsh would be difficult to establish where one had not existed before.

2. *Take into account the surrounding land use and the future plans for the land:* Future land use plans such as abandoning agricultural fields to become old-field ecosystems may obviate the need for a wetland to control nonpoint runoff.

3. *Undertake a detailed hydrologic study of the site, including a determination of the potential interaction of groundwater with the proposed wetland:* Without flooding or saturated soils for at least part of the growing season, a wetland will not develop. For coastal wetlands, the tidal cycle and stages are important.

4. *Find a site where natural inundation is frequent:* Sites should be inspected during flood season and heavy rains, and the annual and extreme event flooding history of the site should be determined as closely as possible.

5. *Inspect and characterize the soils in some detail to determine their permeability, texture, and stratigraphy:* Highly permeable soils are not likely to support a wetland unless water inflow rates are excessive.

6. *Determine the chemistry of the soils, groundwater, surface flows, flooding streams and rivers, and tides that may influence the site water quality:* Even if the wetland is being built primarily for wildlife enhancement, chemicals in the soil and water may be significant either to wetland productivity or to the bioaccumulation of toxic materials.

7. *Evaluate on-site and nearby seed banks to ascertain their viability and response to hydrologic conditions.*

8. *Ascertain the availability of necessary fill material, seed, and plant stocks and access to infrastructure (e.g., roads, electricity):* This is particularly important for the construction phase.

9. *Determine the ownership of the land and, hence, the price:* These are often overriding considerations. Additional lands may need to be purchased in the future to provide a buffer zone and room for expansion.

10. *For wildlife and fisheries enhancement, determine if the wetland site is along ecological corridors such as migratory flyways or spawning runs.*

11. *Assess site access:* Public access to the site will eventually need to be controlled to avoid vandalism and personal injury. A remote site that offers possibilities of fewer mosquito complaints, lower property costs, and less drastic social impact is often preferable to an urban one. Urban wetlands, however, offer many more possibilities for programs on wetland education for school groups and the public.

12. *Ensure that an adequate amount of land is available to meet the objectives:* If "aging" of a wetland, defined as an impairment of wetland function after several years of perturbation, is anticipated because of the inputs of sediments, nutrients, or other materials, then larger land parcels to build additional wetlands in the future should be considered.

13. *Evaluate the position of the proposed wetland in the landscape:* Landscapes have natural patterns that maximize the value and function of individual habitats. For example, isolated wetlands function in ways that are quite different from wetlands adjacent to rivers. A forested wetland island created in an otherwise grassy or agricultural landscape will support far different species than those that inhabit a similar wetland created as part of a large forest tract.

Introducing Vegetation

The species of vegetation types to be introduced to created and restored wetlands depend on the type of wetland desired, the region, and the climate as well as the design characteristics described previously. Table 19-3 summarizes some of the plant species used for wetland creation and restoration projects.

Freshwater Marshes

Common plants used for freshwater marshes include bulrush (*Scirpus* spp. and *Schoenoplectus* spp.), cattails (*Typha* spp.), sedges (*Carex* spp.), and floating-leaved aquatic plants such as white water lilies (*Nymphaea* spp.) and spatterdock (*Nuphar* spp.). Submerged plants are not common in wetland design, and their propagation

Table 19-3 Selected plant species planted in created and restored wetlands

Scientific Name	Common Name	Scientific Name	Common Name
FRESHWATER MARSH—EMERGENT			
Acorus calamus	sweet flag	*Pontederia cordata*	pickerelweed
Cladium jamaicense	sawgrass	*Sagittaria rigida*	duck potato
Carex spp.	sedges	*Sagittaria latifolia*	duck potato; arrowhe
Eleocharis spp.	spike rush	*Saururus cernuus*	lizard's tail
Glyceria spp.	manna grass	*Schoenoplectus*	soft-stem bulrush
Hibiscus spp.	rose mallow	*tabernaemontani**	
Iris pseudacorus	yellow iris	*Scirpus acutus*	hard-stem bulrush
Iris versicolor	blue iris	*Scirpus americanus*	three-square bulrush
Juncus effusus	soft rush	*Scirpus cyperinus**	woolgrass
Leersia oryzoides	rice cutgrass	*Scirpus fluviatilis*	river bulrush
Panicum virgatum	switchgrass	*Sparganium eurycarpum*	giant bur reed
Peltandra virginica	arrow arum	*Spartina pectinata*	prairie cordgrass
Phalaris arundinacea	reed canary grass	*Typha angustifolia**	narrow-leaved cattail
*Phragmites australis**	giant reed	*Typha latifolia**	wide-leaved cattail
Polygonum spp.	smartweed	*Zizania aquatica*	wild rice
FRESHWATER MARSH—SUBMERGED			
Ceratophyllum demersum	coontail	*Potamogeton pectinatus*	Sago pondweed
Elodea nuttallii	waterweed	*Vallisneria* spp.	wild celery; tape gras
Myriophyllum aquaticum	milfoil		
FRESHWATER MARSH—FLOATING			
Azolla caroliniana	water fern	*Nuphar luteum*	spatterdock
Eichhornia crassipes	water hyacinth	*Pistia stratiotes*	water lettuce
Hydrocotyle umbellata	water pennywort	*Salvinia rotundifolia*	floating moss
Lemna spp.	duckweed	*Wolffia* sp.	water meal
Nymphaea odorata	fragrant white water lily		
BOTTOMLANDS/FORESTED WETLAND			
Acer rubrum	red maple	*Gordonia lasianthus*	loblolly bay
Acer floridanum	Florida maple	*Liquidambar styracifula*	sweetgum
Acer saccharinum	silver maple	*Platanus occidentalis*	sycamore
Alnus spp.	alder	*Populus deltoides*	cottonwood
Carya illinoensis	pecan	*Quercus falcata* var.	cherrybark oak
Celtis occidentalis	hackberry	*pagodifolia*	
Cephalanthus occidentalis	buttonbush	*Quercus nigra*	water oak
Cornus stolonifera	red-osier dogwood	*Quercus nuttallii*	Nuttall oak
Fraxinus caroliniana	water ash	*Quercus phellos*	willow oak
Fraxinus pennsylvanica	green ash	*Salix* spp.	willow
		Ulmus americana	American elm
DEEPWATER SWAMP			
Nyssa aquatica	swamp tupelo	*Taxodium distichum*	bald cypress
Nyssa sylvatica var.*biflora*	black gum	*Taxodium distichum* var *nutans*	pond cypress
SALT MARSH			
Distichlis spicata	spike grass	*Spartina foliosa*	cordgrass (Western U
Salicornia sp.	saltwort	*Spartina patens*	salt meadow grass
Spartina alterniflora	cordgrass (Eastern U.S.)	*Spartina townsendii*	cordgrass (Europe)
Spartina anglica	cordgrass (Europe; China)		
MANGROVE			
Rhizophera mangle	red mangrove		
Avicennia germinans	black mangrove		
Laguncularia racemosa	white mangrove		

*Commonly planted in wastewater wetlands

is often hampered by turbidity and algal growth in the early years of wetland development.

Coastal Marshes

For coastal salt marshes, *Spartina alterniflora* is the primary choice for coastal marsh restoration in the eastern United States (Broome et al., 1988). Both *Spartina townsendii* and *S. anglica* have been used to restore salt marshes in Europe (Ranwell, 1967) and in China (Chung, 1982, 1989). The details of successful coastal wetland creation are, of course, site specific, but several generalizations seem to be valid in most situations:

1. Sediment elevation seems to be the most critical factor determining the successful establishment of vegetation and the plant species that will survive. The site must be intertidal and, in general, the upper half of the intertidal zone is more rapidly vegetated than lower elevations. Sediment composition does not seem to be a critical factor in colonization by plants unless the deposits are almost pure sand that is subject to rapid desiccation at the upper elevations. Another important requirement is protection of the site from high wave energy. It is difficult or impossible to establish vegetation at high-energy sites.

2. Most deposits seem to revegetate naturally if the elevation is appropriate and the wave energy is moderate. Sprigging live plants has been accomplished successfully in a number of cases, and seeding also has been successful in the upper half of the intertidal zone.

3. Good stands can be established during the first season of growth, although sediment stabilization does not occur until after two seasons. Within 4 years, successfully planted sites are indistinguishable superficially from natural marshes.

Sinicrope et al. (1990) described the natural restoration of a 20-ha impounded salt marsh in Connecticut over 10 years (see Fig. 19-10). In that case, a wetland had been isolated from tidal flushing for many years and had become dominated by *Typha angustifolia*. In the late 1970s and early 1980s, several major culverts were installed to reintroduce tidal flushing, leading to a significant decline in *T. angustifolia* from 74 to 16 percent cover and a recovery of *Spartina alterniflora* from less than 1 percent cover to 45 percent cover. *Phragmites australis,* which tolerates brackish conditions, did not decrease as expected but increased from 6 to 17 percent cover and was generally found along the edges of the marsh in a stunted (0.3–1.5 m tall) condition. A relatively simple alteration of the hydrologic conditions reestablished the salt marsh.

Forested Wetlands

Forested wetland restoration and creation usually involve the establishment of seedlings. In the Southeast, deciduous hardwood species typical of bottomland forests are planted. They include Nuttall oak (*Quercus nuttallii*), cherrybark oak (*Q. falcata* var. *pagodifolia*), willow oak (*Q. phellos*), water oak (*Q. nigra*), cottonwood (*Populus deltoides*), sycamore (*Platanus occidentalis*), green ash (*Fraxinus pennsylvanica*), sweetgum (*Liquidambar styracifula*), and pecan (*Carya illinoensis*). There is less use of deepwater plants such as bald cypress (*Taxodium distichum*) and water tupelo (*Nyssa aquatica*) although *Taxodium* spp. was the dominant genera of introduced species

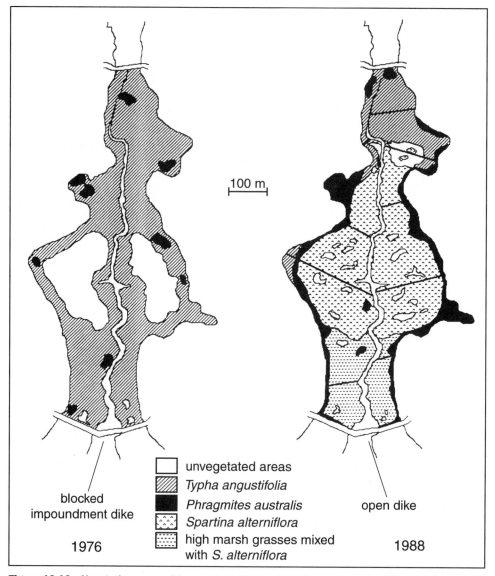

Figure 19-10 Vegetation maps of impounded salt marsh on Connecticut shoreline showing restoration to salt marsh with reintroduction of tidal flushing. 1976 map is prior to opening to tides, while 1988 map indicates return of *Spartina alterniflora* after opening to tidal circulation in late 1970s and early 1980s. Lines in 1988 map indicate vegetation transects. (*After Sinicrope et al., 1990*)

in the long-studied Halls Branch restoration in Florida (Clewell, 1999). In Florida, a wide variety of wetland oaks, bays, gums, ashes, and pines are also used in forested wetland restoration (Erwin et al., 1984; M. T. Brown et al., 1992).

Exotic or Undesirable Plant Species

In some cases, certain plants are viewed as desirable or undesirable because of their value to wildlife or their aesthetics. Reed grass (*Phragmites australis*) is often favored in constructed wetlands in Europe where there is a real concern for "reed dieback" around lakes and ponds. However, reed grass is considered an invasive undesirable plant in much of eastern North America, particularly in coastal freshwater and brackish marshes. Some plants are considered undesirable in wetlands because they are aggressive competitors. In many parts of the tropics and subtropics, the floating aquatic plant water hyacinth (*Eichhornia crassipes*) and alligator weed (*Alternanthera philoxeroides*) are considered undesirable; in eastern North America, particularly around the Great Lakes, purple loosestrife (*Lythrum salicaria*) is considered an undesirable exotic plant in wetlands. Throughout the United States, cattail (*Typha* spp.) is championed by some and disdained by others, for it is a rapid colonizer but of limited wildlife value (W. E. Odum, 1987). In other parts of the world, *Typha* is considered a perfectly acceptable plant in restored wetlands. In New Zealand, several species of willow (*Salix*) are invading marshes and other wetlands, and programs to eradicate them are common.

Natural Succession versus Horticulture

An important general consideration of wetland design is whether plant material is going to be allowed to develop naturally from some initial seeding and planting or whether continuous horticultural selection for desired plants will be imposed. To develop a wetland that will ultimately be low maintenance, natural successional processes need to be allowed to proceed. This may mean some initial period of invasion by undesirable species, but if proper hydrologic conditions are imposed, these invasions will be temporary. The best strategy is to introduce, by seeding and planting, as many choices as possible to allow natural processes to sort out the species and communities in a timely fashion. Wetlands created or restored by this approach are called *self-design wetlands* (see Chapter 8). Providing some help to this selection process, for example, selective weeding, may be necessary in the beginning, but ultimately the system needs to survive with its own successional patterns unless significant labor-intensive management is possible. A somewhat different approach, called *designer wetlands,* occurs when specified plant species are introduced and the success or failure of those plants are used as indicators of success or failure of that wetland.

To Plant or Not to Plant

W. E. Odum (1987) distinguished freshwater wetland succession from coastal saltwater wetland succession by stating that "in many freshwater wetland sites it may be an expensive waste of time to plant species which are of high value to wildlife. . . . It may be wiser to simply accept the establishment of disturbance species as a cheaper although somewhat less attractive solution."

Reinartz and Warne (1993) found that the way vegetation is established can affect the diversity and value of the mitigation wetland system. Their study showed that early introduction of a diversity of wetland plants may enhance the long-term diversity of vegetation in created wetlands. The study examined the natural colonization of plants in 11 created wetlands in southeastern Wisconsin. The wetlands under study were small, isolated, depressional wetlands. A 2-year sampling program was conducted for the created wetlands, aged 1 to 3 years. Colonization was compared to 5 seeded wetlands where 22 species were introduced. The diversity and richness of plants in the colonized wetlands increased with age, size, and proximity to the nearest wetland source. In the colonized sites, *Typha* spp. comprised 15 percent of the vegetation for 1-year wetlands, and 55 percent for 3-year wetlands, with the possibility of monocultures of *Typha* spp. developing over time in colonized wetlands. The seeded wetlands had a high species diversity and richness after 2 years. *Typha* cover in these sites was lower than in the colonized sites after 2 years.

In a multiyear study on the effect of plant introduction on ecosystem function, researchers at the Olentangy River Wetland Research Park in Ohio found that both planted and unplanted wetlands converged in most functions (eight biological measures; eight biophysiochemical measures) in 3 years (Table 8-4; Fig. 19-11). During the year that the planted wetland basin had considerably more vegetation than the

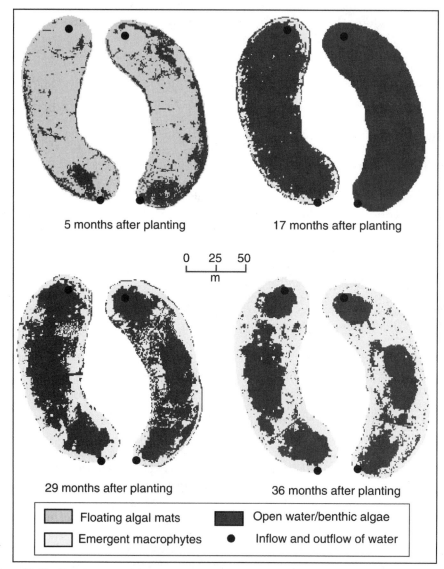

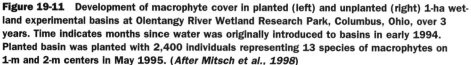

Figure 19-11 Development of macrophyte cover in planted (left) and unplanted (right) 1-ha wetland experimental basins at Olentangy River Wetland Research Park, Columbus, Ohio, over 3 years. Time indicates months since water was originally introduced to basins in early 1994. Planted basin was planted with 2,400 individuals representing 13 species of macrophytes on 1-m and 2-m centers in May 1995. (*After Mitsch et al., 1998*)

unplanted wetland basin, only 12 percent of the indices were similar (see details in Chapter 8). Continued studies at that site subsequent to those 3 years showed a persistence of the planted vegetation in the planted basin but a dominance of colonizing *Typha* in the unplanted basin; functional differences are starting to reemerge. Residual effects of planting included a more spatially diverse macrophyte community but about 20 percent lower net primary productivity 6 years after planting in the planted basin.

Planting Techniques

Plants can be introduced to a wetland by transplanting roots, rhizomes, tubers, seedlings, or mature plants; by broadcasting seeds obtained commercially or from other sites; by importing substrate and its seed bank from nearby wetlands; or by relying completely on the seed bank of the original and surrounding site. If planting stocks rather than site seed banks are used, it is more desirable to choose plants from wild stock rather than nurseries because the former are generally better adapted to the environmental conditions that they will face in constructed wetlands. The plants should come from nearby if possible and should be planted within 36 hours of collection. If nursery plants are used, they should be from the same general climatic conditions and should be shipped by express service to minimize losses. M. T. Brown (1987) suggested that marshes be planted at densities to ensure rapid colonization, adequate seed source, and effective competition with *Typha* spp. Specifically, this could mean introducing from 2,000 to 5,000 plants/ha.

For emergent plants, the use of planting materials with at least 20- to 30-cm stems is recommended; whole plants, rhizomes, or tubers rather than seeds have been most successful (Fig. 19-12). In temperate climates, both fall and spring planting

Figure 19-12 Planting of tubers and rhizomes is labor intensive but is one of the best ways to introduce wetland plants to newly created wetlands. (*Photographs by W. J. Mitsch*)

times are possible for certain species, but spring plantings are generally more successful. Garbisch (1989) suggested that spring planting is desirable to minimize the destructive grazing of plants in the winter by migratory animals and to avoid the uprooting of the new plants by ice.

Transplanting plugs or cores (8–10 cm in diameter) from existing wetlands is another technique that has been used with success, for it brings seeds, shoots, and roots of a variety of wetland plants to the newly restored or created wetland.

If seeds and seed banks are used for wetland vegetation, several precautions must be taken. The seed bank should be evaluated for seed viability and species present (van der Valk, 1981). The use of seed banks from other nearby sites can be an effective way to develop wetland plants in a constructed wetland if the hydrologic conditions in the new wetland are similar. Weller (1981) stated that seed bank transplants were successful for many different species, including sedges (*Carex* spp.), *Sagittaria* sp., *Scirpus acutus, Schoenoplectus tabernaemontani,* and *Typha* spp. The disruption of the wetland site where the seed bank is obtained must also be considered.

When seeds are used directly to vegetate a wetland, they must be collected when they are ripe and stratified if necessary. If commercial stocks are used, the purity of the seed stock should be determined. The seeds can be added with commercial drills or by broadcasting from the ground, watercraft, or aircraft. Seed broadcasting is most effective when there is little to no standing water in the wetland.

Estimating Success

There are few satisfactory methods available to determine the "success" of a created or restored wetland or even a mitigation wetland created to replace the functions lost with the original wetland. Figure 19-13 illustrates conceptually how it should be done for replacement wetlands. On the one hand, *legal success* involves a comparison of the lost wetland function and area with that which is gained in the replacement wetland. *Ecological success* should involve a comparison of the replacement wetland with a reference wetland (natural wetlands of the same type that may occur in the same setting or generally accepted "standards" of regional wetland function) (Wilson and Mitsch, 1996). Overall success would then be gauged by a combination of the legal and ecological comparisons. While this model represents an ideal, the comparison involving both standards is rarely done.

Studies of Wetland Mitigation Success

Contrary to the situation just a few years ago, there have been a considerable number of studies where replacement wetlands have been compared either to the initial goals or to the lost wetlands. Most studies suggest that there is much room for improvement in the building of mitigation wetlands. In the mid-1980s, research focused on the implementation of mitigation for wetland losses; that is, was it ever done at all? Although some scientists involved in wetland restoration and mitigation believed that mitigation was working (Harvey and Josselyn, 1986), others suggested the need for more research (Kusler and Groman, 1986; Race, 1986). Erwin (1991) found that of

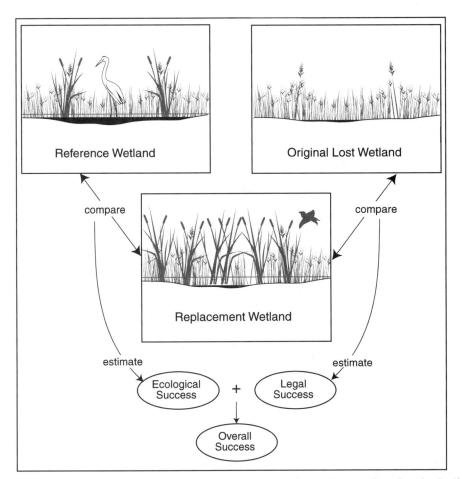

Figure 19-13 In mitigating the loss of wetlands through creation and restoration of wetlands, the proper procedure should involve comparison with both what has been lost (legal success) but also with regional reference wetlands (ecological success). (*After Wilson and Mitsch, 1996*)

40 mitigation projects in south Florida involving wetland creation and restoration, only about half of the required 430 ha of wetlands had been constructed and that 24 of the 40 projects (60 percent) were judged to be incomplete or failures (Fig. 19-14a). The most significant problems identified with the constructed wetlands were improper water levels and hydroperiod.

More recent reviews of the process have suggested mixed results on the efficacy of the process. Sifneos et al. (1992) conducted a study by collecting information from permits on Louisiana wetland mitigation progress and found that only 8 percent of the area for which Section 404 drainage permits were issued in the state was compensated for, and over 50 percent of the permits were for areas less than 0.4 ha. Additionally, only 10 percent of the replacement wetlands were monitored by at least one site visit. A similar study by Kentula et al. (1992a), investigating compensatory mitigation, found net loss in wetland area of 43 percent for Oregon and 26 percent for Washington. The study was difficult as readily available data were either incomplete or of poor quality.

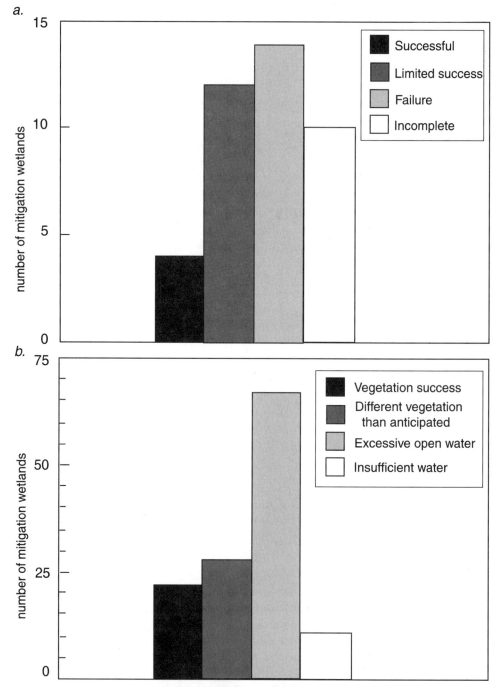

Figure 19-14 Evaluation of wetland mitigation projects in two regions of the United States: **a.** 40 mitigation projects in south Florida involving wetland creation, mitigation, and preservation. The average age of the projects was less than 3 years. Successful meant that the project met all of its stated goals, while failure meant that few goals were met or the created/restored wetland did not have functional equivalency to a reference wetland (Erwin, 1991); **b.** 128 wetland mitigation sites required by 61 permits in the six-county region around Chicago, Illinois. The permits were issued between 1990 and 1994 and this study began in 1996. (*Gallihugh and Rogner, 1998*)

King and Bohlen (1994) illustrated another problem with wetland mitigation efforts. Poorly qualified "experts" often bid for restoration contracts at the low end of the scale to get a contract. This appeals to those who are required to mitigate a lost wetland because of low costs, but the quality of the replacement product suffers. Mitigation is viewed as a means to an end; price, not quality, becomes the major issue in restoration projects.

In most cases, the loss and gain of wetlands is measured in area. Hydrology, soils, wildlife, and water quality have not been evaluated in most cases, nor are any functional measures used. Although there are many factors that could be used to measure the success of mitigation wetlands, vegetation is considered the easiest way to monitor progress. Atkinson et al. (1993) support the use of weighted averages to measure created wetland success. However, they indicate that vegetation should be the main criterion for short-term study of success. This is because vegetation is easily accessible and establishes quickly in a wet environment. On the other hand, Reinartz and Warne (1993) found that while vegetation may be an easy measure of success, it is a poor indicator of function.

Wilson and Mitsch (1996) evaluated five replacement wetlands in detail in Ohio to estimate their ecological and legal success. Only two of the five mitigation cases showed the replacement wetlands in full legal compliance (Table 19-4), but four of the five wetlands were determined to be on a trajectory toward legal compliance with permit requirements. The same four wetlands showed medium to high ecological success. Overall, 24.4 ha of wetlands were lost, and about 16 ha were actually created or restored. Because of the failure of one large site, only 38 percent of the desired wetland area was established at the time of this study. For the four wetlands that were successful, there was an overall mitigation ratio of 1.4 : 1.

Galatowitsch and van der Valk (1996) investigated 2,700 ha of prairie potholes that were restored in Iowa, Minnesota, and South Dakota, most as a result of the Conservation Reserve Program described earlier in this chapter. Using the Stewart and Kantrud (1971) classification of prairie potholes, they found that the potholes were mostly ephemeral or temporarily/seasonally flooded before they had been drained but that most were seasonal to semipermanently flooded after they were "restored." In other words, the restored wetlands were generally wetter. They also found that 18 percent of the restorations in the region were carried out by breaking drainage tiles and that 20 percent of the restorations were hydrologic failures.

Table 19-4 **Permit requirements and compliance for five replacement wetlands investigated in Ohio**

Location, County	Wetland Area (ha)			Percentage of Required Area Replaced
	Lost	Required	Happened	
Portage	0.4	0.6	0.6	100
Delaware	3.7	5.4	~4.0	74
Franklin	15.0	28.0	3.2	11
Jackson	4.8	7.2	7.5	105
Gallia	0.5	0.8	0.7	88
Total	24.4	42.0	~16.0	38

Source: Wilson and Mitsch (1996).

In one of the most comprehensive investigations of the success of wetland mitigation in the Section 404 permit program, 61 permits involving 128 wetland mitigation sites were investigated in the six-county region around Chicago (Gallihugh and Rogner, 1998; Fig. 19-14b). This study found that in 17 percent of the wetlands the vegetation proposed for the project was established, and an additional 22 percent had established wetlands but with vegetation other than that proposed. In what has often been the case in wetland creation and restoration, 52 percent of the wetlands had excessive or unplanned open water, while 9 percent had insufficient hydrology. Over this area, the lost wetland area was 117 ha and the approved wetland mitigation amounted to 144 ha. So, in theory, there was a mitigation ratio of 1.2 : 1, or a net gain of 27 ha. In actuality, the study found that 29 ha were not established and at least 99 ha were found to have unsatisfactory hydrology (too wet or too dry).

Designing for Success

It is clear from the many studies described previously for created and restored wetlands that some cases are successes, while there are still far too many examples of failure of created and restored wetlands to meet expectations. In some cases, expectations were unreasonable, as when endangered-species habitat was to be established in a heavily urbanized environment (Malakoff, 1998). In such cases, the original wetland should not have been lost to begin with (Zedler, 1991). Where expectations are ecologically reasonable, there is optimism that wetlands can be created and restored and that wetland function can be replaced (Mitsch and Wilson, 1996). The spotty record to date is due, in our opinion, to three factors:

1. Little understanding of wetland function by those constructing the wetlands
2. Insufficient time for the wetlands to develop
3. Lack of recognition or underestimation of the self-design capacity of nature

Understanding wetlands enough to be able to create and restore them requires substantial training in plants, soils, wildlife, hydrology, water quality, and engineering. Replacement projects and other restorations involving freshwater marshes need enough time, closer to 15 or 20 years than to 5 years, before success is apparent. Restoration and creation of forested wetlands, coastal wetlands, or peatlands may require even more time. For example, the restoration of certain coastal salt marshes has been suggested to require at least 50 years (Frenkel and Morlan, 1991). Peatland restoration could take decades or more. And, of course, forested wetland restoration generally takes a lifetime. Finally, we should recognize that Nature remains the chief agent of self-design, ecosystem development, and ecosystem maintenance; humans are not the only participants in these processes. Sometimes we refer to these self-design and time requirements for successful ecosystem restoration and creation as invoking "Mother Nature and Father Time."

Wetland science will continue to make significant contributions to the process of reducing our uncertainty about predicting wetland success. Wetland creation and restoration need to become part of an applied ecological science, not a technique without theoretical underpinnings. Scientists need to make the connections between structure (such as vegetation density, diversity) and functions (such as wildlife use, organic sediment accretion, or nutrient retention) in quantitative and carefully designed experiments. Engineers and managers need to recognize that designing systems that emphasize the role of self-design and sustainable structures are more ecologically viable in the long run than are heavily managed systems.

Mitigation Banks

One of the more interesting strategies that the private sector and government agencies have developed to deal with the piecemeal approach to mitigation of wetland loss is the concept of a *mitigation bank*. A mitigation bank is defined as "a site where wetlands and/or other aquatic resources are restored, created, enhanced, or in exceptional circumstances, preserved expressly for the purpose of providing compensatory mitigation in advance of authorized impacts to similar resources" (*Federal Register,* November 28, 1995, "Federal Guidance for the Establishment, Use, and Operation of Mitigation Banks"). In this approach, wetlands are built in advance of development activities that cause wetland loss and credits of wetland area can be sold to those who are in need of mitigation for wetland loss. Banks are seen as a way of streamlining the process of mitigating wetland loss and, in many cases, providing a large, fully functional wetland rather than small, questionable wetlands near the site of wetland loss. The mitigation bank can be set up with bonds ensuring compliance. Arrangements are easier for wetland mitigation banks to be managed "in perpetuity" through conservation easements or transfer of titles to resource agencies. Financial resources can be arranged ahead of time for proper monitoring of the wetland bank. Mitigation banks can be publicly or privately owned, although there is a potentially serious conflict of interest if public agencies run mitigation banks. Public agencies could be involved in enforcing regulations on mitigating wetland loss and then steer permitees toward their own banks rather than to private banks.

Currently, there are hundreds of mitigation banks, both private and public, in the United States. There are 62 formal mitigation banks (proposed and operating) and hundreds of quasi-mitigation banks in Florida alone (Ann Redmond, personal communication). It appears that, if mitigation of wetland loss continues to be the nation's policy and if regulation of mitigation banks can be developed that is fair and uncomplicated, the use of mitigation banks to solve this "wetland trading" issue will continue to increase well into the 21st century.

Summary Recommendations

Robin Lewis and Kevin Erwin, two wetland consultants from Florida, have spent about 5 decades restoring and creating wetlands around the world. Their experience has led to the following 15 recommendations (Lewis et al., 1995). They serve as a fitting summary of the application of principles and practices in effective restoration and creation of wetlands.

1. Wetland restoration and creation proposals must be viewed with great care, particularly when promises are made to restore or recreate a natural system in exchange for a permit.
2. Multidisciplinary expertise in planning and careful project supervision at all project levels is needed.
3. Clear, site-specific measurable goals should be established.

4. A relatively detailed plan concerning all phases of the project should be prepared in advance to help evaluate the probability of success.

5. Site-specific studies should be carried out in the original system prior to wetland alteration if wetlands are being lost in the project.

6. Careful attention to wetland hydrology is needed in design.

7. Wetlands should, in general, be designed to be self-sustaining systems and persistent features of the landscape.

8. Wetland design should consider relationships of the wetland to the watershed, water sources, other wetlands in the watershed, and adjacent upland and deepwater habitat.

9. Buffers, barriers, and other protective measures are often needed.

10. Restoration should be favored over creation.

11. The capability for monitoring and mid-course corrections is needed.

12. The capability for long-term management is needed for some types of systems.

13. Risks inherent in restoration and creation, and the probability of success for restoring or creating particular wetland types and functions, should be reflected in standards and criteria for projects and project design.

14. Restoration for artificial or already altered systems requires special treatment.

15. Emphasis on ecological restoration of watersheds and landscape ecosystem management requires advanced planning.

Recommended Readings

Galatowitsch, S. M., and A. G. van der Valk. 1994. *Restoring Prairie Wetlands: An Ecological Approach.* Iowa State University Press, Ames, IA. 246 pp.

Hammer, D. A. 1997. *Creating Freshwater Wetlands,* 2nd ed. CRC/Lewis Publishers, Boca Raton, FL. 406 pp.

Middleton, B. A. 1999. *Wetland Restoration: Flood Pulsing and Disturbance Dynamics.* John Wiley & Sons, New York. 388 pp.

Mitsch, W. J., X. Wu, R. W. Nairn, P. E. Weihe, N. Wang, R. Deal, and C. E. Boucher. 1998. Creating and restoring wetlands: A whole-ecosystem experiment in self-design. BioScience 48:1019–1030.

National Research Council (NRC). 1992. *Restoration of Aquatic Ecosystems.* National Academy Press, Washington, DC. 552 pp.

Thayer, G. W., ed. 1992. *Restoring the Nation's Marine Environment.* Maryland Sea Grant College, College Park, MD. 716 pp.

Zedler, J. B. 1996. *Tidal Wetland Restoration: A Scientific Perspective and Southern California Focus.* Report T-038. California Sea Grant College System, University of California, La Jolla, CA. 129 pp.

Chapter **20**

Treatment Wetlands

Treatment wetlands have the singular goal of improving water quality. There are three types of treatment wetlands: natural wetlands and two types of constructed wetlands—surface-flow wetlands and subsurface-flow wetlands. Each type has advantages and disadvantages. Studies on the use of natural wetlands to treat wastewater in Florida and Michigan pioneered the use of surface-flow systems; subsurface-flow systems were started and developed in Europe and remain the dominant treatment wetland system there. Because wetlands can be sinks for almost any chemical, applications of treatment wetlands are quite varied, with thousands of applications worldwide to treat domestic wastewater, mine drainage, nonpoint source pollution, stormwater runoff, landfill leachate, and confined livestock operations. The design of treatment wetlands requires particular attention to hydrology, chemical loading, soil physics and chemistry, and wetland vegetation. After the wetlands are completed and wastewater is applied, management can include plant harvesting, wildlife habitat, mosquito and pathogen control, perturbation and water-level management, and sediment dredging. Treatment wetlands are not inexpensive to build and operate but they usually cost less than chemical and physical treatment processes.

To some the idea involves wastewater treatment, to others wastewater disposal, to still others wastewater recycling. Regardless of what it is called, wastewater and polluted-water treatment by wetlands is an intriguing concept involving the forging of a partnership between humanity (our wastes) and an ecosystem (wetlands).

The idea that many wetlands are sinks for chemicals, particularly nutrients, is illustrated in most of the ecosystem chapters (Chapters 9–15) in this book. This tendency for wetlands to serve as sinks for chemicals of all kinds encouraged researchers in the United States in the 1970s to investigate the role of natural wetlands, particularly in regions where they are found in abundance, to treat wastewater and thus recycle clean water back to groundwater and surface water. Earlier than this in Europe, German scientists investigated the use of constructed basins with macrophytes (*Höhere Pflanzen*) for purification of wastewater. The two different approaches, one utilizing natural wetlands and the other using artificial systems, have converged into the general field of *treatment wetlands*. The field now encompasses the construction and/or use of wetlands for a myriad of water quality applications including municipal wastewater, small-scale rural wastewater, acid mine drainage, landfill leachate, and nonpoint source pollution from both urban and agricultural runoff. While water quality improvement by treatment wetlands is the primary goal of treatment wetlands, they also provide habitat for a wide diversity of plants and animals, and can support many of the other wetland functions described in this book.

Much has been written about treatment wetlands, with a tome by Kadlec and Knight (1996) providing the most comprehensive coverage. There are also many books and journal special issues covering the subject of treatment wetlands in general (Godfrey et al., 1985; Reddy and Smith, 1987; Hammer, 1989; Cooper and Findlater, 1990; Knight, 1990; Johnston, 1991; U.S. EPA, 1993; Moshiri, 1993; Reed et al., 1995; Etnier and Guterstam, 1997; Tanner et al., 1999), nonpoint source wetlands (Olson, 1992; Mitsch et al., 2000b), and landfill leachate (Mulamoottil et al., 1999). Papers featuring such systems are frequently published in the journal *Ecological Engineering*.

General Approaches

Three types of wetlands are used to treat wastewater. The first are *natural wetlands* whereby wastewater is purposefully introduced to existing wetlands such as those described in earlier chapters (Fig. 20-1a). In the 1970s, studies involving application of wastewater to natural wetlands were carried out in Michigan and Florida where there were abundant wetlands. At the time, legal protection of wetlands had not been institutionalized. These pioneering studies elevated the importance of wetlands as "nature's kidneys" in the general public's and the government's view, as ecosystems of particular value. This importance was then translated, appropriately, into laws that protected wetlands. However, these same laws also generally prohibited adding wastewater or polluted water to natural wetlands. Only recently has enough research data been gathered concerning the safety and value of natural treatment wetlands, so that the EPA, which enforces the Clean Water Act, has begun to approve their use under carefully controlled conditions.

The reluctance on the part of regulators to allow pollution into natural wetlands, plus the fact that wastewater flows are often continuous and therefore downstream wetland wetlands are continuously flooded, limits the application of wastewater to natural wetlands. *Constructed wetlands* are an alternative. While constructed to mimic natural wetlands, they represent created ecosystems that are not part of the original wetland resources of a region and, therefore, continuous flooding with polluted water is more acceptable. There are two types of constructed wetlands. *Surface-flow constructed wetlands* (Fig. 20-1b) mimic natural wetlands and can be a better habitat for certain wetland species because of standing water throughout most, if not all, of the year. The second type of constructed wetlands, *subsurface-flow constructed wetlands* (Fig. 20-1c) more closely resemble wastewater treatment plants than wetlands. In these systems, the water flows through a porous medium, usually sand or gravel, supporting one or two of a relatively narrow list of macrophytes. There is no standing water in these systems as the wastewater passes laterally through the medium.

Treatment wetlands can also be classified based on the life form of their vegetation. In this case, there are five systems based on herbaceous macrophytes:

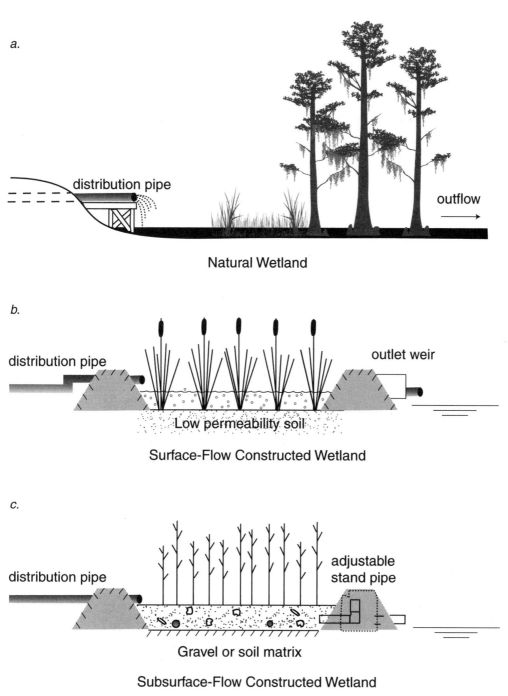

a.

distribution pipe

outflow

Natural Wetland

b.

distribution pipe

outlet weir

Low permeability soil

Surface-Flow Constructed Wetland

c.

distribution pipe

adjustable
stand pipe

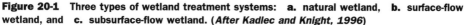

Gravel or soil matrix

Subsurface-Flow Constructed Wetland

Figure 20-1   Three types of wetland treatment systems: **a.** natural wetland,   **b.** surface-flow wetland, and   **c.** subsurface-flow wetland. (*After Kadlec and Knight, 1996*)

1. Free-floating macrophyte systems, e.g., water hyacinths (*Eichhornia crassipes*), duckweed (*Lemna* spp.)

2. Emergent macrophyte systems, e.g., reed grass (*Phragmites australis*), cattails (*Typha* spp.)

3. Submerged macrophyte systems

4. Forested wetland systems

5. Multispecies algal systems, particularly algal-scrubber systems

Subsurface-flow constructed wetlands are limited to emergent macrophyes, whereas surface-flow constructed wetlands often utilize a combination of free-floating, emergent, and submerged macrophytes. Forested wetland treatment systems are generally not constructed wetlands at all, but are natural wetlands to which wastewater is applied. They will often develop extensive communities of all of the other vegetation types described in this classification.

Early Treatment Wetland Studies

Much of the interest in using surface-flow wetlands for water quality management was sparked by two studies begun in the early 1970s. In one of those studies, a peatland in Michigan was investigated by researchers from the University of Michigan for the wetland's capacity to treat wastewater (Fig. 20-2). A pilot operation for disposing of up to 380 m^3/day (100,000 gal/day) of secondarily treated wastewater in a rich fen at Houghton Lake led to significant reductions in ammonia nitrogen and total dissolved phosphorus as the wastewater passed from the point of discharge through the wetland. Inert materials such as chloride did not change as the wastewater passed through the wetland. In 1978, the flow was increased to approximately 5,000 m^3/day over a much larger area, essentially all the wastewater from the local treatment plant. Data after more than 22 years of operation at this high flow show that, although the area of influence of the wastewater on the peatland has grown from 23 to 77 ha (Fig. 20-2b), the effectiveness of the wetland in removing both inorganic nitrogen and total phosphorus remained extremely high (Fig. 20-2c and d).

In another major research effort in the early 1970s, to investigate water quality management in natural wetlands, wastewater was applied to several cypress domes in north-central Florida by a team of researchers from the University of Florida (Ewel, 1976; H. T. Odum et al., 1977a; Ewel and Odum, 1978, 1979, 1984; Dierberg and Brezonik, 1983a, b, 1985). After 5 years of experimentation in which secondarily treated wastewater was added to the cypress domes at a rate of approximately 2.5 cm/week, the results indicated that the wetland filtered nutrients, heavy metals, microbes, and viruses from the water. The productivity of the canopy pond cypress trees also increased (Nessel et al., 1982; Lemlich and Ewel, 1984). The uptake of nutrients in these systems was enhanced by a continuous cover of duckweed on the water, by the retention of nutrients in the cypress wood and litter, and by the adsorption of phosphorus onto clay and organic peat in sediments.

Subsurface treatment wetlands had their start in the Max-Planck Institute in Germany in the 1950s. Dr. Käthe Seidel performed many experiments with emergent macrophytes, *Schoenoplectus lacustris* in particular, and found that the plants contributed to the reduction of bacteria and organic and inorganic chemicals (Seidel, 1964, 1966). This process was translated into a gravel bed macrophyte system that became

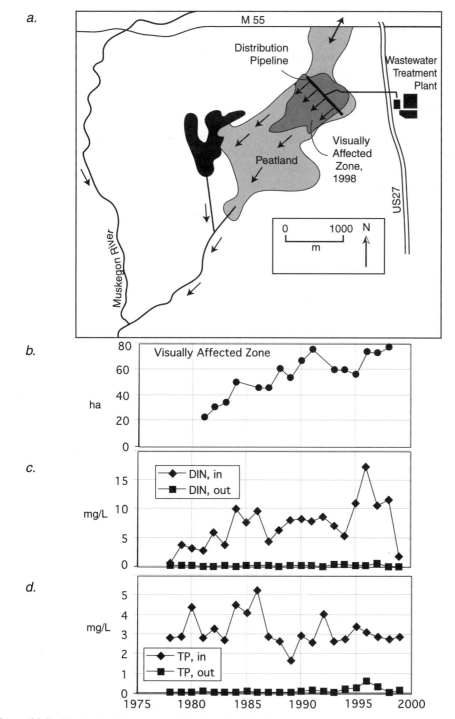

Figure 20-2 Houghton Lake treatment wetland in Michigan, where treated sewage effluent has been applied to a natural peatland since 1978: **a.** map of site showing visually affected area in 1998; **b.** area of visually affected zone in peatland, primarily where vegetation changes have occurred, 1981–1998; **c.** dissolved inorganic nitrogen of influent and outlet stream, 1978–1999; **d.** total phosphorus of influent and outlet stream. (*After Kadlec and Knight, 1996; R. H. Kadlec, personal communication, January 2000*)

known as the *Max-Planck Institute process,* or the *Krefeld system* (Seidel and Happl, 1981, as cited in Brix, 1994). The development of subsurface wetlands continued in Europe through the work of Kickuth (1970, 1982) in Göttingen, Germany, who used a system of subsurface flow basins planted with *Phragmites australis.* These systems were called the *root zone method* (Wurzelraumentsorgung). Subsurface wetland systems continued to be studied and refined through the work of DeJong (1976) in Holland, Brix (1987) in Denmark, and many other scientists in Europe. The appeal of these more "artificial" types of wetlands in Europe (as opposed to North America where free-water surface wetlands) is due to two factors: (1) There are fewer natural wetlands remaining in Europe and those that are left are protected for nature, and (2) space is much more at a premium in Europe and subsurface wetlands require less land area.

Types of Treatment Wetlands by Sources

Treatment wetlands are often classified by the general type of wastewater they are treating. Although many of these systems are used for municipal wastewater, there has been much interest in the use of wetlands to treat acid mine drainage from coal mines, nonpoint source pollution in rural and urban landscapes, livestock and aquacultural wastewaters, and an array of industrial wastewaters.

Municipal Wastewater Wetlands

The use of wetlands for municipal wastewater treatment (i.e., *constructed wetlands*), stimulated by the studies described previously in Florida and Michigan, demonstrated the ability of natural wetlands to remove suspended sediments and nutrients, particularly nitrogen and phosphorus, from domestic wastewater (see also Nichols, 1983; Godfrey et al., 1985; Knight, 1990; U.S. EPA, 1993). Today, these natural wetland systems have been supplanted with constructed wetlands. In Europe, most of the development of subsurface constructed wetlands was to replace both primary and secondary treatment to remove BOD and suspended solids as well as inorganic nutrients. Sanitary engineering manuals have been describing wetlands as alternative natural treatment systems since the early 1990s (Water Pollution Control Federation, 1990; Tchobanoglous and Burton, 1991). Wastewater wetlands that have been studied in some detail include sites in Florida (Knight et al., 1987; J. Jackson, 1989), California (Gersberg et al., 1983, 1984; Gerheart et al., 1989; Gerheart, 1992; Sartoris et al., 2000), Louisiana (Boustany et al., 1997), Arizona (Wilhelm et al., 1989), Kentucky (Steiner et al., 1987), Pennsylvania (Conway and Murtha, 1989), Ohio (Spieles and Mitsch, 2000), North Dakota (Litchfield and Schatz, 1989), and Alberta (White et al., 2000). Hundreds of subsurface wetland treatment systems for municipal wastewater have been constructed in Europe (Vymazal et al., 1998), particularly in the United Kingdom (Cooper and Findlater, 1990), Denmark (Brix and Schierup, 1989a, b; Brix, 1998), and the Czech Republic (Vymazal, 1995b, 1998). There are also many applications of this technology in Australia (Mitchell et al., 1995) and New Zealand (Cooke, 1992; Nguyen et al., 1997; Nguyen, 2000), some going back several decades. Created wetlands for processing water pollution from wastewater have been most effective for controlling organic matter (BOD), suspended sediments, and nutrients. Their value for controlling trace metals and other toxic materials is more controversial, not because these chemicals are not retained in the wetlands but because of concerns that they might concentrate in wetland substrate and fauna.

Mine Drainage Wetlands

Wetlands have been frequently used as downstream treatment systems for mineral mines. Acid mine drainage water, with its low pH and high concentrations of iron, sulfate, aluminum, and trace metals, is a major water pollution problem in many coal-mining regions of the world and constructed wetlands are a viable treatment option (Fig. 20-3). The use of wetlands for coal mine drainage control was probably first considered when volunteer *Typha* wetlands were observed near acid seeps in a harsh environment where no other vegetation could grow. By the 1980s, more than 140 wetlands had been constructed in the eastern United States alone to treat mine drainage water (Wieder, 1989). The most common goal of these systems is usually the removal of iron from the water column to avoid its discharge downstream, but sulfate reduction and the alleviation of extremely acidic conditions are also appropriate goals (Wieder and Lang, 1984; Brodie et al., 1988; Fennessy and Mitsch, 1989; Mitsch and Wise, 1998; Tarutis et al., 1999). Design criteria for these wetlands have been developed, but they are neither consistent from site to site nor generally accepted. Stark and Williams (1995) found design features that enhanced iron removal and decreased acidity included broad drainage basins, nonchannelized flow patterns, high plant diversity, southern exposure, low flow rates and loadings, and shallow depths. It is not always cost effective to construct wetlands when extremely high (>85–90 percent) iron removal efficiencies are necessary or when the pH of the mine drainage water is less than 4. Furthermore, effluent from constructed wetlands does not always meet strict regulatory requirements. Nevertheless, where no other alternative is feasible, the use of wetlands to reduce this type of water pollution is viewed as a low-cost alternative to costly chemical treatment or to downstream water pollution (Baker et al., 1991).

Stormwater and Nonpoint Source Wetlands

The control of stormwater and nonpoint source pollution has been proposed as a valid application of the ecological engineering of wetlands. Unlike municipal wastewater, stormwater and nonpoint source pollution are seasonal, often quite sporadic, and variable in quality, depending on the season and recent land use. Wetlands have been designed for capturing stormwater in urban areas in Florida (Johengen and LaRock, 1993), Washington (Reinelt and Horner, 1995), and England (Shutes et al., 1997) and for preventing pollution from nonpoint runoff in agricultural watersheds in southeastern Australia (Raisin and Mitchell, 1995; Raisin et al., 1997), northeastern Spain (Comin et al., 1997), Illinois (Kadlec and Hey, 1994; Phipps and Crumpton, 1994; Mitsch et al., 1995), Florida (Moustafa, 1999), Ohio (Spieles and Mitsch, 2000; Nairn and Mitsch, 2000), and Sweden (Leonardson et al., 1994; Jacks et al., 1994; Arheimer and Wittgren, 1994). One of the features of these systems is that severe storms have a dramatic effect on treatment efficiency. High flows usually result in less nutrient retention as a percentage of inflow and sometimes even cause a net release of nutrients.

Despite the importance of controlling agricultural runoff, few careful studies of the effects of treatment wetlands on agricultural nonpoint sources have been published

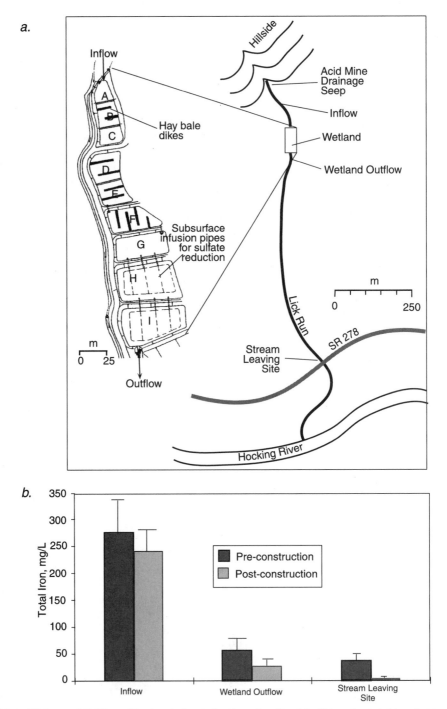

Figure 20-3 **a.** A 0.4-ha acid mine drainage treatment wetland in Ohio; **b.** total iron in stream before wetland was built and after it was constructed. (*After Mitsch and Wise, 1998*)

in peer-reviewed literature in the United States. Several wetland sites have received the equivalent of nonpoint source pollution but under somewhat controlled hydrologic conditions (e.g., river overflow to riparian basins) over several years of study. Bony Marsh, a constructed wetland located along the Kissimmee River in southern Florida, was investigated for nutrient retention of river water for 9 years (1978–1986) by the South Florida Water Management District (Moustafa et al., 1996) and found to be a consistent sink of nitrogen and phosphorus but at relatively low levels. At constructed riparian wetlands at the Des Plaines River Wetlands Demonstration Project in north-eastern Illinois (Fig. 20-4; Hey et al., 1994 a, b; Kadlec and Hey, 1994; Phipps and Crumpton, 1994; Mitsch et al., 1995) and the Olentangy River Wetland Research Park in central Ohio (Fig. 20-5; Mitsch et al., 1998; Spieles and Mitsch, 2000; Nairn and Mitsch, 2000), consistent patterns of nutrient and sediment retention have been observed over multiple years of study of these systems, which both received pumped river water, thus simulating wetlands receiving dilute nonpoint source pollution. One of the largest wetlands constructed for the control of nutrients in stormwater, the Everglades Nutrient Removal (ENR) Project, a 1,545-ha constructed marsh, removed 82 percent of the phosphorus and 55 percent of the total nitrogen applied to it over 3 years (Moustafa, 1999).

Landfill Leachate Wetlands

A more recent application of constructed wetlands has been to treat leachate from landfills. Impermeable liners are used to collect groundwater that has passed through

Figure 20-4   Aerial photo of parts of Des Plaines River Wetlands Demonstration Project in mid-1990s. Four wetland basins used in initial hydrologic experiments are in bottom of photograph. Des Plaines River shown flowing from north (left) to south (right) of photograph. (*Courtesy of The Wetlands Research Initiative, Chicago, reprinted with permission*)

a.

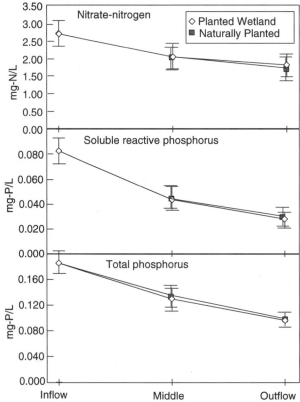

b.

Figure 20-5 Experimental wetlands at Olentangy River Wetland Research Park, The Ohio State University: **a.** aerial photo, showing two kidney-shaped experimental wetlands; **b.** reduction of nutrients flowing through the planted and unplanted wetlands in 1998, four years after one basin was planted with macrophytes while second basin was left to natural colonization of plants.

the landfill. This leachate is often quite variable in water quality but generally has very high concentrations of ammonium nitrogen and chemical oxygen demand (Kadlec, 1999). This wastewater has always presented a problem to landfill operators, and stricter water quality standards are making it necessary for advanced treatment. Wetlands are one of several options for management of leachate; other options include spray irrigation, physical/chemical treatment, biological treatment, and piping to a wastewater treatment plant. A summary of results from several dozen constructed wetlands that are treating landfill leachate in Canada, the United States, and Europe is presented by Mulamoottil et al. (1999).

Agricultural Wastewater Wetlands

In addition to the nonpoint sources from agriculture discussed previously, serious water pollution problems occur in many parts of the world due to runoff from confined animals, particularly dairy, cattle, and swine operations (Tanner et al., 1995; Cronk, 1996; CH2M-Hill and Payne Engineering, 1997). As more animals are concentrated per unit area to increase food production, the concentrations and volumes of effluents are becoming more noticeable, both by the public and by water pollution control authorities. Concentrations of organic matter, organic nitrogen, ammonia nitrogen, phosphorus, and fecal coliforms from animal feed lots far exceed concentrations in most municipal sewer systems. Two examples from the eastern United States of the effectiveness of wetlands for treating wastewater from dairy milkhouses are illustrated in Table 20-1. In both cases, there were significant reductions in most pollutants in the treated water, although ammonia nitrogen increased substantially in the Connecticut wetland and nitrate nitrogen increased by 80 percent in the Maryland case.

In addition to livestock waste from land-based agriculture, constructed wetlands have been used to treat effluent from a number of aquaculture operations, including shrimp ponds in Thailand (Sansanayuth et al., 1996) and tilapia fishponds in the United Kingdom (Redding et al., 1997).

Table 20-1 Concentrations and effectiveness of two wetlands constructed to deal with dairy milkhouse effluent

	Connecticut[a]		Maryland[b]	
	Inflow	Outflow	Inflow	Outflow
Wetland area (m^2)	400		1,160	
Flow (m^3/wk)	18.8		—	
Retention time (day)	41		—	
BOD (mg/L)	2,680	611	1,914	59
Total nitrogen (mg/L)	103	74	170	13
Ammonia nitrogen (mg/L)	8	52	72	32
Nitrate nitrogen (mg/L)	0.3	0.1	5.5	10.0
Total phosphorus (mg/L)	26	14	53	2.2
TSS (mg/L)	1,284	130	1,645	65
Coliform (#/100 mL)	557,000	13,700	—	—

[a] Newman et al. (2000).
[b] Schaafsma et al. (2000).

Wetland Design

The need for rigor in designing a wetland varies widely, depending on the site and application. In general, a design that uses natural processes to achieve the objectives yields a less expensive and more satisfactory solution in the long run. On the other hand, "naturally" designed wetlands may not develop as predictably as more tightly designed systems should. The choice of design is strongly affected by the site and the objectives. In Europe and many parts of North America, subsurface wetlands are designed in rectangular basins to very specific design criteria. In coastal Louisiana, by contrast, there are now several projects where natural wetlands are being used as tertiary treatment systems for the removal of nutrients from wastewater. In the following sections, we focus on rigidly designed wetlands in part because this kind of wetland creation requires much greater ecotechnological sophistication.

Hydrology

Hydrology is the most important variable in wetland design. If the proper hydrologic conditions are developed, chemical and biological conditions will respond accordingly (see Chapters 5–7). The hydrologic conditions, in turn, depend on climate, seasonal patterns of streamflow and runoff, tides (for coastal wetlands), and possible groundwater influences. Improper hydrology leads to the failure of many created wetlands (D'Avanzo, 1989). Improper hydrologic conditions will not always correct themselves as will the more forgiving biological components of the system. Ultimately, hydrologic conditions determine wetland function. Several parameters used to describe the hydrologic conditions of treatment wetlands include hydroperiod, depth, seasonal pulses, hydraulic loading rates, and detention time.

Hydroperiod and Depth

One of the most basic design parameters for constructed wetlands is the hydroperiod, that is, the pattern of water depth over time, including the seasonal pattern of the depth and the frequency of flooding (see Chapter 5). Wetlands that have a seasonal fluctuation of water depth have the most potential for developing a diversity of plants, animals, and biogeochemical processes.

In a constructed wastewater wetland with a similar inflow of wastewater every day, water levels vary little unless several wetlands can be managed in tandem. Water levels are controlled by outflow structures that provide appropriate water depths, for example, for planting and controlling undesirable plants. During the start-up period of constructed wetlands, low water levels are needed to avoid flooding newly emerged plants, whereas continuous flooding is necessary for floating-leaved and submerged plants. Specific periods of drawdown may be necessary if wetland plants are to be started from germinating seeds. Start-up periods for the establishment of plants may take 2 to 3 years and development of an adequate litter sediment compartment may take another 2 to 3 years after that.

Seasonal and Year-to-Year Pulses

Storms and seasonal patterns of floods rarely affect constructed wastewater wetlands, but they can significantly affect the performance of wetlands designed for the control of nonpoint source runoff. A variable hydroperiod, exhibiting dry periods interspersed with flooding, is a natural cycle in prairie pothole freshwater marshes, and fluctuating water levels should be accepted if not encouraged. A fluctuating water level could provide needed oxidation of organic sediments and can, in some cases, rejuvenate a system to higher levels of chemical retention.

Most of the loading of phosphorus to wetlands in rural settings comes adsorbed to sediments, so high phosphorus loading results when erosive forces are highest—during storms and in the winter and early spring when there is less vegetation cover in the watershed. In agricultural watersheds, high pulses of nitrate nitrogen occur during the first storms after fertilizer application to farm fields. A good wetland design should both take advantage of these pulses for system replenishment and provide for excess wet weather storage if nutrient retention is a primary objective. Seasonal pulses are themselves not constant but vary in intensity from year to year. The rarer events, for example, the 100-year flood, could be the streamflow for which the wetland has been designed. Infrequent and nonperiodic floods and droughts are also important for dispersing biological species to the wetland and adjusting resident species composition.

Hydraulic Loading Rate

The hydraulic loading rate is defined as

$$q = Q/A \tag{20.1}$$

where q = inflowing hydraulic loading rate (HLR), the volume per unit time per unit area, which is equivalent to the depth of flooding over the treatment area per unit time (m day^{-1} or m yr^{-1})

Q = flow rate (m^3 day^{-1} or m^3 yr^{-1})

A = wetland surface area (m^2)

Table 20-2 lists the hydrologic loading rates (HLRs) of some of the many surface flow and subsurface flow wastewater wetlands in North America and Europe. Loading rates to surface flow constructed wetlands for wastewater treatment from small municipalities ranged from 1.4 to 22 cm day^{-1} (average = 5.4 cm day^{-1}), while rates to subsurface flow constructed wetlands varied between 1.3 and 26 cm day^{-1} (average = 7.5 cm day^{-1}). Knight (1990) reviewed several dozen wetlands constructed for wastewater treatment and found loading rates to vary between 0.7 and 5 cm day^{-1}. He recommended a rate of 2.5 to 5 cm day^{-1} for surface flow constructed wetlands and 6 to 8 cm day^{-1} for subsurface flow wetlands.

The previous loading rate limitations are probably too restrictive for wetlands used for the control of nonpoint pollution and stormwater runoff, but few studies have been undertaken to determine the optimum design rates for such wetlands. The Des Plaines River Wetlands Demonstration Project in northeastern Illinois had initial experiments designed for loading rates of river water from 1 to 8 cm day^{-1} with pumps delivering the water from the Des Plaines River. These rates, however, were estimated from rates for comparable wastewater wetlands and may be too low for riparian wetlands receiving floodwaters.

Hydraulic loading rates as high as 29 cm day^{-1} have been suggested for wetlands designed for acid mine drainage (Watson et al., 1989), although Fennessy and Mitsch (1989b) recommended 5 cm day^{-1} as a conservative loading rate for this type of wetland (Table 20-3).

Detention Time

As with loading rate, most of the experience with detention time of wetlands (see Chapter 5) is based on wetlands designed to treat wastewater. Detention time is calculated as

Table 20-2 Hydrologic loading for wastewater wetlands in North America and Europe

Wetland Project	Type of Flow	Loading Rate (cm/day) Design	Actual
NORTH AMERICA			
Ann Arundel County, MD			
Freshwater wetland	Surface-flow	15.1	
Peat wetland	Percolation	10.2	
Offshore wetland	Surface-flow	1.1	
Arcata, CA	Surface-flow	24.0	2.7
Benton, KY			
Cell 1	Surface-flow	9.5	5.3
Cell 2	Surface-flow	7.1	4.1
Cell 3	Subsurface[a]	7.1	4.3
Brookhaven, NY			
Marsh	Surface-flow		2.3
Pond	Surface-flow		4.5
Cannon Beach, OR	Surface-flow		5.7
Cobalt, Ontario	Surface-flow		2–10
Collins, MS	Surface	3.8	
East Lansing, MI	Surface-flow marsh/pond		1.8
Emmitsburg, MD	Subsurface		16.4
Foothills Pointe, TN	Subsurface	4.7	
Fort Deposit, AL	Surface-flow	1.5	
Gustine, CA	Surface-flow	3.8	
Hardin, KY	Subsurface	5.9	
Houghton Lake, MI	Surface-flow		0.1
Incline Village, NV	Surface-flow	1.3	
Iron Bridge, FL	Surface-flow	1.6	0.6
Iselin, PA			
Marsh	Subsurface[a]	4.7	5.3
Meadow	Subsurface[a]	9.4	10.7
Lake Buena Vista, FL	Surface-flow		4.0
Lakeland, FL	Surface-flow	0.9	0.5
Listowel, Ontario	Surface-flow		1.3–2.0
Mountain View Sanitary, CA	Surface-flow marsh/pond	7.8	3.5
Neshaminy Falls, PA			
Marsh	Subsurface[a]	4.6	2.4
Meadow	Subsurface[a]	9.3	4.8
Orange County, FL	Surface-flow		1.5
Orlando, FL	Surface-flow	1.6	0.6
Paris Landing, TN	Subsurface	15	
Pembrook, KY	Surface/subsurface	2.2	
Phillips High School, Bear Creek, AL	Subsurface	3.7	
Santee, CA	Subsurface	4.7	4.7
Silver Springs Shores, FL	Surface-flow		1.2
Vermontville, MI	Surface-flow	1.4	
EUROPE			
Gravesend, England	Subsurface–gravel	8.2	
Marnhell, England	Subsurface–soil	4.5–6.9	
Holtby, England	Subsurface–soil	4.9	
Castleroe, England	Subsurface–gravel and soil	4.3	
Middleton, England	Subsurface–sand/gravel	8.9	
Bluther Burn, England	Subsurface–flyash	6.2–10.8	
	Subsurface–gravel	9.9–10.1	

Table 20-2 *(Continued)*.

| Wetland Project | Type of Flow | Loading Rate (cm/day) | |
		Design	Actual
Little Stretton, England	Subsurface–gravel	26	
Ringsted, Denmark	Subsurface–gravel	5.7	
	Subsurface–clay	1.7	
Surface-flow wetlands—average ± standard error		5.4 ± 1.7 (*n* = 15)	
Subsurface-flow wetlands—average ± standard error		7.5 ± 1.0 (*n* = 23)	

[a] Plugging resulted in substantial surface flow.
Source: Watson et al. (1989).

$$t = Vp/Q \qquad\qquad (20.2)$$

where t = theoretical detention time (day)

V = volume of wetland basin (m^3)

    = volume of water for surface flow wetlands

    = volume of medium through which the wastewater flows for subsurface flow

p = porosity of medium (e.g., sand or gravel for subsurface flow wetlands)

    $\sim$ 1.0 for surface flow wetlands

Q = flow rate through wetland (m^3 day^{-1})

    = $(Q_i + Q_o)/2$, where Q_i is inflow and Q_o is outflow

The optimum detention time (or nominal residence time) ranges from 5 to 14 days for treatment of municipal wastewater. Florida regulations on wetlands require that the volume in the permanent pools of the wetland must provide for a residence time of at least 14 days. M. T. Brown (1987) suggested a detention time of a riparian treatment wetland system in Florida of 21 days during the dry season and more than 7 days during the wet season. Fennessy and Mitsch (1989b) recommended a minimum

Table 20-3 Average design parameters for marsh-type wetlands used to treat coal mine drainage

Parameter	Suggested Design
Hydrology	
Loading	5 cm/day
Retention time	>24 hours
Iron loading (g m^{-2} day^{-1})	2–10 (for 90% Fe removal)
	20–40 (for 50% Fe removal)
Basin characteristics	
Depth	<0.3 m
Cells	Multiple (3 or more)
Planting material	*Typha* sp.
Substrate material	Organic peat over clay seal

Source: Fennessy and Mistch (1989b).

detention time of 1 day for acid mine drainage wetlands (Table 20-3) and much longer periods for more effective iron removal.

Calculations of detention time or nominal residence time are not always realistic because of short-circuiting and the ineffective spreading of the waters as they pass through the wetland (see Chapter 5). Tracer studies of flowthrough wetlands (see, e.g., Fig. 20-6) illustrate the importance of not overrelying on the theoretical detention time to design treatment wetlands. Not all parcels of water that enter at a certain time leave the wetland at the same time. In some instances, water will "short-circuit" through the wetland; in other instances, it will remain in backwater locations for considerably more time than the theoretical detention time.

Basin Morphology

Several aspects related to the morphology of constructed wetland basins need to be considered when designing wetlands. Several states have developed guidelines for bottom profiles of constructed wetlands, particularly freshwater marshes. Florida regulations for the Orlando area require, for littoral zones, a shelf with a gentle slope of

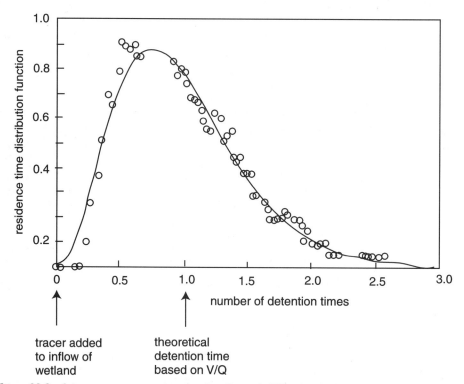

Figure 20-6 A tracer response curve, showing the probability density function of the residence times of water in a constructed wetland basin at the Des Plaines River Wetlands Demonstration Project, northeast Illinois. Data points represent results of a lithium tracer study. Note that while some of the water spent exactly 1 theoretical detention time (V/Q) in the wetland, there were considerable quantities of inflowing water that passed through the wetland much more quickly, and some inflowing water that stayed in the wetland up to double the theoretical detention time. (*After Kadlec and Knight, 1996*)

6 : 1 or flatter to a point 60 to 77 cm below the water surface. Slopes of 10 : 1 or flatter are even better. A flat littoral zone maximizes the area of appropriate water depth for emergent plants, thus allowing more wetland plants to develop more quickly and allowing wider bands of different plant communities. Plants will also have room to move "uphill" if water levels are raised in the basins due to flows being higher than predicted or to enhance treatment. Bottom slopes of less than 1 percent are recommended for wetlands built to control runoff, whereas a substrate slope, from inlet to outlet, of 0.5 percent or less has been recommended for surface flow wetlands used to treat wastewater.

Flow conditions should be designed so that the entire wetland is effective in nutrient and sediment retention if these are desired objectives. This may necessitate several inflow locations and a wetland configuration to avoid channelization of flows. Steiner and Freeman (1989) suggested a length-to-width (L/W) ratio (called the *aspect ratio*) of at least 10 : 1 if water is purposely introduced to the system. A minimum aspect ratio of 2 : 1 to 3 : 1 has been recommended for surface flow wastewater wetlands.

Providing a variety of deep and shallow areas is optimum. Deep areas (>50 cm), while probably too deep for emergent vegetation, offer habitat for fish, increase the capacity of the wetland to retain sediments, enhance nitrification as a prelude to later denitrification if nitrogen removal is desired, and provide low-velocity areas where water flow can be redistributed. Shallow depths (<50 cm) provide maximum soil–water contact for certain chemical reactions such as denitrification and can accommodate a greater variety of emergent vascular plants.

Individual wetland cells, placed in series or parallel, often offer an effective design to create different habitats or establish different functions. Cells can be parallel so that alternate drawdowns can be accomplished for mosquito control or redox enhancement, or they can be in a series to enhance biological processes.

Chemical Loadings

When water flows into a wetland, it brings chemicals that may be beneficial or possibly detrimental to the functioning of that wetland. In an agricultural watershed, this inflow will include nutrients such as nitrogen and phosphorus as well as sediments and possibly pesticides. Wetlands in urban areas can have all of these chemicals plus other contaminants such as oils and salts. Wastewater, when added to wetlands, has high concentrations of nutrients and, with incomplete primary treatment, high concentrations of organic matter (BOD) and suspended solids. At one time or another, wetlands have been subjected to all of these chemicals, and they often serve as effective sinks.

Loading Graphs

The simplest model available to estimate the retention of nutrients or other chemicals by wetlands is to graph some measure of that retention versus chemical loading, either areal (e.g., g m^{-2} yr^{-1}) or volumetric (e.g., g m^{-3} yr^{-1}). If a wetland is designed to retain nutrients, for example, it would be desirable to know how well that retention would occur for various nutrient inflows. Data compiled from a large number of wetland sites in North America and Europe provide an indication of the nutrient retention of wetlands. Figure 20-7, compiled from some of those data sets, illustrates the percentage removal of nitrogen and phosphorus for various areal loading rates of

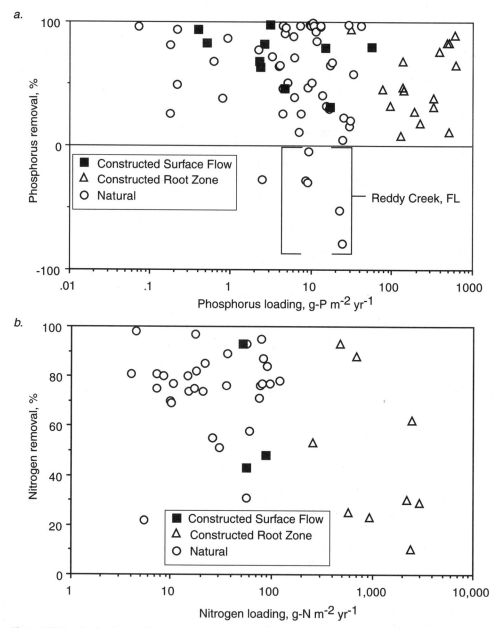

Figure 20-7 Nutrient retention as a function of loading in wetlands receiving wastewater or river water as a function of loading for **a.** phosphorus and **b.** nitrogen. Data are from Knight (1990) and Mitsch et al. (1995). Reddy Creek data in (a) were for a wetland in Florida that consistently exported phosphorus after wastewater was applied (Knight et al., 1987). Data are shown for surface-flow constructed wetlands, subsurface-flow constructed wetlands (root zone), and natural wetlands. Multiple data are given for wetlands with multiple-year monitoring.

wastewater and nonpoint source wetlands. With much scatter, the data show generally decreasing percentage retention with greater loading rates. Phosphorus retention is more erratic and site specific (Fig. 20-7a) and is more dependent on site characteristics (e.g., the chemical characteristics of the wetland substrate; see the following section). One set of phosphorus data in Figure 20-7a shows a wetland in Florida with consistent phosphorus export (negative removal percentage); wastewater was applied to phosphorus-saturated peat soils with insufficient retention capacity to accomplish additional phosphorus removal (R. Knight, personal communication); as a result, that wetland served as a source rather than a sink of phosphorus. Nitrogen retention in constructed wetlands has a more consistent pattern of lower percentage removal at higher loading rates, but there is considerable scatter of data (Fig. 20-7b).

Figure 20-8 shows a similar type of analysis for nonpoint source control of nitrate nitrogen, but here the nitrate–nitrogen retention is expressed in the three ways: (1) mass retention per unit area, (2) percentage retention by mass, and (3) percentage retention by concentration. Each of the data points is based on one year's data at one wetland basin at either the Olentangy River Wetland Research Park in Ohio or the Des Plaines River Wetlands Demonstration Project in Illinois. In other words, each data point represents one wetland-year for wetlands in the midwestern United States receiving generally dilute concentrations of nitrate nitrogen. Data appear to converge and give good design information on the capability of these types of low-nutrient wetlands to retain nitrate nitrogen.

Retention Rates

Another approach to estimating the retention of nutrients is to simply compare a number of studies and estimate the chemical retention that consistently happens in wetlands. For example, the nitrate–nitrogen retention capability of freshwater marshes receiving nonpoint source pollution shows a range of nitrogen retention from 3 to 69 g N m^{-2} yr^{-1} and phosphorus retention of 0.4 to 5.6 g P m^{-2} yr^{-1} (Table 20-4). A "rule of thumb" is that wetlands, when not overloaded, can consistently retain phosphorus in amounts of 1 to 2 g P m^{-2} yr^{-1} and nitrogen in amounts of about 10 to 20 g N m^{-2} yr^{-1}. The problem with these estimates is that the data really do not show if these wetlands were receiving an amount of nutrients for optimum retention. For example, the river-fed, low-flow constructed wetlands in northeast Illinois in Table 20-4 were not loaded to nearly their capacity for removing nutrients (Hey et al., 1994a, b; Mitsch et al., 1995). Figure 20-8 normalizes nitrate–nitrogen data from both the Illinois and the Ohio studies by showing retention for a number of loading rates. When data are viewed in this way, there is a clearer, well-correlated relationship between loading and nutrient removal.

Averages from data from many constructed wastewater wetlands are shown in (Table 20-5). The much higher average mass retention of nitrate nitrogen in subsurface wetlands is due more to very high loading rates of nitrate nitrogen in some of the subsurface flow wetlands in the database than it is to any ability of these systems to sequester more nitrate nitrogen. Note that the percentage nitrate–nitrogen retention is much higher in the surface flow than in the subsurface flow wetlands, partially because of the excessively high loading rates for the latter.

Empirical Models

A third method for estimating the ability of wetlands to retain chemicals is to use equations that either have a theoretical base or are empirically determined from large

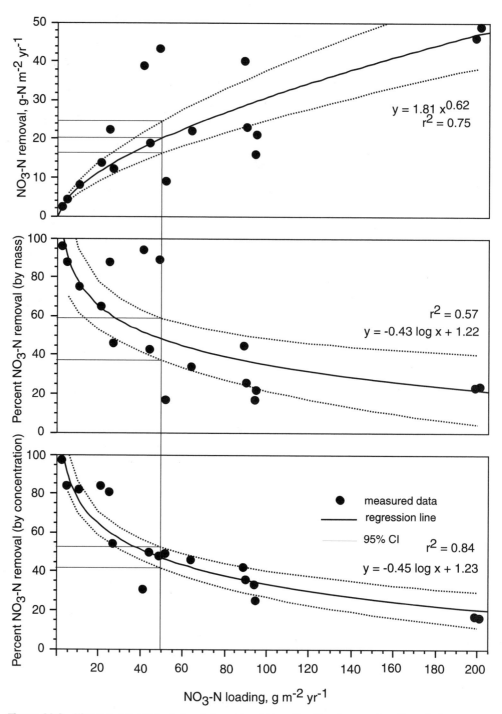

Figure 20-8 Nitrogen retention of river-fed constructed wetland basins as a function of nitrate loading. Nitrogen retention is shown as areal-based removal, percentage nitrate decrease by mass, and percentage nitrate–nitrogen decrease by concentration. Each data point represents one full year of data in one wetland basin at Des Plaines River wetlands in Illinois and Olentangy River wetlands in Ohio. (*After Mitsch et al., 1999*)

Table 20-4 Estimated nutrient retention in constructed and natural wetlands receiving low-concentration (i.e., non-wastewater) nutrient loading from rivers, overflows, or nonpoint source pollution

Wetland Location/type	Wetland Size (ha)	Nitrogen ($g\ N\ m^{-2}\ yr^{-1}$)	Phosphorus ($g\ P\ m^{-2}\ yr^{-1}$)	Reference
WARM CLIMATE				
Everglades marsh, south Florida	8,000	—	0.4–0.6[a]	Richardson and Craft (1993), Richardson et al. (1997)
Boney Marsh, south Florida	49	4.9	0.36	Moustafa et al. (1996)
Everglades Nutrient Removal Project, south Florida	1,545	10.8	0.94	Moustafa (1999)
Restored marshes, Mediterranean delta, Spain	3.5	69	—	Comin et al. (1997)
Constructed rural wetland, Victoria, Australia	0.045	23	2.8	Raisin et al. (1997)
COLD CLIMATE				
Constructed wetlands, northeast Illinois				Mitsch (1992a), Phipps and Crumpton (1994), Mitsch et al. (1995)
River-fed and high-flow	2	11–38[b]	1.4–2.9	
River-fed and low-flow	2–3	3–13[b]	0.4–1.7	
Artificially flooded meadows, southern Sweden	180	43–46	—	Leonardson et al. (1994)
Palustrine freshwater wetlands, northwest Washington				Reinelt and Horner (1995)
Urban area	2	—	0.44	
Rural area	15	—	3.0	
Constructed urban instream wetland, central Ohio	6	—	2.9	Niswander and Mitsch (1995)
Constructed riparian wetlands, central Ohio				Mitsch et al. (1998), Spieles and Mitsch (2000), Nairn and Mitsch (2000)
Planted marsh	1	58[b]	5.6	
Naturally colonizing marsh	1	66[b]	5.2	
Natural marsh, Alberta, Canada	360	—	0.43[a]	White et al., 2000

[a] Estimated by phosphorus accumulation in sediment.
[b] Nitrate nitrogen only.

Table 20-5 Nutrient and sediment removal rates and efficiency in constructed wastewater wetlands

Wetland Type Parameter	Loading (g m^{-2} yr^{-1})	Retention (g m^{-2} yr^{-1}	Percentage Retention
SURFACE-FLOW CONSTRUCTED WETLANDS			
Nitrate + nitrate nitrogen	29	13	44.4
Total nitrogen	277	126	45.6
Total phosphorus	4.7–56	2.1–45	46–80
Suspended solids	107–6,520	65–5,570	61–98
SUBSURFACE-FLOW CONSTRUCTED WETLANDS			
Nitrate + nitrate nitrogen	5,767	547	9.4
Total nitrogen	1,058	569	53.8
Total phosphorus	131–631	11–540	8–89
Suspended solids	1,500–5,880	1,100–4,930	49–89

Source: Knight (1990) and Kadlec and Knight (1996).

databases of existing wastewater wetlands. One such general model, developed by Robert Kadlec and others (Kadlec and Knight, 1996) and based on a mass-balance approach, is called the "k–C^* model" and is given as

$$q\frac{dC}{dy} = k_A (C - C^*) \tag{20.3}$$

where y = fractional distance from inlet to outlet (unitless)
C = chemical concentration (g m^{-3})
k_A = areal removal rate constant (m yr^{-1})
C^* = residual or background chemical concentration (g m^{-3})

This equation is based on the assumption that many processes occur on an areal basis in wetlands. Thus, the coefficient k_A has units of velocity and can be recognized as being similar to a "settling velocity" coefficient used in sedimentation models. C^* represents a background concentration of a chemical or constituent, below which it is generally agreed that treatment wetlands cannot go. Integrating this equation over the entire length of the wetland, the solution can be expressed as a first-order areal model:

$$\frac{C_o - C^*}{C_i - C^*} = \exp\left(\frac{-k_A}{q}\right) \tag{20.4}$$

where C_o = outflow concentration (g m^{-3})
C_i = inflow concentration (g m^{-3})
q = hydraulic loading rate (m yr^{-1})

Estimates of the two parameters needed for this model, C^* and k_A, are listed in Table 20-6. This equation does not work equally well for all parameters but does provide a way of estimating the area of a wetland necessary for achieving a certain removal. Rearranging Equations 20-4 and 20-1 gives the following calculation of wetland area for given results:

Table 20-6 Parameters for first-order areal model given in equations 20-3 to 20-5 for several constituents of wastewater wetlands. Subsurface-flow constructed wetlands and surface-flow constructed wetlands are given as wetland type where appropriate

Constituent and wetland type	k_A (m yr^{-1})	C^* (g m^{-3})
BOD, surface-flow	34	$3.5 + 0.053C_i$
BOD, subsurface-flow	180	$3.5 + 0.053C_i$
Suspended solids, surface-flow	1,000	$5.1 + 0.16C_i$
Total phosphorus, surface and subsurface-flow	12	0.02
Total nitrogen, surface-flow	22	1.5
Total nitrogen, subsurface-flow	27	1.5
Ammonia nitrogen, surface-flow	18	0
Ammonia nitrogen, subsurface-flow	34	0
Nitrate nitrogen, surface-flow	35	0
Nitrate nitrogen, subsurface-flow	50	0

Source: Kadlec and Knight (1996).

$$A = Q \ln \left[\frac{C_o - C^*}{C_i - C^*} \right] / k_A \qquad (20.5)$$

where Q = flow rate through wetland (m^3 yr^{-1})

Where this model is insufficiently supported with good data or does not work properly, strictly empirical relationships of the outflow concentration C_o as a function of the inflow concentration C_i and the HLR (q) have been developed. For example, the equation estimating total phosphorus efflux in a surface flow wastewater marsh is:

$$C_o = 0.195 C_i^{0.91} \, q^{0.53} \qquad (20.6)$$

where C_o and C_i = outflow and inflow concentrations, respectively (g m^{-3})
q = hydraulic loading rate (cm day^{-1})

Table 20-7 illustrates several regression equations of this nature that could be used, with other approaches, to estimate outflow concentrations and, in one case, wetland area.

Other Chemicals

Although most evaluations of the efficiency of wetlands have been concerned with their capacity to remove nutrients, sediments, and organic carbon (BOD), there is some literature on other chemicals such as iron, cadmium, manganese, chromium, copper, lead, mercury, nickel, and zinc (reviewed by Richardson and Nichols, 1985; Kadlec and Knight, 1996). Metals are often easily sequestered by wetland soils or biota or both. B. H. McArthur (1989) reported on a case study in Florida where lead was reduced by 83 percent, zinc by 70 percent, and total solids by 55 percent as stormwater passed through a 0.4-ha pond–wetland complex.

The accumulation of selenium became an issue in marshes in the Kesterson National Wildlife Refuge in California when it began to accumulate in biota there as a result of the inflow of spent irrigation waters for several years. Concentrations of selenium in these agricultural irrigation return flows were as high as 300 μg/L (Ohlendorf et al., 1986, 1990) compared to an average of 0.1 μg/L in the world's

Table 20-7 Empirical equations for the estimation of outflow concentrations or wetland area based on inflow concentrations and hydraulic retention time. Correlation coefficient (r^2) and number of wetlands used in analysis (n) are also given

Constituent	Equation[a]	r^2 (n)
BOD		
Surface-flow wetlands	$C_o = 4.7 + 0.173C_i$	0.62 (440)
Subsurface-flow, soil	$C_o = 1.87 + 0.11C_i$	0.74 (73)
Subsurface-flow, gravel	$C_o = 1.4 + 0.33C_i$	0.48 (100)
Suspended solids		
Surface-flow wetlands	$C_o = 5.1 + 0.158C_i$	0.23 (1,582)
Subsurface-flow wetlands	$C_o = 4.7 + 0.09C_i$	0.67 (77)
Ammonia nitrogen		
Surface-flow wetlands	$A = 0.01Q/\exp[1.527 \ln C_o$ $- 1.05 \ln C_i + 1.69]$	
Surface-flow marshes	$C_o = 0.336C_i^{0.728}q^{0.456}$	0.44 (542)
Subsurface-flow wetlands	$C_o = 3.3 + 0.46C_i$	0.63 (92)
Nitrate nitrogen		
Surface-flow marshes	$C_o = 0.093C_i^{0.474}q^{0.745}$	0.35 (553)
Subsurface-flow wetlands	$C_o = 0.62C_i$	0.80 (95)
Total nitrogen		
Surface-flow marshes	$C_o = 0.409C_i + 0.122q$	0.48 (408)
Subsurface-flow wetlands	$C_o = 2.6 + 0.46C_i +$ $0.124q$	0.45 (135)
Total phosphorus		
Surface-flow marshes	$C_o = 0.195C_i^{0.91}q^{0.53}$	0.77 (373)
Surface-flow swamps	$C_o = 0.37C_i^{0.70}q^{0.53}$	0.33 (166)
Subsurface-flow wetlands	$C_o = 0.51C_i^{1.10}$	0.64 (90)

[a] C_i, inflow concentration (g m^{-3}); C_o, outflow concentration (g m^{-3}); A, area of wetland (ha); Q, wetland inflow, (m^3/day); q, hydraulic retention time, (cm/day).
Source: Kadlec and Knight (1996).

rivers. As a result, the Kesterson marshes were found to be a threat to fish and wildlife in the region and were dewatered.

Substrate/Soils

The substrate is important to the overall function of a constructed wetland and is the primary medium supporting rooted vegetation. The soil or subsoil must also have permeability low enough to retain standing water or saturated soils. If a wetland is designed to improve water quality, the substrate plays a significant role in its ability to retain certain chemicals. The sediments retain certain chemicals and provide a habitat for micro- and macroflora and fauna that are involved in chemical transformations. The substrate is not as significant for the retention of suspended organic matter and solids (along with the chemicals adsorbed to sediment particles), because their retention is based primarily on net deposition (sedimentation less resuspension) that results from the slow water velocities characteristic of wetlands. Some of the characteristics of substrates that do play a role in the design of wetlands are reviewed here.

Organic Content

The organic content of soils has some significance for the retention of chemicals in a wetland. Mineral soils generally have lower cation exchange capacity than organic soils do; the former are dominated by various metal cations, and the latter are dominated by the hydrogen ion (see Chapter 6). Organic soils can, therefore, remove some contaminants (e.g., certain metals) through ion exchange and can enhance nitrogen removal by providing an energy source and anaerobic conditions appropriate for denitrification. Organic matter in wetland soils generally varies between 5 and 75 percent, with higher concentrations in peat-building systems such as bogs and fens and lower concentrations in mineral soil wetlands such as riparian bottomland wetlands subject to mineral sedimentation or erosion. When wetlands are constructed, especially subsurface flow wetlands, organic matter such as composted mushrooms, peat, or detritus is often added in one of the layers. For construction of many wetlands, however, organic soils are avoided because they are low in nutrients, can cause low pH, and often provide inadequate support for rooted aquatic plants.

Soil Texture

Constructed wetland soil texture depends on whether surface flow over the substrate or subsurface flow through the substrate is being considered (Fig. 20-9). Surface flow systems are generally less effective in removing pollutants per unit area but are closer in design to natural wetlands. Clay material, although often favored for surface flow wetlands to prevent percolation of water to groundwater, may also limit root and rhizome penetration and may be impermeable to water for plant roots. In that case, loam soils are preferable. Sandy soils, although generally low in nutrients, do anchor plants and allow water to reach the plant roots readily. The use of local soils, underlain with impermeable clay or bentonite, to prevent downward percolation, is often the best design. If clay is not available on site, it may be advisable to add a layer to minimize percolation. If on-site topsoils are to be returned to the wetland, adequate temporary storage should be provided.

Subsurface flow through artificial wetlands can be through soil media (*root zone method*) or through rocks, gravel, or sand (*rock* reed filters). Flow in both cases is 15 to 30 cm below the surface. Gravel is sometimes added to the substrate of subsurface flow wetlands (*gravel bed*) to provide a relatively high permeability that allows water to percolate into the root zone of the plants where microbial activity is high. In a survey of several hundreds of wetlands built in Europe for sewage treatment in rural settings, Cooper and Hobson (1989) reported that gravel had been used in combination with soil but that the substrate remained the greatest uncertainty in artificial reed (*Phragmites*) wetlands in Europe. Gravel can be silica based or limestone based; the former has much less capacity for phosphorus retention. Another evaluation of European-designed subsurface wetlands indicated that they decreased in hydrologic conductivity after a few years and became essentially overloaded surface flow wetlands (Steiner and Freeman, 1989).

Depth and Layering of Substrate

The depth of substrate is an important design consideration for both wastewater and mine drainage wetlands, particularly those that use subsurface flow. The depth of substrate is of less concern for surface flow wetlands than for subsurface flow wetlands. All wetlands should have an adequate layer of clay materials at an appropriate depth

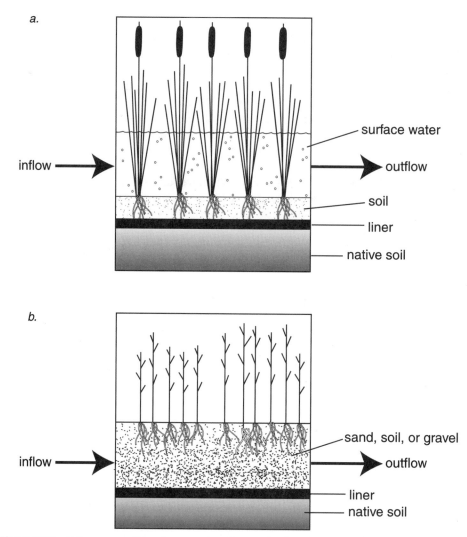

Figure 20-9 Soil cross section of **a.** surface-flow wetland, and **b.** subsurface-flow wetland. (*After Knight, 1990*)

if downward percolation is not desired. Meyer (1985) described a layered substrate in wetlands to control stormwater runoff as having the following materials from the base upward: 60 cm of 1.9-cm limestone; 30 cm of 2-mm crushed limestone to raise pH and aid in the precipitation of dissolved heavy metals and phosphate; 60 cm of coarse-to-medium sand as filter, and 50 cm of organic soil. A common substrate depth for subsurface flow wetlands is 60 cm. The depth of suitable substrate should be adequate to support and hold vegetation roots (see Vegetation later in this chapter).

Nutrients

Although exact specifications of nutrient conditions in a substrate necessary to support aquatic plants are not well known, low nutrient levels characteristic of organic, clay, or sandy soils can cause problems for initial plant growth. Although fertilization may be necessary in some cases to establish plants and enhance growth, it should be avoided if possible in wetlands that eventually will be used as sinks for macronutrients. When fertilization is required to get plants started in constructed wetlands, slow-release granular and tablet fertilizers are often useful.

Iron and Aluminum

When soils are submerged and anoxic conditions result, iron is reduced from the ferric (Fe^{3+}) to the ferrous (Fe^{2+}) state, releasing phosphorus that was previously held as insoluble ferric phosphate compounds. The Fe–P compound can be a significant source of phosphorus to overlying and interstitial waters after flooding and anaerobic conditions occur, particularly if the wetland was constructed on former agricultural land. After an initial pulse of released phosphorus in such constructed wetlands, the iron and aluminum contents of a wetland soil exert significant influences on the ability of that wetland to retain phosphorus. All things being equal, soils with higher aluminum and iron concentrations are more desirable because their affinity for phosphorus is higher.

Vegetation

Just as the question "What plants should be used?" arises for creating and restoring wetlands as discussed in Chapter 19, it is also a consideration for treatment wetlands. However, there is at least one significant difference for treatment wetlands. While creation and restoration of wetlands are principally done to develop a diverse vegetation cover and replace habitat, treatment wetlands are constructed with the main goal of improving water quality. The plants in created and restored wetlands are the solution; in treatment wetlands, they are part of the solution. Furthermore, treatment wetlands invariably have higher concentrations of chemicals in the water, which limit the number of plants that will survive in those wetlands.

In theory, any of the plants listed in Table 19-3 can be used in treatment wetlands. Experience has shown, however, that relatively few plants thrive in the high-nutrient, high-BOD, high-sediment waters of treatment wetlands. Among those plants are cattails (*Typha* spp.), the bulrushes (*Schoenoplectus* spp., *Scirpus* spp.), and reed grass (*Phragmites australis*). The last is the preferred plant in subsurface-flow wetlands around the world but is not favored in many parts of North America because of its aggressive behavior in freshwater and brackish marshes. When water is deeper than 30 cm, emergent plants often have difficulty growing and surface-flow wetlands can become covered with duckweed (*Lemna* spp.) in temperate zones and water hyacinths (*Eichhornia crassipes*) in the Tropics. While rooted floating aquatics such as *Nymphaea*, *Nuphar*, and *Nelumbo* are favored for their aesthetics, they thrive only in rare instances in treatment wetlands where, under high-nutrient conditions, they are easily overwhelmed by duckweed and filamentous algae.

The number of plant species that might be suitable for treatment wetlands is large, and evaluation studies are few to date. Species of choice are certain to vary with the design and purpose of the wetland and with the inflowing water quality. Two studies designed to find the most appropriate plants for treatment wetlands are illustrative. Esry and Cairns (1989) described three cells of a marsh designed for stormwater runoff. They were planted differently to determine the effectiveness and hardiness of different plant covers. The first cell was planted with locally obtained sawgrass (*Cladium*), and the second cell was planted with bulrush (*Scirpus*) obtained in nurseries. Both of these species were hand planted in rows. The third cell was planted with pickerelweed (*Pontederia*) that had been gathered in a nearby farm pond. The plants of the first and the third cells have taken hold and thrived. The second cell still contains some bulrushes, but their overall survival was limited. In this cell, there was rapid volunteer growth of duckweed (*Lemna*) and emergent plants.

Tanner (1996) compared nutrient uptake and pollutant removal of eight macrophytes in gravel bed wetland mesocosms fed by dairy wastes in New Zealand (Fig. 20-10). Greatest above-ground biomass was seen in this highly polluted wastewater by *Glyceria maxima* and *Zizania latifolia,* while greatest below-ground biomass was seen with *Bolboschoenus fluviatilis* (*Scirpus fluviatilis* in the United States), which had below-ground biomass 3.3 times its above-ground biomass (Fig. 20-10a). Total nitrogen removed from these mesocosms was linearly correlated with total plant biomass (Fig. 20-10b). Overall, Tanner (1996) suggested that, based on key growth characteristics of the plants in this wastewater, three productive graminoids (*Zizania latifolia, Glyceria maxima,* and *Phragmites australis*) had the highest overall scores. *Baumea articulata, Cyperus involucratus,* and *Schoenoplectus validus* had medium scores, while *Scirpus fluviatilis* and *Juncus effusus* had the lowest scores and are least likely to be effective plants in wastewater wetlands.

Establishing Vegetation

Vegetation can be established through the same general procedures outlined in Chapter 19, that is, by planting roots and rhizomes directly or by seeding. Because these wetlands are usually constructed on former upland with no connection to rivers or streams, reliance on nature bringing in plant propagules generally does not work. Field-harvested plants or nursery-grown stock can be used for plantings. The former have the advantage of establishing vegetation cover more quickly than would smaller nursery stock. Also these plants, if harvested nearby, are adapted to the local climate and may be from the proper genotype for the region (Kadlec and Knight, 1996). On the other hand, harvesting plants in large numbers from natural wetlands may threaten those wetlands. Nursery plants are easier to plant because of their generally small size, and a greater diversity and number of plants can be obtained from good nurseries. However, it is often unclear what genetic stock was used to start these plants and they may not be from stock adapted to the local climate.

Water, either too much or too little, is the major reason why macrophytes do not become well established in wetlands constructed for wastewater treatment. When plants are first establishing themselves, the optimum conditions are moist soils or very shallow (<5 cm) water depth. If water is too deep, the new plants will be flooded out. If there is inadequate water and topsoil dries out, the plants will not survive. Of course, if the wastewater itself can be used in measured amounts to irrigate the plants, this is optimum. If not, artificial irrigation might be required to make sure that plants are successful.

Figure 20-10 Results of a study comparing eight macrophytes commonly used in wastewater wetlands in New Zealand after 124 days of culture in dairy farm wastewater. **a.** Mean above-ground and below-ground biomass accumulation of the eight macrophytes. Different letters indicate significant differences. **b.** Relationship between total nitrogen removal from ammonium-rich dairy farm wastewater and total biomass. Regression coefficient $r^2 = 0.66$. (*After Tanner, 1996*)

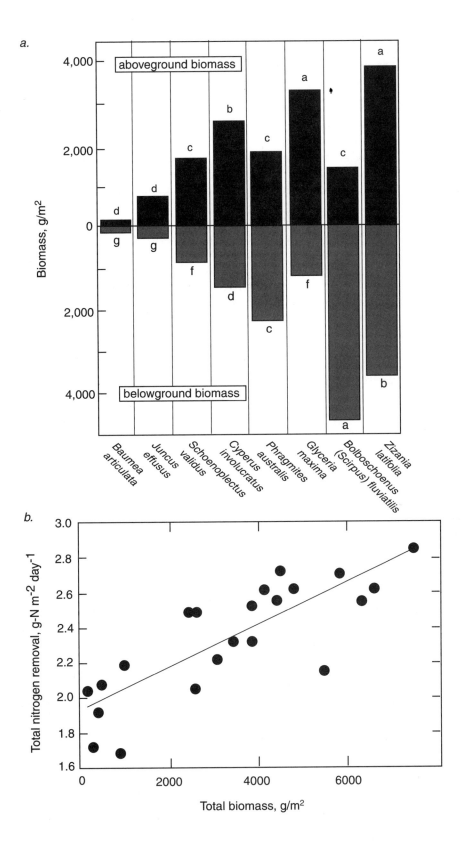

Management After Construction

Plant Harvesting

Nutrients removed from the inflowing water in treatment wetlands are sequestered in the sediments, released to the atmosphere (nitrogen), or taken up into plant roots and above-ground shoots and foliage. The harvesting of plants generally does not result in removal of a large quantity of chemicals, for example, nutrients, from the system unless the plants are harvested several times in a growing season. For example, harvesting *Phragmites* twice during the growing season—once at peak nutrient content and again at the end of plant growth—leads to a maximum removal of nitrogen and phosphorus (Suzuki et al., 1989). In some cases, plant harvesting may be advisable as part of the routine maintenance of the constructed wetland. Wieder et al. (1989) suggested plant harvesting as a mechanism to alter the effect that plants have on the ecosystem, usually by putting it back into an earlier stage of succession when net growth may be greater. Plant harvesting may also be necessary to control mosquitoes, reduce congestion in the water, change the residence time of the basin, and allow for greater plant diversity. The practice in China is to harvest the shoots of emergent plants regularly, thereby increasing the size, strength, and number of remaining shoots and encouraging vegetative reproduction (J. Yan, personal communication).

Sometimes other plant management techniques such as drawdowns followed by burning are used to control the invasion of woody vegetation if that invasion is considered undesirable (Warners, 1987). When the controlled burning of wetlands is used, the wetland manager needs to consider the impact on wildlife (Willard et al., 1989) as well as the potential reintroduction of inorganic nutrients to the water column from the ash and sediments when water is again added.

Wildlife Management

Although the development of wildlife is a welcomed and often desired aspect of treatment wetlands, beaver and muskrat can create obstructions to flow, destroy vegetation, and burrow into dikes; sand, gravel, or wire screening can be used to discourage burrowing. In other cases, animals (e.g., Canada Geese) grazing on newly planted perennial herbs and seedlings are particularly destructive. The timing of planting is important, especially when migratory animals are involved in destructive grazing in the winter. Similarly, deeper wetlands often become havens for undesirable fish such as carp that can cause excessive turbidity and uproot vegetation.

Trapping of muskrats and beavers can alleviate their impact on constructed wetland basins, but it can also be time consuming and often ineffective. Grazing by geese is more difficult to control, although their impact is most significant soon after vegetation is planted. If vegetation can become established, the grazing effects of geese can be minimal.

Weller (1994) recommends a 50 : 50 ratio of open water to vegetation cover in marshes to attract waterfowl, and Brinson et al. (1981b) suggest creating diverse habitats with live and dead vegetation, islands, and floating structures. In many cases of wetland construction, wildlife enhancement begins soon after construction. At a constructed wetland at Pintail Lake in Arizona, the area's waterfowl population increased dramatically by the second year of use; duck nest density increased 97 percent over the first year (Wilhelm et al., 1989). Considerable increase in avian activity was also noted at the Des Plaines River Wetlands Demonstration Project in northeastern Illinois. Migrating waterfowl increased from 3 to 15 species and from 13 to 617 individuals between 1985 (preconstruction) and 1990 (one year after water was introduced to the wetlands). The number of wetland-dependent breeding birds increased from 8 to 17 species and two state-designated endangered birds, the Least Bittern and the Yellow-headed Blackbird, nested at the site after wetland construction (Hickman and Mosca, 1991).

Mosquito Control

Mosquitoes can be controlled in constructed wetlands by changing the hydrologic conditions of the wetland to inhibit mosquito larvae development or by using chemical or biological control. Many have proposed mosquito control by fish, especially the air-gulping mosquito fish (*Gambusia affinis*). One reason to maintain some deeper areas in wetlands in temperate zones is to allow fish such as *Gambusia* and other top minnows to survive the winter and feed on mosquito larvae. Apparently little is known about the role that water quality may perform in controlling mosquitoes, but flowthrough wetlands are less likely to develop mosquito populations than standing-water systems. This is one of the many considerations that should go into the decision on wetland design and turnover time. Bacterial insecticides (e.g., *Bacillus sphaericus*) and the fungus *Lagenidium giganteum* are known pathogens of mosquito larvae, but they have not been extensively tested. Constructing boxes to encourage nesting by swallows (Hirundinidae), swifts (Apodidae), and bats has also been used to control adult mosquito populations at constructed wetlands.

Perturbation Management

The average conditions used in the design of a wetland do not reflect the actual conditions, where seasonal and less frequent perturbations are uncertain in frequency and magnitude but may require some response. Wetlands should be designed and planned for the worst case conditions of perturbations while maintaining a balance among form, function, and persistence. It is useful to remember that wetland design is an inexact science and that perturbations may change the original design (e.g., selected plant species) to something else. If a treatment wetland continues to function according to its main goal—that is, improving water quality—changes in species and forms will not be so significant.

Pathogen Transmission

Because many treatment wetlands are specifically built to deal with human and animal wastewater, proper sanitary engineering techniques should be used to minimize human exposure to pathogens. Treatment wetlands are meant to be biologically rich systems and microbial activity is a major part of the treatment process. Measurements of indicator organisms such as fecal and total coliforms should be part of the monitoring of municipal wastewater treatment wetlands. Nearby wells should also be sampled, as water seeping from a wastewater wetland near potable water supplies should be carefully monitored. If a wetland is being used as tertiary treatment to a conventional treatment plant, consideration in the design of the conventional treatment plant has to be given to the disinfection system. Chlorine disinfection and the resulting chlorine residual would cause significant problems in the downstream wetland so other means of disinfection [ozonation or ultraviolet (UV) radiation] should be used.

Water-Level Management

The water level of surface flow treatment wetlands is the key to both water quality enhancement as well as vegetation success. Most constructed municipal wastewater wetlands have little control on the inflow (unless authorities have given permission to bypass the wetland if needed, or if a wastewater storage lagoon is built upstream of the wetland). However, most constructed wetlands have a control structure such as a flume to control outflows and these structures can often be manipulated to balance inflows and control water depth. For example, macrophytes are stressed by too much water as much as they are stressed by too little water. Water depths of about 30 cm are optimum for many herbaceous macrophytes used in treatment wetlands. When forested wetlands are used as treatment systems, although some such as *Taxodium* will tolerate continuous flooding, most will survive only in moist-soil conditions.

Compounding the effect that water level has on vegetation is the effect that it has on wastewater treatment itself. High water favors treatment of phosphorus that is associated with sedimentation and similar processes; it leads to less resuspension and longer retention time. Shallow water leads to closer proximity of sediments and overlying water, often causing anaerobic or near anaerobic conditions in the aquatic system during the growing season. Low water thus favors the reduction of nitrate nitrogen through denitrification.

Sediment Dredging

This is a rarely used management technique in constructed wetlands. Its use depends on whether the wetlands are filling in with sediments, thus shortening their effective lives, and whether sediment accumulation is viewed as an undesirable feature relative to the objectives of the wetland. Dredging is generally an expensive operation and one that should not be attempted frequently in the life of a constructed wetland. Dredging not only removes sediments but also removes the seed bank and rooted plants. The process of dredging sediments from a constructed wetland may require a regulatory permit even if it is done to "improve" wetland function. The best approach, unless dredging is unavoidable, is to "accept the [sediment] accumulation as a natural part of wetland dynamics" (Willard et al., 1989), and to design the system to minimize its influence on the efficiency of the treatment wetland.

Other Benefits of Treatment Wetlands

There can be other long-term benefits of using wetlands for water quality management. Subsurface treatment wetlands provide little additional benefit beyond the water quality improvement they were designed to provide. However, surface-flow treatment wetlands have a variety of additional benefits. Certainly the habitat that is created can be a major ancillary benefit of these systems. In addition to providing habitat for mammals such as nutria, beavers, muskrats, and voles, surface-flow treatment wetlands are often a haven for waterfowl and wading birds. Human uses such as trapping and hunting are not incompatible with some wastewater wetlands. Wetlands, if designed properly in an urban area, are locations where the public can visit and learn about their important water quality role. This message is a powerful one to the uninitiated, who will often become ardent wetland conservationists as a result of seeing "wetlands at work." Another benefit of using both natural and constructed wetlands for water quality improvement is related to areas where "land-building" is needed, such as in the Mississippi River delta of Louisiana. In the subsiding environment of Louisiana's Gulf Coast (see Chapter 9), nutrients are permanently retained in the peat of wetlands receiving high-nutrient wastewater as the wetland aggrades to match subsidence. In this case, wastewater discharge into a wetland can occur without saturating the system and simultaneously helps counteract the deleterious effects of land subsidence (Conner and Day, 1989).

Economics of Treatment Wetlands

Construction Costs

The construction of a new wetland involves careful consideration of a number of criteria, including a realistic look at cost. The amount of funding available, the period of time for which it is available, and the limits and rules concerning its expenditure are questions to be dealt with early in the planning stages of a constructed wetland. It is generally believed that treatment wetlands are less expensive to build and maintain than conventional wastewater treatment and that is the appeal of these systems to many. However, cost comparisons should be carefully made before investing in these systems. Any estimate of the cost of a new wetland's development should include the following items:

1. Engineering plan
2. Preconstruction site preparation
3. Construction costs (labor, equipment, materials, supervision, indirect and overhead charges)
4. Cost of land

The cost of wetland construction varies widely, depending on the location, type, size, and objectives of the wetland (Table 20-8). Other factors that add to the cost variation include access to the site, substrate characteristics, liner requirements, cost of protective structures, local labor rates, and the availability of equipment. When the data in Table 20-8 are normalized for wetland size, a strong relationship is seen between wetland cost per unit area and wetland size (Fig. 20-11):

$$C_A = \$196,336 A^{-0.511} \qquad (20.7)$$

where C_A = capital cost of wetland construction per unit area ($ ha^{-1})
A = wetland area (ha)

This relationship suggests that a 1-ha wetland would cost almost $200,000, a 10-ha wetland would cost $60,000 per ha, and a 100-ha wetland would cost $19,000 per ha. The data clearly suggest that there is an economy of scale involved in wetland construction.

Some small wastewater wetlands that require human and technological intervention such as the Santee Marsh in California and the Iselin wastewater wetland plant in Pennsylvania were extremely costly to construct. Pintail Lake and Jacques Marsh in Arizona were fairly inexpensive to build because they were constructed in preexisting but dry lake basins and they involved large areas. Digging and basin formation were not necessary at those sites, and the natural formations helped to bring down the construction costs (Wilhelm et al., 1989). Projects where existing land features are used can cost considerably less. Projects that have extensive plumbing and/or pumps cost considerably more.

Operating and Maintenance Costs

Operating and maintenance costs vary according to the wetland's use and to the amount and complexity of mechanical parts and plumbing that the wetland contains.

Table 20-8 Construction costs of treatment wetlands for various uses in the United States

Wetland	State	Use	Area (ha)	Cost ($/ha)	Cost ($/acre)	Reference
Greenwood Wetland	Florida	Stormwater runoff	11.0	$51,500	$20,800	Palmer and Hunt (1989)
Lake Jackson	Florida	Urban runoff	4.0	$199,500[a]	$80,700[a]	Esry and Cairns (1989)
Santee Marsh	California	Wastewater treatment	0.1	$1,820,000[b]	$737,000[b]	Gersberg et al. (1989)
Iselin Marsh/Pond	Pennsylvania	Wastewater treatment	0.2	$2,080,000[b]	$842,000[b]	Conway and Murtha (1989)
Pintail Lake	Arizona	Wastewater treatment	20.2	$73,800	$30,000	Wilhelm et al. (1989)
Jacques Marsh	Arizona	Wastewater treatment	18.0	$75,300	$30,500	Wilhelm et al. (1989)
Kash Creek	Alabama	Acid mine drainage	0.4	$84,200	$34,000	Brodie et al. (1989a)
SIMCO Mine	Ohio	Acid mine drainage	0.2	$480,000	$194,000	R. Kolbash, personal communication
Widows Creek	Alabama	Ash pond seepage	0.5	$69,800	$28,200	Brodie et al. (1989b)
Kingston	Alabama	Ash pond seepage	0.9	$142,100	$57,500	Brodie et al. (1989b)
Bolivar Peninsula	Texas	Disposal site for dredge	8.0	$34,100	$13,800	Newling (1982)
Windmill Point	Virginia	Disposal site for dredge	8.0	$25,300	$10,300	Newling (1982)
Incline Village	Nevada	Wastewater treatment	175	$23,700	$9,600	Kadlec and Knight (1996)
Olentangy River	Ohio	Experimental wetlands with pump	2.0	$155,000[b]	$62,000[b]	W. J. Mitsch, unpublished data (1994)
Licking County	Ohio	Tertiary wastewater treatment	6.4	$59,500	$23,800	E. Bischoff, personal communication (1998)
Average				$358,000	$145,000	
Median				$75,300	$30,500	

[a] Includes area on which an impoundment and filter were built.
[b] Cost reflects wetland excavation as well as a good deal of plumbing.

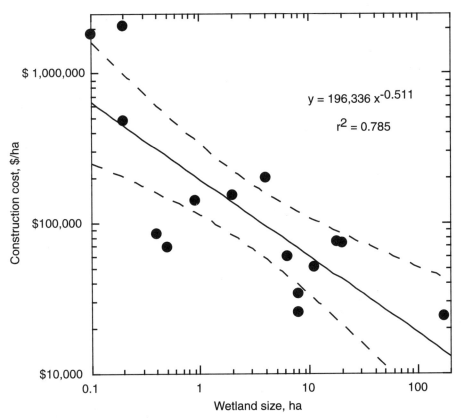

Figure 20-11 **Construction cost per hectare versus wetland size based on data of constructed wetlands in Table 20-8. Dotted lines are 95% confidence interval.**

Fewer data on these operational costs are available. A pump, filter, impoundment tank, and piping add considerably to both the construction and the maintenance costs of a wetland. Kadlec and Knight (1996) estimate the operation and maintenance costs for a wastewater wetland at Incline Village to be about $85,500 per year. That estimate included $50,000 per year for personnel to be in charge of the 175-ha wetland. A wide range of $5,000 to $50,000 per year of operating and maintenance costs was estimated by Kadlec and Knight (1996) for smaller wetlands. Gravity-fed wetlands are far less expensive to maintain than highly mechanized wetlands that need significant plumbing and pumps.

Comparing Wetlands and Conventional Technology

A comparison of the construction and operating costs of a proposed large (>2,000 ha) wetland that was to be constructed in the Florida Everglades with conventional chemical treatment is illustrated in Table 20-9. In this example, if land costs are not considered, the wetland alternative has an 11 percent lower capital cost and 56 percent lower operating costs than the chemical treatment alternative. Although land costs can be significant for treatment wetlands, particularly in urban areas (in essence solar energy is being substituted for fossil fuel energy), it is generally not appropriate to use the cost of land in the comparison with technological solutions

Table 20-9 Estimated cost comparison for phosphorus control in 760,000 m³ day⁻¹ agricultural runoff in Florida

	Treatment Wetland	Chemical Treatment
Land cost	$34,434,000	$2,140,000
Capital costs (land free)	$95,836,000	$108,260,000
Total annual operation/maintenance	$1,094,000	$2,490,000
O&M, present worth	$33,443,000	$76,153,000
Total present worth, without land cost	$129,279,000	$185,637,000
Total present worth, with land cost	$163,713,000	$187,777,000

Source: Kadlec and Knight (1996).

that require little land. This is because the land being used by the wetland can be sold after the life of the wetland is completed, while the salvage value of the worn-out equipment used for conventional treatment alternatives is generally zero (Kadlec and Knight, 1996).

One of the more clever calculations of the difference between using wetlands versus conventional mechanical systems for wastewater treatment is an illustration of the relative impact on the emission of the greenhouse gas CO_2. Normalizing estimates for a 3,800 m³ day⁻¹ (1 million gal/day) flow of wastewater, mechanical treatment leads to 27 times more emission of carbon dioxide to the atmosphere than does a treatment wetland (Table 20-10). The wetland system, in fact, has the additional benefit of sequestering a small amount of carbon. Conventional wastewater treatment uses 3.9 kg of fossil fuel carbon to remove 1 kg of carbon; a wetland treatment system uses 0.16 kg of fossil fuel carbon to remove 1 kg of carbon (Ogden, 1999).

Critical Technical and Institutional Considerations

There are many considerations, both technical and institutional, that must be considered as treatment wetlands are designed and built.

Technical Considerations

1. Values of the wetlands such as wildlife habitat should be considered in any treatment wetland development.

2. Acceptable pollutant and hydrologic loadings must be determined for the use of wetlands in wastewater management. Appropriate loadings, in turn,

Table 20-10 Net atmospheric generation of carbon for a 3,800 m³ day⁻¹ wastewater treatment facility using treatment wetlands or conventional mechanical treatment

Carbon Flow	Treatment Wetland (metric tons C day⁻¹)	Conventional Treatment (metric tons C day⁻¹)
Atmospheric carbon from power generation	53	1,350
Carbon sequestration	−3	0
Net atmospheric carbon	50	1,350

Source: Ogden (1999).

determine the size of the wetland to be constructed. Overloading a constructed wetland is worse than not building it at all.

3. All existing characteristics of local wetlands, including vegetation, geomorphology, hydrology, and water quality, should be well understood so that natural wetlands can be "copied" in the construction of treatment wetlands.

4. Particular care should be taken in the wetland design to address mosquito control and protection of groundwater resources.

Institutional Considerations

5. Potential conflicts over the protection and use of wetlands may arise among state agencies, federal agencies, and local groups. Some agencies view treatment wetlands as a beneficial use. Others view them as polluting a wetland habitat. Still others bring up questions of disease vectors and potential groundwater contamination.

6. Wastewater treatment by wetlands can often serve the dual purposes of both wetland habitat development and wastewater treatment and recycling. The creation of treatment wetlands as mitigation for lost wetlands is still generally unacceptable because of the lack of sustainability and the high level of pollution of treatment wetlands compared to restored wetlands.

7. Many permit processes, administered by local, state, and federal agencies, do not recognize treatment wetland systems as alternatives for wastewater treatment. In these cases, experimental systems should first be established for a given region. The modification of requirements for granting permits is needed to make use of this effective method of wastewater management.

Recommended Readings

Brix, H. 1994. Use of constructed wetlands in water pollution control: Historical development, present status, and future perspectives. Water Science and Technology 30:209–223.

Kadlec, R. H., and R. L. Knight. 1996. Treatment Wetlands. CRC/Lewis Press, Boca Raton, FL. 893 pp.

Mitsch, W. J., A. J. Horne, and R. W. Nairn, eds. 2000. Nitrogen and Phosphorus Retention in Wetlands. Special Issue of Ecological Engineering 14:1–206.

Olson, R. K., ed. 1992. The Role of Created and Natural Wetlands in Controlling Nonpoint Source Pollution. Special Issue of Ecological Engineering 1:1–170.

Reed, S. C., R. W. Crites, and E. J. Middlebrooks. 1995. Natural Systems for Waste Management and Treatment, 2nd ed. McGraw-Hill, New York. 433 pp.

Tanner, C. C., G. Raisin, G. Ho, and W. J. Mitsch, eds. 1999. Constructed and Natural Wetlands for Pollution Control. Special Issue of Ecological Engineering 12:1–170.

Chapter **21**

Classification, Inventory, and Delineation of Wetlands

Wetlands have been classified since the early 1900s, beginning with the peatland classifications of Europe and North America. Regional wetland classifications and inventories have been developed for several states. Some of these have classified wetlands according to their vegetative life forms, whereas others also use the hydrologic regime.

The U.S. Fish and Wildlife Service has been involved in two major wetland classifications and inventories, one completed in 1954, and one begun in the mid-1970s but not yet completed for the entire United States. The early classification described 20 wetland types based on flooding depth, dominant forms of vegetation, and salinity regimes. "Classification of Wetlands and Deepwater Habitats of the United States," a more recent classification, uses a hierarchical approach based on systems, subsystems, classes, subclasses, dominance types, and special modifiers to define wetlands and deepwater habitats precisely.

Classifications based on environmental forcing functions, particularly hydrologic flow, offer a promising approach. More recently, classifications based on wetland function and value have been explored including a functionally based approach called the hydrogeomorphic classification.

Wetland inventories are carried out at many different scales with several different imageries and with both aircraft and satellite platforms. Eventually, wetlands in the entire United States will be mapped at a scale of 1 : 24,000 for most areas and at 1 : 100,000 or coarser for a few areas.

The determination of the exact edge of wetlands, referred to as wetland delineation, has become an important practice in the United States because of the various legal protection regulations regarding wetlands. The general approach uses hydrology, vegetation, and soils to estimate a wetland's boundary.

In order to deal realistically with wetlands on a regional scale, wetland managers have found it necessary both to define the different types of wetlands that exist and to determine their extent and distribution. Those activities, the first called *wetland classification*, and the second called a *wetland inventory*, are valuable undertakings for both the wetland scientist and the wetland manager. Classifications and inventories of wetlands have been developed for the entire United States, for Canada, and in individual states, provinces, and regions for many purposes over the past century. Some of the earliest efforts were undertaken for the purpose of finding wetlands that could be drained for human use; later classifications and inventories centered on the desire to compare different types of wetlands in a given region, often for their value to waterfowl. The protection of multiple ecological values of wetlands is the most recent purpose and now the most common reason for wetland classification and inventory. Recognition of wetland "value" has led some to now seek wetland classifications based on priorities for protection, with highest protection afforded to those wetlands with the greatest value. Like other techniques, classifications and inventories are valuable only when the user is familiar with their scope and limitations.

Goals of Wetland Classification

Several attempts have been made to classify wetlands into categories that follow their structural and functional characteristics. These classifications depend on a well-understood general definition of wetlands (see Chapter 2), although a classification contains definitions of individual wetland types. A primary goal of wetland classifications, according to Cowardin et al. (1979), "is to impose boundaries on natural ecosystems for the purposes of inventory, evaluation, and management." These authors identified four major objectives of a classification system:

1. To describe ecological units that have certain homogeneous natural attributes
2. To arrange these units in a unified framework for the characterization and description of wetlands, that will aid decisions about resource management
3. To identify classification units for inventory and mapping
4. To provide uniformity in concepts and terminology

The first objective deals with the important task of grouping ecosystems that have similar characteristics in much the same way that taxonomists categorize species in taxonomic groupings. The wetland attributes that are frequently used to group and compare wetlands include the geomorphic and hydrologic regime, vegetation physiognomic type, and plant and/or animal species.

The second objective, to aid wetland managers, can be met in several ways when wetlands are classified. Classifications (which are definitions of different types of wetlands) enable the wetland manager to deal with wetland regulation and protection in a consistent manner from region to region and from one time to the next. Classifications also enable the wetland manager to pay selectively more attention to those types of wetlands that are most threatened or functionally the most valuable to a given region. It should be emphasized that recent classification systems [e.g., the U.S. Wetland Inventory system (Cowardin et al., 1979), the Canadian system (Warner and Rubec, 1997), the international Ramsar Convention system, and the hydrogeomorphic approach (Brinson, 1993a)] are scientific classifications based on natural properties, not evaluation systems developed for regulatory purposes. Thus, they do not focus on factors relating to environmental, social, or economic importance.

Although these classification systems are useful planning tools, they are not structured by the requirements of management, that is, the need to make choices about relative social priorities and values.

The third and fourth objectives, to provide consistency in the formulation and use of inventories, mapping, concepts, and terminology, are also important in wetland management. The use of consistent terms to define particular types of wetlands is needed in the field of wetland science (see Chapter 3). These terms should then be applied uniformly to wetland inventories and mapping so that different regions can be compared and so that there will be a common understanding of wetland types among wetland scientists, wetland managers, and wetland owners.

History of Wetland Classification

Peatland Classifications

Many of the earliest wetland classifications were undertaken for the northern peatlands of Europe and North America. An early peatland classification in the United States, developed by Davis (1907), described Michigan bogs according to three criteria: (1) the landform on which the bog was established such as shallow lake basins or deltas of streams, (2) the method by which the bog was developed such as from the bottom up or from the shores inward, and (3) the surface vegetation such as tamarack or mosses. Based on the work of Weber (1907), Potonie (1908), Kulzynski (1949), and others in Europe, Moore and Bellamy (1974) described seven types of peatlands based on flowthrough conditions (Table 21-1). Three general categories, called rheophilous, transition, and ombrophilous, describe the degree to which peatlands are influenced by outside drainage. These categories are discussed in more detail in Chapter 12 in the more modern terminology of minerotrophic, transition, and ombrotrophic peatlands.

Most peatlands, of course, are limited to northern temperate climes and do not include all or even most types of wetlands in North America. These classifications, however, served as models for more inclusive classifications. They are significant because they combined the chemical and physical conditions of the wetland with the vegetation description to present a balanced approach to wetland classification.

Table 21-1   Hydrologic classification of European peatlands

A. Rheophilous mire—Peatland influenced by groundwater derived from outside the immmediate watershed
   Type 1—Continuously flowing water that inundates the peatland surface
   Type 2—Continuously flowing water beneath a floating mat of vegetation
   Type 3—Intermittent flow inundating the mire surface
   Type 4—Intermittent flow of water beneath a floating mat of vegetation

B. Transition mire—Peatland influenced by groundwater derived solely from the immediate watershed
   Type 5—Continuous flow of water
   Type 6—Intermittent flow of water

C. Ombrophilous mire
   Type 7—Peatland never subject for flowing groundwater

Source: Bellamy (1968) and Moore and Bellamy (1974).

Circular 39 Classification

In the early 1950s, the U.S. Fish and Wildlife Service recognized the need for a national wetlands inventory to determine "the distribution, extent, and quality of the remaining wetlands in relation to their value as wildlife habitat" (Shaw and Fredine, 1956). A classification was developed for that inventory (Martin et al., 1953), and the results of both the inventory and the classification scheme were published in U.S. Fish and Wildlife Circular 39 (Shaw and Fredine, 1956). Twenty types of wetlands were described under four major categories:

1. Inland fresh areas
2. Inland saline areas
3. Coastal fresh areas
4. Coastal saline areas

In each of the four categories, the wetlands were arranged in order of increasing water depth or frequency of inundation. A brief description of the site characteristics of the 20 wetland types is given in Table 21-2.

Types 1 through 8 are freshwater wetlands that include bottomland hardwood forests (type 1), infrequently flooded meadows (type 2), freshwater nontidal marshes (types 3 and 4), open water less than 2 m deep (type 5), shrub–scrub swamps (type 6), forested swamps (type 7), and bogs (type 8). Types 9 through 11 are inland wetlands that have saline soils. They are defined according to the degree of flooding. Types 12 through 14 are wetlands that, although freshwater, are close enough to the coast to be influenced by tides. Types 15 through 20 are coastal saline wetlands that are influenced by both salt water and tidal action. These include salt flats and meadows (types 15 and 16), true salt marshes (types 17 and 18), open bays (type 19), and mangrove swamps (type 20).

This wetland classification was the most widely used in the United States until 1979, when the present National Wetlands Inventory classification was adopted. It is still referred to today by some wetland managers and is regarded by many as elegantly simple compared with its successor. It primarily used the physiognomy (life forms) of vegetation and the depth of flooding to identify the wetland type. Salinity was the only chemical parameter used, and although wetland soils were addressed in the Circular 39 publication, they were not used to define wetland types.

Two additional attempts to classify wetlands using categories similar to those in Circular 39 are noteworthy. The prairie potholes of the upper Midwest were included in a classification by Stewart and Kantrud (1971, 1972), who used seven classes, all based on vegetation zones as defined according to their hydrologic characteristics (e.g., temporary pond). Golet and Larson (1974) detailed a freshwater wetland classification for the glaciated northeastern United States based on categories similar to those in Circular 39, types 1 through 8. This classification introduced 24 subclasses based on 18 possible life forms, as shown in Figure 21-1.

Functional Classification Systems

Although virtually all classification systems use some aspect of hydrology as a defining characteristic, early systems generally identified wetlands classes by their general vegetation type, for example, forested or nonforested, evergreen or deciduous, grass domi-

Table 21-2 Circular 39 wetland classification by U.S. Fish and Wildlife Service

Type Number	Wetland Type	Site Characteristics
INLAND FRESH AREAS		
1.	Seasonally flooded basins or flats	Soil covered with water or waterlogged during variable periods, but well drained during much of the growing season; in upland depressions and bottomlands
2.	Fresh meadows	Without standing water during growing season; waterlogged to within a few centimeters of surface
3.	Shallow fresh marshes	Soil waterlogged during growing season; often covered with 15 cm or more of water
4.	Deep fresh marshes	Soil covered with 15 cm to 1 m of water
5.	Open fresh water	Water less than 2 m deep
6.	Shrub swamps	Soil waterlogged; often covered with 15 cm or more of water
7.	Wooded swamps	Soil waterlogged; often covered with 30 cm of water; along sluggish streams, flat uplands, shallow lake basins
8.	Bogs	Soil waterlogged; spongy covering of mosses
INLAND SALINE AREAS		
9.	Saline flats	Flooded after periods of heavy precipitation; waterlogged within few centimeters of surface during the growing season
10.	Saline marshes	Soil waterlogged during growing season; often covered with 0.7 to 1 m of water; shallow lake basins
11.	Open saline water	Permanent areas of shallow saline water; depth variable
COASTAL FRESH AREAS		
12.	Shallow fresh marshes	Soil waterlogged during growing season; at high tide, as much as 15 cm of water; on landward side, deep marshes along tidal rivers, sounds, deltas
13.	Deep fresh marshes	At high tide, covered with 15 cm to 1 m water; along tidal rivers and boys
14.	Open fresh water	Shallow portions of open water along fresh tidal rivers and sounds
COASTAL SALINE AREAS		
15.	Salt flats	Soil waterlogged during growing season; sites occasionally to fairly regularly covered by high tide; landward sides or islands within salt meadows and marshes
16.	Salt meadows	Soil waterlogged during growing season; rarely covered with tide water; landward side of salt marshes
17.	Irregularly flooded salt marshes	Covered by wind tides at irregular intervals during the growing season; along shores of nearly enclosed bays, sounds etc.
18.	Regularly flooded salt marshes	Covered at average high tide with 15 cm or more of wate; along open ocean and along sounds
19.	Sounds and bays	Portions of saltwater sounds and bays shallow enough to be diked and filled; all water landward from average lowtide line
20.	Mangrove swamps	Soil covered at average high tide with 15 cm to 1 m of water; along coast of southern Florida

Source: Shaw and Fredine (1956).

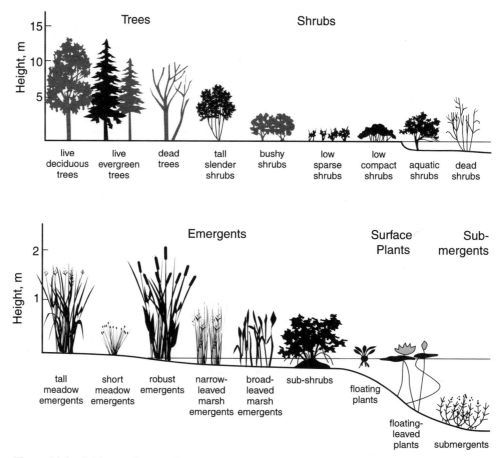

Figure 21-1 Subforms of vegetation used to classify wetlands of glaciated northeastern United States. (*After Golet and Larson, 1974*)

nated or herbaceous. These types of classes are still used as a part of every major classification system because of their utility in mapping and easy identification. During the 1970s, however, several wetland classification systems were proposed, based primarily on functional properties.

Coastal Wetland Classification

H. T. Odum et al. (1974) described coastal ecosystems by their major forcing functions (e.g., seasonal programming of sunlight and temperature) and stresses (e.g., ice) (Fig. 21-2). Coastal wetland types in this classification include salt marshes and mangrove swamps. Salt marshes, found in the type C category of natural temperate ecosystems with seasonal programming, have "light tidal regimes" and "winter cold" as forcing function and stress, respectively. Mangrove swamps are classified as type B (natural tropical ecosystems) because they have abundant light, show little stress, and reflect little seasonal programming. Three additional classes, type A (naturally stressed systems of wide latitudinal range), type D (natural arctic ecosystems with ice stress), and type E (emerging new systems associated with human activity), were included in this classification. The last class, which includes new systems formed by pollution such as

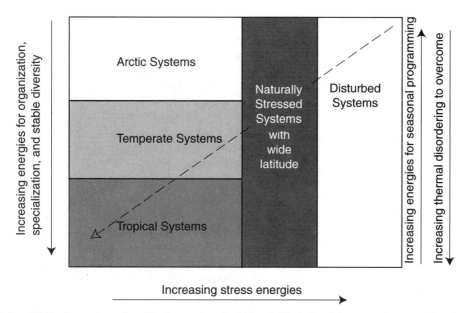

Figure 21-2 Ecosystem classification system based on latitude (and, hence, solar energy) and major stresses. (*After H. T. Odum et al., 1974*)

pesticides and oil spills, is still an interesting concept that could be applied to other wetland classifications.

Florida Wetland Classification

Wharton et al. (1976) described the forested swamps of Florida according to their hydrologic inputs. The wetlands are arranged according to water flow. The first two classes, "cypress ponds" and "other nonstream swamps," involve wetlands that experience little water inflow except precipitation and groundwater in some cases. The third class, "cypress strands," includes slowly flowing water typical of the southern Florida river basins, whereas "river swamps and floodplains" are more typical of continuously flooded alluvial swamps and periodically flooded riparian forests. The fifth class, "saltwater swamps—mangroves," are those forested wetlands that are affected by tides and salt water.

Hydrodynamic Energy Gradient Classification

A similar approach to classifying wetlands according to their source and velocity of water flow was developed by Gosselink and Turner (1978). As described by the authors, "In general flow rate, or other indications of hydrologic energy such as renewal time or frequency of flooding, increases from raised convex (ombrotrophic) wetlands to lentic and tidal wetlands." This classification, although applied only to nonforested freshwater wetlands, has applicability to wetlands in general.

Hydrogeomorphic Wetland Classification System

Brinson (1993a) described a classification system designed to be used for evaluation of wetland functions. It is being used increasingly as a means of assessing the physical, chemical, and biological functions of wetlands, and is extremely useful for comparing the level of functional integrity of wetlands within a functional class, or of evaluating the impact of proposed human activities on wetlands and mitigation alternatives (see Chapter 16). The classification received its inspiration from a number of earlier classification schemes that use hydrodynamics as a primary classifier: Gosselink and Turner (1978) for freshwater marshes; O'Brien and Motts (1980) and Hollands (1987) for glaciated northeastern U.S. wetlands; Novitzki (1979) for Wisconsin wetlands; Gilvear et al. (1989) for East Anglian fens; Prance (1979) for Amazon forests; National Wetlands Working Group (1988) for Canadian wetlands; Brinson (1989) for tidal wetlands; Winter (1977) for north-central U.S. lakes; and Leopold et al. (1964) and Rosgen (1985) for floodplains and riparian systems. Especially pertinent are the coastal classification of H. T. Odum et al. (1974) described earlier; Lugo and Snedaker's (1974) mangrove classification; and variations of the mangrove classification described by Cintròn et al. (1985) and Lugo et al. (1990a) (Fig. 21-3).

The classification is based primarily on hydrodynamic differences as they function within four geomorphic settings. Thus, the three core components of the classification system are geomorphology, water source, and hydrodynamics (Table 21-3). Geomorphic setting is the topographic location of a wetland in the surrounding landscape. Four geomorphic settings are identified: depressional, riverine, and fringe, and extensive peatlands. The first three are clearly related to the hydrologic setting. Extensive peatlands are different because the dominant influence on hydrology is biogenic accretion. Water sources are precipitation, surface or near-surface flow, and groundwater discharge (into a wetland). Hydrodynamics refers to the direction and strength of water movement within a wetland. The three core features are heavily interdepen-

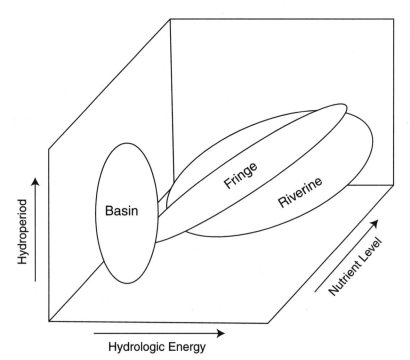

Figure 21-3 Basis of the hydrogeomorphic classification system. Geomorphic settings (basin, fringe, and riverine) are arranged around three core factors. (*After Lugo et al., 1990; Brinson, 1993a*)

Table 21-3 **Functional classification of wetlands by geomorphology, water source, and hydrodynamics**

Core Component	Description	Example
GEOMORPHIC SETTING	Topographic location of a wetland in the surrounding landscape	
Depressional	Wetlands in depressions that typically receive most moisture from precipitation, hence often ombrotrophic; found in dry and moist climates	Kettles, potholes, vernal pools, Carolina bays groundwater slope wetlands
Extensive peatlands	Peat substrate isolates wetland from mineral substrate; peat dominates movement and storage of water and chemicals	Blanket bogs, tussock tundra
Riverine	Linear strips in landscape; subject predominantly to unidirectional surface flow	Riparian wetlands along rivers, streams
Fringe	Estuarine and lacustrine wetlands with bidirectional surface flow	Estuarine tidal wetlands, lacustrine fringes subject to winds, waves, and seiches
WATER SOURCE	Relative importance of three main sources of water to a wetland	
Precipitation	Wetlands dominated by precipitation as the primary water source; water level may be variable because of evapotranspiration	Ombrotrophic bogs, pocosins
Groundwater discharge	Primary water source from regional or perched mineral groundwater sources	Fens, groundwater slope wetlands
Surface inflow	Water source dominated by surface inflow	Alluvial swamps, tidal wetlands, montane streamside wetlands
HYDRODYNAMICS	Motion of water and its capacity to do work	
Vertical fluctuation	Vertical fluctuation of the water table resulting from evapotranspiration and replacement by precipitation or groundwater discharge	Usually depressional wetlands, bogs (annual), prairie potholes (multiyear)
Unidirectional flow	Unidirectional surface or near-surface flow; velocity corresponds to gradient	Usually riverine wetlands
Bidirectional flow	Occurrence in wetlands dominated by tidal and wind-generated water-level fluctuations	Usually fringe wetlands

Source: Brinson (1993a).

dent, so it is difficult to describe any one without the other two. Taken as a group, the three core features may be combined in 36 combinations, but because of the interdependence not all combinations are found in nature.

It becomes clear when examining Table 21-3 how interrelated the three core features are. The water entering a wetland is seldom from only one of the three sources—precipitation, groundwater discharge, and surface inflow. Figure 21-4 shows characteristic mixtures of water sources for some major wetland types. Ombrotrophic peat wetlands are typically dominated by precipitation; mineral fens and seep wetlands by groundwater discharge; and riverine and fringe wetlands by surface flows.

The core component *hydrodynamics* is an expression of the fluvial energy that drives the system. This ranges from low-energy water table fluctuations typical of depressional wetlands to unidirectional flows found in riverine wetlands to bidirectional flows of tidal and high-energy lacustrine wetland systems. Uni- and bidirectional surface flows range widely in energy from hardly perceptible movement to strong erosive currents. Combined with the geographic setting, hydrodynamics can result in a range of different wetland types.

This classification system was designed to be independent of plant communities, since it depends on the geomorphic and hydrologic properties of the wetlands. In practice, however, vegetation often provides important clues to the hydrogeomorphic forces at work, and since most modern classification systems developed for inventory purposes have some basis in hydrogeomorphology, their classes often give important clues as to function. The reverse is not true. Classification schemes based on functional

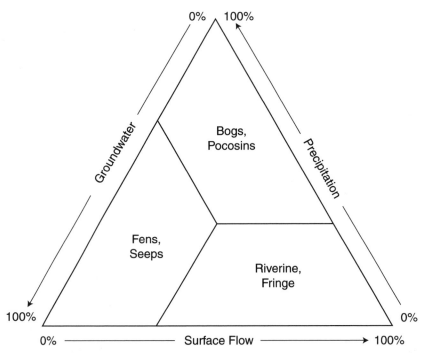

Figure 21-4 The relative contribution of a combination of three water sources—precipitation, groundwater discharge, and surface inflow—to wetlands. The location of major wetland types within the triangle shows the relative importance of different water sources. (*After Brinson, 1987*)

processes, such as the hydrogeomorphic system, are difficult to delimit spatially because the classes are not necessarily identified by different vegetation associations or other easily mapped features. Hence, the major classification schemes used for inventory purposes rely heavily on vegetation life forms, although all include some aspects of geomorphology and hydrology.

Wetland Classification for Value

Since the introduction of legislation aimed at wetland conservation and regulation, there has been considerable interest in classifying wetlands for their value to society in order to simplify the issuance of permits for wetlands activities; that is, high-value wetlands would presumably receive more protection than low-value wetlands. Figure 21-5 illustrates some of the difficulties of using this approach to classify a bottomland forest riparian system. We pose the question of whether one can classify this floodplain into high-value wetlands and wetlands of lesser value. This is not a trivial question because wetland status and, hence, the protection of the upper end of the bottomland zone is an issue of serious concern to developers, farmers, and agency regulators. Floodplains are multiple-value resources; as Figure 21-5 shows, different ecological processes peak in different zones, and these peaks are not directly related to their value to humans. For example, although most floodwater is stored at low elevations in the floodplain, the highest zone is important because it helps moderate large infrequent floods that do the most damage. Because tree growth rates are highest in the seasonally flooded zone, that zone might be considered most valuable for timber harvest. In fact, however, bald cypress, which is found primarily in a permanently flooded zone, is a more valuable timber crop because of its superior rot and insect resistance. The example illustrates some of the difficulties of classifying wetlands according to value:

1. A wetland is a multiple-value system (see Chapter 16). Any categorization by relative value implies that priorities have been set that trade off one value for another. Who makes these decisions? They are not technical decisions; rather, they involve human and societal judgments.

2. Value is not linearly related to function. It is influenced by human perceptions that may change as development pressure, population density, political interests, and other human factors change.

3. Wetland value classification involves, either implicitly or explicitly, some sort of risk assessment, for example, the value weight to give to a wetland's ability to moderate rare but severe floods.

4. Categorization by value also involves decisions about scale. Conventional economics places a higher value on 1 ha in a 10-ha forest than on 1 ha in a 100,000-ha forest. For a bear, however, a 10-ha forest patch is useless: It cannot survive in such a small area. If a bear population is a desired component of a wetland system, it is necessary to elevate the priority assigned to large, unbroken forest tracts and to linear "bridges" that connect large natural areas.

5. The classification of wetlands by value involves trade-offs among competing social priorities that imply an earlier commitment to certain goals (e.g., to farming over conservation, or to hunting over fishing).

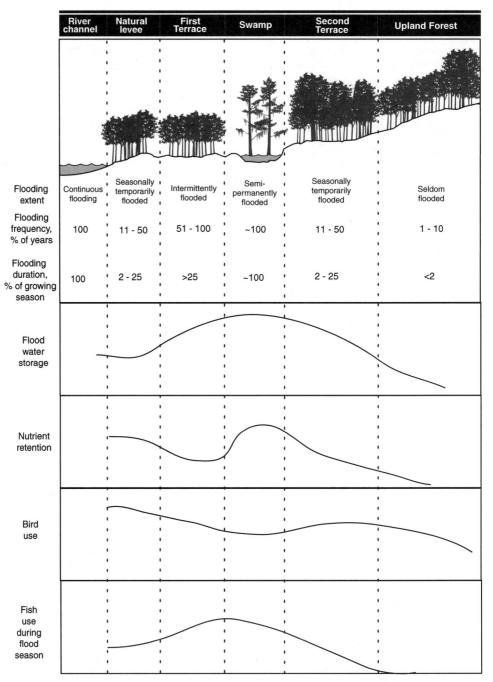

Figure 21-5 A simplified representation of a section across a bottomland hardwood forest from stream to upland, showing how various functions of interest to humans change across the transect. In actuality, the area is a complex spatial pattern of intermixed zones, not a linear gradient. This is a multiple-value resource; different ecological processes peak in different zones, and these peaks are not directly related to their value to humans.

Widely Used Wetland Classification Systems

The United States Classification of Wetlands and Deepwater Habitats

The U.S. Fish and Wildlife Service began a rigorous wetland inventory of the nation's wetlands in 1974. Because this inventory was designed to fulfill several scientific and management objectives, a new classification scheme, broader than the Circular 39 classification, was developed and finally published in 1979 as a "Classification of Wetlands and Deepwater Habitats of the United States" (Cowardin et al., 1979). Because wetlands were found to be continuous with deepwater ecosystems, both categories were addressed in this classification. It is thus a comprehensive classification of all continental aquatic and semiaquatic ecosystems. As described in that publication,

> This classification, to be used in a new inventory of wetlands and deepwater habitats of the United States, is intended to describe ecological taxa, arrange them in a system useful to resource managers, furnish units for mapping, and provide uniformity of concepts and terms. Wetlands are defined by plants (hydrophytes), soils (hydric soils), and frequency of flooding. Ecologically related areas of deep water, traditionally not considered wetlands, are included in the classification as deepwater habitats.

This classification is based on a hierarchical approach analogous to taxonomic classifications used to identify plant and animal species. The first three levels of the classification hierarchy are given in Figure 21-6. The broadest level is *systems:* "a complex of wetlands and deepwater habitats that share the influence of similar hydrologic, geomorphologic, chemical, or biological factors." Thus, the systems, subsystems, and classes are based primarily on geologic and, to some extent, hydrologic considerations. Broad vegetation types are included primarily at the class level, and, even here, the vegetation types are generic, that is, perennial, emergent, forested, scrub–shrub, or moss–lichen. Systems include the following:

1. *Marine:* Open ocean overlying the continental shelf and its associated high-energy coastline.
2. *Estuarine:* Deepwater tidal habitats and adjacent tidal wetlands that are usually semi-enclosed by land but have open, partially obstructed, or sporadic access to the ocean and in which ocean water is at least occasionally diluted by freshwater runoff from the land.
3. *Riverine:* Wetlands and deepwater habitats contained within a channel with two exceptions: (1) wetlands dominated by trees, shrubs, persistent emergents, emergent mosses, or lichens; and (2) deepwater habitats with water containing ocean-derived salts in excess of 0.5 parts per thousand (ppt).
4. *Lacustrine:* Wetlands and deepwater habitats with all of the following characteristics: (1) situated in a topographic depression or a dammed river channel; (2) lacking trees, shrubs, persistent emergents, emergent mosses, or lichens with greater than 30 percent areal coverage; and (3) total area in excess of 8 ha. Similar wetland and deepwater habitats totaling less than 8 ha are also included in the lacustrine system when an active wave-formed or bedrock shoreline feature makes up all or part of the boundary or when the depth in the deepest part of the basin exceeds 2 m at low water.

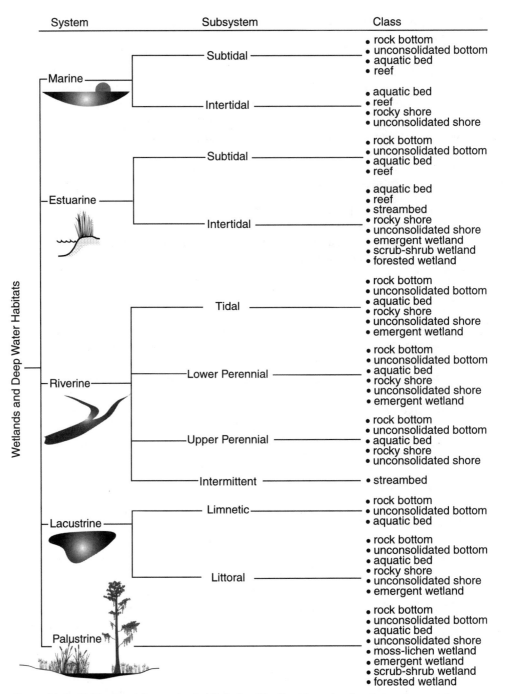

System	Subsystem	Class

Marine
- Subtidal
  - rock bottom
  - unconsolidated bottom
  - aquatic bed
  - reef
- Intertidal
  - aquatic bed
  - reef
  - rocky shore
  - unconsolidated shore

Estuarine
- Subtidal
  - rock bottom
  - unconsolidated bottom
  - aquatic bed
  - reef
- Intertidal
  - aquatic bed
  - reef
  - streambed
  - rocky shore
  - unconsolidated shore
  - emergent wetland
  - scrub-shrub wetland
  - forested wetland

Riverine
- Tidal
  - rock bottom
  - unconsolidated bottom
  - aquatic bed
  - rocky shore
  - unconsolidated shore
  - emergent wetland
- Lower Perennial
  - rock bottom
  - unconsolidated bottom
  - aquatic bed
  - rocky shore
  - unconsolidated shore
  - emergent wetland
- Upper Perennial
  - rock bottom
  - unconsolidated bottom
  - aquatic bed
  - rocky shore
  - unconsolidated shore
- Intermittent
  - streambed

Lacustrine
- Limnetic
  - rock bottom
  - unconsolidated bottom
  - aquatic bed
- Littoral
  - rock bottom
  - unconsolidated bottom
  - aquatic bed
  - rocky shore
  - unconsolidated shore
  - emergent wetland

Palustrine
- rock bottom
- unconsolidated bottom
- aquatic bed
- unconsolidated shore
- moss-lichen wetland
- emergent wetland
- scrub-shrub wetland
- forested wetland

(Vertical label: Wetlands and Deep Water Habitats)

Figure 21-6 Wetland and deepwater habitat classification hierarchy showing systems, subsystems, and classes. (*After Cowardin et al., 1979*)

738

5. *Palustrine:* All nontidal wetlands dominated by trees, shrubs, persistent emergents, emergent mosses, or lichens, and all such wetlands that occur in tidal areas where salinity stemming from ocean-derived salts is below 0.5 ppt. It also includes wetlands lacking such vegetation but with all of the following characteristics: (1) area less than 8 ha; (2) lack of active wave-formed or bedrock shoreline features; (3) water depth in the deepest part of the basin of less than 2 m at low water; and (4) salinity stemming from ocean-derived salts of less than 0.5 ppt.

Subsystems, as shown in Figure 21-6, give further definition to the systems. These include the following:

1. *Subtidal:* Substrate continuously submerged
2. *Intertidal:* Substrate exposed and flooded by tides, including the splash zone
3. *Tidal:* For riverine systems, gradient low and water velocity fluctuates under tidal influence
4. *Lower Perennial:* Riverine systems with continuous flow, low gradient, and no tidal influence
5. *Upper perennial:* Riverine systems with continuous flow, high gradient, and no tidal influence
6. *Intermittent:* Riverine systems in which water does not flow for part of the year
7. *Limnetic:* All deepwater habitats in lakes
8. *Littoral:* Wetland habitats of a lacustrine system that extends from shore to a depth of 2 m below low water or to the maximum extent of nonpersistent emergent plants

The *class* of a particular wetland or deepwater habitat describes the general appearance of the ecosystem in terms of either the dominant vegetation life form or the substrate type. When more than 30 percent cover by vegetation is present, a vegetation class is used (e.g., shrub–scrub wetland). When less than 30 percent of the substrate is covered by vegetation, then a substrate class is used (e.g., unconsolidated bottom). Definitions and examples of most of the classes in this classification system are given in Table 21-4. The typical demarcation of many of the classes of the palustrine system is shown in Figure 21-7.

Most inland wetlands fall into the Palustrine system, in the classes moss–lichen, emergent, scrub–shrub, or forested wetland. Coastal wetlands are classified in the same classes within the estuarine system and intertidal subsystem. Only nonpersistent emergent wetlands are classified into other systems.

Further descriptions of the wetlands and deepwater habitats are possible through the use of *subclasses, dominance types,* and *modifiers.* Subclasses such as "persistent" and "nonpersistent" give further definition to a class such as emergent vegetation (Table 21-4). Type refers to a particular dominant plant species (e.g., bald cypress, *Taxodium distichum,* for a needle-leaved deciduous forested wetland) or a dominant sedentary or sessile animal species (e.g., eastern oyster, *Crassostrea virginica,* for a

Table 21-4 Classes, subclasses, and examples of dominance types for wetland and deepwater habitat classification by U.S. fish and wildlife service

Class/Subclass	Definition	Examples of Dominance Type		
		Marine/Estuarine	Lacustrine/Riverine	Palustrine
ROCK BOTTOM				
Bedrock	Bedrock covers 75% or more of surface	Lobster (*Homarus*)	Brook leech (*Helobdella*)	—
Rubble	Stones and boulders cover more than 75% of surface	Sponge (*Hippospongia*)	Chironomids	Water penny (*Psephenus*)
UNCONSOLIDATED BOTTOM				
Cobble-gravel	At least 25% of particles smaller than stones and less than 30% vegetation cover	Clam (*Mya*)	Mayfly (*Baetis*)	Oligochaete worms
Sand	At least 25% sand cover and less than 30% vegetation cover	Wedge shell (*Donax*)	Mayfly (*Ephemerella*)	Sponge (*Eunapius*)
Mud	At least 25% silt and clay, although coarser sediments can be intermixed; less than 25% vegetation	Scallop (*Placopecten*)	Freshwater mollusk (*Anodonta*)	Fingernail calm (*Pisidium*)
Organic	Unconsolidated material predominantly organic matter and less than 25% vegetation cover	Clam (*Mya*)	Sewage worm (*Tubifex*)	Oligochaete worms
AQUATIC BED				
Algal	Algae growing on or below surface of water	Kelp (*Macrocystis*)	Stonewort (*Chara*)	Stonewort (*Chara*)
Aquatic moss	Aquatic moss growing at or below the surface	—	Moss (*Fissidens*)	—
Rooted vascular	Rooted vascular plants growing submerged or as floating-leaved	Turtlegrass (*Thalassia*)	Water lily (*Nymphaea*)	Ditch grasses (*Ruppia*)
Floating vascular	Floating vascular plants growing on water surface	—	Water hyacinth (*Eichhornia crassipes*)	Duckweed (*Lemna*)

REEF		Ridgelike or moundlike structures formed by sedentary invertebrates			
	Coral		Coral (*Porites*)	—	—
	Mollusk		Eastern oyster (*Crassostrea virginica*)	—	—
	Worm		Reefworm (*Sabellaria*)	—	
STREAMBED		Intermittent streams (riverine system) or systems completely dewatered at low tide			
	Bedrock	Bedrock covers 75% or more of surface	—	Mayfly (*Ephemerella*)	—
	Rubble	Stones, boulders, and bedrock cover more than 75% of channel	—	Fingernail clam (*Pisidium*)	—
	Cobble-gravel	At least 25% of substrate smaller than stones	Blue mussel (*Mytilus*)	Snail (*Physa*)	—
	Sand	Sand particles predominate	Ghost shrimp (*Callianassa*)	Snail (*Lymnea*)	—
	Mud	Silt and clay predominate	Mud snail (*Nassarius*)	Crayfish (*Procambarus*)	—
	Organic	Peat or muck predominates	Mussel (*Modiolus*)	Oligochaete worms	—
ROCKY SHORE		High-energy habitats that lie exposed to wind-driven waves or strong currents			
	Bedrock	Bedrock covers 75% or more of surface	Acorn barnacle (*Chthamalus*)	Liverwort (*Marsupella*)	—
	Rubble	Stones, boulders, and bedrock cover more than 75% of surface	Mussel (*Mytilus*)	Lichens	—
UNCONSOLIDATED SHORE		Landforms such as beaches, bars, and flats that have less than 30% vegetation and are found adjacent to unconsolidated bottoms			
	Cobble-gravel	At least 25% of particles smaller than stones	Periwinkle (*Littorina*)	Mollusk (*Ellipia*)	—

(cont'd)

Table 21-4 *(Continued)*.

Class/Subclass	Definition	Examples of Dominance Type		
		Marine/Estuarine	Lacustrine/Riverine	Palustrine
Sand	At least 25% sand	Wedge shell (*Donax*)	Fingernail clam (*Pisidium*)	—
Mud	At least 25% silt and clay	Fiddler crab (*Uca*)	Fingernail clam (*Pisidium*)	—
Organic	Unconsolidated material, predominantly organic matter	Fiddler crab (*Uca*)	Chironomids	—
Vegetated	Nontidal shores exposed for sufficient time to colonize annuals or perennials	—	Cocklebur (*Xanthium*)	Summer cypress (*Kochia*)
MOSS–LICHEN WETLAND				
Moss	Mosses cover substrate other than rock; emergents, shrubs, and trees cover less than 30% of area	—	—	Peatmoss (*Sphagnum*)
Lichen	Lichens cover substrate other than rock; emergents, shrubs, and trees cover less than 30% of area	—	—	Reindeer moss (*Cladonia*)
EMERGENT WETLAND				
	Erect, rooted, herbaceous aquatic plants			
Persistent	Species that normally remain standing until the beginning of the next growing season	Cordgrass (*Spartina*)	—	Cattail (*Typha*)

742

		Samphire (*Salicornia*)	Wild rice (*Zizania*)	Pickerelweed (*Pontederia*)
Nonpersistent	No obvious sign of emergent vegetation at certain seasons			
SCRUB–SHRUB WETLAND	Dominated by wood vegetation less than 6 m tall			
Broad-leaved deciduous		Marsh elder (*Iva*)	—	Buttonbush (*Cephalanthus*)
Needle-leaved deciduous		—	—	Dwarf cypress (*Taxodium*)
Broad-leaved evergreen		Mangrove (*Rhizophora*)	—	Fetterbush (*Lyonia*)
Needle-leaved evergreen		—	—	Stunted pond pine (*Pinus serotina*)
Dead		—	—	—
FORESTED WETLAND	Woody vegetation 6 m or taller			
Broad-leaved deciduous		—	—	Red maple (*Acer rubrum*)
Needle-leaved deciduous		—	—	Bald cypress (*Taxodium distichum*)
Broad-leaved evergreen		Mangrove (*Rhizophora*)	—	Red bay (*Persea*)
Needle-leaved evergreen		—	—	Northern white cedar (*Thuia occidentalis*)
Dead		—	—	—

Source: Cowardin et al. (1979).

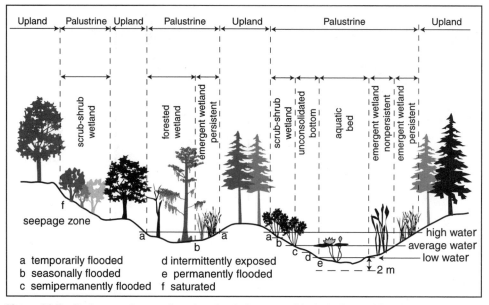

Figure 21-7 **Features and examples of wetland classes and hydrologic modifiers in the palustrine system. (*After Cowardin et al., 1979*)**

mollusk reef). Modifiers (Table 21-5) are used after classes and subclasses to describe more precisely the water regime, the salinity, the pH, and the soil. For many wetlands, the description of the environmental modifiers adds a great deal of information about their physical and chemical characteristics. Unfortunately, those parameters are difficult to measure consistently in large-scale surveys such as inventories.

Canadian Wetlands Classification System

The Canadian Wetland Classification System (Warner and Rubec, 1997) is designed to be practical. It is hierarchical as shown below:

> *Classes* are based on natural features of the wetlands, rather than on interpretation for various uses. They have direct application to large wetland regions. Wetland classes are recognized on the basis of properties that reflect the overall "genetic origin" of the wetland system and the nature of the wetland environment. Division into classes allows ready identification in the field and delineation on maps. They are also convenient groupings for data storage, retrieval, and interpretation.

> *Forms* are subdivisions of wetland classes based on surface morphology, water type, and morphology characteristics of the underlying mineral soil. Some forms are further subdivided into subforms. Forms are easily recognized features of the landscape and are the basic wetland-mapping unit.

> *Types* are subdivisions of wetland forms and subforms based on physiognomic characteristics of the vegetation communities. They appear to be comparable to the modifiers used in the U.S. Fish and Wildlife Service classification system. Types are most useful for evaluation of wetland values and benefits,

Table 21-5 Modifiers used in wetland and deepwater habitat classification by U.S. Fish and Wildlife Service

WATER REGIME MODIFIERS (TIDAL)

Subtidal—substrate permanently flooded with tidal water
Irregularly exposed—land suface exposed by tides less often than daily
Regularly flooded—alternately floods and exposes land surfaces at least daily
Irregularly flooded—land surface flooded less often than daily

WATER REGIME MODIFIERS (NONTIDAL)

Permanently flooded—water covers land surface throughout year in all years
Intermittently exposed—surface water present throughout year except in years of extreme drought
Semipermanently flooded—surface water persists throughout growing season in most years; when
  surface water is absent, water table is at or near surface
Seasonally flooded—surface water is present for extended periods, especially in early growing
  season but is absent by the end of the season
Saturated—substrate is saturated for extended periods during growing season but surface water is
  seldom present
Temporarily flooded—surface water is present for brief periods during growing season but water table
  is otherwise well below the soil surface
Intermittently flooded—substrate is usually exposed but surface water is present for variable periods
  with no seasonal periodicity

SALINITY MODIFIERS

Marine and Estuarine	Riverine, Lacustrine, and Palustrine	Salinity (ppt)
Hyperhaline	Hypersaline	>40
Euhaline	Eusaline	30–40
Mixohaline (brackish)	Mixosaline	0.5–30
Polyhaline	Polysaline	18.0–30
Mesohaline	Mesosaline	5.0–18
Oligohaline	Oligosaline	0.5–5
Fresh	Fresh	<0.5

pH MODIFIERS

Acid	pH less than 5.5	
Circumneutral	pH 5.5–7.4	
Alkaline	pH greater than 7.4	

SOIL MATERIAL MODIFIERS

Mineral	(1) Less than 20% organic carbon and never saturated with water for more than a few days, or
	(2) Saturated or artificially drained and has
	(a) less than 18% organic carbon if 60% or more is clay
	(b) less than 12% organic carbon if no clay
	(c) a proportional content of organic carbon between 12 and 18% if clay content is between 0 and 60%
Organic	Other than mineral as described above

Source: Cowardin et al. (1979).

management for wetland hydrology and wildlife habitat, and conservation and protection of rare and endangered species.

Table 21-6 shows the classes, forms, and subforms of the Canadian Wetland Classification System. Currently, the system recognizes 49 wetland forms and 72 subforms, based on 5 wetland classes. Although geomorphological, hydrologic, and chemical characteristics do not appear in the classification, Figure 21-8 shows the ontogenetic development of wetland forms in Canada. The similarity to the hydrogeomorphic (HGM) classification is clear. Two basic types of wetlands occur—ombrogenous and minerogenous. Ombrogenous systems are similar to HGM extensive peatlands. Minerogenous wetlands can be broken down further into terrigenous—wetlands similar to HGM depressional wetlands—and littogenous, which encompass HGM riverine and fringe wetlands. The five wetland classes—bog, fen, swamp, marsh, and shallow-water marsh—flow from this ontogenetic development, but there is no one-to-one correlation.

Ramsar Wetland Classification System

As a part of its mission to identify and conserve wetlands of international importance for biodiversity and wildlife (see Chapter 18), the International Union for the Conservation of Nature and Natural Resources (IUCN) has developed a "Classification System for Wetland Types." As with both the U.S. Fish and Wildlife Service classification system and the Canadian Wetland Classification System, most of the wetland types are readily recognized by their location in the landscape combined with their vegetation life form. Table 21-7 compares the Canadian and U.S. Fish and Wildlife Service systems to the Ramsar Convention system. The Ramsar system has 32 classes, divided into a marine/coastal group and an inland group. Because it attempts to be global, it has categories that neither the U.S. nor the Canadian system have, for example, underground karst systems and oases. The U.S. system, because it is hierarchical, has fewer classes at the system and subsystem level, but uses a number of modifiers to zero in on specific wetland types. The Canadian system also has only five classes, but with forms and subforms reaches a total of over 70 different categories by which to classify wetlands.

Wetland Inventories

One of the major objectives of wetland classification is to be able to inventory the location, extent, and type of wetlands in a region of concern. An inventory can be made of a small watershed, a county or a parish, an entire state or a province, or an entire nation. Whatever the size of the area to be surveyed, the inventory must be based on some previously defined classification and should be constructed to meet the needs of specific users of information on wetlands. Generally, inventories require not only information about the types and extent of wetlands, but also documentation of their geographical locations and boundaries. To accomplish this, *remote platforms*—aircraft and/or satellites—produce *imagery*—photographs or digital information that can be used to make images. This imagery must be *interpreted* to identify the locations, boundaries, and types of wetlands on the image. The interpreted images are typically overlain on topographic maps, such as U.S. Geological Survey (USGS) quadrangle maps, to obtain wetland *maps* (paper or digital) of the scene.

Table 21-6   Classes, forms, and subforms in the Canadian Wetland Classification System

Class/Form/Subform	Class/Form/Subform	Class/Form/Subform
Bog	**Swamp**	Riparian meltwater
Paalsa bog	Tidal swamp	Channel marsh
Peat mound bog	Tidal saltwater swamp	Lacustrine marsh
Mound bog	Tidal freshwater swamp	Lacustrine shore
		marsh
Domed bog	Inland salt swamp	Lacustrine bay marsh
Polygonal peat plateau bog	Flat swamp	Lacustrine lagoon
		marsh
Lowland polygon bog	Basin swamp	Basin marsh
Peat plateau bog	Swale swamp	Linked basin marsh
Plateau bog	Unconfined flat swamp	Isolated basin marsh
Northern plateau bog	Riparian swamp	Discharge basin
		marsh
Atlantic plateau bog	Lacustrine swamp	Hummock marsh
Collapse scar bog	Riverine swamp	Spring marsh
Riparian bog	Floodplain swamp	Slope marsh
Floating bog	Channel swamp	
Shore bog	Slope swamp	**Shallow water marsh**
Basin bog	Unconfined slope swamp	Tidal water
Flat bog	Peat margin swamp	Tidal bay water
String bog	Lagg swamp	Tidal shore water
Blanket bog	Drainageway swamp	Tidal channel water
Slope bog	Discharge swamp	Tidal lagoon water
Veneer bog	Spring swamp	Tidal basin water
	Seepage swamp	Estuarine water
Fen	Mineral-rise swamp	Estuarine delta water
String fen	Island swamp	Estuarine bay water
Northern ribbed fen	Levee swamp	Estuarine shore water
Atlantic ribbed fen	Beach ridge swamp	Estuarine channel
		water
Ladder fen	Mound swamp	Estuarine basin water
Net fen	Raised peatland swamp	Estuarine lagoon
		water
Paalsa fen		Riparian water
Snowpatch fen	**Marsh**	Riparian stream water
Spring fen	Tidal marsh	Riparian meltwater
Feather fen	Tidal bay Marsh	Channel water
Slope fen	Tidal lagoon marsh	Riparian floodplain
		water
Lowland polygon fen	Tidal channel marsh	Riparian delta water
Riparian fen	Tidal basin marsh	Lacustrine water
Floating fen	Estuarine marsh	Lacustrine shore
		water
Stream fen	Estuarine delta marsh	Lacustrine lagoon
		water
Shore fen	Estuarine bay marsh	Basin water
Collapse scar fen	Estuarine lagoon marsh	Discharge basin water
Horizontal fen	Estuarine shore marsh	Linked basin water
Channel fen	Riparian marsh	Isolated basin water
Basin fen	Riparian stream marsh	Polygon basin water
	Riparian floodplain marsh	Tundra basin water
	Riparian delta marsh	Thermokarst basin
		water

Source: Warner and Rubec (1997).

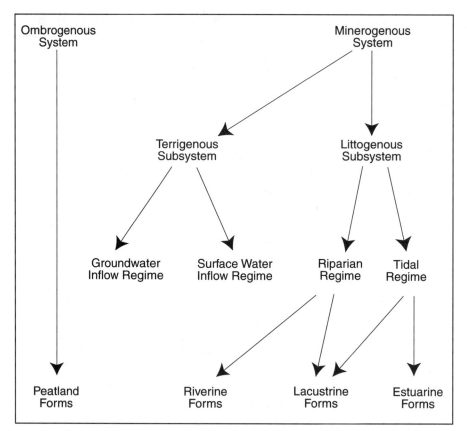

Figure 21-8 **Hydrological systems and wetland development as the basis of the Canadian Wetland Classification System. (*After Warner and Rubec, 1997*)**

Remote Sensing of Wetlands

Platform

In the early days of wetland classification, wetlands were mapped from surveyors' records and from boats, as O'Neil (1949) did for the first map of Louisiana wetlands. Later, interpretation of aerial photographs combined with verification in the field made the process both faster and more accurate. Today, high-altitude imagery from aircraft, especially the U-2 "spy plane," is used extensively, but satellite imagery is rapidly becoming the norm. These remote platforms are an effective way of gathering data for large-scale wetland surveys. The choice of which platform to use depends on the resolution required, the area to be covered, and the cost of the data collection. Low-altitude aircraft surveys offer a relatively inexpensive and fairly effective way to survey small areas. High-altitude aircraft offer much greater coverage in each scene (photograph) and may be less expensive per unit area than low-altitude aircraft when costs of photo interpretation are included.

Orbiting satellites have been providing data for Earth resources classification since the launching of the first of the Landsat satellites in 1972. Today, a number of highly effective satellites orbit the Earth (Table 21-8). One problem of early satellites was

Table 21-7 Comparison of wetland types among the Ramsar Convention system, the Canadian wetland classification system, and the U.S. Fish and Wildlife wetland and deepwater habitat classification

Ramsar Convention Code Name		U.S. Fish and Wildlife System[a]	Canadian System[b]
MARINE/COASTAL WETLANDS			
A	Marine water <6 m	Marine Subtidal	Shallow (<2 m) water marsh
B	Marine subtidal aquatic beds	Marine subtidal aquatic bed	
C	Coral reefs	Marine subtidal reef	—
D	Rocky marine shores	Marine intertidal rock bottom	—
E	Sandy shore or dune	Marine intertidal unconsolidated	—
F	Estuarine waters	Estuarine subtidal	Estuarine marsh, water
G	Intertidal flats	Estuarine intertidal unconsolidated bottom	Estuarine water, tidal water
H	Intertidal marshes	Estuarine intertidal emergent wetland	Tidal marsh
I	Intertidal forested wetland	Estuarine intertidal forested wetland	Tidal swamp
J	Coastal saline lagoon	Estuarine subtidal unconsolidated, saline	Estuarine water
K	Coastal fresh lagoon	Estuarine subtidal unconsolidated, fresh	Estuarine water
Zk(a)	Marine/coastal karst	Estuarine subtidal rocky shore	—
INLAND WETLANDS			
L	Permanent inland deltas	Riverine perennial	Riparian delta marsh
		Estuarine delta marsh	
		Shallow riparian delta water	
M	Permanent rivers/streams	Riverine perennial swamp, marsh	Shallow riparian water
N	Intermittent rivers/streams	Riverine intermittent	—
O	Permanent fresh lakes Riparian water (oxbows)	Lacustrine littoral or limnetic	Shallow lacustrine water
P	Intermittent fresh lakes	Lacustrine or riparian littoral	—
Q	Permanent saline lakes	Lacustrine littoral unconsolidated, saline	—
R	Intermittent saline lakes	Lacustrine littoral intermittent, saline	—
Sp	Permanent saline marshes/pools	Palustrine emergent wetland or unconsolidated bottom, saline	Estuarine marsh, inland salt swamp
Ss	Intermittent saline marshes/pools	Palustrine emergent wetland or unconsolidated bottom, intermittent	Spring, slope, or basin marsh
Tp	Permanent fresh marsh/pools (<8 ha)	Palustrine emergent wetland or unconsolidated bottom, fresh	Shallow basin water, lacustrine marsh
Ts	Intermittent fresh marsh/pools, inorganic soils	Palustrine emergent wetland, intermittently flooded	Shallow basin water
U	Nonforested peatlands	Palustrine emergent wetland, persistent	Bogs, fens
Va	Alpine wetlands	Palustrine emergent wetland, persistent	—

Table 21-7 *(Continued).*

Ramsar Convention Code Name	U.S. Fish and Wildlife System[a]	Canadian System[b]
Vt Tundra wetlands	Palustrine emergent wetland, persistent	Bogs, fens, shallow basin water
W Shrub-dominated wetlands		Riparian, flat, slope, discharge or mineral-rise swamp
	Palustrine scrub–shrub wetland	
Xf Fresh forested wetlands on inorganic soils	Palustrine forested wetland	Riparian swamp
Xp Forested peatlands	Palustrine forested or scrub–shrub wetland	Flat bog, flat or raised peatland swamp
Y Freshwater springs, oases	—?	Hummock marsh
Zg Geothermal wetlands	—	—
Zg(b) Inland karst systems, underground	—	—

[a] Ramsar class can be further approximated with additional modifiers.
[b] Class and form indicated only; subforms can approximate Ramsar more closely. The term "shallow" in the wetland type indicates the shallow marsh class.

the poor resolution (Landsat has a resolution of 30 m). Wetland scientists working on the National Wetlands Inventory (see discussion later) found that Landsat 7 could not provide the desired level of detail without what appeared to be an excessive amount of collateral data such as aerial photographs and field work (Nyc and Brooks, 1979). Today's satellites can resolve features of the Earth's surface to as little as 1 m, but high resolution is not an unmixed blessing. It requires the ability to transmit and

Table 21-8 **Summary of a few of the many government-operated orbiting satellites capable of imaging land cover. A number of privately owned satellites are also in orbit**

Country	Program	Date (year)	Instrument Type[a]	Resolution (m) P	M	R	Number of Color Bands
Europe	ENVISAT	1999	R			30	
France	Spot 5A	1999	P/M	5	10		4
India	IRS-1D	1999	P/M	10	20		4
United States	GDE	1998	P	1			
United States	Orb View	1998	P/M	1, 2	8		4
United States	Resource 21	1998	M		10, 100		6
Korea	KOMPSAT	1998	P/M	10	10		3
United States/ Japan	EOS AM-1	1998	M	15	15		14
United States	Landsat 7	1998	P/M	15	30		7
United States	Earth Watch	1997	P/M	1	4		4
Russia	Almaz2	1997	R			5	
United States	TRW Lewis	1997	P/M	5	30		384
China/Brazil	CBERS	1997	P/M	20	20		7
Canada	Radarsat	1996	R			9	

[a] M, multispectral only; P/M, panchromatic and multispectral; P, panchromatic only; R, radar.
Source: Wulder (2000).

process enormous amounts of data, since the data generated for a given surface area quadruples every time the resolution doubles.

Imagery

In addition to choosing the remote-sensing platform, the wetland scientist or manager has the choice of several types of imagery from different types of sensors. Color photography and color–infrared photography have been popular for many years for wetland inventories from aircraft, although black-and-white photography has been used with some success. Color–infrared film provides good definition of plant communities and is the film of choice. Satellites, and some aircraft, gather digital data in one or more electromagnetic spectral bands. For example, *Landsat 7* has panchromatic and multispectral capability in seven color bands, including infrared. It can resolve the Earth's surface to 30 m in multispectral mode and 15 m in panchromatic mode.

Interpretation

Interpretation of wetland areas from imagery is typically the most difficult and time consuming part of the mapping procedure, and the accuracy of the resulting maps is determined at this step. Much interpretation is still done visually by a photointerpreter, identifying and delimiting image colors correlated with field observation of the wetland types that correspond to different colors. The ability to program computers to recognize these colors (or their equivalent digital signal), and thus to interpret the imagery directly, is improving as computer power and image resolution increases. In particular, the use of several spectral bands at once allows for much improved resolution of the spectral signature of different surface features. Depending on the expected use of the classified map products, supervised computer interpretation is frequently used today, usually with satisfactory results. In addition, satellite data can be used in conjunction with other databases such as hydric soil maps and on-site evaluation, as in the example of Figure 21-9.

Wetland Maps

The product of image interpretation is a map delineating the locations, boundaries, and types of wetlands in the mapped scene. These typically appear as an overlay on a USGS topographic map or as a standalone map that has been georectified using established coordinates and procedures. Maps are generally created for specific purposes, and none serves all management requirements equally well. Coarse-scale maps covering large areas are often useful for inventories addressing regional wetland types, areas, and changes through time (e.g., the Status and Trends Reports of the U.S. Fish and Wildlife Service). Finer scale resolution maps generally cover much smaller areas because of the much greater data requirements per unit area, especially when computerized interpretation of multispectral, high-resolution satellite data is used. These maps tend to be much more accurate at the local level than regional maps.

The National Wetlands Inventory

The U.S. National Wetlands Inventory (NWI) is a good example of a major wetlands mapping project, and illustrates some of the problems encountered in any mapping enterprise. The Cowardin et al. (1979) classification scheme has provided the basic mapping units for the NWI being carried out by the U.S. Fish and Wildlife Service. For the NWI, aerial photography at scales ranging from 1 : 60,000 to 1 : 130,000

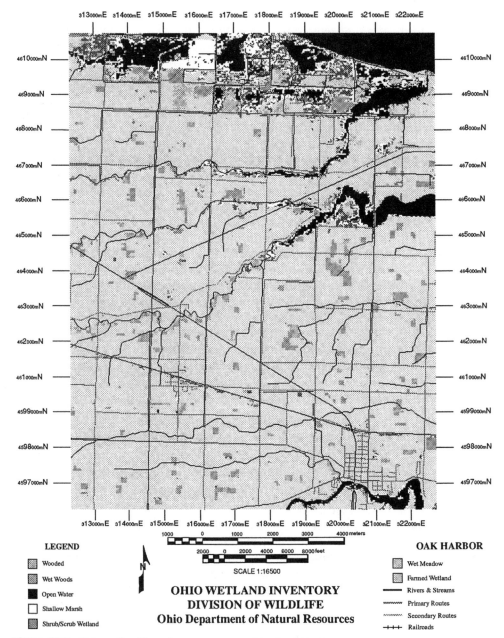

Figure 21-9 Ohio wetland inventory map prepared by the Ohio Department of Natural Resources, Division of Wildlife, from supervised classification of *Landsat 5* Thematic Mapper (TM) satellite imagery (30-m resolution), hydric soil data, and limited on-site evaluation. (*Courtesy of Ohio Department of Natural Resources, Division of Wildlife*)

is the primary source of data, with color–infrared photography providing the best delineation of wetlands (Wilen and Pywell, 1981; Tiner and Wilen, 1983). In the 1970s, maps were mostly created from 1 : 80,000-scale black-and-white photography. Now 1 : 60,000 color–infrared photography is used for most of the mapping. Manual photointerpretation and field reconnaissance are then used to define wetland boundaries according to the wetland classification system. The information is summarized on 1 : 24,000 and 1 : 100,000 maps, using an alphanumeric system as illustrated in Figure 21-10.

Although the NWI currently still relies primarily on manual photo or image interpretation, recent research suggests that computerized interpretation of satellite imagery could be superior for delineating wetland hydrology, particularly for agricultural areas (National Research Council, 1995). The Federal Geographic Data Committee has published a review of the use of satellite data for mapping and monitoring wetlands (FGDC, 1992).

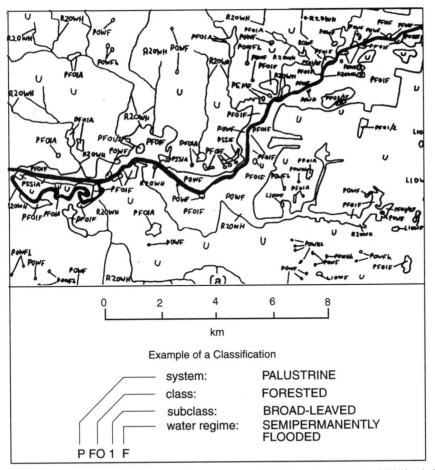

Figure 21-10 Sample of mapping technique used at 1 : 100,000 scale by National Wetlands Inventory, showing a portion of map (redrawn) and an example of a classification notation. (*After Dyersburg, Tennessee, 1 : 100,000 wetland map provided courtesy of National Wetlands Inventory, U.S. Fish and Wildlife Service*)

By late 1999, 89 percent of the lower 48 states, 31 percent of Alaska, and all of Hawaii had been mapped. Initial areas for mapping were selected based on agency needs and requests from states and other government units. Thirty nine percent of U.S. wetlands maps have been computerized into digital geographic information systems (GIS), including New Jersey, Delaware, Illinois, Maryland, Washington, and Indiana and parts of Virginia, Minnesota, and South Carolina, and 11 percent of Alaska.

The NWI program is now also producing "status and trends" reports on U.S. wetlands, the most recent documenting wetlands losses in the past 200 years (Dahl, 1990) and wetlands status and trends from the mid-1970s to the mid-1980s (Dahl and Johnson, 1991; see Chapter 4). Data for these reports were derived from a sample of U.S. wetlands and will be revised every 10 years, with updates due early in the 21st century. Other products from the NWI include "map reports" for each 1 : 100,000-scale wetland map, a wetland plant database, and state wetland reports. The entire wetland inventory, at the level of detail envisioned for the NWI, will take many years to complete. Once it is completed and the database is computerized, a continuous and more vigilant monitoring of our nation's wetland resources will be possible.

Wetland Delineation

Although NWI wetland maps can provide much important information about the general location of wetlands, the determination of wetland boundaries to within a tolerance of less than a meter for regulatory purposes (termed *jurisdictional wetlands*) will probably always require on-site evaluation. In part, this simply reflects the limit of resolution of the 1 : 24,000-scale maps currently used by the NWI. The width of a fine line on these maps represents about 5 m on the ground. Improving the scale to 1 : 12,000 would theoretically double the resolution but would require four times as many maps to cover the same area.

In a comparison of wetlands delineated by the 1987 Wetlands Delineation Manual with wetlands mapped by the NWI and hydric soils mapped by the Soil Conservation Service, Rolband (1995) found that NWI maps typically significantly understated, and Soil Conservation Service hydric soils maps significantly overstated the areas of jurisdictional wetlands on the examined site.

The requirement of a permit for dredge-and-fill activities in wetlands (Clean Water Act, Section 404) spawned a new industry that specializes in determining the locations and boundaries of wetlands, as well as the impact of the proposed activities. In order to standardize, for regulatory purposes, the way in which boundaries were determined the U.S. Army Corps of Engineers wrote several manuals on wetland delineation. The 1987 Wetlands Delineation Manual (U.S. Army Corps of Engineers, 1987) is currently the manual in official use (see Chapter 18).

The 1987 delineation manual differs from the U.S. Fish and Wildlife (FWS) Service Classification of Wetlands and Deepwater Habitats (Cowardin et al., 1979) in two principal ways: (1) Not all wetlands identified in the FWS classification are included in the delineation manual; the FWS classification includes all six categories of special aquatic sites that are covered under Section 404(b) guidelines (sanctuaries and refuges, wetlands, mud flats, vegetated shallows, coral reefs, and riffle and pool complexes). TR Y-87-1 considers only wetlands. (2) The FWS system requires that a positive indicator of wetlands be present for any *one* of the three parameters— vegetation, soils, and hydrology—in the Section 404 definition of wetlands. TR Y-

87-1 requires a positive wetland indicator be present for each of these parameters, with some exceptions.

Technical Guidelines

Guidelines follow from the Section 404(b)(1) definition of wetlands (*Federal Register*, 1980; Federal Register, 1982): "those areas that are inundated or saturated by surface or ground water [*hydrology*] at a frequency and duration sufficient to support, and that under normal circumstances do support, a prevalence of vegetation [*vegetation*] typically adapted for life in saturated soil conditions [*soil*]" (bracketed words added). The definition refers to wetland-adapted vegetation, hydric soil, and flooding or saturating hydrology, and wetland delineation in TR Y-87-1 depends on determining the boundaries of the area for which these three parameters are met.

Vegetation

Wetland vegetation is defined as macrophytes typically adapted to inundated or saturated conditions. Plants are grouped into five categories (Table 21-9): obligate wetland plants (OBL), facultative wetland plants (FACW), facultative plants (FAC), facultative upland plants (FACU), and obligate upland plants (UPL). To meet the wetland vegetation requirement, more than 50 percent of the dominant species must be OBL,

Table 21-9 Plant indicator status categories used in wetland delineation

Indicator Category	Indicator Symbol	Definition
Obligate wetland plants	OBL	Plants that occur almost always (estimated probability >99%) in wetlands under natural conditions, but which may also occur rarely (estimated probability < 1%) in nonwetlands. Examples: *Spartina alterniflora*, *Taxodium distichum*.
Facultative wetland plants	FACW	Plants that occur usually (estimated probability >67–99%) in wetlands, but also occur (estimated probability 1–33% in nonwetlands). Examples: *Fraxinus pennsylvanica*, *Cornus stolonifera*.
Facultative plants	FAC	Plants with a similar likelihood (estimated probability 33–67%) of occurring in both wetlands and nonwetlands. Examples: *Gleditsia triaconthos*, *Smilax rotundifolia*.
Facultative upland plants	FACU	Plants that occur sometimes (estimated probability 1–<33%) in wetlands, but occur more often (estimated probability >67–99%) in nonwetlands. Examples: *Quercus rubra*, *Potentilla arguta*.
Obligate upland plants	UPL	Plants that occur rarely (estimated probability <1%) in wetlands, but occur almost always (estimated probability >99%) in nonwetlands under natural conditions. Examples: *Pinus echinata*, *Bromus mollis*.

Source: U.S. Army Corps of Engineers (1987).

FACW, or FAC. Species lists of plants in these categories are available from a number of sources (U.S. Army Corps of Engineers, 1987). Other indicators of wetland plants may also be used, including morphological, physiological, and reproductive adaptations such as buttressed tree trunks, pneumatophores, adventitious roots, and enlarged lenticels. Furthermore, the technical literature may provide additional information about the ability of plants to endure saturated soils.

Hydric Soils

A hydric soil is a soil that is saturated, flooded, or ponded long enough during the growing season to develop anaerobic conditions that favor the growth and regeneration of hydrophytic vegetation (see Chapter 6). All histosols (organic soils) except folists are hydric, as well as soils in a few other groups, particularly aquic soils that are poorly drained, are saturated, or have shallow (typically less than 15 cm) water tables for a significant period (usually more than 1 week) during the growing season. When a hydric soil is drained, it may not be referred to as hydric, unless the vegetation is hydrophytic and indicators of hydrology support the designation as a hydric soil. Hydric soil designation can be supported by additional indicators (defined in detail in the manual) such as low permeability, appropriate soil chroma, development of mottles, and iron or manganese concretions.

Wetland Hydrology

Areas with evident characteristics of wetland hydrology are those in which the presence of water has an overriding influence on characteristics of vegetation and soils due to anaerobic and reducing conditions, respectively. Generally, determination of hydrology as wetland depends on the frequency, timing, and duration of inundation, or soil saturation, as presented in Table 21-10 for nontidal areas. Zone I is aquatic and Zone VI is upland. Zones II, III, and IV are wetlands. Zone V may or may not be considered wetland, depending on other indicators. Additional indicators of wetland hydrology use recorded data from stream, lake, or tidal gages, flood predictions, and historical data on flooding. Visual observations are also indicators, such as soil saturation,

Table 21-10 Hydrologic zones for nontidal areas

Zone	Name	Duration[a]	Comments
I	Permanently inundated	100%	Inundation >2 m mean water depth. Aquatic, not wetlands
II	Semipermanently to nearly permanently inundated or saturated	>75%–<100%	Inundation defined as <2 m mean water depth
III	Regularly inundated or saturated	>25%–75%	
IV	Seasonally inundated or saturated	>12.5%–25%	
V	Irregularly inundated or saturated	≥5%–12.5%	Many areas having these hydrologic characteristics are not wetlands
VI	Intermittently or never inundated or saturated	<5%	Areas with these hydrologic characteristics are not wetlands

[a] Refers to duration of inundation and/or soil saturation during the growing season.
Source: U.S. Army Corps of Engineers, 1987

watermarks on trees or other structures, drift lines, sediment deposits, and drainage patterns.

Delineation Procedure

Routine delineation methods require a combination of office gathering and synthesis of available data on the site, combined, when necessary, with on-site inspection and additional data generation. Comprehensive delineation methods are reserved for particularly sensitive cases and usually require significant time and effort to obtain the needed quantitative data.

All methods begin with accumulation of available data on the site to be delineated, data such as USGS quadrangle maps, NWI wetland maps, plant surveys, soil surveys, gage data, environmental assessments or impact statements, remotely sensed data,

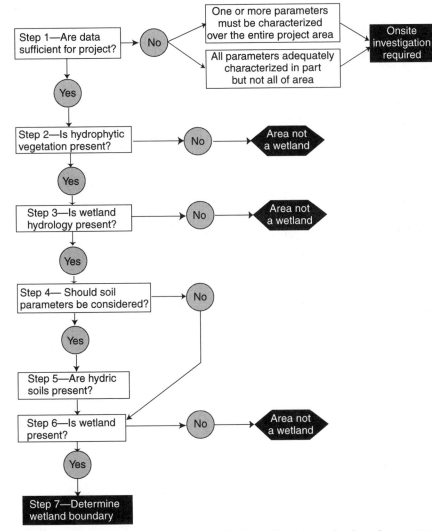

Figure 21-11 Flowchart of steps involved in making a wetland determination when an on-site inspection is unnecessary. (*U.S. Army Corps of Engineers, 1987*)

local expertise, and the applicant's survey plans and engineering designs (often with topographic surveys). These data are synthesized into a preliminary determination of whether the information is adequate to make a wetland delineation of the entire tract in question. A flowchart (Fig. 21-11) shows the steps to determine first if on-site inspection is necessary, and second, if unnecessary, to determine whether the area is a jurisdictional wetland.

TR Y-87-1 also details methods, depending on the size of the project area, for on-site evaluation when available data are inadequate. These methods may include, for example, the use of transects when the area is too large to survey in its entirety. The intensity of the on-site investigation is determined by the available information on the site, the type of project anticipated, the ecological sensitivity of the area, and other factors. The objective of on-site investigations is, of course, to obtain adequate data to determine whether all or part of the project area fits the criteria for wetlands and, if so, where the wetland boundaries lie.

Recommended Readings

Brinson, M. M. 1993. A Hydrogeomorphic Classification for Wetlands. Wetlands Research Program Technical Report WRP-DE-4. U.S. Army Corps of Engineers Waterways Experiment Station, Vicksburg, MS.

Cowardin, L. M., V. Carter, F. C. Golet, and E. T. LaRoe. 1979. Classification of Wetlands and Deepwater Habitats of the United States. U.S. Fish and Wildlife Service, FWS/OBS-79/31, Washington, DC. 103 pp.

Tiner, R. W. 1999. Wetland Indicators: A Guide to Wetland Identification, Delineation, Classification, and Mapping. CRC Press, Boca Raton, FL. 392 pp.

U.S. Army Corps of Engineers. 1987. Corps of Engineers Wetlands Delineation Manual. Technical Report Y-87-1. U.S. Army Corps of Engineers Waterways Experiment Station, Vicksburg, MS. 100 pp. and appendices.

Warner, B. G., and C. D. A. Rubec, eds. 1997. The Canadian Wetland Classification System. National Wetlands Working Group, Wetlands Research Centre, University of Waterloo, Waterloo, Ontario.

Selected Wetland Books and Similar Publications

Selected Wetland Books and Similar Publications

Category and Background	Subject	Reference
GENERAL SINGLE-AUTHOR BOOKS		
Salt Marshes and Salt Deserts of the World	Ecology	Chapman (1960)
Peatlands	Ecology	Moore and Bellamy (1974)
Coastal Vegetation	Ecology	Chapman (1976a)
Mangrove Vegetation	Ecology	Chapman (1976b)
Wet Coastal Ecosystems	Ecology	Chapman (1977)
Freshwater Marshes, 1st, 2nd, 3rd eds.	Freshwater marshes/wildlife management	Weller (1981, 1987, 1994)
Wetlands, 1st and 2nd eds.	Ecology and management	Mitsch and Gosselink (1986, 1993)
Waterlogged Wealth	World's wetlands and their losses	Maltby (1986)
Coastal Marshes	Ecology/wildlife management	Chabreck (1988)
Creating Freshwater Wetlands, 1st and 2nd eds.	Creation and restoration	Hammer (1992, 1997)
Wetland Identification and Delineation	Handbook for wetland delineation	Lyon (1994)
Restoring Prairie Wetlands	Restoration of U.S. prairie wetlands	Galatowitsch and van der Valk (1994)
Algae and Element Cycling in Wetlands	Algae and wetlands	Vymazal (1995a)
Treatment Wetlands	Wastewater wetlands	Kadlec and Knight (1996)
Wetlands of the American Midwest	Wetland drainage history	Prince (1997)
In Search of Swampland	Wetland primer and identification guide	Tiner (1998)
Wetland Indicators	Wetland delineation, classification, mapping	Tiner (1999)
Wetland Restoration	Flood pulsing and disturbance dynamics	Middleton (1999)
Constructed Wetlands in the Sustainable Landscape	Wastewater wetlands	Campbell and Ogden (1999)
Wetland Birds	Habitat resources	Weller (1999)
Heavy Metals in the Environment	Using wetlands for metal removal	H. T. Odum (2000)
INTECOL WETLAND CONFERENCES		
1980 (New Delhi, India)	Ecology and management	Gopal et al. (1982)
1984 (Trebon, Czechoslovakia)	Waterplants	Pokorny et al. (1987)
	Modeling	Mitsch et al. (1988)
1988 (Rennes, France)	Inventory, ecology, management	Whigham et al. (1993)
	Ecology and Management	Lefeuvre (1989)
	Conservation and development	Maltby et al. (1992)

Event	Topic	Reference
1992 (Columbus, Ohio)	Ecology and management	Mitsch (1994)
	Wetlands—general	Wetzel et al. (1994)
	Constructed wetlands	Gopal and Mitsch (1995)
	Inventory of world's wetlands	Finlayson and van der Valk (1995)
	Wetland interactions with lakes	Jørgensen (1995)
1996 (Perth, Australia)	Ecology and Management	McComb and Davis (1998)
	Treatment wetlands	Tanner et al. (1999)

OTHER EDITED CONFERENCE PROCEEDINGS, BOOKS, AND JOURNAL SPECIAL ISSUES

Event	Topic	Reference
1977 wetland symposium	Ecology and management	Good et al. (1978)
1977 wetland symposium		Kusler and Montanari (1978)
1978 wetland symposium		Greeson et al. (1979)
1978 symposium on floodplain and riparian wetlands		Johnson and McCormick (1979)
1978 workshop on bottomland hardwood forest		Clark and Benforado (1981)
1981 USSR wetlands workshop	Ecology and management	Logofet and Luckyanov (1982)
Mires of the world (2 volumes)	Ecology, management and inventory	Gore (1983a, b)
Canadian wetlands	Classification and ecology	National Wetlands Working Group (1998)
1986 anaerobiosis symposium (Charleston, South Carolina)	Ecology and management	Hook et al. (1988)
1985 cumulative loss workshop	Cumulative effects on wetlands	Bedford and Preston (1988)
1985 regional symposium (Jamestown, Virginia)	Prairie potholes	van der Valk (1989)
1986 wetland meeting (Charleston, South Carolina)		Sharitz and Gibbons (1989)
Chinese wetland inventory (in Chinese)		Lu (1990)
Wetland creation and restoration		Kusler and Kentula (1990)
1983 SCOPE Conference (Tallinn, Estonia)		Patten (1990, 1994b)
Wetlands of the world	Ecology and management	Williams (1990)
Forested wetlands of the World	Ecology and distribution	Lugo et al. (1990a)
Bottomland hardwood forest ecosystems	Management and cumulative loss	Gosselink et al. (1990b)
1989 Great Lakes wetlands workshop	Great Lakes wetlands	Krieger et al. (1992)
1990 conference (Cambridge, UK)	Wastewater wetlands	Cooper and Findlater (1990)
1991 Workshop—Salt Marsh dynamics (UK)	Salt marsh physics	Allen and Pye (1992)
Fens and bogs of the Netherlands		Verhoeven (1992)
Wetlands and nonpoint source pollution	Wetlands and water quality	Olson (1992)
Wetland delineation manual discussion	U.S. wetlands	Environmental Defense Fund and World Wildlife Fund (1992)
1989 symposium on Netherland wetlands (Arnhem, The Netherlands)	Wetlands in the Netherlands	Best and Bakker (1993)

Category and Background	Subject	Reference
1991 Land–water Interactions Conference (New Delhi, India)		Gopal et al. (1993)
1991 conference (Pensacola, Florida)	Wastewater wetlands	Moshiri (1993)
The world's wetlands	Ecology and management	Whigham et al. (1993)
1989 symposium on hydrogeology of wetlands	Hydrology	Winter and Llamas (1993)
1992 SWS conference (New Orleans, USA)	Ecology and management	Landin (1993)
Wetlands: Guide to science, law and technology	Wetland management	Dennison and Berry (1993)
Everglades, Florida	Ecology and management	Davis and Ogden (1994)
Des Plaines River wetlands, Illinois	Wetland creation	Sanville and Mitsch (1994)
Plants and processes in wetlands	Wetland vegetation and their processes	Brock et al. (1994)
Applied wetlands science and technology	Ecology and management	Kent (1994)
The Pantanal of Mato Grosso		Por (1995)
Wetlands: Characteristics and Boundaries	U.S. wetland characterization	National Research Council (1995)
1994 symposium on wetland gradients	Environmental gradients, boundaries and buffers	Mulamoottil et al. (1996)
Tidal wetland restoration	Southern California emphasis	Zedler (1996)
Ecology and management of tidal marshes	Gulf of Mexico model	Coultas and Hsieh (1997)
Northern forested wetlands	Ecology and management—North America	Trettin et al. (1997)
1990-91 workshops on coastal forests	Ecology and management	Laderman (1998)
1994 international symposium on wetland gradients	Wetland gradients	Mulamoottil et al. (1996)
Southern forested wetlands	Ecology and management-southern USA	Messina and Conner (1998)
1996 Great Lakes wetlands workshop (Columbus, Ohio)	Wetland restoration	Mitsch and Bouchard (1998)
1997 SWS meeting, Bozeman, MT	Functions/values of prairie wetlands	Murkin (1998)
1997 symposium on wetlands and landfill leachate	Treatment wetlands	Mulamoottil et al. (1999)
Nitrogen and phosphorus retention in wetlands	Water quality and wetlands	Mitsch et al. (2000b)
1997 wetland values symposium (Stockholm, Sweden)	Wetland values: landscape/institutional	Söderqvist et al. (2000)
1999 conference on wetlands and remediation (Salt Lake City, Utah)	Remediation of contaminated wetlands	Means and Hinchee (2000)

POPULAR DESCRIPTIONS OF WETLANDS W/ COLOR PHOTOGRAPHS

Swamps of the United States		Thomas (1976)
Wetlands	Field guide	Niering (1985)
Cypress Swamps		Dennis (1988)
Australian Wetlands	Inventory	McComb and Lake (1990)
Wetlands	General ecology and management	Finlayson and Moser (1991)

Wetlands of North America	Littlehales and Niering (1991)
National Geographic article	Mitchell et al. (1992)
Wetlands in Danger	Dugan (1993)
Scientific American article	Kusler et al. (1994)
Wetlands: The Web of Life	Rezendes and Roy (1996)

SELECTED U.S. GOVERNMENT REPORTS

Wetland inventory	Shaw and Fredine (1956)
Wetlands and construction activities	Darnell (1976)
Wetland definition and classification	Cowardin et al. (1979)
Wetland community profiles	Many authors
Wetland management	Zinn and Copeland (1982)
Wetland inventory	Frayer et al. (1983
Wastewater wetlands	U.S. Environmental Protection Agency (1983)
Wetland use and regulation	Office of Technology Assessment (1984)
Wetland inventory	Dahl (1990)
Wetland indicators	Adamus and Brandt (1990)
Wetland inventory	Dahl and Johnson (1991)
Wetland creation and restoration	Kentula et al. (1992b)
Hydrogeomorphic classification	Brinson (1993a)

INTERNATIONAL UNION FOR THE CONSERVATION OF NATURE AND NATURAL RESOURCES (IUCN) AND THE RAMSAR CONVENTION BUREAU PUBLICATIONS

Wetlands of Africa	Burgis and Symoens (1987)
	Davies and Gasse (1988)
1989 Wetlands of Greece Conference	Gerakis (1992)
Wetlands of Africa	Hughes and Hughes (1992)
Wetlands of Africa	Jones (1993a)
Wetlands of Asia and Oceania	Jones (1993b)
Wetlands of Europe	Jones (1993c)
Wetlands of Neotropics and North America	Jones (1993d)
The Ramsar Convention	Matthews (1993)
Wetlands of Central and Eastern Europe	IUCN (1993)
Wetlands of Oceania	Scott (1993)
Wetlands of the Middle East	Scott (1995)
Wetlands of New Zealand	Cromarty and Scott (1996)
Wetlands of Mediterrania	Costa et al. (1996)
Wetlands of Russia	Krivenko (1999)
Ramsar Sites Overview	Frazier (1999)

Note: The left-column labels for the IUCN section correspond as follows: Inventory and directory; Bibliography; Inventory and directory; Directory; Directory; Directory; Directory; General description; Description and inventory; Description; Principal wetlands; Description; Inventory; Most valuable wetlands.

Useful Wetland
Web Sites

Useful wetland web sites

Address	Organization Responsible	Remarks
http://swamp.ag.ohio-state. edu	The Ohio State University	Olentangy River Wetland Research Park
http://iucn.org/themes/ramsar	IUCN	Ramsar Convention home page
http://www.wetlands.agro.nl	Wetlands International—Africa, Europe, Middle East (AEME)	Includes Ramsar data base
http://www.wetlands.org/	Wetlands International—the Americans	
http://www.nwi.fws.gov/	U.S. Fish and Wildlife Service	National Wetland Inventory
http://www.epa.gov/OWOW	U.S. Environmental Protection Agency	Office of Wetlands, Oceans, and Watersheds
http://www.sws.org	Society of Wetland Scientists	Membership, meetings, journal *Wetlands*
http://www.ducks.org	Ducks Unlimited	Membership, meetings, publications, news
http://www.env.duke.edu/wetland	Duke University	Wetland program
http://www.wetlands.cas.psu.edu/cwhome.htm	Pennsylvania State University	Cooperative wetland center
http://www.wes.army.mil/el/wetlands/wetlands.html	U.S. Army Corps of Engineers	Waterways Experiment Station (WES)
http://www.pwrc.usgs.gov/wli	U.S. Department of Agriculture, Natural Resources Conservation Service (USDA/NRCS)	Environmental Laboratory—Wetlands: publications, delineation manual, research
http://www.nwrc.usgs.gov	U.S. Geological Survey	Wetland Science Center
http://www.ifas.ufl.edu/~wbl	University of Florida	National Wetlands Center: publications, news
http://www.enveng.ufl.edu/wetlands	University of Florida	Wetland Biogeochemistry Laboratory
http://www.statlab.iastate.edu:80/soils/hydric	USDA/NRCS	Center for Wetlands
http://www.wl.fb-net.org	USDA/NRCS	Hydric soils list for the United States
http://www.umanitoba.ca/faculties/science/delta_marsh/	University of Manitoba	Wetland Reserve Program (WRP)
http://ibs.uel.ac.uk/imcg/	International Mire Conservation Group	Delta Marsh Field Station

Useful Conversion Factors

Multiply	By	To Obtain
LENGTH		
centimeters (cm)	0.3937	inches
feet	0.3048	meters (m)
inches	2.54	centimeters (cm)
kilometers (km)	0.6214	miles
meters (m)	3.2808	feet
meters (m)	39.37	inches
meters (m)	1.0936	yards
miles	1.6093	kilometers (km)
yards	0.9144	meters (m)
AREA		
acres	0.4047	hectares (ha)
hectares (ha)	2.47	acres
hectares (ha)	10,000	square meters (m^2)
acres	4047	square meters (m^2)
hectares (ha)	0.01	square kilometers (km^2)
square kilometers (km^2)	100	hectares (ha)
square meters (m^2)	0.0001	hectares (ha)
VOLUME		
cubic feet	0.02834	cubic meters (m^3)
cubic meters (m^3)	35.31	cubic feet
cubic centimeters (cm^3)	10^{-3}	liters (L)
acre-feet	1223.5	cubic meters (m^3)
gallons	3.785	liters (L)
gallons	0.003785	cubic meters (m^3)
cubic meters (m^3)	264.2	gallons
liters (L)	0.2642	gallons
FLOW		
cubic feet per second (cfs)	0.002832	cubic meters per second ($m^3 \ s^{-1}$)
cubic feet per second (cfs)	10.1952	cubic meters per hour ($m^3 \ hr^{-1}$)
cubic feet per second (cfs)	448.86	gallons per minute (gpm)
cubic meters per second ($m^3 \ s^{-1}$)	35.31	cubic feet per second (cfs)
cubic meters per second ($m^3 \ s^{-1}$)	3600	cubic meters per hour ($m^3 \ hr^{-1}$)
gallons per minute (gpm)	0.002228	cubic feet per second (cfs)
gallons per minute (gpm)	0.06308	liters per second (L s^{-1})
gallons per minute (gpm)	0.00379	cubic meters per minute ($m^3 \ min^{-1}$)
MASS		
grams (g)	0.002205	pounds
grams (g)	0.001	kilograms (kg)
kilograms (kg)	2.2046	pounds
kilograms (kg)	1000	grams (g)
pounds	453.6	grams (g)
pounds	0.4536	kilograms (kg)
metric tons (t)	2205	pounds
metric tons (t)	1000	kilograms (kg)
PRESSURE		
Atmosphere (atm)	1.01325×10^5	pascal (Pa)
Atmosphere (atm)	760	millimeters of mercury (mm Hg)

Multiply	By	To Obtain
FLUX OF MASS		
grams per square meter per year ($g\ m^{-2}\ yr^{-1}$)	10	kilograms per hectare per year ($kg\ ha^{-1}\ yr^{-1}$)
grams per square meter per year ($g\ m^{-2}\ yr^{-1}$)	8.924	pounds per acre per year
kilograms per hectare per year ($kg\ ha^{-1}\ yr^{-1}$)	0.1	grams per square meter per year ($g\ m^{-2}\ yr^{-1}$)
pound per acre per year	1.12	kilograms per hectare per year ($kg\ ha^{-1}\ yr^{-1}$)
pounds per acre per year	0.112	grams per square meter per year ($g\ m^{-2}\ yr^{-1}$)
ENERGY		
British Thermal Units (BTU)	0.2530	kilocalories (kcal)
British Thermal Units (BTU)	1054	joules (J)
calories (cal)	4.1869	joules (J)
calories (cal)	0.001	kilocalories (kcal)
joules (J)	0.239	calories (cal)
joules (J)	2.390×10^{-4}	kilocalories (kcal)
kilocalories (kcal)	1000	calories (cal)
kilocalories (kcal)	3.968	British Thermal Units (BTU)
kilocalories (kcal)	4183	joules (J)
kilocalories (kcal)	4.183	kilojoules (kJ)
kilocalories (kcal)	0.001162	kilowatt-hours (kWhr)
kilojoule (kJ)	0.239	kilocalories (kcal)
kilowatt-hours (kWhr)	860.5	kilocalories (kcal)
kilowatt-hours (kWhr)	3.6×10^{6}	joules (J)
langley (ly)*	1	calories per square centimeter ($cal\ cm^{-2}$)
langley (ly)	10	kilocalories per square meter ($kcal\ m^{-2}$)
POWER		
horsepower	0.7457	kilowatts (kW)
horsepower	10.70	kilocalories per minute (kcal/min)
kilocalories/day (kcal/day)	6.4937×10^{-5}	horsepower
kilocalories/day (kcal/day)	4.8417×10^{-5}	kilowatts (kW)
kilowatts (kW)	1.341	horsepower
kilowatts (kW)	14.34	kilocalories per minute (kcal/min)
kilowatts (kW)	1000	watts (W)
watt (W)	1	joule per second (J/sec)
PRIMARY PRODUCTIVITY/ENERGY FLOW		
grams dry weight (g-dw)	4.5	kilocalories (kcal)
grams dry weight (g-dw)	0.45	grams C (g-C)
grams O_2 (g-O_2)**	3.7	kilocalories (kcal)
grams O_2 (g-O_2)**	0.375	grams C (g-C)
grams C (g-C)**	10	kilocalories (kcal)
grams C (g-C)**	2.67	grams O_2 (g-O_2)
kilocalories (kcal)	4.18	kilojoules (kJ)
kilocalories (kcal)**	0.1	grams C (g-C)
STOICHIOMETRY OF ORGANIC MATTER*		
molar ratio	106C : 16N : 1P	
weight ratio	41C : 7.2N : 1P	

Multiply	By	To Obtain
CONCENTRATIONS IN WATER		
part per thousand (ppt)	1	grams per liter (g L^{-1})
part per million (ppm)	1	milligrams per liter (mg L^{-1})
parts per million (ppm)	1	grams per cubic meter (g m^{-3})
parts per million (ppm)	1000	parts per billion (ppb)
parts per billion (ppb)	1	micrograms per liter (μg L^{-1})
milligrams per liter (mg L^{-1})	1000	micrograms per liter (μg L^{-1})
millimolarity (m mole/L)	molecular weight	milligrams per liter (mg L^{-1})
micromolarity (μ mole/L)	molecular weight	micrograms per liter (μg L^{-1})
microgram-atoms per liter (μg-atom L^{-1})	molecular weight	micrograms per liter (μg L^{-1})
milligrams per liter (mg L^{-1})	$\left[\dfrac{\text{ionic charge}}{\text{molecular weight}}\right]$	milliequivatents per liter (meq L^{-1})
milliequivatents per liter (meq L^{-1})	$\left[\dfrac{\text{molecular weight}}{\text{ionic charge}}\right]$	milligrams per liter (mg L^{-1})
micromhos person (μmho cm^{-1})	1	microSiemens per centimeter (μS cm^{-1})

* solar constant = radiant energy at outer limit of earth's atmosphere ~2.00 langleys per minute (ly min^{-1})

** based generally on the production of glucose: $6CO_2 + 12 H_2O + (118 \times 6)$ kcal $\rightarrow C_6H_{12}O_6 + 6 O_2 + 6 H_2O$

*** based on Redfield (1958) molecule of plantkon organic matter: $[CH_2O)_{106}(NH_3)_{16}(H_2PO_4)]$

Useful atomic weights

H	1	Al	27	Ca	40
C	12	Si	28	Mn	55
N	14	P	31	Fe	56
O	16	S	32	Ni	59
Na	23	Cl	35.5	Cu	63.5
Mg	24	K	39	Se	79

Glossary

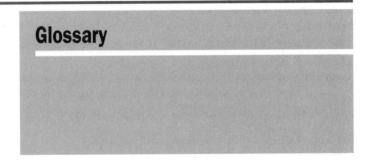

Aapa peatlands—Also called string bogs and patterned fens; peatlands identified by watertracks of long, narrow alignment of the high peat hummocks (strings) that form ridges perpendicular to the slope of the peatland and are separated by deep pools (flarks).

ADH—Alcohol dehydrogenase, the enzyme that catalyzes the reduction of acetaldehyde to ethanol in fermentation.

Adventitious roots—Roots that develop from some part of a vascular plant other than the seed. Usually they originate from the stem, and while common in most plants, also develop as adaptations to anoxia in both flood-tolerant trees (e.g., *Salix* and *Alnus*) and herbaceous species, and flood-intolerant (e.g., tomato) plants just above the anaerobic zone when these plants are flooded.

Aerenchyma—Large air spaces in roots and stems of some wetland plants that allow the diffusion of oxygen from the aerial portions of the plant into the roots.

Alcohol dehydrogenase—*see* ADH.

Allochthonous—Pertains to material that is imported into a ecological system of interest from outside that system; usually refers to organic material and/or nutrients and minerals.

Allogenic succession—Ecosystem development whereby the distribution of species is governed by their individual responses to their environment with little or no feedback from organisms to their environment. Also called individualistic hypothesis, continuum concept, and Gleasonian succession.

Alluvial plain—The floodplain of a river, where the soils are alluvial deposits carried in by the overflowing river.

Ammonia volatilization—NH_3 released to the atmosphere as a gas.

Anadromous—Refers to marine species that spawn in freshwater streams.

Anaerobic—Refers to oxygenless conditions.

Anoxia—Waters or soils with no dissolved oxygen.

Artificial wetland—*See* Constructed wetland.

Assimilatory nitrate reduction—Nitrate (NO_3) assimilated by plants or microbes and converted into biomass.

Assimilatory sulfate reduction—Process in the sulfur cycle whereby sulfur-reducing obligate anaerobes such as *Desulfovibrio* bacteria utilize sulfates as terminal electron acceptors in anaerobic respiration.

Autochthonous—Pertains to material that is produced within the ecological system of interest (e.g. organic material produced by photosynthesis). *See also* Allochthonous.

Autogenic succession—Clementian theory of succession of ecosystems whereby vegetation occurs in recognizable and characteristic communities; community change through time is brought about by the biota; changes are linear and directed toward a mature stable climax ecosystem.

Bankfull discharge—Streamflow at which a river begins to overflow onto its floodplain.

Basin wetland—A wetland that is hydrologically isolated with little or no flooding from streams, rivers, or tides.

Billabong—Australian term for a riparian wetland that is periodically flooded by the adjacent stream or river.

Biogeochemical cycling—The transport and transformation of chemicals in ecosystems.

Blanket bogs—In humid climates, peat that blankets large areas far from the site of the original peat accumulation, through the process of paludification.

BOD—Biochemical oxygen demand, a biological test for degradable organic matter in water.

Bog—A peat-accumulating wetland that has no significant inflows or outflows and supports acidophilic mosses, particularly *Sphagnum*.

Bottomland—Lowland along streams and rivers, usually on alluvial floodplains, that is periodically flooded.

Bottomland hardwood forest—Term used principally in southeastern and eastern United States to mean a mesic riparian forested ecosystem along a higher order stream or river that is subject to intermittent to frequent flooding from that stream or river; dominated by oaks and other deciduous hardwood tree species.

Bulk density—Dry weight of a known volume of soil, divided by that volume.

Buttress—Swollen bases of tree trunks growing in water.

Carr—Term used in Europe for forested wetlands characterized by alders (*Alnus*) and willows (*Salix*).

Cat clays—When coastal wetlands are drained, soil sulfides often oxidize to sulfuric acid, and the soils become too acidic to support plant growth.

Cation exchange capacity—The sum of exchangeable cations (positive ions) that a soil can hold.

Cheia—Annual period of flooding from March through May in the Pantanal region of South America that supports luxurious aquatic plant and animal life. *See also* Enchente, Seca, and Vazante.

Clay depletions—Clay is selectively removed along root channels after iron and manganese oxides have been depleted in wetland soils, only to redeposit as clay coatings on soil particles below the clay depletions.

Coliforms—A quantitative biological test for the presence of colon bacteria or related forms; because of their ubiquitous presence, they are used as a presumptive index of general bacterial contamination of water.

Concentric domed bog—A concentric pattern of pools and peat communities formed around the most elevated part of a bog.

Constructed wetland—A wetland developed on a former uplands to create poorly drained soils and wetland flora and fauna for the primary purpose of contaminant or pollution removal from wastewater or runoff.

Consumer surplus—In economics, the net benefit of a good to the consumer.

Continuum concept—*See* Allogenic succession.

Created wetland—A wetland constructed where one did not exist before.

Cumbungi swamp—Cattail (*Typha*) marsh in Australia.

Cumulative loss—When ecosystems such as wetlands are lost, usually due to human development, one small piece at a time, with the cumulative loss being substantial.

Cypress domes—Also called cypress ponds or cypress heads; poorly drained to permanently wet depressions dominated by pond cypress (*Taxodium distichum* var. *nutans*). Called domes because the cypress grows more vigorously in the center than around the perimeter of the dome, giving it a domed appearance from a distance.

Cypress strand—A diffuse freshwater stream flowing through a shallow forested depression (dominated by *Taxodium*) on a gently sloping plain.

Dalton's law—Flux is proportional to a pressure gradient. An example of a process that follows Dalton's law is evaporation, which is proportional to the difference between the vapor pressure at the water surface and the vapor pressure in the overlying air.

Darcy's law—Groundwater equation that states that flow of groundwater is proportional to a hydraulic gradient and the hydraulic conductivity, or permeability, of the soil or substrate.

Delineation—Technique of determining an exact boundary of a wetland. Used for identifying jurisdictional wetlands in United States.

Delta—Location where rivers meet the sea and deposit sediments, often in a broad alluvial fan.

Demand curve—Economist's estimate of consumer benefits; see Chapter 16.

Denitrification—Process in the nitrogen cycle carried out by microorganisms in anaerobic conditions, where nitrate acts as a terminal electron acceptor, resulting in the loss of nitrogen as it is converted to nitrous oxide (N_2O) and nitrogen gas (N_2).

Designer wetland—Created or restored wetland in which certain plant species or other organisms are introduced and the success or failure of those plants or organisms is used as the indicator of success or failure of that wetland.

Detention time—A measure of the length of time a parcel of water stays in a wetland; equivalent to the turnover time or retention time and the inverse of the turnover rate. *See also* Retention time. Detention time is the term used most frequently for designing treatment wetlands.

Discharge wetland—Wetland that has surface water (or groundwater) level lower hydrologically than the surrounding water table, leading to an inflow of groundwater.

Dissimilatory nitrogen reduction—Several pathways of nitrate reduction, particularly nitrate reduction to ammonia and denitrification.

DMS—Dimethyl sulfide, one of the gases given off by wetlands.

Drop roots—*See* Prop roots.

Duck stamps—Stamps sold in several countries to hunters to help pay for the protection of waterfowl habitat. The Duck Stamp program in the United States started in 1934.

Eat-out—A major wetland vegetation removal by herbivory, often by geese or muskrats.

Ebullitive flux—Flux of gases from wetland soils as bubbles or diffusion to the surface of the water and then to the atmosphere.

Ecological engineering—The design, creation, and restoration of ecosystems for the benefit of humans and nature.

Ecosystem engineers—Plants, animals, and microbes that carry out essential biological feedbacks in ecosystems. Examples in wetlands are beavers and muskrats.

Embodied energy—The total energy required to produce a commodity.

Emergy—Calculation of total energy requirement for any product in nature or humanity based on using transformities. Short for "energy memory." See Odum (1996).

Enchente—Period of rising waters from December through February in Pantanal region of South America. *See also* Cheia, Seca, and Vanzante.

Ericaceous plants—Flowering plants of the family Ericaceae, which, as a group, are acid-loving or acid-tolerant plants that often dominate bogs and other sites with acidic substrates.

Estuary—General location where rivers meet the sea and freshwater mixes with saltwater.

Eutrophic—Nutrient rich; generally used in lake classification, but is also applicable to peatlands.

Eutrophication—Process of aquatic ecosystem development whereby an ecosystem such as a lake, estuary, or wetland goes from an oligotrophic (nutrient poor) to eutrophic (nutrient rich) condition. If caused by humans, it is called cultural eutrophication.

Excentric raised bogs—Bogs that form from previously separate basins on sloping land and form elongated hummocks and pools aligned perpendicular to the slope.

Facultative—Adapted equally to either wet or dry condition. Usually used in the context of vegetation adapted to growing in saturated soils or upland soils.

Fen—A peat-accumulating wetland that receives some drainage from surrounding mineral soil and usually supports marshlike vegetation.

Fermentation—Partial oxidation of organic matter, when organic matter itself is the terminal electron acceptor in anaerobic respiration by microorganisms; forms various low-molecular-weight acids and alcohols and CO_2. Also called glycolysis.

Fibrists—*See* Peat.

Flarks—*See* Aapa peatlands.

Flood duration—The amount of time that a wetland is in standing water.

Flood frequency—The average number of times that a wetland is flooded during a given period.

Flood peak—Peak runoff into a wetland caused by a specific rainfall event.

Flood pulse concept (FPC)—Pulsing river discharge as the major force controlling biota in river floodplains, including the lateral exchange between floodplains and river channels.

Folists—Organic soils caused by excessive moisture (precipitation > evapotranspiration) that accumulate in tropical and boreal mountains; these soils are not classified as hydric soils as saturated conditions are the exception rather than the rule.

Gardians—"Cowboys" who ride horses through the wetlands of southern France's Camargue.

Gator holes—Deep sloughs and solution holes that hold water during the dry season and that serve as wildlife refuges; term mostly used in the Florida Everglades.

Geogenous—Peatland subject to external flows.

Gleying—Development of black, gray, or sometimes greenish or blue–gray color in soils when flooded.

Glycolysis—*See* Fermentation.

Guild—A group of functionally similar species in a community.

Halophiles—"Salt-loving" organisms.

Halophytes—Salt-tolerant plants.

Hammock—Slightly raised tree islands, such as tree island freshwater hammocks or mangrove islands in the Florida Everglades.

Hatch–Slack–Kortschak pathway—Biochemical pathway of photosynthesis for C_4 plants.

Hemists—Mucky peat or peaty muck; conditions between saprist and fibrist soil.

HGM—*See* Hydrogeomorphic classification.

High marsh—Upper zone of a salt marsh that is flooded irregularly and generally is located between mean high water and extreme high water. Called inland salt marsh in Gulf of Mexico coastline.

Histosols—Organic soils that have organic soil material in more than half of the upper 80 cm, or that are of any thickness if they overlie rock or fragmental materials that have interstices filled with organic soil material.

HLR—*See* Hydraulic loading rate.

Hydrarch succession—Development of a terrestrial forested climax community from a shallow lake with wetland as an intermediate sere. In this view, lakes gradually fill in as organic material from dying plants accumulates and minerals are carried in from upslope.

Hydraulic conductivity—*See* Permeability.

Hydraulic loading rate (HLR)—Amount of water added to a wetland, generally described as the depth of water (volume of flooding per wetland area) per unit time; generally used for treatment wetlands.

Hydric soils—Soils that formed under conditions of saturation, flooding, or ponding long enough during the growing season to develop anaerobic conditions in the upper part.

Hydrochory—Seed dispersal by water.

Hydrodynamics—An expression of the fluvial energy that drives a system.

Hydrogeomorphic classification (HGM)—Wetland classification system based on type and direction of hydrologic conditions, local geomorphology and climate.

Hydrogeomorphology—Combination of climate, basin geomorphology, and hydrology that collectively influences a wetland's function.

Hydroperiod—The seasonal pattern of the water level of a wetland. This approximates the hydrologic signature of each wetland type.

Hydrophyte—Plant adapted to the wet conditions.

Hydrophytic vegetation—Plant community dominated by hydrophytes.

Hypoxia—Waters with dissolved oxygen less than 2 mg/L.

Interception—Precipitation that is retained in the overlying vegetation canopy.

Intermittently exposed—Refers to nontidal wetlands that are flooded throughout the year, except during periods of extreme drought.

Intermittently flooded—Refers to nontidal wetlands that are usually exposed, with surface water present for variable periods without detectable seasonal patterns.

Intertidal—Part of coastal wetland flooded periodically with tidal water.

Intrariparian continuum—The structure and function of riparian communities along a river system.

Irregularly exposed—Refers to coastal wetlands with surface exposed by tides less often than daily.

Irregularly flooded—Refers to coastal wetlands with surface flooded by tides less often than daily.

Jurisdictional wetland—Term used in United States to refer to wetlands that fall under the jurisdiction of federal laws for the purpose of permit issuance or other legal matters.

Kahikatea—Refers to both the tree (*Dacrycarpus dacrydiodes*) and the forested wetlands found throughout New Zealand. Referred to as "white pine" forests by locals.

Karst—A topography formed over limestone, dolomite, or gypsum.

Krefeld system or Max-Planck-Institute process—Gravel bed macrophyte subsurface flow treatment wetlands.

Lacustrine—Pertaining to lakes or lake shores.

Lagoon—Term frequently used in Europe to denote deepwater enclosed or partially opened aquatic system, especially in coastal delta regions.

Lentic—Related to slow-moving or standing water systems; usually refers to lake (lacustrine) and stagnant swamp systems.

Lenticels—Small pores found on mangrove tree prop roots and pneumatophores above low tide and presumed to be sites of oxygen influx for anaerobic roots survival.

Limnogenous peatland—Geogenous peatland that develops along a slow-flowing stream or a lake.

Littoral—Zone between high and low tide in coastal waters or the shoreline of a freshwater lake.

Loading rate—The amount of a material (e.g., a chemical) applied to a wetland, measured either per unit area (e.g., $g\ m^{-2}\ yr^{-1}$) or volumetrically (e.g., $g\ m^{-3}\ yr^{-1}$).

Lotic—Pertaining to running water (i.e., rivers and streams).

Low marsh—Intertidal or lower marsh in salt marsh that is located in the intertidal zone and is flooded daily. Called streamside salt marsh in coastal Gulf of Mexico.

Mangal—Same as mangrove.

Mangrove—Subtropical and tropical coastal ecosystem dominated by halophytic trees, shrubs, and other plants growing in brackish to saline tidal waters. The word "mangrove" also refers to the dozens of tree and shrub species that dominate mangrove wetlands.

Marginal value—The value of an additional increment of a commodity in a free market.

Marsh—A frequently or continually inundated wetland characterized by emergent herbaceous vegetation adapted to saturated soil conditions. In European terminology, a marsh has a mineral soil substrate and does not accumulate peat. *See also* Tidal freshwater marsh and Salt marsh.

Mesotrophic peatlands—Also called transition or poor fens. Peatlands intermediate between minerotrophic and ombrotrophic.

Methanogenesis—Carbon process under extremely reduced conditions when certain bacteria (methanogens) use CO_2 or low-molecular-weight organic compounds as electron acceptors for the production of gaseous methane (CH_4).

Methanogens—Bacteria that carry out methanogenesis.

Mineral soil—Soil that has less than 20 to 35 percent organic matter.

Minerotrophic peatlands—Also called rheotrophic peatlands or rich fens; peatlands that receive water that has passed through mineral soil.

Mire—Synonymous with any peat-accumulating wetland (European definition); from the Norse word "myrr." The Danish and Swedish word for peatland is now "mose."

Mitigate—To lessen or compensate for an impact. Used here in the context of mitigating wetland loss by restoring or creating wetlands.

Mitigation ratio—The ratio of restored or created wetland to wetland lost to development.

Mitigation wetland—*See* Replacement wetland.

Moor—Synonymous with peatland (European definition). A highmoor is a raised bog; a lowmoor is a peatland in a basin or depression that is not elevated above its perimeter. The primitive sense of the Old Norse root is "dead" or barren land.

Mottles (or redox concentrations)—Orange/reddish-brown (because of iron oxides) or dark reddish-brown/black (because of manganese oxides) accumulations in hydric soils throughout an otherwise gray (gleyed) soil matrix. Mottles suggest intermittently exposed soils and are relatively insoluble, enabling them to remain in soil long after it has been drained.

Muck—Sapric organic soil material with virtually all of the organic matter decomposed, not allowing for the identification of plant forms. Bulk density generally greater than 0.2 g/cm^3 (more than peat).

Munsell soil color chart—Book of standard color chips for determining soil color value and chroma. Used to identify hydric soils.

Muskeg—Large expanse of peatlands or bogs; particularly used in Canada and Alaska.

NAD—Nicotinamide adenine dinucleotide, an enzyme that accumulates in anaerobic conditions.

NADP—NAD phosphate.

Nitrification—Ammonium nitrogen oxidized by microbes to nitrite nitrogen and nitrate nitrogen.

Nernst equation—Equation based on a hydrogen scale showing how redox potential is related to the concentrations of oxidants {ox} and reductants {red} in a redox reaction.

Nitrogen fixation—Process in the nitrogen cycle whereby N_2 gas is converted to organic nitrogen through the activity of certain organisms in the presence of the enzyme nitrogenase.

No net loss—Wetland policy in the United States that began in the late 1980s and means that if wetlands are lost they must be replaced so that there is no "net loss" of wetlands overall.

Nutrient budget—Mass balance of a nutrient in an ecosystem.

Nutrient spiraling—The process whereby resources (organic carbon, nutrients, etc.) are temporarily stored, then released as they "spiral" downstream from organic to inorganic form and back again.

Obligate—Requiring a specific environment to grow, as in adapted only a wet environment. In the context of wetlands, obligate generally refers to plants requiring saturated soils.

Oligotrophic—Nutrient rich; generally used in lake classification, but is also applicable to peatlands.

Oligotrophication—Often the process of peatland development whereby a peatland eventually elevates itself above the surrounding landscape and goes from eutrophic (nutrient rich) to oligotrophic (nutrient poor).

Ombrogenous—Peatland with inflow from precipitation only; also called ombrotrophic.

Ombrotrophic—Literally rain fed, referring to wetlands that depend on precipitation as the sole source of water.

Opportunity cost—The net worth of a non–free market resource in its best alternative use; that is, the net benefit of the area in its best alternative use that has to be foregone in order to keep it in its natural state.

Organic soil—Soil that has more than 12 to 18 percent organic carbon, depending on clay content (see Fig. 6-1).

Osmoconformers—Marine animals in which the internal cell environment follows closely the osmotic concentration of the external medium.

Osmoregulators—Marine animals that control their internal cell environment despite a different osmotic concentration of the external medium.

Outwelling—Function of coastal wetlands as "primary production pumps" that feed large areas of adjacent waters with organic material and nutrients; analogous to upwelling of deep ocean water, which supplies nutrients to some coastal waters from deep water.

Overland flow—Nonchannelized sheet flow that usually occurs during and immediately following rainfall or a spring thaw, or as tides rise in coastal wetlands.

Oxbow—Abandoned river channel, often developing into a swamp or marsh, on a river floodplain.

Oxidation—Chemical process of giving up an electron (e.g., $Fe^{2+} \rightarrow Fe^{3+} + e^-$). Special cases involve uptake of oxygen or removal of hydrogen (e.g., $H_2S \rightarrow S^{2-} + 2H^+$).

Oxidized pore linings—*See* Oxidized rhizosphere.

Oxidized rhizosphere (also called oxidized pore linings)—Thin traces of oxidized soils through an otherwise dark matrix indicating where roots of hydrophytes were once found.

Paalsa peatlands—Peatlands found in the southern limit of the tundra biome; large plateaus of peat (20–100 m in breadth and length and 3 m high) generally underlain by frozen peat and silt.

Pakihi—Peatland in southwestern New Zealand dominated by sedges, rushes, ferns, and scattered shrubs. Most pakihi form on terraces or plains of glacial or fluvial outwash origin and are acid and exceedingly infertile.

Palmer Drought Severity Index (PDSI)—A relative measure of climatic "wetness." Used primarily to estimate the severity of droughts.

Paludification—The blanketing of terrestrial ecosystems by overgrowth of bog vegetation. *See also* Blanket bog.

Palustrine—Nontidal wetlands.

Panne—Bare, exposed, or water-filled depression in a salt marsh.

Patterned fens—*See* Aapa peatlands.

Peat—Fibric organic soil material with virtually all of the organic matter allowing for the identification of plant forms. Bulk density generally less than 0.1 g/cm^3 (less than muck).

Peatland—A generic term of any wetland that accumulates partially decayed plant matter (peat).

Penman equation—Empirical equation for estimating evapotranspiration using an energy budget approach.

Perched wetland—Wetland that holds water well above the groundwater table.

Permanently flooded—Refers to nontidal wetlands that are flooded throughout the year in all years.

Permeability—The capacity of soil to conduct water flow. Also known as hydraulic conductivity. *See also* Darcy's law.

Phreatophytes—Plants that obtain their water from phreatic sources (i.e., groundwater or the capillary fringe of the groundwater table).

Physiognomy—The appearance or life form of vegetation.

Piezometers—Groundwater wells that are only partially screended and thus measure the piezometric head of an isolated part of the groundwater.

Playa—An arid- to semiarid-region wetland that has distinct wet and dry seasons. Term used in the southwestern United States for marshlike ponds similar to potholes, but with a different geologic origin.

Pneumatophores—"Air roots" that protrude out of the mud from the main roots of wetland plants such as black mangroves (*Avicennia*) and cypress (*Taxodium distichum*) and are thought to be organs for transport of oxygen and other gases to and from the roots of the plant. Called "knees" for cypress.

Pocosin—Peat-accumulating, nonriparian freshwater wetland, generally dominated by evergreen shrubs and trees and found on the southeastern Coastal Plain of the United States. The term comes from the Algonquin for "swamp on a hill."

Porosity—Total pore space in soil, generally expressed as a percentage.

Pothole—Shallow marshlike pond, particularly as found in the Dakotas and central Canadian provinces; the so-called "prairie pothole" region.

Producer surplus or economic rent—The area over a good's supply curve bounded by price.

Prop roots—Above-ground arched roots that aid in support of some wetland trees such as the mangrove *Rhizophora*.

Pulse stability concept—Concept that pulses can be both a subsidy and a stress to an ecosystem, depending on their strength, with subsidies occurring with moderate pulses, while both weak and excessive pulses can result in stress responses.

Quaking bog—Schwingmoor in German. Bog in which the peat layer and plant cover is only partially attached in the basin bottom or is floating like a raft.

Quickflow—Direct runoff component of streamflow during a storm that causes an immediate increase in streamflow.

Raised bogs—Peat deposits that fill entire basins, are raised above groundwater levels, and receive their major inputs of nutrients from precipitation. *See* Ombrogenous and Ombrotrophic.

Ramsar Convention—International treaty originally started in Ramsar, Iran, in the early 1970s to protect wetland habitat around the world, especially for migratory waterfowl. The Convention has enrolled over 1,000 wetland sites around the world in 116 countries as "Wetlands of International Importance."

Raupo swamp—Cattail (*Typha*) marsh in New Zealand.

Recharge wetland—Wetland that has surfacewater (or groundwater) level higher hydrologically than the surrounding water table, leading to an outflow of groundwater.

Recurrence interval—The average interval between the recurrence of floods at a given or greater magnitude.

Redox concentrations—Bodies of accumulated iron and manganese oxides in wetland soils such as nodules and concretions, masses (formerly called "reddish mottles"), and pore linings (formerly called "oxidized rhizospheres").

Redox depletions—Bodies of low chroma (2 or less) where the natural (gray or black) color of the parent sand, silt, or clay results when soluble forms of iron, manganese, or clay are leached out of the soil. Generally have Munsell color values of 4 or greater. *See also* Clay depletions.

Redoximorphic features—Features formed by the reduction, translocation, and/or oxidation of iron and manganese oxides; used to identify hydric soils. Formerly called mottles and low-chroma colors.

Redox potential—Reduction–oxidation potential, a measure of the electron pressure (or availability) in a solution or measure of the tendency of soil solution to oxidize or reduce substances. Low redox potential indicates reduced conditions; high redox potential indicates oxidized conditions.

Reduced matrix—Soil that has low chroma and high value but whose color changes in hue or chroma when exposed to air.

Reduction—Chemical process of gaining an electron (e.g., $Fe^{3+} + e^- \rightarrow Fe^{2+}$). Special cases involve releasing oxygen or gaining hydrogen (hydrogenation) (e.g., $S^{2-} + 2H^+ \rightarrow H_2S$).

Reedmace swamp—Cattail (*Typha*) marsh in the United Kingdom.

Reedswamp—Marsh dominated by *Phragmites* (common reed); term used particularly in Europe.

Reference wetland—Natural wetland used as a reference or control site to judge the condition of another created, restored, or impacted wetland.

Regularly flooded—Refers to coastal wetlands with surface flooded and exposed by tides at least once daily.

Regulators (or avoiders)—In reference to biological adaptations to stress, organisms that actively avoid stress or modify it to minimize its effects.

Rehabilitation—Less than full restoration of an ecosystem to its predisturbance condition.

Renewal rate—*See* Turnover rate.

Replacement value—The sum of the cheapest way of replacing all the various services performed by a natural ecosystem area.

Replacement wetland—A wetland constructed to replace the functions lost by human development, usually in the same or an adjacent watershed.

Residence time—See Retention time.

Resource spiraling—*See* Nutrient spiraling.

Restoration—To return a site to an approximation of it condition before alteration. *See also* Wetland restoration.

Retention rate—The amount of a material retained in a wetland per unit time and area; usually refers to material more or less removed from water flowing over or through a wetland, as contrasted to *detention,* which is transitory.

Retention time—A measure of the average time that water remains in the wetland. Nominal residence time or retention time refers to the theoretical time that water stays in a wetland as calculated from the flowthrough and the water volume in the wetland.

Rheotrophic peatlands—*See* Minerotrophic peatlands.

Riparian—Pertaining to the bank of a body of flowing water; the land adjacent to a river or stream that is, at least periodically, influenced by flooding.

Riparian ecosystem—Ecosystem with a high water table because of proximity to an aquatic ecosystem, usually a stream or river. Also called bottomland hardwood forest, floodplain forest, bosque, riparian buffer, and streamside vegetation strip.

River continuum concept (RCC)—Theory to describe the longitudinal patterns of biota found in streams and rivers.

Root zone method (Wurzelraumentsorgung)—Subsurface flow wetland basins, almost always found in Europe, and generally planted with *Phragmites australis.*

Runoff—Nonchannelized surfacewater flow.

Salt exclusion—A salinity adaptation by some wetland plants by which plants prevent salt from entering the plant at the roots.

Salt marsh—A halophytic grassland on alluvial sediments bordering saline water bodies where water level fluctuates either tidally or nontidally.

Salt secretion—A salinity adaptation by which some wetland plants excrete salt from specialized organs in the leaves.

Saprists—*See* Muck.

Saturated soils—Refers to nontidal wetlands where the soil or substrate is saturated for extended periods in the growing season, but standing water is rarely present.

Sclerophylly—Refers to the thickening of the plant epidermis.

Seasonally flooded—Refers to nontidal wetlands that are flooded for extended periods in the growing season, but with no surface water by the end of the growing season.

Seca—Dry period in Pantanal region of South America from September through November when the wetland reverts to vegetation typical of dry savannas. *See also* Cheia, Enchente, and Vazante.

Secondary treatment—Treatment of wastewater to remove organic material.

Sedge meadow—Very shallow wetland dominated by several species of sedges (e.g., *Carex, Scirpus, Cyperus*).

Seed bank—Seeds stored in soils, often for many years. In wetlands changing hydroperiod, as in wetland restoration or wetland drainage, can often lead to germination.

Self-design—The process of ecosystem development whereby the continual or periodic introduction of species propagules (plants, animals, microbes) and their subsequent survival or nonsurvial provide the essence of the successional and functional development of an ecosystem.

Semipermanently flooded—Refers to nontidal wetlands that are flooded in the growing season in most years.

Serial discontinuity concept—Describes the effects that floodplains, dams, and the transverse dimension in general has on the functioning of a river system.

Shrub–scrub—Wetlands dominated by woody, low-stature vegetation such as freshwater buttonwood (*Cephalanthus*) or saltwater dwarf mangrove (*Rhizophora*) swamps.

Sink—Term used in the context of wetland nutrient budgets to define a wetland that imports more of a certain nutrient than it exports.

Slough—An elongated swamp or shallow lake system, often adjacent to a river or stream. A slowly flowing shallow swamp or marsh in the southeastern United States (e.g., cypress slough). From the Old English word ''sloh'' meaning a watercourse running in a hollow. *See also* Cypress strand.

Soligenous peatland—Geogenous peatland that develops with regional interflow and surface runoff.

Source—Term used in the context of wetland nutrient budgets to define a wetland that exports more of a certain nutrient than it imports.

Spit—A neck of land along a coastline behind which coastal wetlands sometimes develop.

SRP—Soluble reactive phosphorus, similar to orthophosphate; a measure of biologically available phosphorus.

Stemflow—Precipitation that passes down the stems of vegetation. Used generally in connection with forests and forested wetlands.

Streamflow—Channelized surfacewater flow.

Stream order—Numerical system that classifies stream and river segments by size according to the order of its tributaries.

String bogs—*See* Aapa peatlands.

Strings—*See* Aapa peatlands.

Subsidence—Sinking of ground level, caused by natural and artificial settling of sediments over time.

Subsurface-flow constructed wetlands—Constructed wetlands through which water flows beneath the surface rather than over the surface. *See also* Root zone method.

Subtidal—Coastal wetland permanently flooded with tidal water.

Supply curve—Economist's estimate of producer benefits; see Chapter 16.

Surface-flow constructed wetlands—Constructed wetlands that mimic many natural wetlands with flow on surface rather than below the surface.

Swamp—Wetland dominated by trees or shrubs (U.S. definition). In Europe, forested fens and wetlands dominated by reed grass (*Phragmites*) are also called swamps (*see* Reedswamp).

Swampbuster—Provision of the U.S. Food Security Act that encourages farmers not to drain wetlands and thereby lose their farm subsidies.

Swamp gas (or marsh gas)—Methane.

Taking—The legal denial of an individual's right to use all or part of the area or structure (trees, wildlife, etc.) of his or her property.

Telmatology—A term originally coined to mean "bog science." From the Greek word "telma" for bog.

Temporarily flooded—Refers to nontidal wetlands that are flooded for brief periods in the growing season, but otherwise the water table is well below the surface.

Terrestrialization—Generally in reference to succession of peatlands, the infilling of shallow lakes until they become, in appearance, a peat basin supporting terrestrial vegetation.

Tertiary treatment—Advanced treatment of wastewater after secondary treatment to remove inorganic nutrients and other trace materials. Wetlands are often used for this purpose.

Thornthwaite equation—Empirical equation for estimating potential evapotranspiration as a function of air temperature.

Throughfall—Precipitation that passes through vegetation cover to the water or substrate below. Used particularly in forests and forested wetlands.

Tidal creeks—Small streams that serve as important conduits for material and energy transfer between salt marshes and adjacent bodies of water.

Tidal freshwater marsh—Marsh along rivers and estuaries close enough to the coastline to experience significant tides by nonsaline water. Vegetation is often similar to nontidal freshwater marshes.

Tolerators (also called resisters)—In reference to biological adaptations to stress, organisms that have functional modifications that enable them to survive and often to function efficiently in the presence of stress.

Topogenous—Refers to peatland development when the peatland modifies the pattern of surface water flow.

Total suspended solids—*See* TSS.

Transformer—Term used in the context of wetland nutrient budgets to define a wetland that imports and exports the same amount of a certain nutrient but changes it from one form to another.

Transitional ecosystems—Subirrigated riparian sites such as inactive floodplains, terraces, toeslopes, and meadows that have seasonally high water tables that recede to below the rooting zone in late summer.

Treatment wetland—Wetland constructed to treat wastewater or polluted runoff. *See also* Constructed wetland.

TSS—Total suspended solids, a measure of the sediments in a unit volume of water.

Turnover rate—Ratio of throughput of water to average volume of water within a wetland. This is the inverse of turnover time, residence time, or retention time of a wetland.

Turnover time—*See* Retention time.

Value—Something worthy, desirable, or useful to humanity; although the term is used often in ecology to refer to processes (e.g., primary production) or ecological structures (e.g., trees) as they are "valuable" to the way an ecosystem functions, the term generally should be limited to an anthropocentric connotation. Humans decide what is of "value" in an ecosystem.

Vazante—Period of declining water in Pantanal region of South America from June through August. *See also* Cheia, Enchente, and Seca.

Vernal pool—Shallow, intermittently flooded wet meadow, generally typical of Mediterranean-type climate with dry season for most of the summer and fall. Term is now used to indicate wetlands temporarily flooded during the spring throughout the United States.

Viviparity—The production of young in a living state.

Viviparous seedlings—Seedlings of trees germinate while still attached to the tree canopy, as with the mangrove genera *Rhizophora*. A specific case of viviparity.

Wad (pl. Wadden)—Unvegetated tidal flat originally referring to the northern Netherlands and northwestern German coastline. Now used throughout the world for coastal areas.

Watertracks—*See* Aapa peatlands.

Wetland—See various wetland definitions in Chapter 2. Generally wetlands have the presence of shallow water or flooded soils for part of the growing season, have organisms adapted to this wet environment, and have soil indicators of this flooding such as hydric soils.

Wetland creation—The conversion of a persistent upland or shallow water area into a wetland by human activity.

Wetland delineation—The demarcation of wetland boundaries for legal purposes. *See* Jurisdictional wetlands.

Wetlanders—People who live in proximity to wetlands and whose culture is linked to the wetlands.

Wetland restoration—The return of a wetland from a condition disturbed or altered by human activity to a previously existing condition.

Wet meadow—Grassland with waterlogged soil near the surface but without standing water for most of the year.

Wet prairie—Similar to a marsh, but with water levels usually intermediate between a marsh and a wet meadow.

Willingness-to-pay, or net willingness-to-pay—A hypothetical market that establishes the amount society would be willing to pay to produce and/or use a good beyond that which it actually does pay.

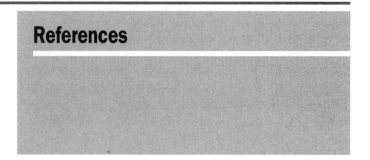

References

Abell, D. L., ed. 1984. California Riparian Systems—Protection, Management and Restoration for the 1990s. General Technical Report PSW-110, Pacific Southwest Forest and Range Experiment Station, Forest Service, U.S. Department of Agriculture, Berkeley, CA.

Abernethy, R. K. 1986. Environmental Conditions and Waterfowl Use of a Backfilled Pipeline Canal. Master's Thesis, Louisiana State University, Baton Rouge. 125 pp.

Abernethy, R. K., and J. G. Gosselink. 1988. Environmental conditions of a backfilled pipeline canal four years after construction. Wetlands 8:109–121.

Abernethy, V., and R. E. Turner. 1987. U.S. forested wetlands: 1940–1980. BioScience 37:721–727.

Adam, P. 1990. Saltmarsh Ecology. Cambridge University Press, Cambridge, UK. 461 pp.

Adam, P. 1994. Saltmarsh and mangrove. In R. H. Groves, ed. Australian Vegetation. Cambridge University Press, Cambridge, UK, pp. 97–105.

Adam, P. 1998. Australian saltmarshes: A review. In A. J. McComb and J. A. Davis, eds. Wetlands for the Future. Contributions from INTECOL's Fifth International Wetlands Conference. Gleneagles Publishing, Adelaide, Australia, pp. 287–295.

Adams, D. A. 1963. Factors influencing vascular plant zonation in North Carolina salt marshes. Ecology 44:445–456.

Adams, D. D. 1978. Habitat Development Field Investigations: Windmill Point Marsh Development Site, James River, Virginia; App. F: Environmental Impacts of Marsh Development with Dredged Material: Sediment and Water Quality, Vol. II: Substrate and Chemical Flux Characteristics of a Dredged Material Marsh. Technical Report D-77-23, U.S. Army Corps of Engineers Waterways Experiment Station, Vicksburg, MS. 72 pp.

Adams, D. F., S. O. Farwell, E. Robinson, M. R. Pack, and W. L. Bamesberger. 1981. Biogenic sulfur source strengths. Environmental Science and Technology 15:1493–1498.

Adams, G. D. 1988. Wetlands of the prairies of Canada. In National Wetlands Working Group, ed. Wetlands of Canada. Environment Canada, Ottawa, Ontario, and Polyscience Publications, Montreal, Quebec, pp. 155–198.

Adamus, P. R. 1983. A Method for Wetland Functional Assessment, Vol. I: Critical Review and Evaluation Concepts, and Vol. II: FHWA Assessment Method. Federal Highway Reports FHWA-IP-82-83 and FHWA-IP-82-84, U.S. Department of Transportation, Washington, DC. 16 pp. and 134 pp.

Adamus, P. R., E. J. Clairain, R. D. Smith, and R. E. Young. 1987. Wetland Evaluation Technique (WET), Vol. II: Methodology. U.S. Army Corps of Engineers Waterways Experiment Station, Vicksburg, MS.

Alexandrov, G. A., N. I. Bazilevich, D. A. Logofet, A. A. Tishkov, and T. E. Shytikova. 1994. Conceptual and mathematical modeling of matter cycling in Tajozhny Log Bog ecosystem (Russia). In B. C. Patten, ed. Wetlands and Shallow Continental Water Bodies, Vol. 2. SPB Academic Publishing, The Hague, pp. 45–93.

Allam, A. I., and J. P. Hollis. 1972. Sulfide inhibition of oxidases in rice roots. Phytopathology 62:634–639.

Allen, J. A. 1998. Mangroves as alien species: The case of Hawaii. Global Ecology and Biogeography Letters 7:61–71.

Allen, J. R. L. 1992. Tidally influenced marshes in the Severn Estuary, southwest Britain. In J. R. L. Allen and K. Pye, eds. Saltmarshes: Morphodynamics, Conservation and Engineering Significance. Cambridge University Press, Cambridge, UK, pp. 123–147.

Allen, J. R. L., and K. Pye, eds. 1992. Saltmarshes: Morphodynamics, Conservation and Engineering Significance. Cambridge University Press, Cambridge, UK. 184 pp.

Alongi, D. M. 1998. Coastal Ecosystem Processes. CRC Press, Boca Raton, FL.

Alper, J. 1998. Ecosystem "engineers" shape habitats for other species. Science 280:1195–1196.

Anderson, F. O. 1976. Primary production in a shallow water lake with special reference to a reed swamp. Oikos 27:243–250.

Anderson, R. C., and J. White. 1970. A cypress swamp outlier in southern Illinois. Illinois State Academy of Science Transactions 63:6–13.

Andreae, M. O., and H. Raembonck. 1983. Di-methyl sulfide in the surface ocean and the marine atmosphere: A global view. Science 221:744–747.

Aneja, V. P., J. H. Overton, Jr., and A. P. Aneja. 1981. Emission survey of biogenic sulfur flux from terrestrial surfaces. Journal of the Air Pollution Control Association 31:256–258.

Anisfeld, S. C., M. J. Tobin, and G. Benoit. 1999. Sedimentation rates in flow-restricted and restored salt marshes in Long Island Sound. Estuaries 22:231–244.

Antlfinger, A. E., and E. L. Dunn. 1979. Seasonal patterns of CO_2 and water vapor exchange of three salt-marsh succulents. Oecologia (Berlin) 43:249–260.

Arheimer, B., and H. B. Wittgren. 1994. Modelling the effects of wetlands on regional nitrogen transport. Ambio 23:378–386.

Armentano, T. V. 1990. Soils and ecology: Tropical wetlands. In M. Williams, ed. Wetlands: A Threatened Landscape. Basil Blackwell, Oxford, UK, pp. 115–144.

Armstrong, W. 1975. Waterlogged soils. In J. R. Etherington, ed. Environment and Plant Ecology. John Wiley & Sons, London, pp. 181–218.

Armstrong, W., M. E. Strange, S. Cringle, and P. M. Beckett. 1994. Microelectrode and modelling study of oxygen distribution in roots. Annals of Botany 74:287–299.

Aselmann, I., and P. J. Crutzen. 1989. Global distribution of natural freshwater wetlands and rice paddies, their net primary productivity, seasonality and possible methane emissions. Journal of Atmospheric Chemistry 8:307–358.

Aselmann, I., and P. J. Crutzen. 1990. A global inventory of wetland distribution and seasonality, net primary productivity, and estimated methane emissions. In A. F. Bouwman, ed. Soils and the Greenhouse Effect. John Wiley & Sons, New York, pp. 441–449.

Atchue, J. A., III, H. G. Marshall, and F. P. Day, Jr. 1982. Observations of phytoplankton composition from standing water in the Great Dismal Swamp. Castanea 47:308–312.

Atchue, J. A., III, F. P. Day, Jr., and H. G. Marshall. 1983. Algae dynamics and nitrogen and phosphorus cycling in a cypress stand in the seasonally flooded Great Dismal Swamp. Hydrobiologia 106:115–122.

Atkinson, R. B., J. E. Perry, E. Smith, and J. Cairns, Jr. 1993. Use of created wetlands delineation and weighted averages as a component of assessment. Wetlands 13:185–193.

Auble, G. T. 1991. Modeling wetland and riparian vegetation change. In J. A. Kusler and S. Daly, eds. Wetlands and River Corridor Management. Association of State Wetland Managers, Berne, NY, pp. 399–403.

Auble, G. T., B. C. Patten, R. W. Bosserman, and D. B. Hamilton. 1982. A hierarchical model to organize integrated research on the Okefenokee Swamp. In D. O. Logofet and N. K. Luckyanov, eds. Ecosystem Dynamics in Freshwater Wetlands and Shallow Water Bodies, Vol. II. SCOPE and UNEP Workshop, Center of International Projects, Moscow, pp. 203–217.

Auble, G. T., and M. L. Scott. 1998. Fluvial disturbance patches and cottonwood recruitment along the upper Missouri River, Montana. Wetlands 18:546–556.

Auble, G. T., J. Friedman, and M. L. Scott. in press. Relating riparian vegetation to existing and future streamflows. Ecological Applications.

Aust, W. M., R. Lea, and J. D. Gregory. 1991. Removal of floodwater sediments by a clearcut tupelo–cypress wetland. Water Resources Bulletin 27:111–116.

Australian Nature Conservation Agency. 1996. Wetlands Are Important. 2-page flyer, National Wetlands Program, ANCA, Canberra, Australia.

Babcock, H. M. 1968. The phreatophyte problem in Arizona. Arizona Watershed Symposium Proceedings 12:34–36.

Bachand, P. A. M., and A. J. Horne. 2000a. Denitrification in constructed free-water surface wetlands. I. Very high nitrate removal rates in a macrocosm study. Ecological Engineering 14:9–15.

Bachand, P. A. M., and A. J. Horne. 2000b. Denitrification in constructed free-water surface wetlands. II. Vegetation community effects. Ecological Engineering 14:17–32.

Baker, J. M. 1973. Recovery of salt marsh vegetation from successive oil spillages. Environmental Pollution 4:223–230.

Baker, K., M. S. Fennessy, and W. J. Mitsch. 1991. Designing wetlands for controlling coal mine drainage: An ecologic–economic modelling approach. Ecological Economics 3:1–24.

Baker, L. A. 1992. Introduction to nonpoint source pollution in the United States and prospects for wetland use. Ecological Engineering 1:1–26.

Baker, W. L. 1989. Macro- and micro-scale influences on riparian vegetation in western Colorado. Annals of the Association of American Geographers 79:65–78.

Baker-Blocker, A., T. M. Donahue, and K. H. Mancy. 1977. Methane flux from wetlands. Tellus 29:245–250.

Baldwin, A. H., K. L. McKee, and I. A. Mendelssohn. 1996. The influence of vegetation, salinity, and inundation on seed banks of oligohaline coastal marshes. American Journal of Botany 83:470–479.

Baldwin, A. H., and I. A. Mendelssohn. 1998. Response of two oligohaline marsh comunities to lethal and nonlethal disturbance. Oecologia 116:543–555.

Ball, M. C. 1980. Patterns of secondary succession in a mangrove forest in south Florida. Oecologia 44:226–235.

Balling, S. S., and V. H. Resh. 1983. The influence of mosquito control recirculation ditches on plant biomass, production, and composition in two San Francisco Bay salt marshes. Estuarine, Coastal and Shelf Science 16:151–161.

Balogh, G. R., and T. A. Bookhout. 1989. Purple loosestrife (*Lythrum salicaria*) in Ohio's Lake Erie marshes. Ohio Journal of Science 89:62–64.

Barbier, E. B. 1994. Valuing environmental functions: Tropical wetlands. Land Economics 70:155–173.

Barbier, E. B., J. C. Burgess, and C. Folke. 1994. Paradise Lost? The Ecological Economics of Biodiversity. Earthscan Publications, London. 267 pp.

Barbier, E. B., M. Acreman, and D. Knowler. 1997. Economic Valuation of Wetlands: A Guide for Policy Makers and Planners. Ramsar Convention Bureau, Gland, Switzerland.

Bardecki, M. J. 1987. Wetland Evaluation: Methodology Development and Pilot Area Selection. Report 1, Canadian Wildlife Service and Wildlife Habitat Canada, Toronto.

Barry, J. M. 1997. Rising Tide: The Great Mississippi Flood of 1927 and How It Changed America. Simon & Schuster, New York.

Barsdate, R. J., and V. Alexander. 1975. The nitrogen balance of Arctic tundra: Pathways, rates, and environmental implications. Journal of Environmental Quality 4:111–117.

Bartlett, C. H. 1904. Tales of Kankakee Land. Charles Scribner's Sons, New York. 232 pp.

Bartsch, I., and T. R. Moore. 1985. A preliminary investigation of primary production and decomposition in four peatlands near Schefferville, Quebéc. Canadian Journal of Botany 63:1241–1248.

Batie, S. S., and J. R. Wilson. 1978. Economic values attributable to Virginia's coastal wetlands as inputs in oyster production. Southern Journal of Agricultural Economics 10:111–117.

Batt, B. D. J., M. G. Anderson, C. D. Anderson, and F. D. Caswell. 1989. The use of prairie potholes by North American ducks. In A. G. van der Valk, ed. Northern Prairie Wetlands. Iowa State University Press, Ames, pp. 204–227.

Baumann, R. H., J. W. Day, Jr., and C. A. Miller. 1984. Mississippi deltaic wetland survival: Sedimentation versus coastal submergence. Science 224:1093–1095.

Baumann, R. H., and R. E. Turner. 1990. Direct impacts of outer continental shelf activities on wetland loss in the Central Gulf of Mexico. Environmental Geology and Water Resources 15:189–198.

Bay, R. R. 1967. Groundwater and vegetation in two peat bogs in northern Minnesota. Ecology 48:308–310.

Bay, R. R. 1969. Runoff from small peatland watersheds. Journal of Hydrology 9:90–102.

Bayley, P. B. 1995. Understanding large river-floodplain ecosystems. BioScience 45:153–158.

Bazilevich, N. I., L. Ye. Rodin, and N. N. Rozov. 1971. Geophysical aspects of biological productivity. Soviet Geography 12:293–317.

Bazilevich, N. I., and A. A. Tishkov. 1982. Conceptual balance model of chemical element cycles in a mesotrophic bog ecosystem. In D. O. Logofet and N. K. Luckyanov, eds. Ecosystem Dynamics in Freshwater Wetlands and Shallow Water Bodies, Vol. II. SCOPE and UNEP Workshop, Center of International Projects, Moscow, pp. 236–272.

Beadle, L. C. 1974. The Inland Waters of Tropical Africa. Longman, London.

Beaumont, P. 1975. Hydrology. In B. Whitton, ed. River Ecology. Basil Blackwell, Oxford, UK, pp. 1–38.

Beck, L. T. 1977. Distribution and Relative Abundance of Freshwater Macroinvertebrates of the Lower Atchafalaya River Basin, Louisiana. Master's Thesis, Louisiana State University, Baton Rouge.

Bedford, B. L., and E. M. Preston. 1988. Developing the scientific basis for assessing cumulative effects of wetland loss and degradation on landscape functions: Status, perspectives and prospects. Environmental Management 12:751–771.

Bedinger, M. S. 1981. Hydrology of bottomland hardwood forests of the Mississippi Embayment. In J. R. Clark and J. Benforado, eds. Wetlands of Bottomland Hardwood Forests. Elsevier, Amsterdam, pp. 161–176.

Beeftink, W. G. 1977a. Salt marshes. In R. S. K. Barnes, ed. The Coastline. John Wiley & Sons, New York, pp. 93–121.

Beeftink, W. G. 1977b. The coastal salt marshes of Western and Northern Europe: An ecological and phytosociological approach. In V. J. Chapman, ed. Ecosystems of the World 1: Wet Coastal Ecosystems. Elsevier Science Publishing, New York, pp. 109–149.

Beeftink, W. G., and J. M. Gehu. 1973. Spartinetea maritimae. In R. Tüxen, ed. Prodrome des Groupements Végétaux d'Europe, lieferung. 1. J. Cramer Verlag, Lehre, pp. 1–43.

Beilfuss, R. D., and J. A. Barzen. 1994. Hydrological wetland restoration in the Mekong Delta, Vietnam. In W. J. Mitsch, ed. Global Wetlands: Old World and New. Elsevier, Amsterdam, pp. 453–468.

Bell, F. W. 1997. The economic value of saltwater marsh supporting marine recreational fishing in the southeastern United States. Ecological Economics 21:243–254.

Bellamy, D. J. 1968. An Ecological Approach to the Classification of the Lowland Mires in Europe. In Proceedings of the Third International Peat Congress, Quebec, Canada, pp. 74–79.

Bellamy, D. J., and J. Rieley. 1967. Some ecological statistics of a "miniature bog." Oikos 18:33–40.

Belt, C. B., Jr. 1975. The 1973 flood and man's constriction of the Mississippi River. Science 189:681–684.

Berggren, T. J., and J. T. Lieberman. 1977. Relative contributions of Hudson, Chesapeake, and Roanoke striped bass, *Morone saxatilis*, to the Atlantic coast fishery. Fisheries Bulletin 76:335–345.

Berkeley, E., and D. Berkeley. 1976. Man and the Great Dismal. Virginia Journal of Science 27:141–171.

Bernard, J. M., and B. A. Solsky. 1977. Nutrient cycling in a *Carex lacustris* wetland. Canadian Journal of Botany 55:630–638.

Bernatowicz, S., S. Leszczynski, and S. Tyczynska. 1976. The influence of transpiration by emergent plants on the water balance of lakes. Aquatic Botany 2:275–288.

Bertani, A., I. Bramblila, and F. Menegus. 1980. Effect of anaerobiosis on rice seedlings: Growth, metabolic rate, and rate of fermentation products. Journal of Experimental Botany 3:325–331.

Bertness, M. D. 1992. The ecology of a New England salt marsh. American Scientist 80:260–268.

Bertness, M. D., and R. Callaway. 1994. Positive interactions in communities. Trends in Ecology and Evolution 9:191–193.

Bertness, M. D., and S. D. Hacker. 1994. Physical stress and positive associations among marsh plants. American Naturalist 144:363–372.

Beschel, R. E., and P. J. Webber. 1962. Gradient analysis in swamp forests. Nature 194:207–209.

Best, E. P. H., and J. P. Bakker, eds. 1993. Netherlands—Wetlands. Kluwer Academic Publishers, Dordrecht, The Netherlands. 328 pp.

Bhowmik, N. G., A. P. Bonini, W. C. Bogner, and R. P. Byrne. 1980. Hydraulics of flow and sediment transport to the Kankakee River in Illinois. Illinois State Water Survey Report of Investigation 98, Champaign, IL. 170 pp.

Bishop, J. M., J. G. Gosselink, and J. M. Stone. 1980. Oxygen consumption and hemolymph osmolality of brown shrimp, *Penaeus aztecus*. Fisheries Bulletin 78:741–757.

Black, C. C., Jr. 1973. Photosynthetic carbon fixation in relation to net CO_2 uptake. Annual Review of Plant Physiology 24:253–286.

Blaustein, A. R., and D. B. Wake. 1990. Declining amphibian populations: A global phenomenon? Trends in Ecology and Evolution 5:203–204.

Blem, C. R., and L. B. Blem. 1975. Density, biomass, and energetics of the bird and mammal populations of an Illinois deciduous forest. Illinois Academy of Science Transactions 68:156–184.

Blom, C. W. P. M., G. M. Bögemann, P. Laan, A. J. M. van der Sman, H. M. van de Steeg, and L. A. C. J. Voesenek. 1990. Adaptations to flooding in plants from river areas. Aquatic Botany 38:29–47.

Blom, C. W. P. M., and L. A. C. J. Voesenek. 1996. Flooding: The survival strategies in plants. Trends in Ecology and Evolution 11:290–295.

Blum, J. L. 1968. Salt marsh *Spartinas* and associated algae. Ecological Monographs 38:199–221.

Boatman, D. J., and P. M. Lark. 1971. Inorganic nutrition of the protonemata of *Sphagnum papillosum* Lindb., *S. magellanicum* Brid., and *S. cuspidatum* Ehrh. New Phytologist 70:1053–1059.

Boelter, D. H., and E. S. Verry. 1977. Peatland and Water in the Northern Lake States. General Technical Report NC-31, North Central Experiment Station, Forest Service, U.S. Department of Agriculture, St. Paul, MN.

Bolin, E. G., and F. S. Guthery. 1982. Playa, irrigation and wildlife in West Texas. Transactions of the 47th North American Wildlife Natural Resource Conference 47:528–541.

Bolin, E. G., L. H. Smith, and H. L. Scramm, Jr. 1989. Playa lakes: Prairie wetlands of the Southern High Plains. BioScience 39:615–623.

Bonasera, J., J. Lynch, and M. A. Leck. 1979. Comparison of the allelopathic potential of four marsh species. Torrey Botanical Club Bulletin 106:217–222.

Boon, P. I. 1999. Carbon cycling in Australian wetlands: The importance of methane. Verhandlungen Internationale Vereinigung für Limnologie 27:1–14.

Boon, P. I., and B. K. Sorrell. 1991. Biogeochemistry of billabong sediments. I. The effect of macrophytes. Freshwater Biology 26:209–226.

Boon, P. I., and A. Mitchell. 1995. Methanogenesis in the sediments of an Australian freshwater wetland: Comparison with aerobic decay and factors controlling methanogenesis. FEMS Microbiology Ecology 18:174–190.

Boon, P. I., and Sorrell, B. K. 1995. Methane fluxes from an Australian floodplain wetland: The importance of emergent macrophytes. Journal of North American Benthological Society 14:582–598.

Bormann, F. H., G. E. Likens, and J. S. Eaton. 1969. Biotic regulation of particulate and solution losses from a forested ecosystem. BioScience 19:600–610.

Bosselman, F. P. 1996. Limitations inherent in the title to wetlands at common law. Stanford Law Journal 15:247–337.

Bosserman, R. W. 1983a. Dynamics of physical and chemical parameters in Okefenokee Swamp. Journal of Freshwater Ecology 2:129–140.

Bosserman, R. W. 1983b. Elemental composition of *Utricularia*—Periphyton ecosystems from Okefenokee Swamp. Ecology 64:1637–1645.

Botch, M. S., K. I. Kobak, T. S. Vinson, and T. P. Kolchugina. 1995. Carbon pools and accumulation in peatlands of the former Soviet Union. Global Biogeochemical Cycles 9:37–46.

Boto, K. G., and J. T. Wellington. 1984. Soil characteristics and nutrient status in a northern Australian mangrove forest. Estuaries 7:61–69.

Boulé, M. E. 1988. Wetland Creation and Enhancement in the Pacific Northwest. In J. Zelazny and J. S. Feierabend, eds. Proceedings of the Conference on Wetlands: Wetlands: Increasing Our Wetland Resources. Corporate

Conservation Council, National Wildlife Federation, Washington, DC, pp. 130–136.

Boustany, R. G., C. R. Crozier, J. M. Rybczyk, and R. R. Twilley. 1997. Denitrification in a south Louisiana wetland forest receiving treated sewage effluent. Wetlands Ecology and Management 4:273–283.

Boutin, C., and P. A. Keddy. 1993. A functional classification of wetland plants. Journal of Vegetation Science 4:591–600.

Bowden, W. B. 1984. Nitrogen and phosphorous in the sediments of a tidal freshwater marsh in Massachusetts (USA). Estuaries 7:108–118.

Bowden, W. B. 1987. The biogeochemistry of nitrogen in freshwater wetlands. Biogeochemistry 4:313–348.

Bowden, W. B., C. J. Vorosmarty, J. T. Morris, B. J. Peterson, J. E. Hobbie, P. A. Steudler, and B. Moore. 1991. Transport and processing of nitrogen in a tidal freshwater wetland. Water Resources Research 27:389–408.

Boyd, C. E. 1970. Losses of mineral nutrients during decomposition of *Typha latifolia*. Archiv für Hydrobiologie 66:511–517.

Boyd, C. E. 1971. The dynamics of dry matter and chemical substances in a *Juncus effusus* population. American Midland Naturalist 86:28–45.

Boyt, F. L., S. E. Bayley, and J. Zoltek. 1977. Removal of nutrients from treated wastewater by wetland vegetation. Journal of the Water Pollution Control Federation 49:789–799.

Brackke, F. H., ed. 1976. Impact of Acid Precipitation on Forests and Freshwater Ecosystems in Norway. SNSF Report 6, Oslo-As. 111 pp.

Bradbury, I. K., and J. Grace. 1983. Primary production in wetlands. In A. J. P. Gore, ed. Ecosystems of the World, Vol. 4A: Mires: Swamp, Bog, Fen, and Moor. Elsevier, Amsterdam, pp. 285–310.

Bradford, K. J., and S. F. Yang. 1980. Xylem transport of 1-amino-cyclopropane-1-carboxylic acid, an ethylene precursor in waterlogged tomato plants. Plant Physiology 65:322–326.

Bradley, C. E., and D. G. Smith. 1986. Plains cottonwood recruitment and survival on a prairie meandering river floodplain, Milk River, southern Alberta and northern Montana. Canadian Journal of Botany 64:1433–1442.

Bradley, P. M., and E. L. Dunn. 1989. Effects of sulfide on the growth of three salt marsh halophytes of the southeastern United States. American Journal of Botany 76:1707–1713.

Bradshaw, A. D. 1996. Underlying principles of restoration. Canadian Journal of Fisheries and Aquatic Sciences 53 (Suppl. 1):3–9.

Brakenridge, G. R. 1988. River flood regime and floodplain stratigraphy. In V. R. Baker, R. C. Kochel, and P. C. Patton, eds. Flood Geomorphology. John Wiley & Sons, New York, pp. 139–156.

Brandt, K., and K. C. Ewel. 1989. Ecology and Management of Cypress Swamps: A Review. Florida Cooperative Extension Service, University of Florida, Gainesville, FL.

Bricker-Urso, S., S. W. Nixon, J. K. Cochran, D. J. Hirschberg, and C. Hunt. 1989. Accretion rates and sediment accumulation in Rhode Island salt marshes. Estuaries 12:300–317.

Bridgewater, P. B., and I. D. Cresswell. 1993. Phytosociology and phytogeography of Western Australia salt marshes. Fragmentum Floristica et Geobotanica Suppl. 2:609–629.

Bridgham, S. D. 1991. Cellulose decay in natural and disturbed peatlands in North Carolina. Journal of Environmental Quality 20:695–701.

Bridgham, S. D., S. P. Faulkner, and C. J. Richardson. 1991. Steel rod oxidation as a hydrologic indicator in wetland soils. Soil Science Society of America Journal 55:856–862.

Bridgham, S. D., J. Pastor, J. A. Janssens, C. Chapin, and T. J. Malterer. 1996. Multiple limiting gradients in peatlands: A call for a new paradigm. Wetlands 16:45–65.

Bridgham, S. D., K. Updegraff, and J. Pastor. 1998. Carbon, nitrogen, and phosphorus mineralization in northern wetlands. Ecology 79:1545–1561.

Bridgham, S. D., C.-L. Ping, J. L. Richardson, and K. Updegraff. in press. Soils of northern peatlands: Histosols and gelisols. In J. L. Richardson and M. J. Vepraskas, eds. Wetland Soils: Their Genesis, Morphology, Hydrology, Landscapes, and Classification. CRC Press, Boca Raton, FL.

Brinson, M. M. 1977. Decomposition and nutrient exchange of litter in an alluvial swamp forest. Ecology 58:601–609.

Brinson, M. M. 1987. Cumulative Increases in Water Table as a Dimension for Quantifying Hydroperiod in Wetlands. In Estuarine Research Federation Meeting, October 26, 1987, New Orleans.

Brinson, M. M. 1989. Fringe Wetlands in Albemarle and Pamlico Sounds: Landscape Position, Fringe Swamp Structure, and Response to Rising Sea Level. In Albemarle–Pamlico Estuarine Study, Raleigh, NC.

Brinson, M. M. 1993a. A Hydrogeomorphic Classification for Wetlands. Wetlands Research Program Technical Report WRP-DE-4, U.S. Army Corps of Engineers Waterways Experiment Station, Vicksburg, MS.

Brinson, M. M. 1993b. Changes in the functioning of wetlands along environmental gradients. Wetlands 13:65–74.

Brinson, M. M., H. D. Bradshaw, R. N. Holmes, and J. B. Elkins, Jr. 1980. Litterfall, stemflow, and throughfall nutrient fluxes in an alluvial swamp forest. Ecology 61:827–835.

Brinson, M. M., A. E. Lugo, and S. Brown. 1981a. Primary productivity, decomposition and consumer activity in freshwater wetlands. Annual Review of Ecology and Systematics 12:123–161.

Brinson, M. M., B. L. Swift, R. C. Plantico, and J. S. Barclay. 1981b. Riparian Ecosystems: Their Ecology and Status. FWS/OBS-81/17, U.S. Fish and Wildlife Service, Washington, DC. 151 pp.

Brinson, M. M., W. Kruczynski, L. C. Lee, W. L. Nutter, R. D. Smith, and D. F. Whigham. 1994. Developing an approach for assessing the functions of wetlands. In W. J. Mitsch, ed. Global Wetlands: Old World and New. Elsevier, Amsterdam, pp. 615–624.

Britsch, L. D., and E. B. Kemp. 1990. Land Loss Rates, Report 1, Mississippi River Deltaic Plain. Technical Report GL-90-2, U. S. Army Corps of Engineers Waterways Experiment Station, Vicksburg, MS.

Brix, H. 1987. Treatment of wastewater in the rhizosphere of wetland plants: The root-zone method. Water Science and Technology 19:107–118.

Brix, H. 1989. Gas exchange through dead culms of reed, *Phragmites australis* (Cav.) Trin. ex Steudel. Aquatic Botany 35:81–98.

Brix, H. 1994. Use of constructed wetlands in water pollution control: Historical development, present status, and future perspectives. Water Science and Technology 30:209–223.

Brix, H. 1998. Denmark. In J. Vymazal, H. Brix, P. F. Cooper, M. D. Green, and R. Haberl, eds. Constructed Wetlands for Wastewater Treatment in Europe. Backhuys Publishers, Leiden, Netherlands, pp. 123–152.

Brix, H., and H.-H. Schierup. 1989a. The use of aquatic macrophytes in water-pollution control. Ambio 18:100–107.

Brix, H., and H.-H. Schierup. 1989b. Sewage treatment in constructed reed beds—Danish experiences. Water Science and Technology 21:1655–1668.

Brix, H., B. K. Sorrell, and P. T. Orr. 1992. Internal pressurization and convective gas flow in some emergent freshwater macrophytes. Limnology and Oceanography 37:1420–1433.

Brock, J. H. 1985. Physical characteristics and pedogenesis of soils in riparian habitats along the upper Gila River basin. In R. R. Johnson, C. D. Ziebell, D. R. Patton, P. F. Fiolliott, and R. Hamre, eds. Riparian Ecosystems and Their Management: Reconciling Conflicting Uses. General Technical Report RM-120, Rocky Mountain Forest and Range Experiment Station, Forest Service, U.S. Department of Agriculture, Fort Collins, CO, pp. 49–53.

Brock, M. A., P. I. Boon, and A. Grant, eds. 1994. Plants and Processes in Wetlands. Special Issue of Australian Journal of Marine and Freshwater Research 45:1369–1564.

Brodie, G. A., D. A. Hammer, and D. A. Tomljanovich. 1988. An evaluation of substrate types in constructed wetland drainage treatment systems. In Mine Drainage and Surface Mine Reclamation, Vol. I: Mine Water and Mine Waste. U.S. Department of the Interior, Pittsburgh, pp. 389–398.

Brodie, G. A., D. A. Hammer, and D. A Tomljanovich. 1989a. Treatment of acid drainage with a constructed wetland at the Tennessee Valley Authority 950 coal mine. In D. A. Hammer, ed. Constructed Wetlands for Wastewater Treatment. Lewis Publishers, Chelsea, MI, pp. 201–210.

Brodie, G. A., D. A. Hammer, and D. A. Tomljanovich. 1989b. Constructed wetlands for treatment of ash pond seepage. In D. A. Hammer, ed. Constructed Wetlands for Wastewater Treatment. Lewis Publishers, Chelsea, MI, pp. 211–220.

Brody, M., and E. Pendleton. 1987. FORFLO: A Model to Predict Changes in Bottomland Hardwood Forests. Office of Information Transfer, U.S. Fish and Wildlife Service, Fort Collins, CO.

Brooks, R. P., D. E. Samuel, and J. B. Hill, eds. 1985. Wetlands and Water Management on Mined Lands. Proceedings of a Conference October 23–24, 1985. Pennsylvania State University Press, University Park. 393 pp.

Broome, S. W. 1990. Creation and restoration of tidal wetlands of the southeastern United States. In J. A. Kusler and M. E. Kentula, eds. Wetland Creation and Restoration. Island Press, Washington, DC, pp. 37–72.

Broome, S. W., E. D. Seneca, and W. W. Woodhouse, Jr. 1988. Tidal salt marsh restoration. Aquatic Botany 32:1–22.

Browder, J. A., H. A. Bartley, and K. S. Davis. 1989. Probabilistic model of the relationship between marshland-water interface and marsh disintegration. Ecological Modelling 29:245–260.

Brown, C. A. 1984. Morphology and biology of cypress trees. In K. C. Ewel and H. T. Odum, eds. Cypress Swamps. University Presses of Florida, Gainesville, pp. 16–24.

Brown, D. J., W. A. Hubert, and S. H. Anderson. 1996. Beaver ponds create wetland habitat for birds in mountains of southeastern Wyoming. Wetlands 16:127–133.

Brown, M. T. 1987. Conceptual Design for a Constructed Wetlands System for the Renovation of Treated Effluent. Report from the Center for Wetlands, University of Florida, Gainesville. 18 pp.

Brown, M. T. 1988. A simulation model of hydrology and nutrient dynamics in wetlands. Computers, Environment and Urban Systems 12(4):221–237.

Brown, M. T., R. E. Tighe, T. R. McClanahan, and R. W. Wolfe. 1992. Landscape reclamation at a central Florida phosphate mine. Ecological Engineering 1:323–354.

Brown, S. L. 1978. A Comparison of Cypress Ecosystems in the Landscape of Florida. Ph.D. Dissertation, University of Florida, Gainesville. 569 pp.

Brown, S. L. 1981. A comparison of the structure, primary productivity, and transpiration of cypress ecosystems in Florida. Ecological Monographs 51:403–427.

Brown, S. L. 1990. Structure and dynamics of basin forested wetlands in North America. In A. E. Lugo, M. M. Brinson, and S. L. Brown, eds. Forested Wetlands: Ecosystems of the World. Elsevier, Amsterdam, pp. 171–199.

Brown, S. L., and A. E. Lugo. 1982. A comparison of structural and functional characteristics of saltwater and freshwater forested wetlands. In B. Gopal, R. E. Turner, R. G. Wetzel, and D. F. Whigham, eds. Wetlands: Ecology and Management. National Institute of Ecology and International Scientific Publications, Jaipur, India, pp. 109–130.

Brown, S. L., and D. L. Peterson. 1983. Structural characteristics and biomass production of two Illinois bottomland forests. American Midland Naturalist 110:107–117.

Brown, S. L., E. W. Flohrschutz, and H. T. Odum. 1984. Structure, productivity, and phosphorus exchange of the scrub cypress ecosystem. In K. C. Ewel and H. T. Odum, eds. Cypress Swamps. University Presses of Florida, Gainesville, pp. 304–317.

Bruins, R. J. F., S. Cai, S. Chen, and W. J. Mitsch. 1998. Ecological engineering strategies to reduce flooding damage to wetland crops in central China. Ecological Engineering 11:231–259.

Brundage, H. M., III, and R. E. Meadows. 1982. Occurrence of the endangered shortnose sturgeon, Acipenser brevirostrum, in the Delaware River estuary. Estuaries 5:203–208.

Bryan, C. F., D. J. DeMost, D. S. Sabins, and J. P. Newman, Jr. 1976. A Limnological Survey of the Atchafalaya Basin. Annual Report, Louisiana Cooperative Fishery Research Unit, School of Forestry and Wildlife Management, Louisiana State University, Baton Rouge. 285 pp.

Burdick, D. M., and I. A. Mendelssohn. 1990. Relationship between anatomical and metabolic responses to soil waterlogging in the coastal grass *Spartina patens.* Journal of Experimental Botany 41:223–228.

Buresh, R. J., M. E. Casselman, and W. H. Patrick, Jr. 1980a. Nitrogen fixation in flooded soil systems: A review. Advances in Agronomy 33:149–192.

Buresh, R. J., R. D. Delaune, and W. H. Patrick, Jr. 1980b. Nitrogen and phosphorus distribution and utilization by *Spartina alterniflora* in a Louisiana Gulf Coast marsh. Estuaries 3:111–121.

Burger, J. 1991. Coastal landscapes, coastal colonies, and seabirds. Reviews in Aquatic Sciences 4:23–43.

Burgis, M. J., and J. J. Symoens, eds. 1987. African Wetlands and Shallow Water Bodies: Directory. ORSTOM, Paris, 650 pp.

Burns, L. A. 1978. Productivity, Biomass, and Water Relations in a Florida Cypress Forest. Ph.D. Dissertation, University of North Carolina.

Burton, J. D., and P. S. Liss. 1976. Estuarine Chemistry. Academic Press, London. 229 pp.

Busch, D. E., N. L. Ingraham, and S. D. Smith. 1992. Water uptake in woody riparian phreatophytes of the southwestern United States: A stable isotope study. Ecological Applications 2:450–459.

Busch, D. E., and S. D. Smith. 1993. Effects of fire on water and salinity relations of riparian woody taxa. Oecologia 94:186–194.

Busch, D. E., and S. D. Smith. 1995. Mechanisms associated with decline of woody species in riparian ecosystems of the southwestern U.S. Ecological Monographs 65:347–370.

Buxton, R. P., P. N. Johnson, and P. R. Espie. 1996. Sphagnum Research Programme: The Ecological Effects of Commercial Harvesting. Science for Conservation: 25. New Zealand Department of Conservation, Wellington. 33 pp.

Caldwell, R., and T. E. Crow. 1992. A floristic and vegetation analysis of a tidal freshwater marsh on the Merrimac River, West Newberry, MA. Rhodora 94:63–97.

Callaway, J. C., R. D. Delaune, and W. H. Patrick, Jr. 1997. Sediment accretion rates from four coastal wetlands along the Gulf of Mexico. Journal of Coastal Research 13:181–191.

Cameron, C. C. 1970. Peat deposits of northeastern Pennsylvania. Bulletin 1317-A, U.S. Geological Survey, Washington, DC. 90 pp.

Camilleri, J. C. 1992. Leaf-litter processing by invertebrates in a mangrove forest in Queensland. Marine Biology 114:139–145.

Campbell, C. S., and M. H. Ogden. 1999. Constructed Wetlands in the Suitable Landscape. John Wiley & Sons, Inc., New York, 270 pp.

Campeau, S., H. R. Murkin, and R. D. Titman. 1994. Relative importance of algae and emergent plant litter to freshwater marsh invertebrates. Canadian Journal of Fisheries and Aquatic Sciences 51:681–692.

Capone, D. G., and R. P. Kiene. 1988. Comparison of microbial dynamics in marine and freshwater sediments: Contrasts in anaerobic carbon metabolism. Limnology and Oceanography 33:725–749.

Carey, O. L. 1965. The economics of recreation: Progress and problems. Western Economics Journal 8:169–173.

Carpenter, E. J., C. D. Van Raalte, and I. Valiela. 1978. Nitrogen fixation by algae in a Massachusetts salt marsh. Limnology and Oceanography 23:318–327.

Carroll, P., and P. Crill. 1997. Carbon balance of a temperate poor fen. Global Biogeochemical Cycles 11:349–356.

Carter, M. R., L. A. Burns, T. R. Cavinder, K. R. Dugger, P. L. Fore, D. B. Hicks, H. L. Revells, T. W. Schmidt. 1973. Ecosystem Analysis of the Big Cypress Swamp and Estuaries. EPA-904/9-74-002, U.S. Environmental Protection Agency, Region IV, Atlanta.

Carter, N. 1932. A comparative study of the algal flora of two salt marshes. I. Journal of Ecology 20:341–370.

Carter, N. 1933a. A comparative study of the algal flora of two salt marshes. II. Journal of Ecology 21:128–208.

Carter, N. 1933b. A comparative study of the algal flora of two salt marshes. III. Journal of Ecology 21:385–403.

Carter, V. 1986. An overview of the hydrologic concerns related to wetlands in the United States. Canadian Journal of Botany 64:364–374.

Carter, V., M. S. Bedinger, R. P. Novitzki, and W. O. Wilen. 1979. Water resources and wetlands. In P. E. Greeson, J. R. Clark, and J. E. Clark, eds. Wetland Functions and Values: The State of Our Understanding. American Water Resources Association, Minneapolis, pp. 344–376.

Carter, V., and R. P. Novitzki. 1988. Some comments on the relation between ground water and wetlands. In D. D. Hook, W. H. McKee, Jr., H. K. Smith, J. Gregory, V. G. Burrell, M. R. DeVoe, R. E. Sojka, S. Gilbert, R. Banks, L. G. Stolzy, C. Brooks, T. D. Matthews, and T. H. Shear, eds. The Ecology and Management of Wetlands, Vol. 1: Ecology of Wetlands. Timber Press, Portland, OR, pp. 68–86.

Casparie, W. A., and J. G. Streefkerk. 1992. Climatological, stratigraphic and paleoecological aspects of mire development. In J. T. A. Verhoeven, ed. Fens and Bogs in the Netherlands: Vegetation, History, Nutrient Dynamics and Conservation. Kluwer Academic Publishers, Dordrecht, pp. 81–129.

Castro, M. S., and F. E. Dierberg. 1987. Biogenic hydrogen sulfide emissions from selected Florida wetlands. Water, Air, and Soil Pollution 33:1–13.

Catallo, W. J., W. Henk, L. Younger, O. Mills, D. J. Thiele, and S. P. Meyers. 1996. Trace metal uptake by Pichia spartinae, an endosymbiotic yeast in the salt marsh cord grass *Spartina alterniflora*. Chemistry and Ecology 13:113–131.

Cely, J. 1974. Is the Beidler Tract in Congaree Swamp virgin? In Congaree Swamp: Greatest Unprotected Forest on the Continent. Sierra Club Publication, Columbia, SC.

CH2M-Hill and Payne Engineering. 1997. Constructed Wetlands for Livestock Wastewater Management. Prepared for Gulf of Mexico Program, Nutrient Enrichment Committee, CH2M-Hill, Gainesville, FL.

Chabreck, R. H. 1972. Vegetation, water and soil characteristics of the Louisiana coastal region, Louisiana State University Agriculture Experiment Station Bulletin 664, Baton Rouge. 72 pp.

Chabreck, R. H. 1979. Wildlife harvest in wetlands of the United States. In P. E. Greeson, J. R. Clark, and J. E. Clark, eds. Wetland Functions and Values: The State of Our Understanding, American Water Resources Association, Minneapolis, pp. 618–631.

Chabreck, R. H. 1988. Coastal Marshes: Ecology and Wildlife Management. University of Minnesota Press, Minneapolis. 138 pp.

Chalmers, A. G., R. G. Wiegert, and P. L. Wolf. 1985. Carbon balance in a salt marsh: Interactions of diffusive export, tidal deposition and rainfall-caused erosion. Estuarine, Coastal and Shelf Science 21:757–771.

Chambers, J. M., and A. J. McComb. 1994. Establishment of wetland ecosystems in lakes created by mining in Western Australia. In W. J. Mitsch, ed. Global Wetlands: Old World and New. Elsevier, Amsterdam, pp. 431–441.

Chamie, J. P., and C. J. Richardson. 1978. Decomposition in northern wetlands. In R. E. Good, D. F. Whigham, and R. L. Simpson, eds. Freshwater Wetlands: Ecological Processes and Management Potential. Academic Press, New York, pp. 115–130.

Chapin, C. T., and J. Pastor. 1995. Nutrient limitations in the northern pitcher plant *Sarracenia purpurea*. Canadian Journal of Botany 73:728–734.

Chapman, S. B. 1965. The ecology of Coom Rigg Moss, Northumberland: Some water relations of the bog system. Journal of Ecology 53:371–384.

Chapman, V. J. 1938. Studies in salt marsh ecology. I–III. Journal of Ecology 26:144–221.

Chapman, V. J. 1940. Studies in salt marsh ecology. VI–VII. Journal of Ecology 28:118–179.

Chapman, V. J. 1960. Salt Marshes and Salt Deserts of the World. Interscience, New York. 392 pp.

Chapman, V. J. 1974. Salt marshes and salt deserts of the world. In R. J. Reimold and W. H. Queen, eds. Ecology of Halophytes. Academic Press, New York, pp. 3–19.

Chapman, V. J. 1975. The salinity problem in general: Its importance and distribution with special reference to natural halophytes. In A. Poljakoff-Mayber and J. Gale, eds. Plants in Saline Environments. Ecological Studies 15. Springer-Verlag, New York, pp. 7–24.

Chapman, V. J. 1976a. Coastal Vegetation, 2nd ed. Pergamon Press, Oxford, UK. 292 pp.

Chapman, V. J. 1976b. Mangrove Vegetation. J. Cramer, Vaduz, Germany. 447 pp.

Chapman, V. J. 1977. Wet Coastal Ecosystems. Elsevier, Amsterdam. 428 pp.

Chen, R., and R. R. Twilley. 1998. A gap dynamic model of mangrove forest development along gradients of soil salinity and nutrient resources. Journal of Ecology 86:37–51.

Chen, Y. 1995. Study of Wetlands in China. Jilin Sciences Technology Press, Changchun, China. [in Chinese with English summaries]

Childers, D. L. 1994. Fifteen years of marsh flumes: A review of marsh–water column interactions in southeastern USA estuaries. In W. J. Mitsch, ed. Global Wetlands: Old World and New. Elsevier, Amsterdam, pp. 277–293.

Childers, D. L., H. N. McKellar, Jr., R. Dame, F. Sklar, and E. Blood. 1993. A dynamic nutrient budget of subsystem interactions in a salt marsh estuary. Estuarine, Coastal and Shelf Science 36:105–131.

Childers, D. L., J. W. Day, Jr., and H. N. McKellar, Jr. 2000. Twenty more years of marsh and estuarine flux studies: Revisiting Nixon (1980). In M. P. Weinstein and D. A. Kreeger, eds. International Symposium: Concepts and Controversies in Tidal Marsh Ecology. Kluwer Academic Publishers, Dordrecht, The Netherlands, pp. 389–421.

Chow, V. T., ed. 1964. Handbook of Applied Hydrology. McGraw-Hill, New York. 1453 pp.

Christensen, T. 1991. Arctic and sub-Arctic soil emissions: Possible implications for global climate change. The Polar Record 27:205–210.

Chung, C. H. 1982. Low marshes, China. In R. R. Lewis III, ed. Creation and Restoration of Coastal Plant Communities. CRC Press, Boca Raton, FL, pp. 131–145.

Chung, C. H. 1983. Geographical distribution of *Spartina anglica* Hubbard in China. Bulletin of Marine Science 33:753–758.

Chung, C. H. 1985. The effects of introduced *Spartina* grass on coastal morphology in China. Zeitschrift für Geomorphologie N.F. Suppl. Bd. 57:169–174.

Chung, C. H. 1989. Ecological engineering of coastlines with salt marsh plantations. In W. J. Mitsch and S. E. Jørgensen, eds. Ecological Engineering: An Introduction to Ecotechnology. John Wiley & Sons, New York, pp. 255–289.

Chung, C. H. 1994. Creation of *Spartina* plantations as an effective measure for reducing coastal erosion in China. In W. J. Mitsch, ed. Global Wetlands: Old World and New. Elsevier, Amsterdam, pp. 443–452.

Cicchetti, G., and R. J. Diaz. 2000. Types of salt marsh edge and export of trophic energy from marshes to deeper habitats. In M. P. Weinstein and D. A. Kreeger, eds. International Symposium: Concepts and Controversies in Tidal Marsh Ecology. Kluwer Academic Publishers, Dordrecht, The Netherlands, pp. 513–539.

Cintrón, G., A. E. Lugo, D. J. Pool, and G. Morris. 1978. Mangroves of arid environments in Puerto Rico and adjacent islands. Biotropica 10:110–121.

Cintrón, G., A. E. Lugo, and R. Martinez. 1985. Structural and functional properties of mangrove forests. In W. G. D'Arcy and M. D. Corma, eds. The Botany and Natural History of Panama, IV Series: Monographs in Systematic Botany, Vol. 10. Missouri Botanical Garden, St. Louis, pp. 53–66.

Clark, J. E. 1979. Fresh water wetlands: Habitats for aquatic invertebrates, amphibians, reptiles, and fish. In P. E. Greeson, J. R. Clark, and J. E. Clark, eds. Wetland Functions and Values: The State of Our Understanding. American Water Resources Association, Minneapolis, pp. 330–343.

Clark, J. R., and J. Benforado, eds. 1981. Wetlands of Bottomland Hardwood Forests. Elsevier, Amsterdam. 401 pp.

Clarkson, B. R. 1997. Vegetation recovery following fire in two Waikato peatlands at Whangamarino and Moanatuatua. New Zealand Journal of Botany 35:167–179.

Clarkson, B. R., K. Thompson, L. A. Schipper, and M. McLeod. 1999. Moanatuatua Bog—Proposed restoration of a New Zealand restiad peat bog ecosystem. In W. Streever, ed. An International Perspective on Wetland Rehabilitation. Kluwer Academic Publishers, Dordrecht, pp. 127–137.

Clements, F. E. 1916. Plant Succession. Publication 242, Carnegie Institution of Washington. 512 pp.

Cleverly, J. R., S. D. Smith, A. Sala, and D. A. Devitt. 1997. Invasive capacity of *Tamarix ramosissima* in a Mojave Desert floodplain: The role of drought. Oecologia 111:12–18.

Clewell, A. F. 1999. Restoration of riverine forest at Hall Branch on phosphate-mined land, Florida. Restoration Ecology 7:1–14.

Clewell, A. F., L. F. Ganey, Jr., D. P. Harlos, and E. R. Tobi. 1976. Biological Effects of Fill Roads across Salt Marshes. Report FL-E.R-1-76, Florida Department of Transportation, Tallahassee.

Clewell, A. F., and D. B. Ward. 1987. White Cedar in Florida and along the northern Gulf Coast. In A. D. Laderman, ed. Atlantic White Cedar Wetlands. Westview Press, Boulder, CO, pp. 69–82.

Clewell, A. F., and R. Lea. 1990. Creation and restoration of forested wetland vegetation in the southeastern United States. In J. A. Kusler and M. E. Kentula, eds. Wetland Creation and Restoration. Island Press, Washington, DC, pp. 195–230.

Clymo, R. S. 1963. Ion exchange in *Sphagnum* and its relation to bog ecology. Annals of Botany (London) New Series 27:309–324.

Clymo, R. S. 1965. Experiments on breakdown of *Sphagnum* in two bogs. Journal of Ecology 53:747–758.

Clymo, R. S. 1970. The growth of *Sphagnum:* Methods of measurement. Journal of Ecology 58:13–49.

Clymo, R. S. 1983. Peat. In A. J. P. Gore, ed. Ecosystems of the World, Vol. 4A: Mires: Swamp, Bog, Fen, and Moor. Elsevier, Amsterdam, pp. 159–224.

Clymo, R. S., and E. J. F. Reddaway. 1971. Productivity of *Sphagnum* (bog-moss) and peat accumulation. Hydrobiologia (Bucharest) 12:181–192.

Clymo, R. S., and P. M. Hayward. 1982. The ecology of *Sphagnum.* In A. J. E. Smith, ed. Bryophyte Ecology. Chapman & Hall, London, pp. 229–289.

Cohen, A. D., D. J. Casagrande, M. J. Andrejko, and G. R. Best, eds. 1984. The Okefenokee Swamp: Its Natural History, Geology, and Geochemistry. Wetland Surveys, Los Alamos, NM. 709 pp.

Cole, T. G., K. C. Ewel, and N. N. Devoe. 1999. Structure of mangrove trees and forests in Micronesia. Forest Ecology and Management 117:95–109.

Coleman, J. M., and H. H. Roberts. 1989. Deltaic coastal wetlands. Geologie en Mijnbouw 68:1–24.

Coles, B., and J. Coles. 1989. People of the Wetlands, Bogs, Bodies and Lake-Dwellers. Thames & Hudson, New York. 215 pp.

Comin, F. A., J. A. Romero, V. Astorga, and C. Garcia. 1997. Nitrogen removal and cycling in restored wetlands used as filters of nutrients for agricultural runoff. Water Science and Technology 35:255–261.

Conner, W. H., and J. W. Day, Jr. 1976. Productivity and composition of a bald cypress–water tupelo site and a bottomland hardwood site in a Louisiana swamp. American Journal of Botany 63:1354–1364.

Conner, W. H., J. G. Gosselink, and R. T. Parrondo. 1981. Comparison of the vegetation of three Louisiana swamp sites with different flooding regimes. American Journal of Botany 68:320–331.

Conner, W. H., and J. W. Day, Jr. 1982. The ecology of forested wetlands in the southeastern United States. In B. Gopal, R. E. Turner, R. G. Wetzel, and D. F. Whigham, eds. Wetlands: Ecology and Management. National Institute of Ecology and International Scientific Publications, Jaipur, India, pp. 69–87.

Conner, W. H., and J. W. Day, eds. 1987. The Ecology of Barataria Basin, Louisiana: An Estuarine Profile. Biological Report 85(7.13), U.S. Fish and Wildlife Service, Washington, DC. 165 pp.

Conner, W. H., and J. W. Day, Jr. 1989. A Use Attainability Analysis of Wetlands for Receiving Treated Municipal and Small Industry Wastewater: A Feasibility Study Using Baseline Data from Thibodaux La. Prepared for Louisiana Department of Environmental Quality, Center for Wetland Resources, Louisiana State University, Baton Rouge.

Conner, W. H., and J. R. Toliver. 1990. Long-term trends in the bald cypress (*Taxodium distichum*) resource in Louisiana (U.S.A.). Forest Ecology and Management 33/34:543–557.

Conner, W. H., and M. A. Buford. 1998. Southern deepwater swamps. In M. G. Messina and W. H. Conner, eds. Southern Forested Wetlands: Ecology and Management. Lewis Publishers, Boca Raton, FL, pp. 263–289.

Conway, T. E., and J. M. Murtha. 1989. The Iselin Marsh Pond Meadow. In D. A. Hammer, ed. Constructed Wetlands for Wastewater Treatment. Lewis Publishers, Chelsea, MI, pp. 139–144.

Cooke, J. G. 1992. Phosphorus removal processes in a wetland after a decade of receiving a sewage effluent. Journal of Environmental Quality 21:733–739.

Cooper, A. W. 1974. Salt marshes. In H. T. Odum, B. J. Copeland, and E. A. McMahan, eds. Coastal Ecological Systems of the United States, Vol. II. Conservation Foundation, Washington, DC, pp. 55–96.

Cooper, J. R., J. W. Gilliam, and T. C. Jacobs. 1986. Riparian areas as a control of nonpoint pollutants. In D. L. Correll, ed. Watershed Research Perspectives. Smithsonian Institution Press, Washington, DC, pp. 166–192.

Cooper, J. R., and J. W. Gilliam. 1987. Phosphorous redistribution from cultivated fields into riparian areas. Soil Science Society of America Journal 51:1600–1604.

Cooper, J. R., J. W. Gilliam, R. B. Daniels, and W. P. Robarge. 1987. Riparian areas as filters for agricultural sediment. Soil Science Society of America Journal 51:416–420.

Cooper, P. F., and J. A. Hobson. 1989. Sewage treatment by reed bed systems: The present situation in the United Kingdom. In D. A. Hammer, ed. Constructed Wetlands for Wastewater Treatment. Lewis Publishers, Chelsea, MI, pp. 153–172.

Cooper, P. F., and B. C. Findlater, eds. 1990. Constructed Wetlands in Water Pollution Control. Pergamon Press, Oxford, UK. 605 pp.

Cooper, W. S. 1913. The climax forest of Isle Royale, Lake Superior, and its development. Botanical Gazette 55:1–44, 115–140, 189–235.

Corlett, R. T. 1986. The mangrove understory: some additional observations. Journal of Tropical Ecology 2:93–94.

Costa, L. T., J. C. Farinha, N. Hecker, and P. Tomàs Vives. 1996. Mediterranean Wetland Inventory, Vol 1: A Reference Manual. Wetlands International-AEME, Wageningen, The Netherlands, 111 pp.

Costanza, R. 1980. Embodied energy and economic evaluation. Science 210:1219–1224.

Costanza, R. 1984. Natural resource valuation and management: Toward ecological economics. In A. M. Jansson, ed. Integration of Economy and Ecology—An Outlook for the Eighties. University of Stockholm Press, Stockholm, pp. 7–18.

Costanza, R., S. C. Farber, and J. Maxwell. 1989. Valuation and management of wetland ecosystems. Ecological Economics 1:335–361.

Costanza, R., R. d'Arge, R. de Groot, S. Farber, M. Grasso, B. Hannon, K. Limburg, S. Naeem, R. V. O'Neill, J. Paruelo, R. G. Raskin, P. Sutton, and M. van den Belt. 1997. The value of the world's ecosystem services and natural capital. Nature 387:253–260.

Costlow, J. D., C. G. Boakout, and R. Monroe. 1960. The effect of salinity and temperature on larval development of *Sesarma cinereum* (Bosc.) reared in the laboratory. Biological Bulletin 118:183–202.

Coultas, C. L., and Y.-P. Hsieh, eds. 1997. Ecology and Management of Tidal Marshes: A Model from the Gulf of Mexico. St. Lucie Press, Delray Beach, FL. 355 pp.

Cowardin, L. M., V. Carter, F. C. Golet, and E. T. LaRoe. 1979. Classification of Wetlands and Deepwater Habitats of the United States. FWS/OBS-79/31, U.S. Fish and Wildlife Service, Washington, DC. 103 pp.

Cowles, H. C. 1899. The ecological relations of the vegetation on the sand dunes of Lake Michigan. Botanical Gazette 27:95–117, 167–202, 281–308, 361–369.

Cowles, H. C. 1901. The physiographic ecology of Chicago and vicinity. Botanical Gazette 31:73–108, 145–182.

Cowles, H. C. 1911. The causes of vegetative cycles. Botanical Gazette 51:161–183.

Cowles, S. 1975. Metabolism Measurements in a Cypress Dome. Master's Thesis, University of Florida, Gainesville. 275 pp.

Craft, C. B., and C. J. Richardson. 1993. Peat accretion and phosphorus accumulation along a eutrophication gradient in the northern Everglades. Biogeochemistry 22:133–156.

Craft, C. B., J. Vymazal, and C. J. Richardson. 1995. Response of Everglades plant communities to nitrogen and phosphorus additions. Wetlands 15:258–271.

Craft, C. B., and C. J. Richardson. 1997. Relationships between soil nutrients and plant species composition in Everglades peatlands. Journal of Environmental Quality 26:224–232.

Cragg, J. B. 1961. Some aspects of the ecology of moorland animals. Journal of Animal Ecology 30:205–234.

Craighead, F. C. 1971. The Trees of South Florida. University of Miami Press, Coral Gables. 212 pp.

Cresswell, I. D., and P. B. Bridgewater. 1998. Major plant communities of coastal saltmarsh vegetation in Western Australia. In A. J. McComb and J. A. Davis, eds. Wetlands for the Future. Contributions from INTECOL's Fifth International Wetlands Conference. Gleneagles Publishing, Adelaide, Australia, pp. 297–325.

Crill, P. M., K. B. Bartlett, R. C. Harriss, E. Gorham, E. S. Verry, D. I. Sebacher, L. Madzar, and W. Sanner. 1988. Methane flux from Minnesota peatlands. Global Biogeochemical Cycles 2:371–384.

Crisp, D. T. 1966. Input and output of minerals for an area of Pennine moorland: The importance of precipitation, drainage, peat erosion, and animals. Journal of Applied Ecology 3:327–348.

Cromarty, D., and D. Scott. 1996. A Directory of Wetlands in New Zealand. Department of Conservation, Wellington, New Zealand.

Cronk, J. K. 1996. Constructed wetlands to treat wastewater from dairy and swine operations: A review. Agriculture, Ecosystems and Environment 58:97–114.

Cronk, J. K., and W. J. Mitsch. 1994. Aquatic metabolism in four newly constructed freshwater wetlands with different hydrologic inputs. Ecological Engineering 3:449–468.

Crozier, C. R., and R. D. Delaune. 1996. Methane production by soils from different Louisiana marsh vegetation types. Wetlands 16:121–126.

Crum, H. 1988. A Focus on Peatlands and Peat Mosses. University of Michigan Press, Ann Arbor. 306 pp.

Cullis, C. F., and M. M. Hirschler. 1980. Atmospheric sulfur natural and man-made sources. Atmospheric Environment 14:1263–1278.

Curtis, J. T. 1959. The Vegetation of Wisconsin. University of Wisconsin Press, Madison. 657 pp.

Cushing, E. J. 1963. Late-Wisconsin Pollen Stratigraphy in East-Central Minnesota. Ph.D. Dissertation, University of Minnesota, Minneapolis.

Cypert, E. 1961. The effect of fires in the Okefenokee Swamp in 1954 and 1955. American Midland Naturalist 66:485–503.

Cypert, E. 1972. The origin of houses in the Okefenokee prairies. American Midland Naturalist 87:448–458.

Dabel, C. V., and F. P. Day, Jr. 1977. Structural composition of four plant communities in the Great Dismal Swamp, Virginia. Torrey Botanical Club Bulletin 104:352–360.

Dacey, J. W. H. 1980. Internal winds in water lilies: An adaption for life in anaerobic sediments. Science 210:1017–1019.

Dacey, J. W. H. 1981. Pressurized ventilation in the yellow waterlily. Ecology 62:1137–1147.

Dachnowski-Stokes, A. P. 1935. Peat land as a conserver of rainfall and water supplies. Ecology 16:173–177.

Dahl, T. E. 1990. Wetlands Losses in the United States, 1780s to 1980s. U.S. Department of Interior, Fish and Wildlife Service, Washington, DC. 21 pp.

Dahl, T. E., and C. E. Johnson. 1991. Wetlands Status and Trends in the Conterminous United States Mid-1970s to Mid-1980s. U.S. Department of Interior, Fish and Wildlife Service, Washington, DC. 28 pp.

Dahm, C. N., ed. 1995. Kissimmee River Restoration. Special Issue of Restoration Ecology 3:145–283.

Daiber, F. C. 1977. Salt marsh animals: Distributions related to tidal flooding, salinity and vegetation. In V. J. Chapman, ed. Wet Coastal Ecosystems. Elsevier, Amsterdam, pp. 79–108.

Daiber, F. C. 1982. Animals of the Tidal Marsh. Van Nostrand Reinhold, New York. 442 pp.

Dalkey, N. C. 1972. Studies in the Quality of Life: Delphi and Decision Making. Lexington Books, Lexington, MA. 161 pp.

Damman, A. W. H. 1978. Distribution and movement of elements in ombrotrophic peat bogs. Oikos 30:480–495.

Damman, A. W. H. 1979. Geographic patterns in peatland development in eastern North America. In Proceedings of the International Symposium of Classification of Peat and Peatlands, Hyytiala, Finland. International Peat Society, pp. 42–57.

Damman, A. W. H. 1986. Hydrology, development, and biogeochemistry of ombrogenous peat bogs with special reference to nutrient relocation in a western Newfoundland bog. Canadian Journal of Botany 64:384–394.

Damman, A. W. H. 1990. Nutrient status of ombrotrophic peat bogs. Aquilo, Ser. Bot. 28:5–14.

Damman, A. W. H., and T. W. French. 1987. The Ecology of Peat Bogs of the Glaciated Northeastern United States: A Community Profile. Biological Report 85(7.16), U.S. Fish and Wildlife Service, Washington, DC. 100 pp.

Darnell, R. 1976. Impacts of Construction Activities in Wetlands of the United States. EPA-600/3-76-045, U.S. Environmental Protection Agency, Corvallis, OR. 393 pp.

D'Avanzo, C. 1989. Long-term evaluation of wetland creation projects. In J. A. Kusler and M. E. Kentula, eds. Wetland Creation and Restoration: The Status of the Science. U.S. Environmental Protection Agency, Corvallis, OR, pp. 75–84.

Davies, B., and F. Gasse, eds. 1988. African Wetlands and Shallow Water Bodies. ORSTOM, Paris, 502 pp.

Davis, A. M. 1984. Ombrotrophic peatlands in Newfoundland, Canada: Their origins, development and trans-Atlantic affinities. Chemical Geology 44:287–309.

Davis, C. A. 1907. Peat: Essays on its origin, uses, and distribution in Michigan. In Report State Board Geological Survey Michigan for 1906, pp. 95–395.

Davis, C. B., and A. G. van der Valk. 1978a. The decomposition of standing and fallen litter of *Typha glauca* and *Scirpus fluviatilis*. Canadian Journal of Botany 56:662–675.

Davis, C. B., and A. G. van der Valk. 1978b. Litter decomposition in prairie glacial marshes. In R. E. Good, D. F. Whigham, and R. L. Simpson, eds. Freshwater Wetlands: Ecological Processes and Management Potential. Academic Press, New York, pp. 99–113.

Davis, D. G. 1989. No net loss of the nation's wetlands: A goal and a challenge. Water Environment and Technology 4:513–514.

Davis, J. H. 1940. The ecology and geologic role of mangroves in Florida. Publication 517, Carnegie Institution of Washington, pp. 303–412.

Davis, J. H. 1943. The natural features of southern Florida, especially the vegetation and the Everglades. Florida Geological Survey Bulletin 25. 311 pp.

Davis, L. V., and I. E. Gray. 1966. Zonal and seasonal distribution of resects in North Carolina salt marshes. Ecological Monographs 36:275–295.

Davis, S. M., and J. C. Ogden, eds. 1994. Everglades: The Ecosystem and Its Restoration. St. Lucie Press, Delray Beach, FL. 826 pp.

Davis, S. N., and R. J. M. DeWiest. 1966. Hydrogeology. John Wiley & Sons, Inc., New York. 463 pp.

Day, F. P., Jr. 1984. Biomass and litter accumulation in the Great Dismal Swamp. In K. C. Ewel and H. T. Odum, eds. Cypress Swamps. Univ. Florida Press, Gainesville, pp. 386–392.

Day, F. P., Jr. 1982. Litter decomposition rates in the seasonally flooded Great Dismal Swamp. Ecology 63:670–678.

Day, F. P. 1987. Production and decay in a *Chamaecyparis thyoides* swamp in southeastern Virginia. In A. Laderman, ed. Atlantic White Cedar Wetlands. Westview Press, Boulder, CO, pp. 123–132.

Day, F. P., Jr., and C. V. Dabel. 1978. Phytomass budgets for the Dismal Swamp ecosystem. Virginia Journal of Science 29:220–224.

Day, F. P., Jr., J. P. Megonigal, and L. C. Lee. 1989. Cypress root decomposition in experimental wetland mesocosms. Wetlands 9:263–282.

Day, J. H. 1981. Estuarine Ecology: With Particular Reference to Southern Africa. A. A. Balkema, Rotterdam. 411 pp.

Day, J. W., Jr., W. G. Smith, P. Wagner, and W. Stowe. 1973. Community Structure and Carbon Budget in a Salt Marsh and Shallow Bay Estuarine System in Louisiana. Sea Grant Publication LSU-SG-72-04, Center for Wetland Resources, Louisiana State University, Baton Rouge, 30 pp.

Day, J. W., Jr., T. J. Butler, and W. G. Conner. 1977. Productivity and nutrient export studies in a cypress swamp and lake system in Louisiana. In M. Wiley, ed. Estuarine Processes, Vol. II. Academic Press, New York, pp. 255–269.

Day, J. W., Jr., W. H. Conner, F. Ley-Lou, R. H. Day, and A. M. Navarro. 1987. The productivity and composition of mangrove forests, Laguna de Términos, Mexico. Aquatic Botany 27:267–284.

Day, J. W., Jr., C. A. S. Hall, W. M. Kemp, and A. Yánez-Arancibia. 1989. Estuarine Ecology. John Wiley & Sons, New York. 558 pp.

Day, J. W., Jr., and P. H. Templet. 1989. Consequences of sea level rise: Implications from the Mississippi Delta. Coastal Management 17:241–257.

Day, J. W., Jr., D. Reed, J. Syhayda, P. Kemp, D. Cahoon, R. M. Boumans, and N. Latif. 1994. Physical processes of marsh deterioration. In H. H. Roberts, ed. Critical Physical Processes of Wetland Loss, 1988–1994. Final Report. Louisiana State University, Baton Rouge. Prepared for U.S. Geological Survey, Reston, VA.

Day, J. W., Jr., C. Coronado-Molina, F. R. Vera-Herrera, R. Twilley, V. H. Rivera-Monroy, H. Alvarez-Guillen, R. Day, and W. Conner. 1996. A 7 year record of above-ground net primary production in a southeastern Mexican mangrove forest. Aquatic Botany 55:39–60.

Day, R. T., P. A. Keddy, J. M. McNeil, and T. J. Carleton. 1988. Fertility and disturbance gradients: a summary model for riverine marsh vegetation. Ecology 69:1044–1054.

Deegan, L. A., J. E. Hughes, and R. A. Rountree. 2000. Salt marsh ecosystem support of marine transient species. In M. P. Weinstein and D. A. Kreeger, eds. International Symposium: Concepts and Controversies in Tidal Marsh Ecology. Kluwer Academic Publishers, Dordrecht, The Netherlands, pp. 331–363.

Deghi, G. S., K. C. Ewel, and W. J. Mitsch. 1980. Effects of sewage effluent application on litterfall and litter decomposition in cypress swamps. Journal of Applied Ecology 17:397–408.

DeGraaf, R. M., and D. D. Rudis. 1986. New England Wildlife: Habitat Natural History, and Distribution. General Technical Report NE-108, Northeastern Forest Experiment Station, Forest Service, U.S. Department of Agriculture. 491 pp.

DeGraaf, R. M., and D. D. Rudis. 1990. Herpetofaunal species composition and relative abundance among three New England forest types. Forest Ecology and Management 32:155–165.

De Jong, J. 1976. The purification of wastewater with the aid of rush or reed ponds. In J. Tourbier and R. W. Pierson, Jr., eds. Biological Control of Water Pollution. University of Pennsylvania, Philadelphia, pp. 133–139.

de la Cruz, A. A. 1974. Primary productivity of coastal marshes in Mississippi. Gulf Research Reports 4:351–356.

de la Cruz, A. A. 1978. Primary production processes: Summary and recommendations. In R. E. Good, D. F. Whigham, and R. L. Simpson, eds. Freshwater Wetlands: Ecological Processes and Management Potential. Academic Press, New York, pp. 79–86.

Delaune, R. D., and C. W. Lindau. 1990. Fate of added 15N labelled nitrogen in a *Sagittaria lancifolia* L. Gulf Coast marsh. Journal of Freshwater Ecology 5(3):429–431.

Delaune, R. D., and W. H. Patrick, Jr. 1979. Nitrogen and phosphorus cycling in a Gulf Coast salt marsh. In V. S. Kennedy, ed. Estuarine Perspectives. Academic Press, New York, pp. 143–151.

Delaune, R. D., C. J. Smith, and W. H. Patrick. 1983a. Methane release from Gulf coast wetlands. Tellus 35B:8–15.

Delaune, R. D., R. H. Baumann, and J. G. Gosselink. 1983b. Relationships among vertical accretion, coastal submergence, and erosion in a Louisiana Gulf Coast marsh. Journal of Sedimentary Petrology 53:147–157.

Delaune, R. D., J. A. Nyman, and W. H. Patrick, Jr. 1994. Peat collapse, ponding and wetland loss in a rapidly submerging coastal marsh. Journal of Coastal Research 10:1021–1030.

Delmas, R., J. Baudey, J. Servant, and Y. Baziard. 1980. Emissions and concentrations of hydrogen sulfide in the air of the tropical forest of the Ivory Coast and of temperate regions in France. Journal of Geophysical Research 85:4468–4474.

Demaree, D. 1932. Submerging experiments with *Taxodium*. Ecology 13:258–262.

Dennis, J. G., and P. L. Johnson. 1970. Shoot and rhizome-root standing crops of tundra vegetation at Barrow, Alaska. Arctic Alpine Research 2:253–266.

Dennis, J. V. 1988. The Great Cypress Swamps. Louisiana State University Press, Baton Rouge. 142 pp.

Dennis, W., and W. Batson. 1974. The floating log and stump communities in the Santee Swamp of South Carolina. Castanea 39:166–170.

Dennison, M. S., and J. F. Berry, eds. 1993. Wetlands: Guide to Science, Law, and Technology. Noyes Publishing, Park Ridge, NJ. 4439 pp.

Denny, P., ed. 1985a. The Ecology and Management of African Wetland Vegetation. Junk, Dordrecht. 344 pp.

Denny, P. 1985b. Wetland plants and associated plant life-forms. In P. Denny, ed. The Ecology and Management of African Wetland Vegetation. Junk, Dordrecht, pp. 1–18.

Denny, P. 1993. Wetlands of Africa: Introduction. In D. F. Whigham, D. Dykyjová, and S. Hejny, eds. Wetlands of the World, I: Inventory, Ecology, and Management. Kluwer Academic Publishers, Dordrecht, pp. 1–31.

Dent, D. L. 1986. Acid Sulphate Soils: A Baseline for Research and Development. ILRI Publication 39, Wageningen, Netherlands.

Dent, D. L. 1992. Reclamation of acid sulphate soils. In R. Lal and B. A. Stewart, eds. Soil Restoration: Advances in Soil Science, Vol. 17. Springer-Verlag, New York, pp. 29–122.

Derksen, A. J. 1989. Autumn movements of underyearling northern pike, *Esox lucius,* from a large Manitoba marsh. Canadian Field Naturalist 103:429–431.

DeRoia, D. M., and T. A. Bookhout. 1989. Spring feeding ecology of teal on the Lake Erie marshes (abstract). Ohio Journal of Science 89(2):3.

DeSylva, D. P. 1969. Trends in marine sport fisheries research. American Fisheries Society Transactions 98:151–169.

deSzalay, F. A., and V. H. Resh. 1996. Spatial and temporal variability of trophic relationships among aquatic macrophytes in a seasonal marsh. Wetlands 16:458–466.

deSzalay, F. A., and V. H. Resh. 1997. Responses of wetland invertebrates and plants important in waterfowl diets to burning and mowing of emergent vegetation. Wetlands 17:149–156.

Devitt, D. A., J. M. Piorkowski, S. C. Smith, J. R. Cleverly, and A. Sala. 1997. Plant water relations of *Tamarix ramosissima* in response to the imposition and alleviation of soil moisture stress. Journal of Arid Environments 36:527–540.

Devol, A. H., Richey, J. E., B. R. Forsberg, and L. A. Martinelli. 1994. Environmental methane in the Amazon River floodplain. In W. J. Mitsch, ed. Global Wetlands: Old World and New. Elsevier, Amsterdam, pp. 151–165.

Diaz, R. J. 1977. The Effects of Pollution on Benthic Communities of the Tidal James River, Virginia. Ph.D. Dissertation, University of Virginia. 149 pp.

Dierberg, F. E., and P. L. Brezonik. 1983a. Tertiary treatment of municipal wastewater by cypress domes. Water Research 17:1027–1040.

Dierberg, F. E., and P. L. Brezonik. 1983b. Nitrogen and phosphorus mass balances in natural and sewage-enriched cypress domes. Journal of Applied Ecology 20:323–337.

Dierberg, F. E., and P. L. Brezonik. 1984. Nitrogen and phosphorus mass balances in a cypress dome receiving wastewater. In K. C. Ewel and H. T. Odum, eds. Cypress Swamps. University Presses of Florida, Gainesville, pp. 112–118.

Dierberg, F. E., and P. L. Brezonik. 1985. Nitrogen and phosphorus removal by cypress swamp sediments. Water, Air and Soil Pollution 24:207–213.

Dolan, T. J., S. E. Bayley, J. Zoltek, Jr., and A. Hermann. 1981. Phosphorus dynamics of a Florida freshwater marsh receiving treated wastewater. Journal of Applied Ecology 18:205–219.

Donovan, L. A., and J. Gallagher. 1984. Anaerobic substrate tolerance in *Sporobolus virginicum* (L.) Kunth. American Journal of Botany 71:1424–1431.

Dopson, J. R., P. W. Greenwood, and R. L. Jones. 1986. Holocene forest and wetland vegetation dynamics at Barrington Tops, New South Wales. Journal of Biogeography 13:561–585.

Dorffling, K., D. Tietz, J. Streich, and M. Ludewig. 1980. Studies on the role of abscisic acid in stomatal movements. In F. Skoog, ed. Plant Growth Substances 1979. Springer-Verlag, Berlin, pp. 274–285.

Dorge, C. L., W. J. Mitsch, and J. R. Wiemhoff. 1984. Cypress wetlands in southern Illinois. In K. C. Ewel and H. T. Odum, eds. Cypress Swamps. University Presses of Florida, Gainesville, pp. 393–404.

Dortch, M. S. 1996. Removal of solids, nitrogen, and phosphorus in the Cache River wetland. Wetlands 16:358–365.

Doss, P. K. 1993. The nature of a dynamic water table in a system of non-tidal, freshwater coastal wetlands. Journal of Hydrology 141:107–126.

Doss, P. K. 1995. Physical-hydrogeologic processes in wetlands. Natural Areas Journal 15:216–266.

Douglas, M. S. 1947. The Everglades: River of Grass. Ballantine, New York. 308 pp.

Doyle, G. J. 1973. Primary production estimates of native blanket bog and meadow vegetation growing on reclaimed peat at Glenamoy, Ireland. In L. C. Bliss and F. E. Wielgolaski, eds. Primary Production and Production Processes, Tundra Biome. Tundra Biome Steering Committee, Edmonton and Stockholm, pp. 141–151.

Dreyer, G. D., and W. A. Niering, eds. 1995. Tidal Marshes of Long Island Sound: Ecology, History and Restoration. Bulletin 34, Connecticut College Arboretum, New London, CT.

Driscoll, C. T., G. E. Likens, L. O. Hedin, J. S. Eaton, and F. H. Bormann. 1989. Changes in the chemistry of surface waters. Environmental Science & Technology 23:137–143.

Drum, R. W., and E. Webber. 1966. Diatoms from a Massachusetts salt marsh. Botanica Marina 9:70–77.

DuBarry, A. P., Jr. 1963. Germination of bottomland tree seed while immersed in water. Journal of Forestry 61:225–226.

Dubinski, B. J., R. L. Simpson, and R. E. Good. 1986. The retention of heavy metals in sewage sludge applied to a freshwater tidal wetland. Estuaries 9:102–111.

Duever, M. J. 1988. Hydrologic processes for models of freshwater wetlands. In W. J. Mitsch, M. Straskraba, and S. E. Jørgensen, eds. Wetland Modelling. Elsevier, Amsterdam, pp. 9–39.

Duever, M. J. 1990. Hydrology. In B. C. Patten, ed. Wetlands and Shallow Continental Water Bodies, Vol. 1. SPB Academic Publishing, The Hague, pp. 61–89.

Duever, M. J., J. E. Carlson, and L. A. Riopelle. 1984. Corkscrew Swamp: A virgin cypress strand. In K. C. Ewel and H. T. Odum, eds. Cypress Swamps. University Presses of Florida, Gainesville, pp. 334–348.

Duever, M. J., J. E. Carlson, J. F. Meeder, L. C. Duever, L. H. Gunderson, L. A. Riopelle, T. R. Alexander, R. L. Myers, and D. P. Spangler. 1986. The Big Cypress National Preserve. Research Report 8, National Audubon Society, New York.

Dugan, P. 1993. Wetlands in Danger. Michael Beasley, Reed International Books, London. 192 pp.

Dunbar, J. B. 1990. Land Loss Rates, Report 2, Louisiana Chenier Plain. Technical Report GL-90-2, U.S. Army Corps of Engineers Waterways Experiment Station, Vicksburg, MS.

Dunbar, J. B., L. D. Britsch, and E. B. Kemp. 1992. Land Loss Rates, Report 3, Louisiana Coastal Plain. Technical Report GL-90-2, U.S. Army Corps of Engineers Waterways Experiment Station, Vicksburg, MS.

Dunn, M. L. 1978. Breakdown of Freshwater Tidal Marsh Plants. Master's Thesis, University of Virginia, Charlottesville.

Dunne, T., and L. B. Leopold. 1978. Water in Environmental Planning. W. H. Freeman and Company, New York. 818 pp.

Du Rietz, G. E. 1949. Huvudenheter och huvudgränser i Svensk myrvegetation. Svensk Botanisk Tidkrift 43:274–309.

Du Rietz, G. E. 1954. Die Mineralbodenwasserzeigergrenze als Grundlage Einer Natürlichen Zweigleiderung der Nord-und Mitteleuropäischen Moore. Vegetatio 5–6:571–585.

Durno, S. E. 1961. Evidence regarding the rate of peat growth. Journal of Ecology 49:347–351.

Dvorak, J. 1978. Macrofauna of invertebrates in helohyte communities. In D. Dykyjová and J. Kvet, eds. Pond Littoral Ecosystems. Springer-Verlag, Berlin, pp. 389–392.

Dykyjová, D., and J. Kvet, eds. 1978. Pond Littoral Ecosystems. Springer-Verlag, Berlin. 464 pp.

Economic Research Service. 1998. Wetlands and Agriculture: Private Interests and Public Benefits. AER-765, U.S. Department of Agriculture, Washington, DC.

Eggelsmann, R. 1963. Die Potentielle und Aktuelle Evaporation eines Seeklimathochmoores. Publication 62, International Association of Hydrological Sciences, pp. 88–97.

Eggelsmann, R., and J. Schwar. 1979. Regeneration, recreation, and renascence of peatlands in northwestern Germany. In Proceedings of the International Symposium on Classification of Peat and Peatlands. Hyytiälä, Finland, pp. 251–260.

Egler, F. E. 1952. Southeast saline Everglades vegetation, Florida, and its management. Vegetatio Acta Geobotica 3:213–265.

Eisenlohr, W. S. 1972. Hydrologic Investigations of Prairie Potholes in North Dakota, 1958–1969. Paper 585, U.S. Geological Survey, Washington, DC.

Eisenlohr, W. S. 1976. Water loss from a natural pond through transpiration by hydrophytes. Water Resources Research 2:443–453.

Elder, J. F. 1985. Nitrogen and phosphorus speciation and flux in a large Florida river–wetland system. Water Resources Research 21:724–732.

Elder, J. F., and H. C. Mattraw, Jr. 1982. Riverine transport of nutrient and detritus to the Apalachicola Bay estuary, Florida. Water Resources Bulletin 18:849–856.

Ellanna, L. J., and P. C. Wheeler. 1986. Subsistence use of wetlands in Alaska. In A. G. van der Valk and J. Hall, eds. Alaska: Regional Wetland Functions. Proceedings of a Workshop Held at Anchorage, Alaska, May 28–29, 1986. Publication 90-1, Environmental Institute, University of Massachusetts, Amherst, pp. 85–103.

Ellison, A. M., E. J. Farnsworth, and R. R. Twilley. 1996. Facultative mutualism between red mangroves and root-fouling sponges in Belizean mangal. Ecology 77:2431–2444.

Ellison, R. L., and M. M. Nichols. 1976. Modern and holocene foraminifera in the Chesapeake Bay region. Mar. Sed. Spec. Publ. 1:131–151.

Emery, K. O., and E. Uchupi. 1972. Western North Atlantic Ocean: Topography, rocks, structure, water, life, and sediments. Memoirs of the American Association of Petroleum Geologists 17. 1532 pp.

Engelaar, W. M. H. G., M. W. Vanbruggen, W. P. M. Vandenhoek, M. A. Huyser, and C. W. P. M. Blom. 1993. Root porosities and radial oxygen losses of *Rumex* and *Plantago* species as influenced by soil pore diameter and soil aeration. New Phytologist 125:565–574.

Environmental Defense Fund and World Wildlife Fund. 1992. How Wet Is a Wetland? The Impact of the Proposed Revisions to the Federal Wetlands Delineation Manual. Environmental Defense Fund and World Wildlife Fund, Washington, DC. 175 pp.

Ernst, J. P., and V. Brown. 1989. Conserving endangered species on southern forested wetlands. In D. D. Hook and R. Lea, eds. Proceedings of the Symposium: The Forested Wetlands of the Southern United States. Southeastern Forest Experiment Station, Forest Service, U.S. Department of Agriculture, Asheville, NC, pp. 135–145.

Ernst, W. H. O. 1990. Ecophysiology of plants in waterlogged and flooded environments. Aquatic Botany 38:73–90.

Erwin, K. L. 1990. Freshwater marsh creation and restoration in the Southeast. In J. A. Kusler and M. E. Kentula, eds. Wetland Creation and Restoration. Island Press, Washington, DC, pp. 233–265.

Erwin, K. L. 1991. An Evaluation of Wetland Mitigation in the South Florida Water Management District, Vol. I. Final Report to South Florida Water Management District, West Palm Beach, FL. 124 pp.

Erwin, K. L., G. R. Best, W. J. Dunn, and P. M. Wallace. 1984. Marsh and forested wetland reclamation of a central Florida phosphate mine. Wetlands 4:87–104.

Esry, D. H., and D. J. Cairns. 1989. Overview of the Lake Jackson Restoration Project with artificially created wetlands for treatment of urban runoff. In D. W. Fisk, ed. Wetlands Concerns and Successes. Proceedings of the Conference American Water Resources Association: Wetlands Concerns and Successes, Tampa, Florida. American Water Resources Association, Tampa, pp. 247–257.

Eswaran, H., and P. Reich. 1996. Global Wetlands: A Fragile Ecosystem Vulnerable to Degradation. Unpublished Mimeo, Natural Resources Conservation Service, U.S. Department of Agriculture, Washington, DC.

Etherington, J. R. 1983. Wetland Ecology. Edward Arnold, London. 67 pp.

Etnier, C., and B. Guterstam. 1997. Ecological Engineering for Wastewater Treatment. CRC Press/Lewis Publishers, Boca Raton, FL. 451 pp.

Evers, D. E., J. G. Gossselink, C. E. Sasser, and J. M. Hill. 1992. Wetland loss dynamics in southwestern Barataria basin, Louisiana (USA), 1945–1985. Wetlands Ecology and Management 2:103–118.

Evers, D. E., C. E. Sasser, J. G. Gosselink, D. A. Fuller, and J. M. Visser. 1998. The impact of vertebrate herbivores on wetland vegetation in Atchafalaya Bay, Louisiana. Estuaries 21:1–13.

Evink, G. L. 1980. Studies of Causeways in the Indian River, Florida. Report FL-ER-7-80, Florida Department of Transportation, Tallahassee. 140 pp.

Ewel, K. C. 1976. Effects of sewage effluent on ecosystem dynamics in cypress domes. In D. L. Tilton, R. H. Kadlec, and C. J. Richardson, eds. Freshwater Wetlands and Sewage Effluent Disposal. University of Michigan Press, Ann Arbor, pp. 169–195.

Ewel, K. C. 1990. Swamps. In R. L. Myers and J. J. Ewel, eds. Ecosystems of Florida. University of Central Florida Press, Orlando, pp. 281–323.

Ewel, K. C., and W. J. Mitsch. 1978. The effects of fire on species composition in cypress dome ecosystems. Florida Science 41:25–31.

Ewel, K. C., and H. T. Odum. 1978. Cypress swamps for nutrient removal and wastewater recycling. In M. B. Wanielista and W. W. Eckenfelder, Jr., eds. Advances in Water and Wastewater Treatment Biological Nutrient Removal. Ann Arbor Science, Ann Arbor, MI, pp. 181–198.

Ewel, K. C., and H. T. Odum. 1979. Cypress domes: Nature's tertiary treatment filter. In W. E. Sopper and S. N. Kerr, eds. Utilization of Municipal Sewage Effluent and Sludge on Forest and Disturbed Land. Pennsylvania State University Press, University Park, pp. 103–114.

Ewel, K. C., and Odum H. T., eds. 1984. Cypress Swamps. University Presses of Florida, Gainesville. 472 pp.

Ewel, K. C., and L. P. Wickenheiser. 1988. Effects of swamp size on growth rates of cypress (*Taxodium distichum*) trees. American Midland Naturalist 120:362–370.

Ewel, K. C., and J. E. Smith. 1992. Evapotranspiration from Florida pond cypress swamps. Water Resources Bulletin 28:299–304.

Ewel, K. C., R. R. Twilley, and J. E. Ong. 1998. Different kinds of mangrove forests provide different goods and services. Global Ecology and Biogeography Letters 7:83–94.

Ewing, K., and K. A. Kershaw. 1986. Vegetation patterns in James Bay coastal marshes. I. Environmental factors on the south coast. Canadian Journal of Botany 64:217–226.

Faber, P. A., E. Keller, A. Sands, and B. M. Masser. 1989. The Ecology of Riparian Habitats of the Southern California Coastal Region: A Community Profile. Biological Report 85(7.27), U.S. Fish and Wildlife Service, Washington, DC. 152 pp.

Fail, J. L., M. N. Hamzah, B. L. Haines, and R. L. Todd. 1989. Above and belowground biomass, production, and element accumulation in riparian forests of an agricultural watershed. In D. D. Hook and R. Lea, eds. Proceedings of the Symposium: The Forested Wetlands of the Southern

United States. Southeastern Forest Experiment Station, Forest Service, U.S. Department of Agriculture, Asheville, NC, pp. 193–223.

Farber, S. 1987. The value of coastal wetlands for protection of property against hurricane wind damage. Journal of Environmental Economics and Management 14:143–151.

Farber, S., and R. Costanza. 1987. The economic value of wetlands systems. Journal of Environmental Management 24:41–51.

Farrar, J., and R. Gersib. 1991. Nebraska Salt Marshes: Last of the Least. Nebraska Game and Park Commission, Lincoln, NE, 23 pp.

Faulkner, S. P., and C. J. Richardson. 1989. Physical and chemical characteristics of freshwater wetland soils. In D. A. Hammer, ed. Constructed Wetlands for Wastewater Treatment. Lewis Publishers, Chelsea, MI, pp. 41–72.

Faulkner, S. P., W. H. Patrick, Jr., and R. P. Gambrell. 1989. Field techniques for measuring wetland soil parameters. Soil Science Society of America Journal 53:883–890.

Federal Geographic Data Committee (FGDC). 1992. Application of Satellite Data for Mapping and Monitoring Wetlands. Technical Report 1, Wetlands Subcommittee, FGDC, Washington, DC.

Feierabend, S. J., and J. M. Zelazny. 1987. Status Report on Our Nation's Wetlands. National Wildlife Federation, Washington, DC. 50 pp.

Feminella, J. W., and V. H. Resh. 1989. Submersed macrophytes and grazing crayfish: An experimental study of herbivory in a California freshwater marsh. Holarctic Ecology 12:1–8.

Fenchel, T. 1969. The ecology of marine microbenthos. IV. Structure and function of the benthic ecosystem, its chemical and physical factors and the microfauna communities with special reference to the ciliated protozoa. Ophelia 6:1–182.

Fennessy, M. S., and W. J. Mitsch. 1989. Treating coal mine drainage with an artificial wetland. Research Journal of the Water Pollution Control Federation 61:1691–1701.

Fetherston, K. L., R. J. Naiman, and R. E. Bilby. 1995. Large woody debris, physical processes, and riparian forest development in montane river networks of the Pacific Northwest. Geomorphology 13:133–144.

Fetter, C. W., Jr., W. E. Sloey, and F. L. Spangler. 1978. Use of a natural marsh for wastewater polishing. Journal of the Water Pollution Control Federation 50:290–307.

Field, D. W., A. J. Reyer, P. V. Genovese, and B. D. Shearer. 1991. Coastal Wetlands of the United States. National Oceanic and Atmospheric Administration and U.S. Fish and Wildlife Service, Washington, DC. 59 pp.

Finlayson, C. M., and J. Volz. 1994. A strategic action plan for the conservation of the wetlands of the lower Volga, Russia. In W. J. Mitsch, ed. Global Wetlands: Old World and New. Elsevier, Amsterdam, pp. 729–736.

Finlayson, C. M., and A. G. van der Valk, eds. 1995. Classification and Inventory of the World's Wetlands. Special Issue of Vegetatio 118:1–192.

Finlayson, M., and M. Moser, eds. 1991. Wetlands. Facts on File, Oxford, UK. 224 pp.

Fitter, A. H., and R. K. M. Hay. 1987. Environmental Physiology of Plants. Academic Press, New York. 423 pp.

Fleischer, S., L. Stibe, and L. Leonardsson. 1991. Restoration of wetlands as a means of reducing nitrogen transport to coastal waters. Ambio 20:271–272.

Fleischer, S., A. Gustafson, A. Joelsson, C. Johansson, and L. Stibe. 1994. Restoration of wetlands to counteract coastal eutrophication in Sweden. In W. J. Mitsch, ed. Global Wetlands: Old World and New. Elsevier, Amsterdam, pp. 901–907.

Fleming, M., G. Lin, and L. da S. L. Sternberg. 1990. Influence of mangrove detritus in an estuarine ecosystem. Bulletin of Marine Science 47:663–669.

Flores-Verdugo, F., J. W. Day, Jr., and R. Briseno-Duenas. 1987. Structure, litter fall, decomposition, and detritus dynamics of mangroves in a Mexican coastal lagoon with an ephemeral inlet. Marine Ecology—Progress Series 35:83–90.

Flores-Verdugo, F., F. Gonzáles-Farías, O. Ramírez-Flores, F. Amezcua-Linares, A. Yánez-Arancibia, M. Alvarez-Rubio, and J. W. Day, Jr. 1990. Mangrove ecology, aquatic primary productivity, and fish community dynamics in the Teacapan–Agua Brava lagoon–estuarine system (Mexican Pacific). Estuaries 13:219–230.

Folke, C. 1991. The societal value of wetland life-support. In C. Folke and T. Kaberger, eds. Linking the Natural Environment and the Economy. Kluwer Academic Publishers, Dordrecht, pp. 141–171.

Folke, C., and T. Kaberger, eds. 1991. Linking the Natural Environment and the Economy. Kluwer Academic Publishers, Dordrecht. 305 pp.

Ford, J., and B. L. Bedford. 1987. The hydrology of Alaskan wetlands, U.S.A.: A review. Arctic and Alpine Research 19:209–229.

Forman, R. T. T., and M. Godron. 1986. Landscape Ecology. John Wiley & Sons, New York.

Forrest, G. I., and R. A. H. Smith. 1975. The productivity of a range of blanket bog types in the northern Pennines. Journal of Ecology 63:173–202.

Forsyth, J. L. 1960. The Black Swamp. Ohio Department of Natural Resources, Division of Geological Survey, Columbus, OH. 1 p.

Foster, D. R., G. A. King, P. H. Glaser, and H. E. Wright. 1983. Origin of string patterns in boreal peatlands. Nature 306:256–258.

Fox, T. C., R. A. Kennedy, and A. A. Alani. 1988. Biochemical adaptations to anoxia in barnyard grass. In D. D. Hook et al., eds. The Ecology and Management of Wetlands, Vol. 1: Ecology of Wetlands. Timber Press, Portland, OR, pp. 359–372.

Frayer, W. E., T. J. Monahan, D. C. Bowden, and F. A. Graybill. 1983. Status and Trends of Wetlands and Deepwater Habitat in the Conterminous United States, 1950s to 1970s. Department of Forest and Wood Sciences, Colorado State University, Fort Collins. 32 pp.

Frazer, C., J. R. Longcore, and D. G. McAuley. 1990a. Habitat use by postfledgling American black ducks in Maine and New Brunswick. Journal of Wildlife Management 54(3):451–459.

Frazer, C., J. R. Longcore, and D. G. McAuley. 1990b. Home range and movements of postfledgling American ducks in eastern Maine. Canadian Journal of Zoology 68:1228–1291.

Frazier, S. 1999. Ramsar Sites Overview: A Synopsis of the World's Wetlands of International Importance. Wetlands International, Wageningen, The Netherlands. 48 pp.

Fredrickson, L. H. 1979. Lowland hardwood wetlands: Current status and habitat values for wildlife. In P. E. Greeson, J. R. Clark, and J. E. Clark, eds. Wetland Functions and Values: The State of Our Understanding. American Water Resources Association, Minneapolis, pp. 296–306.

Fredrickson, L. H., and F. A. Reid. 1990. Impacts of hydrologic alteration on management of freshwater wetlands. In J. M. Sweeney, ed. Management of Dynamic Ecosystems, North Central Section, Wildlife Society, West Lafayette, IN, pp. 71–90.

Freiberger, H. J. 1972. Streamflow Variation and Distribution in the Big Cypress Watershed During Wet and Dry Periods. Map Series 45, Bureau of Geology, Florida Department of Natural Resources, Tallahassee.

Frenkel, R. E., and J. C. Morlan. 1991. Can we restore our salt marshes? Lessons from the Salmon River, Oregon. Northwest Environmental Journal 7:119–135.

Frey, R. W., and P. B. Basan. 1985. Coastal salt marshes. In R. A. Davis, Jr., ed. Coastal Sedimentary Environments. Springer-Verlag, New York, pp. 225–301.

Friedman, J. M., M. L. Scott, and D. T. Patten, eds. 1998a. Semiarid Riparian Ecosystems. Special Issue of Wetlands 18:497–696.

Friedman, J. M., W. R. Osterkamp, M. L. Scott, and G. T. Auble. 1998b. Downstream effects of dams on channel geometry and bottomland vegetation: Regional patterns in the Great Plains. Wetlands 18:619–633.

Frissell, C. A., W. J. Liss, C. E. Warren, and M. D. Hurley. 1986. A hierarchical framework for stream habitat classification: Viewing streams in a watershed context. Environmental Management 10:199–214.

Fuller, D. A., G. W. Peterson, R. K. Abernethy, and M. A. LeBlanc. 1988. The distribution and habitat use of waterfowl in Atchafalaya Bay, Louisiana. In C. E. Sasser and D. A. Fuller, eds. Vegetation and Waterfowl Use of Islands in Atchafalaya Bay. Final Report Submitted to Louisiana Board of Regents, Coastal Ecology Institute, Louisiana State University, Baton Rouge, pp. 73–103.

Gaddy, L. L. 1978. Congaree: Forest of giants. American Forests 84:51–53.

Galatowitsch, S. M., and A. G. van der Valk. 1994. Restoring Prairie Wetlands: An Ecological Approach. Iowa State University Press, Ames. 246 pp.

Galatowitsch, S. M., and A. G. van der Valk. 1996. Characteristics of recently restored wetlands in the prairie pothole region. Wetlands 16:75–83.

Galatowitsch, S. M., N. O. Anderson, and P. D. Ascher. 1999. Invasiveness in wetland plants in temperate North America. Wetlands 19:733–755.

Gallagher, J. L. 1978. Decomposition processes: Summary and recommendations. In R. E. Good, D. F. Whigham, and R. L. Simpson, eds. Freshwater Wetlands: Ecological Processes and Management Potential. Academic Press, New York, pp. 145–151.

Gallagher, J. L., and F. C. Daiber. 1974. Primary production of edaphic algae communities in a Delaware salt marsh. Limnology and Oceanography 19:390–395.

Gallagher, J. L., G. F. Somers, D. M. Grant, and D. M. Seliskar. 1988. Persistent differences in two forms of *Spartina alterniflora:* A common garden experiment. Ecology 69:1005–1008.

Gallihugh, J. L., and J. D. Rogner. 1998. Wetland Mitigation and 404 Permit Compliance Study, Vol. I. U.S. Fish and Wildlife Service, Region III, Burlington, IL, and U.S. Environmental Protection Agency, Region V, Chicago. 161 pp.

Gambrell, R. P., and W. H. Patrick, Jr. 1978. Chemical and microbiological properties of anaerobic soils and sediments. In D. D. Hook and R. M. M. Crawford, eds. Plant Life in Anaerobic Environments. Ann Arbor Science, Ann Arbor, pp. 375–423.

Garbisch, E. W. 1977. Recent and Planned Marsh Establishment Work Throughout the Contiguous United States: A Survey and Basic Guidelines. CR D-77-3, U.S. Army Corps of Engineers Waterways Experiment Station, Vicksburg, MS.

Garbisch, E. W. 1989. Wetland enhancement, restoration, and construction. In S. K. Majumdar, ed. Wetlands Ecology and Conservation: Emphasis in Pennsylvania. Pennsylvania Academy of Science. Easton, pp. 261–275.

Garbisch, E. W., P. B. Woller, and R. J. McCallum. 1975. Salt Marsh Establishment and Development. Technical Memo 52, U.S. Army Coastal Engineering Research Center, Fort Belvoir, VA.

Gardner, L. R. 1975. Runoff from an intertidal marsh during tidal exposure: regression curves and chemical characteristics. Limnology and Oceanography 20:81–89.

Gaudet, J. J. 1977. Natural drawdown of Lake Naivasha and the formation of papyrus swamps. Aquatic Botany 3:1–47.

Gaudet, J. J. 1979. Seasonal changes in the nutrients in a tropical swamp: North Swamp, Lake Naivasha. Journal of Ecology 67:953–981.

Gemborys, S. R., and E. J. Hodgkins. 1971. Forests of small stream bottoms in the coastal plain of southwestern Alabama. Ecology 52:70–84.

Gerakis, P. A., ed. 1992. Conservation and Management of Greek Wetlands. IUCN, Gland, Switzerland.

Gerheart, R. A. 1992. Use of constructed wetlands to treat domestic wastewater, city of Arcata, California. Water Science and Technology 26:1625–1637.

Gerheart, R. A., F. Klopp, and G. Allen. 1989. Constructed free surface wetlands to treat and receive wastewater: Pilot project to full scale. In D. A. Hammer, ed. Constructed Wetlands for Wastewater Treatment. Lewis Publishers, Chelsea, MI, pp. 121–137.

Gerritsen, J., and H. S. Greening. 1989. Marsh seed banks of the Okefenokee Swamp: Effects of hydrologic regime and nutrients. Ecology 70:750–763.

Gersberg, R. M., B. V. Elkins, and C. R. Goldman. 1983. Nitrogen removal in artificial wetlands. Water Resources 17(9):1009–1014.

Gersberg, R. M., B. V. Elkins, and C. R. Goldman. 1984. Use of artificial wetlands to remove nitrogen from wastewater. Journal of the Water Pollution Control Federation 56(2):152–156.

Gersberg, R. M., S. R. Lyon, R. Brenner, and B. V. Elkins. 1989. Integrated wastewater treatment using artificial wetlands: A gravel marsh case study. In D. A. Hammer, ed. Constructed Wetlands for Wastewater Treatment. Lewis Publishers, Chelsea, MI, pp. 145–152.

Gibbons, A. 1997. Did birds fly through the K–T extinction with flying colors? Science 275:1068.

Gilman, K. 1982. Nature conservation in wetlands: Two small fen basins in western Britain. In D. O. Logofet and N. K. Luckyanov, eds. Ecosystem Dynamics in Freshwater Wetlands and Shallow Water Bodies, Vol. I. SCOPE and UNEP Workshop, Center of International Projects, Moscow, pp. 290–310.

Gilvear, D. J., J. H. Tellam, J. W. Lloyd, and D. N. Lerner. 1989. The Hydrodynamics of East Anglian Fen Systems. University of Birmingham Press, Edgbaston, UK.

Giurgevich, J. R., and E. L. Dunn. 1978. Seasonal patterns of CO_2 and water vapor exchange of *Juncus roemerianus* Scheele in a Georgia salt marsh. American Journal of Botany 65:502–510.

Giurgevich, J. R., and E. L. Dunn. 1979. Seasonal patterns of CO_2 and water vapor exchange of the tall and short forms of *Spartina alterniflora* Loisel in a Georgia salt marsh. Oecologia 43:139–156.

Glaser, P. H. 1983a. *Carex exilis* and *Scirpus cespitosus* var. *callosus* in patterned fens in northern Minnesota. Michigan Botany 22:22–26.

Glaser, P. H. 1983b. *Eleocharis rostellata* and its relation to spring fens in Minnesota. Michigan Botany 22:19–21.

Glaser, P. H. 1983c. Vegetation patterns in the North Black River peatland, northern Minnesota. Canadian Journal of Botany 61:2085–2104.

Glaser, P. H. 1987. The Ecology of Patterned Boreal Peatlands of Northern Minnesota: A Community Profile. Biological Report 85(7.14), U.S. Fish and Wildlife Service, Washington, DC. 98 pp.

Glaser, P. H., and G. A. Wheeler. 1980. The development of surface patterns in the Red Lake peatland, northern Minnesota. In Proceedings of the Sixth International Peat Congress, Duluth, MN, pp. 31–35.

Glaser, P. H., G. A. Wheeler, E. Gorham, and H. E. Wright, Jr. 1981. The patterned mires of the Red Lake peatland, northern Minnesota: Vegetation, water chemistry, and landforms. Journal of Ecology 69:575–599.

Glaser, P. H., J. A. Janssens, and D. I. Siegel. 1990. The response of vegetation to chemical and hydrological gradients in the Lost River peatland, northern Minnesota. Journal of Ecology 78:1021–1048.

Glaser, P. H., D. I. Siegel, E. A. Romanowicz, and Y. P. Shen. 1997a. Regional linkages between raised bogs and the climate, groundwater, and landscape of north-western Minnesota. Journal of Ecology 85:3–16.

Glaser, P. H., P. C. Bennett, D. I. Siegel, and E. A. Romanowicz. 1997b. Palaeo-reversals in groundwater flow and peatland development at Lost River, Minnesota, USA. Holocene 6:413–421.

Gleason, P. J., ed. 1974. Environments of South Florida: Present and Past, Memoir No. 2, Miami Geological Society, Miami. 452 pp.

Gleason, P. J., ed. 1984. Environments of South Florida: Present and Past II. Memoir No. 2, 2nd ed. Miami Geological Society, Miami. 551 pp.

Gleason, H. A. 1917. The structure and development of the plant association. Torrey Botanical Club Bulletin 44:463–481.

Glob, P. V. 1969. The Bog People: Iron Age Man Preserved. Translated by R. Bruce-Mitford. Cornell University Press, Ithaca, NY. 200 pp.

Glooschenko, V., and P. Grondin. 1988. Wetlands of eastern temperate Canada. In National Wetlands Working Group, ed. Wetlands of Canada. Environment

Canada, Ottawa, Ontario, and Polyscience Publications, Montreal, Quebec, pp. 199–248.

Godfrey, P. J., E. R. Kaynor, S. Pelczarski, and J. Benforado, eds. 1985. Ecological Considerations in Wetlands Treatment of Municipal Wastewaters. Van Nostrand Reinhold, New York. 474 pp.

Godwin, H. 1981. The Archives of the Peat Bogs. Cambridge University Press, Cambridge, UK. 229 pp.

Goldman, C. R. 1961. The contribution of alder trees (*Alnus tenuifolia*) to the primary production of Castle Lake, California. Ecology 42:282–288.

Golet, F. C., and J. S. Larson. 1974. Classification of Freshwater Wetlands in the Glaciated Northeast. Resources Publication 116, U.S. Fish and Wildlife Service, Washington, DC. 56 pp.

Golet, F. C., and D. J. Lowry. 1987. Water regimes and tree growth in Rhode Island Atlantic white cedar swamps. In A. D. Laderman ed. Atlantic White Cedar Wetlands. Westview Press, Boulder, CO. pp. 91–110.

Golet, F. C., A. J. K. Calhoun, W. R. DeRagon, D. J. Lowry, and A. J. Gold. 1993. Ecology of Red Maple Swamps in the Glaciated Northeast: A Community Profile. Biological Report 12, U.S. Fish and Wildlife Service, Washington, DC. 151 pp.

Golley, F. B., H. T. Odum, and R. F. Wilson. 1962. The structure and metabolism of a Puerto Rican red mangrove forest in May. Ecology 43:9–19.

Gomez, M. M., and F. P Day, Jr. 1982. Litter nutrient content and production in the Great Dismal Swamp. American Journal of Botany 69:1314–1321.

Gonthier, G. J. 1996. Ground-water-flow conditions within a bottomland hardwood wetland, Eastern Arkansas. Wetlands 16:334-346.

Good, R. E., D. F. Whigham, and R. L. Simpson, eds. 1978. Freshwater Wetlands: Ecological Processes and Management Potential. Academic Press, New York. 378 pp.

Good, R. E., N. F. Good, and B. R. Frasco. 1982. A review of primary production and decomposition dynamics of the belowground marsh component. In V. S. Kennedy, ed. Estuarine Comparisons. Academic Press, New York, pp. 139–157.

Goodman, G. T., and D. F. Perkins. 1968. The role of mineral nutrients in *Eriophorum* communities. IV. Potassium supply as a limiting factor in an *E. vaginatum* community. Journal of Ecology 56:685–696.

Goodman, P. J., and W. T. Williams. 1961. Investigations into "die-back." Journal of Ecology 49:391–398.

Gopal, B., R. E. Turner, R. G. Wetzel, and D. F. Whigham, eds. 1982. Wetlands: Ecology and Management. National Institute of Ecology and International Scientific Publications, Jaipur, India. 514 pp.

Gopal, B., and V. Masing. 1990. Biology and ecology. In B. C. Patten, ed. Wetlands and Shallow Continental Water Bodies. SPB Academic Publishing, The Hague, pp. 91–239.

Gopal, B., A. Hillbricht-Ilkowska, and R. G. Wetzel, eds. 1993. Wetlands and Ecotones: Studies on Land–Water Interactions. National Institute of Ecology, New Delhi, India. 301 pp.

Gopal, B., and W. J. Mitsch, eds. 1995. The Role of Vegetation in Created and Restored Wetlands. Special Issue of Ecological Engineering 5:1–121.

Gore, A. J. P. ed. 1983a. Ecosystems of the World, Vol. 4A: Mires: Swamp, Bog, Fen, and Moor. Elsevier, Amsterdam. 440 pp.

Gore, A. J. P. 1983b. Introduction. In A. J. P. Gore, ed. Ecosystems of the World, Vol. 4B: Mires: Swamp, Bog, Fen, and Moor. Elsevier, Amsterdam, pp. 1–34.

Gorham, E. 1956. The ionic composition of some bogs and fen waters in the English lake district. Journal of Ecology 44:142–152.

Gorham, E. 1961. Factors influencing supply of major ions to inland waters, with special references to the atmosphere. Geological Society of America Bulletin 72:795–840.

Gorham, E. 1967. Some Chemical Aspects of Wetland Ecology. Technical Memorandum 90, Committee on Geotechnical Research, National Research Council of Canada, pp. 2–38.

Gorham, E. 1974. The relationship between standing crop in sedge meadows and summer temperature. Journal of Ecology 62:487–491.

Gorham, E. 1991. Northern peatlands: Role in the carbon cycle and probable responses to climatic warming. Ecological Applications 1:182–195.

Gorham, E., S. E. Bayley, and D. W. Schindler. 1984a. Ecological effects of acid deposition upon peatlands: A neglected field in "acid-rain" research. Canadian Journal of Fisheries and Aquatic Sciences 41:1256–1268.

Gorham, E., S. J. Eisenreich, J. Ford, and M. V. Santelmann. 1984b. The chemistry of bog waters. In W. Strum, ed. Chemical Processes in Lakes. John Wiley & Sons, New York, pp. 339–363.

Gorham, E., and J. A. Janssens. 1992. Concepts of fen and bog reexamined in relation to bryophyte cover and the acidity of surface waters. Acta Societatis Botanicorum Poloniae 61:7–20.

Gosselink, J. G. 1984. The Ecology of Delta Marshes of Coastal Louisiana: A Community Profile. FWS/OBS-84/09, U.S. Fish and Wildlife Service, Washington, DC. 134 pp.

Gosselink, J. G., E. P. Odum, and R. M. Pope. 1974. The Value of the Tidal Marsh. Publication LSU-SG-74-03, Center for Wetland Resources, Louisiana State University, Baton Rouge. 30 pp.

Gosselink, J. G., and R. E. Turner. 1978. The role of hydrology in freshwater wetland ecosystems. In R. E. Good, D. F. Whigham, and R. L. Simpson, eds. Freshwater Wetlands: Ecological Processes and Management Potential. Academic Press, New York, pp. 63–78.

Gosselink, J. G., and R. H. Baumann. 1980. Wetland inventories: Wetland loss along the United States coast. Zeitschrift für Geomorphologie N.F. Suppl. Bd. 34:173–187.

Gosselink, J. G., W. H. Conner, J. W. Day, Jr., and R. E. Turner. 1981a. Classification of wetland resources: Land, timber, and ecology. In B. D. Jackson and J. L. Chambers, eds. Timber Harvesting in Wetlands. Division of Continuing Education, Louisiana State University, Baton Rouge, pp. 28–48.

Gosselink, J. G., S. E. Bayley, W. H. Conner, and R. E. Turner. 1981b. Ecological factors in the determination of riparian wetland boundaries. In J. R. Clark and J. Benforado, eds. Wetlands of Bottomland Hardwood Forests. Elsevier, Amsterdam, pp. 197–219.

Gosselink, J. G., and E. Maltby. 1990. Wetland losses and gains. In M. Williams, ed. Wetlands: A Threatened Landscape. Basil Blackwell, Oxford, UK, pp. 296–322.

Gosselink, J. G., G. P. Shaffer, L. C. Lee, D. M. Burdick, D. L. Childers, N. C. Leibowitz, S. C. Hamilton, R. Boumans, D. Cushman, S. Fields, M. Koch, and J. M. Visser. 1990a. Landscape conservation in a forested wetland watershed. BioScience 40:588–600.

Gosselink, J. G., L. C. Lee, and T. A. Muir, eds. 1990b. Ecological Processes and Cumulative Impacts: Illustrated by Bottomland Hardwood Wetland Ecosystems. Lewis Publishers, Chelsea, MI. 708 pp.

Gosselink, J. G., M. M. Brinson, L. C. Lee, and G. T. Auble. 1990c. Human activities and ecological processes in bottomland hardwood ecosystems: The report of the ecosystem workgroup. In J. G. Gosselink, L. C. Lee, and T. A. Muir, eds. Ecological Processes and Cumulative Impacts: Illustrated by Bottomland Hardwood Wetland Ecosystems. Lewis Publishers, Chelsea, MI, pp. 549–598.

Gosselink, J. G., and C. E. Sasser. 1995. Causes of wetland loss. In D. J. Reed, ed. Status and Historical Trends of Hydrologic Modification, Reduction in Sediment Availability, and Habitat Loss/Modification in the Barataria–Terrebonne Estuarine System. BT/NEP Publication 20, Barataria–Terrebonne National Estuary Program, Thibodaux, LA, pp. 203–236.

Gosselink, J. G., Coleman, J. M., and R. E. Stewart, Jr. 1998. Coastal Louisiana. In M. J. Mac, P. A. Opler, C. E. Puckett Haecker, and P. D. Doran, eds. Status and Trends of the Nation's Biological Resources, Vol. 1. U.S. Department of Interior, U.S. Geological Survey, Reston, VA, pp. 385–436.

Gottgens, J. F., B. P. Swartz, R. W. Kroll, and M. Eboch. 1998. Long-term GIS-based records of habitat changes in a Lake Erie coastal marsh. Wetlands Ecology and Management 6:5–17.

Gottgens, J. F., R. H. Fortney, J. Meyer, J. E. Perry, and B. E. Rood. In press. Impacts of the Paraguay–Paraná Hidrovia on the Pantanal of Brazil: Large-scale channelization or a "tyranny of small decisions"? BioScience.

Grace, J. B., and M. A. Ford. 1996. The potential impact of herbivores on the susceptibility of the marsh plant *Sagittaria lancifolia* to saltwater intrusion in coastal wetlands. Estuaries 19:13–20.

Graf, W. L. 1988. Definition of flood plains along arid-region rivers. In V. R. Baker, R. C. Kochel, and P. C. Patton, eds. Flood Geomorphology. John Wiley & Sons, New York, pp. 231–259.

Grant, J., and U. V. Bathmann. 1987. Swept away: Resuspension of bacterial mats regulates benthic–pelagic exchange of sulfur. Science 236:1472–1474.

Grant, M. 1962. Myths of the Greeks and Romans. Mentor/New American Library, New York.

Grant, R. R., and R. Patrick. 1970. Tinicum Marsh as a water purifier. In J. McCormick, R. R. Grant, Jr., and R. Patrick, eds. Two Studies of Tinicum Marsh, Delaware and Philadelphia Counties, Pa. Conservation Foundation, Washington, DC, pp. 105–131.

Gray, A. J. 1992. Saltmarshes plant ecology: Zonation and succession revisited. In Saltmarshes: Morphodynamics, Conservation and Engineering Significance. Cambridge University Press, Cambridge, UK, pp. 63–79.

Gray, L. C., O. E. Baker, F. J. Marschner, B. O. Weitz, W. R. Chapline, W. Shepard, and R. Zon. 1924. The Utilization of Our Lands for Crops, Pasture, and Forests. In U.S. Department of Agriculture Yearbook, 1923. Government Printing Office, Washington. DC, pp. 415–506.

Greenwood, D. J. 1961. The effect of oxygen concentration on the decomposition of organic materials in soil. Plant and Soil 14:360–376.

Greeson, P. E., J. R. Clark, and J. E. Clark, eds. 1979. Wetland Functions and Values: The State of Our Understanding. American Water Resources Association, Minneapolis. 674 pp.

Gregory, S. V., F. J. Swanson, W. A. McKee, and K. W. Cummins. 1991. An ecosystem perspective of riparian zones. BioScience 41:540–551.

Gren, I.-M. 1994. Costs and benefits of restoring wetlands: Two Swedish case studies. Ecological Engineering 4:153–162.

Gren, I.-M., C. Folke, K. Turner, and I. Batemen. 1994. Primary and secondary values of wetland ecosystems. Environmental and Resource Economics 4:55–74.

Grigal, D. F. 1985. *Sphagnum* production in forested bogs of northern Minnesota. Canadian Journal of Botany 63:1204–1207.

Grigal, D. F., C. C. Buttlema, and L. K. Kernik. 1985. Productivity of the woody strata of forested bogs in northern Minnesota. Canadian Journal of Botany 63:2416–2424.

Grime, H. H. 1979. Plant Strategies and Vegetation Processes. John Wiley & Sons, New York.

Groffman, P. M. 1994. Denitrification in freshwater wetlands. Current Topics in Wetland Biogeochemistry 1:15–35.

Groffman, P. M., E. A. Axelrod, J. L. Lemunyon, and W. M. Sullivan. 1991. Denitrification in grass and forest vegetated filter strips. Journal of Environmental Quality 20:571–674.

Groffman, P. M., and G. C. Hanson. 1997. Wetland denitrification: Influence of site quality and repationships with wetland delineation protocols. Soil Science Society of America Journal 61:323–329.

Gross, W. J. 1964. Trends in water and salt regulation among aquatic and amphibious crabs. Biological Bulletin 127:447–466.

Grosse, W., and P. Schröder. 1984. Oxygen supply of roots by gas transport in alder trees. Zeitschrift für Naturforsch. 39C:1186–1188.

Grosse, W., and J. Mevi-Schütz. 1987. A beneficial gas transport system in *Nymphoides peltata*. American Journal of Botany 74:941–952.

Grosse, W., S. Sika, and S. Lattermann. 1990. Oxygen supply of roots by thermo-osmotic gas transport in *Alnus glutinosa* and other wetland trees. In D. Werner and P. Müller, eds. Fast Growing Trees and Nitrogen Fixing Trees. Gustav Fischer Verlag, New York, pp. 246–249.

Grosse, W., H. B. Büchel, and H. Tiebel. 1991. Pressurized ventilation in wetland plants. Aquatic Botany 39:89–98.

Grosse, W., J. Frye, and S. Lattermann. 1992. Root aeration in wetland trees by pressurized gas transport. Tree Physiology 10:285–295.

Grosse, W., and D. Meyer. 1992. The effect of pressurized gas transport on nutrient uptake during hypoxia of alder roots. Botany Acta 105:223–226.

Grosse, W., H. B. Büchel, and S. Lattermann. 1998. Root aeration in wetland trees and its ecophysiological significance. In A. D. Laderman, ed. Coastally Restricted Forests. Oxford University Press, New York, pp. 293–305.

Gum, R. L., and W. E. Martin. 1975. Problems and solutions in estimating the demand for and value of rural outdoor recreation. American Journal of Agricultural Economics 57:558–566.

Gunderson, L. H., and W. T. Loftus. 1993. The Everglades. In W. H. Martin, S. G. Boyce, and A. C. E. Echternacht, eds. Biodiversity of the Southeastern United States: Lowland Terrestrial Communities. John Wiley & Sons, New York, pp. 199–255.

Guthery, F. S., J. M. Pates, and F. A. Stormer. 1982. Characterization of playas of the north-central Llano Estacado in Texas. Transactions of the 47th American Wildlife and Natural Resources Conference 47:516–527.

Hall, F. R., R. J. Rutherford, and G. L. Byers. 1972. The Influence of a New England Wetland on Water Quantity and Quality. New Hampshire Water Resource Center Research Report 4, University of New Hampshire, Durham. 51 pp.

Hall, H. D., and V. W. Lambou. 1990. The ecological significance to fisheries of bottomland hardwood ecosystems: Values, detrimental impacts, and assessment: The report of the fisheries workgroup. In J. G. Gosselink, L. C. Lee, and T. A. Muir, eds. Ecological Processes and Cumulative Impacts: Illustrated by Bottomland Hardwood Wetland Ecosystems. Lewis Publishers, Chelsea, MI, pp. 481–531.

Hall, J. V., W. E. Frayer, and B. O. Wilen. 1994. Status of Alaska Wetlands. U.S. Fish and Wildlife Service, Alaska Region, Anchorage. 32 pp.

Hall, S. L., and F. M. Fisher, Jr. 1985. Annual productivity and extracellular release of dissolved organic compounds by the epibenthic algal community of a brackish marsh. Journal of Phycology 21:277–281.

Hall, T. F., and W. T. Penfound. 1939. A phytosociological study of a cypress–gum swamp in southern Louisiana. American Midland Naturalist 21:378–395.

Hamilton, S. K., S. J. Sippel, and J. M. Melack. 1996. Inundation patterns in the Pantanal wetland of South America determined from passive microwave remote sensing. Archiv für Hydrobiologie 137:1–23.

Hammack, J., and G. M. Brown. 1974. Waterfowl and Wetlands: Toward Bioeconomic Analysis. Johns Hopkins University Press, Baltimore.

Hammer, D. A., ed. 1989. Constructed Wetlands for Wastewater Treatment. Lewis Publishers, Chelsea, MI. 831 ppHammer, D. A. 1992. Creating Freshwater Wetlands. Lewis Publishers, Chelsea, MI. 298 pp.

Hammer, D. A. 1997. Creating Freshwater Wetlands, 2nd ed. CRC Press/Lewis Publishers, Boca Raton, FL. 406 pp.

Hammer, D. E., and R. H. Kadlec. 1983. Design Principles for Wetland Treatment Systems. EPA-600/2-83-26, U.S. Environmental Protection Agency, Ada, OK. 244 pp.

Hanby, J., and D. Bygott. 1998. Ngorongoro Conservation Area. Kibuyu Partners, Karatu, Tanzania.

Hanson, G. C., P. M. Groffman, and A. J. Gold. 1994a. Denitrification in riparian wetlands receiving high and low groundwater nitrate inputs. Journal of Environmental Quality 23:917–922.

Hanson, G. C., P. M. Groffman, and A. J. Gold. 1994b. Symptoms of nitrogen saturation in a riparian wetland. Ecological Applications 4:750–756.

Hare, F. K. 1980. Long-term annual surface heat and water balances over Canada and the United States south of 60°N: Reconciliation of precipitation, runoff, and temperature fields. Atmosphere—Ocean 18:127–153.

Harper, D. M. 1992. The ecological relationships of aquatic plants at Lake Naivasha, Kenya. Hydrobiologia 232:65–71.

Harper, D. M., C. Adams, and K. Mavuti. 1995. The aquatic plant communities of the Lake Naivasha wetland, Kenya: Pattern, dynamics, and conservation. Wetlands Ecology and Management 3:111–123.

Harris, L. D. 1989. The faunal significance of fragmentation of southeastern bottomland forests. In D. D. Hook and R. Lea, eds. Proceedings of the Symposium: The Forested Wetlands of the Southern United States. Southeastern Forest Experiment Station, Forest Service, U.S. Department of Agriculture, Asheville, NC, pp. 126–134.

Harris, L. D., and J. G. Gosselink. 1990. Cumulative impacts of bottomland hardwood forest conversion on hydrology, water quality, and terrestrial wildlife. In J. G. Gosselink, L. C. Lee, and T. A. Muir, eds. Ecological Processes and Cumulative Impacts: Illustrated by Bottomland Hardwood Wetland Ecosystems. Lewis Publishers, Chelsea, MI, pp. 259–322.

Harris, R. R. 1988. Associations between stream valley geomorphology and riparian vegetation as a basis for landscape analysis in the Eastern Sierra Nevada, California, USA. Environmental Management 12:219–228.

Harris, T. 1991. Death in the Marsh. Island Press, Washington, DC. 245 pp.

Harrison, A. F., P. M. Latter, and D. W. H. Walton. 1988. Cotton strip assay: An index of decomposition in soils. In Proceedings of the International Terrestrial Ecology Symposium, Grange-Over-Sands, UK. 176 pp.

Harriss, R. C., D. I. Sebacher, and F. P. Day, Jr. 1982. Methane flux in the Great Dismal Swamp. Nature 297:673–674.

Harvey, H. T., and M. N. Josselyn. 1986. Wetlands restoration and mitigation policies: Comment. Environmental Management 10:567–569.

Harvey, J. W., and W. E. Odum. 1990. The influence of tidal marshes on upland groundwater discharge to estuaries. Biogeochemistry 12:217–236.

Harvey, J. W., and W. K. Nuttle. 1995. Fluxes of water and solute in a coastal wetland sediment. 2. Effect of macropores on solute exchange with surface water. Journal of Hydroloogy 164:109–125.

Hatton, R. S. 1981. Aspects of Marsh Accretion and Geochemistry: Barataria Basin, La. Master's Thesis, Louisiana State University, Baton Rouge.

Havill, D. C., A. Ingold, and J. Pearson. 1985. Sulfide tolerance in coastal halophytes. Vegetatio 62:279–285.

Heal, O. W., H. E. Jones, and J. B. Whittaker. 1975. Moore House, U.K. In T. Rosswall and O. W. Heal, eds. Structure and Function of Tundra Ecosystems. Ecological Bulletin 20, Swedish Natural Science Research Council, Stockholm, pp. 295–320.

Heald, E. J. 1969. The Production of Organic Detritus in a South Florida Estuary. Ph.D. Dissertation, University of Miami. 110 pp.

Heald, E. J. 1971. The Production of Organic Detritus in a South Florida Estuary. University of Miami Sea Grant Technical Bulletin 6, Coral Gables, FL. 110 pp.

Healey, M. C. 1994. Effects of dams and dikes on fish habitat in two Canadian river deltas. In W. J. Mitsch, ed. Global Wetlands: Old World and New. Elsevier, Amsterdam, pp. 385–398.

Heard, R. W. 1982. Guide to Common Tidal Marsh Invertebrates of the Northeastern Gulf of Mexico. MASGP-79-004, Mississippi-Alabama Sea Grant Consortium, University of South Alabama, Mobile.

Heckman, C. W. 1994. The seasonal succession of biotic components in wetlands of the tropical wet-and-dry climatic zone. I. Physical and chemical causes and biological effects in the Pantanal of Mato Grosso, Brazil. Internationale Revue Gesamten Hydrobiologie 79:397–421.

Hedmon, C. W., D. Van Lear, and W. T. Swank. 1996. Instream large woody debris loading and riparian forest seral stage associations in the southern Appalachian mountains. Canadian Journal of Forest Research 26:1218–1227.

Hefner, J. M., and J. D. Brown. 1985. Wetland trends in the southeastern United States. Wetlands 4:1–11.

Heilman, P. E. 1968. Relationship of availability of phosphorus and cations to forest succession and bog formation in interior Alaska. Ecology 49:331–336.

Heimburg, K. 1984. Hydrology of north-central Florida cypress domes. In K. C. Ewel and H. T. Odum, eds. Cypress Swamps. University Presses of Florida, Gainesville, pp. 72–82.

Heinselman, M. L. 1963. Forest sites, bog processes, and peatland types in the glacial Lake Agassiz Region, Minnesota. Ecological Monographs 33:327–374.

Heinselman, M. L. 1970. Landscape evolution and peatland types, and the Lake Agassiz Peatlands Natural Area, Minnesota. Ecological Monographs 40:235–261.

Hem, J. D. 1970. Study and Interpretation of the Chemical Characteristics of Natural Water. U.S. Geological Survey Water Supply Paper 1473, Washington, DC.

Hemond, H. F. 1980. Biogeochemistry of Thoreau's Bog, Concord, Mass. Ecological Monographs 50:507–526.

Hemond, H. F. 1983. The nitrogen budget of Thoreau's Bog. Ecology 64:99–109.

Hemond, H. F., and J. L. Fifield. 1982. Subsurface flow in salt marsh peat: A model and field study. Limnology and Oceanography 27:126–136.

Hemond, H. F., and J. C. Goldman. 1985. On non-Darcian water flow in peat. Journal of Ecology 73:579–584.

Hemond, H. F., T. P. Army, W. K. Nuttle, and D. G. Chen. 1987. Element cycling in wetlands: Interactions with physical mass transport. In R. A. Hites and S. J. Eisenreich, eds. Sources and Fates of Aquatic Pollutants. American Chemical Society, Chicago, pp. 519–540.

Hensel, B. R., and M. V. Miller. 1991. Effects of wetlands creation on groundwater flow. Journal of Hydrology 126:293–314.

Herdendorf, C. E. 1987. The Ecology of the Coastal Marshes of Western Lake Erie: A Community Profile. Biological Report 85(7.9), U.S. Fish and Wildlife Service, Washington, DC. 171 pp. plus microfiche appendices.

Hern, S. C., V. W. Lambou, and J. R. Butch. 1980. Descriptive Water Quality for the Atchafalaya Basin, Louisiana. EPA-600/4-80-614, U.S. Environmental Protection Agency, Las Vegas. 168 pp.

Hester, M. W., I. A. Mendelssohn, and K. L. McKee. 1996. Intraspecific variation in salt tolerance and morphology in the coastal grass *Spartina patens* (Poaceae). American Journal of Botany 83:1521–1527.

Hewlett, J. D., and A. R. Hibbert. 1967. Factors affecting the response of small watersheds to precipitation in humid areas. In W. E. Sopper and H. W. Lull, eds. International Symposium on Forest Hydrology. Pergamon Press, New York, pp. 275–290.

Hey, D. L., M. A. Cardamone, J. H. Sather, and W. J. Mitsch. 1989. Restoration of riverine wetlands: The Des Plaines River wetlands demonstration project. In W. J. Mitsch and S. E. Jørgensen, eds. Ecological Engineering: An Introduction to Ecotechnology. John Wiley & Sons, New York, pp. 159–183.

Hey, D. L., K. R. Barrett, and C. Biegen. 1994a. The hydrology of four experimental constructed marshes. Ecological Engineering 3:319–343.

Hey, D. L., A. L. Kenimer, and K. R. Barrett. 1994b. Water quality improvement by four experimental wetlands. Ecological Engineering 3:381–398.

Hey, D. L., and N. S. Philippi. 1995. Flood reduction through wetland restoration: The Upper Mississippi River Basin as a case study. Restoration Ecology 3:4–17.

Hibbard, B. H. 1965. A History of the Public Land Policies. University of Wisconsin Press, Madison.

Hickman, S. C., and V. J. Mosca. 1991. Improving Habitat Quality for Migratory Waterfowl and Nesting Birds: Assessing the Effectiveness of the Des Plaines River Wetlands Demonstration Project. Technical Paper 1, Wetlands Research, Chicago. 13 pp.

Hill, B. H. 1985. The breakdown of macrophytes in a reservoir wetland. Aquatic Botany 21:23–31.

Hill, F. B., V. P. Aneja, and R. M. Felder. 1978. A technique for measurement of biogenic sulfur emission fluxes. Environmental Science Health 13:199–225.

Hill, M. O., P. M. Latter, and G. Bancroft. 1985. A standard curve for inter-site comparison of cellulose degradation using the cotton strip method. Canadian Journal of Soil Science 65:609–619.

Hill, P. L. 1983. Wetland–Stream Ecosystems of the Western Kentucky Coalfield: Environmental Disturbance and the Shaping of Aquatic Community Structure. Ph.D. Dissertation, University of Louisville, Louisville, KY. 290 pp.

Ho, C. L., and S. Schneider. 1976. Water and sediment chemistry. In J. G. Gosselink, R. Miller, M. Hood, and L. M. Bahr, Jr., eds. Louisiana Offshore Oil Port: Environmental Baseline Study, Vol. IV. LOOP, Inc., New Orleans.

Hofstetter, R. H. 1983. Wetlands in the United States. In A. J. P. Gore, ed. Ecosystems of the World, Vol. 4B: Mires: Swamp, Bog, Fen, and Moor. Elsevier, Amsterdam, pp. 201–244.

Hogg, E. H., and R. W. Wein. 1987. Growth dynamics of floating Typha mats: Seasonal transformation and internal deposition of organic material. Oikos 50:197–205.

Hollands, G. G. 1987. Hydrogeologic classification of wetlands in glaciated regions. In J. Kusler, ed. Wetland Hydrology. Proceedings of a National Wetland Symposium. Association of State Wetland Managers, Berne, NY.

Hollis, J. 1967. Toxicant Diseases of Rice. Bulletin 614, Louisiana Agriculture Experiment Station, Louisiana State University, Baton Rouge.

Holm, G. O., Jr. 1998. Master's Thesis. Louisiana State University, Baton Rouge.

Hook, D. D., W. H. McKee, Jr., H. K. Smith, J. Gregory, V. G. Burrell, M. R. DeVoe, R. E. Sojka, S. Gilbert, R. Banks, L. G. Stolzy, C. Brooks, T. D. Matthews, and T. H. Shear, eds. 1988. The Ecology and Management of Wetlands, Vols. 1 and 2. Croom Helm, London, and Timber Press, Portland OR. 592 pp. and 394 pp.

Hook, D. D., and R. Lea, eds. 1989. Proceedings of the Symposium: The Forested Wetlands of the Southern United States. General Technical Report SE-50, Southeastern Forest Experiment Station, Forest Service, U.S. Department of Agriculture, Asheville, NC. 168 pp.

Hopkinson, C. S. 1985. Shallow water benthic and pelagic metabolism: Evidence for heterotrophy in the nearshore. Marine Biology 87:19–32.

Hopkinson, C. S., Jr., and J. W. Day, Jr. 1977. A model of the Barataria Bay salt marsh ecosystem. In C. A. J. Hall and J. W. Day, Jr., eds. Ecosystem Modeling in Theory and Practice. John Wiley & Sons, New York, pp. 235–265.

Hopkinson, C. S., Jr., J. G. Gosselink, and F. T. Parrondo. 1980. Production of coastal Louisiana marsh plants calculated from phenometric techniques. Ecology 61:1091–1098.

Howarth, R. W., and A. Giblin. 1983. Sulfate reduction in the saltmarshes at Sapelo Island, Georgia. Limnology and Oceanography 28:70–82.

Howes, B. L., R. W. Howarth, J. M. Teal, and I. Valiela. 1981. Oxidation–reduction potentials in a salt marsh: Spatial patterns and interactions with primary production. Limnology and Oceanography 26:350–360.

Howes, B. L., J. W. H. Dacey, and G. M. King. 1984. Carbon flow through oxygen and sulfate reduction pathways in salt marsh sediments. Limnology and Oceanography 29:1037–1051.

Howes, B. L., J. W. Dacey, and J. M. Teal. 1985. Annual carbon mineralization and below-ground production of *Spartina alterniflora* in a New England salt marsh. Ecology 66:595–605.

Hsu, S.-A., M. E. C. Giglioli, P. Reiter, and J. Davies. 1972. Heat and water balance studies on Grand Cayman. Caribbean Journal of Science 12(1–2):9–22.

Hudec, K., and K. Stastny. 1978. Birds in the reedswamp ecosystem. In D. Dykyjová and J. Kvet, eds. Pond Littoral Ecosystems. Springer-Verlag, Berlin, pp. 366–375.

Huenneke, L. F., and R. R. Sharitz. 1986. Microsite abundance and distribution of woody seedlings in a South Carolina cypress–tupelo swamp. American Midland Naturalist 115:328–335.

Huffman, R. T., and R. E. Lonard. 1983. Successional patterns on floating vegetation mats in a southwestern Arkansas bald cypress swamp. Castanea 48:73–78.

Hughes, R. H., and J. S. Hughes. 1992. A Directory of African Wetlands. IUCN, Gland, Switzerland and UNEP, Nairobi, Kenya, 820 pp.

Hunt, R. J., D. P. Krabbenhoft, and M. P. Anderson. 1996. Groundwater inflow measurements in wetland systems. Water Resources Research 32:495–507.

Hupp, C. R., and E. E. Morris. 1990. A dendrogeomorphic approach to measurement of sedimentation in a forested wetland, Black Swamp, Arkansas. Wetlands 10:107–124.

Hupp, C. R., and D. E. Bazemore. 1993. Temporal and spatial patterns of wetland sedimentation, West Tennessee. Journal of Hydrology 141:179–196.

Hutchinson, G. E. 1973. Eutrophication: the scientific background of a contemporary practical problem. American Scientist 61:269–279.

Hyatt, R. A., and G. A. Brook. 1984. Groundwater flow in the Okefenokee Swamp and hydrologic and nutrient budgets for the period August, 1981 through July, 1982. In A. D. Cohen, D. J. Casagrande, M. J. Andrejko, and G. R. Best, eds. The Okefenokee Swamp: Its Natural History, Geology, and Geochemistry. Wetland Surveys, Los Alamos, NM, pp. 229–245.

Immirzi, C. P., E. Maltby, and R. S. Clymo. 1992. The Global Status of Peatlands and Their Role in Carbon Cycling. Wetland Ecosystems Research Group, Department of Geography, University of Exeter. Prepared for Friends of the Earth, London. 145 pp.

Ingram, H. A. P. 1967. Problems of hydrology and plant distribution in mires. Journal of Ecology 55:711–724.

Ingram, H. A. P. 1983. Hydrology. In A. J. P. Gore, ed. Ecosystems of the World, Vol. 4A: Mires: Swamp, Bog, Fen, and Moor. Elsevier, Amsterdam, pp. 67–158.

Ingvorsen, K., and T. D. Brock. 1982. Electron flow via sulfate reduction and methanogenesis in the anaerobic hypolimnion of Lake Mendota. Limnology and Oceanography 27:559–564.

IUCN. 1993. The Wetlands of Central and Eastern Europe. IUCN, Gland, Switzerland, and Cambridge, UK. 83 pp.

Ivanov, K. E. 1981. Water Movement in Mirelands. Translated from Russian by A. Thomson and H. A. P. Ingram. Academic Press, London. 276 pp.

Jacks, G., A. Joelsson, and S. Fleischer. 1994. Nitrogen retention in forested wetlands. Ambio 23:358–362.

Jackson, J. 1989. Man-made wetlands for wastewater treatment: Two case studies. In D. A. Hammer, ed. Constructed Wetlands for Wastewater Treatment. Lewis Publishers, Chelsea, MI, pp. 574–580.

Jackson, M. B. 1985. Ethylene and responses of plants to soil waterlogging and submergence. Annual Review of Plant Physiology 36:145–174.

Jackson, M. B. 1988. Involvement of the hormones ethylene and abscisic acid in some adaptive responses of plants to submergence, soil waterlogging and oxygen shortage. In D. D. Hook, W. H. McKee, Jr., H. K. Smith, J. Gregory, V. G. Burrell, M. R. DeVoe, R. E. Sojka, S. Gilbert, R. Banks, L. G. Stolzy, C. Brooks, T. D. Matthews, and T. H. Shear, eds. The Ecology and Management of Wetlands, Vol. 1: Ecology of Wetlands. Timber Press, Portland, OR, pp. 373–382.

Jackson, M. B. 1990. Hormones and developmental change in plants subjected to submergence or soil waterlogging. Aquatic Botany 38:49–71.

Jackson, M. B. 1993. Are plant hormones involved in root to shoot communication? Advances in Botanical Research 19:103–187.

Jackson, M. B., and C. R. Black, eds. 1993. Interacting Stresses on Plants in a Changing Climate. Springer-Verlag, New York.

Jackson, S. T., R. P. Rutyma, and D. A. Wilcox. 1988. A paleoecological test of a classical hydrosere in the Lake Michigan dunes. Ecology 69:928–936.

Jacobs, T. C., and J. W. Gilliam. 1985. Riparian losses of nitrate from agricultural drainage waters. Journal of Environmental Quality 14:472–478.

James, P. S. B. R. 1999. Shrimp farming development in India—An overview of environmental, socio-economic, legal and other implications. Aquaculture Magazine, online edition. www.aquaculturemag.com.

Janzen, D. H. 1985. Mangroves: Where's the understory? Journal of Tropical Ecology 1:89–92.

Jasinski, S. M. 1999. Peat. In Minerals Yearbook 1999: Volume I—Metals and Minerals. Minerals and Information, U.S. Geological Survey, Reston, VA.

Jiménez, J. A., A. E. Lugo, and G. Cintrón. 1985. Tree mortality in mangrove forests. Biotropica 17:177–185.

Johengen, T. H., and P. A. LaRock. 1993. Quantifying nutrient removal processes within a constructed wetland designed to treat urban stormwater runoff. Ecological Engineering 2:347–366.

Johnson, B. L., W. B. Richardson, and T. J. Naimo. 1995. Past, present, and future concepts in large river ecology. BioScience 45:134–141.

Johnson, C. W. 1985. Bogs of the Northeast. University Press of New England, Hanover, NH. 269 pp.

Johnson, C. W. in press. Dams and riparian forests: Case study from the Upper Missouri River. Rivers.

Johnson, D. C. 1942. The Origin of the Carolina Bays. Columbia University Press, New York. 341 pp.

Johnson, F. L., and D. T. Bell. 1976. Plant biomass and net primary production along a flood-frequency gradient in a streamside forest. Castanea 41:156–165.

Johnson, R. L. 1979. Timber harvests from wetlands. In P. E. Greeson, J. R. Clark, and J. E. Clark, eds. Wetland Functions and Values: The State of Our Understanding. American Water Resources Association, Minneapolis, pp. 598–605.

Johnson, R. R. 1979. The lower Colorado River, a western system. In R. R. Johnson and J. F. McCormick, tech. coords. Strategies for the Protection and Management of Floodplain Wetlands and Other Riparian Ecosystems. Proceedings of the Symposium, Callaway Gardens, Georgia, December 11–13, 1978. General Technical Report WO-12, U.S. Forest Service, Washington, DC, pp. 41–55.

Johnson, R. R., and J. F. McCormick, tech. coords. 1979. Strategies for the Protection and Management of Floodplain Wetlands and Other Riparian Ecosystems. Proceedings of the Symposium, Callaway Gardens, Georgia, December 11–13, 1978. General Technical Report WO-12, U.S. Forest Service, Washington, DC. 410 pp.

Johnson, R. R., and C. H. Lowe. 1985. On the development of riparian ecology. In R. R. Johnson, C. D. Zieball, D. R. Patton, P. F. Ffolliott, and R. H. Hamre, eds. Riparian Ecosystems and Their Management: Reconciling Conflicting Uses. General Technical Report RM-120, Forest Service, U.S. Department of Agriculture, Washington, DC, pp. 112–116.

Johnson, R. R., C. D. Ziebell, D. R. Patton, P. F. Ffolliott, and R. H. Hamre, eds. 1985. Riparian Ecosystems and Their Management: Reconciling

Conflicting Uses. General Technical Report RM-120, Forest Service, U.S. Department of Agriculture, Washington, DC.

Johnson, W. C. 1998. Adjustment of riparian vegetation to river regulation in the Great Plains, USA. Wetlands 18:608–618.

Johnson, W. C., R. L. Burgess, and W. R. Keammerer. 1976. Forest overstory vegetation and environment on the Missouri River floodplain in North Dakota. Ecological Monographs 46:59–84.

Johnson, W. C., T. L. Sharik, R. A. Mayes, and E. P. Smith. 1987. Natural and cause of zonation discreteness around glacial prairie marshes. Canadian Journal of Botany 65(8):1622–1632.

Johnston, C. A. 1991. Sediment and nutrient retention by freshwater wetlands: Effects on surface water quality. Critical Reviews in Environmental Control 21:491–565.

Johnston, C. A. 1994. Ecological engineering of wetlands by beavers. In W. J. Mitsch, ed. Global Wetlands: Old World and New. Elsevier, Amsterdam, pp. 379–384.

Johnston, C. A., G. D. Bubenzer, G. B. Lee, F. W. Madison, and J. R. McHenry. 1984. Nutrient trapping by sediment deposition in a seasonally flooded lakeside wetland. Journal of Environmental Quality 13:283–290.

Johnston, C. A., N. E. Detenbeck, and G. J. Niemi. 1990. The cumulative effect of wetlands on stream water quality and quantity: A landscape approach. Biogeochemistry 10:105–141.

Johnston, C. A., and R. J. Naiman. 1990. Aquatic patch creation in relation to beaver population trends. Ecology 71:1617–1621.

Jones, C. G., J. H. Lawton, and M. Shachak. 1994. Organisms as ecosystem engineers. Oikos 69:373–386.

Jones, C. G., J. H. Lawton, and M. Shachak. 1997. Positive and negative effects of organisms as physical ecosystem engineers. Ecology 78:1946–1957.

Jones, D. A. 1984. Crabs of the mangal ecosystem. In F. D. Por and I. Dor, eds. Hydrobiology of the Mangal. Junk, The Hague, pp. 89–110.

Jones, T. A., comp. 1993a. A Directory of Wetlands of International Importance, Part I: Africa. Ramsar Convention Bureau, Gland, Switzerland. 97 pp.

Jones, T. A., comp. 1993b. A Directory of Wetlands of International Importance, Part II: Asia and Oceania. Ramsar Convention Bureau, Gland, Switzerland. 120 pp.

Jones, T. A., comp. 1993c. A Directory of Wetlands of International Importance, Part III: Europe. Ramsar Convention Bureau, Gland, Switzerland. 375 pp.

Jones, T. A., comp. 1993d. A Directory of Wetlands of International Importance, Part IV: Neotropics and North America. Ramsar Convention Bureau, Gland, Switzerland. 109 pp.

Jordan, T. E., D. L. Correll, W. T. Peterjohn, and D. E. Weller. 1986. Nutrient flux in a landscape: The Rhode River watershed and receiving waters. In D. L. Correll, ed. Watershed Research Perspectives. Smithsonian Institution Press, Washington, DC, pp. 57–76.

Jørgensen, S. E. 1995. Interactions of Watersheds, Lakes, and Riparian Zones. Special Issue of Wetlands Ecology and Management. 3(2):79–137.

Josephson, J. 1992. Status of wetlands. Environmental Science and Technology 26:422.

Josselyn, M. 1983. The Ecology of San Francisco Bay Tidal Marshes: A Community Profile. FWS/OBS-83/23, U.S. Fish and Wildlife Service, Slidell, LA. 102 pp.

Josselyn, M., J. Zedler, and T. Griswold. 1990. Wetland mitigation along the Pacific coast of the United States. In J. A. Kusler and M. E. Kentula eds. Wetland Creation and Restoration. Island Press, Washington, DC, pp. 3–36.

Junk, W. J. 1970. Investigations on the ecology and production biology of the "floating meadows" (*Paspalo echinochloetum*) on the middle Amazon. I. The floating vegetation and its ecology. Amazonia 2:449–495.

Junk, W. J. 1982. Amazonian floodplains: Their ecology, present and potential use. In D. O. Logofet and N. K. Luckyanov, eds. Ecosystem Dynamics in Freshwater Wetlands and Shallow Water Bodies, Vol. I. SCOPE and UNEP Workshop, Center of International Projects, Moscow, pp. 98–126.

Junk, W. J. 1993. Wetlands of tropical South America. In D. F. Whigham, D. Dykyjová, and S. Hejny, eds. Wetlands of the World, I: Inventory, Ecology, and Management. Kluwer Academic Publishers, Dordrecht, pp. 679–739.

Junk, W. J., P. B. Bayley, and R. E. Sparks. 1989. The flood pulse concept in river-floodplain systems. In D. P. Dodge, ed. Proceedings of the International Large River Symposium. Special Issue of Journal of Canadian Fisheries and Aquatic Sciences 106:11–127.

Kaatz, M. R. 1955. The Black Swamp: A study in historical geography. Annals of the Association of American Geographers 35:1–35.

Kadlec, R. H. 1983. The Bellaire wetlands: Wastewater alteration and recovery. Wetlands 3:44–63.

Kadlec, R. H. 1989. Hydrologic factors in wetland water treatment. In D. A. Hammer, ed. Constructed Wetlands for Wastewater Treatment. Lewis Publishers, Chelsea, MI, pp. 21–40.

Kadlec, R. H. 1999. Constructed wetlands for treating landfill leachate. In G. Mulamoottil, E. A. McBean, and F. Rovers, eds. Constructed Wetlands for the Treatment of Landfill Leachates. Lewis Publishers, Boca Raton, FL, pp. 17–31.

Kadlec, R. H., and D. L. Tilton. 1979. The use of freshwater wetlands as a tertiary wastewater treatment alternative. CRC Critical Reviews in Environmental Control 9:185–212.

Kadlec, R. H., R. B. Williams, and R. D. Scheffe. 1988. Wetland evapotranspiration in temperate and arid climates. In D. D. Hook, W. H. McKee, Jr., H. K. Smith, J. Gregory, V. G. Burrell, M. R. DeVoe, R. E. Sojka, S. Gilbert, R. Banks, L. G. Stolzy, C. Brooks, T. D. Matthews, and T. H. Shear, eds. The Ecology and Management of Wetlands, Vol. 1: Ecology of Wetlands. Timber Press, Portland, OR, pp. 146–160.

Kadlec, R. H., and X.-M. Li. 1990. Peatland ice/water quality. Wetlands 10:93–106.

Kadlec, R. H., and D. L. Hey. 1994. Constructed wetlands for river water quality improvement. Water Science and Technology 29:159–168.

Kadlec, R. H., and R. L. Knight. 1996. Treatment Wetlands. CRC Press/Lewis Publishers, Boca Raton, FL. 893 pp.

Kale, H. W. 1965. Ecology and Bioenergetics of the Long-billed Marsh Wren, *Telmatodytes palustris griseus* (Brewster) in Georgia Salt Marshes. Publication 5, Nuttall Ornithological Club, Cambridge, UK.

Kangas, P. C. 1990. An energy theory of landscape for classifying wetlands. In A. E. Lugo, M. Brinson, and S. Brown, eds. Forested Wetlands. Elsevier, Amsterdam, pp. 15–23.

Kangas, P. C., and A. E. Lugo. 1990. The distribution of mangroves and saltmarsh in Florida. Tropical Ecology 31:32–39.

Kantrud, H. A., and R. E. Stewart. 1977. Use of natural basin wetlands by breeding waterfowl in North Dakota. Journal of Wildlife Management 41:243–253.

Kantrud, H. A., J. B. Millar, and A. G. van der Valk. 1989. Vegetation of wetlands of the prairie pothole region. In A. G. van der Valk, ed. Northern Prairie Wetlands. Iowa State University Press, Ames, pp. 132–187.

Kaplan, W., I. Valiela, and J. M. Teal. 1979. Denitrification in a salt marsh ecosystem. Limnology and Oceanography 24:726–734.

Kaswadji, R. F., J. G. Gosselink, and R. E. Turner. 1990. Estimation of primary production using five different methods in a *Spartina alterniflora* salt marsh. Wetlands Ecology and Management 1:57–64.

Kauffman, J. B., and W. C. Kreuger. 1984. Livestock impacts on riparian ecosystems and streamside management implications—A review. Journal of Range Management 37:685–691.

Kawase, M. 1981. Effects of ethylene on aerenchyma development. American Journal of Botany 68:61–65.

Keammerer, W. R., W. C. Johnson, and R. L. Burgess. 1975. Floristic analysis of the Missouri River bottomland forests in North Dakota. Canadian Field Naturalist 89:5–19.

Keddy, P. A. 1983. Freshwater wetland human-induced changes: Indirect effects must also be considered. Environmental Management 7:299–302.

Keddy, P. A. 1992a. Assembly and response rules: Two goals for predictive community ecology. Journal of Vegetation Science 3:157–164.

Keddy, P. A. 1992b. Water level fluctuations and wetland conservation. In J. Kusler and R. Smandon, eds. Wetlands of the Great Lakes. Proceedings of an International Symposium. Association of State Wetland Managers, Berne, NY, pp. 79–91.

Keddy, P. A., and A. A. Reznicek. 1986. Great Lakes vegetation dynamics: The role of fluctuating water levels and buried seeds. Journal of Great Lakes Research 12:25–36.

Kemp, G. P., and J. W. Day, Jr. 1984. Nutrient dynamics in a Louisiana swamp receiving agricultural runoff. In K. C. Ewel and H. T. Odum, eds. Cypress Swamps. University Presses of Florida, Gainesville, pp. 286–293.

Kennedy, H. E., Jr. 1982. Growth and survival of water tupelo coppice regeneration after six growing seasons. Southern Journal of Applied Forestry 6:133–135.

Kent, D. M. 1994. Applied Wetlands Science and Technology. Lewis Publishers, Boca Raton, FL, 436 pp.

Kentula, M. E., J. C. Sifneos, J. W. Good, M. Rylko, and K. Kuntz. 1992a. Trends and patterns in Section 404 permitting requiring compensatory mitigation on Oregon and Washington, USA. Environmental Management 16:109–119.

Kentula, M. E., R. P. Brooks, S. E. Gwin, C. C. Holland, A. D. Sherman, and J. C. Sifness. 1992b. Wetlands: An Approach to Improving Decision Making in Wetland Restoration and Creation. Island Press, Washington, DC. 151 pp.

Keough, J. R., T. A. Thompson, G. R. Guntenspergen, and D. A. Wilcox. 1999. Hydrogeomorphid factors and ecosystem responses in coastal wetlands of the Great Lakes. Wetlands 19:821–834.

Kesel, R., and D. J. Reed. 1995. Status and trends in Mississippi River sediment regime and its role in Louisiana wetland development. In D. J. Reed, ed. Status and Historical Trends of Hydrologic Modification, Reduction in Sediment Availability, and Habitat Loss/Modification in the Barataria–Terrebonne Estuarine System. BT/NEP Publication 20, Barataria–Terrebonne National Estuary Program, Thibodaux, LA, pp. 80–98.

Kessler, E., and M. Jansson, eds. 1994. Wetlands and Lakes as Nitrogen Traps. Special Issue of Ambio 23:319–386.

Kickuth, R. 1970. Ökochemische Leistungen höhere Pflanzen. Naturwissenschaften 57:55–61.

Kickuth, R. 1982. A low-cost process for purification of municipal and industrial waste water. Der Tropenlandwirt 83:141–154.

Kilham, P. 1982. The biogeochemistry of bog ecosystems and the chemical ecology of *Sphagnum*. Michigan Botany 21:159–168.

Killgore, K. J., and J. A. Baker. 1996. Patterns of larval fish abundance in a bottomland hardwood wetland. Wetlands 16:288–295.

King, D. M., and C. Bohlen. 1994. Estimating the costs of restoration. National Wetlands Newsletter 16(3):3–5.

King, D. M., and L. W. Herbert. 1997. The fungibility of wetlands. National Wetlands Newsletter 19:10–13.

King, D. R., and G. S. Hunt. 1967. Effect of carp on vegetation in a Lake Erie marsh. Journal of Wildlife Management 31(1):181–188.

King, S. L., and B. D. Keeland. 1999. Evaluation of reforestation in the Lower Mississippi River alluvial valley. Restoration Ecology 7:348–359.

Kinler, N. W., G. Linscombe, and P. R. Ramsey. 1987. Nutria. In M. Novak, J. A. Balen, M. E. Obbard, and B. Mallocheds, eds. Wild Furbearer Management and Conservation in North America. Ministry of Natural Resources, Ontario, Canada, pp. 326–343.

Kirk, P. W., Jr. 1979. The Great Dismal Swamp. University Press of Virginia, Charlottesville. 427 pp.

Kitchens, W. M., Jr., J. M. Dean, L. H. Stevenson, and J. M. Cooper. 1975. The Santee Swamp as a nutrient sink. In F. G. Howell, J. B. Gentry, and M. H. Smith, eds. Mineral Cycling in Southeastern Ecosystems. ERDA Symposium Series 740513, Government Printing Office, Washington, DC, pp. 349–366.

Kivinen, E., and P. Pakarinen. 1981. Geographical distribution of peat resources and major peatland complex types in the world. Annals Acad. Sciencia Fennicae, Series A, Geology–Geography 132:1–28.

Klarer, D. M., and D. F. Millie. 1989. Amelioration of storm-water quality by a freshwater estuary. Archiv für Hydrobiologie 116:375–389.

Kleiss, B. A., ed. 1996a. Bottomland Hardwoods of the Cache River, Arkansas. Special Issue of Wetlands 16:255–396.

Kleiss, B. 1996b. Sediment retention in a bottomland hardwood wetland in eastern Arkansas. Wetlands 16:321–333.

Klopatek, J. M. 1974. Production of Emergent Macrophytes and Their Role in Mineral Cycling Within a Freshwater Marsh. Master's Thesis, University of Wisconsin, Milwaukee.

Klopatek, J. M. 1978. Nutrient dynamics of freshwater riverine marshes and the role of emergent macrophytes. In R. E. Good, D. F. Whigham, and R. L. Simpson, eds. Freshwater Wetlands: Ecological Processes and Management Potential. Academic Press, New York, pp. 195–216.

Knetsch, J. L., and R. K. Davis. 1966. Comparisons of methods for recreation evaluation. In A. Kneese and S. Smith, eds. Water Research. Johns Hopkins University Press, Baltimore, pp. 125–142.

Knight, A. W., and R. L. Bottorff. 1984. The importance of riparian vegetation to stream ecosystems. In R. F. Warner and K. M. Hendrix, eds. California Riparian Systems: Ecology, Conservation, and Productive Management. University of California Press, Berkeley, pp. 160–167.

Knight, R. L. 1990. Wetland Systems. In Natural Systems for Wastewater Treatment, Manual of Practice FD-16. Water Pollution Control Federation, Alexandria, VA, pp. 211–260.

Knight, R. L., B. H. Winchester, and J. C. Higman. 1984. Carolina Bays—Feasibility for effluent advanced treatment and disposal. Wetlands 4:177–204.

Knight, R. L., T. W. McKim, and H. R. Kohl. 1987. Performance of a natural wetland treatment system for wastewater management. Journal of the Water Pollution Control Federation 59:746–754.

Knopf, F. L., and M. L. Scott. 1990. Altered flows and created landscapes. In J. M. Sweeney, ed. Management of Dynamic Ecosystems. North Central Section, Wildlife Society, West Lafayette, IN, pp. 47–70.

Koch, M. S., and I. A. Mendelssohn. 1989. Sulphide as a soil phytotoxin: Differential responses in two marsh species. Journal of Ecology 77:565–578.

Koch, M. S., I. A. Mendelssohn, and K. L. McKee. 1990. Mechanism for the hydrogen sulfide-induced growth limitation in wetland macrophytes. Limnology and Oceanography 35:399–408.

Koch, M. S., and K. R. Reddy. 1992. Distribution of soil and plant nutrients along a trophic gradient in the Florida Everglades. Soil Science Society of America Journal 56:1492–1499.

Koerselman, W., M. B. Van Kerkhoven, and J. T. A. Verhoeven. 1993. Release of inorganic N, P, and K in peat soils: Effect of temperature, water chemistry, and water level. Biogeochemistry 20:63–81.

Koerselman, W., and A. F. M. Meuleman. 1996. The vegetation N : P ratio: A new tool to detect the nature of nutrient limitation. Journal of Applied Ecology 33:1441–1450.

Kolb, C. R., and J. R. Van Lopik. 1958. Geology of the Mississippi Deltaic Plain—Southeastern Louisiana. U.S. Army Corps of Engineers Waterways Experiment Station, Vicksburg, MS. 482 pp.

Kologiski, R. L. 1977. The Phytosociology of the Green Swamp, North Carolina. Technical Bulletin 20, North Carolina Agricultural Experiment Station, Raleigh. 101 pp.

Koreny, J. S., W. J. Mitsch, E. S. Bair, and X. Wu. 1999. Regional and local hydrology of a constructed riparian wetland system. Wetlands 19:182–193.

Kortekaas, W. M., E. Van der Maarel, and W. G. Beeftink. 1976. A numerical classification of European *Spartina* communities. Vegetatio 33:51–60.

Kovacic, D. A., M. B. David, and L. E. Gentry. 1996. Grassed detention buffer strips for reducing agricultural nonpoint-source pollution from tile drainage systems. In Research on Agricultural Chemicals in Illinois Groundwater: Status and Future Directions VI. Presented at Illinois Groundwater Consortium, Sixth Annual Conference Proceedings, pp. 88–97.

Kovalchik, B. L. 1987. Riparian Zone Associations of the Deschutes, Ochoco, Fremont, and Winema National Forests. Ecology Technical Paper 279-287, Pacific Northwest Region, Region 6, Forest Service, U.S. Department of Agriculture.

Kramer, P. J., W. S. Riley, and T. T. Bannister. 1952. Gas exchange of cypress knees. Ecology 33:117–121.

Kratz, T. K., and C. B. DeWitt. 1986. Internal factors controlling peatland–lake ecosystem development. Ecology 67:100–107.

Kreeger, D. A., and R. I. E. Newell. 2000. Trophic complexity between producers and invertebrate consumers in salt marshes. In M. P. Weinstein and D. A. Kreeger, eds. International Symposium: Concepts and Controversies in Tidal Marsh Ecology. Kluwer Academic Publishers, Dordrecht, pp. 187–220.

Kreitler, C. W. 1977. Faulting and land subsidence from groundwater and hydrocarbon production, Houston/Galveston, Texas. In Proceedings of the Second International Symposium on Land Subsidence. Publication 121, International Association of Hydrological Sciences, pp. 435–446.

Krieger, K., D. M. Klarer, and R. Heath, eds. 1992. Coastal wetlands of the Laurentian Great Lakes: Current knowledge and research needs. Journal of Great Lakes Research 18(4):525–699.

Krivenko, V. G., ed. 1999. Wetlands in Russia. Wetlands International, Wageningen, The Netherlands. 194 pp.

Kroll, R. W., J. F. Gottgens, and B. P. Swartz. 1997. Wild rice to rip-rap: 120 years of habitat changes and management of a Lake Erie coastal marsh. Transactions of the 62nd North American Wildlife and Natural Resources Conference 62:490–500.

Kuenzler, E. J. 1974. Mangrove swamp systems. In H. T. Odum, B. J. Copeland, and E. A. McMahan, eds. Coastal Ecological Systems of the United States, Vol. 1. Conservation Foundation, Washington, DC, pp. 346–371.

Kuenzler, E. J., P. J. Mulholland, L. A. Yarbro, and L. A. Smock. 1980. Distributions and Budgets of Carbon, Phosphorus, Iron and Manganese in a Floodplain Swamp Ecosystem. Report 17, Water Resources Research Institute of North Carolina, Raleigh. 234 pp.

Kuenzler, E. J., and N. J. Craig. 1986. Land use and nutrient yields of the Chowan River watershed. In D. L. Correll, ed. Watershed Research Perspectives. Smithsonian Institution Press, Washington, DC, pp. 57–76.

Kulzynski, S. 1949. Peat bogs of Polesie. Acad. Pol. Sci. Mem., Ser. B, No. 15. 356 pp.

Kurz, H. 1928. Influence of *Sphagnum* and other mosses on bog reactions. Ecology 9:56–69.

Kurz, H., and D. Demaree. 1934. Cypress buttresses in relation to water and air. Ecology 15:36–41.

Kurz, H., and K. A. Wagner. 1953. Factors in cypress dome development. Ecology 34:17–164.

Kushlan, J. A. 1989. Avian use of fluctuating wetlands. In R. R. Sharitz and J. W. Gibbons, eds. Freshwater Wetlands and Wildlife. Department of Energy Symposium Series 61. Office of Scientific and Technical Information, Department of Energy, Oak Ridge, TN, pp. 593–604.

Kushlan, J. A. 1990. Freshwater marshes. In R. L. Myers and J. J. Ewel, eds. Ecosystems of Florida. University of Central Florida Press, Orlando, pp. 324–363.

Kushlan, J. A. 1991. The Everglades. In R. J. Livingston, ed. The Rivers of Florida. Springer-Verlag, New York, pp. 121–142.

Kushner, D. J. 1978. Life in high salt and solute concentrations: Halophilic bacteria. In D. J. Kushner, ed. Microbial Life in Extreme Environments. Academic Press, New York, pp. 318–357.

Kusler, J. A. 1983. Our National Wetland Heritage: A Protection Guidebook. Environmental Law Institute, Washington, DC. 167 pp.

Kusler, J. A., and J. Montanari, eds. 1978. Proceedings of the National Wetland Protection Symposium. FWS/OBS-78/97, U.S. Fish and Wildlife Service, Washington, DC. 55 pp.

Kusler, J., and H. Groman. 1986. Mitigation: An introduction. National Wetlands Newsletter 8:2–3.

Kusler, J. A., and M. E. Kentula, eds. 1990. Wetland Creation and Restoration: The Status of the Science. Island Press, Washington, DC. 594 pp.

Kusler, J., W. J. Mitsch, and J. S. Larson. 1994. Wetlands. Scientific American 270(1):64–70.

Kvet, J., and S. Husak. 1978. Primary data on biomass and production estimates in typical stands of fishpond littoral plant communities. In D. Dykyjová and J. Kvet, eds. Pond Littoral Ecosystems. Springer-Verlag, Berlin, pp. 211–216.

Kwak, T. J. 1988. Lateral movement and the use of floodplain habitat by fishes of the Kankakee River, Illinois. American Midland Naturalist 120:241–249.

Laanbroek, H. J. 1990. Bacterial cycling of minerals that affect plant growth in waterlogged soils: A review. Aquatic Botany 38:109–125.

LaBaugh, J. W. 1986. Wetland ecosystem studies from a hydrologic perspective. Water Resources Bulletin 22:1–10.

LaBaugh, J. W. 1989. Chemical characteristics of water in northern prairie wetlands. In A. van der Valk, ed. Northern Prairie Wetlands. Iowa State University Press, Ames, pp. 56–90.

LaBaugh, J. W., T. C. Winter, G. A. Swanson, D. O. Rosenberry, R. D. Nelson, and N. H. Euliss. 1996. Changes in atmospheric circulation patterns affect midcontinent wetlands sensitive to climate. Limnology and Oceanography 41:864–870.

LaBaugh, J. W., T. C. Winter, and D .O. Rosenberry. 1998. Hydrologic functions of prairie wetlands. Great Plains Research 8:17–37.

Laderman, A. D., ed. 1987. Atlantic White Cedar Wetlands. Westview Press, Boulder, CO. 401 pp.

Laderman, A. D. 1989. The Ecology of the Atlantic White Cedar Wetlands: A Community Profile. Biological Report 85(7.21), U.S. Fish and Wildlife Service, Washington, DC. 114 pp.

Laderman, A. D., ed. 1998. Coastally Restricted Forests. Oxford University Press, Oxford, UK. 334 pp.

Lambers, H., E. Steingrover, and G. Smakman. 1978. The significance of oxygen transport and of metabolic adaptation in flood tolerance of *Senecio* species. Physiologia Plantarum 43:277–281.

Lambert, J. M. 1964. The *Spartina* story. Nature 204:1136–1138.

Lambou, V. W. 1965. The commercial and sport fisheries of the Atchafalaya Basin floodway. Proceedings of the 17th Annual Conference of S.E. Associated Game and Fish Commissioners, pp. 256–281.

Lambou, V. W. 1990. Importance of bottomland forest zones to fishes and fisheries: A case history. In J. G. Gosselink, L. C. Lee, and T. A. Muir, eds. Ecological Processes and Cumulative Impacts: Illustrated by Bottomland Hardwood Wetland Ecosystems. Lewis Publishers, Chelsea, MI, pp. 125–193.

Landin, M., ed. 1993. Wetlands. Proceedings of the 13th Annual Conference of Society of Wetland Scientists. SWS South Central Chapter, Utica, MS. 990 pp.

Larcher, W. 1995. Physiological Plant Ecology, 3rd ed. Springer-Verlag, New York. 522 pp.

Larsen, J. A. 1982. Ecology of the Northern Lowland Bogs and Conifer Forests. Academic Press, New York. 307 pp.

Larson, J. S. 1982. Wetland value assessment—State of the art. In B. Gopal, R. E. Turner. R. G. Wetzel, and D. F. Whigham, eds. Wetlands: Ecology and Management. National Institute of Ecology and International Scientific Publications, Jaipur, India, pp. 417–424.

Larson, J. S., and J. A. Kusler. 1979. Preface. In P. E. Greeson, J. R. Clark, and J. E. Clark, eds. Wetland Functions and Values: The State of Our Understanding. American Water Resources Association, Minneapolis.

Lashof, D. A., and D. R. Ahuja. 1990. Relative contributions of greenhouse gas emissions to global warming. Nature 344:529–531.

Latchford, J. A. 1998. The impacts of runnelling on saltmarsh vegetation and substrate. In A. J. McComb and J. A. Davis, eds. Wetlands for the Future. Contributions from INTECOL's Fifth International Wetlands Conference. Gleneagles Publishing, Adelaide, Australia, pp. 327–334.

Latter, P. M., and G. Howson. 1977. The use of cotton strips to indicate cellulose decomposition in the field. Pedobiologia 17:145–155.

Lavoie, C., and L. Rochefort. 1996. The natural revegetation of a harvested peatland in southern Quebec: A spatial and dentroecological analysis. Ecoscience 3:101–111.

Leck, M. A. 1979. Germination behavior of *Impatiens capensis* Meerb. (Balsaminaceae). Bartonia 46:1–11.

Leck, M. A., and K. J. Graveline. 1979. The seed bank of a freshwater tidal marsh. American Journal of Botany 66:1006–1015.

Leck, M. A., and R. L. Simpson. 1987. Seed bank of a freshwater tidal wetland: Turnover and relationship to vegetation change. American Journal of Botany 74:360–370.

Leck, M. A., and R. L. Simpson. 1995. The-year seed bank and vegetation dynamics of a tidal freshwater marsh. American Journal of Botany 82:1547–1557.

Lee, D. R. 1977. A device for measuring seepage flux in lakes and estuaries. Limnology and Oceanography 22:155–163.

Lee, D. R., and J. A. Cherry. 1979. A field exercise on groundwater flow using seepage meters and mini-piezometers. Journal of Geological Education 27:6–10.

Lee, G. F., E. Bentley, and R. Amundson. 1975. Effect of marshes on water quality. In A. D. Hasler, ed. Coupling of Land and Water Systems. Springer-Verlag, New York, pp. 105–127.

Lee, J. K., R. A. Park, and P. W. Mausel. 1991. Application of geoprocessing and simulation modeling to estimate impacts of sea level rise on northeastern coast of Florida. Photogrammetric Engineering and Remote Sensing 58:1579–1586.

Lee, L. C. 1983. The Floodplain and Wetland Vegetation of Two Pacific Northwest River Ecosystems. Ph.D. Dissertation, University of Washington, Seattle.

Lee, R. 1980. Forest Hydrology. Columbia University Press, New York. 349 pp.

Lee, R. W., D. W. Kraus, and J. E. Doeller. 1999. Oxidation of sulfide by *Spartina alterniflora* roots. Limnology and Oceanography 44:1155–1159.

Lee, S. Y. 1989. Litter production and turnover of the mangrove *Kandelia candel* (L.) Druce in a Hong Kong tidal shrimp pond. Estuarine, Coastal and Shelf Science 29:75–87.

Lee, S. Y. 1990. Primary productivity and particulate organic matter flow in an estuarine mangrove-wetland in Hong Kong. Marine Biology 106:453–463.

Lee, S. Y. 1991. Herbivory as an ecological process in a *Kandelia candel* (Rhizophoraceae) mangal in Hong Kong. Journal of Tropical Ecology 7:337–348.

Lee, S. Y. 1995. Mangrove outwelling: A review. Hydrobiologia 295:203–212.

Lefeuvre, J. C., ed. 1989. Conservation and Development: The Sustainable Use of Wetland Resources. Published Abstracts of the Third INTECOL Wetlands Conference. Rennes, France, September 1988. Laboratoire d'Evolution des Systems Naturels et Modifies, Paris.

Lefeuvre, J. C. 1990. Ecological impact of sea level rise on coastal ecosystems of Mont-Saint-Michel Bay. In J. J. Beukema, W. J. Wolff, & J. W. M. Brouns, eds. Expected Effects of Climatic Change on Marine Coastal Ecosystems. Kluwer Academic Publishers, Dordrecht, The Netherlands. pp. 139–153.

Lefeuvre, J. C., and R. F. Dame. 1994. Comparative studies of salt marsh processes in the New and Old Worlds: An introduction. In W. J. Mitsch, ed. Global Wetlands: Old World and New. Elsevier, Amsterdam, pp. 169–179.

Leitch, J. A. 1986. Economics of prairie wetland values. North Dakota Academy of Science 40:44.

Leitch, J. A. 1989. Politicoeconomic overview of prairie potholes. In A. G. van der Valk, ed. Northern Prairie Wetlands. Iowa State University Press, Ames, pp. 3–14.

Leitch, J. A., and L. E. Danielson. 1979. Social, Economic, and Institutional Incentives to Drain or Preserve Prairie Wetlands. Department of Agricultural and Applied Economics, University of Minnesota, St. Paul. 78 pp.

Leitch, J. A., and B. L. Ekstrom. 1989. Wetland Economics and Assessment: An Annotated Bibliography. Garland Publishing, New York.

Lemlich, S. K., and K. C. Ewel. 1984. Effects of wastewater disposal on growth rates of cypress trees. Journal of Environmental Quality 13:602–604.

Leonardson, L., L. Bengtsson, T. Davidsson, T. Persson, and U. Emanuelsson. 1994. Nitrogen retention in artificially flooded meadows. Ambio 23:332–341.

Leopold, L. B., M. G. Wolman, and J. E. Miller. 1964. Fluvial Processes in Geomorphology. W. H. Freeman, San Francisco. 522 pp.

Levitt, J. 1980. Responses of Plants to Environmental Stresses, Vol. II: Water, Radiation, Salt, and Other Stresses. Academic Press, New York. 607 pp.

Lewis, R. R.1990a. Wetland restoration/creation/enhancement terminology: Suggestions for standardization. In J. A. Kusler and M. E. Kentula, eds. Wetland Creation and Restoration. Island Press, Washington, DC, pp. 1–7.

Lewis, R. R. 1990b. Creation and restoration of coastal plain wetlands in Florida. In J. A. Kusler and M. E. Kentula, eds. Wetland Creation and Restoration. Island Press, Washington, DC, pp. 73–101.

Lewis, R. R. 1990c. Creation and restoration of coastal plain wetlands in Puerto Rico and the U.S. Virgin Islands. In J. A. Kusler and M. E. Kentula, eds. Wetland Creation and Restoration. Island Press, Washington, DC, pp. 103–123.

Lewis, R. R., J. A. Kusler, and K. L. Erwin. 1995. Lessons learned from five decades of wetland restoration and creation in North America. In C. Montes, G. Oliver, F. Molina, and J. Cobos, eds. Bases Ecológicas para la Restauracion de Humedales en la Cuenca Mediterránea. Consejeria de Medio Ambiente, Junta de Andalucía, Andalucía, Spain.

Lide, R. F., V. G. Meentemeyer, J. E. Pinder, and L. M. Beatty 1995. Hydrology of a Carolina Bay located on the Upper Coastal Plain of western South Carolina. Wetlands 15:47–57.

Lieth, H. 1975. Primary production of the major units of the world. In H. Lieth and R. H. Whitaker, eds. Primary Productivity of the Biosphere. Springer-Verlag, New York, pp. 203–215.

Light, S. S., and J. W. Dineen. 1994. Water control in the Everglades: A historical perspective. In S. M . Davis and J. C. Ogden, eds. Everglades: The Ecosystem and Its Restoration. St. Lucie Press, Delray Beach, FL, pp. 47–84.

Likens, G. E., F. H. Bormann, R. S. Pierce, and J. S. Eaton. 1985. The Hubbard Brook Valley. In G. E. Likens, ed. An Ecosystem Approach to Aquatic Ecology: Mirror Lake and Its Environment. Springer-Verlag, New York, pp. 9–39.

Linacre, E. 1976. Swamps. In J. L. Monteith, ed. Vegetation and the Atmosphere, Vol. 2: Case Studies. Academic Press, London, pp. 329–347.

Lindeman, R. L. 1941. The developmental history of Cedar Creek Lake, Minnesota. American Midland Naturalist 25:101–112.

Lindeman, R. L. 1942. The trophic-dynamic aspect of ecology. Ecology 23:399–418.

Linsley, R. K., and J. B. Franzini. 1979. Water Resources Engineering, 3rd ed. McGraw-Hill, New York. 716 pp.

Linthurst, R. A., and R. J. Reimold. 1978. An evaluation of methods for estimating the net aerial primary productivity of estuarine angiosperms. Journal of Applied Ecology 15:919–931.

Lippson, M. A. J., M. S. Haire, A. F. Holland, F. Jacobs, J. Jensen, R. L. Moran-Johnson, T. T. Polgar, and W. A. Richkus. 1979. Environmental Atlas of the Potomac Estuary. Williams and Heintz Map Corporation, Washington, DC.

Lipschultz, F. 1981. Methane release from a brackish intertidal salt-marsh embayment of Chesapeake Bay, Maryland. Estuaries 4:143–145.

Litchfield, D. K., and D. D. Schatz. 1989. Constructed wetlands for wastewater treatment at Amoco Oil Company's Mandan, North Dakota, refinery. In D. A. Hammer, ed. Constructed Wetlands for Wastewater Treatment. Lewis Publishers, Chelsea, MI, pp. 101–119.

Little, E. L., Jr. 1971. Atlas of United States Trees, Vol. 1: Conifers and Important Hardwoods. Miscellaneous Publication 1146, Forest Service, U.S. Department of Agriculture. Government Printing Office, Washington, DC.

Littlehales, B., and W. A. Niering. 1991. Wetlands of North America. Thomasson-Grant, Charlottesville, VA. 160 pp.

Livingston, D. A. 1963. Chemical composition of rivers and lakes. Professional Paper 440G, U.S. Geological Survey, Washington, DC. 64 pp.

Livingston, E. H. 1989. Use of wetlands for urban stormwater management. In D. A. Hammer, ed. Constructed Wetlands for Wastewater Treatment. Lewis Publishers, Chelsea, MI, pp. 253–264.

Lockaby, B. G., and M. R. Walbridge. 1998. Biogeochemistry. In M. G. Messina and W. H. Conner, eds. Southern Forested Wetlands: Ecology and Management. Lewis Publishers, Boca Raton, FL, pp. 149–172.

Logofet, D. O., and N. K. Luckyanov, eds. 1982. Ecosystem Dynamics in Freshwater Wetlands and Shallow Water Bodies. Proceedings Workshop Minsk, Pinsk, and Tskhaltoubo, USSR, July 12–26, 1981, Vols. I and II. Scientific Committee on Problems of the Environment and United Nations Environment Program, Center of International Projects, Moscow. 312 pp. and 424 pp.

Logofet, D. O., and G. A. Alexandrov. 1984. Modelling of matter cycle in a mesotrophic bog ecosystem. 1. Linear analysis of carbon environs. Ecological Modelling 21:247–258.

Logofet, D. O., and G. A. Alexandrov. 1988. Interference between mosses and trees in the framework of a dynamic model of carbon and nitrogen cycling in a mesotrophic bog ecosystem. In W. J. Mitsch, M. Straskraba, and S. E. Jørgensen, eds. Wetland Modelling. Elsevier, Amsterdam, pp. 55–66.

Loucks, O. L. 1990. Restoration of the pulse control function of wetlands and its relationship to water quality objectives. In J. A. Kusler and M. E. Kentula, eds. Wetland Creation and Restoration. Island Press, Washington, DC, pp. 467–477.

Louisiana Coastal Wetlands Conservation and Restoration Task Force (LCWCR) and the Wetlands Conservation and Restoration Authority. 1998. Coast 2050:

Toward a Sustainable Coastal Louisiana, an Executive Summary. Louisiana Department of Natural Resources, Baton Rouge. 12 pp.

Lowrance, R., R. Todd, J. Fail, Jr., O. Hendrickson, Jr., R. Leonard, and L. Asmussen. 1984. Riparian forests as nutrient filters in agricultural watersheds. BioScience 34:374–377.

Lowrance, R., R. Leonard, and J. Sheridan. 1985. Managing riparian ecosystems to control nonpoint pollution. Journal of Soil and Water Conservation 40:87–97.

Lowrance, R., L. Altier, J. D. Newbold, R. Schnabel, P. M. Groffman, J. M. Denver, D. L. Correll, J. W. Gilliam, J. M. Robinson, R. B. Brinsfield, K. W. Staver, W. Lucas, and A. H. Todd. 1997. Water quality functions of riparian forest buffers in Chesapeake Bay watersheds. Environmental Management 21:687–712.

Lu, J., ed. 1990. Wetlands in China. East China Normal University Press, Shanghai. 177 pp. [in Chinese]

Lu, J. 1995. Ecological significance and classification of Chinese wetlands. Vegetatio 118:49–56.

Lugo, A. E. 1980. Mangrove ecosystems: Successional or steady state? Biotropica (Supplement) 12:65–72.

Lugo, A. E. 1981. The island mangroves of Inagua. Journal of Natural History 15:845–852.

Lugo, A. E. 1986. Mangrove understory: An expensive luxury? Journal of Tropical Ecology 2:287–288.

Lugo, A. E. 1988. The mangroves of Puerto Rico are in trouble. Acta Cientifica 2:124.

Lugo, A. E. 1990a. Mangroves of the Pacific Islands: Research Opportunities. General Technical Report PSW-118, Pacific Southwest Forest and Range Experiment Station, Forest Service, U.S. Department of Agriculture, Berkeley, CA. 13 pp.

Lugo, A. E. 1990b. Fringe wetlands. In A. E. Lugo, M. Brinson, and S. Brown, eds. Forested Wetlands: Ecosystems of the World 15. Elsevier, Amsterdam, pp. 143–169.

Lugo, A. E., and S. C. Snedaker. 1974. The ecology of mangroves. Annual Review of Ecology and Systematics 5:39–64.

Lugo, A. E., G. Evink, M. M. Brinson, A. Broce, and J. C. Snedaker. 1975. Diurnal rates of photosynthesis, respiration, and transpiration in mangrove forests in South Florida. In F. B. Golley and E. Medina, eds. Tropical Ecological Systems—Trends in Terrestrial and Aquatic Research. Springer-Verlag, New York, pp. 335–350.

Lugo, A. E., M. Sell, and S. C. Snedaker. 1976. Mangrove ecosystem analysis. In B. C. Patten, ed. Systems Analysis and Simulation in Ecology, Vol. IV. Academic Press, New York, pp. 113–145.

Lugo, A. E., and C. Patterson-Zucca. 1977. The impact of low temperature stress on mangrove structure and growth. Tropical Ecology 18:149–161.

Lugo, A. E., G. Cintrón, and C. Goenaga. 1981. Mangrove ecosystems under stress. In G. W. Barrett and R. Rosenberg, eds. Stress Effects on Natural Ecosystems. John Wiley & Sons, New York, pp. 129–153.

Lugo, A. E., M. M. Brinson, and S. L. Brown, eds. 1990a. Forested Wetlands: Ecosystems of the World 15. Elsevier, Amsterdam. 527 pp.

Lugo, A. E., S. Brown, and M. M. Brinson. 1990b. Concepts in wetland ecology. In A. E. Lugo, M. M. Brinson, and S. Brown, eds. Forested Wetlands: Ecosystems of the World 15. Elsevier, Amsterdam, pp. 53–85.

Lunz, J. D., T. W. Zweigler, R. T. Huffman, R. J. Diaz, E. J. Clairain, and L. J. Hunt. 1978. Habitat Development Field Investigations Windmill Point Marsh Development Site, James River Virginia, Summary Report. Technical Report D-79-23, U.S. Army Corps of Engineers Waterways Experiment Station, Vicksburg, MS. 116 pp.

Lynn, G. D., P. D. Conroy, and F. J. Prochaska. 1981. Economic valuation of marsh areas for marine production processes (Florida). Journal of Environmental Economic Management 8:175–186.

Lyon, J. G. 1994. Practical Handbook for Wetland Identification and Delineation. Lewis Publishers, Boca Raton, FL. 157 pp.

Ma, S., and J. Yan. 1989. Ecological engineering for treatment and utilization of wastewater. In W. J. Mitsch and S. E. Jørgensen, eds. Ecological Engineering: An Introduction to Ecotechnology. John Wiley & Sons, New York, pp. 185–218.

Ma, X., X. Liu, and R. Wang. 1993. China's wetlands and agro-ecological engineering. Ecological Engineering 2:291–330.

MacDonald, K. B. 1977. Plant and animal communities of Pacific North American salt marshes. In B. J. Chapman, ed. Ecosystems of the World, Vol. 1: Wet Coastal Ecosystems. Elsevier, Amsterdam, pp. 167–191.

MacIntyre, H. L., and J. J. Cullen. 1995. Fine-scale vertical resolution of chlorophyll and photosynthetic parameters in shallow-water benthos. Marine Ecology Progress Series 122:227–237.

MacIntyre, H. L., R. J. Geider, and D. C. Miller. 1996. Microphytobenthos: The ecological role of the "secret garden" of unvegetated, shallow-water marine habitats. I. Distribution, abundance and primary production. Estuaries 19:186–201.

MacMannon, M., and R. M. M. Crawford. 1971. A metabolic theory of flooding tolerance: The significance of enzyme distribution and behavior. New Phytologist 10:299–306.

Macnae, W. 1963. Mangrove swamps in South Africa. Journal of Ecology 51:1–25.

Maguire, C., and P. O. S. Boaden. 1975. Energy and evolution in the thiobios: An extrapolation from the marine gastrotrich *Thiodasys sterreri*. Cahiers de Biol. Mar. 16:635–646.

Mahall, B. E., and R. B. Park. 1976a. The ecotone between *Spartina foliosa* Trin. and *Salicornia virginica* L. in salt marshes of northern San Francisco Bay. I. Biomass and production. Journal of Ecology 64:421–433.

Mahall, B. E., and R. B. Park. 1976b. The ecotone between *Spartina foliosa* Trin. and *Salicornia virginica* L. in salt marshes of northern San Francisco Bay. II. Soil water and salinity. Journal of Ecology 64:793–809.

Mahall, B. E., and R. B. Park. 1976c. The ecotone between *Spartina foliosa* Trin. and *Salicornia virginica* L. in salt marshes of northern San Francisco Bay. III. Soil aeration and tidal immersion. Journal of Ecology 64:811–819.

Mahoney, J. M., and S. B. Rood. 1998. Streamflow requirements for cottonwood seedling recruitment—An integrative model. Wetlands 18:634–645.

Malakoff, D. 1998. Restored wetlands flunk real-world test. Science 280:371–372.

Malley, D. F. 1978. Degradation of mangrove leaf litter by the tropical sesarmid crab *Chiromanthes onychophorum*. Marine Biology 49:377–386.

Malmer, N. 1975. Development of bog mires. In A. D. Hasler, ed. Coupling of Land and Water Systems. Ecology Studies 10. Springer-Verlag, New York, pp. 85–92.

Malone, M., and I. Ridge. 1983. Ethylene-induced growth and proton excretions in the aquatic plant *Nymphoides peltata*. Planta 157:71–73.

Maltby, E. 1986. Waterlogged Wealth: Why Waste the World's Best Wet Places? Earthscan Publications, Washington, DC. 200 pp.

Maltby, E., and R. E. Turner. 1983. Wetlands of the world. Geographic Magazine 55:12–17.

Maltby, E., P. J. Dugan, and J. C. Lefeuvre, eds. 1992. Conservation and Development: The Sustainable Use of Wetland Resources. IUCN, Gland, Switzerland. 219 pp.

Manning, E., M. Bardecki, and W. Bond. 1990. Measuring the value of renewable resources: The case of wetlands. In The Common Property Conference, Duke University, Durham, NC, September 26–28, 1990.

Mantel, L. H. 1968. The foregut of *Gecarcinus lateralis* as an organ of salt and water balance. American Zoologist 8:433–442.

Marble, A. D. 1992. A Guide to Wetland Functional Design. Lewis Publishers, Chelsea, MI, 222 pp.

Marois, K. C., and K. C. Ewel. 1983. Natural and management-related variation in cypress domes. Forest Science 29:627–640.

Martin, A. C., N. Hutchkiss, F. M. Uhler, and W. S. Bourn. 1953. Classification of Wetlands of the United States. Special Science Report—Wildlife 20, U.S. Fish and Wildlife Service, Washington, DC. 14 pp.

Martini, I. P., D. W. Cowell, and G. M. Wickware. 1980. Geomorphology of southwestern James Bay: A low energy, emergent coast. In S. B. McCann, ed. The Coastline of Canada. Geological Survey of Canada, pp. 293–301.

Mason, C. F., and R. J. Bryant. 1975. Production, nutrient content and decomposition of *Phragmites communis* Trin. and *Typha angustifolia* L. Ecology 63:71–95.

Matthews, E. 1990. Global distribution of forested wetlands. In A. E. Lugo, M. Brinson, and S. Brown, eds. Addendum to Forested Wetlands. Elsevier, Amsterdam.

Matthews, E., and I. Fung. 1987. Methane emissions from natural wetlands: Global distribution, area, and environmental characteristics of sources. Global Biogeochemical Cycles 1:61–86.

Matthews, E., I. Fung, and J. Lerner. 1991. Methane emission from rice cultivation: Geographic and seasonal distribution of cultivated areas and emissions. Global Biogeochemical Cycles 5:3–24.

Matthews, G. V. T. 1993. The Ramsar Convention on Wetlands: Its History and Development. Ramsar Convention, Gland, Switzerland, 120 pp.

Mattoon, W. R. 1915. The Southern Cypress. Bulletin 272, U.S. Department of Agriculture, Washington, DC.

Mazda, Y., Y. Sato, S. Sawamoto, H. Yokochi, and E. Wolanski. 1990. Links between physical, chemical and biological processes in Bashita-minato, a mangrove swamp in Japan. Estuarine, Coastal and Shelf Science 31:817–833.

McArthur, B. H. 1989. The use of isolated wetlands in Florida for stormwater treatment. In D. W. Fisk, ed., Wetlands Concerns and Successes. Proceedings of the Conference, Tampa, FL. American Water Resources Association, Bethesda, MD, pp. 185–194.

McArthur, J. V. 1989. Aquatic and terrestrial linkages: Floodplain functions. In D. D. Hook and R. Lea, eds. Proceedings of the Symposium: The Forested Wetlands of the Southern United States. Southeastern Forest Experiment Station, Forest Service, U.S. Department of Agriculture, Asheville, NC, pp. 107–116.

McArthur, J. V., and G. R. Marzolf. 1987. Changes in soluble nutrients of prairie riparian vegetation during decomposition on a floodplain. American Midland Naturalist 117:26–34.

McCaffrey, R. J. 1977. A Record of the Accumulation of Sediment and Trace Metals in a Connecticut, U.S.A. Salt Marsh. Ph.D. Dissertation, Yale University, New Haven, CT. 156 pp.

McComb, A. J., and P. S. Lake. 1990. Australian Wetlands. Angus and Robertson, London. 258 pp.

McComb, A. J., and J. A. Davis, eds. 1998. Wetlands for the Future. Contributions from INTECOL's Fifth International Wetlands Conference. Gleneagles Publishing, Adelaide, Australia. 780 pp.

McCoy, M. B., and J. M. Rodriguez. 1994. Cattail (*Typha domingensis*) eradication methods in the restoration of the tropical seasonal freshwater marsh. In W. J. Mitsch, ed. Global Wetlands: Old World and New. Elsevier, Amsterdam, pp. 469–482.

McIntosh, R. P. 1980. The background and some current problems of theoretical ecology. Synthese 43:195–255.

McIntosh, R. P. 1985. The Background of Ecology, Concept and Theory. Cambridge University Press, Cambridge, UK. 383 pp.

McJannet, C. L., P. A. Keddy, and F. R. Pick. 1995. Nitrogen and phosphorus tissue concentrations in 41 wetland plants: A comparison across habitats and functional groups. Functional Ecology 9:231–238.

McKee, K. L., I. A. Mendelssohn, and M. W. Hester. 1988. Reexamination of pore water sulfide concentrations and redox potentials near the aerial roots of *Rhizophora mangle* and *Avicennia germinans*. American Journal of Botany 75:1352–1359.

McKee, K. L., and I. A. Mendelssohn. 1989. Response of a freshwater marsh plant community to increased salinity and increased water level. Aquatic Botany 34(4):301–316.

McKinley, C. E., and F. E. Day, Jr. 1979. Herbaceous production in cut-burned, uncut-burned, and control areas of a *Chamaecyparis thyoides* (L.) BSP (Cupressaceae) stand in the Great Dismal Swamp. Torrey Botanical Club Bulletin 106:20–28.

McKnight, D., E. Thurman, R. Wershaw, and H. Hemond. 1985. Biogeochemistry of aquatic humic substances in Thoreau's Bog, Concord, Mass. Ecology 66:1339–1352.

McKnight, J. S., D. D. Hook, O. G. Langdon, and R. L. Johnson. 1981. Flood tolerance and related characteristics of trees of the bottomland forests of the southern United States. In J. R. Clark and J. Benforado, eds. Wetlands of Bottomland Hardwood Forests. Elsevier, Amsterdam, pp. 29–69.

McLaughlin, D. B., and H. J. Harris. 1990. Aquatic insect emergence in two Great Lakes marshes. Wetlands Ecology and Management 1:111–121.

McLeod, K. W., L. S. Donovan, and N. J. Stumpff. 1988. Responses of woody seedlings to elevated flood water temperatures. In D. D. Hook et al., eds. The Ecology and Management of Wetlands, Vol. 1: Ecology of Wetlands. Timber Press, Portland, OR, pp. 441–451.

McNaughton, S. J. 1966. Ecotype function in the *Typha* community-type. Ecological Monographs 36:297–325.

McNaughton, S. J. 1968. Autotoxic feedback in relation to germination and seedling growth in *Typha latifolia*. Ecology 49:367–369.

Meade, R. 1992. Some early changes following the rewetting of a vegetated cutover peatland surface at Danes Moss, Cheshire, UK, and their relevance to conservation management. Biological Conservation 61:31–34.

Meade, R., and R. Parker. 1985. Sediment in Rivers of the United States. In National Water Summary. Water Supply Paper 2275, U.S. Geological Survey, Washington, DC, pp. 49–60.

Means, J. L., and R. E. Hinchee. 2000. Wetlands and Remediation. Battelle Press, Columbus. 445 pp.

Medina, E., E. Cuevas, M. Popp, and A. E. Lugo. 1990. Soil salinity, sun exposure, and growth of *Acrostichum aureum*, the mangrove fern. Botanical Gazette 151:41–49.

Meeks, G., and L. C. Runyon. 1990. Wetlands Protection and the States. National Conference of State Legislatures, Denver. 26 pp.

Megonigal, J. P., and F. P. Day, Jr. 1988. Organic matter dynamics in four seasonally flooded forest communities of the Dismal Swamp. American Journal of Botany 75:1334–1343.

Megonigal, J. P., W. H. Conner, S. Kroeger, and R. R. Sharitz. 1997. Aboveground production in southeastern floodplain forests: A test of the subsidy–stress hypothesis. Ecology 78:370–384.

Meijer, L. E., and Y. Avnimelech. 1999. On the use of micro-electrodes in fish pond sediments. Aquaculture Engineering 21:71–83.

Mendelssohn, I. A. 1979. Nitrogen metabolism in the height forms of *Spartina alterniflora* in North Carolina. Ecology 60:574–584.

Mendelssohn, I. A., K. L. McKee, and M. L. Postek. 1982. Sublethal stresses controlling *Spartina alterniflora* productivity. In B. Gopal, R. E. Turner, R. G. Wetzel, and D. F. Whigham, eds. Wetlands: Ecology and Management. National Institute of Ecology and International Science Publications, Jaipur, India, pp. 223–242.

Mendelssohn, I. A., and M. L. Postek. 1982. Elemental analysis of deposits on the roots of *Spartina alterniflora* Loisel. American Journal of Botany 69:904–912.

Mendelssohn, I. A., and D. M. Burdick. 1988. The relationship of soil parameters and root metabolism to primary production in periodically inundated soils. In D. D. Hook, W. H. McKee, Jr., H. K. Smith, J. Gregory, V. G. Burrell, M. R.

DeVoe, R. E. Sojka, S. Gilbert, R. Banks, L. G. Stolzy, C. Brooks, T. D. Matthews, and T. H. Shear, eds. The Ecology and Management of Wetlands, Vol. 1: Ecology of Wetlands. Timber Press, Portland, OR, pp. 398–428.

Mendelssohn, I. A., and J. T. Morris. 2000. Eco-physiological controls on the productivity of *Spartina alterniflora* Loisel. In M. P. Weinstein and D. A. Kreeger, eds. International Symposium: Concepts and Controversies in Tidal Marsh Ecology. Kluwer Academic Publishers, Dordrecht, The Netherlands, pp. 59–80.

Mesléard, P., Grillas, L. T. Ham. 1995. Restoration of seasonally flooded marshes in abandoned ricefields in the Camargue (southern France)—Preliminary results on vegetation and use by ducks. Ecological Engineering 5:95–106.

Messina, M. G., and W. H. Conner, eds. 1998. Southern Forested Wetlands. Lewis Publishers, Boca Raton, FL. 616 pp.

Metzker, K. D., and W. J. Mitsch. 1997. Modelling self-design of the aquatic community in a newly created freshwater wetland. Ecological Modelling 100:61–86.

Metzler, K., and R. Rosza. 1982. Vegetation of fresh and brackish tidal marshes in Connecticut. Connecticut Botanical Society Newsletter 10:2–4.

Mevi-Schütz, J., and W. Grosse. 1988. A two-way gas transport system in *Nelumbo mucifera*. Plant Cell Environment 11:27–34.

Meyer, A. H. 1935. The Kankakee "Marsh" of northern Indiana and Illinois. Michigan Academy of Science, Arts, and Letters Papers 21:359–396.

Meyer, J. L. 1985. A detention basin/artificial wetland treatment system to renovate stormwater runoff from urban, highway, and industrial areas. Wetlands 5:135–145.

Micheli, F., F. Gherardi, and M. Vannini. 1991. Feeding and burrowing ecology of two East African mangrove crabs. Marine Biology 111:247–254.

Middleton, B. A. 1999. Wetland Restoration: Flood Pulsing and Disturbance Dynamics. John Wiley & Sons, New York. 388 pp.

Millar, J. B. 1971. Shoreline-area as a factor in rate of water loss from small sloughs. Journal of Hydrology 14:259–284.

Miller, D. C., R. J. Geider, and H. L. MacIntyre. 1996. Microphytobenthos: The ecological role of the "secret garden" of unvegetated, shallow-water marine habitats. II. Role in sediment stability and shallow-water food webs. Estuaries 19:202–212.

Miller, R. S., D. B. Botkin, and R. Mendelssohn. 1974. The whooping crane (*Grus americana*) population of North America. Biological Conservation 6:106–111.

Minnesota Department of Natural Resources. 1984. Recommendations for the Protection of Ecologically Significant Peatlands in Minnesota. Minnesota Department of Natural Resources, Minneapolis, MN. 55 pp.

Minshall, G. W., R. C. Peterson, K. W. Cummins, T. L. Bott, J. R. Sedall, C. E. Cushing, and R. L. Vannote. 1983. Interbiome comparison of stream ecosystem dynamics. Ecological Monographs 53:1–25.

Minshall, G. W., K. W. Cummins, R. C. Peterson, C. E. Cushing, D. A. Bruins, J. R. Sedall, and R. L. Vannote. 1985. Development in stream ecosystem theory. Canadian Journal of Fisheries and Aquatic Sciences 42:1045–1055.

Minshall, G. W., W. S. E. Jensen, and W. S. Platts. 1989. The Ecology of Stream and Riparian Habitats of the Great Basin Region: A Community Profile. Biological Report 85(7.24), U.S. Fish and Wildlife Service, Washington, DC. 142 pp.

Mitchell, D. S., and B. Gopal. 1991. Invasion of tropical freshwaters by alien aquatic plants. In P. S. Ramakrishnan, ed. Ecology of Biological Invasion in the Tropics, pp. 139–154.

Mitchell, D. S., A. J. Chick, and G. W. Rasin. 1995. The use of wetlands for water pollution control in Australia: An ecological perspective. Water Science and Technology 32:365–373.

Mitchell, J. G., R. Gehman, and J. Richardson. 1992. Our disappearing wetlands. National Geographic 182(4):3–45.

Mitchell, R. C., and R. T. Carson. 1989. Using Surveys to Value Public Goods: The Contingent Valuation Method. Resources for the Future, Washington, DC.

Mitsch, W. J. 1977a. Water hyacinth (*Eichhornia crassipes*) nutrient uptake and metabolism in a north-central Florida marsh. Archiv für Hydrobiologie 81:188–210.

Mitsch, W. J. 1977b. Energy conservation through interface ecosystems. In Proceedings of the International Conference on Energy Use Management. Pergamon Press, Oxford, UK, pp. 875–881.

Mitsch, W. J. 1979. Interactions between a riparian swamp and a river in southern Illinois. In R. R. Johnson and J. F. McCormick, tech. coords. Strategies for the Protection and Management of Floodplain Wetlands and Other Riparian Ecosystems. Proceedings of the Symposium, Calaway Gardens, Georgia, December 11–13, 1978. General Technical Report WO-12, U.S. Forest Service, Washington, DC, pp. 63–72.

Mitsch, W. J. 1983. Ecological models for management of freshwater wetlands. In S. E. Jørgensen and W. J. Mitsch, eds. Application of Ecological Modeling in Environmental Management, Part B. Elsevier, Amsterdam, pp. 283–310.

Mitsch, W. J. 1984. Seasonal patterns of a cypress dome pond in Florida. In K. C. Ewel and H. T. Odum, eds. Cypress Swamps. University Presses of Florida, Gainesville, pp. 25–33.

Mitsch, W. J. 1988. Productivity–hydrology–nutrient models of forested wetlands. In W. J. Mitsch, M. Straskraba, and S. E. Jørgensen, eds. Wetland Modelling. Elsevier, Amsterdam, pp. 115–132.

Mitsch, W. J., ed. 1989. Wetlands of Ohio's Coastal Lake Erie: A Hierarchy of Systems. NTIS, OHSU-BS-007, Ohio Sea Grant Program, Columbus. 186 pp.

Mitsch, W. J. 1991a. Ecological engineering: Approaches to sustainability and biodiversity in the U.S. and China. In R. Costanza, ed. Ecological Economics: The Science and Management of Sustainability. Columbia University Press, New York, pp. 428–448.

Mitsch, W. J. 1991b. Ecological engineering: The roots and rationale of a new ecological paradigm. In C. Etnier and B. Guterstam, eds. Ecological Engineering for Wastewater Treatment. Bokskogen, Godlenburg, Sweden, pp. 19–37.

Mitsch, W. J. 1992a. Landscape design and the role of created, restored, and natural riparian wetlands in controlling nonpoint source pollution. Ecological Engineering 1:27–47.

Mitsch, W. J. 1992b. Combining ecosystem and landscape approaches to Great Lakes wetlands. Journal of Great Lakes Research 18:552–570.

Mitsch, W. J. 1993. Ecological engineering—A cooperative role with the planetary life-support systems. Environmental Science & Technology 27:438–445.

Mitsch, W. J., ed. 1994. Global Wetlands: Old World and New. Elsevier, Amsterdam. 967 pp.

Mitsch, W. J. 1996. Managing the world's wetlands—Preserving and enhancing their ecological functions. Verhandlungen Internationale Vereinigung für Limnologie 26:139–147.

Mitsch, W. J. 1998a. Ecological engineering—The seven-year itch. Ecological Engineering 10:119–138.

Mitsch, W. J. 1998b. Protecting the world's wetlands: Threats and opportunities in the 21st century. In A. J. McComb and J. A. Davis, eds. Wetlands for the Future. Proceedings of INTECOL's Fifth International Wetlands Conference. Gleneagles Publishing, Adelaide, Australia, pp. 19–31.

Mitsch, W. J. 1998c. Self-design and wetland creation: Early results of a freshwater marsh experiment. In A. J. McComb and J. A. Davis, eds. Wetlands for the Future. Contributions from INTECOL's Fifth International Wetlands Conference. Gleneagles Publishing, Adelaide, Australia, pp. 635–655.

Mitsch, W. J. 1999. Hypoxia solution through wetland restoration in America's breadbasket. National Wetlands Newsletter 21(6):9–10, 14.

Mitsch, W. J., C. L. Dorge, and J. R. Wiemhoff. 1977. Forested Wetlands for Water Resource Management in Southern Illinois. Research Report 132, Illinois University Water Resources Center, Urbana. 225 pp.

Mitsch, W. J., and K. C. Ewel. 1979. Comparative biomass and growth of cypress in Florida wetlands. American Midland Naturalist 101:417–426.

Mitsch, W. J., C. L. Dorge, and J. R. Wiemhoff. 1979a. Ecosystem dynamics and a phosphorus budget of an alluvial cypress swamp in southern Illinois. Ecology 60:1116–1124.

Mitsch, W. J., W. Rust, A. Behnke, and L. Lai. 1979b. Environmental Observations of a Riparian Ecosystem during Flood Season. Research Report 142, Illinois University Water Resources Center, Urbana. 64 pp.

Mitsch, W. J., M. D. Hutchison, and G. A. Paulson. 1979c. The Momence Wetlands of the Kankakee River in Illinois—An Assessment of Their Value. Document 79/17, Illinois Institute of Natural Resources, Chicago. 55 pp.

Mitsch, W. J., J. W. Day, Jr., J. R. Taylor, and C. Madden. 1982. Models of North American freshwater wetlands. International Journal of Ecological and Environmental Science 8:109–140.

Mitsch, W. J., J. R. Taylor, and K. B. Benson. 1983a. Classification, modelling and management of wetlands—A case study in western Kentucky. In W. K. Lauenroth, G. V. Skogerboe, and M. Flug, eds. Analysis of Ecological Systems: State-of-the-Art in Ecological Modelling. Elsevier, Amsterdam, pp. 761–769.

Mitsch, W. J., J. R. Taylor, K. B. Benson, and P. L. Hill, Jr. 1983b. Atlas of Wetlands in the Principal Coal Surface Mine Region of Western Kentucky. FWS/OBS-82/72, U.S. Fish and Wildlife Service, Washington, DC. 135 pp.

Mitsch, W. J., J. R. Taylor, K. B. Benson, and P. L. Hill, Jr. 1983c. Wetlands and coal surface mining in western Kentucky—A regional impact assessment. Wetlands 3:161–179.

Mitsch, W. J., and W. G. Rust. 1984. Tree growth responses to flooding in a bottomland forest in northeastern Illinois. Forest Science 30:499–510.

Mitsch, W. J., and J. G. Gosselink. 1986. Wetlands. Van Nostrand Reinhold, New York.

Mitsch, W. J., M. Straskraba, and S. E. Jørgensen, eds. 1988. Wetland Modelling. Elsevier, Amsterdam. 227 pp.

Mitsch, W. J., and B. C. Reeder. 1991. Modelling nutrient retention of a freshwater coastal wetland: Estimating the roles of primary productivity, sedimentation, resuspension and hydrology. Ecological Modelling 54:151–187.

Mitsch, W. J., J. R. Taylor, and K. B. Benson. 1991. Estimating primary productivity of forested wetland communities in different hydrologic landscapes. Landscape Ecology 5:75–92.

Mitsch, W. J., and J. K. Cronk. 1992. Creation and restoration of wetlands: Some design consideration for ecological engineering. In R. Lal and B. A. Stewart, eds. Advances in Soil Science, Vol. 17: Soil Restoration. Springer-Verlag, New York, pp. 217–259.

Mitsch, W. J., and B. C. Reeder. 1992. Nutrient and hydrologic budgets of a Great Lakes coastal freshwater wetland during a drought year. Wetlands Ecology and Management 1(4):211–223.

Mitsch, W. J., and J. G. Gosselink. 1993. Wetlands, 2nd ed. Van Nostrand Reinhold, New York. 722 pp.

Mitsch, W. J., R. H. Mitsch, and R. E. Turner. 1994a. Wetlands of the Old and New Worlds—Ecology and management. In W. J. Mitsch, ed. Global Wetlands: Old World and New. Elsevier, Amsterdam, pp. 3–56.

Mitsch, W. J., B. C. Reeder, and D. M. Robb. 1994b. Modelling ecosystem and landscape scales of Lake Erie coastal wetlands. In W. J. Mitsch, ed. Global Wetlands: Old World and New. Elsevier, Amssterdam, pp. 563–574.

Mitsch, W. J., J. K. Cronk, X. Wu, R. W. Nairn, and D. L. Hey. 1995. Phosphorus retention in constructed freshwater riparian marshes. Ecological Applications 5:830–845.

Mitsch, W. J., and X. Wu. 1995. Wetlands and global change. In R. Lal, J. Kimble, E. Levine, and B. A. Stewart, eds. Advances in Soil Science: Soil Management and Greenhouse Effect. CRC Press/Lewis Publishers, Boca Raton, FL, pp. 205–230.

Mitsch, W. J., and R. F. Wilson. 1996. Improving the success of wetland creation and restoration with know-how, time, and self-design. Ecological Applications 6:77–83.

Mitsch, W. J., and V. Bouchard, eds. 1998. Great Lakes Coastal Wetlands: Their Potential for Restoration. Special Issue of Wetlands Ecology and Management 6:1–82.

Mitsch, W. J., and K. M. Wise. 1998. Water quality, fate of metals, and predictive model validation of a constructed wetland treating acid mine drainage. Water Research 32:1888–1900.

Mitsch, W. J., X. Wu, R. W. Nairn, P. E. Weihe, N. Wang, R. Deal, and C. E. Boucher. 1998. Creating and restoring wetlands: A whole-ecosystem experiment in self-design. BioScience 48:1019–1030.

Mitsch, W. J., J. W. Day, Jr., J. W. Gilliam, P. M. Groffman, D. L. Hey, G. W. Randall, and N. Wang. 1999. Reducing Nutrient Loads, Especially Nitrate-Nitrogen, to Surface Water, Ground Water, and the Gulf of Mexico. Topic 5 Report to the National Oceanic and Atmospheric Administration Coastal Ocean Program, Silver Spring, MD. 111 pp.

Mitsch, W. J., A. Horne, and R. W. Nairn, eds. 2000a. Nitrogen and Phosphorus Retention in Wetlands. Special Issue of Ecological Engineering 14:1–206.

Mitsch, W. J., N. Wang, and V. Bouchard. 2000b. Fringe wetlands of the Laurentian Great Lakes: Effects of dikes, water level fluctuations, and climate change. Verhandlungen Internationale Vereinigung für Limnologie, in press.

Mitsch, W. J., and J. G. Gosselink. 2000. The value of wetlands: Importance of scale and landscape setting. Ecological Economics 36 (in press).

Mohanty, S. K., and R. N. Dash. 1982. The chemistry of waterlogged soils. In B. Gopal, R. E. Turner, R. G. Wetzel, and D. F. Whigham, eds. Wetlands: Ecology and Management. National Institute of Ecology and International Scientific Publications, Jaipur, India, pp. 389–396.

Monk, C. D., and T. W. Brown. 1965. Ecological consideration of cypress heads in north-central Florida. American Midland Naturalist 74:126–140.

Montague, C. L., S. M. Bunker, E. B. Haines, M. L. Pace, and R. L. Wetzel. 1981. Aquatic macroconsumers. In L. R. Pomeroy and R. G. Wiegert, eds. The Ecology of a Salt Marsh. Springer-Verlag, New York, pp. 69–85.

Montague, C. L., and R. G. Wiegert. 1990. Salt marshes. In R. L. Myers and J. J. Ewel, eds. Ecosystems of Florida. University of Central Florida Press, Orlando, pp. 481–516.

Montz, G. N., and A. Cherubini. 1973. An ecological study of a bald cypress swamp in St. Charles Parish, Louisiana. Castanea 38:378–386.

Moon, G. J., B. F. Clough, C. A. Peterson, and W. G. Allaway. 1986. Apoplastic and symplastic pathways in *Avicennia marina* (Forsk.) Vierh. roots revealed by fluorescent tracer dyes. Australian Journal of Plant Physiology 13:637–648.

Moore, D. R. J., and P. A. Keddy. 1989. The relationship between species richness and standing crop in wetlands: The importance of scale. Vegetatio 79:99–106.

Moore, D. R. J., P. A. Keddy, C. L. Gaudet, and I. C. Wisheu. 1989. Conservation of wetlands: Do infertile wetlands deserve a higher priority? Biological Conservation 47:203–217.

Moore, P. D., ed. 1984. European Mires. Academic Press, London. 367 pp.

Moore, P. D., and D. J. Bellamy. 1974. Peatlands. Springer-Verlag, New York. 221 pp.

Moore, T. R. 1989. Growth and net production of *Sphagnum* at five fen sites, subarctic eastern Canada. Canadian Journal of Botany 67:1203–1207.

Moore, T. R., and R. Knowles. 1989. The influence of water table levels on methane and carbon dioxide emissions from peatland soils. Canadian Journal of Soil Science 69:33–38.

Moore, T. R., and N. T. Roulet. 1995. Methane emissions from Canadian peatlands. In R. Lal, J. Kimble, E. Levine, and B. A. Stewart, eds. Advances in Soil Science: Soils and Global Change. CRC Press, Boca Raton, FL, pp. 153–164.

Morris, J. T. 1982. A model of growth responses by *Spartina alterniflora* to nitrogen limitation. Journal of Ecology 70:25–42.

Morris, J. T. 1984. Effects of oxygen and salinity on ammonium uptake by *Spartina alterniflora* Loisel and *Spartina patens* (Aiton). Journal of Experimental Marine Biology and Ecology 78:87–98.

Morris, J. T., and J. W. H. Dacey. 1984. Effects of O_2 on ammonium uptake and root respiration by *Spartina alterniflora*. American Journal of Botany 71:979–985.

Morris, J. T., and W. B. Bowden. 1986. A mechanistic, numerical model of sedimentation, mineralization, and decomposition for marsh sediments. Journal of Soil Science Society of America 50:96–105.

Morris, J. T., B. Kjerfve, and J. M. Dean. 1990. Dependence of estuarine productivity on anomalies in mean sea level. Limnology and Oceanography 35:926–930.

Moshiri, G. A., ed. 1993. Constructed Wetlands for Water Quality Improvement. Lewis Publishers, Boca Raton, FL.

Motzkin, G. 1991. Atlantic white cedar wetlands of Massachusetts. Research Bulletin 731, Massachusetts Agricultural Experiment Station, Amherst.

Motzkin, G., W. A. Patterson III, and N. E. R. Drake. 1993. Fire history and vegetation dynamics of a *Chamaecyparis thyoides* wetland on Cape Cod, Massachusetts. Journal of Ecology 81:391–402.

Moustafa, M. Z. 1999. Nutrient retention dynamics of the Everglades nutrient removal project. Wetlands 19:689–704.

Moustafa, M. Z., M. J. Chimney, T. D. Fontaine, G. Shih, and S. Davis. 1996. The response of a freshwater wetland to long-term "low level" nutrient loads—Marsh efficiency. Ecological Engineering 7:15–33.

Mulamoottil, G., B. G. Warner, and E. A. McBean. 1996. Wetlands: Environmental Gradients, Boundaries, and Buffers. CRC Press, Boca Raton, FL. 298 pp.

Mulamoottil, G., E. A. McBean, and F. Rovers, eds. 1999. Constructed Wetlands for the Treatment of Landfill Leachates. Lewis Publishers, Boca Raton, FL. 281 pp.

Mulholland, P. J. 1979. Organic Carbon in a Swamp–Stream Ecosystem and Export by Streams in Eastern North Carolina. Ph.D. Dissertation, University of North Carolina.

Mulholland, P. J., and E. J. Kuenzler. 1979. Organic carbon export from upland and forested wetland watersheds. Limnology and Oceanography 24:960–966.

Murkin, H. R. ed. 1998. Freshwater Functions and Values of Prairie Watlands. Special Issue of Great Plains Research 8:1–182.

Naiman, R. J. 1988. Animal influences on ecosystem dynamics. BioScience 38:750–752.

Naiman, R. J., T. Manning, and C. A. Johnston. 1991. Beaver population fluctuations and tropospheric methane emissions in boreal wetlands. Biogeochemistry 12:1–15.

Naiman, R. J., and H. Decamps. 1997. The ecology of interfaces: Riparian zones. Annual Review of Ecology and Systematics 28:621–658.

Naiman, R. J., and K. H. Rogers. 1997. Large animals and system-level characteristics in river corridors: Implications for river management. BioScience 47:521–529.

Nairn, R. W., and W. J. Mitsch. 2000. Phosphorus removal in created wetland ponds receiving river overflow. Ecological Engineering 14:107–126.

Nakamura, T., T. Murakami, M. Saotome, K. Tomikta, T. Kitsuwa, and S. P. Meyers. 1991. Identification of indole-3-acetic acid in *Pichia spartinae,* an ascosporogenous yeast from *Spartina alterniflora* marshland environments. Mycologia 83:662–664.

National Research Council (NRC). 1985. Glaciers, Ice Sheets and Sea Level. Effect of a CO_2-Induced Climatic Change. Committee on the Relationship Between Land Ice and Sea Level. National Academy Press, Washington, DC.

National Research Council (NRC). 1992. Restoration of Aquatic Ecosystems. National Academy Press, Washington, DC. 552 pp.

National Research Council (NRC). 1995. Wetlands: Characteristics and Boundaries. National Academy Press, Washington, DC. 306 pp.

National Wetlands Policy Forum. 1988. Protecting America's Wetlands: An Action Agenda, Conservation Foundation, Washington, DC. 69 pp.

National Wetlands Working Group. 1988. Wetlands of Canada. Ecological Land Classification Series 24, Environment Canada, Ottawa, Ontario, and Polyscience Publications, Montreal, Quebec. 452 pp.

Natural Resources Conservation Service (NRCS). 1998. Field Indicators of Hydric Soils in the United States, Version 4.0. G. W. Gurt, P. M. Whited, and R. F. Pringle, eds. USDA, NRCS, Ft. Worth, TX. 30 pp.

The Nature Conservancy. 1992. The Forested Wetlands of the Mississippi River: An Ecosystem in Crisis. Nature Conservancy, Baton Rouge. 25 pp.

Naumann, E. 1919. Nagra sypunkte angaende planktons ökilogi. Med. sarskild hänsyn till fytoplankton. Svensk Botanisk Tidkrift 13:129–158.

Navid, D. 1989. The international law of migratory species: The Ramsar Convention. Natural Resources Journal 29:1001–1016.

Neill, C. 1990a. Effects of nutrients and water levels on emergent macrophytes biomass in a prairie marsh. Canadian Journal of Botany 68:1007–1014.

Neill, C. 1990b. Effects of nutrients and water levels on species composition in prairie whitetop (*Scolochloa festucacea*) marshes. Canadian Journal of Botany 68:1015–1020.

Neill, C. 1990c. Nutrient limitation of hardstem bulrush (*Scirpus acutus* Muhl.) in a Manitoba interlake region marsh. Wetlands 10(1):69–75.

Nelson, J. W., J. A. Kadlec, and H. R. Murkin. 1990a. Seasonal comparisons of weight loss of two types of *Typha glauca* Godr. leaf litter. Aquatic Botany 37:299–314.

Nelson, J. W., J. A. Kadlec, and H. R. Murkin. 1990b. Response by macroinvertebrates to cattail litter quality and timing of litter submergence in a northern prairie marsh. Wetlands 10:47–60.

Nelson, U. C. 1947. Woodland caribou in Minnesota. Journal of Wildlife Management 11:283–284.

Nessel, J. K. 1978. Distribution and Dynamics of Organic Matter and Phosphorus in a Sewage Enriched Cypress Strand. Master's Thesis, University of Florida, Gainesville. 159 pp.

Nessel, J. K., K. C. Ewel, and M. S. Burnett. 1982. Wastewater enrichment increases mature pond cypress growth rates. Forest Science 28:400–403.

Nessel, J. K., and S. E. Bayley. 1984. Distribution and dynamics of organic matter and phosphorus in a sewage-enriched cypress swamp. In K. C. Ewel and H. T. Odum, eds. Cypress Swamps. University Presses of Florida, Gainesville, pp. 262–278.

Newell, S. Y., and D. Porter. 2000. Microbial secondary production from saltmarsh grass shoots, and its known and potential fates. In M. P. Weinstein and D. A. Kreeger, eds. International Symposium: Concepts and Controversies in Tidal Marsh Ecology. Kluwer Academic Publishers, Dordrecht, pp. 159–185.

Newling, C. J. 1982. Feasibility report on a Santa Ana River marsh restoration and habitat development project. In M. C. Landin, ed. Habitat Development at Eight Corps of Engineers Sites: Feasibility and Assessment. U.S. Army Corps of Engineers Waterways Experiment Station, Vicksburg, MS, pp. 45–84.

Newman, J. M., J. C. Clausen, and J. A. Neafsey. 2000. Seasonal performance of a wetland constructed to process dairy milkhouse wastewater in Connecticut. Ecological Engineering 14:181–198.

Newman, S., J. B. Grace, and J. W. Kooebel. 1996. The effects of nutrients and hydroperiod on mixtures of *Typha domingensis, Cladium jamaicense,* and *Eleocharis interstincta:* Implications for Everglades restoration. Ecological Applications 6:774–783.

Nguyen, L. M. 2000. Phosphate incorporation and transformation in surface sediments of a sewage-impacted wetland as influenced by sediment sites, sediment pH and added phosphate concentration. Ecological Engineering 14:139–155.

Nguyen, L. M., J. G. Cooke, and G. B. McBride. 1997. Phosphorus retention and release characteristics of sewage-impacted wetland sediments. Water, Air, and Soil Pollution 100:163–179.

Nicholas, G. P. 1998. Wetlands and hunter-gatherers: A global perspective. Current Anthropology 39:720–731.

Nichols, D. S. 1983. Capacity of natural wetlands to remove nutrients from wastewater. Journal of the Water Pollution Control Federation 55:495–505.

Nicholson, B. J., L. D. Gignac, and S. E. Bayley. 1996. Peatland distribution along a north-south transect in the Mackenzie River Basin in relation to climatic and environmental gradients. Vegetatio 126:119–133.

Niering, W. A. 1985. Wetlands. Alfred A. Knopf, New York. 638 pp.

Niering, W. A. 1988. Endangered, threatened and rare wetland plants and animals of the continental United States. In D. D. Hook et al., eds. The Ecology and Management of Wetlands, Vol. 1: Ecology of Wetlands. Timber Press, Portland, OR, pp. 227–238.

Niering, W. A. 1989. Wetland vegetation development. In S. K. Majumdar, R. P. Brooks, F. J. Brenner, and J. R. W. Tiner, eds. Wetlands Ecology and Conservation: Emphasis in Pennsylvania. Pennsylvania Academy of Science, Easton, pp. 103–113.

Niswander, S. F., and W. J. Mitsch. 1995. Functional analysis of a two-year-old created in-stream wetland: Hydrology, phosphorus retention, and vegetation survival and growth. Wetlands 15:212–225.

Nixon, S. W. 1980. Between coastal marshes and coastal waters—A review of twenty years of speculation and research on the role of salt marshes in estuarine productivity and water chemistry. In P. Hamilton and K. B. MacDonald, eds. Estuarine and Wetland Processes. Plenum Press, New York, pp. 437–525.

Nixon, S. W. 1982. The Ecology of New England High Salt Marshes: A Community Profile. FWS/OBS-81/55, U.S. Fish and Wildlife Service, Washington, DC.

Nixon, S. W., and C. A. Oviatt. 1973. Ecology of a New England salt marsh. Ecological Monographs. 43:463–498.

Nixon, S. W., and V. Lee. 1986. Wetlands and Water Quality. Technical Report Y-86-2, U.S. Army Corps of Engineers Waterways Experiment Station, Vicksburg, MS.

Nordquist, G. E., and E. C. Birney. 1980. The Importance of Peatland Habitats to Small Mammals in Minnesota. Peat Program, Minnesota Department of Natural Resources, Minneapolis.

Norgress, R. E. 1947. The history of the cypress lumber industry in Louisiana. Louisiana Historical Quarterly 30:979–1059.

Novacek, J. M. 1989. The water and wetland resources of the Nebraska Sandhills. In A. G. van der Valk, ed. Northern Prairie Wetlands. Iowa State University Press, Ames, pp. 340–384.

Novak, M., J. A. Balen, M. E. Obbard, and B. Mallocheds, eds. 1987. Wildlife Furbearer Management and Conservation in North America. Ontario Trappers Association, Ontario. 1150 pp.

Novák, M., and R. K. Wieder. 1992. Inorganic and organic sulfur profiles in nine *Sphagnum* peat bogs in the United States and Czechoslovakia. Water, Air, and Soil Pollution 65:353–369.

Novitzki, R. P. 1979. Hydrologic characteristics of Wisconsin's wetlands and their influence on floods, stream flow, and sediment. In P. E. Greeson, J. R. Clark, and J. E. Clark, eds. Wetland Functions and Values: The State of Our Understanding. American Water Resources Association, Minneapolis, MN, pp. 377–388.

Novitzki, R. P. 1982. Hydrology of Wisconsin Wetlands. University of Wisconsin Extension Geological Natural History Survey Circular 40, University of Wisconsin, Madison. 22 pp.

Novitzki, R. P. 1985. The effects of lakes and wetlands on flood flows and base flows in selected northern and eastern states. In H. A. Groman et al., eds. Proceedings of a Conference—Wetlands of the Chesapeake. Environmental Law Institute, Washington, DC, pp. 143–154.

Nuttle, W. K., and H. F. Hemond. 1988. Salt marsh hydrology: Implications of biogeochemical fluxes to the atmosphere and estuaries. Global Biogeochemical Cycles 2:91–114.

Nuttle, W. K., and J. W. Harvey. 1995. Fluxes of water and solute in a coastal wetland sediment. I. The contribution of regional groundwater discharge. Journal of Hydrology 164:89–107.

Nyc, R., and P. Brooks. 1979. National Wetlands Inventory Project: Inventorying the Nation's Wetlands with Remote Sensing. Paper Presented at U.S. Army Corps of Engineers Remote Sensing Symposium, Reston, VA, October 29–31, 1979. 11 pp.

Nyman, J. A., R. D. Delaune, and W. H. Patrick, Jr. 1990. Wetland soil formation in the rapidly subsiding Mississippi River Deltaic Plain: Mineral and organic matter relationships. Estuarine, Coastal and Shelf Science 31:57–69.

Oaks, R. Q. Jr., and D. R. Whitehead. 1979. Geologic setting origin of the Dismal Swamp, southeastern Virginia and northeastern North Carolina. In: P. W. Kirk, ed. The Great Dismal Swamp. Univ Press of Virginia Charlottesville, VA, pp. 1–24.

O'Brien, A. J., and W. S. Motts. 1980. Hydrogeologic evaluation of wetland basins for land use planning. Water Resources Bulletin 16:785–789.

O'Brien, A. L. 1988. Evaluating the cumulative effects of alteration on New England wetlands. Environmental Management 12:627–636.

O'Neil, T. 1949. The Muskrat in the Louisiana Coastal Marsh. Louisiana Department of Wildlife and Fisheries, New Orleans. 152 pp.

Odum, E. P. 1961. The role of tidal marshes in estuarine production. New York State Conservation 15(6):12–15.

Odum, E. P. 1968. A research challenge: Evaluating the productivity of coastal and estuarine water. Proceedings of the Second Sea Grant Conference, University of Rhode Island, pp. 63–64.

Odum, E. P. 1969. The strategy of ecosystem development. Science 164:262–270.

Odum, E. P. 1971. Fundamentals of Ecology, 3rd ed. W. B. Saunders, Philadelphia. 544 pp.

Odum, E. P. 1979a. Ecological importance of the riparian zone. In R. R. Johnson and J. F. McCormick, tech. coords. Strategies for Protection and Management of Floodplain Wetlands and Other Riparian Ecosystems. Proceedings of the Symposium, Callaway Gardens, Georgia, December 11–13, 1978. General Technical Report WO-12, U.S. Forest Service, Washington, DC, pp. 2–4.

Odum, E. P. 1979b. The value of wetlands: A hierarchical approach. In P. E. Greeson, J. R. Clark, and J. E. Clark, eds. Wetland Functions and Values: The State of Our Understanding. American Water Resources Association, Minneapolis, pp. 1–25.

Odum, E. P. 1980. The status of three ecosystem-level hypotheses regarding salt marsh estuaries: Tidal subsidy, outwelling, and detritus-based food chains. In V. S. Kennedy, ed. Estuarine Perspectives. Academic Press, New York, pp. 485–495.

Odum, E. P. 1981. Foreword. In J. R. Clark and J. Benforado, eds. Wetlands of Bottomland Hardwood Forests. Elsevier, Amsterdam, pp. xi–xiii.

Odum, E. P. 2000. Tidal marshes as outwelling/pulsing systems. In M. P. Weinstein and D. A. Kreeger, eds. International Symposium: Concepts and Controversies in Tidal Marsh Ecology. Kluwer Academic Publishers, Dordrecht, The Netherlands (in press).

Odum, E. P., and A. A. de la Cruz. 1967. Particulate organic detritus in a Georgia salt marsh–estuarine ecosystem. In G. H. Lauer, ed. Estuaries. American Association for the Advancement of Science, Washington, DC, pp. 383–388.

Odum, H. T. 1951. The Carolina Bays and a Pleistocene weather map. American Journal of Science 250:262–270.

Odum, H. T. 1971. Environment, Power and Society. John Wiley & Sons, Inc., New York.

Odum, H. T. 1982. Role of wetland ecosystems in the landscape of Florida. In D. O. Logofet and N. K. Luckyanov, eds. Ecosystem Dynamics in Freshwater Wetlands and Shallow Water Bodies, Vol. II. SCOPE and UNEP Workshop, Center of International Projects, Moscow, pp. 33–72.

Odum, H. T. 1983. Systems Ecology—An Introduction. John Wiley & Sons, New York. 644 pp.

Odum, H. T. 1984. Summary: Cypress swamps and their regional role. In K. C. Ewel and H. T. Odum, eds. Cypress Swamps. University Presses of Florida, Gainesville, pp. 416–443.

Odum, H. T. 1988. Self-organization, transformity, and information. Science 242:1132–1139.

Odum, H. T. 1989. Ecological engineering and self-organization. In W. J. Mitsch and S. E. Jørgensen, eds. Ecological Engineering. John Wiley & Sons, New York, pp. 79–101.

Odum, H. T. 1996. Environmental Accounting: Energy and Environmental Decision Making. John Wiley & Sons, New York. 370 pp.

Odum, H. T. 2000. Heavy Metals in the Environment: Using Wetlands for Their Removal. CRC Press, Boca Raton, FL.

Odum, H. T., B. J. Copeland, and E. A. McMahan, eds. 1974. Coastal Ecological Systems of the United States. Conservation Foundation, Washington, DC. 4 vols.

Odum, H. T., K. C. Ewel, W. J. Mitsch, and J. W. Ordway. 1977a. Recycling Treated Sewage Through Cypress Wetlands in Florida. In F. M. D'Itri, ed. Wastewater Renovation and Reuse. Marcel Dekker, New York, pp. 35–67.

Odum, H. T., W. M. Kemp, M. Sell, W. Boynton, and M. Lehman. 1977b. Energy analysis and coupling of man and estuaries. Environmental Management 1:297–315.

Odum, H. T., P. Kangas, G. R. Best, B. T. Rushton, S. Leibowitz, J. R. Butner, and T. Oxford. 1981. Studies on Phosphate Mining, Reclamation, and Energy. Center for Wetlands, University of Florida, Gainesville. 142 pp.

Odum, W. E. 1970. Pathways of Energy Flow in a South Florida Estuary. Ph.D. Dissertation, University of Miami, Coral Gables. 62 pp.

Odum, W. E. 1987. Predicting ecosystem development following creation and restoration of wetlands. In J. Zelazny and J. S. Feierabend, eds. Wetlands: Increasing Our Wetland Resources. Proceedings of the Conference Wetlands: Increasing Our Wetland Resources. Corporate Conservation Council, National Wildlife Federation, Washington, DC, pp. 67–70.

Odum, W. E. 1988. Comparative ecology of tidal freshwater and salt marshes. Annual Review of Ecology and Systematics 19:147–176.

Odum, W. E., and E. J. Heald. 1972. Trophic analyses of an estuarine mangrove community. Bulletin of Marine Science 22:671–738.

Odum, W. E., and M. A. Heywood. 1978. Decomposition of intertidal freshwater marsh plants. In R. E. Good, D. G. Whigham, and R. L. Simpson, eds. Freshwater Wetlands: Ecological Processes and Management Potential. Academic Press, New York, pp. 89–97.

Odum, W. E., J. S. Fisher, and J. C. Pickral. 1979. Factors controlling the flux of particulate organic carbon from estuarine wetlands. In R. J. Livingston, ed.

Ecological Processes in Coastal and Marine Systems. Plenum Press, New York, pp. 69–80.

Odum, W. E., C. C. McIvor, and T. J. Smith III. 1982. The Ecology of the Mangroves of South Florida: A Community Profile. FWS/OBS-81/24, U.S. Fish and Wildlife Service, Washington, DC.

Odum, W. E., T. J. Smith III, J. K. Hoover, and C. C. McIvor. 1984. The Ecology of Tidal Freshwater Marshes of the United States East Coast: A Community Profile. FWS/OBS-87/17, U.S. Fish and Wildlife Service, Washington, DC. 177 pp.

Odum, W. E., and C. C. McIvor. 1990. Mangroves. In R. L. Myers and J. J. Ewel, eds. Ecosystems of Florida. University of Central Florida Press, Orlando, pp. 517–548.

Odum, W. E., E. P. Odum, and Odum, H. T. 1995. Nature's pulsing paradigm. Estuaries 18:547–555.

Office of Technology Assessment. 1984. Wetlands: Their Use and Regulation. Report O-206, Office of Technology Assessment, U.S. Congress, Washington, DC. 208 pp.

Ogaard, L. A., J. A. Leitch, D. F. Scott, and W. C. Nelson. 1981. The Fauna of the Prairie Wetlands: Research Methods and Annotated Bibliography. Research Report 86, Agricultural Experiment Station, North Dakota State University, Fargo. 2 pp.

Ogawa, H., and J. W. Male. 1983. The Flood Mitigation Potential of Inland Wetlands. Publication 138, Water Resources Research Center, University of Massachusetts, Amherst. 164 pp.

Ogawa, H., and J. W. Male. 1986. Simulating the flood mitigation role of wetlands. Journal of Water Resource Planning and Management 112:114–128.

Ogden, M. H. 1999. Constructed wetlands for small community wastewater treatment. Paper Presented at Wetlands for Wastewater Recycling Conference, November 3, 1999, Baltimore. Environmental Concern, Inc., St. Michaels, MD.

Ohlendorf, H. M., D. J. Hoffman, M. K. Saiki, and T. W. Aldrich. 1986. Embryonic mortality and abnormalities of aquatic birds: Apparent impacts of selenium from irrigation drainwater. Science of the Total Environment 52:49–63.

Ohlendorf, H. M., R. L. Hothem, C. M. Bunck, and K. C. Marois. 1990. Bioaccumulation of selenium in birds at Kesterson Reservoir, California. Archives of Environmental Contamination and Toxicology 19:495–507.

Oliver, J. D., and T. Legovic. 1988. Okefenokee marshland before, during, and after nutrient enrichment by a bird rookery. Ecological Modelling 43:195–223.

Olson, R. K., ed. 1992. The Role of Created and Natural Wetlands in Controlling Nonpoint Source Pollution. Special Issue of Ecological Engineering 1:1–170.

Omernik, J. M. 1977. Nonpoint Source-Stream Nutrient Level Relationships: A Nationwide Study. EPA-600/3-79-105, Corvallis Environmental Research Laboratory, U.S. Environmental Protection Agency, Corvallis, OR.

Orozco-Segovia, A. D. L. 1980. One option for the use of marshes of Tabasco, Mexico. In D. P. Cole, ed. Wetlands Restoration and Creation: Proceedings of the Seventh Annual Conference, May 16–17, 1979, Tampa, FL, pp. 209–218. Available from Hillsborough Community College, Tampa, FL.

Orson, R. A., R. L. Simpson, and R. E. Good. 1990. Rates of sediment accumulation in a tidal freshwater marsh. Journal of Sedimentology and Petrology 60:859–869.

Osborne, K., and T. J. Smith. 1990. Differential predation on mangrove propagules in open and closed canopy forest habitats. Vegetatio 89:1–6.

Ostendorp, W. 1989. "Die-back" of reeds in Europe—A critical review of literature. Aquatic Botany 35:5–26.

Ostendorp, W., C. Iseli, M. Krauss, P. Krumscheid-Plankert, J.-L. Moret, M. Rollier, and F. Schanz. 1995. Lake shore deterioration, reed management and bank restoration in some Central European lakes. Ecological Engineering 5:51–75.

Otte, L. J. 1981. Origin, Development and Maintenance of Pocosin Wetlands of North Carolina. North Carolina Department of Natural Resources and Community Development, Raleigh. Unpublished Report to the North Carolina Natural Heritage Program.

Overbeck, F., and H. Happach. 1957. Uber das Wachstum und den Wasserhaushalt einiger Hochmoor Sphagnum, *Flora*. Jena 144:335–402.

Özesmi, U., and W. J. Mitsch. 1997. A spatial model for the marsh-breeding red-winged blackbird (*Agelaius phoeniceus* L.) in coastal Lake Erie wetlands. Ecological Modelling 101:139–152.

Padgett, W. G., A. P. Youngblood, and A. H. Winward. 1989. Riparian Community Type Classification of Utah and Southeastern Idaho. R4-Ecol-89-01, Intermountain Region, Forest Service, U.S. Department of Agriculture.

Pakarinen, P. 1978. Production and nutrient ecology of three *Sphagnum* species in southern Finnish raised bogs. Annales Botanici Fennici 15:15–26.

Pakarinen, P., and E. Gorham. 1983. Mineral element composition of *Sphagnum fuscum* peats collected from Minnesota, Manitoba, and Ontario. In Proceedings of the International Symposium on Peat Utilization, Bemidji, Minnesota. Bemidji State University Center for Environmental Studies, Bemidji, MN, pp. 417–429.

Pallis, M. 1915. The structural history of Plav: The floating fen of the delta of the Danube. J. Linn. Soc. Bot. 43:233–290.

Palmer, C. N., and J. D. Hunt. 1989. Greenwood urban wetland a manmade stormwater treatment facility. In D. W. Fisk, ed. Wetlands Concerns and Successes. Proceedings of the American Water Resources Association Conference, Tampa, FL, pp. 1–10.

Parish, D., and C. Elliott. 1990. Foreword. In J. Lu. Wetlands in China, East China Normal University, Shanghai, p. x.

Park, R. A., J. K. Lee, P. W. Mausel, and R. C. Howe. 1991. Using remote sensing for modeling the impacts of sea level rise. World Resource Review 3:184–205.

Patrick, R., J. Cairns, Jr., and S. S. Roback. 1967. An ecosystematic study of the fauna and flora of the Savannah River. Academy of Natural Sciences of Philadelphia Proceedings 118:109–407.

Patrick, W. H., Jr. 1981. Bottomland soils. In J. R. Clark and J. Benforado, eds. Wetlands of Bottomland Hardwood Forests. Elsevier, Amsterdam, pp. 177–185.

Patrick, W. H., Jr., and R. D. Delaune. 1972. Characterization of the oxidized and reduced zones in flooded soil. Proceedings of the Soil Science Society of America 36:573–576.

Patrick, W. H., Jr., and R. A. Khalid. 1974. Phosphate release and sorption by soils and sediments: Effect of aerobic and anaerobic conditions. Science 186:53–55.

Patrick, W. H., Jr., and R. D. Delaune. 1990. Subsidence, accretion, and sea level rise in south San Francisco Bay marshes. Limnology and Oceanography 35:1389–1395.

Patten, B. C., ed. 1990. Wetlands and Shallow Continental Water Bodies, Vol. 1: Natural and Human Relationships. SPB Academic Publishing, The Hague. 759 pp.

Patten, B. C. 1994a. Systems ecology of Okefenokee Swamp (USA). In B. C. Patten, ed. Wetlands and Shallow Continential Water Bodies. SPB Academic Publishing, The Hague, pp. 129–180.

Patten, B. C. 1994b. Wetlands and Shallow Continental Water Bodies, Vol. 2: Case Studies. SPB Academic Publishing, The Hague. 732 pp.

Patten, D. T. 1998. Riparian ecosystems of semi-arid North America: Diversity and human impacts. Wetlands 18:498–512.

Patterson, S. G. 1986. Mangrove Community Boundary Interpretation and Detection of Areal Changes in Marco Island, Florida: Application of Digital Image Processing and Remote Sensing Techniques. Biological Services Report 86(10), U.S. Fish and Wildlife Service, Washington, DC.

Pearlstein, L., H. McKellar, and W. Kitchens. 1985. Modelling the impacts of a river diversion on bottomland forest communities in the Santee River floodplain, South Carolina. Ecological Modelling 29:283–302.

Pearsall, W. H. 1920. The aquatic vegetation of the English lakes. Journal of Ecology 8:163–201.

Pearson, J., and D. C. Havill. 1988. The effect of hypoxia and sulfide on culture grown wetland and nonwetland plants. 2. Metabolic and physiological changes. Journal of Experimental Botany 39:431–439.

Pedersen, A. 1975. Growth measurements of five *Sphagnum* species in South Norway. Norwegian Journal of Botany 22:277–284.

Pederson, R. L., and L. M. Smith. 1988. Implications of wetland seed bank research: A review of Great Britain and prairie marsh studies. In D. A. Wilcox, ed. Interdisciplinary Approaches to Freshwater Wetlands Research. Michigan State University Press, East Lansing, pp. 81–95.

Pelikan, J. 1978. Mammals in the reedswamp ecosystem. In D. Dykyjová and J. Kvet, eds. Pond Littoral Ecosystems. Springer-Verlag, Berlin, pp. 357–365.

Penfound, W. T. 1952. Southern swamps and marshes. Botanical Review 18:413–446.

Penfound, W. T., and E. S. Hathaway. 1938. Plant communities in the marshlands of southeastern Louisiana. Ecological Monographs 8:1–56.

Penfound, W. T., and T. T. Earle. 1948. The biology of the water hyacinth. Ecological Monographs 18:447–472.

Penland, S., K. E. Ramsey, R. A. McBride, J. T. Mestayer, and K. A. Westphal. 1988. Relative sea-level rise and subsidence in Louisiana and the Gulf of Mexico: 1908–1988. Coastal Geology Technical Report 4, Louisiana Geological Survey, Baton Rouge, pp. 20–21.

Penman, H. L. 1948. Natural evaporation from open water, bare soil and grass. Proceedings of the Royal Society of London 93:120–145.

Perry, J. E., and C. H. Hershner. 1999. Temporal changes in the vegetation pattern in a tidal freshwater marsh. Wetlands 19:90–99.

Peterjohn, W. T., and D. L. Correll. 1984. Nutrient dynamics in an agricultural watershed: Observations on the role of a riparian forest. Ecology 65:1466–1475.

Peverly, J. H. 1982. Stream transport of nutrients through a wetland. Journal of Environmental Quality 11:38–43.

Pezeshki, S. R., S. W. Matthews, and R. D. Delaune. 1991. Root cortex structure and metabolic responses of *Spartina patens* to soil redox conditions. Environmental and Experimental Botany 31:91–97.

Pfeiffer, W. J., and R. G. Wiegert. 1981. Grazers on *Spartina* and their predators. In L. R. Pomeroy and R. G. Wiegert, eds. The Ecology of a Salt Marsh. Springer-Verlag, New York, pp. 87–112.

Phillips, J. D. 1989. Fluvial sediment storage in wetlands. Water Resources Bulletin 25:867–873.

Phipps, R. G., and W. G. Crumpton. 1994. Factors affecting nitrogen loss in experimental wetlands with different hydrologic loads. Ecological Engineering 3:399–408.

Phipps, R. L. 1979. Simulation of wetlands forest vegetation dynamics. Ecological Modelling 7:257–288.

Phipps, R. L., and L. H. Applegate. 1983. Simulation of management alternatives in wetland forests. In S. E. Jørgensen and W. J. Mitsch, eds. Application of Ecological Modelling in Environmental Management, Part B. Elsevier, Amsterdam, pp. 311–339.

Pianka, E. R. 1983. Evolutionary Ecology. McGraw-Hill, New York.

Pinckney, J., and R. G. Zingmark. 1993. Modeling the annual production of intertidal benthic microalgae in estuarine ecosystems. Journal of Phycology 29:396–407.

Pjavchenko, N. J. 1982. Bog ecosystems and their importance in nature. In D. O. Logofet and N. K. Luckyanov, eds. Ecosystem Dynamics in Freshwater Wetlands and Shallow Water Bodies, Vol. I. SCOPE and UNEP Workshop, Center of International Projects, Moscow, pp. 7–21.

Pokorny, I., O. Lhotsky, P. Denny, and E. G. Turner, eds. 1987. Waterplants and wetland processes. Archiv für Hydrobiologie, Heft 27. E. Schweizerbartische Verlagsbuchhandlung, Stuttgart, Germany. 265 pp.

Polderman, P. J. 1975. The algal communities of the northeastern part of the saltmarsh "DeMok" on Texel (the Netherlands). Acta Botanica Neerlandica 24:361–378.

Polderman, P. J., and R. A. Polderman-Hall. 1980. Algal communities in Scottish saltmarshes. British Phycological Journal 15:59–71.

Pomeroy, L. R. 1959. Algae productivity in salt marshes of Georgia. Limnology and Oceanography 4:386–397.

Pomeroy, L. R., L. R. Shenton, R. D. Jones, and R. J. Reimold. 1972. Nutrient flux in estuaries. In G. E. Likens, ed. Nutrients and Eutrophication. American Society of Limnology and Oceanography Special Symposium. Allen Press, Lawrence, KS, pp. 274–291.

Pomeroy, L. R., and R. G. Wiegert, eds. 1981. The Ecology of a Salt Marsh. Springer-Verlag, New York. 271 pp.

Pomeroy, L. R., W. M. Darley, E. L. Dunn, J. L. Gallagher, E. B. Haines, and D. M. Witney. 1981. Primary production. In L. R. Pomeroy and R. G. Wiegert, eds. Ecology of a Salt Marsh. Springer-Verlag, New York, pp. 39–67.

Ponnamperuma, F. N. 1972. The chemistry of submerged soils. Advances in Agronomy 24:29–96.

Por, F. D. 1995. The Pantanal of Mato Grosso (Brazil). Kluwer Academic Publishers, Dordrecht. 122 pp.

Porcher, R. 1981. The vascular flora of the Francis Beidler Forest in Four Holes Swamp, Berkeley and Dorchester Counties, South Carolina. Castanea 46:248–280.

Post, R. A. 1996. Functional Profile of Black Spruce Wetlands in Alaska. EPA-910/R-96-006, U.S. Environmental Protection Agency, Seattle. 170 pp.

Post, W. M. 1990. Report of a Workshop on Climate Feedbacks and the Role of Peatlands, Tundra, and Boreal Ecosystems in the Global Carbon Cycle. Publication 3289, Environmental Science Division, Oak Ridge National Laboratory, Oak Ridge, TN.

Potonie, R. 1908. Aufbau und Vegetation der Moore Norddeutschlands. Englers. Bot. Jahrb. 90. Leipzig.

Powell, S. W., and F. P. Day. 1991. Root production in four communities in the Great Dismal Swamp. American Journal of Botany 78:288–297.

Prance, G. T. 1979. Notes on the vegetation of Amazonia. III. The terminology of Amazonian forest types subject to inundation. Brittonia 31:26–38.

Prasad, V. P., D. Mason, J. E. Marburger, and C. R. A. Kumar. 1996. Illustrated Flora of Keoladeo National Park, Bharatpur, Rajasthan. Bombay Natural History Society, Mumbai, India. 435 pp.

Prentki, R. T., T. D. Gustafson, and M. S. Adams. 1978. Nutrient movements in lakeshore marshes. In R. E. Good, D. F. Whigham, and R. L. Simpson, eds. Freshwater Wetlands: Ecological Processes and Management Potential. Academic Press, New York, pp. 169–194.

Presser, T. S., and H. M. Ohlendorf. 1987. Biogeochemical cycling of selenium in the San Joaquin Valley. Environmental Management 11:805–821.

Price, J. S., and M. K. Woo. 1988. Origin of salt in coastal marshes of Hudson and James bays. Canadian Journal of Earth Sciences 25:145–147.

Price, J., L. Rochefort, and F. Quinty. 1998. Energy and moisture considerations on cutover peatlands: Surface microtopography, mulch cover and *Sphagnum* regeneration. Ecological Engineering 10:293–312.

Pride, R. W., F. W. Meyer, and R. N. Cherry. 1966. Hydrology of Green Swamp Area in Central Florida. Report 42, Florida Division of Geology, Tallahassee. 137 pp.

Prince, H. 1997. Wetlands of the American Midwest. University of Chicago Press, Chicago. 395 pp.

Prince, H. H., and F. M. D'Itri, eds. 1985. Coastal Wetlands. Lewis Publishers, Chelsea, MI. 286 pp.

Prouty, W. F. 1952. Carolina Bays and their origin. Geological Society of America Bulletin 63:167–224.

Puriveth, P. 1980. Decomposition of emergent macrophytes in a Wisconsin marsh. Hydrobiology 72:231–242.

Putnam, J. A., G. M. Furnival, and J. S. McKnight. 1960. Management and Inventory of Southern Hardwoods. USDA Agricultural Handbook 181, U.S. Department of Agriculture, Washington, DC. 102 pp.

Puustjarvi, V. 1957. On the base status of peat soils. Acta Agriculturae Scandinavica 7:190–223.

Qin, P., M. Xie, and Y. Jiang. 1998. *Spartina* green food ecological engineering. Ecological Engineering 11:147–156.

Qualls, R. G., and C. J. Richardson. 1995. Forms of soil phosphorus along a nutrient enrichment gradient in the northern Everglades. Soil Science 160:183–198.

Quinty, F., and L. Rochefort. 1997. Plant reintroduction on a harvested peat bog. In C. C. Trettin, M. F. Jurgensen, D. F. Grigal, M. R. Gale, and J. K. Jeglum, eds. Northern Forested Wetlands: Ecology and Management. CRC Press/Lewis Publishers, Boca Raton, FL, pp. 133–145.

Rabalais, N. N., W. J. Wiseman, R. E. Turner, B. K. Sengupta, and Q. Dortch. 1996. Nutrient changes in the Mississippi River and system responses on the adjacent continental shelf. Estuaries 19:386–407.

Rabalais, N. N., R. E. Turner, W. J. Wiseman, and Q. Dortch. 1998. Consequences of the 1993 Mississippi River flood in the Gulf of Mexico. Regulated Rivers 14:161–177.

Rabinowitz, D. 1978. Dispersal properties of mangrove propagules. Biotropica 10:47–57.

Race, M. S. 1986. Wetlands restoration and mitigation policies: Reply. Environmental Management 10:571–572.

Radforth, N. W., and C. O. Brawner, eds. 1977. Muskeg and the Northern Environment in Canada. University of Toronto Press, Toronto. 399 pp.

Ragotzkie, R. A., L. R. Pomeroy, J. M. Teal, and D. C. Scott, eds. 1959. Proceedings of the Salt Marsh Conference, Marine Institute, University of Georgia, Sapelo Island, Georgia. Marine Institute, University of Georgia, Athens. 133 pp.

Rainey, G. B. 1979. Factors Affecting Nutrient Chemistry Distribution in Louisiana Coastal Marshes. Master's Thesis, Louisiana State University, Baton Rouge.

Raisin, G. W., and D. S. Mitchell. 1995. The use of wetlands for the control of non-point source pollution. Water Science and Technology 32:177–186.

Raisin, G. W., D. S. Mitchell, and R. L. Croome. 1997. The effectiveness of a small constructed wetland in ameliorating diffuse nutrient loadings from an Australian rural catchment. Ecological Engineering 9:19–35.

Ranwell, D. S. 1967. World resources of *Spartina townsendii* and economic use of *Spartina* marshland. Coastal Zone Management Journal 1:65–74.

Ranwell, D. S. 1972. Ecology of Salt Marshes and Sand Dunes. Chapman & Hall, London. 258 pp.

Raskin, I., and H. Kende. 1983. Regulation of growth in rice seedlings. Journal of Plant Growth Regulation 2:193–203.

Reader, R. J. 1978. Primary production in northern bog marshes. In R. E. Good, D. F. Whigham, and R. L. Simpson, eds. Freshwater Wetlands: Ecological Processes and Management Potential. Academic Press, New York, pp. 53–62.

Reader, R. J., and J. M. Stewart. 1971. Net primary productivity of the bog vegetation in southeastern Manitoba. Canadian Journal of Botany 49:1471–1477.

Reader, R. J., and J. M. Stewart. 1972. The relationship between net primary production and accumulation for a peatland in southeastern Manitoba. Ecology 53:1024–1037.

Redding, T., S. Todd, and A. Midlen. 1997. The treatment of aquaculture wastewaters—A botanical approach. Journal of Environmental Management 50:283–299.

Reddy, K. R., and W. H. Patrick, Jr. 1984. Nitrogen transformations and loss in flooded soils and sediments. CRC Critical Reviews in Environmental Control 13:273–309.

Reddy, K. R., and W. H. Smith, eds. 1987. Aquatic Plants for Water Treatment and Resource Recovery. Magnolia Publishing, Orlando, FL.

Reddy, K. R., and D. A. Graetz. 1988. Carbon and nitrogen dynamics in wetland soils. In D. D. Hook, W. H. McKee, Jr., H. K. Smith, J. Gregory, V. G. Burrell, M. R. DeVoe, R. E. Sojka, S. Gilbert, R. Banks, L. G. Stolzy, C. Brooks, T. D. Matthews, and T. H. Shear, eds. The Ecology and Management of Wetlands, Vol. 1: The Ecology of Wetlands. Timber Press, Portland, OR, pp. 307–318.

Reddy, K. R., and E. M. D'Angelo. 1994. Soil processes regulating water quality in wetlands. In W. J. Mitsch, ed. Global Wetlands: Old World and New. Elsevier, Amsterdam, pp. 309–324.

Reddy, K. R., and E. G. Flaig, eds. 1995. Phosphorus Dynamics in the Lake Okeechobee Watershed, Florida. Special Issue of Ecological Engineering 5:127–414.

Reddy, K. R., R. H. Kadlec, E. Flaig, and P. M. Gale. 1999. Phosphorus retention in streams and wetlands: A review. Critical Reviews in Environmental Science and Technology 29:83–146.

Redfield, A. C. 1958. The biological control of chemical factors in the environment. American Scientist 46:206–226.

Redfield, A. C. 1965. Ontogeny of a salt marsh estuary. Science 147:50–55.

Redfield, A. C. 1972. Development of a New England salt marsh. Ecological Monographs 42:201–237.

Redman, F. H., and W. H. Patrick, Jr. 1965. Effect of Submergence on Several Biological and Chemical Soil Properties. Bulletin 592, Louisiana Agricultural Experiment Station, Louisiana State University, Baton Rouge.

Redmond, A. M. 1981. Considerations for design of an artificial marsh for use in stormwater renovation. In R. H. Stovall, eds. Wetlands Restoration and Creation. Proceedings of the Eighth Annual Conference on Wetlands

Restoration and Creation, Hillsborough Community College, Tampa, FL, pp. 189–199.

Reed, S. C., R. W. Crites, and E. J. Middlebrooks. 1995. Natural Systems for Waste Management and Treatment, 2nd ed. McGraw-Hill, New York. 433 pp.

Reeve, A. S., D. I. Siegel, and P. H. Glaser. 1996. Geochemical controls on peatland pore water from the Hudson Bay Lowland: A multivariate statistical approach. Journal of Hydrology 181:285–304.

Reilly, J. F., A. J. Horne, and C. D. Miller. 2000. Nitrate removal from a drinking water supply with a large free-surface constructed wetland prior to groundwater recharge. Ecological Engineering 14:33–47.

Reimold, R. J. 1972. The movement of phosphorus through the marsh cord grass, *Spartina alterniflora* Loisel. Limnology and Oceanography 17:606–611.

Reimold, R. J. 1974. Mathematical modeling—Spartina. In R. J. Reimold and W. M. Queen, eds. Ecology of Halophytes. Academic Press, New York, pp. 393–406.

Reimold, R. J., and F. C. Daiber. 1970. Dissolved phosphorus concentrations in a natural salt marsh of Delaware. Hydrobiologia 36:361–371.

Reinartz, J. A., and E. L. Warne. 1993. Development of vegetation in small created wetlands in southeast Wisconsin. Wetlands 13:153–164.

Reinelt, L. E., and R. R. Horner. 1995. Pollutant removal from stormwater runoff by palustrine wetlands based on comprehensive budgets. Ecological Engineering 4:77–97.

Reiners, W. A. 1972. Structure and energetics of three Minnesota forests. Ecological Monographs 42:71–94.

Rezendes, P., and P. Roy. 1996. Wetlands: The Web of Life. Sierra Club Books, San Francisco. 156 pp.

Rheinhardt, R. D., M. M. Brinson, and P. M. Farley. 1997. Applying wetland reference data to functional assessment, mitigation, and restoration. Wetlands 17:195–215.

Rheinhardt, R. D., M. C. Rheinhardt, M. M. Brinson, and K. Faser. 1998. Forested wetlands of low order streams in the inner coastal plain of North Carolina, USA. Wetlands 18:365–378.

Richardson, C. J. 1979. Primary productivity values in freshwater wetlands. In P. E. Greeson, J. R. Clark, and J. E. Clark, eds. Wetland Functions and Values: The State of Our Understanding. American Water Resources Association, Minneapolis, pp. 131–145.

Richardson, C. J., ed. 1981. Pocosin Wetlands. Hutchinson Ross Publishing, Stroudsburg, PA. 364 pp.

Richardson, C. J. 1983. Pocosins: Vanishing wastelands or valuable wetlands? BioScience 33:626–633.

Richardson, C. J. 1985. Mechanisms controlling phosphorus retention capacity in freshwater wetlands. Science 228:1424–1427.

Richardson, C. J., W. A. Wentz, J. P. M. Chamie, J. A. Kadlec, and D. L. Tilton. 1976. Plant growth, nutrient accumulation and decomposition in a central Michigan peatland used for effluent treatment. In D. L. Tilton, R. H. Kadlec, and C. J. Richardson, eds. Freshwater Wetlands and Sewage Effluent Disposal. University of Michigan Press, Ann Arbor, pp. 77–117.

Richardson, C. J., D. L. Tilton, J. A. Kadlec, J. P. M. Chamie, and W. A. Wentz. 1978. Nutrient dynamics of northern wetland ecosystems. In R. E. Good, D. F. Whigham, and R. L. Simpson, eds. Freshwater Wetlands—Ecological Processes and Management Potential. Academic Press, New York, pp. 217–241.

Richardson, C. J., R. Evans, and D. Carr. 1981. Pocosins: An ecosystem in transition. In C. J. Richardson, ed. Pocosin Wetlands. Hutchinson Ross Publishing, Stroudsburg, PA, pp. 3–19.

Richardson, C. J., and D. S. Nichols. 1985. Ecological analysis of wastewater management criteria in wetland ecosystems. In P. J. Godfrey, E. R. Kaynor, S. Pelczarski, and J. Benforado, eds. Ecological Considerations in Wetlands Treatment of Municipal Wastewaters. Van Nostrand Reinhold, New York, pp. 351–391.

Richardson, C. J., and C. B. Craft. 1993. Effective phosphorus retention in wetlands—Fact or fiction? In G. A. Moshiri, ed. Constructed Wetlands for Water Quality Improvement. CRC Press, Boca Raton, FL, pp. 271–282.

Richardson, C. J., S. Qian, C. B. Craft, and R. G. Qualls. 1997. Predictive models for phosphorus retention in wetlands. Wetlands Ecology and Management 4:159–175.

Richardson, J., P. A. Straub, K. C. Ewel, and H. T. Odum. 1983. Sulfate-enriched water effects on a floodplain forest in Florida. Environmental Management 7:321–326.

Richter, K. O., and A. L. Azous. 1995. Amphibian occurrence and wetland characteristics in the Puget Sound Basin. Wetlands 15:305–312

Riley, J. P., and G. Skirrow. 1975. Chemical Oceanography, 2nd ed., Vol. 2. Academic Press, New York. 647 pp.

Risotto, S., and R. E. Turner. 1985. Annual fluctuations in the abundance of the commercial fisheries of the Mississippi River and tributaries. North American Journal of Fisheries Management 4:557–574.

Rivera-Monroy, V. H., J. W. Day, R. R. Twilley, R. Vera-Herrera, and C. Coronado-Molina. 1995. Flux of nitrogen and sediment in a fringe mangrove forest in Términos Lagoon, Mexico. Estuarine, Coastal and Shelf Science 40:139–160.

Rivers, J. S., D. I. Siegel, L. S. Chasar, J. P. Chanton, P. H. Glaser, N. T. Roulet, and J. M. McKenzie. 1998. A stochastic appraisal of the annual carbon budget of a large circumboreal peatland, Rapid River Watershed, northern Minnesota. Global Biogeochemical Cycles 12:715–727.

Robb, D. M. 1989. Diked and Undiked Freshwater Coastal Marshes of Western Lake Erie. Master's Thesis, Ohio State University, Columbus. 145 pp.

Roberts, H. H., and L. Van Heerden. 1992. Atchafalaya–Wax Lake Delta Complex: The New Mississippi River Delta Lobe. Research Report CSA-IARP, Report 1, First Annual CSI Industrial Association Research Program, Coastal Studies Institute, Louisiana State University, Baton Rouge.

Roberts, J. K. M. 1988. Cytoplasmic acidosis and flooding in crop plants. In Hook, D. D., W. H. McKee, Jr., H. K. Smith, J. Gregory, V. G. Burrell, M. R. DeVoe, R. E. Sojka, S. Gilbert, R. Banks, L. G. Stolzy, C. Brooks, T. D. Matthews, and T. H. Shear, eds. The Ecology and Management of

Wetlands, Vol. 1: Ecology of Wetlands. Timber Press, Portland, OR, pp. 392–397.

Roberts, L. 1993. Wetlands trading is a loser's game, say ecologists. Science 260:1890–1892.

Robertson, A. I. 1986. Leaf-burying crabs: Their influence on energy flow and export from mixed mangrove forests (*Rhizophora* spp.) in northeastern Australia. Journal of Experimental Marine Biology and Ecology 102:237–248.

Robertson, A. I., and P. A. Daniel. 1989. The influence of crabs on litter processing in high intertidal mangrove forests in tropical Australia. Oecologia 78:191–198.

Robertson, A. I., and N. C. Duke. 1990. Mangrove fish-communities in tropical Queensland, Australia: Spatial and temporal patterns in densities, biomass and community structure. Marine Biology 104:369–379.

Robertson, P. A. 1987. The woody vegetation of Little Black Slough: An undisturbed upland–swamp forest in southern Illinois. In R. L. Hays, F. W. Woods, and H. deSelm, eds. Proceedings of the Sixth Central Hardwood Forest Conference, Knoxville, TN. Department of Forestry, Wildlife and Fisheries, Department of Botany, University of Tennessee; Southern Forest Experiment Station, Forest Service, U.S. Department of Agriculture, pp. 353–367.

Rochefort, L., D. H. Vitt, and S. E. Bayley. 1990. Growth, production, and decomposition dynamics of *Sphagnum* under natural and experimentally acidified conditions. Ecology 71:1986–2000.

Rochefort, L., and S. Campeau. 1997. Rehabilitation work on post-harvested bogs in south eastern Canada. In L. Parkyn, R. E. Stoneman, and H. A. P. Ingram, eds. Conserving Peatlands. CAB International, Walingford, UK, pp. 287–284.

Rodin, L. E., N. I. Bazilevich, and N. N. Rozov. 1975. Productivity of the world's ecosystems. In D. E. Reichle, J. Franklin, and D. W. Goodal, eds. Productivity of the World's Ecosystems. National Academy of Sciences, Washington, DC.

Roe, H. B., and Q. C. Ayres. 1954. Engineering for Agricultural Drainage. McGraw-Hill, New York. 501 pp.

Rolband, M. S. 1995. A comparison of field delineated wetland areas vs. NWI mapping and SCS hydric soils mapping. In M. Landin, ed. National Interagency Workshop on Wetlands: Technology Advances for Wetlands Science. U.S. Army Corps of Engineers Waterways Experiment Station, Vicksburg, MS.

Romanov, V. V. 1968. Hydrophysics of Bogs. Translated from Russian by N. Kaner; edited by Prof. Heimann. Israel Program for Scientific Translation, Jerusalem. Available from Clearinghouse for Federal Scientific and Technical Information, Springfield, VA. 299 pp.

Rood, S. B., and J. M. Mahoney. 1990. Collapse of riparian poplar forests downstream from dams in western prairies: Probable causes and prospects for mitigation. Environmental Management 14:451–464.

Rosgen, D. L. 1985. A stream classification system. In R. R. Johnson, C. D. Ziebell, D. R. Patton, P. F. Ffolliott, and R. H. Hamre, eds. Riparian Ecosystems and Their Management: Reconciling Conflicting Uses. General Technical Report RM-120, Rocky Mountain Forest and Range Experiment Station, Forest Service, U.S. Department of Agriculture, Fort Collins, CO.

Rosswall, T. and O. W. Heal. 1975. Structure and function of tundra ecosystems. Ecological Bulletin (Sweden) 20:265–294.

Roulet, N. T., R. Ash, and T. R. Moore. 1992a. Low boreal wetlands as a source of atmospheric methane. Journal of Geophysical Research 97:3739–3749.

Roulet, N. T., T. R. Moore, J. Bubier, and P. Lafleur. 1992b. Northern fens: CH_4 flux and climate change. Tellus 44B:100–105.

Round, F. E. 1960. The diatom flora of a salt marsh on the River Dee. New Phytologist 59:332–348.

Rowe, J. S. 1972. Forest Regions of Canada. Publication 1300, Canadian Forest Service. 172 pp.

Rubec, C. D. A. 1994. Canada's federal policy on wetland conservation: A global model. In W. J. Mitsch, ed. Global Wetlands: Old World and New. Elsevier, Amsterdam, pp. 909–917.

Ruebsamen, R. N. 1972. Some Ecological Aspects of the Fish Fauna of a Louisiana Intertidal Pond System. Master's Thesis, Louisiana State University, Baton Rouge. 80 pp.

Rundquist, D. C. 1983. Wetland Inventories of Nebraska's Sandhills. Resource Report 9, Nebraska Remote Sensing Center, University of Nebraska, Lincoln. 46 pp.

Runyon, L. C. 1993. The Lucas Court Case and Land-Use Planning. National Conference of State Legislators, Denver, CO, Supplement to State Legislatures, Vol. 1, No. 10 (March).

Russell, H. S. 1976. A Long, Deep Furrow: Three Centuries of Farming in New England. University Press of New England, Hanover, NH. 671 pp.

Ruttner, F. 1963. Fundamentals of Limnology, 3rd ed. University of Toronto Press, Toronto. 295 pp.

Rützler, K., and I. C. Feller. 1996. Caribbean mangrove swamps. Scientific American (March 1996):94–99.

Rycroft, D. W., D. J. A. Williams, and H. A. E. Ingram. 1975. The transmission of water through peat. I. Review. Journal of Ecology 63:535–556.

Rykiel, E. J., Jr. 1984. General hydrology and mineral budgets for Okefenokee Swamp: Ecological significance. In A. D. Cohen, D. J. Casagrande, M. J. Andrejko, and G. R. Best, eds. The Okefenokee Swamp: Its Natural History, Geology, and Geochemistry. Wetland Surveys, Los Alamos, NM, pp. 212–228.

Rymal, D. E., and G. W. Folkerts. 1982. Insects associated with pitcher plants (*Sarracenia*, Sarraceniaceae), and their relationship to pitcher plant conservation: A review. Journal of Alabama Academy of Science 53:131–151.

Ryther, J. H., R. A. Debusk, M. D. Hanisak, and L. D. Williams. 1979. Freshwater macrophytes for energy and waste water treatment. In P. E. Greeson, J. R. Clark, and J. E. Clark, eds. Wetland Functions and Values: The State of Our Understanding. American Water Resources Association, Minneapolis, pp. 652–660.

Sala, A., D. Smith, and D. A. Devitt. 1996. Water use by *Tamarix ramosissima* and associated phreatophytes in a Mojave Desert floodplain. Ecological Applications 6:888–898.

Salonen, V. 1987. Relationship between the seed rain and the establishment of vegetation in two areas abandoned after peat harvesting. Holarctic Ecology 10:171–174.

Salonen, V. 1990. Early plant succession in two abandoned cutover peatland areas. Holarctic Ecology 13:217–223.

Sanders, M. D., and M. J. Winterbourn. 1993. Effect of *Sphagnum* harvesting on invertebrate species diversity and community size. New Zealand Department of Conservation Science and Research Series 57, Wellington.

Sansanayuth, P., A. Phadungchep, S. Ngammontha, S. Ngdngam, P. Sukasem, H. Hoshino, and M. S. Ttabucanon. 1996. Shrimp pond effluent: Pollution problems and treatment by constructed wetlands. Water Science and Technology 34:93–98.

Sanville, W., and W. J. Mitsch, eds. 1994. Creating Freshwater Marshes in a Riparian Landscape: Research at the Des Plaines River Wetland Demonstration Project. Special Issue of Ecological Engineering 3(4):315–521.

Sartoris, J. J., J. S. Thullen, L. B. Barber, and D. E. Salas. 2000. Investigation of nitrogen transformations in a southern California constructed wastewater treatment wetland. Ecological Engineering 14:49–65.

Sass, R. L., F. M. Fisher, and P. A. Harcombe. 1990. Methane production and emission in a Texas rice field. Global Biogeochemical Cycles 4:47–68.

Sass, R. L., F. M. Fisher, Y. B. Wang, F. T. Turner, and M. F. Jund. 1992. Methane emission from rice fields: The effect of floodwater management. Global Biogeochemical Cycles 6:249–262.

Sasser, C. E. 1994. Vegetation Dynamics in Relation to Nutrients in Floating Marshes in Louisiana, USA. Ph.D. Dissertation, Utrecht University, Utrecht, Netherlands. 193 pp.

Sasser, C. E., G. W. Peterson, D. A. Fuller, R. K. Abernethy, and J. G. Gosselink. 1982. Environmental Monitoring Program, Louisiana Offshore Oil Port Pipeline. 1981 Annual Report, Coastal Ecology Laboratory, Center for Wetland Resources, Louisiana State University, Baton Rouge. 299 pp.

Sasser, C. E., and J. G. Gosselink. 1984. Vegetation and primary production in a floating freshwater marsh in Louisiana. Aquatic Botany 20:245–255.

Sasser, C. E., M. D. Dozler, J. G. Gosselink, and J. M. Hill. 1986. Spatial and temporal changes in Louisiana's Barataria Basin marshes. Environmental Management 10:671–680.

Sasser, C. E., J. G. Gosselink, and G. P. Shaffer. 1991. Distribution of nitrogen and phosphorus in a Louisiana freshwater floating marsh. Aquatic Botany 41:317–331.

Sasser, C. E., J. M. Visser, D. E. Evers, and J. G. Gosselink. 1995. The role of environmental variables on interannual variation in species composition and biomass in a subtropical minerotrophic floating marsh. Canadian Journal of Botany 73:413–424.

Sasser, C. E., J. G. Gosselink, E. M. Swenson, C. M. Swarzenski, and N. C. Leibowitz. 1996. Vegetation, substrate and hydrology in floating marshes in the Mississippi river delta plain wetlands, USA. Vegetatio 122:129–142.

Sather, J. H., and R. D. Smith. 1984. An Overview of Major Wetland Functions and Values. FWS/OBS-84/18, Western Energy and Land Use Team, U.S. Fish and Wildlife Service, Washington, DC. 68 pp.

Savage, H. 1956. River of the Carolinas: The Santee. Rinehart, New York. 435 pp.

Savage, H. 1983. The Mysterious Carolina Bays. University of South Carolina Press, Columbia. 121 pp.

Schaafsma, J. A., A. H. Baldwin, and C. A. Streb. 2000. An evaluation of a constructed wetland to treat wastewater from a dairy farm in Maryland, USA. Ecological Engineering 14:199–206.

Schaeffer-Novelli, Y., G. Cintrón-Molero, R. R. Adaime, and T. M. de Camargo. 1990. Variability of mangrove ecosystems along the Brazilian coast. Estuaries 13:204–218.

Schamberger, M. L., C. Short, and A. Farmer. 1979. Evaluation wetlands as a wildlife habitat. In P. E. Greeson, J. R. Clark, and J. E. Clark, eds. Wetland Functions and Values: The State of Our Understanding. American Water Resources Association, Minneapolis, pp. 74–83.

Schat, H. 1984. A comparative ecophysiological study on the effects of waterlogging and submergence on dune slack plants: Growth, survival and mineral nutrition in sand culture experiments. Oecologia (Berlin) 62:279–286.

Scheffe, R. D. 1978. Estimation and Prediction of Summer Evapotranspiration from a Northern Wetland. Master's Thesis, University of Michigan, Ann Arbor. 69 pp.

Schlesinger, W. H. 1978. Community structure, dyes, and nutrient ecology in the Okefenokee Cypress Swamp-Forest. Ecological Monographs 48:43–65.

Schlesinger, W. H. 1991. Biogeochemistry: An Analysis of Global Change. Academic Press, San Diego. 443 pp.

Schlesinger, W. H., and B. F. Chabot. 1977. The use of water and minerals by evergreen and deciduous shrubs in Okefenokee Swamp. Botanical Gazette 138:490–497.

Schlosser, I. J., and J. R. Karr. 1981a. Water quality in agricultural watersheds: Impact of riparian vegetation during base flow. Water Resources Bulletin 17:233–240.

Schlosser, I. J., and J. R. Karr. 1981b. Riparian vegetation and channel morphology impact on spatial patterns of water quality in agricultural watersheds. Environmental Management 5:233–243.

Schneider, R. L., and R. R. Sharitz. 1986. Seed bank dynamics in a southeastern riverine swamp. American Journal of Botany 73:1022–1030.

Schneider, R. L., and R. R. Sharitz. 1988. Hydrochory and regeneration in a bald cypress–water tupelo swamp forest. Ecology 69:1055–1063.

Scholander, P. F. 1968. How mangroves desalinate seawater. Physiologia Plantarum 21:251–261.

Scholander, P. F., L. van Dam, and S. I. Scholander. 1955. Gas exchange in the roots of mangroves. American Journal of Botany 42:92–98.

Scholander, P. F., H. T. Hammel, E. D. Bradstreet, and E. A. Hemmingsen. 1965. Sap pressure in vascular plants. Science 148:339–346.

Scholander, P. F., E. D. Bradstreet, H. T. Hammel, and E. A. Hemmingsen. 1966. Sap concentrations in naiophytes and some other plants. Plant Physiology 41:529–532.

Schouwenaars, J. M. 1993. Hydrological differences between bogs and bog-relicts and consequences for bog restoration. Hydrobiologia 265:217–224.

Schröder, P. 1989. Characterization of a thermo-osmotic gas transport mechanism in *Alnus glutinosa* (L.) Gaertn. Trees 3:38–44.

Schubel, J. R., and D. W. Pritchard. 1990. Great Lakes estuaries—Phooey. Estuaries 13:508–509.

Schutz, H., A. Holzapfel-Pschorn, R. Conrad, H. Rennenberg, and W. Seiler. 1989. A three year continuous record on the influence of daytime, season and fertilizer treatment on methane emission rates from an Italian rice paddy field. Journal of Geophysical Research 94:16405–16416.

Scodari, P. F. 1990. Wetlands Protection: The Role of Economics. Environmental Law Institute, Washington, DC. 89 pp.

Scott, D. A. 1993. A Directory of Wetlands in Oceania. Wetlands International, Wageningen, The Netherlands, 444 pp.

Scott, D. A. 1995. A Directory of Wetlands in the Middle East. Wetlands International, Wageningen, The Netherlands, 560 pp.

Scott, D. A., and T. A. Jones. 1995. Classification and inventory of wetlands: A global overview. Vegetatio 118:3–16.

Scott, M. L., G. T. Auble, and J. M. Friedman. 1997. Flood dependency of cottonwood establishment along the Missouri River, MT, USA. Ecological Applications 7:677–690.

Sedell, J. R., P. A. Bisson, and J. A. June. 1980. Ecology and habitat requirements of fish populations in South Fork Hoh River, Olympic National Park. In Proceedings of the Second Conference on Scientific Research in National Parks, NPS/ST-80/02/7 7:47–63.

Sedell, J. R., F. H. Everest, and F. J. Swanson. 1982. Fish habitat and streamside management: Past and present. In Proceedings of the Society of American Foresters Annual Meeting, 1981, Bethesda, MD. Society of American Foresters, pp. 244–255.

Sedell, J. R., and K. J. Luchessa. 1982. Using the historical record as an aid to salmonid habitat enhancement. In N. B. Armantrout, ed. Proceedings of a Symposium on Acquisition and Utilization of Aquatic Habitat Inventory Information. Western Division, American Fisheries Society, Portland, OR, pp. 210–223.

Sedell, J. R., and J. L. Froggatt. 1984. Importance of streamside forests to large rrivers: The isolation of the Willamette River, Oregon, USA, from its floodplain by snagging and streamside forest removal. Verhandlungen Internationale Vereinigung für Theoretische und Augewantre Limnologie 22:1828–1834.

Sedell, J. R., and W. S. Duval. 1985. Influence of Forest and Rangeland Management on Anadromous Fish Habitat in Western North America. 5. Water Transportation and Storage of Logs. General Technical Report PNW-186, Pacific Northwest Forest and Range Experiment Station, Forest Service, U.S. Department of Agriculture, Portland, OR.

Seidel, K. 1964. Abbau von Bacterium Coli durch höhere Wasserpflanzen. Naturwissenschaften 51:395.

Seidel, K. 1966. Reinigung von Gewässern durch höhere Pflanzen. Naturwissenschaften 53:289–297.

Seidel, K., and H. Happl. 1981. Pflanzenkläranlage "Krefelder system." Sicherheit in Chemic und Umbelt 1:127–129.

Seiler, W. A., A. Holzapfel-Pschorn, R. Conrad, and D. Scharffe. 1984. Methane emission from rice paddies. Journal of Atmospheric Chemistry 1:214–268.

Sell, M. G. 1977. Modeling the Respse of Mangrove Ecosystems to Herbicide Spraying, Hurricanes, Nutrient Enrichment and Economic Development. Ph.D. Dissertation, University of Florida, Gainesville. 389 pp.

Shabman, L. 1986. The contribution of economics to wetlands valuation and management. In J. A. Kusler and P. Reixinger, eds. Proceedings of the National Wetlands Assessment Symposium. Association of State Wetland Managers Technical Report, pp. 9–13.

Shaffer, G. W., C. E. Sasser, J. G. Gosselink, and M. Rejmanek. 1992. Vegetation dynamics in the emergent Atchafalaya delta, Louisiana, USA. Journal of Ecology 80.

Sharitz, R. R., and J. W. Gibbons, eds. 1982. The Ecology of Southeastern Shrub Bogs (Pocosins) and Carolina Bays: A Community Profile. FWS/OBS-82/04, U.S. Fish and Wildlife Service, Washington, DC.

Sharitz, R. R., and J. W. Gibbons. 1989. Freshwater Wetlands and Wildlife. U.S. Department of Energy, NTIS, Springfield, VA. 1265 pp.

Sharitz, R. R., and W. J. Mitsch. 1993. Southern floodplain forests. In W. H. Martin, S. G. Boyce, and A. C. E. Echternacht, eds. Biodiversity of the Southeastern United States: Lowland Terrestrial Communities. John Wiley & Sons, New York, pp. 311–372.

Shaver, G. R., and J. M. Melillo. 1984. Nutrient budgets of marsh plants: Efficiency concepts and relation to availability. Ecology 65:1491–1510.

Shaw, S. P., and C. G. Fredine. 1956. Wetlands of the United States, Their Extent, and Their Value for Waterfowl and Other Wildlife. Circular 39, U.S. Fish and Wildlife Service, U.S. Department of Interior, Washington, DC. 67 pp.

Shearer, J. C., and B. R. Clarkson. 1998. Whangamarino Wetland: Effects of lowered river levels on peat and vegetation. International Peat Journal 8:52–65.

Shedlock, R. J., D. A. Wilcox, T. A. Thompson, and D. A. Cohen. 1993. Interactions between ground water and wetlands, southern shore of Lake Michigan, USA. Journal of Hydrology 141:127–155.

Sheffield, R. M., T. W. Birch, W. H. McWilliams, and J. B. Tansey. 1998. *Chamaecyparis thyoides* (Atlantic white cedar) in the United States. In A. D. Laderman, ed. Coastally Restricted Forests. Oxford University Press, New York, pp. 111–123.

Shelford, V. E. 1907. Preliminary note on the distribution of the tiger beetle (Cicindela) and its relation to plant succession. Biological Bulletin 14.

Shelford, V. E. 1911. Ecological succession. II. Pond fishes. Biological Bulletin 21:127–151.

Shelford, V. E. 1913. Animal Communities in Temperate America as Illustrated in the Chicago Region. University of Chicago Press, Chicago.

Shepard, J. P., S. J. Brady, N. D. Cost, and C. G. Storrs. 1998. Classification and inventory. In M. G. Messina and W. H. Conner, eds. Southern Forested Wetlands. Lewis Publishers, Boca Raton, FL, pp. 3–28.

Shew, D. M., R. A. Linthurst, and E. D. Seneca. 1981. Comparison of production computation methods in a southeastern North Carolina *Spartina alterniflora* salt marsh. Estuaries 4:97–109.

Shiel, R. J. 1994. Death and life of the billabong. In X. Collier, ed. Restoration of Aquatic Habitats. Selected Papers from New Zealand Limnological Society 1993 Annual Conference, Department of Conservation, pp. 19–37.

Shisler, J. K. 1990. Creation and restoration of coastal wetlands of the northeastern United States. In J. A. Kusler and M. E. Kentula, eds. Wetland Creation and Restoration. Island Press, Washington, DC, pp. 143–170.

Shjeflo, J. B. 1968. Evapotranspiration and the water budget of prairie potholes in North Dakota. Professional Paper 585-B, U.S. Geological Survey, Washington, DC. 49 pp.

Shugart, H. H., T. M. Smith, and W. M. Post. 1992. The potential for application of individual-based simulation models for assessing the effects of global change. Annual Review of Ecology and Systematics 23:15–38.

Shutes, R. B. E., D. M. Revitt, A. S. Mungar, and L. N. L. Scholes. 1997. The design of wetland systems for the treatment of urban runoff. Water Science and Technology 35:19–25.

Siegel, D. I. 1983. Ground water and evolution of patterned mires, glacial Lake Agassiz peatlands, northern Minnesota. Journal of Ecology 71:913–921.

Siegel, D. I. 1988a. A review of the recharge–discharge function of wetlands. In D. D. Hook, W. H. McKee, Jr., H. K. Smith, J. Gregory, V. G. Burrell, M. R. DeVoe, R. E. Sojka, S. Gilbert, R. Banks, L. G. Stolzy, C. Brooks, T. D. Matthews, and T. H. Shear, eds. The Ecology and Management of Wetlands, Vol. 1: Ecology of Wetlands. Timber Press, Portland, OR, pp. 59–67.

Siegel, D. I. 1988b. Evaluating cumulative effects of disturbance on the hydrologic function of bogs, fens, and mires. Environmental Management 12:621–626.

Siegel, D. I. 1988c. The recharge–discharge function of wetlands near Juneau, Alaska. I. Hydrogeological investigations. Groundwater 26:427–434.

Siegel, D. I. 1988d. The recharge–discharge function of wetlands near Juneau, Alaska. II. Geochemical studies. Groundwater 26:526–544.

Siegel, D. I., and P. H. Glaser. 1987. Groundwater flow in the bog/fen complex, Lost River Peatland, northern Minnesota. Journal of Ecology 75:743–754.

Siegel, D. I., A. S. Reeve, P. H. Glaser, and E. A. Romanowicz. 1995. Climate-driven flushing of pore water in peatlands. Nature 531–533.

Siegley, C. E., R. E. J. Boerner, and J. M. Reutter. 1988. Role of the seed bank in the development of vegetation on a freshwater marsh created from dredge spoil. Journal of Great Lakes Research 14:267–276.

Sifneos, J. C., E. W. Cake, Jr., and M. E. Kentula. 1992. Effects of Section 404 permitting on freshwater wetlands in Louisiana, Alabama, and Mississippi. Wetlands 12:28–36.

Sikora, J. P., W. B. Sikora, C. W. Erkenbrecher, and B. C. Coull. 1977. Significance of ATP, carbon and caloric content of meiobenthic nematodes in partitioning benthic biomass. Marine Biology 44:7–14.

Sikora, L. J., and D. R. Keeney. 1983. Further aspects of soil chemistry under anaerobic conditions. In A. J. P. Gore, ed. Ecosystems of the World, Vol. 4A: Mires: Swamp, Bog, Fen, and Moor. Elsevier, Amsterdam, pp. 247–256.

Sikora, W. B. 1977. The Ecology of *Palaemonetes pugio* in a Southeastern Salt Marsh Ecosystem with Particular Emphasis on Production and Trophic Relationship. Ph.D. Dissertation, University of South Carolina. 122 pp.

Silander, J. A. 1984. The genetic basis of the ecological amplitude of *Spartina patens*. III. Allozyme variation. Botanical Gazette 145:569–577.

Silvola, J., and I. Hanski. 1979. Carbon accumulation in a raised bog. Oecologia (Berlin) 37:285–295.

Simberloff, D., and T. Dayan. 1991. The guild concept and the structure of ecological communities. Annual Review of Ecology and Systematics 22:115–143.

Simpson, R. L., D. F. Whigham, and R. Walker. 1978. Seasonal patterns of nutrient movement in a freshwater tidal marsh. In R. E. Good, D. F. Whigham, and R. L. Simpson, eds. Freshwater Wetlands: Ecological Processes and Management Potential. Academic Press, New York, pp. 243–257.

Simpson, R. L., R. E. Good, M. A. Leck, and D. F. Whigham. 1983. The ecology of freshwater tidal wetlands. BioScience 33:255–259.

Singer, D. K., S. T. Jackson, B. J. Madsen, and D. A. Wilcox. 1996. Differentiating climatic and successional influences on long-term development of a marsh. Ecology 77:1765–1778.

Singer, P. C., and W. Stumm. 1970. Acidic mine drainage: The rate-determining step. Science 167:1121–1123.

Sinicrope, T. L., P. G. Hine, R. S. Warren, and W. A. Niering. 1990. Restoring of an impounded salt marsh in New England. Estuaries 13:25–30.

Sjörs, H. 1948. Myrvegetation i bergslagen. Acta Phytogeographica Suecica 21:1–299.

Sjörs, H. 1950. On the relationship between vegetation and electrolytes in North Swedish mire waters. Oikos 2:239–258.

Sjörs, H. 1961a. Bogs and fens on Attawapiskat River, Northern Ontario. National Museum of Canada Bulletin 186. 133 pp.

Sjörs, H. 1961b. Surface patterns in boreal peatlands. Endeavor 20:217–224.

Skaggs, R. W., J. W. Gilliam, and R.O. Evans. 1991. A computer simulation study of pocosin hydrology. Wetlands 11:399–416.

Skaggs, R. W., D. Amatya, R. O. Evans, and J. E. Parsons. 1994. Characterization and evaluation of proposed hydrologic criteria for wetlands. Journal of Soil and Water Conservation 49:501–510.

Sklar, F. H., and W. H. Conner. 1979. Effects of altered hydrology on primary production and aquatic animal populations in a Louisiana swamp forest. In J. W. Day, Jr., D. D. Culley, Jr., R. E. Turner, and A. T. Humphrey, Jr. eds. Proceedings of the Third Coastal Marsh and Estuary Management Symposium. Division of Continuing Education, Louisiana State University, Baton Rouge, pp. 101–208.

Smalley, A. E. 1960. Energy flow of a salt marsh grasshopper population. Ecology 41:672–677.

Smart, R. M., and J. W. Barko. 1980. Nitrogen nutrition and salinity tolerance of *Distichlis spicata* and *Spartina alterniflora*. Ecology 61:630–638.

Smith, C. J., R. D. Delaune, and W. H. Patrick, Jr. 1982. Carbon and nitrogen cycling in a *Spartina alterniflora* salt marsh. In J. R. Freney and I. E. Galvally, eds. The Cycling of Carbon, Nitrogen, Sulfur and Phosphorus in Terrestrial and Aquatic Ecosystems. Springer-Verlag, New York, pp. 97–104.

Smith, C. S., M. S. Adams, and T. D. Gustafson. 1988. The importance of belowground mineral element states in cattails (*Typha latifolia* L.). Aquatic Botany 30:343–352.

Smith, K. J., G. L. Taghon, and K. W. Able. 2000. Trophic linkages in marshes: Ontogenetic changes in diet for young-of-the-year mummichog, *Fundulus heteroclitus*. In M. Weinstein and D. A. Kreeger, eds. International Symposium: Concepts and Controversies in Tidal Marsh Ecology. Kluwer Academic Publishers, Dordrecht, The Netherlands, pp. 221–236.

Smith, L. M., and J. A. Kadlec. 1985. Predictions of vegetation change following fire in a Great Salt Lake marsh. Aquatic Botany 21:43–51.

Smith, R. C. 1975. Hydrogeology of the experimental cypress swamps. In H. T. Odum and K. C. Ewel, eds. Cypress Wetlands for Water Management, Recycling and Conservation. Second Annual Report to NSF and Rockefeller Foundation, Center for Wetlands, University of Florida, Gainesville, pp. 114–138.

Smith, R. L. 1980. Ecology and Field Biology, 3rd ed. Harper & Row, New York. 835 pp.

Smith, R. L., and M. J. Klug. 1981. Electron donors utilized by sulfate-reducing bacteria in eutrophic lake sediments. Applied and Environmental Microbiology 42:116–121.

Smith, S. D., D. A. Devitt, A. Sala, J. R. Cleverly, and D. E. Busch. 1998. Water relations of riparian plants from warm desert regions. Wetlands 18:687–696.

Smith, T. J. 1983. Alteration of salt marsh community composition by grazing snow geese. Holarctic Ecology 6:204–210.

Smith, T. J. 1987. Seed predation in relation to tree dominance and distribution in mangrove forests. Ecology 68:266–273.

Smith, T. J. I., and W. E. Odum. 1981. The effects of grazing by snow geese on coastal salt marshes. Ecology 62:98–106.

Smith, T. J., H.-T. Chan, C. C. McIvor, and M. B. Robblee. 1989. Comparisons of seed predation in tropical, tidal forests on three continents. Ecology 70:146–151.

Smith, T. J., K. G. Boto, S. D. Frusher, and R. L. Giddins. 1991. Keystone species and mangrove forest dynamics: The influence of burrowing by crabs on soil nutrient status and forest productivity. Estuarine, Coastal and Shelf Science 33:419–432.

Smith, T. J., M. B. Robblee, H. R. Wanless, and T. W. Doyle. 1994. Mangroves, hurricanes, and lightning strikes. BioScience 44:256–262.

Smits, A. J. M., R. M. J. C. Kleukers, C. J. Kok, and A. G. van der Velde. 1990a. Alcohol dehydrogenase isozymes in the roots of some nymphaeid and isoetid macrophytes: Adaptations to hypoxic sediment conditions? Aquatic Botany 38:19–27.

Smits, A. J. M., P. Laan, R. H. Thier, and A. G. van der Velde. 1990b. Root aerenchyma, oxygen leakage patterns and alcohol fermentation ability of the roots of some nymphaeid and isoetid macrophytes in relation to the sediment type of their habitat. Aquatic Botany 38:3–17.

Smock, L. A., and K. L. Harlowe. 1983. Utilization and processing of freshwater wetland macrophytes by the detritivore Asellus forbesi. Ecology 64:1556–1565.

Snedaker, S. C. 1989. Overview of ecology of mangroves and information needs for Florida Bay. Bulletin of Marine Science 44:341–347.

Söderqvist, T., W. J. Mitsch, and R. K. Turner, eds, 2000. The Values of Wetlands: Landscape and Institutional Perspectives. Special Issue of Ecological Economics. 36 (in press).

Solem, T. 1986. Age, origin and development of blanket mires in soer-Troendelag, central Norway. Boreas 15:101–115.

Soper, E. K. 1919. The peat deposits of Minnesota. Minnesota Geological Survey Bulletin 16:1–261.

Sorrell, B. K., and P. I. Boon. 1992. Biogeochemistry of billabong sediments. II. Seasonal variations in methane production. Freshwater Biology 27:435–445.

Sorrell, B. K., and P. I. Boon. 1994. Convective gas flow in *Eleocharis sphacelata* R. Br.: Methane transport and release from wetlands. Aquatic Botany 47:197–212.

Sorrell, B. K., H. Brix, and P. I. Boon. 1994. Modelling of *in situ* oxygen transport and aerobic metabolism in the hydrophyte *Eleocharis sphacelata* R. Br. Proceedings of the Royal Society of Edinburgh 102B:367–372.

Spieles, D. J., and W. J. Mitsch. 2000. The effects of season and hydrologic and chemical loading on nitrate retention in constructed wetlands: A comparison of low- and high-nutrient wetlands. Ecological Engineering 14:77–91.

Standford, J. A., J. V. Ward, W. J. Liss, C. A. Frissell, R. N. Williams, J. A. Lichatowich, and C. C. Coutant. 1996. A general protocol for restoration of regulated rivers. Regulated Rivers Research and Management 12:391–413.

Stark, L. R., and F. M. Williams. 1995. Assessing the performance indices and design parameters of treatment wetlands for H, Fe, and Mn retention. Ecological Engineering 5:433–444.

Stavins, R. 1987. Conversion of Forested Wetlands to Agricultural Uses: Executive Summary. Environmental Defense Fund, New York. 72 pp.

Stearns, F. 1978. Management potential: Summary and recommendations. In R. E. Good, D. F. Whigham, and R. L. Simpson, eds. Freshwater Wetlands: Ecological Processes and Management Potential. Academic Press, New York, pp. 357–363.

Steever, E. Z., R. S. Warren, and W. A. Niering. 1976. Tidal energy subsidy and standing crop production of *Spartina alterniflora*. Estuarine, Coastal and Marine Science 4:473–478.

Steiner, G. R., J. T. Watson, D. Hammer, and D. F. Harker, Jr. 1987. Municipal wastewater treatment with artificial wetlands—A TVA/Kentucky demonstration. In K. R. Reddy and W. H. Smith, eds. Aquatic Plants for Wastewater Treatment and Resource Recovery. Magnolia Publishing, Orlando, FL, p. 923.

Steiner, G. R., and R. J. Freeman, Jr. 1989. Configuration and substrate design considerations for constructed wetlands for wastewater treatment. In D. A. Hammer, ed. Constructed Wetlands for Wastewater Treatment. Lewis Publishers, Chelsea, MI, pp. 363–378.

Stelzer, R., and A. Lauchli. 1978. Salt and flooding tolerance of *Puccinellia peisonis*. III. Distribution and localization of ions in the plant. Zeitschrift für Pflanzenphysiologie 88:437–448.

Stephenson, T. D. 1990. Fish reproductive utilization of coastal marshes of Lake Ontario near Toronto. Journal of Great Lakes Research 16:71–81.

Steudler, P. A., and B. J. Peterson. 1984. Contribution of gaseous sulfur from salt marshes to the global sulfur cycle. Nature 311:455–457.

Stevenson, J. S., D. R. Heinle, D. A. Flemer, R. J. Small, R. A. Rowland, and J. F. Ustach. 1977. Nutrient exchanges between brackish water marshes and the estuary. In M. Wiley, ed. Estuarine Processes, Vol. II. Academic Press, New York, pp. 219–240.

Steward, K. K. 1990. Aquatic weed problems and management in the eastern United States. In A. H. Pietersen and K. J. Murphy, eds. Aquatic Weeds: The Ecology and Management of Nuisance Aquatic Vegetation. Oxford University Press, New York, pp. 391–405.

Stewart, G. R., and M. Popp. 1987. The ecophysiology of mangroves. In R. M. M. Crawford, ed. Plant Life in Aquatic and Amphibious Habitats. Special Publication of the British Ecological Society, Vol. 5, pp. 333–345.

Stewart, R. E. 1962. Waterfowl Populations in the Upper Chesapeake Region. Special Science Report on Wildlife, Research Publication 65, U.S. Fish and Wildlife Service, Washington, DC. 208 pp.

Stewart, R. E., and C. S. Robbins. 1958. Birds of Maryland and the District of Columbia. North America Fauna Service Research Publication 62, U.S. Fish and Wildlife Service, Washington, DC. 401 pp.

Stewart, R. E., and H. A. Kantrud. 1971. Classification of Natural Ponds and Lakes in the Glaciated Prairie Region. Research Publication 92, U.S. Fish and Wildlife Service, Washington, DC. 57 pp.

Stewart, R. E., and H. A. Kantrud. 1972. Vegetation of the Prairie Potholes, North Dakota, in Relation to Quality of Water and Other Environmental Factors. Professional Paper 85-D, U.S. Geological Survey, Washington, DC. 36 pp.

Stewart, R. E., and H. A. Kantrud. 1973. Ecological distribution of breeding water-fowl populations in North Dakota. Journal of Wildlife Management 37:39–50.

Stockwell, S. S., and M. L. Hunter. 1985. Distribution and Abundance of Birds, Amphibians and Reptiles, and Small Mammals in Peatlands of Central Maine. Maine Department of Inland Fisheries and Wildlife, Augusta. 89 pp.

Stout, J. P. 1978. An Analysis of Annual Growth and Productivity of *Juncus roemerianus* Scheele and *Spartina alterniflora* Loisel in Coastal Alabama. Ph.D. Dissertation, University of Alabama, Tuscaloosa.

Stout, J. P. 1984. The Ecology of Irregularly Flooded Salt Marshes of the Northeastern Gulf of Mexico: A Community Profile. Biological Report 85(7.1), U.S. Fish and Wildlife Service, Washington, DC. 98 pp.

Stromberg, J. S., and D. C. Patten. 1990. Riparian vegetation instream flow requirements: A case study from a diverted stream in the eastern Sierra Nevada, California, USA. Environmental Management 14:185–194.

Stuckey, R. L. 1980. Distributional history of *Lythrum salicaria* (purple loosestrife) in North America. Bartonia 47:3–20.

Stumm, W., and J. J. Morgan. 1970. Aquatic Chemistry: An Introduction Emphasizing Chemical Equilibria in Natural Waters. John Wiley & Sons, New York. 583 pp.

Stumm, W., and J. J. Morgan. 1996. Aquatic Chemistry: Chemical Equilibria and Rates in Natural Waters, 3rd ed. John Wiley & Sons, New York. 1022 pp.

Sullivan, M. J. 1975. Diatom communities from a Delaware salt marsh. Journal of Phycology 11:384–390.

Sullivan, M. J. 1977. Edaphic diatom communities associated with *Spartina alterniflora* and *S. patens* in New Jersey. Hydrobiologia 52:207–211.

Sullivan, M. J. 1978. Diatom community structure: Taxonomic and statistical analyses of a Mississippi salt marsh. Journal of Phycology 14:468–475.

Sullivan, M. J., and C. A. Moncreiff. 1988. Primary production of edaphic algal communities in a Mississippi salt marsh. Journal of Phycology 24:49–58.

Sullivan, M. J., and C. A. Currin. 2000. Community structure and functional dynamics of benthic microalgae in salt marshes. In M. P. Weinstein and D. A. Kreeger, eds. International Symposium: Concepts and Controversies in Tidal Marsh Ecology. Kluwer Academic Publishers, Dordrecht, The Netherlands, pp. 81–106.

Suzuki, T., W. G. A. Nissanka, and Y. Kurihara. 1989. Amplification of total dry matter, nitrogen and phosphorus removal from stands of *Phragmites australis* by harvesting and reharvesting regenerated shoots. In D. A. Hammer, ed. Constructed Wetlands for Wastewater Treatment. Lewis Publishers, Chelsea, MI, pp. 530–535.

Svengsouk, L., and W. J. Mitsch. in review. *Typha latifolia* and *Schoenoplectus tabernaemontani* mixture dynamics and nutrient limitation in nutrient-enrichment wetland experiments.

Svensson, B. H., and T. Rosswall. 1980. Energy flow through the subarctic mire at Stordalen. Ecological Bulletin (Stockholm) 30:283–301.

Swarzenski, C., E. M. Swenson, C. E. Sasser, and J. G. Gosselink. 1991. Marsh mat flotation in the Louisiana Delta Plain. Journal of Ecology 79:999–1011.

Swerhone, G. D. W., J. R. Lawrence, J. G. Richards, and M. J. Hendry. 1999. Construction and testing of a durable platinum wire electrode for *in situ* redox measurements in the subsurface. Ground Water Monitoring and Remediation 19(2):132–136.

Szaro, R. C. 1989. Riparian forest and scrubland community types of Arizona and New Mexico. Desert Plants 9:69–139.

Sze, N. D. 1977. Anthropogenic CO_2 emissions: Implications for the atmospheric CO_2–OH–CH_4 cycle. Science 195:673–675.

Szumigalski, A. R., and S. E. Bayley. 1996a. Net above-ground primary production along a bog-rich fen gradient in central Alberta, Canada. Wetlands 16:467–476.

Szumigalski, A. R., and S. E. Bayley. 1996b. Decomposition along a bog to rich fen gradient in central Alberta, Canada. Canadian Journal of Botany 74:573–581.

Tanner, C. C. 1996. Plants for constructed wetland treatment systems—A comparison of the growth and nutrient uptake of eight emergent species. Ecological Engineering 7:59–83.

Tanner, C. C., J. S. Clayton, and M. P. Upsdell. 1995. Effect of loading rate and planting on treatment of dairy farm wastewaters in constructed wetlands. II. Removal of nitrogen and phosphorus. Water Research 29:27–34.

Tanner, C. C., G. Raisin, G. Ho, and W. J. Mitsch, eds. 1999. Constructed and Natural Wetlands for Pollution Control. Special Issue of Ecological Engineering 12:1–170.

Tarnocai, C. 1979. Canadian Wetland Registry. In C. D. A. Rubec and F. C. Pollett, eds. Proceedings of a Workshop on Canadian Wetlands Environment. Ecological Land Classification Series 12, Canada Land Directorate, pp. 9–38.

Tarnocai, C., G. D. Adams, V. Glooschenko, W. A. Glooschenko, P. Grondin, H. E. Hirvonen, P. Lynch-Stewart, G. F. Mills, E. T. Oswald, F. C. Pollett, C. D. A. Rubec, E. D. Wells, and S. C. Zoltai. 1988. The Canadian wetland classification system. In National Wetlands Working Group, ed. Wetlands of Canada. Ecological Land Classification Series 24, Environment Canada, Ottawa, Ontario, and Polyscience Publications, Montreal, Quebec, pp. 413–427.

Tarutis, W. J., L. R. Stark, and F. M. Williams. 1999. Sizing and performance estimation of coal mine drainage wetlands. Ecological Engineering 12:353–372.

Taylor, J. R. 1985. Community Structure and Primary Productivity of Forested Wetlands in Western Kentucky. Ph.D. Dissertation, University of Louisville, Louisville, KY. 139 pp.

Taylor, J. R., M. A. Cardamone, and W. J. Mitsch. 1990. Bottomland hardwood forests: Their functions and values. In J. G. Gosselink, L. C. Lee, and T. A. Muir, eds. Ecological Processes and Cumulative Impacts: Illustrated by Bottomland Hardwood Wetland Ecosystems. Lewis Publishers, Chelsea, MI, pp. 13–86.

Taylor, K. L., J. B. Grace, G. R. Guntenspergen, and A. L. Foote. 1994. The interactive effects of herbivory and fire on an oligohaline marsh, Little Lake, Louisiana, USA. Wetlands 14:82–87.

Tchobanoglous, G., and F. L. Burton. 1991. Wastewater Engineering: Treatment, Disposal and Reuse, 3rd ed. McGraw-Hill, New York. 1334 pp.

Teal, J. M. 1958. Distribution of fiddler crabs in Georgia salt marshes. Ecology 39:18–19.

Teal, J. M. 1962. Energy flow in the salt marsh ecosystem of Georgia. Ecology 43:614–624.

Teal, J. M. 1986. The Ecology of Regularly Flooded Salt Marshes of New England: A Community Profile. Biological Report 85(7.4), U.S. Fish and Wildlife Service, Washington, DC. 61 pp.

Teal, J. M., and J. W. Kanwisher. 1966. Gas transport in the marsh grass *Spartina alterniflora*. Journal of Experimental Botany 17:355–361.

Teal, J. M., and M. Teal. 1969. Life and Death of the Salt Marsh. Little, Brown, Boston. 278 pp.

Teal, J. M., I. Valiela, and D. Berla. 1979. Nitrogen fixation by rhizosphere and free-living bacteria in salt marsh sediments. Limnology and Oceanography 24:126–132.

Teal, J. M., and B. L. Howes. 2000. Salt marsh values: Retrospection from the end of the century. In M. P. Weinstein and D. A. Kreeger, eds. International Symposium: Concepts and Controversies in Tidal Marsh Ecology. Kluwer Academic Publishers, Dordrecht, pp. 9–19.

Teas, H. J., and R. J. McEwan. 1982. An epidemic dieback gall disease of *Rhizophora* mangroves in Gambia, West Africa. Plant Disease 66:522–523.

Thayer, G. W., ed. 1992. Restoring the Nation's Marine Environment. Maryland Sea Grant College, College Park, MD. 716 pp.

Thibodeau, F. R., and N. H. Nickerson. 1986. Differential oxidation of mangrove substrate by *Avicennia germinans* and *Rhizophora mangle*. American Journal of Botany 73:512–516.

Thom, B. G. 1982. Mangrove ecology: A geomorphological perspective. In B. F. Clough, ed. Mangrove Ecosystems in Australia. Australian National University Press, Canberra, pp. 3–17.

Thomas, B. 1976. The Swamp. Norton, New York. 223 pp.

Thompson, E. 1983. Origin of Surface Patterns in a Subarctic Peatland. In C. H. Fuchsman and S. A. Spigarelli, eds. Proceedings of the International Symposium on Peat Utilization, Bemidji, MN.

Thormann, M. N., and S. E. Bayley. 1997. Aboveground net primary productivity along a bog–fen–marsh gradient in southern boreal Alberta, Canada. Ecoscience 4:374–384.

Thorp, J. H., E. M. McEwan, M. F. Flynn, and F. R. Hauer. 1985. Invertebrate colonization of submerged wood in a cypress–tupelo swamp and blackwater stream. American Midland Naturalist 113:56–68.

Tilman, D. 1982. Resource Competition and Community Structure. Princeton University Press, Princeton, NJ. 296 pp.

Tilman, D. 1988. Plant Strategies and the Dynamics and Structure of Plant Communities. Princeton University Press, Princeton, NJ. 376 pp.

Tiner, R. W. 1984. Wetlands of the United States: Current Status and Recent Trends. National Wetlands Inventory, U.S. Fish and Wildlife Service, Washington, DC. 58 pp.

Tiner, R. W. 1998. In Search of Swampland: A Wetland Sourcebook and Field Guide. Rutgers University Press, New Brunswick, NJ. 264 pp.

Tiner, R. W. 1999. Wetland Indicators: A Guide to Wetland Identification, Delineation, Classification, and Mapping. Lewis Publishers, Boca Raton, FL. 392 pp.

Tiner, R. W., and B. O. Wilen. 1983. The U.S. Fish and Wildlife Services National Wetlands Inventory Project. Unpublished Report, U.S. Fish and Wildlife Service, Washington, DC. 19 pp.

Titus, J. G. 1991. Greenhouse effect and coastal wetland policy: How Americans could abandon an area the size of Massachusetts at minimum cost. Environmental Management 15:39–58.

Titus, J. G., R. A. Park, S. P. Leatherman, J. R. Weggel, M. S. Greene, P. W. Mausel, S. Brown, C. Gaunt, M. Trehan, And G. Yohe. 1991. Greenhouse effect and sea level rise: The cost of holding back the sea. Coastal Management 19:171–204.

Todd, D. K. 1964. Groundwater. In V. T. Chow, ed. Handbook of Applied Hydrology. McGraw-Hill, New York, pp. 13-1–13-55.

Tomlinson, P. B. 1986 The Botany of Mangroves. Cambridge University Press, London. 413 pp.

Toth, L. A., D. A. Arrington, M. A. Brady, and D. A. Muszick. 1995. Conceptual evaluation of factors potentially affecting restoration of habitat structure within the channelized Kissimmee River ecosystem. Restoration Ecology 3:160–180.

Toth, L. A., S. L. Melvin, D. A. Arrington, and J. Chamberlain. 1998. Hydrologic manipulations of the channelized Kissimmee River. BioScience 48:757–764.

Tourbier, J. T., and R. Westmacott. 1989. Looking good: The use of natural methods to control urban runoff. Urban Land (April 1989):32–35.

Train, E., and F. P. Day, Jr. 1982. Population age structure of tree species in four communities in the Great Dismal Swamp. Castanea 47:1–16.

Trettin, C. C., M. F. Jurgensen, D. F. Grigal, M. R. Gale, and J. K. Jeglum. 1997. Northern Forested Wetlands: Ecology and Management. Lewis Publishers, Boca Raton, FL. 486 pp.

Tupacz, E. G., and F. P. Day. 1900. Decomposition of roots in a seasonally flooded swamp ecosystem. Aquatic Botany 37:199–214.

Turner, K. 1991. Economics and wetland management. Ambio 20:59–63.

Turner, M. G. 1989. Landscape ecology: The effect of pattern of process. Annual Review of Ecology and Systematics 20:171–197.

Turner, R. E. 1976. Geographic variations in salt marsh macrophyte production: A review. Contributions to Marine Science 20:47–68.

Turner, R. E. 1977. Intertidal vegetation and commercial yields of penaeid shrimp. American Fisheries Society Transactions 106:411–416.

Turner, R. E. 1982. Protein yields from wetlands. In B. Gopal, R. E. Turner, R. G. Wetzel, and D. F. Whigham, eds. Wetlands: Ecology and Management. National Institute of Ecology and International Scientific Publications, Jaipur, India, pp. 405–415.

Turner, R. E. 1988a. Secondary production in riparian wetlands. Transactions of the 53rd North American Wildlife and Natural Resource Conference, pp. 491–501.

Turner, R. E. 1988b. Fish and fisheries of inland wetlands. Water Quality Bulletin 13:7–9, 13.

Turner, R. E. 1997. Wetland loss in the northern Gulf of Mexico: Multiple working hypotheses. Estuaries 20:1–13.

Turner, R. E., W. Woo, and H. R. Jitts. 1979. Estuarine influences on a continental shelf plankton comunity. Science 206:218–220.

Turner, R. E., S. W. Forsythe, and N. J. Craig. 1981. Bottomland hardwood forest land resources of the southeastern United States. In J. R. Clark and J. Benforado, eds. Wetlands of Bottomland Hardwood Forests. Elsevier, Amsterdam, pp. 13–28.

Turner, R. E., and M. E. Boyer. 1997. Mississippi River diversions, coastal wetland restoration/creation and an economy of scale. Ecological Engineering 8:117–128.

Twilley, R. R. 1982. Litter Dynamics and Organic Carbon Exchange in Black Mangrove (*Avicennia germinans*) Basin Forests in a Southwest Florida Estuary. Ph.D. Dissertation, University of Florida, Gainesville.

Twilley, R. R. 1985. The exchange of organic carbon in basin mangrove forests in a southwest Florida estuary. Estuarine, Coastal and Shelf Science 20:543–557.

Twilley, R. R. 1988. Coupling of mangroves to the productivity of estuarine and coastal waters. In B. O. Jansson, ed. Coastal/Offshore Ecosystem: Interactions. Lecture Notes on Coastal and Estuarine Studies, Vol. 22. Springer-Verlag, Berlin, pp. 155–180.

Twilley, R. R. 1995. Properties of mangrove ecosystems related to the energy signature of coastal environments. In C. A. S. Hall, ed. Maximum Power: The Ideas and Applications of H. T. Odum. University Press of Colorado, Niwot, pp. 43–62.

Twilley, R. R. 1998. Mangrove wetlands. In M. G. Messina and W. H. Conner, eds. Southern Forested Wetlands: Ecology and Management. Lewis Publishers, Boca Raton, FL, pp. 445–473.

Twilley, R. R., A. E. Lugo, and C. Patterson-Zucca. 1986. Litter production and turnover in basin mangrove forests in southwest Florida. Ecology 67:670–683.

Twilley, R. R., R. H. Chen, and T. Hargis. 1992. Carbon sinks in mangroves and their implications to carbon budget of tropical coastal ecosystems. Water, Air, and Soil Pollution 64:265–288.

Twilley, R. R., S. C. Snedaker, A. Yáñez-Arancibia, and E. Medina. 1996. Biodiversity and ecosystem processes in tropical estuaries: Perspectives on mangrove ecosystems. In H. A. Mooney, J. H. Cushman, E. Medina, O. E. Sala, and E.-D. Schulze, eds. Functional Role of Biodiversity: A Global Perspective. John Wiley & Sons, London, pp. 327–370.

Twilley, R. R., M. Pozo, V. H. Garcia, V. H. Rivera-Monroy, R. Zambrano, and A. Bodero. 1997. Litter dynamics in riverine mangrove forests in the Guayas River estuary, Ecuador. Oecologia 111:109–122.

Tyler, P. A. 1976. Lagoon of Islands, Tasmania—Deathknell for a unique ecosystem? Biological Conservation 9:1–11.

Urban, N. R., S. J. Eisenreich, and E. Gorham. 1985. Proton cycling in bogs: Geographic variation in northeastern North America. In T. C. Hutchinson and K. Meema, eds. Proceedings of NATO Advanced Research Workshop on the Effects of Acid Deposition on Forest, Wetland, and Agricultural Ecosystems, Toronto, May 13–17, 1985. Springer-Verlag, New York, pp. 577–598.

Urban, N. R., and S. J. Eisenreich. 1988. Nitrogen cycling in a forested Minnesota bog. Canadian Journal of Botany 66:435–449.

U.S. Army Corps of Engineers. 1972. Charles River Watershed, Massachusetts. New England Division, Waltham, MA. 65 pp.

U.S. Army Corps of Engineers. 1987. Corps of Engineers Wetlands Delineation Manual. Technical Report Y-87-1, U.S. Army Corps of Engineers Waterways Experiment Station, Vicksburg, MS. 100 pp. and appendices.

U.S. Department of Interior, Bureau of Reclamation. 1984. Water Measurement Manual, 2nd ed., revised reprint. U.S. Government Printing Office, Washington, DC. 327 pp.

U.S. Environmental Protection Agency. 1983. Freshwater Wetlands for Wastewater Management: Environmental Impact Statement—Phase I Report. EPA-90419-83-107, U.S. Environmental Protection Agency, Region IV, Atlanta. 380 pp.

U.S. Environmental Protection Agency. 1993. Constructed Wetlands for Wastewater Treatment and Wildlife Habitat: 17 Case Studies.

EPA-832-R-93-005, U.S. Environmental Protection Agency, Washington, DC. 174 pp.

U.S. Soil Conservation Service. 1975. Soil Taxonomy: A Basic System of Soil Classification for Making and Interpreting Soil Surveys. Agricultural Handbook 436, U.S. Soil Conservation Service, Washington, DC. 754 pp.

Vaithiyanathan, R., and C. J. Richardson. 1997. Nutrient profiles in the Everglades: Examination along the eutrophication gradient. Science of Total Environment 205:81–95.

Valiela, I. 1984. Marine Ecological Processes. Springer-Verlag, New York. 546 pp.

Valiela, I., J. M. Teal, and W. Sass. 1973. Nutrient retention in salt marsh plots experimentally fertilized with sewage sludge. Estuarine Coastal Marine Science 1:261–269.

Valiela, I., and J. M. Teal. 1974. Nutrient limitation in salt marsh vegetation. In R. J. Reimold and W. H. Queen, eds. Ecology of Halophytes. Academic Press, New York, pp. 547–563.

Valiela, I., J. M. Teal, S. Volkmann, D. Shafer, and E. J. Carpenter. 1978. Nutrient and particulate fluxes in a salt marsh ecosystem: Tidal exchanges and inputs by precipitation and groundwater. Limnology and Oceanography 23:798–812.

Valiela, I., and J. M. Teal. 1979. The nitrogen budget of a salt marsh ecosystem. Nature 280:652–656.

van der Valk, A. G. 1981. Succession in wetlands: A Gleasonian approach. Ecology 62:688–696.

van der Valk, A. G. 1982. Succession in temperate North American wetlands. In B. Gopal, R. E. Turner, R. G. Wetzel, and D. F. Whigham, eds. Wetlands: Ecology and Management. National Institute for Ecology and International Science Publications, Jaipur, India, pp. 169–179.

van der Valk, A. G., ed. 1989. Northern Prairie Wetlands. Iowa State University Press, Ames. 400 pp.

van der Valk, A. G. 1998. Succession theory and wetland restoration. In A. J. McComb and J. A. Davis, eds. Wetlands for the Future. Contributions from INTECOL's Fifth International Wetland Conference. Gleneagles Publishing, Adelaide, Australia, pp. 657–667.

van der Valk, A. G., and C. B. Davis. 1978a. Primary production of prairie glacial marshes. In R. E. Good, D. F. Whigham, and R. L. Simpson, eds. Freshwater Wetlands: Ecological Processes and Management Potential. Academic Press, New York, pp. 21–37.

van der Valk, A. G., and C. B. Davis. 1978b. The role of seed banks in the vegetation dynamics of prairie glacial marshes. Ecology 59:322–335.

van der Valk, A. G., J. M. Rhymer, and H. R. Murkin. 1991. Flooding and the decomposition of litter of four emergent plant species in a prairie wetland. Wetlands 11:1–16.

van Eck, G. T. M. 1982. Forms of phosphorus in particulate matter from the Holland's Diep/Haringvliet, the Netherlands. Hydrobiologia 92:665–681.

Van Engel, W. A., and E. B. Joseph. 1968. Characterization of Coastal and Estuarine Fish Nursery Grounds as Natural Communities. Final Report, Bureau of Commercial Fisheries, Virginia Institute of Marine Science, Gloucester Point, VA. 43 pp.

Vannote, R. L., G. W. Minshall, K. W. Cummins, J. R. Sedell, and C. E. Cushing. 1980. The river continuum concept. Canadian Journal of Fisheries and Aquatic Sciences 37:130–137.

Van Raalte, C. D. 1976. Production of epibenthic salt marsh algae: Light and nutrient limitation. Limnology and Oceanography 21:862–872.

Vartapetian, B. B. 1988. Ultrastructure studies as a means of evaluating plant tolerance to flooding. In D. D. Hook, W. H. McKee, Jr., H. K. Smith, J. Gregory, V. G. Burrell, M. R. DeVoe, R. E. Sojka, S. Gilbert, R. Banks, L. G. Stolzy, C. Brooks, T. D. Matthews, and T. H. Shear, eds. The Ecology and Management of Wetlands, Vol. 1: Ecology of Wetlands. Timber Press, Portland, OR, pp. 452–466.

Vega, A., and K. C. Ewel. 1981. Wastewater effects on a water hyacinth marsh and adjacent impoundment. Environmental Management 5:537–541.

Vepraskas, M. J. 1995. Redoximorphic Features for Identifying Aquic Conditions. Technical Bulletin 301, North Carolina Agricultural Research Service, North Carolina State University, Raleigh. 33 pp.

Verhoeven, J. T. A. 1986. Nutrient dynamics in minerotrophic peat mires. Aquatic Botany 25:117–137.

Verhoeven, J. T. A., ed. 1992. Fens and Bogs in the Netherlands: Vegetation, History, Nutrient Dynamics and Conservation. Kluwer Academic Publishers, Dordrecht. 490 pp.

Verhoeven, J. T. A., E. Maltby, and M. B. Schmitz. 1990. Nitrogen and phosphorus mineralization in fens and bogs. Journal of Ecology 78:713–726.

Verhoeven, J. T. A., and M. B. Schmitz. 1991. Control of plant growth by nitrogen and phosphorus in mesotrophic fens. Biogeochemistry 12:135–148.

Verhoeven, J. T. A., D. F. Whigham, M. van Kerkhoven, J. O'Neill, and E. Maltby. 1994. Comparative study of nutrient-related processes in geographically separated wetlands: Toward a science base for functional assessment procedures. In W. J. Mitsch, ed. Global Wetlands: Old World and New. Elsevier, Amsterdam, pp. 91–106.

Vermeer, J. G., and F. Berendse. 1983. The relationship between nutrient availability, shoot biomass and species richness in grassland and wetland communities. Vegetatio 53:121–126.

Vernberg, F. J. 1981. Benthic macrofauna. In F. J. Vernberg and W. B. Vernberg, eds. Functional Adaptations of Marine Organisms. Academic Press, New York, pp. 179–230.

Vernberg, F. J., and W. B. Vernberg, eds. 1981. Functional Adaptations of Marine Organisms. Academic Press, New York.

Vernberg, W. B., and F. J. Vernberg. 1972. Environmental Physiology of Marine Animals. Springer-Verlag, New York. 346 pp.

Vernberg, W. B., and B. C. Coull. 1981. Meiofauna. In F. J. Vernberg and W. B. Vernberg, eds. Functional Adaptations of Marine Organisms. Academic Press, New York, pp. 147–177.

Vernon, R. O. 1947. Cypress domes. Science 105:97–99.

Verry, E. S., and D. H. Boelter. 1979. Peatland hydrology. In P. E. Greeson, J. R. Clark, and J. E. Clark, eds. Wetland Functions and Values: The State of Our Understanding. American Water Resources Association, Minneapolis, pp. 389–402.

Viereck, L. A., and E. L. Little, Jr. 1972. Alaska trees and shrubs. Handbook 410, U.S. Department of Agriculture, Washington, DC. 265 pp.

Visser, J. M. 1989. The Impact of Vertebrate Herbivores on the Primary Production of *Sagittaria* Marshes in the Wax Lake Delta, Atchafalaya Bay, Louisiana. Ph.D. Dissertation, Louisiana State University, Baton Rouge. 88 pp.

Visser, J. M., C. E. Sasser, R. G. Chabreck, and R. G. Linscombe. 1998. Marsh vegetation types of the Mississippi River deltaic plain. Estuaries 21:818–828.

Visser, J. M., C. E. Sasser, R. H. Chabreck, and R. G. Linscombe. 1999. Long-term vegetation change in Louisiana tidal marshes. Wetlands 19:168–175.

Vitousek, P. M., J. Abner, R. W. Howarth, G. E. Likens, P. A. Matson, D. W. Schindler, W. H. Schlesinger, and G. D. Tilman. 1997. Human alteration of the global nitrogen cycle: Causes and consequences. Issues in Ecology 1. Ecological Society of America, Washington, DC.

Vitt, D. H., P. Achuff, and R. E. Andrus. 1975a. The vegetation and chemical properties of patterned fens in the Swan Hills, north central Alberta. Canadian Journal of Botany 53:2776–2795.

Vitt, D. H., H. Crum, and J. A. Snider. 1975b. The vertical zonation of *Sphagnum* species in hummock–hollow complexes in northern Michigan. Michigan Botany 14:190–200.

Voesenek, L. A. C. J. 1990. Adaptation of *Rumex* in Flooding Gradients. Thesis, Catholic University of Nijmegen, Nijmegen, Netherlands. 159 pp.

Voigts, D. K. 1976. Aquatic invertebrate abundance in relation to changing marsh vegetation. American Midland Naturalist 95:312–322.

von Post, L., and E. Granlund. 1926. Södra Sveriges Torvtillgängar. Sveriges Geologiska Undersökning Ser. C Avhandlingar och uppsater, No. 355. Arsbok 19(2):1–127.

Vorosmarty, C. J., B. Moore, W. B. Bowden, J. E. Hobbie, B. J. Peterson, and J. Morris. 1982. The transport and processing of nitrogen in a tidal, freshwater marsh and river ecosystem: Modeling the roles of water movement and biotic activity in determining water quality. In W. K. Lauenroth, G. V. Skogerboe, and M. Flug, eds. Analysis of Ecological Systems: State of the Art in Ecological Modelling. Elsevier Science Publishing, New York, pp. 689–698.

Vymazal, J. 1995a. Algae and Element Cycling in Wetlands. Lewis Publishers, Boca Raton, FL. 689 pp.

Vymazal, J. 1995b. Constructed wetlands for wastewater treatment in the Czech Republic—State of the art. Water Science and Technology 32:357–364.

Vymazal, J. 1998. Czech Republic. In J. Vymazal, H. Brix, P. F. Cooper, M. B. Green, and R. Haberl, eds. Constructed Wetlands for Wastewater Treatment in Europe. Backhuys Publishers, Leiden, Netherlands, pp. 95–121.

Vymazal, J., H. Brix, P. F. Cooper, M. B. Green, and R. Haberl, eds. 1998. Constructed Wetlands for Wastewater Treatment in Europe. Backhuys Publishers, Leiden, Netherlands.

Waddington, J. M., and N. T. Roulet. 1997. Groundwater flow and dissolved carbon movement in a boreal peatland. Journal of Hydrology 191:122–138.

Wadsworth, J. R., Jr. 1979. Duplin River Tidal System: Sapelo Island, Georgia. Map reprinted in January 1982 by the University of Georgia Marine Institute, Sapelo Island, GA.

Wafar, S., A. G. Untawale, and M. Wafar. 1997. Litter fall and energy flux in a mangrove ecosystem. Estuarine, Coastal and Shelf Science 44:111–124.

Wainscott, V. J., C. Bardey, and P. Kangas. 1990. Effect of muskrat mounds on microbial density on plant litter. American Midland Naturalist 123:399–401.

Walker, D. 1970. Direction and rate in some British post-glacial hydroseres. In D. Walker and R. G. West, eds. Studies in the Vegetational History of the British Isles. Cambridge University Press, Cambridge, UK, pp. 117–139.

Walker, D. J. 1998. Modelling residence time in stormwater ponds. Ecological Engineering 10:247–262.

Walmsley, M. E. 1977. Physical and chemical properties of peat. In N. W. Radforth and C. O. Brawner, eds. Muskeg and the Northern Environment of Canada. University of Toronto Press, Toronto, pp. 82–129.

Walter, H. 1973. Vegetation of the Earth. Springer-Verlag, New York. 237 pp.

Walters, C., L. Gunderson, and C. S. Holling. 1992. Experimental policies for water management in the Everglades. Ecological Applications 2:189–202.

Walton, R., J. E. Davis, T. H. Martin, and R. S. Chapman. 1996. Hydrology of the Black Swamp wetlands on the Cache River, AR. Wetlands 16:279–287.

Wample, R. L., and D. M. Reid. 1979. The role of endogenous auxins and ethylene in the formation of adventitious roots and hypocotyl hypertrophy in flooded sunflower plants (*Helianthus annuus* L.). Physiologia Plantarum 45:219–226.

Wang, F. C., T. Lu, and W. B. Sikora. 1993. Intertidal marsh suspended sediment transport processes, Terrebonne Bay, Louisiana, U.S.A. Journal of Coastal Research 9:209–220.

Wang, N., and W. J. Mitsch. in press. A detailed ecosystem model of phosphorus dynamics in created riparian wetlands. Ecological Modelling.

Want, W. L. 1994. Law of Wetlands Regulation. Clark-Boardman Company, New York.

Ward, J. V. 1998. Riverine landscapes: Biodiversity patterns, disturbance regimes, and aquatic conservation. Biological Conservation 83:269–278.

Ward, J. V., and J. A. Stanford. 1983. The immediate-disturbance hypothesis: An explanation for biotic diversity patterns in lotic ecosystems. In T. D. Fontaine and S. M. Bartell, eds. Dynamics of Lotic Systems. Ann Arbor Science, Ann Arbor, MI, pp. 347–356.

Ward, J. V., and J. A. Stanford. 1995. The serial discontinuity concept: Extending the model to floodplain rivers. Regulated Rivers 10:1598.

Warner, B. G., and C. D. A. Rubec, eds. 1997. The Canadian Wetland Classification System. National Wetlands Working Group, Wetlands Research Centre, University of Waterloo, Waterloo, Ontario.

Warner, D., and D. Wells. 1980. Bird Population Structure and Seasonal Habitat Use as Indicators of Environment Quality of Peatlands. Minnesota Peat Program, Minnesota Department of Natural Resources. 84 pp.

Warners, D. P. 1987. Effects of burning on sedge meadow studied. Restoration and Management Notes 5(2):90–91.

Wassen, M. J., A. Barendregt, P. P. Schot, and B. Beltman. 1990. Dependence of local mesotrophic fens on a regional groundwater flow system in a poldered river plain in the Netherlands. Landscape Ecology 5:21–38.

Water Pollution Control Federation. 1990. Natural Systems for Wastewater Treatment. Manual of Practice FD-16, Water Environment Federation, Alexandria, VA.

Watershed. In D. L. Correll, ed. Watershed Research Perspectives. Smithsonian Institution Press, Washington, DC, pp. 57–76.

Watson, J. T., and J. A. Hobson. 1989. Hydraulic design considerations and control structures for constructed wetlands for wastewater treatment. In D. A. Hammer, ed. Constructed Wetlands for Wastewater Treatment. Lewis Publishers, Chelsea, MI, pp. 379–392.

Watson, J. T., S. C. Reed, R. H. Kadlec, R. L. Knight, and A. E. Whitehouse. 1989. Performance expectations and loading rates for constructed wetlands. In D. A. Hammer, ed. Constructed Wetlands for Wastewater Treatment. Lewis Publishers, Chelsea, MI, pp. 319–361.

Watzin, M. C., and J. G. Gosselink. 1992. Coastal Wetlands of the United States. Louisiana Sea Grant College Program, Baton Rouge, and U.S. Fish and Wildlife Service, Lafayette, LA. 15 pp.

Waughman, G. J. 1980. Chemical aspects of the ecology of some south German peatlands. Journal of Ecology 68:1025–1046.

Webb, J., and M. B. Jackson. 1986. A transmission and cryo-scanning electron microscopy study of the formation of aerenchyma (cortical gas-filled space) in adventitious roots of rice (*Oryza sativa*). Journal of Experimental Botany 37:832–841.

Webber, E. E. 1967. Bluegreen algae from a Massachusetts salt marsh. Torrey Botanical Club Bulletin 94:99–106.

Webber, E. E. 1968. Systematics and ecology of benthic salt marsh algae at Ipswich, Massachusetts. Ph.D. Dissertation, University of Massachusetts, Amherst.

Weber, C. A. 1907. Aufbau und Vegetation der Moore Norddutschlands. Beibl. Bot. Jahrb. 90:19–34.

Webster, J. R., and G. M. Simmons. 1978. Leaf breakdown and invertebrate colonization on a reservoir bottom. Verhandlungen Internationale Vereinigung für Limnologie 20:1587–1596.

Weinstein, M. P., J. H. Balletto, J. M. Teal, and D. F. Ludwig. 1997. Success criteria and adaptive management for a large-scale wetland restoration project. Wetlands Ecology and Management 4:111–127.

Weinstein, M. P., and D. A. Kreeger, eds. 2000. International Symposium: Concepts and Controversies in Tidal Marsh Ecology. Kluwer Academic Publishers, Dordrecht, 864 pp.

Welcomme, R. L. 1976. The role of African flood plains in fisheries. In M. Smart, ed. Proceedings of the International Conference on Conservation of Wetlands and Waterfowl. International Waterfowl Research Bureau, Slimbridge, UK, pp. 332–335.

Welcomme, R. L. 1979. Fisheries Ecology of Floodplain Rivers. Longman, New York.

Weller, M. W. 1978. Management of freshwater marshes for wildlife. In R. E. Good, D. F. Whigham, and R. L. Simpson, eds. Freshwater Wetlands: Ecological Processes and Management Potential. Academic Press, New York, pp. 267–284.

Weller, M. W. 1981. Freshwater Marshes. University of Minnesota Press, Minneapolis. 146 pp.

Weller, M. W. 1987. Freshwater Marshes, Ecology and Wildlife Management, 2nd ed. University of Minnesota Press, Minneapolis. 165 pp.

Weller, M. W. 1994. Freshwater Marshes, 3rd ed. University of Minnesota Press, Minneapolis. 192 pp.

Weller, M.W. 1999. Wetland Birds. Cambridge University Press, Cambridge, UK.

Weller, M. W., and C. S. Spatcher. 1965. Role of habitat in the distribution and abundance of marsh birds. Special Report 43, Agriculture and Home Economics Experiment Station, Iowa State University, Ames. 31 pp.

Welling, C. H., R. L. Pederson, and A. G. van der Valk. 1988a. Recruitment from the seed bank and the development of zonation of emergent vegetation during a drawdown in a prairie wetland. Journal of Ecology 76:483–496.

Welling, C. H., R. L. Pederson, and A. G. van der Valk. 1988b. Temporal patterns in recruitment from the seed bank during drawdowns in a prairie wetland. Journal of Applied Ecology 25:99–107.

Wells, B. W. 1928. Plant communities of the coastal plain of North Carolina and their successional relations. Ecology 9:230–242.

Werme, C. E. 1981. Resource Partitioning in the Salt Marsh Fish Community. Ph.D. Dissertation, Boston University, Boston. 126 pp.

Werner, T. M., and R. H. Kadlec. 1996. Application of residence time distributions to stormwater treatment systems. Ecological Engineering 7:213–234.

West, R. G. 1964. Inter-relations of ecology and quaternary paleobotany. Journal of Ecology (Supplement) 52:47–57.

Wetzel, R. G., and D. F. Westlake. 1969. Periphyton. In R. A. Vollenweider, ed. A Manual on Methods for Measuring Primary Productivity in Aquatic Environments. IBP Handbook 12. Basil Blackwell, Oxford, UK, pp. 33–40.

Wetzel, R. G., A. van der Valk, R. E. Turner, W. J. Mitsch, and B. Gopal, eds. 1994. Recent Studies on Ecology and Management of Wetlands. Special Issue of International Journal of Ecology and Environmental Sciences 20:1–246.

Wetzel, R. L., and S. Powers. 1978. Habitat Development Field Investigations, Windmill Point Marsh Development Site, James River, Virginia: App. D: Environmental Impacts of Marsh Development with Dredged Material: Botany, Soil, Aquatic Biology, and Wildlife. Technical Report D-77-2, U.S. Army Corps of Engineers Waterways Experiment Station Vicksburg, MS. 292 pp.

Whalen, M. N., and W. S. Reeburgh. 1990. Consumption of atmospheric methane by tundra soils. Nature 346:160–162.

Wharton, C. H. 1970. The Southern River Swamp—A Multiple-Use Environment. Bureau of Business and Economic Research, Georgia State University, Atlanta. 48 pp.

Wharton, C. H. 1978. The Natural Environments of Georgia. Georgia Department of Natural Resources, Atlanta. 227 pp.

Wharton, C. H., H. T. Odum, K. Ewel, M. Duever, A. Lugo, R. Boyt, J. Bartholomew, E. DeBellevue, S. Brown, M. Brown, and L. Duever. 1976.

Forested Wetlands of Florida—Their Management and Use. Center for Wetlands, University of Florida, Gainesville. 421 pp.

Wharton, C. H., V. W. Lambou, J. Newsom, P. V. Winger, L. L. Gaddy, and R. Mancke. 1981. The fauna of bottomland hardwoods in southeastern United States. In J. R. Clark and J. Benforado, eds. Wetlands of Bottomland Hardwood Forests. Elsevier, Amsterdam, pp. 87–100.

Wharton, C. H., W. M. Kitchens, E. C. Pendleton, and T. W. Sipe. 1982. The Ecology of Bottomland Hardwood Swamps of the Southeast: A Community Profile. FWS/OBS-81/37, U.S. Fish and Wildlife Service, Washington, DC. 133 pp.

Wheeler, B. D. 1980. Plant communities of rich-fen systems in England and Wales. Journal of Ecology 68:365–395.

Wheeler, B. D., and K. E. Giller. 1982. Species richness of herbaceous fen vegetation in Broadland, Norfolk, in relation to the quantity of above-ground plant material. Journal of Ecology 70:179–200.

Whigham, D. F., and R. L. Simpson. 1975. Ecological Studies of the Hamilton Marshes. Progress Report for the period June 1974–January 1975, Biology Department, Rider College, Lawrenceville, NJ.

Whigham, D. F., and R. L. Simpson. 1977. Growth, mortality, and biomass partitioning in freshwater tidal wetland populations of wild rice (*Zizania aquatica* var. *aquatica*). Torrey Botanical Club Bulletin 104:347–351.

Whigham, D. F., R. L. Simpson, and M. A. Leck. 1979. The distribution of seeds, seedlings, and established plants of arrow arum (*Peltandra virginica* (L.) Kunth) in a freshwater tidal wetland. Torrey Botanical Club Bulletin 106:193–199.

Whigham, D. F., R. L. Simpson, and K. Lee. 1980. The Effect of Sewage Effluent on the Structure and Function of a Freshwater Tidal Wetland. Water Resources Research Institute Report, Rutgers University, New Brunswick, NJ. 160 pp.

Whigham, D. F., C. Chitterling, B. Palmer, and J. O'Neill. 1986. Modification of runoff from upland watersheds—The influence of a diverse riparian ecosystem. In D. L. Correll, ed. Watershed Research Perspectives. Smithsonian Institution Press, Washington, DC, pp. 305–332.

Whigham, D. F., and C. J. Richardson. 1988. Soil and plant chemistry of an Atlantic white cedar wetland on the Inner Coastal Plain of Maryland. Canadian J. Botany 66:568–576.

Whigham, D. F., D. Dykyjová, and S. Hejny, eds. 1993. Wetlands of the World, I: Inventory, Ecology, and Management. Kluwer Academic Publishers, Dordrecht. 768 pp.

White, D. A. 1993. Vascular plant community development on mudflats in the Mississippi River delta, Louisiana, USA. Aquatic Botany 45:171–194.

White, D. A., T. E. Weiss, J. M. Trapani, and L. B. Thien. 1978. Productivity and decomposition of the dominant salt marsh plants in Louisiana. Ecology 59:751–759.

White, J. S., S. E. Bayley, and P. J. Curtis. 2000. Sediment storage of phosphorus in a northern prairie wetland receiving municipal and agro-industrial wastewater. Ecological Engineering 14:127–138.

Whitehead, D. R. 1972. Developmental and environmental history of the Dismal Swamp. Ecological Monographs 42:301–315.

Whitehead, J. C. 1992. Economic valuation of wetland resources: A review of value estimates. Journal of Environmental Systems 22:151.

Whitney, D. E., G. M. Woodwell, and R. W. Howarth. 1975. Nitrogen fixation in Flax Pond: A Long Island salt marsh. Limnology and Oceanography 20:640–643.

Whitney, D. M., A. G. Chalmers, E. B. Haines, R. B. Hanson, L. R. Pomeroy, and B. Sherr. 1981. The cycles of nitrogen and phosphorus. In L. R. Pomeroy and R. G. Wiegert, eds. The Ecology of a Salt Marsh. Springer-Verlag, New York, pp. 163–181.

Whittaker, R. H. 1967. Gradient analysis of vegetation. Biological Reviews 42:207–264.

Whooten, H. H., and M. R. Purcell. 1949. Farm Land Development: Present and Future by Clearing, Drainage, and Irrigation. Circular 825, U.S. Department of Agriculture, Washington, DC.

Wicker, K. M., D. Davis, and D. Roberts. 1983. Rockefeller State Wildlife Refuge and Game Preserve: Evaluation of Wetland Management Techniques. Coastal Environments, Inc., Baton Rouge, LA.

Widdows, J., B. L. Bayne, D. R. Livingstone. R. I. E. Newell, and E. Donkin. 1979. Physiological and biochemical responses of bivalve mollusks to exposure to air. Comparative Biochemistry and Physiology 62A(2):301–308.

Wiebe, W. J., R. R. Christian, J. A. Hansen, G. King, B. Sherr, and G. Skyring. 1981. Anaerobic respiration and fermentation. In L. R. Pomeroy and R. G. Wiegert, eds. The Ecology of a Salt Marsh. Springer-Verlag, New York, pp. 137–159.

Wieder, R. K. 1989. A survey of constructed wetlands for acid coal mine drainage treatment in the eastern United States. Wetlands 9:299–315.

Wieder, R. K., and G. E. Lang. 1983. Net primary production of the dominant bryophytes in a *Sphagnum*-dominated wetland in West Virginia. Bryologist 86:280–286.

Wieder, R. K., and G. E. Lang. 1984. Influence of wetlands and coal mining on stream water chemistry. Water, Air, and Soil Pollution 23:381–396.

Wieder, R. K., G. Tchobanoglous, and R. W. Tuttle. 1989. Preliminary considerations regarding constructed wetlands for wastewater treatment. In D. A. Hammer, ed. Constructed Wetlands for Wastewater Treatment. Lewis Publishers, Chelsea, MI, pp. 297–306.

Wiegert, R. G. 1986. Modeling spatial and temporal variability in a salt marsh: Sensitivity to rates of primary production, tidal migration and microbial degradation. In D. A. Wolfe, ed. Estuarine Variability. Academic Press, New York, pp. 405–426.

Wiegert, R. G., R. R. Christian, J. L. Gallagher, J. R. Hall, R. D. H. Jones, and R. L. Wetzel. 1975. A preliminary ecosystem model of a Georgia salt marsh. In L. E. Cronin, ed. Estuarine Research, Vol. 1. Academic Press, New York, pp. 583–601.

Wiegert, R. G., and R. L. Wetzel. 1979. Simulation experiments with a fourteen compartment model of a *Spartina* salt marsh. In R. F. Dame, ed.

Marsh-Estuarine Systems Simulations. University of South Carolina Press, Columbia, pp. 7–39.

Wiegert, R. G., R. R. Christian, and R. L. Wetzel. 1981. A model view of the marsh. In L. R. Pomeroy and R. G. Wiegert, eds. The Ecology of a Salt Marsh. Springer-Verlag, New York, pp. 183–218.

Wiegert, R. G., and B. J. Freeman. 1990. Tidal Salt Marshes of the Southeast Atlantic Coast: A Community Profile. Biological Report 85(7.29), U.S. Fish and Wildlife Service, Washington, DC.

Wijte, A. H. B. M., and J. L. Gallagher. 1996a. Effect of oxygen availability and salinity on early life history stages of salt marsh plants. I. Different germination strategies of *Spartina alterniflora* and *Phragmites australis* (Poaceae). American Journal of Botany 83:1337–1342.

Wijte, A. H. B. M., and J. L. Gallagher. 1996b. Effect of oxygen availability and salinity on early life history stages of salt marsh plants. II. Early seedling development advantage of *Spartina alterniflora* over *Phragmites australis* (Poaceae). American Journal of Botany 83:1343–1350.

Wilcox, D. A., R. A. Shedlock, and W. H. Hendrickson. 1986. Hydrology, water chemistry, and ecological relations in the raised mound of Cowles Bog. Journal of Ecology 74:1103–1117.

Wilcox, D. A., and H. A. Simonin. 1987. A chronosequence of aquatic macrophyte communities in dune ponds. Aquatic Botany 28:227–242.

Wilcox, D. A., and T. H. Whillans. 1999. Techniques for restoration of disturbed coastal wetlands of the Great Lakes. Wetlands 19:835–857.

Wilen, B. O., and H. R. Pywell. 1981. The National Wetlands Inventory. Paper Presented at In-Place Resource Inventories: Principles and Practices—A National Workshop, Orono, ME, August 9–14. 10 pp

Wilhelm, M., S. R. Lawry, and D. D. Hardy. 1989. Creation and management of wetlands using municipal wastewater in northern Arizona: A status report. In D. A. Hammer, ed. Constructed Wetlands for Wastewater Treatment. Lewis Publishers, Chelsea, MI, pp. 179–185.

Willard, D. E., V. M. Finn, D. A. Levine, and J. E. Klarquist. 1989. Creation and restoration of riparian wetlands in the agricultural Midwest. In J. A. Kusler and M. E. Kentula, eds. Wetland Creation and Restoration. Island Press, Washington, DC, pp. 327–337.

Williams, G. P., and M. G. Wolman. 1984. Downstream Effects of Dams on Alluvial Rivers. Professional Paper 1286, U.S. Geological Survey, Washington, DC.

Williams, J. D., and C. K. Dodd, Jr. 1979. Importance of wetlands to endangered and threatened species. In P. E. Greeson, J. R. Clark, and J. E. Clark, eds. Wetland Functions and Values: The State of Our Understanding. American Water Resources Association, Minneapolis, pp. 565–575.

Williams, M., ed. 1990. Wetlands: A Threatened Landscape. Basil Blackwell, Oxford, UK. 419 pp.

Williams, R. B. 1962. The Ecology of Diatom Populations in a Georgia Salt Marsh. Ph.D. Dissertation, Harvard University, Cambridge, MA.

Williams, R. B., and M. B. Murdock. 1972. Compartmental analysis of the production of *Juncus roemerianus* in a North Carolina salt marsh. Chesapeake Science 13:69–79.

Willis, C., and W. J. Mitsch. 1995. Effects of hydrology and nutrients on seedling emergence and biomass of aquatic macrophytes from natural and artificial seed banks. Ecological Engineering 4:65–76.

Williston, H. L., F. W. Shropshire, and W. E. Balmer. 1980. Cypress Management: A Forgotten Opportunity. U.S. D. A. Forest Service, Forestry Report SA-FR8, Atlanta, GA. 1980.

Wilson, C. L., and W. E. Loomis. 1967. Botany, 4th ed. Holt, Rinehart and Winston, New York. 626 pp.

Wilson, J. O., R, Buchsbaum, I. Valiela, and T. Swain. 1986. Decomposition in salt marsh ecosystems: Phenolic dynamics during decay of litter of *Spartina alterniflora*. Marine Ecology Progress Series 29:177–187.

Wilson, K. A. 1962. North Carolina Wetlands: Their Distribution and Management. North Carolina Wildlife Resources Commission, Raleigh. 169 pp.

Wilson, L. R. 1935. Lake development and plant succession in Vilas County, Wisconsin. 1. The medium hard water lakes. Ecological Monographs 5:207–247.

Wilson, R. F., and W. J. Mitsch. 1996. Functional assessment of five wetlands constructed to mitigate wetland loss in Ohio, USA. Wetlands 16:436–451.

Wind-Mulder, H. L., L. Rochefort, and D. H. Vitt. 1996. Water and peat chemistry comparisons of natural and post-harvested peatlands across Canada and their relevance to peatland restoration. Ecological Engineering 7:161–181.

Winter, T. C. 1977. Classification of the hydrologic settings of lakes in the north-central United States. Water Resources Research 13:753–767.

Winter, T. C. 1986. Effect of ground-water recharge on configuration of the water table beneath sand dunes and on seepage in lakes in the sandhills of Nebraska, U.S.A. Journal of Hydrology 86:221–237.

Winter, T. C. 1988. A conceptual framework for assessing cumulative impacts on the hydrology of nontidal wetlands. Environmental Management 12:605–620.

Winter, T. C. 1989. Hydrologic studies of wetlands in the northern prairie. In A. G. van der Valk, ed. Northern Prairie Wetlands. Iowa State University Press, Ames, pp. 16–54.

Winter, T. C. 1992. A physiographic and climatic framework for hydrologic studies of wetlands. In R. D. Robarts and M. L. Bothwell, eds. N.H.R.I. Symposium Series 7. Environment Canada, Saskatoon, Saskatchewan, pp. 127–148.

Winter, T. C., and M.-K. Woo. 1990. Hydrology of lakes and wetlands. In M. G. Wolman and H. C. Riggs, eds. Surface Water Hydrology: The Geology of North America, Vol. 0-1. Geological Society of America, Boulder, CO, pp. 159–187.

Winter, T. C., and M. R. Llamas, eds. 1993. Hydrogeology of Wetlands. Special Issue of Journal of Hydrology 141:1–269.

Winter, T. C., and D. O. Rosenberry. 1995. The interaction of groundwater with prairie pothole wetlands in the Cottonwood Lake area, east-central North Dakota, 1979–1990. Wetlands 15:193–211.

Wisheu, I. C., and P. A. Keddy. 1992. Competition and centrifugal organization of plant communities: Theory and tests. Journal of Vegetation Science 3:147–156.

Wolff, W. J. 1993. Netherlands—Wetlands. In E. P. H. Best and J. P. Bakker, eds. Netherlands—Wetlands. Kluwer Academic Publishers, Dordrecht, pp. 1–14.

Woo, M.-K., and T. C. Winter. 1993. The role of permafrost and seasonal frost in the hydrology of nothern wetlands in North America. Journal of Hydrology 141:5–31.

Woodhouse, W. W., Jr. 1979. Building Salt Marshes Along the Coasts of the Continental United States. Special Report 4, U.S. Army Coastal Engineering Research Center, Fort Belvoir, VA.

Woodwell, G. M. 1956. Phytosociology of Coastal Plain Wetlands of the Carolinas. Master's Thesis, Duke University, Durham, NC. 52 pp.

Woodwell, G. M., and D. E. Whitney. 1977. Flax Pond ecosystem study: Exchanges of phosphorus between a salt marsh and the coastal waters of Long Island Sound. Marine Biology 41:1–6.

Woodwell, G. M., C. A. S. Hall, D. E. Whitney, and R. A. Houghton. 1979. The Flax Pond ecosystem study: Exchanges of inorganic nitrogen between an estuarine marsh and Long Island Sound. Ecology 60:695–702.

Wright, A. H., and A. A. Wright. 1932. The habits and composition of vegetation of Okefenokee Swamp, Georgia. Ecological Monographs 2:109–232.

Wright, J. O. 1907. Swamp and Overflow Lands in the United States. Circular 76, U.S. Department of Agriculture, Washington, DC.

Wu, Y., F. H. Sklar, K. Gopu, and K. Rutchey. 1996. Fire simulations in the Everglades landscape using parallel programming. Ecological Modelling 93:113–124.

Wu, Y., F. H. Sklar, and K. Rutchey. 1997. Analysis and simulations of fragmentation patterns in the Everglades. Ecological Applications 7:268–276.

Wulder, M. 2000. High Resolution Optical Resource Satellites: A Review. Pacific Forestry Centre, Victoria, British Columbia. 9 pp.

Yagi, K., and K. Minami. 1990. Effect of organic matter application on methane emission from some Japanese paddy fields. Soil Science and Plant Nutrition 36:599–610.

Yan, J. 1992. Ecological techniques and their applications with some case studies in China. Ecological Engineering 1:261–285.

Yáñez-Arancibia, A., A. L. Lara-Dominguez, J. L. Rojan-Galaviz, P. Sánchez-Gil, J. W. Day, and C. J. Madden. 1988. Seasonal biomass and diversity of estuarine fishes coupled with tropical habitat heterogeneity (southern Gulf of Mexico). Journal of Fish Biology 33(Suppl. A):191–200.

Yáñez-Arancibia, A., A. L. Lara-Dominguez, and J. W. Day. 1993. Interactions between mangrove and seagrass habitats mediated by estuarine nekton assemblages: Coupling of primary and secondary production. Hydrobiologia 264:1–12.

Yarbro, L. A. 1979. Phosphorus Cycling in the Creeping Swamp Floodplain Ecosystem and Exports from the Creeping Swamp Watershed. Ph.D. Dissertation, University of North Carolina, Chapel Hill.

Yarbro, L. A. 1983. The influence of hydrologic variations on phosphorus cycling and retention in a swamp stream ecosystem. In T. D. Fontaine and S. M. Bartell, eds. Dynamics of Lotic Ecosystems. Ann Arbor Science, Ann Arbor, MI, pp. 223–245.

Yates, R. F. K., and F. R. Day, Jr. 1983. Decay rates and nutrient dynamics in confined and unconfined leaf litter in the Great Dismal Swamp. American Midand Naturalist 110:37–45.

Young, P. 1996. The "new science" of wetland restoration. Environmental Science & Technology 30:292–296.

Zedler, J. B. 1980. Algae mat productivity: Comparisons in a salt marsh. Estuaries 3:122–131.

Zedler, J. B. 1982a. The Ecology of Southern California Coastal Salt Marshes: A Community Profile. FWS/OBS-81/54, U.S. Fish and Wildlife Service, Washington, DC.

Zedler, J. B. 1982b. Salt marsh algal mat composition: Spatial and temporal comparisons. Bulletin Southern California Academy of Sciences 81:41–50.

Zedler, J. B. 1988. Salt marsh restoration: Lessons from California. In J. Cairns, ed. Rehabilitating Damaged Ecosystems, Vol. I. CRC Press, Boca Raton, FL, pp. 123–138.

Zedler, J. B. 1991. The challenge of protecting endangered species habitat along the southern California coast. Coastal Management 19:35–53.

Zedler, J. B. 1996. Tidal Wetland Restoration: A Scientific Perspective and Southern California Focus. Report T-038, California Sea Grant College System, University of California, La Jolla. 129 pp.

Zedler, P. H. 1987. The Ecology of Southern California Vernal Pools: A Community Profile. Biological Report 85(7.11), U.S. Fish and Wildlife Service, Washington, DC.

Zieman, J. C., and W. E. Odum. 1977. Modeling of Ecological Succession and Production in Estuarine Marshes. Technical Report D-77-35, U.S. Army Corps of Engineers Waterways Experiment Station, Vicksburg, MS.

Zimmerman, R. J., T. J. Minellos, D. L. Smith, and J. Kostera. 1990. The Use of *Juncus* and *Spartina* Marshes by Fisheries Species in Lavaca Bay, Texas, with Reference to Effects of Floods. NOAA Technical Memo NMFS-SEFC-25-1, National Marine Fisheries Service, National Oceanographic and Atmospheric Agency, Washington, DC.

Zinn, J. A., and C. Copeland. 1982. Wetland Management. Congressional Research Service, Library of Congress, Washington, DC. 149 pp.

Ziser, S. W. 1978. Seasonal variations in water chemistry and diversity of the phytophilia macroinvertebrates of three swamp communities in southeastern Louisiana. Southwest Naturalist 23:545–562.

Zobel, M., and V. Masing. 1987. Bog changing in time and space. Archiv für Hydrobiologie, Beiheft: Ergebnisse der Limnologie 27:41–55.

Zoltai, S. C. 1979. An outline of the wetland regions of Canada. In C. D. A. Rubec and F. C. Pollett, eds. Proceedings of a Workshop on Canadian Wetlands. Ecological Land Classification Series 12, Lands Directorate, Environment Canada, Saskatoon, Saskatchewan, pp. 1–8.

Zoltai, S. C. 1988. Wetland environments and classification. In National Wetlands Working Group, ed. Wetlands of Canada. Ecological Land Classification Series 24, Environment Canada, Ottawa, Ontario, and Polyscience Publications, Montreal, Quebec, pp. 1–26.

Zoltai, S. C., and F. C. Pollet. 1983. Wetlands in Canada: Their classification, distribution, and use. In A. J. P. Gore, ed. Ecosystems of the World, Vol. 4A: Mires: Swamp, Bog, Fen, and Moor. Elsevier, Amsterdam, pp. 245–268.

Zoltai, S. C., S. Taylor, J. K. Jeglum, G. F. Mills, and J. D. Johnson. 1988. Wetlands of boreal Canada. In National Wetlands Working Group, ed. Wetlands of Canada. Ecological Land Classification Series 24, Environment Canada, Ottawa, Ontario, and Polyscience Publications, Montreal, Quebec, pp. 97–154.

Zoltai, S. C., and D. H. Vitt. 1995. Canadian wetlands: Environmental gradients and classification. Vegetatio 118:131–137.

Zucker, L. A., and L. C. Brown, eds. 1998. Agricultural Drainge: Water Quality Impacts and Subsurface Drainage Studies in the Midwest. Ohio State University Extension Bulletin 871, Ohio State University, Columbus. 40 pp. 36

General Index

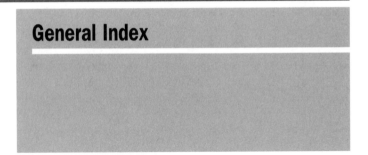

Organism Index

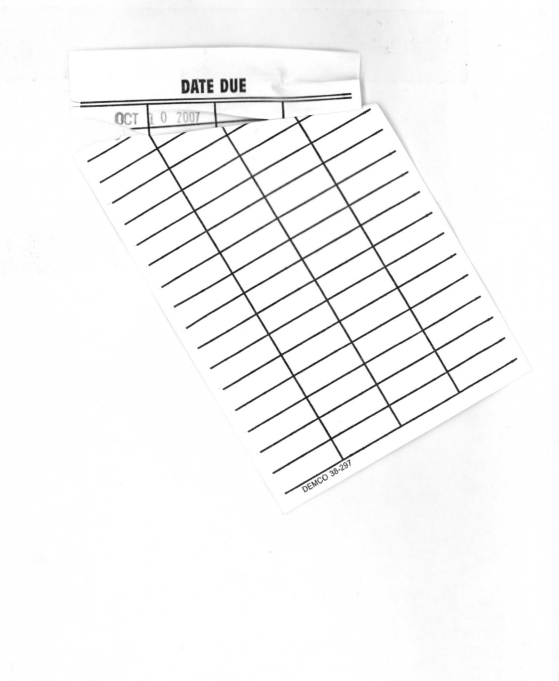

DATE DUE

OCT 1 0 2007

DEMCO 38-297